Homogeneous Flows, Moduli Spaces and Arithmetic

Clay Mathematics Proceedings
Volume 10

Homogeneous Flows, Moduli Spaces and Arithmetic

Proceedings of the Clay Mathematics Institute Summer School
Centro di Recerca Matematica Ennio De Giorgi, Pisa, Italy
June 11–July 6, 2007

Manfred Leopold Einsiedler
David Alexandre Ellwood
Alex Eskin
Dmitry Kleinbock
Elon Lindenstrauss
Gregory Margulis
Stefano Marmi
Jean-Christophe Yoccoz
Editors

American Mathematical Society
Clay Mathematics Institute

Cover photographs courtesy of the Scuola Normale Superiore, Pisa, Italy.

2000 *Mathematics Subject Classification.* Primary 37A17, 37A45, 37A35, 37C85, 37D40, 37E05, 11J13, 11J83, 58J51, 81Q50.

Library of Congress Cataloging-in-Publication Data

Clay Mathematics Institute. Summer School (2007 : Centro di recerca matematica Ennio de Giorgi)

Homogeneous flows, moduli spaces and arithmetic : Clay Mathematics Institute Summer School, June 11–July 6, 2007, Centro di recerca matematica Ennio de Giorgi, Pisa, Italy / Manfred Leopold Einsiedler... [et al.], editors.

p. cm. — (Clay mathematics proceedings ; 10)

Includes bibliographical references.

ISBN 978-0-8218-4742-8 (alk. paper)

1. Ergodic theory—Congresses. 2. Analytic spaces—Congresses. 3. Differentiable dynamical systems—Congresses. I. Einsiedler, Manfred Leopold, 1973– II. Title.

QA313.C53 2007
515′.48—dc22

2010021098

Published by the American Mathematical Society, Providence, RI,
for the Clay Mathematics Institute, Cambridge, MA.
Printed in the United States of America.

∞ The paper used in this book is acid-free and falls within the guidelines
established to ensure permanence and durability.
Visit the AMS home page at http://www.ams.org/
Visit the Clay Mathematics Institute home page at http://www.claymath.org/

10 9 8 7 6 5 4 3 2 1   15 14 13 12 11 10

Contents

Introduction

These are the proceedings of the 2007 Clay Summer School on Homogeneous Flows, Moduli Spaces and Arithmetic, which took place at the Centro di Ricerca Matematica Ennio De Giorgi in Pisa between June 11th and July 6th, 2007. More than 100 young researchers and graduate students attended this intensive four week school, as well as 18 lecturers and other established researchers.

As suggested by the name, the topic of this summer school consisted of two connected but distinct areas of active current research: flows on homogeneous spaces of algebraic groups (or Lie groups), and dynamics on moduli spaces of abelian or quadratic differentials on surfaces. These two subjects have common roots and have several important features in common; most importantly, they give concrete examples of dynamical systems with highly interesting behavior and a rich and powerful theory. Moreover, both have applications whose scope lies well outside that of the theory of dynamical systems.

The first three weeks of the summer school were devoted to the basic theory, and consisted mostly of three long lecture series. Based on these lecture series, the following four sets of notes were written:

[1] *Interval exchange maps and translation surfaces* by J. C. Yoccoz
[2] *Unipotent flows and applications* by A. Eskin
[3] *Quantitative nondivergence and its Diophantine applications* by D. Kleinbock
[4] *Diagonal actions on locally homogeneous spaces* by M. Einsiedler and E. Lindenstrauss

Furthermore, there was a shorter lecture series

[5] *Fuchsian groups, geodesic flows on surfaces of constant negative curvature and symbolic coding of geodesics* by S. Katok.

Extensive notes for all the lecture series given in the first three weeks of the school are included in this proceedings volume (the content of the course by Eskin and Kleinbock has been separated into two different sets of notes). These papers were written to be read independently, and any of the five papers [1]–[5] could serve as a good starting point for the interested reader. More advanced topics were covered by several lecture series and individual lectures mostly given in the last week of the summer school; it was left to the discretion of the lecturers in these shorter courses whether to provide notes for these proceedings (though they were strongly encouraged to contribute). A list of these lecture notes with some additional details is given below.

The common root of both main topics of the summer school mentioned above lie (at least in part) in the theory of flows on surfaces of constant negative curvature, particularly the modular surface $\mathrm{SL}(2,\mathbb{Z})\backslash\mathbb{H}$, where pioneering work was done in the early 20th century by mathematicians such as Artin, Hedlund, Morse and others, and this theory has been developed much further in the times since. One highlight was the discovery that the geodesic flow on the modular surface is intimately connected to the continued fraction expansion of real numbers; indeed, when things are properly set up, one can view the continued fraction expansion as a symbolic coding of trajectories of the geodesic flow. These flows and their symbolic codings are carefully explained in Katok's notes; in later sections of that work, recent extensions of this classical result are also discussed.

One can view the modular surface $\mathrm{SL}(2,\mathbb{Z})\backslash\mathbb{H}$ in two ways: firstly, it can be viewed as the locally homogeneous space $\mathrm{SL}(2,\mathbb{Z})\backslash\mathrm{SL}(2,\mathbb{R})/\mathrm{SO}(2,\mathbb{R})$, in which case the geodesic flow as well as another important geometric flow — the horocycle flow — can be viewed as in the projection of trajectories of the one parameter groups

$$g_t = \begin{pmatrix} e^{t/2} & 0 \\ 0 & e^{-t/2} \end{pmatrix} \quad \text{and} \quad u_t = \begin{pmatrix} 1 & t \\ 0 & 1 \end{pmatrix} \tag{1}$$

on the quotient space $\mathrm{SL}(2,\mathbb{Z})\backslash\mathrm{SL}(2,\mathbb{R})$. Another way to view $\mathrm{SL}(2,\mathbb{Z})\backslash\mathbb{H}$ is as a moduli space of flat structure (up to rotations) on a two-dimensional torus. These two different points of view generalize to the two main themes of this Clay Summer School: flows on homogeneous spaces, and flows on moduli spaces of abelian or quadratic differentials (which are essentially fancy names for flat structures in two related but slightly different senses).

Flows on moduli spaces of flat structures. The torus is the only surface admitting a flat structure with no singularities. When one considers flat structures for surfaces of higher genus, one is forced to admit singularities: points where the total angles add up to more than 2π. It turns out that *interval exchange maps* play an important role in studying the analogue of the geodesic flow (sometimes called the Teichmüller geodesic flow) on these moduli spaces of flat structures. We recall that interval exchange maps are the following simple yet intriguing dynamical system: divide the unit interval $[0,1]$ into finitely many intervals $I_1, I_2, \dots, I_d$ and then permute these intervals according to a permutation $\pi \in S_d$. Yoccoz' contribution to this proceedings provides an introduction to this theory, and provides full proofs of the most fundamental theorems (by Keane, Masur, Veech, Zorich) in the first ten sections and an introduction to some more advanced topics (Kontsevich-Zorich cocycle, cohomological equation, connected components of the moduli space, exponential mixing of the Teichmüller flow) in the last four sections. Further advanced topics are provided by notes based on the shorter lecture series

[6] *Chaoticity of the Teichmüller flow* by A. Avila

given in the last week of school; in these notes the interested reader can find surveys of the proof of two recent theorems: the simplicity of the Lyapunov spectrum for the Kontsevich-Zorich cocycle and that a typical interval exchange map with three or more intervals is weak mixing.

Flows on homogeneous spaces and applications to arithmetic. Flows on homogeneous spaces concerns the dynamics of group actions on quotient spaces $\Gamma\backslash G$, where G is usually taken to be either a **(i)** Lie group, or **(ii)** an algebraic group over $\mathbb{R}$, or **(iii)** an algebraic group over the p-adic numbers $\mathbb{Q}_p$, or **(iv)** a product of algebraic groups as in (ii) and (iii) above, involving several different fields (sometimes called an S-algebraic group, where S refers to the set of "primes" p that are used[(1)].)

A simple case is the case of $G = \mathrm{SL}(2, \mathbb{R})$ and Γ a lattice in G, for instance $\Gamma = \mathrm{SL}(2, \mathbb{Z})$. In this case we have discussed (e.g. (1)) the action of two one-parameter subgroups of $\mathrm{SL}(2, \mathbb{R})$: the group g_t corresponding to the geodesic flow on the unit tangent bundle on $\Gamma \backslash \mathbb{H}$ and u_t, which corresponds to the horocycle flow on the same space. These two flows behave very differently: the u_t-flow is very rigid, and one can algebraically classify orbit closures, invariant measures, measurable factors, self joinings, and even the asymptotic distribution of individual orbits. The g_t-flow is very flexible: it is certainly ergodic, but individual orbits can behave very badly. Moreover, the g_t-flow is measure-theoretically equivalent to a Bernoulli shift which has a wealth of measurable factors and self joinings.

The group u_t is an example of a *unipotent group*. In a fundamental series of papers published in 1990; 91, M. Ratner proved that the above mentioned rigidity properties of u_t-flow are shared by all unipotent group actions on homogeneous spaces, in particular establishing in complete generality Raghunathan's conjecture about orbit closures for such actions (some cases of which were known previously, notably in the context of the Oppenheim conjecture discussed below). For the g_t-flow the situation is rather different: while a diagonalizable one-parameter group in general behaves very much like g_t, higher-dimensional diagonalizable groups seem to behave much more rigidly (though not as rigidly as unipotent group actions).

The notes by Eskin discuss in detail unipotent flows, with an emphasis on applications, particularly regarding values attained by indefinite quadratic forms and Oppenheim's Conjecture. This long-standing conjecture was proved by Margulis in the mid-80s using homogeneous dynamics, and in particular unipotent dynamics. In dynamical terms, what Margulis has shown is that any bounded orbit of $\mathrm{SO}(2,1)$ on $\mathrm{SL}(3, \mathbb{Z}) \backslash \mathrm{SL}(3, \mathbb{R})$ is closed. The notes also give a detailed exposition of a more delicate result giving precise asymptotics to the distribution of these values by Eskin, Margulis and Mozes (under certain assumptions on the signature of a quadratic form). Some of the ideas and methods used in the theory of unipotent flows, and in particular some of the ideas used by Ratner in her proof of the Measure Classification Theorem are also described in these notes. Eskin's notes also contain other interesting applications of unipotent rigidity as well as connections to dynamics of rational billiards.

Kleinbock's notes focus on a method originally introduced by Margulis and developed significantly since, to show that orbits of unipotent group actions do not diverge to infinity. In particular, a quantitative version of the non-divergence statement due to S. G. Dani is an important ingredient in the proof of various versions of orbit closure and equidistribution theorems, including Ratner's Orbit Closure Theorem. However these techniques are more widely applicable and, in particular, were used by Kleinbock and Margulis to prove a conjecture of Sprindžuk on Diophantine approximations; this connection is also carefully discussed.

The notes by Einsiedler and Lindenstrauss discuss diagonalizable group actions, based mostly on work by the authors and by A. Katok in various combinations. A crucial role in current analysis of these actions is played by the concept of entropy.

[(1)]For this purpose ∞ is a prime and $\mathbb{Q}_\infty = \mathbb{R}$.

These notes give a detailed and self-contained account of the theory of the entropy in the locally homogeneous context, the construction of leafwise measures on foliations, and the connection between the two. Subsequently an account is given of two rather different and complementary methods to study measures invariant under multidimensional diagonalizability actions under suitable entropy assumptions, which go under the names of the *high entropy method* and the *low entropy method.* Two applications of this theory are also discussed: a partial result toward a conjecture of Littlewood on simultaneous Diophantine approximations, and how these techniques can be used to establish Arithmetic Quantum Unique Ergodicity on compact surfaces.

The material given in these three basic papers about homogeneous dynamics is complemented by the following two more advanced notes:

[7] *Counting and equidistribution on homogeneous spaces, via mixing and unipotent flows* by H. Oh

[8] *Equidistribution of Heegner points and L-functions* by G. Harcos

In Oh's notes, the use of equidistribution of unipotent flows (and the closely related but more quantitative mixing properties of diagonalizable flows) to count integer and rational points on certain varieties, a theme touched upon in Eskin's note, is developed further, and several state-of-the-art applications are explained. The notes by Harcos give some brief background in the theory of L-functions and how it relates to equidistribution of periodic orbits of the diagonal group in SL(2).

Semiclassical analysis and dynamics. One of the applications of the theory of diagonalizable actions discussed in the Einsiedler-Lindenstrauss note is establishing Arithmetic Quantum Unique Ergodicity for compact (arithmetic) surfaces. The Quantum Unique Ergodicity conjecture deals with the asymptotic distribution of eigenfunctions of the Laplacian; the arithmetic case is a very special case where the surface is arithmetic and eigenfunctions of the Laplacian are chosen so as to respect the rich set of symmetries of such surfaces. This question is considered from a completely different point of view in the notes

[9] *Eigenfunctions of the Laplacian on negatively curved manifolds: a semiclassical approach* by N. Anantharaman.

In these notes the basics of semiclassical analysis are reviewed, the connections between eigenvalues of the Laplacian and the geodesic flow, which have been discussed to some extent in the Einsiedler-Lindenstrauss notes, are developed in a much more systematic way, and very recent work relating entropy and limiting distributions of eigenfunctions of the Laplacian in general compact negatively curved manifolds (including the variable curvature case) is exposed.

Acknowledgement. This Clay Summer School was hosted by the Centro di Ricerca Matematica Ennio De Giorgi in Pisa; we are grateful to its director, Mariano Giaquinta, for accepting to host the school in this inspiring venue. The hospitality of this institute was outstanding, and the local staff, particularly Antonella Gregorace, Ilaria Gabbani, and Valentina Giuffra, went out of their way to help this school be a success. The summer school would not come to being without the vision and generosity of the Clay Mathematics Institute, and the hard work put into the school by its president, Jim Carlson, and its program manager, Christa Carter. We would especially like to thank CMI's publication manager Vida Salahi for all her work and dedication in bringing this volume to completion.

In addition to the authors of the notes listed above, the following mathematicians gave one or more lectures during this school: G. Forni, A. Gamburd, Y. Manin, G. Margulis, J. Marklof, M. Mirzakhani, S. Mozes, N. Templier, C. Ulcigrai, and A. Venkatesh. All lecturers and participants contributed to the enthusiastic and stimulating atmosphere at this school, and we thank them warmly for this.

For a variety of reasons, these lecture notes appear almost 3 years after the summer school. Quite a bit of work went into them, and indeed this is one of the reasons for the delay. They contain a substantial amount of material which cannot be found in any textbook, and we hope you, the reader, would find them useful!

Manfred Einsiedler, David Ellwood, Alex Eskin, Dmitry Kleinbock, Elon Lindenstrauss, Gregory Margulis, Stefano Marmi and Jean-Christophe Yoccoz

May 2010

Clay Mathematics Proceedings
Volume 10, 2010

Interval exchange maps and translation surfaces

Jean-Christophe Yoccoz

Introduction

Let T be a 2-dimensional torus equipped with a flat Riemannian metric and a vector field which is unitary and parallel for that metric. Then there exists a unique lattice $\Lambda \subset \mathbb{R}^2$ such that T is isometric to $\mathbb{R}^2/\Lambda$ and the vector field on T corresponds to the vertical vector field $\frac{\partial}{\partial y}$ on $\mathbb{R}^2/\Lambda$. The corresponding "Teichmüller space" (classification modulo diffeomorphisms isotopic to the identity) is thus $GL(2,\mathbb{R})$, viewed as the space of lattices equipped with a basis; the "moduli space" (classification modulo the full diffeomorphism group) is the homogeneous space $GL(2,\mathbb{R})/GL(2,\mathbb{Z})$, viewed as the space of lattices in $\mathbb{R}^2$.

The dynamics of the vertical vector field on $\mathbb{R}^2/\Lambda$ can be analyzed through the return map to a non vertical closed oriented geodesic S on $\mathbb{R}^2/\Lambda$; in the natural parameter on S which identifies S with $\mathbb{T} = \mathbb{R}/\mathbb{Z}$ (after scaling time), the return map is a rotation $x \mapsto x + \alpha$ on $\mathbb{T}$ for some $\alpha \in \mathbb{T}$. When $\alpha \notin \mathbb{Q}/\mathbb{Z}$, all orbits are dense and equidistributed on $\mathbb{R}^2/\Lambda$: the rotation and the vectorfield are uniquely ergodic (which means that they have a unique invariant probability measure, in this case the respective normalized Lebesgue measures on S and $\mathbb{R}^2/\Lambda$).

In the irrational case, an efficient way to analyze the recurrence of orbits is to use the continuous fraction of the angle α. It is well-known that the continuous fraction algorithm is strongly related to the action of the 1-parameter diagonal subgroup in $SL(2,\mathbb{R})$ on the moduli space $SL(2,\mathbb{R})/SL(2,\mathbb{Z})$ of "normalized" lattices in $\mathbb{R}^2$. It is also important in this context that the discrete subgroup $SL(2,\mathbb{Z})$ of $SL(2,\mathbb{R})$ is itself a lattice, i.e. has finite covolume, but is not cocompact.

Our aim is to explain how every feature discussed so far can be generalized to higher genus surfaces. In the first ten sections, we give complete proofs of the basic facts of the theory, which owes a lot to the pionneering work of W. Veech [**Ve1**]-[**Ve5**], with significant contributions by M. Keane [**Kea1, Kea2**], H. Masur [**Ma**], G. Rauzy [**Rau**], A. Zorich [**Zo2**]-[**Zo4**], A. Eskin, G.Forni [**For1**]-[**For3**] and many others. In the last four sections, we present without proofs some more advanced results in different directions.

2010 *Mathematics Subject Classification.* Primary 54C40, 14E20; Secondary 46E25, 20C20.
Key words and phrases. Interval exchange maps, translation surfaces, moduli space, Teichmller flow, Rauzy-Veech algorithm.

The reader is advised to consult [**Zo1**] for an excellent and very complete survey on translation surfaces. See also [**Y1**] for a first and shorter version of these notes.

In Section 1 we give the definition of a translation surface, and introduce the many geometric structures attached to it. Section 2 explains how translation surfaces occur naturally in connection with billiards in rational polygonal tables. In Section 3, we introduce interval exchange maps, which occur as return maps of the vertical flow of a translation surface. We explain in Section 4 Veech's fundamental zippered rectangle construction which allow to obtain a translation surface from an interval exchange map and appropriate suspension data. The relation between interval exchange maps and translation surfaces is further investigated in Section 5, which concludes with Keane's theorem on the minimality of interval exchange maps with no connection. Section 6 introduces the Teichmüller spaces and the moduli spaces; the fundamental theorem of Masur and Veech on the finiteness of the canonical Lebesgue measure in normalized moduli space is stated. In Section 7, we introduce the Rauzy-Veech algorithm for interval exchange maps with no connection, which is a substitute for the continuous fraction algorithm. The basic properties of this algorithm are established. Invariant measures for interval exchange maps with no connection are considered in Section 8. In Section 9, the dynamics in parameter space are introduced, whose study lead ultimately to a proof of the Masur-Veech theorem. Almost sure unique ergodicity of interval exchange maps, a related fundamental result of Masur and Veech, is proven in Section 10.

In Section 11, we introduce the Kontsevich-Zorich cocycle, and present the related results of Forni and Avila-Viana. In section 12, we consider the cohomological equation for an interval exchange map and present the result of Marmi, Moussa and myself, which extend previous fundamental work of Forni. In Section 13, we present the classification of the connected components of the moduli space by Kontsevich and Zorich. In the last section, we discuss the exponential mixing of the Teichmüller flow proved by Avila, Gouezel and myself.

1. Definition of a translation surface

1.1. We start from the following combinational data :

- a compact orientable topological surface M of genus $g \geqslant 1$;
- a non-empty finite subset $\Sigma = \{A_1, \dots, A_s\}$ of M ;
- an associated family $\kappa = (\kappa_1, \dots, \kappa_s)$ of positive integers which should be seen as **ramification indices**.

Moreover we require (for reasons that will be apparent soon) that κ and g are related through

$$2g - 2 = \sum_{i=1}^{s} (\kappa_i - 1) \,. \tag{1.1}$$

The classical setting considered in the introduction corresponds to $g = 1, s = 1, \kappa_1 = 1$.

Definition 1.1. A structure of translation surface on (M, Σ, K) is a maximal atlas ζ for $M - \Sigma$ of charts by open sets of $\mathbb{C} \simeq \mathbb{R}^2$ which satisfies the two following properties:
(i) any coordinate change between two charts of the atlas is locally a translation of $\mathbb{R}^2$;

(ii) for every $1 \leqslant i \leqslant s$, there exists a neighbourhood V_i of A_i, a neighbourhood W_i of 0 in $\mathbb{R}^2$ and a ramified covering $\pi : (V_i, A_i) \to (W_i, 0)$ of degree κ_i such that every injective restriction of π is a chart of ζ.

1.2. Because many structures on $\mathbb{R}^2$ are translation-invariant, a translation surface $(M, \Sigma, \kappa, \zeta)$ is canonically equipped with several auxiliary structures:

- a preferred orientation ; actually, one frequently starts with an **oriented** (rather than orientable) surface M and only considers those translation surface structures which are compatible with the preferred orientation ;
- a structure of Riemann surface ; this is only defined initially by the atlas ζ on $M - \Sigma$, but is easily seen to extend to M in a unique way : if V_i is a small disk around $A_i \in \Sigma, V_i - \{A_i\}$ is the κ_i- fold covering of $W_i - \{0\}$, with W_i a small disk around $0 \in \mathbb{C}$, hence is biholomorphic to $\mathbb{D}^*$;
- a flat metric on $M - \Sigma$; the metric exhibits a true singularity at each A_i such that $\kappa_i > 1$; the total angle around each $A_i \in \Sigma$ is $2\pi\kappa_i$;
- an area form on $M - \Sigma$, extending smoothly to M ; in the neighbourhood of $A_i \in \Sigma$, it takes the form $\kappa_i^2(x^2 + y^2)^{\kappa_i - 1} dx \wedge dy$ in a natural system of coordinates ;
- the geodesic flow of the flat metric on $M - \Sigma$ gives rise to a 1-parameter family of constant unitary directional flows on $M - \Sigma$, containing in particular a vertical flow $\partial/\partial y$ and a horizontal flow $\partial/\partial x$.

We will be interested in the dynamics of these vector fields. By convention (and symmetry) we will generally concentrate on the vertical vector field.

1.3. Together with the complex structure on M, a translation surface structure ζ also provides an holomorphic (w.r.t that complex structure) 1-form ω, characterized by the property that it is written as dz in the charts of ζ. In particular, this holomorphic 1-form does not vanish on $M - \Sigma$. At a point $A_i \in \Sigma$, it follows from condition (ii) that ω has a zero of order $(\kappa_i - 1)$. The relation (1) between g and κ is thus a consequence of the Riemann-Roch formula.

We have just seen that a translation surface structure determine a complex structure on M and a holomorphic 1-form ω with prescribed zeros. Conversely, such data determine a translation surface structure ζ : the charts of ζ are obtained by local integration of the 1-form ω.

The last remark is also a first way to provide explicit examples of translation surfaces. Another very important way, that will be presented in Section 5, is by suspension of one-dimensional maps called interval exchange maps. A third way, which however only gives rise to a restricted family of translation surfaces, is presented in the next section.

2. The translation surface associated to a rational polygonal billiard

2.1. Let U be a bounded connected open subset in $\mathbb{R}^2 \simeq \mathbb{C}$ whose boundary is a finite union of line segments ; we say that U is a polygonal billiard table. We say that U is **rational** if the angle between any two segments in the boundary is commensurate with π.

The billiard flow associated to the billiard table U is governed by the laws of optics (or mechanics) : point particles move linearly at unit speed inside U, and reflect on the smooth parts of the boundary ; the motion is stopped if the boundary is hit at a non smooth point, but this only concerns a codimension one subset of initial conditions.

The best way to study the billiard flow on a rational polygonal billiard table is to view it as the geodesic flow on a translation surface constructed from the table; this is the construction that we now explain.

2.2. Let $\widehat{U}$ be the **prime end compactification** of U : a point of $\widehat{U}$ is determined by a point z_0 in the closure $\overline{U}$ of U in $\mathbb{C}$ and a component of $B(z_0,\varepsilon)\cap U$ with ε small enough (as U is polygonal, this does not depend on ε if ε is small enough).

EXERCISE 2.1. Define the natural topology on $\widehat{U}$; prove that $\widehat{U}$ is compact, and that the natural map from U into $\widehat{U}$ is an homeomorphism onto a dense open subset of $\widehat{U}$.

EXERCISE 2.2. Show that the natural map from $\widehat{U}$ onto $\overline{U}$ is injective (and then a homeomorphism) iff the boundary of U is the disjoint union of finitely many polygonal Jordan curves.

A point in $\widehat{U}-U$ is **regular** if the corresponding sector in $B(z_0,\varepsilon)\cap U$ is flat; the non regular points of $\widehat{U}-U$ are the vertices of $\widehat{U}$.

EXERCISE 2.3. Show that every component of $\widehat{U}-U$ is homeomorphic to a circle and contain at least two vertices. Show that there are only finitely many vertices.

A connected component of regular points in $\widehat{U}-U$ is a **side** of $\widehat{U}$. The closure in $\widehat{U}$ of a side C of $\widehat{U}$ is the union of C and two distinct vertices called the **endpoints** of C. A vertex is the endpoint of exactly two sides.

2.3. The previous considerations only depend on U being a polygonal billiard table ; we now assume that U is rational. For each side C of $\widehat{U}$, let $\sigma_C \in O(2,\mathbb{R})$ the orthogonal symmetry with respect to the direction of the image of C in $\overline{U}\subset\mathbb{R}^2$. Let G be the subgroup of $O(2,\mathbb{R})$ generated by the σ_C.

As U is rational, G is finite. More precisely, if N is the smallest integer such that the angle between any two sides of $\widehat{U}$ can be written as $\pi m/N$ for some integer m, G is a dihedral group of order $2N$, generated by the rotations of order N and a symmetry σ_C.

For any vertex $q\in\widehat{U}$, we denote by G_q the subgroup of G generated by σ_C and $\sigma_{C'}$, where C and C' are the sides of $\widehat{U}$ having q as endpoint ; if the angle of C and C' is $\pi\, m_q/N_q$ with $m_q\wedge N_q=1, G_q$ is dihedral of order $2N_q$.

We now define a topological space M as the quotient of $\widehat{U}\times G$ by the following equivalence relation : two points $(z,g),(z',g')$ are equivalent iff $z=z'$ and moreover

- $g^{-1}g'=\mathbf{1}_G$ if $z\in U$;
- $g^{-1}g'\in\{\mathbf{1}_G,\sigma_C\}$ if z belongs to a side C of $\widehat{U}$;
- $g^{-1}g'\in G_z$ if z is a vertex of $\widehat{U}$.

We also define a finite subset Σ of M as the image in M of the vertices of $\widehat{U}$.

EXERCISE 2.4. Prove that M is a compact topological orientable surface.

To define a structure of translation surface on (M,Σ) (with appropriate ramification indices), we consider the following atlas on $M-\Sigma$.

- for each $g \in G$, we have a chart
$$U \times \{g\} \to \mathbb{R}^2$$
$$(z, g) \mapsto g(z) ;$$
- for each z_0 belonging to a side C of $\widehat{U}$, and each $g \in G$, let $\tilde{z}_0$ be the image of z_0 in $\overline{U}, \varepsilon$ be small enough, V be the component of $B(\tilde{z}_0, \varepsilon) \cap U$ corresponding to $z_0, \widehat{V}$ be interior of the closure of the image of V in $\widehat{U}$; we have a map
$$\widehat{V} \times \{g, g\, \sigma_c\} \to \mathbb{R}^2$$
sending (z, g) to $g(z)$ and $(z, g\sigma_c)$ to $g(\tilde{\sigma}_c(z))$, where $\tilde{\sigma}_c$ is the **affine** orthogonal symmetry with respect to the line containing the image of C in $\mathbb{R}^2$. This map is compatible with the identifications defining M and defines a chart from a neighbourhood of (z, g) in M onto an open subset of $\mathbb{R}^2$.

One checks easily that the coordinate changes between the charts considered above are translations. One then completes this atlas to a maximal one with property (i) of the definition of translation surfaces.

EXERCISE 2.5. Let q be a vertex of $\widehat{U}$, and let $\pi m_q/N_q$ be the angle between the sides at q and G_q the subgroup of G as above. Show that property (ii) in the definition of a translation surface is satisfied at any point $(q, g\, G_q) \in \Sigma$, with ramification index m_q (independent of the coset $g\, G_q$ under consideration).

We have therefore defined the ramification indices κ_i at the points of Σ and constructed a translation surface structure on (M, Σ, κ).

2.4. The relation between the trajectories of the billiard flow on U and the geodesics on $M - \Sigma$ is as follows.

Let $z(t), 0 \leqslant t \leqslant T$ be a billiard trajectory ; let $t_1 < \cdots < t_N$ be the successive times in $(0, T)$ where the trajectory bounces on the sides of $\widehat{U}$ (by hypothesis, the trajectory does not go through a vertex, except perhaps at the endpoints 0 and T). Denote by C_i the side met at time t_i and define inductively $g_0, \ldots, g_N$ by
$$g_0 = \mathbf{1}_G ,$$
$$g_{i+1} = g_i\, \sigma_{C_{i+1}} .$$
For any $g \in G$, the formulas
$$z_g(t) = \begin{cases} (z(t), gg_0), & \text{for } 0 \leqslant t \leqslant t_1, \\ (z(t), gg_i), & \text{for } t_i \leqslant t \leqslant t_{i+1}\ (1 \leqslant i < N), \\ (z(t), gg_N), & \text{for } t_N \leqslant t \leqslant T, \end{cases}$$
define a geodesic path on M. Conversely, every geodesic path on M (contained in $M - \Sigma$ except perhaps for its endpoints) defines by projection on the first coordinate a trajectory of the billiard flow on U.

2.5. The left action

$$g_0(z,g) = (z, g_0\, g)$$

of G on $\widehat{U} \times G$ is compatible with the equivalence relation defining M and therefore defines a left action of G on M. The corresponding transformations of M are isometries of the flat metric of M but not isomorphisms of the translation surface structure (except for the identity !). The existence of such a large group of isometries explain why the translation surfaces constructed from billiard tables are special amongst general translation surfaces.

2.6. On the other hand, when a billiard table admits non trivial symmetries, this gives rise to isomorphisms of the translation surface structures. More precisely, let H be the subgroup of G formed of the $h \in G$ such that $h(U)$ is a translate $U + t_h$ of U. The group H acts on the left on M through the formula

$$h(z,g) = (h(z) - t_h, g\, h^{-1}),$$

which is compatible with the equivalence relation defining M. Each $h \in H$ acts through an isomorphism of the translation surface structure (permuting the points of Σ). This allows to consider the quotient under the action of H to get a reduced translation surface $(M', \Sigma', \kappa', \zeta')$ and a ramified covering from (M, Σ) onto (M', Σ').

2.7. To illustrate all this, consider the case where U is a regular n-gon, $n \geqslant 3$. The angle at each vertex is then $\pi \frac{n-2}{n}$.

EXERCISE 2.6. Show that $G = G_q$ for every vertex q and that G has order n if n is even, $2n$ if n is odd. Show that Σ has n points, each having ramification index $n-2$ if n is odd, $\frac{n-2}{2}$ if n is even. Conclude that the genus of M is $\frac{(n-1)(n-2)}{2}$ if n is odd, $(\frac{n}{2} - 1)^2$ if n is even.

EXERCISE 2.7. Show that the subgroup H of subsection 2.6. is equal to G if n is even, and is of index 2 if n is odd. Show that the reduced translation surface satisfies $\#\Sigma' = 2$ if $N - 2$ is divisible by 4, $\#\Sigma' = 1$ otherwise. Show that the corresponding ramification index is $n-2$ if n odd, $\frac{(n-2)}{2}$ if n is divisible by 4, $\frac{(n-2)}{4}$ if $n - 2$ is divisible by 4. Conclude that the genus g' is $\frac{(n-1)}{2}$ if n is odd, $\frac{n}{4}$ if n is divisible by 4 , $\frac{(n-2)}{4}$ if $n-2$ is divisible by 4.

3. Interval exchange maps : basic definitions

3.1. Let $(M, \Sigma, \kappa, \zeta)$ be a translation surface and let X be one of the non zero constant vector fields on $M - \Sigma$ defined by ζ.

DEFINITION 3.1. An **incoming (resp. outgoing) separatrix** for X is an orbit of X ending (resp. starting) at a marked point in Σ. A **connection** is an orbit of X which is both an incoming and outgoing separatrix.

At a point $A_i \in \Sigma$, there are κ_i incoming separatrices and κ_i outgoing separatrices.

Let S be an open bounded geodesic segment in $M - \Sigma$, parametrized by arc length, and transverse to X. Consider the first return map T_S to S of the flow generated by the vectorfield X.

As X is area-preserving, the Poincaré recurrence theorem guarantees that the map T_S is defined on a subset D_{T_S} of S of full 1-dimensional Lebesgue measure.

The domain D_{T_S} is open because S itself is open and the restriction of T_S to each component of D_{T_S} is a translation (because the flow of X is isometric). Also, the return time is constant on each component of D_{T_S}.

We now show that D_{T_S} has only finitely many components. Indeed, let $x \in S$ be an endpoint of some component J of D_{T_S}, and let t_J the return time to S of points in J. Either there exists $T \in (0, t_J)$ such that the orbit of X starting at x stops at time T at a point of Σ without having crossed S, or the orbit of X starting at x is defined up to time t_J and is at this moment at one of the endpoints of S, also without having crossed S. This leaves only a finite number of possibilities for x, which gives the finiteness assertion.

The return map T_S is thus an interval exchange map according to the following definition.

DEFINITION 3.2. Let $I \subset \mathbb{R}$ be a bounded open interval. An interval exchange map (i.e.m) T on I is a one-to-one map $T : D_T \to D_{T^{-1}}$ such that $D_T \subset I, D_{T^{-1}} \subset I, I - D_T$ and $I - D_{T^{-1}}$ are finite sets (with the same cardinality) and the restriction of T to each component of D_T is a translation onto some component of $D_{T^{-1}}$.

3.2. Markings, combinatorial data. Let $T : D_T \to D_{T^{-1}}$ be an interval exchange map. Let $d = \#\pi_0(D_T) = \#\pi_0(D_{T^{-1}})$. Then T realizes a bijection between $\pi_0(D_T)$ and $\pi_0(D_{T^{-1}})$. To keep track of the combinatorial data, in particular when we will consider below the Rauzy-Veech continuous fraction algorithm for i.e.m, it is convenient to give names to the components of D_T (and therefore through T also to those of $D_{T^{-1}}$). This is formalized as follow.

A **marking** for T is given by an alphabet $\mathcal{A}$ with $\#\mathcal{A} = d$ and a pair $\pi = (\pi_t, \pi_b)$ of one-to-one maps

$$\begin{matrix} \pi_t \\ \\ \pi_b \end{matrix} \quad \mathcal{A} \to \{1, \ldots, d\}$$

such that, for each $\alpha \in \mathcal{A}$, the component of D_T in position $\pi_t(\alpha)$ (counting from the left) is sent by T to the component of $D_{T^{-1}}$ in position $\pi_b(\alpha)$. We summarize these combinatorial data by writing just

$$\begin{pmatrix} \pi_t^{-1}(1) & \ldots & \pi_t^{-1}(d) \\ \pi_b^{-1}(1) & \ldots & \pi_b^{-1}(d) \end{pmatrix}$$

expressing how the intervals which are exchanged appear before and after applying T.

Two markings $(\mathcal{A}, \pi_t, \pi_b), (\mathcal{A}', \pi'_t, \pi'_b)$ are equivalent if there exists a bijection $i : \mathcal{A} \to \mathcal{A}'$ with $\pi_t = \pi'_t \circ i, \pi_b = \pi'_b \circ i$. Clearly T determines the marking up to equivalence.

3.3. Irreducible combinatorial data. We say that combinatorial data $(\mathcal{A}, \pi_t, \pi_b)$ are irreducible if for every $1 \leqslant k < d = \#\mathcal{A}$, we have

$$\pi_t^{-1}(\{1, \ldots, k\}) \neq \pi_b^{-1}(\{1 \ldots k\}).$$

The condition is invariant under equivalence of markings. We will always assume that the i.e.m under consideration satisfy this property. Otherwise, if we have

$$\pi_t^{-1}(\{1, \ldots, k\}) = \pi_b^{-1}(\{1 \ldots k\}).$$

T is the juxtaposition of an i.e.m with k intervals and another with $d - k$, and the dynamics of T reduce to simpler cases.

3.4. Terminology and notations. Let $T : D_T \to D_{T^{-1}}$ be an i.e.m on an interval I ; let $(\mathcal{A}, \pi_t, \pi_b)$ a marking for T.

The points $u_1^t < u_2^t < \cdots < u_{d-1}^t$ of $I - D_T$ are called the **singularities** of T ; the points $u_1^b < u_2^b < \cdots < u_{d-1}^b$ of $I - D_{T^{-1}}$ are called the **singularities** of T^{-1}.

For each $\alpha \in \mathcal{A}$, we denote by I_α^t or just I_α the component of D_T in position $\pi_t(\alpha)$ (counting from the left), and by I_α^b its image by T which is also the component of $D_{T^{-1}}$ in position $\pi_b(\alpha)$.

We denote by λ_α the common length of I_α^t and I_α^b. The vector $\lambda = (\lambda_\alpha)_{\alpha\in\mathcal{A}}$ in $\mathbb{R}^{\mathcal{A}}$ is the **length vector** and will be considered as a **row vector**.

On the other hand, let δ_α be the real number such that $I_\alpha^b = I_\alpha^t + \delta_\alpha$. The vector $\delta = (\delta_\alpha)_{\alpha\in\mathcal{A}}$ is the **translation vector** and will be considered as a column vector.

The length vector and the translation vector are related through the obvious formulas

$$\delta_\alpha = \sum_{\pi_b(\beta)<\pi_b(\alpha)} \lambda_\beta - \sum_{\pi_t(\beta)<\pi_t(\alpha)} \lambda_\beta = \sum_\beta \Omega_{\alpha\beta}\lambda_\beta$$

where the antisymmetric matrix Ω is defined by

$$\Omega_{\alpha\beta} = \begin{cases} +1 & \text{if } \pi_b(\beta) < \pi_b(\alpha) \text{ and } \pi_t(\beta) > \pi_t(\alpha), \\ -1 & \text{if } \pi_b(\beta) > \pi_b(\alpha) \text{ and } \pi_t(\beta) < \pi_t(\alpha), \\ 0 & \text{otherwise.} \end{cases}$$

4. Suspension of i.e.m : the zippered rectangle construction

4.1. We have seen in subsection 3.1 that we come naturally to the definition of an interval exchange map by considering return maps for constant vector fields on translation surfaces.

Conversely, starting from an interval exchange map T, we will construct, following Veech [Ve2] a translation surface for which T appears as a return map of the vertical vector field. However, as the case of the torus for rotations already demonstrates, supplementary data such as return times are needed to specify uniquely the translation surface.

Let $T : D_T \to D_{T^{-1}}$ be an i.e.m on an interval I, equipped with a marking $(\mathcal{A}, \pi_t, \pi_b)$ as above.

A vector $\tau \in \mathbb{R}^{\mathcal{A}}$ is a **suspension vector** if it satisfies the following inequalities

$$(S_\pi) \qquad \sum_{\pi_t(\alpha)<k} \tau_\alpha > 0\,, \qquad \sum_{\pi_b(\alpha)<k} \tau_\alpha < 0 \qquad \text{for all } 1 < k \leqslant d\,.$$

Define

$$\tau_\alpha^{can} = \pi_b(\alpha) - \pi_t(\alpha) \quad , \quad \alpha \in \mathcal{A}\,.$$

Then the vector τ^{can} satisfies (S_π) iff the combinatorial data are irreducible (an hypothesis that we will assume from now on). When the combinatorial data are not irreducible, no vector $\tau \in \mathbb{R}^{\mathcal{A}}$ satisfies (S_π).

4.2. A simple version of the construction. Let T as above ; we assume that the combinatorial data are irreducible and use the notations of subsection 3.4. Let also $\tau \in \mathbb{R}^{\mathcal{A}}$ be a suspension vector.

We will construct from these data a translation surface $(M, \Sigma, \kappa, \zeta)$. We first give a simple version of the construction that unfortunately is not valid for all values of the data. We identify as usual $\mathbb{R}^2$ with $\mathbb{C}$ and set $\zeta_\alpha = \lambda_\alpha + i\tau_\alpha$ for $\alpha \in \mathcal{A}$.

Consider the "top" polygonal line connecting the points 0, $\zeta_{\pi_t^{-1}(1)}$, $\zeta_{\pi_t^{-1}(1)} + \zeta_{\pi_t^{-1}(2)}, \dots, \zeta_{\pi_t^{-1}(1)} + \zeta_{\pi_t^{-1}(2)} + \dots + \zeta_{\pi_t^{-1}(d)}$ and the "bottom" polygonal line connecting the points $0, \zeta_{\pi_b^{-1}(1)}, \zeta_{\pi_b^{-1}(1)} + \zeta_{\pi_b^{-1}(2)}, \dots, \zeta_{\pi_b^{-1}(1)} + \zeta_{\pi_b^{-1}(2)} + \dots + \zeta_{\pi_b^{-1}(d)}$. Observe that both lines have the same endpoints and that, from the suspension condition (S_π), all intermediary points in the top (resp. bottom) line lie in the upper (resp. lower) half-plane.

When the two lines do not intersect except from their endpoints, their union is a Jordan curve and we can construct a translation surface as follows : denoting by W the closed polygonal disk bounded by the two lines, we identify for each $\alpha \in \mathcal{A}$ the ζ_α side of the top line with the ζ_α side of the bottom line through the appropriate translation and define M to be the topological space obtained from W with this identifications. The finite subset Σ is the image of the vertices of W.

EXERCISE 4.1. Check that M is indeed a compact oriented topological surface.

The atlas defining the translation surface structure is obvious : besides the identity map on the interior of W, we use charts defined on neighbourhoods of the interiors of the ζ_α sides which have been identified.

Condition (ii) in the definition of a translation surface and ramification indices will be discussed below.

This construction is very easy to visualize, and the non intersection condition is frequently satisfied : for instance when $\sum_\alpha \tau_\alpha = 0$ (in particular for $\tau = \tau^{can}$), or when $\lambda_{\pi_t^{-1}(d)} = \lambda_{\pi_b^{-1}(d)}$. Unfortunately, it is not always satisfied. For instance, taking for combinatorial data (with $\mathcal{A} = \{A, B, C, D\}$),

$$\pi = (\pi_t, \pi_b) = \begin{pmatrix} A & B & D & C \\ D & A & C & B \end{pmatrix},$$

we may have $\zeta_A = 1 + i$, $\zeta_B = 3 + 3i$, $\zeta_C = \varepsilon + i$, $\zeta_D = 3 - 3i$ with $\varepsilon > 0$. Then the suspension condition (S_π) is satisfied but the two lines intersect non trivially when $0 < \varepsilon < 1$.

4.3. Zippered rectangles. Let $T, \lambda, \tau, \zeta = \lambda + i\tau$ as above. The length vector and the translation vector δ are related through.

$$\delta = \Omega^t \lambda .$$

We define

$$h = -\Omega^t \tau ,$$

$$\theta = \delta - ih = \Omega^t \zeta .$$

We consider here λ, τ as row vectors in $\mathbb{R}^{\mathcal{A}}$, ζ as a row vector in $\mathbb{C}^{\mathcal{A}}$, δ, h as column vectors in $\mathbb{R}^{\mathcal{A}}$ and θ as a column vector in $\mathbb{C}^{\mathcal{A}}$.

EXERCISE 4.2. Check that in the construction of subsection 4.2, the ζ_α side of the "top line" was identified to the ζ_α side of the "bottom line" through a translation by θ_α.

We observe that for all $\alpha \in \mathcal{A}$ we have

$$h_\alpha = \sum_{\pi_t\beta < \pi_t\alpha} \tau_\beta - \sum_{\pi_b\beta < \pi_b\alpha} \tau_\beta$$

and therefore, from the suspension condition (S_π) :

$$h_\alpha > 0 \, .$$

Indeed, the first sum on the right-hand side is > 0 except if $\pi_t\alpha = 1$ when it is 0 and the second sum is < 0 except if $\pi_b\alpha = 1$ when it is 0. By irreducibility, we cannot have both $\pi_t\alpha = 1$ and $\pi_b\alpha = 1$.

Define the rectangles in $\mathbb{R}^2 = \mathbb{C}$:

$$R^t_\alpha = I^t_\alpha \times [0, h_\alpha] \, ,$$

$$R^b_\alpha = I^b_\alpha \times [-h_\alpha, 0] \, ,$$

Let $u^t_1 < u^t_2 < \cdots < u^t_{d-1}$ be the singularities of T, $u^b_1 < u^b_2 < \cdots < u^b_{d-1}$ those of T^{-1}. Write also $I = (u_0, u_d)$. Define, for $1 \leqslant i \leqslant d-1$:

$$S^t_i = \{u^t_i\} \times [0, \sum_{\pi_t\alpha \leqslant i} \tau_\alpha),$$

$$S^b_i = \{u^b_i\} \times (\sum_{\pi_b\alpha \leqslant i} \tau_\alpha, 0].$$

Define the points

$$C_0 = (u_0, 0), \qquad C_d = (u_d, \sum_\alpha \tau_\alpha),$$

$$C^t_i = C_0 + \sum_{\pi_t\alpha \leqslant i} \zeta_\alpha, \qquad C^b_i = C_0 + \sum_{\pi_b\alpha \leqslant i} \zeta_\alpha \, , \qquad \text{for } 0 < i < d.$$

Finally, let S^* be the closed vertical segment whose endpoints are $(u_d, 0)$ and C_d (Figure 1). Let $\widehat{M}$ be the union of all the elements just defined : the R^t_α, R^b_α, $(\alpha \in \mathcal{A})$, S^t_i, S^b_i, $(0 < i < d)$, C_0, C_d, C^t_i, C^b_i, $(0 < i < d)$ and S^*.

We use translations by $\theta_\alpha, \alpha \in \mathcal{A}$ to identify some of these elements :

- We identify R^t_α and $R^b_\alpha = R^t_\alpha + \theta_\alpha$.
- We identify $C^t_{\pi_t(\alpha)}$ and $C^b_{\pi_b(\alpha)} = C^t_{\pi_t(\alpha)} + \theta_\alpha$, and also $C^t_{\pi_t(\alpha)-1}$ and $C^b_{\pi_b(\alpha)-1} = C^t_{\pi_t(\alpha)-1} + \theta_\alpha$; here, we have by convention $C^t_0 = C^b_0 = C_0, C^t_d = C^b_d = C_d$.
- finally, if $\Sigma_\alpha \, \tau_\alpha > 0$, we identify by $\theta_{\pi_b^{-1}(d)}$ the top part of $S^t_{\pi_t\pi_b^{-1}(d)}$ with S^* ; if $\Sigma_\alpha \, \tau_\alpha < 0$, we identify S^* with the bottom part of $S^b_{\pi_b\pi_t^{-1}(d)}$ by $\theta_{\pi_t^{-1}(d)}$.

We denote by M the topological space deduced from $\widehat{M}$ by these identifications. We denote by Σ the part of M which is the image of $\{C_0, C_d, C^t_i, C^b_i\}$.

One easily checks that M is compact and that $M - \Sigma$ is a topological orientable surface. Every point in $M - \Sigma$, except those in the image of S^* when $\Sigma \, \tau_\alpha \neq 0$, has a representative in the interior of $\widehat{M}$; for those points, a local continuous section of the projection from $\widehat{M}$ onto M provides a chart for the atlas defining the translation surface structure. We leave the reader provide charts around points in the image of S^*.

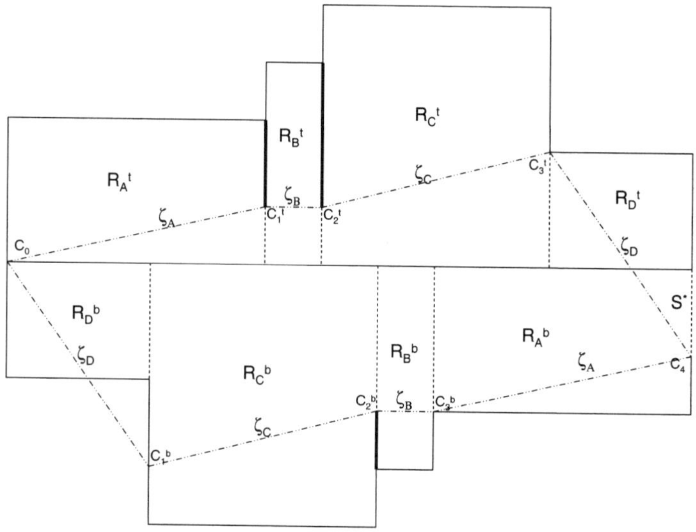

FIGURE 1

In the next section, we complete the construction by investigating the local structure at points in Σ : this means checking that M is indeed a topological surface, that condition (ii) in the definition of translation surfaces is satisfied, and computing the ramification indices.

Let us however observe right now that we have indeed a suspension for the i.e.m. T on I. The return map on the horizontal segment $I \times \{0\}$ (or rather its image in M) of the vertical vector field $\frac{\partial}{\partial y}$ is exactly T. The return time of I^t_α is equal to h_α.

4.4. Ramification indices. Let $\mathcal{C}$ the set $\{C^t_i, C^b_i \ ; 0 < i < d\}$ with $2d-2$ elements ; turning around points of Σ in an anticlockwise manner, we define a "successor" map $\sigma : \mathcal{C} \to \mathcal{C}$:

- $\sigma(C^t_i) = C^b_{\pi_b\pi_t^{-1}(i+1)-1}$, except if $\pi_b\pi_t^{-1}(i+1) = 1$ in which case $\sigma(C^t_i) = C^b_{\pi_b\pi_t^{-1}(1)-1}$;
- $\sigma(C^b_j) = C^t_{\pi_t\pi_b^{-1}(j)}$ except if $\pi_t\pi_b^{-1}(j) = d$ in which case $\sigma(C^b_j) = C^t_{\pi_t\pi_b^{-1}(d)}$.

We see that σ is a permutation of $\mathcal{C}$, exchanging the C^t_i and the C^b_j. Therefore every cycle of σ has even length.

From the very definition of σ, points of Σ are in one-to-one correspondance with the cycles of σ. Moreover, one checks that small neighbourhoods of points of Σ are homeomorphic to disks, and that condition (ii) in the definition of a translation index is satisfied, the ramification index being half the length of the corresponding cycle. Summing up :

- The number s of points in Σ is the number of cycles of the permutation σ.

- The ramification indices κ_j are the half lengths of the cycles ; in particular, we have

$$d - 1 = \sum_{j=1}^{s} \kappa_j \ .$$

If g is the genus of the compact surface M, we also must have

$$2g - 2 = \sum_{i=1}^{s} (\kappa_i - 1) \ .$$

We therefore can relate d, g, s by

$$d = 2g + s - 1 \ .$$

4.5. Homology and cohomology of M. Consider the homology groups $H_1(M, \mathbb{Z}), H_1(M - \Sigma, \mathbb{Z}), H_1(M, \Sigma, \mathbb{Z})$. The first one has rank $2g$, the last two have rank $2g + s - 1 = d$. They are related through maps

$$H_1(M - \Sigma, \mathbb{Z}) \to H_1(M, \mathbb{Z}) \to H_1(M, \Sigma, \mathbb{Z}),$$

where the first map is onto and the second is injective. The zippered rectangle construction provides natural bases for $H_1(M - \Sigma, \mathbb{Z})$ and $H_1(M, \Sigma, \mathbb{Z})$.

For $\alpha \in \mathcal{A}$, let $[\theta_\alpha]$ be the image in $H_1(M - \Sigma, \mathbb{Z})$ of a path joining in the interior of $\widehat{M}$ the center of R^t_α to the center of R^b_α ; and let $[\zeta_\alpha]$ be the image in $H_1(M, \Sigma, \mathbb{Z})$ of a path joining in $R^t_\alpha \cup \{C^t_{\pi_t(\alpha)-1}, C^t_{\pi_t(\alpha)}\}$ the point $C^t_{\pi_t(\alpha)-1}$ to $C^t_{\pi_t(\alpha)}$ (if $\pi_t(\alpha) = d$ and $\Sigma_\alpha \tau_\alpha < 0$, the path should be allowed to go through S^* also).

The intersection form establishes a duality between $H_1(M - \Sigma, \mathbb{Z})$ and $H_1(M, \Sigma, \mathbb{Z})$. Now we clearly have, for $\alpha, \beta \in \mathcal{A}$:

$$< [\theta_\alpha], [\zeta_\beta] >= \delta_{\alpha\beta} \ ,$$

which shows that $([\theta_\alpha])_{\alpha \in \mathcal{A}}, ([\zeta_\beta])_{\beta \in \mathcal{A}}$ are respectively bases of $H_1(M - \Sigma, \mathbb{Z})$, $H_1(M, \Sigma, \mathbb{Z})$ dual to each other.

Considering $[\theta_\alpha]$ as classes in $H_1(M, \mathbb{Z})$, the intersection form now reads :

$$< [\theta_\alpha], [\theta_\beta] >= \Omega_{\beta\alpha} \ ,$$

Indeed, writing $[\overline{\theta_\alpha}]$ for the image of $[\theta_\alpha]$ in $H_1(M, \Sigma, \mathbb{Z})$, we have

$$[\overline{\theta_\alpha}] = \sum_\beta \Omega_{\alpha\beta} [\zeta_\beta]$$

which shows in particular that

$$\text{rk}\, \Omega = 2g \ .$$

Going to cohomology, we have maps

$$H^1(M, \Sigma, \mathbb{Z}) \to H^1(M, \mathbb{Z}) \to H^1(M - \Sigma, \mathbb{Z})$$

(and similar maps with real and complex coefficients) where the first map is onto and the second is injective.

The holomorphic 1−form ω associated to the translation surface structure determines by integration a class $[\omega] \in H^1(M, \Sigma, \mathbb{C})$ (this will be studied in more details and generality in section 6 below). One has

$$< [\omega], [\zeta_\alpha] >= \zeta_\alpha \ ,$$

$$< [\overline{\omega}], [\theta_\alpha] >= \theta_\alpha \ ,$$

where $[\overline{\omega}]$ is the image of $[\omega]$ in $H^1(M-\Sigma,\mathbb{C})$.

Therefore the vectors λ,τ can be considered as elements of $H^1(M,\Sigma,\mathbb{R})$, the vector $\zeta=\lambda+i\tau$ as an element of $H^1(M,\Sigma,\mathbb{C})$. The vectors δ,h can be considered as elements of $H^1(M-\Sigma,\mathbb{R})$; they actually belong to the image of $H^1(M,\mathbb{R})$ into $H^1(M-\Sigma,\mathbb{R})$ because they vanish on the kernel of the map from $H_1(M-\Sigma,\mathbb{Z})$ to $H_1(M,\mathbb{Z})$. Similarly, $\theta=\delta-ih$ belongs to the image of $H^1(M,\mathbb{C})$ into $H^1(M-\Sigma,\mathbb{C})$. Finally, the area of the translation surface M is given by

$$A=\sum_\alpha \lambda_\alpha\, h_\alpha=\tau\,\Omega\,{}^t\lambda\,.$$

5. Representability, minimality, connections

5.1. We have seen in subsection 3.1 that for any translation surface, the return map of the vertical vector field on any horizontal segment is an interval exchange map. In the zippered rectangle construction, the horizontal segment $I\times\{0\}$ is wide enough to intersect all orbits of the vertical vector field.

Already in the case of the torus, when the vertical vectorfield has rational slope with respect to the lattice, it is clear that a short enough horizontal segment will not intersect all orbits. In higher genus, the same can happen even when the vertical vector field has no periodic orbits, as the following construction shows.

Let Λ_1,Λ_2 be two lattices in $\mathbb{R}^2$ with no non zero vertical vectors ; let $T_i=\mathbb{R}^2/\Lambda_i$; choose on each T_i two vertical segments $[A_i,B_i]$ of the same length. Slit T_i along $[A_i,B_i]$ and glue isometrically the left side of $[A_1,B_1]$ to the right side of $[A_2,B_2]$ and vice-versa. We obtain a compact oriented surface M of genus 2, with two marked points A (image of A_1,A_2) and B (image of B_1,B_2) ; the canonical translation surface structures on T_1,T_2 generate a translation surface structure on $(M,\{A,B\})$ with ramification indices $\kappa_A=\kappa_B=2$. The vector field has no periodic orbit in view of the hypothesis on Λ_1,Λ_2 but obviously any small horizontal segment in T_1 not intersecting $[A_1,B_1]$ will only intersect the orbits of the vectorfield contained in T_1.

Even when an horizontal segment intersects all orbits of the vertical vectorfield, the number of intervals in the i.e.m obtained as return map depends on the segment.

EXERCISE 5.1. For a torus with one marked point and a minimal vertical vectorfield, show that the return map on a horizontal segment starting at the marked point is an i.e.m with 2 or 3 intervals. Find necessary and sufficient conditions for the return map to have only 2 intervals.

5.2. In order to understand which translation surfaces can be obtained via the zippered rectangle construction, the following lemma is useful.

Let (M,Σ,κ,ζ) be a translation surface. Denote by (Φ_t^V), resp. (Φ_t^H), the flow of the vertical, resp. horizontal, vectorfield. Let $x_0\in M-\Sigma$ a point of period T for the vertical vectorfield.

LEMMA 5.2. *There exists a maximal open bounded interval J around 0 such that for $s\in J$, the vertical flow $\Phi_t^V(\Phi_s^H(x_0))$ is defined for all times $t\in\mathbb{R}$. One has*

$$\Phi_{t+T}^V(\Phi_s^H(x_0))=\Phi_t^V(\Phi_s^H(x_0)),$$

for $s \in J, t \in \mathbb{R}$, *and the map*

$$J \times \mathbb{R}/T\mathbb{Z} \to M$$

$$(s,t) \mapsto \Phi_t^V(\Phi_s^H(x_0))$$

is injective. The compact set

$$Z^+ = \lim_{s \nearrow \sup J} \{\Phi_{[0,T]}^V(\Phi_s^H(x_0))\}$$

is a finite union of points of Σ *and vertical connections between them. The same holds for*

$$Z^- = \lim_{s \searrow \inf J} \{\Phi_{[0,T]}^V(\Phi_s^H(x_0))\}.$$

The image $\Phi_{[0,T]}^V(\Phi_J^H(x_0))$ is called the **cylinder** around the periodic orbit of x_0. Its boundary in M is $Z^+ \cup Z^-$.

PROOF. Let J be an open bounded interval around 0 such that $\Phi_s^H(x_0)$ is defined for $s \in J$ and $\Phi_t^V(\Phi_s^H(x_0))$ is defined for all $t \in \mathbb{R}, s \in J$. Any J small enough will have this property. Moreover, we must have

$$\Phi_T^V(\Phi_s^H(x_0)) = \Phi_s^H(x_0)$$

for all $s \in J$ because the set of s with this property contains 0 and is open and closed in J. The map

$$J \times \mathbb{R}/T\mathbb{Z} \to M$$
$$(s,t) \mapsto \Phi_t^V(\Phi_s^H(x_0))$$

must be injective : if we had

$$\Phi_{t_0}^V(\Phi_{s_0}^H(x_0)) = \Phi_{t_1}^V(\Phi_{s_1}^H(x_0)),$$

then either $s_0 = s_1$, $0 < t_1 - t_0 < T$ would contradict that T is the minimal period of x_0 or $s_0 < s_1$ would imply that

$$\Phi_{[0,T]}^V(\Phi_{[s_0,s_1]}^H(x_0))$$

is open and closed in M, hence equal to M, contradicting that Σ is non empty. The injectivity gives a bound on the length of J, namely

$$|J| \leqslant AT^{-1}$$

where A is the area of M. This bound means that there exists indeed a maximal bounded open interval with the required properties. The maximality in turn implies that the set Z^+ must meet Σ (otherwise $\Phi_{\sup J}^H(x_0)$ is defined and of period T for the vertical flow), and thus is a finite union of points of Σ and vertical connections between them. Similarly for Z^-. □

PROPOSITION 5.3. *Let* $(M, \Sigma, \kappa, \zeta)$ *be a translation surface, and* S *be an open bounded horizontal segment in* M. *Assume that* S *meets every vertical connection (if any). Then, either every infinite half-orbit of the vertical vectorfield meets* S, *or there is a cylinder containing every (infinite) orbit of the vertical vectorfield not meeting* S.

PROOF. Denote by T_S the return map of the vertical vectorfield to S, by Φ_t the flow of the vertical vectorfield. Let u^t be a singularity of T_S, J the component of the domain of T_S to the left of u^t, t_J the return time to S in J. For $0 \leqslant t \leqslant t^L(u^t) := t_J$, let

$$\Phi_t^L(u^t) = \lim_{x \nearrow u^t} \Phi_t(x).$$

In the same way we define a right-limit $\Phi_t^R(u^t), 0 \leqslant t \leqslant t^R(u^t)$, and, for a singularity u^b of T_S^{-1}, we define left and right limits $\Phi_t^L(u^b), \Phi_t^R(u^b)$ (for negative time intervals $0 \geqslant t \geqslant t^L(u^b), 0 \geqslant t \geqslant t^R(u^b)$ respectively).

CLAIM 5.4. *The sets* $X^L = [\bigcup_{u^t} \Phi^L_{[0,t^L(u^t)]}(u^t)\,] \bigcup [\bigcup_{u^b} \Phi^L_{[t^L(u^b),0]}(u^b)\,]$ *and* $X^R = [\bigcup_{u^t} \Phi^R_{[0,t^R(u^t)]}(u^t)\,] \bigcup [\bigcup_{u^b} \Phi^R_{[t^R(u^b),0]}(u^b)\,]$ *are equal.*

PROOF. Let u^t be a singularity of T_S. We prove that $\Phi^L_{[0,t^L(u^t)]}(u^t)$ is contained in X^R. The claim then follows by symmetry. We distinguish two cases.

- **(a):** Assume first that $\lim_{x \nearrow u^t} T_S(x)$ is not the right endpoint of S. Then, it is a singularity u^b of T_S^{-1}. As S meets every vertical connection, the set $\Phi^L_{[0,t^L(u^t)]}(u^t)$ contains exactly one point of Σ, say $\Phi^L_{t^*}(u^t)$. Then $\Phi^L_{[0,t^*]}(u^t)$ is equal to $\Phi^R_{[0,t^*]}(u^t)$, and $\Phi^L_{[t^*,t^L(u^t)]}(u^t)$ is contained in $\Phi^R_{[t^R(u^b),0]}(u^b)$.
- **(b):** Assume now that $\lim_{x \nearrow u^t} T_S(x) = u^*$ is the right endpoint of S. Then $u^b = \lim_{x \nearrow u^*} T_S(x)$ is a singularity of T_S^{-1}. Again, as S meets every vertical connection, the union $\Phi^L_{[0,t^L(u^t)]}(u^t) \cup \Phi^L_{[t^L(u^b),0]}(u^b)$ contains at most one point of Σ, and it is contained in $\Phi^R_{[0,t^R(u^t)]}(u^t) \cup \Phi^R_{[t^R(u^b),0]}(u^b)$.

□

End of proof of proposition : Let X be the union, over the components J of the domain pf T_S, of the $\Phi_{[0,t_J]}(J)$ (with t_J the return time to S on J) ; let $\widehat{X}$ be the union of X and $X^L = X^R$. As $X^L = X^R, \widehat{X} \cap (M - \Sigma)$ is open in $M - \Sigma$. There are now two possibilities.

- **(a):** the return map T_S does not coincide with the identity in the neighbourhood of either endpoint of S. Then, the set $\widehat{X}$ is easily seen to be also closed in M. Therefore $\widehat{X} = M$ and every infinite half-orbit of the vertical vectorfield meets S.
- **(b):** The return map T_S coincides with the identity in the neighbourhood of at least one of the endpoints of S. Let Y be the cylinder containing the corresponding periodic orbits. As the boundary of Y is made of vertical connections and points of Σ, it is contained in $\widehat{X}$. Then $\widehat{X} \cup Y$ must be equal to M and the second possibility in the statement of the proposition holds.

□

COROLLARY 5.5. *If the vertical vectorfield on a translation surface has no connection, it is minimal : every infinite half orbit is dense.*

PROOF. Otherwise there exists an open bounded horizontal segment S which does not meet every infinite vertical half-orbit. By the proposition, there would exist a cylinder containing these orbits ; but this is also not possible, since the boundary of a cylinder contains a vertical connection. □

COROLLARY 5.6. *Let $(M, \Sigma, \kappa, \zeta)$ be a translation surface and S be an open bounded horizontal segment. Assume that*

(H1): *S meets every vertical connection (if any).*
(H2): *The left endpoint of S is in Σ.*
(H3): *The right endpoint of S either belongs to Σ, or to a vertical separatrix segment which does not meet S.*

Then the translation surface is isomorphic to the one constructed from the return map T_S by the zippered rectangle construction with appropriate suspension data.

PROOF. Applying the Proposition 5.3, we see that the second possibility in the statement of the proposition is forbidden by the hypothesis $(H2)$ and therefore S meets every infinite half-orbit of the vertical vectorfield. Therefore, every ingoing separatrix of the vertical vectorfield meets S ; the intersection point which is closest (on the separatrix) to the marked point is a singularity of T_S and we obtain in this way a one-to-one map between ingoing separatrices of the vertical vectorfield and singularities of T_S ; in the same way, there is a natural one-to-one correspondence between outgoing separatrices and singularities of T_S^{-1}. The vertical lengths of the corresponding separatrices segments determine the suspension data. It is now a direct verification, which we leave to the reader, to check that our translation surface is indeed isomorphic to the one obtained from these suspension data by the zippered rectangle construction. □

PROPOSITION 5.7. *Let $(M, \Sigma, \kappa, \zeta)$ be a translation surface and let S_∞ be an outgoing separatrix of the horizontal vectorfield. If either the horizontal or the vertical vectorfield has no connection, then some initial segment S of S_∞ satisfies the hypotheses $(H1), (H2), (H3)$ of Corollary 5.6*

PROOF. First assume that there is no vertical connection. Then any initial segment S of S_∞ satisfies $(H1)$ and $(H2)$.Let $\widetilde{S}$ be some initial segment of S_∞, and S' be some vertical separatrix ; as there is no vertical connection, S' is dense, and therefore intersects $\widetilde{S}$. Let B be the intersection point closest along S' to the point of Σ at the end of S' ; the initial segment S of S_∞ with right endpoint B satisfies $(H1)$, $(H2)$ and $(H3)$.

Assume now that there is no horizontal connection. Then S_∞ is dense. As there are only finitely many vertical connections, every initial segment S of S_∞ which is long enough satisfies $(H1)$, and also $(H2)$. Let S' be a short enough vertical separatrix segment ; if the initial segment $\widetilde{S}$ of S_∞ is long enough it will intersect S', but only after having met all vertical connections ; again we cut $\widetilde{S}$ at the intersection point with S' which is closest to the marked point at the end of S'. We get an initial segment S of S_∞ which satisfies $(H1)$, $(H2)$ and $(H3)$. □

5.3. We reformulate Corollary 5.5 in the context of i.e.m.

DEFINITION 5.8. A connection for an i.e.m. T on an interval I is a triple (m, u^t, u^b) where m is a non negative integer, u^t is a singularity of T, u^b is a singularity of T^{-1}, such that

$$T^m(u^b) = u^t \ .$$

THEOREM 5.9. *(Keane* [**Kea1**]*) If an i.e.m. has no connection, it is minimal : every half-orbit is dense.*

PROOF. Choose suspension data, construct a translation surface by the zippered rectangle construction ; the vertical vectorfield has no connection because the i.e.m. does not have either ; thus it is minimal and the same holds for the i.e.m. □

5.4. In this context, the following result of Keane is also relevant.

PROPOSITION 5.10. *If the coordinates of the length vector of an i.e.m. are rationally independent, it has no connection.*

PROOF. Choose suspension data, construct a translation surface by the zippered rectangle construction. We use the notations of 4.5. If the i.e.m. had a connection, the vertical vectorfield on the translation surface would have a connection which we could express as a linear combination $\Sigma\, n_\alpha[\zeta_\alpha]$ in $H_1(M, \Sigma, \mathbb{Z})$ with integer coefficients. Integrating against the holomorphic 1-form, we have $\Sigma\, n_\alpha\, \lambda_\alpha = 0$ but $\Sigma\, n_\alpha \tau_\alpha \neq 0$, a contradiction. □

EXERCISE 5.11. For $d = 2$, T is minimal iff there is no connection, anf iff the lengths of the intervals are rationally independent. For $d \geqslant 3$, show that there exists T minimal but having a connection, and also T with no connection but lengths data rationally dependent.

6. The Teichmüller space and the Moduli space

6.1. The Teichmüller space. Let M be a compact orientable topological surface, Σ a finite non-empty subset, κ a set of ramification indices.

We denote by $\mathrm{Diff}(M, \Sigma)$ the group of homeomorphisms of M fixing each point of Σ, by $\mathrm{Diff}^+(M, \Sigma)$ the subgroup of index 2 formed of orientation preserving homeomorphisms, by $\mathrm{Diff}_0(M, \Sigma)$ the neutral component of $\mathrm{Diff}(M, \Sigma)$, by $\mathrm{Mod}(M, \Sigma)$ the **modular group** (or mapping class group) $\mathrm{Diff}(M, \Sigma)/\mathrm{Diff}_0(M, \Sigma)$, and by $\mathrm{Mod}^+(M, \Sigma)$ the subgroup (of index 2) $\mathrm{Diff}^+(M, \Sigma)/\mathrm{Diff}_0\,(M, \Sigma)$.

The group $\mathrm{Diff}(M, \Sigma)$ acts on the set of translation surface structures on (M, Σ, κ): if $\zeta = (\varphi_\alpha)$ is an atlas defining such a structure, $f_*\zeta$ is the atlas $(\varphi_\alpha \circ f^{-1})$ (for $f \in \mathrm{Diff}(M, \Sigma)$).

DEFINITION 6.1. The Teichmüller space $Q(M, \Sigma, \kappa)$ is the set of orbits of the action of $\mathrm{Diff}_0(M, \Sigma)$ on the set of translation surface structures on (M, Σ, κ).

6.2. Topology on $Q(M, \Sigma, \kappa)$. We will fix once and for all a universal cover

$$p : (\widetilde{M}, *) \to (M, A_1)$$

where A_1 is the first point of Σ.

Given a translation surface structure ζ on (M, Σ, κ), we define an associated **developing map**

$$D_\zeta : (\widetilde{M}, *) \to (\mathbb{C}, 0)$$

by integrating from $*$ the 1-form $p^*\omega$, where ω is the holomorphic 1-form determined by ζ.

Conversely, the developing map determines ζ. The set of translation surface structures on (M, ζ, κ) can therefore be considered as a subset of $C(\widetilde{M}, \mathbb{C})$; we equip this set with the compact-open topology, the set of translation surface structures with the induced topology, and the Teichmüller space $Q(M, \Sigma, \kappa)$ with the quotient topology.

6.3. The period map. Let ζ be a translation surface structure on (M, Σ, κ), ω be the associated holomorphic 1-form, γ a relative homology class in $H_1(M, \Sigma, \mathbb{Z})$. As ω is closed, the integral $\int_\gamma \omega$ is well-defined. Moreover, if f is an homeomorphism in $\mathrm{Diff}_0(M, \Sigma)$, f acts trivially on $H_1(M, \Sigma, \mathbb{Z})$, therefore the map

$$\zeta \mapsto (\gamma \to \int_\gamma \omega)$$

is constant on orbits of $\mathrm{Diff}_0(M, \Sigma)$ and defines a map

$$\Theta : Q(M, \Sigma, \kappa) \to \mathrm{Hom}(H_1(M, \Sigma, \mathbb{Z}), \mathbb{C})$$

called the **period map**. Here, we will generally identify in the right-hand side $\mathrm{Hom}(H_1(M, \Sigma, \mathbb{Z}), \mathbb{C})$ with the cohomology group $H^1(M, \Sigma, \mathbb{C})$. The importance of the period map lies in the following property.

PROPOSITION 6.2. *The period map is a local homeomorphism.*

The proposition will be proved in section 6.5

6.4. Action of $GL(2, \mathbb{R})$ on Teichmüller space. Let $\zeta = (\varphi_\alpha)$ be an atlas defining a translation surface structure on (M, Σ, κ), and let g be an element of $GL(2, \mathbb{R})$ acting on $\mathbb{R}^2 \simeq \mathbb{C}$.

Consider the atlas $g_*\zeta = (g \circ \varphi_\alpha)$; because the conjugacy of a translation by an element of $GL(2, \mathbb{R})$ is still a translation, the atlas $g_*\zeta$ defines another translation surface structure on (M, Σ, κ) and we have thus a left action of $GL(2, \mathbb{R})$ on the space of translation surface structures.

It is clear that this action commutes with the action of the group $\mathrm{Diff}(M, \Sigma)$. In particular, it defines a left action of $GL(2, \mathbb{R})$ on the Teichmüller space $Q(M, \Sigma, \kappa)$. One easily checks that this action is continuous.

Regarding the period map Θ, the group $GL(2, \mathbb{R})$ acts on the right-hand side $\mathrm{Hom}(H_1(M, \Sigma, \mathbb{Z}), \mathbb{C})$ by acting on the target $\mathbb{C} = \mathbb{R}^2$. The period map is then covariant with respect to the actions of $GL(2, \mathbb{R})$ on the source and the image.

It is to be noted that the subgroup $SO(2, \mathbb{R})$ preserves some of the auxiliary structures associated to a translation surface structure : the complex structure is invariant, the holomorphic 1-form is replaced by a multiple of modulus 1, the flat metric is preserved as is the associated area. The group $SO(2, \mathbb{R})$ acts transitively on the set of constant unitary vectorfields ; therefore, every result proved for the vertical vectorfield is valid for a non constant unitary vectorfield. Actually, if we use the full action of $GL(2, \mathbb{R})$, we see that in section 5 we can replace the vertical and horizontal vectorfield by any two non-proportional constant vectorfields on the translation surface.

6.5. Proof of proposition 6.2.

PROOF. We first observe that the period map is continuous : this follows immediately from the definition of the topology on Teichmüller space. To study the properties of Θ in the neighbourhood of a point $[\zeta]$ in $Q(M, \Sigma, \kappa)$, we may assume that the translation structure ζ has no vertical connection ; otherwise, we could replace ζ by $R_*\zeta$ for some appropriate $R \in SO(2, \mathbb{R})$ and use the covariance of the period map.

Then we know that the translation surface structure ζ can be obtained by the zippered rectangle construction from some i.e.m. T on some interval I.

Because the conditions on the length data λ and the suspension data τ in the zippered rectangle construction are open, the period map, expressed locally by (λ, τ), is locally onto. It remains to be seen that the period map is locally injective, with continuous inverse.

In the zippered rectangle construction, we will always assume (by choosing the horizontal separatrix S_∞ appropriately in proposition 5.6) that the first marked point A_1 of Σ is the left endpoint of the interval I. The surface M was obtained in section 4.3 from some explicitly defined subset $\widehat{M}$ of $\mathbb{C}$, depending only on π, λ and τ. We can lift $\widehat{M}$ to a (connected) subset $\widehat{M}_\zeta$ of $\widetilde{M}$ (with the left endpoint of I lifted to $*$) with the property that the developing map D_ζ is an homeomorphism from $\widehat{M}_\zeta$ onto $\widehat{M}$.

If ζ_0, ζ_1 are two translation surface structures close to ζ with the same image by the period map, the subset $\widehat{M}$ of $\mathbb{C}$ will be the same for ζ_0 and ζ_1. There will be a unique homeomorphism $h : \widehat{M}_{\zeta_0} \to \widehat{M}_{\zeta_1}$ such that $D_{\zeta_0} = D_{\zeta_1} \circ h$ on $\widehat{M}_{\zeta_0}$. It is easily checked that h extends uniquely as a homeomorphism of $(\widetilde{M}, *)$ still satisfying $D_{\zeta_0} = D_{\zeta_1} \circ h$, and that extension is the lift of an homeomorphism of M. This proves that $[\zeta_0] = [\zeta_1]$ in Teichmüller space. This proves local injectivity of the period map ; the continuity of local inverses is proven along the same lines and left to the reader. □

6.6. Geometric structures on Teichmüller space. First, we can use the locally injective restrictions of the period map as charts defining a structure of complex manifold of complex dimension $d = 2g + s - 1$.

This complex manifold will also be equipped with a canonical volume form. Indeed, we can normalize Lebesgue measure on $\mathrm{Hom}(H_1(M, \Sigma, \mathbb{Z}), \mathbb{R}^2)$ by asking that the lattice $\mathrm{Hom}(H_1(M, \Sigma, \mathbb{Z}), \mathbb{Z}^2)$ has covolume 1. We then lift by the period map this canonical volume to Teichmüller space.

6.7. Examples and remarks. Let us consider the case $g = s = 1$ of the torus T with a single marked point $\{A_1\}$. Fix a basis $[\zeta_1], [\zeta_2]$ for the homology group $H_1(T, \{A_1\}, \mathbb{Z})$.

In this case, the period map is injective and allows to identify the Teichmüller space with its image. The image of the period map is

$$Q(T, \{A_1\}, 1) = \{(\zeta_1, \zeta_2) \in (\mathbb{C}^*)^2, \zeta_2/\zeta_1 \notin \mathbb{R}\} .$$

The two components of $Q(T, \{A_1\}, 1)$ correspond to the two possible orientations. Restricting to $\mathrm{Im}\ \zeta_2/\zeta_1 > 0$, the map $(\zeta_1, \zeta_2) \to \zeta_2/\zeta_1$ presents $Q(T, \{A_1\}, 1)$ as fibered over the upper half-plane $\mathbb{H}$ (representing the classical Teichmüller space of T) with fiber $\mathbb{C}^*$ (representing the choice of a non-zero holomorphic 1-form).

REMARK 6.3. For $g \geqslant 2$, the period map is not injective. Indeed, let γ be a loop on M which is homologous but not homotopic to 0. We assume that $\gamma \cap \Sigma = \emptyset$. Let then f be a Dehn twist along γ ; this can be constructed fixing each point of Σ and thus defining an element of $\mathrm{Diff}(M, \Sigma)$.

If ζ is any translation surface structure on (M, Σ, κ), $f_*\zeta$ and ζ will have the same image by the period map because f induces the identity on $H_1(M, \Sigma, \mathbb{Z})$. On the other hand, ζ and $f_*\zeta$ represent different points in Teichmüller space : indeed, we will see that $[f_*^n\zeta]$ goes to ∞ in Teichmüller space as n goes to $\pm\infty$ in $\mathbb{Z}$.

REMARK 6.4. Regarding the relation to "classical" Teichmüller theory classifying the complex structures on compact surfaces, consider the two extremal cases.

Take first $s = 2g-2, \kappa_1 = \kappa_2 = \cdots = \kappa_s = 2$; this means that the holomorphic 1-form associated with the translation surface structure has only simple zeros, the generic situation for an holomorphic 1-form. The Teichmüller space $Q(M, \Sigma, \kappa)$ of dimension $2g + s - 1 = 4g - 3$ is fibered over the "classical" Teichmüller space of dimension $3g - 3$; the fiber of dimension g corresponds to the choice of the holomorphic 1-form (which form a g-dimensional vector space; however, one has to exclude the zero form and those having multiple zeros).

Consider now the case $s = 1$, $\kappa_1 = 2g - 2$; this means that the holomorphic 1-form has a single zero of maximal multiplicity ; when $g \geqslant 3$, not all Riemann surfaces of genus g admit such an holomorphic 1-form. Indeed the Teichmüller space has dimension $2g+s-1 = 2g$ and the scaling of the holomorphic 1-form corresponds to 1 dimension, hence $Q(M, \Sigma, \kappa)$ is fibered over a subvariety of "classical" Teichmüller space of codimension $\geqslant g - 2$.

6.8. Normalizations.

6.8.1. *Normalization of orientation.* It is generally convenient to fix an orientation of the orientable topological surface M and then to consider only those translation surface structures ζ on (M, Σ, κ) which are compatible with the given orientation. The groups $\mathrm{Diff}^+(M, \Sigma)$ and $GL^+(2, \mathbb{R})$ act on this subset. We denote by $Q^+(M, \Sigma, \kappa)$ the corresponding subset of Teichmüller space.

6.8.2. *Normalization of area.* Given a translation surface structure ζ compatible with a chosen orientation, let $A(\zeta)$ be the surface of M for the area-form (on $M - \Sigma$) induced by ζ. It is clear that the function A is invariant under the action of $\mathrm{Diff}^+(M, \Sigma)$ and therefore induces a function still denoted by A on the Teichmüller space $Q^+(M, \Sigma, \kappa)$.

We will write $Q^{(1)}(M, \Sigma, \kappa)$ for the locus $\{A = 1\}$ in $Q^+(M, \Sigma, \kappa)$). As A is a smooth submersion, $Q^{(1)}(M, \Sigma, \kappa)$ is a codimension 1 real-analytic submanifold of $Q^+(M, \Sigma, \kappa)$.

If $[\zeta] \in Q^+(M, \Sigma, \kappa)$ and $g \in GL^+(2, \mathbb{R})$, we have

$$A(g_*[\zeta]) = \det g\, A([\zeta]) \ .$$

In particular, $Q^{(1)}(M, \Sigma, \kappa)$ is invariant under the action of $\mathrm{Diff}^+(M, \Sigma)$ and $SL(2, \mathbb{R})$.

Let μ be the canonical volume form on $Q^+(M, \Sigma, \kappa)$. We write

$$\mu = \mu_1 \wedge \frac{dA}{A} \ ;$$

then μ_1 induces on $Q^{(1)}(M, \Sigma, \kappa)$ a canonical volume form which is invariant under the action of $\mathrm{Diff}^+(M, \Sigma)$ and $SL(2, \mathbb{R})$.

6.9. he moduli space. The discrete group $\mathrm{Mod}(M, \Sigma)$ acts continuously on the Teichmüller space $Q(M, \Sigma, \kappa)$.

DEFINITION 6.5. The moduli space is the quotient

$$\mathcal{M}(M, \Sigma, \kappa) := Q(M, \Sigma, \kappa) \ / \ \mathrm{Mod}(M, \Sigma) \ .$$

The normalized moduli space is the quotient

$$\mathcal{M}^{(1)}(M, \Sigma, \kappa) := Q^{(1)}(M, \Sigma, \kappa) \ / \ \mathrm{Mod}^+(M, \Sigma) \ .$$

The action of the modular group $\mathrm{Mod}(M, \Sigma)$ on $Q(M, \Sigma, \kappa)$ is proper but not always free, as we explain below. This means that the moduli space is an orbifold (locally the quotient of a manifold by a finite group) but not (always) a manifold.

To see that the action is proper, consider as above a universal cover $p : (\widetilde{M}, *) \to (M, A_1)$. Let $\widetilde{\Sigma} = p^{-1}(\Sigma)$. Given a translation surface structure ζ, we can lift the flat metric defined by ζ to $\widetilde{M}$ and consider the distance d_ζ on $\widetilde{\Sigma}$ induced by this metric (as length of shortest path). It is clear that this distance only depends on the class of ζ in Teichmüller space. If ζ, ζ' are two translation surface structures, the distances $d_\zeta, d_{\zeta'}$ on $\widetilde{\zeta}$ are quasiisometric : there exists $C \geqslant 1$ such that

$$C^{-1} d_\zeta(B, B') \leqslant d_{\zeta'}(B, B') \leqslant C\, d_\zeta(B, B')$$

for all $B, B' \in \widetilde{\Sigma}$. We write $C(\zeta, \zeta')$ for the best constant C.

EXERCISE 6.6. Prove that a subset $X \subset Q(M, \Sigma, \kappa)$ is relatively compact iff (given any $\zeta_0 \in Q(M, \Sigma, \kappa)$) the quantities $C(\zeta, \zeta_0), \zeta \in X$, are bounded.

The distances d_ζ have the property that any ball of finite radius only contain finitely many points. On the other hand, the modular group $\mathrm{Mod}(M, \Sigma)$ acts on $\widetilde{\Sigma}$, and there exists a finite subset $\widetilde{\Sigma}_0$ of $\widetilde{\Sigma}$ such that, for any finite subset $\widetilde{\Sigma}_1$ of $\widetilde{\Sigma}$, the set $\{g \in \mathrm{Mod}(M, \Sigma), g(\widetilde{\Sigma}_0) \subset \widetilde{\Sigma}_1\}$ is finite. Using the compactness criterion given by the exercise, it is easy to conclude that the action is proper.

To see that the action is not always free, it is sufficient to construct a translation surface with a non trivial group of automorphisms. Start with an integer $k \geqslant 2$ and k copies of the same translation torus T with two marked points A, B. Denote by T_i, A_i, B_i the ith copy, $1 \leqslant i \leqslant k$. Slit T_i along a geodesic segment A_iB_i (the same for all i). For each i, glue isometrically the left side of A_iB_i in T_i to the right side of $A_{i+1}B_{i+1}$ (with $(T_{k+1}, A_{k+1}, B_{k+1}) = (T_1, A_1, B_1)$). One obtains a translation surface of genus k with 2 marked points of ramification index k and an obvious automorphism group cyclic of order k.

6.10. Marked translation surfaces and marked moduli space. From the point of view of the zippered rectangles construction, it is more convenient to consider translation surfaces with an additional marking.

Indeed, if the construction starts from an i.e.m. T on an interval I, we have said above that we always take the left endpoint of I as the first marked point A_1 of the set Σ on the surface M. But the interval I itself appears on the surface as an outgoing separatrix of the horizontal vector field.

DEFINITION 6.7. A marked translation surface is a translation surface $(M, \Sigma, \kappa, \zeta)$ with a marked outgoing horizontal separatrix coming out of A_1.

Obviously, we require that an isomorphism between marked translation surfaces should respect the marked horizontal separatrices. We can then define a Teichmüller space $\widetilde{Q}(M, \Sigma, \kappa)$ of marked translation surfaces. It is a κ_1-fold cover of $Q(M, \Sigma, \kappa)$, because there are κ_1 possible choices for an horizontal separatrix out of A_1. In particular, when $\kappa_1 = 1$, the marking is automatic and $\widetilde{Q}(M, \Sigma, \kappa) = Q(M, \Sigma, \kappa)$.

On the other hand, it is quite obvious that a marked translation surface cannot have an automorphism distinct from the identity. Therefore, the modular group $\mathrm{Mod}(M, \Sigma)$ acts freely on $\widetilde{Q}(M, \Sigma, \kappa)$ and the quotient space, that we denote by $\widetilde{\mathcal{M}}(M, \Sigma, \kappa)$, is now a complex manifold. This moduli space is a κ_1-fold ramified covering of the moduli space $\mathcal{M}(M, \Sigma, \kappa)$. Normalizing orientation and area gives a codimension 1 real-analytic submanifold $\widetilde{\mathcal{M}}^{(1)}(M, \Sigma, \kappa)$.

6.11. In the following sections, we will present the proofs of the following results, obtained independently by H. Masur [**Ma**] and W. Veech [**Ve2**].

THEOREM 6.8. *Almost all i.e.m. are uniquely ergodic.*

The combinatorial data are here fixed and "almost all" refer to the choice of length data according to Lebesgue measure.

THEOREM 6.9. *The normalized moduli space $\widetilde{\mathcal{M}}^1(M, \Sigma, \kappa)$ has finite volume. The action of the group $SL(2, \mathbb{R})$ on it is ergodic.*

We will follow the approach of W. Veech [**Ve5**]. The **Teichmüller flow** on the moduli space $\widetilde{\mathcal{M}}^{(1)}(M, \Sigma, \kappa)$ is the restriction of the action of $SL(2, \mathbb{R})$ to the 1-parameter diagonal subgroup $\begin{pmatrix} e^t & 0 \\ 0 & e^{-t} \end{pmatrix}$. The ergodicity of the action will follow from the ergodicity of this flow (stronger properties of this flow will be presented in later sections).

Let us consider what happens in the simple case $g = s = 1$. Then, the normalized Teichmüller space is $Q^{(1)}(M, \Sigma, \kappa) = SL(2, \mathbb{R})$, the modular group $\text{Mod}^+(M, \Sigma)$ is $SL(2, \mathbb{Z})$, the normalized moduli space is the space of normalized lattices $SL(2, \mathbb{R}) \,/\, SL(2, \mathbb{Z})$ which has unit area and on which $SL(2, \mathbb{R})$ obviously acts transitively. The Teichmüller flow is essentially the geodesic flow on the modular surface. It is well known that this flow is closely related to the classical continuous fraction algorithm. G. Rauzy and W. Veech, introduced a renormalization algorithm for i.e.m., later refined by A. Zorich, which plays the role of the classical continuous fraction algorithm for more than 2 intervals. This will be the subject of the next sections.

7. The Rauzy-Veech algorithm

7.1. The aim of the Rauzy-Veech algorithm [**Rau, Ve1, Ve2**], to be defined below, is to understand the dynamics of an i.e.m. by looking at the return map on shorter and shorter intervals. What makes this general "renormalization" method available is the fact that the return maps are still i.e.m. with bounded combinatorial complexity : actually, by choosing the small intervals carefully, they have the same number of singularities than the i.e.m. we started with.

7.2. Definition of one step of the algorithm. Let T be an i.e.m. on an interval I, with irreducible combinatorial data $(\mathcal{A}, \pi_t, \pi_b)$. Let $d = \#\mathcal{A}$; let $u_1^t < \cdots < u_{d-1}^t$ be the singularities of T, $u_1^b < \cdots < u_{d-1}^b$ be the singularities of T^{-1}.

The step of the algorithm is defined for T if $u_{d-1}^t \neq u_{d-1}^b$. Observe that if $u_{d-1}^t = u_{d-1}^t$, then $(0, u_{d-1}^t, u_{d-1}^b)$ is a **connection** for T (see Subsection 5.3).

When $u_{d-1}^t \neq u_{d-1}^b$, we define $\widetilde{I}$ to be the open interval with the same left endpoint than I and right endpoint equal to $\max(u_{d-1}^t, u_{d-1}^b)$.

Let $\widetilde{T}$ be the return map of T to $\widetilde{I}$. To understand $\widetilde{T}$, let us introduce the letters α_t, α_b satisfying $\pi_t(\alpha_t) = \pi_b(\alpha_b) = d$ which correspond to the intervals at the right of I before and after applying T. The hypothesis $u_{d-1}^t \neq u_{d-1}^b$ corresponds to $\lambda_{\alpha_t} \neq \lambda_{\alpha_b}$. We distinguish two cases.

1) $u^b_{d-1} > u^t_{d-1} \Longleftrightarrow \lambda_{\alpha_t} > \lambda_{\alpha_b}$: We say that α_t is the **winner** and α_b is the **loser** of this step of the algorithm, and that the step is of **top** type. We have in this case

$$\widetilde{T}(x) = \begin{cases} T(x) & \text{if} \quad x \notin I^t_{\alpha_b} \\ T^2(x) & \text{if} \quad x \in I^t_{\alpha_b} \end{cases}$$

We use the same alphabet to label the intervals of $\widetilde{T}$; we define :

$$\widetilde{I}^t_\alpha = I^t_\alpha \quad \text{for} \quad \alpha \neq \alpha_t ,$$

$$\widetilde{I}^t_{\alpha_t} = I^t_{\alpha_t} \cap \widetilde{I} = (u^t_{d-1}, u^b_{d-1}) ,$$

$$\widetilde{I}^b_\alpha = I^b_\alpha \quad \text{for} \quad \alpha \neq \alpha_b, \alpha_t ,$$

$$\widetilde{I}^b_{\alpha_b} = T(I^b_{\alpha_b}) ,$$

$$\widetilde{I}^b_{\alpha_t} = I^b_{\alpha_t} / \widetilde{I}^b_{\alpha_b} .$$

The new length data are given by

$$\widetilde{\lambda}_\alpha = \begin{cases} \lambda_\alpha & \text{if} \quad \alpha \neq \alpha_t \\ \lambda_{\alpha_t} - \lambda_{\alpha_b} & \text{if} \quad \alpha = \alpha_t . \end{cases}$$

The new combinatorial data are given by

$$\widetilde{\pi}_t = \pi_t ;$$

$$\widetilde{\pi}_b(\alpha) = \begin{cases} \pi_b(\alpha) & \text{if} \quad \pi_b(\alpha) \leqslant \pi_b(\alpha_t), \\ \pi_b(\alpha_t) + 1 & \text{if} \quad \alpha = \alpha_b , \\ \pi_b(\alpha) + 1 & \text{if} \quad \pi_b(\alpha_t) < \pi_b(\alpha) < d . \end{cases}$$

2) $u^t_{d-1} > u^b_{d-1} \Longleftrightarrow \lambda_{\alpha_b} > \lambda_{\alpha_t}$: We now say that α_b is the winner, α_t the loser, and the step is of **bottom** type. We have

$$\widetilde{T}^{-1}(x) = \begin{cases} T^{-1}(x) & \text{if} \quad x \notin I^b_{\alpha^t} \\ T^{-2}(x) & \text{if} \quad x \in I^b_{\alpha^t} \end{cases}$$

(we could also write the formulas for $\widetilde{T}$; we prefer to write them for $\widetilde{T}^{-1}$ in order to keep more obvious the bottom/top time symmetry of the setting). The new labelling is

$$\widetilde{I}^b_\alpha = I^b_\alpha \quad \text{for} \quad \alpha \neq \alpha_b ,$$

$$\widetilde{I}^b_{\alpha_b} = I^b_{\alpha_b} \cap \widetilde{I} = (u^b_{d-1}, u^t_{d-1}) ,$$

$$\widetilde{I}^t_\alpha = I^t_\alpha \quad \text{for} \quad \alpha \neq \alpha_t, \alpha_b ,$$

$$\widetilde{I}^t_{\alpha_t} = T^{-1}(I^t_{\alpha_t}) ,$$

$$\widetilde{I}^t_{\alpha_b} = I^t_{\alpha_b} / \widetilde{I}^t_{\alpha_t} .$$

The new length data are given by

$$\widetilde{\lambda}_\alpha = \begin{cases} \lambda_\alpha & \text{if} \quad \alpha \neq \alpha_b \\ \lambda_{\alpha_b} - \lambda_{\alpha_t} & \text{if} \quad \alpha = \alpha_b \,. \end{cases}$$

The new combinatorial data are given by

$$\widetilde{\pi}_b = \pi_b \ ;$$

$$\widetilde{\pi}_t(\alpha) = \begin{cases} \pi_t(\alpha) & \text{if} \quad \pi_t(\alpha) \leqslant \pi_t(\alpha_t), \\ \pi_t(\alpha_b) + 1 & \text{if} \quad \alpha = \alpha_t \ , \\ \pi_t(\alpha) + 1 & \text{if} \quad \pi_t(\alpha_b) < \pi_t(\alpha) < d \,. \end{cases}$$

EXERCISE 7.1. Show that the combinatorial data $(\widetilde{\pi}_t, \widetilde{\pi}_b)$ for $\widetilde{T}$ are irreducible.

EXERCISE 7.2. Show that if T has no connection, then $\widetilde{T}$ also has no connection.

This means that for i.e.m. with no connections, it is possible to iterate indefinitely the algorithm ; the converse is also true, see below.

EXERCISE 7.3. Check that the return map of T on an interval I' with the same left endpoint than I and $|\widetilde{I}| < |I'| < |I|$ is an i.e.m with $d+1$ intervals.

In the case of 2 intervals, there is only one possible set of irreducible combinatorial data and the algorithm is given by

$$(\lambda_A, \lambda_B) \longmapsto \begin{cases} (\lambda_A - \lambda_B, \lambda_B) & \text{if} \quad \lambda_A > \lambda_B, \\ (\lambda_A, \lambda_B - \lambda_A) & \text{if} \quad \lambda_B > \lambda_A \,. \end{cases}$$

the iteration of which gives the classical continued fraction algorithm.

7.3. Rauzy diagrams. Let $\mathcal{A}$ be an alphabet. For irreducible combinatorial data $\pi = (\pi_t, \pi_b)$, we have defined in the last section new combinatorial data $\widetilde{\pi} = (\widetilde{\pi}_t, \widetilde{\pi}_b)$ depending only on (π_t, π_b) and the type (top or bottom) of the step ; we write $\widetilde{\pi} = R_t(\pi)$ or $\widetilde{\pi} = R_b(\pi)$ accordingly.

A **Rauzy class** on the alphabet $\mathcal{A}$ is a set of irreducible combinatorial data $\pi = (\pi_t, \pi_s)$ which is invariant under both R_t and R_b and minimal with this property. The associated **Rauzy diagram** has the elements of this set as vertices. The arrows of the diagram join a vertex to its images by R_t and R_b and are of **top** and **bottom** type accordingly.

The **winner** of an arrow of top type (resp. bottom type) starting at (π_t, π_b) is the letter α_t (resp. α_b) such that $\pi_t(\alpha_t) = d$ (resp. $\pi_b(\alpha_b) = d$). The **loser** is the letter α_b (resp. α_t) such that $\pi_b(\alpha_b) = d$ (resp. $\pi_t(\alpha_t) = d$).

EXERCISE 7.4. Show that the maps R_t, R_b are invertible and that each vertex is therefore the endpoint of exactly one arrow of top type and an arrow of bottom type.

EXERCISE 7.5. Let γ, γ' be arrows in a Rauzy diagram of the same type such that the endpoint of γ is the starting point of γ' ; show that γ, γ' have the same winner.

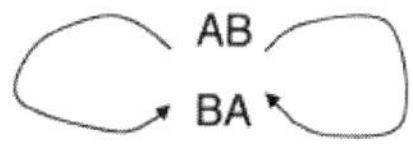

g=1, s=1,d=2

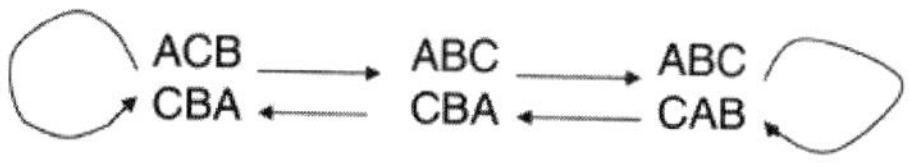

g=1,s=2,d=3

FIGURE 2

For $d = 2$ or 3, there is, up to equivalence, only one Rauzy diagram pictured (Figure 2).

For $d = 4$, there are two non-equivalent Rauzy diagrams pictured (Figure 3). They correspond respectively (see next section) to the cases $g = 2, s = 1$ and $g = 1, s = 3$.

7.4. The basic step for suspensions. Recall from section 4.1 that for combinatorial data $\pi = (\pi_t, \pi_b)$, suspension data $(\tau_\alpha)_{\alpha \in \mathcal{A}}$ must satisfy

$$(S_\pi) \qquad \sum_{\pi_t(\alpha)<k} \tau_\alpha > 0\,, \qquad \sum_{\pi_b(\alpha)<k} \tau_\alpha < 0 \qquad \text{for all } 1 < k \leqslant d\,.$$

We denote by Θ_π the convex open cone in $\mathbb{R}^{\mathcal{A}}$ defined by these inequalities. The main reason to consider Θ_π is the following property. Set $\widetilde{\pi} = R_t(\pi)$. Define also, for $\tau \in \mathbb{R}^{\mathcal{A}}$

$$\widetilde{\tau}_\alpha = \begin{cases} \tau_\alpha & \text{if} \quad \alpha \neq \alpha_t \\ \tau_{\alpha_t} - \tau_{\alpha_b} & \text{if} \quad \alpha = \alpha_t\,. \end{cases}$$

where $\pi_t(\alpha_t) = \pi_b(\alpha_b) = d\,.$

LEMMA 7.6. *The linear map $\tau \to \widetilde{\tau}$ sends Θ_π onto $\Theta_{\widetilde{\pi}} \cap \{\sum_\alpha \widetilde{\tau}_\alpha < 0\}$.*

There is a symmetric statement exchanging top and bottom.

PROOF. Let $\tau \in \Theta_\pi$. As $\widetilde{\pi}_t = \pi_t$, and $\widetilde{\tau}_\alpha = \tau_\alpha$ for $\pi_t(\alpha) < d$, the first half of the conditions for $(S_{\widetilde{\pi}})$ are satisfied. Let $\ell = \pi_b(\alpha_t)$; for $k \leqslant \ell$, we have

$$\sum_{\widetilde{\pi}_b(\alpha)<k} \widetilde{\tau}_\alpha = \sum_{\pi_b(\alpha)<k} \widetilde{\tau}_\alpha = \sum_{\pi_b(\alpha)<k} \tau_\alpha < 0\,.$$

Next we have

$$\sum_{\widetilde{\pi}_b(\alpha)\leqslant\ell} \widetilde{\tau}_\alpha = \sum_{\pi_b(\alpha)<\ell} \tau_\alpha + \tau_{\alpha_t} - \tau_{\alpha_b} = \sum_{\pi_b(\alpha)<\ell} \tau_\alpha - \sum_{\pi_t(\alpha)<d} \tau_\alpha + \sum_{\pi_b(\alpha)<d} \tau_\alpha < 0\,,$$

and for $\ell < k \leqslant d$

$$\sum_{\widetilde{\pi}_b(\alpha)\leqslant k} \widetilde{\tau}_\alpha = \sum_{\pi_b(\alpha)\leqslant k-1} \tau_\alpha < 0\,.$$

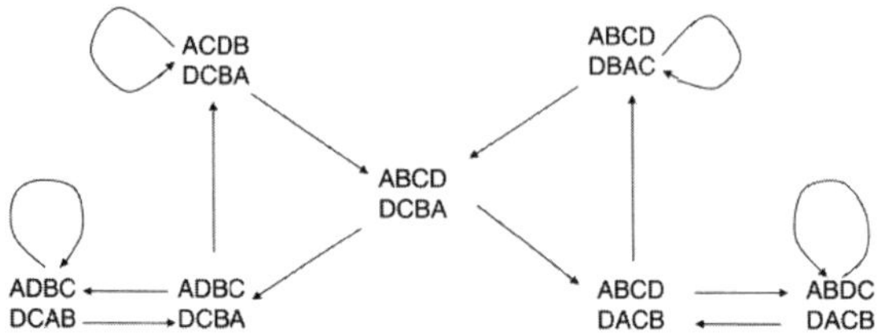

g=2,s=1,d=4

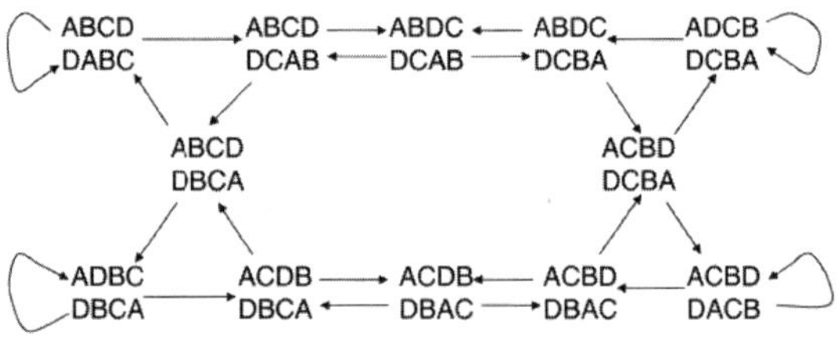

g=1,s=3,d=4

FIGURE 3

Conversely, let $\widetilde{\tau} \in \Theta_{\widetilde{\pi}} \cap \{\sum \widetilde{\tau}_\alpha < 0\}$. Again the first half of (S_π) is satisfied. For the second half, we have

$$\sum_{\pi_b(\alpha)<k} \tau_\alpha = \begin{cases} \sum_{\widetilde{\pi}_b(\alpha)<k} \tau_\alpha & \text{if } \ 1 < k \leqslant \ell, \\ \sum_{\widetilde{\pi}_b(\alpha)\leqslant k} \tau_\alpha & \text{if } \ \ell < k \leqslant d\,. \end{cases}$$

Thus, condition (S_π) is satisfied. □

Let then T be an i.e.m. on an interval I, with (irreducible) combinatorial data $\pi = (\pi_t, \pi_b)$ on an alphabet $\mathcal{A}$. Assume that the condition $\lambda_{\alpha_t} \neq \lambda_{\alpha_b}$ (with $\pi_t(\alpha_t) = \pi_b(\alpha_b) = d$) for one step of the algorithm is satisfied. Let $\tau \in \Theta_\pi$ be suspension data satisfying the required conditions (S_π).

If the step is of top type, we define

$$\widetilde{\tau}_\alpha = \begin{cases} \tau_\alpha & \text{if } \ \alpha \neq \alpha_t \\ \tau_{\alpha_t} - \tau_{\alpha_b} & \text{if } \ \alpha = \alpha_t\,. \end{cases}$$

If the step is of bottom type, we define

$$\widetilde{\tau}_\alpha = \begin{cases} \tau_\alpha & \text{if } \ \alpha \neq \alpha_b \\ \tau_{\alpha_b} - \tau_{\alpha_t} & \text{if } \ \alpha = \alpha_b\,. \end{cases}$$

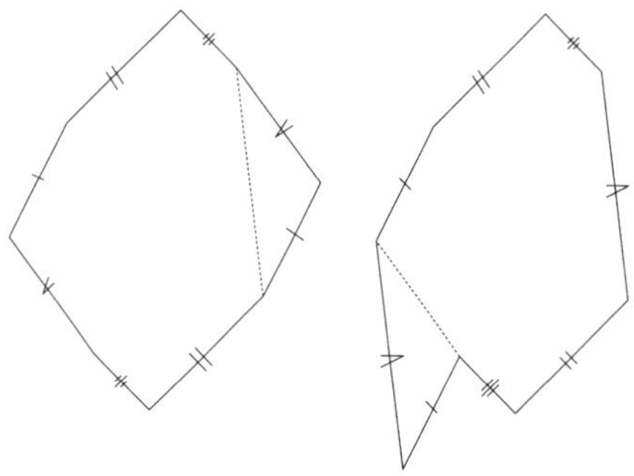

FIGURE 4. $M(\pi, \lambda, \tau)$ $\quad M(\widetilde{\pi}, \widetilde{\lambda}, \widetilde{\tau})$

(The formulas are the same than for the length data).

We have explained in Section 4 how to construct a translation surface $M(\pi, \lambda, \tau)$ from the given data by the zippered rectangle construction. Writing $\widetilde{\pi} = R_t(\pi)$ or $R_b(\pi)$ according to the type of the step and writing $\widetilde{\lambda}$ for the length data of $\widetilde{T}$ as above, we construct another translation surface $M(\widetilde{\pi}, \widetilde{\lambda}, \widetilde{\tau})$ from these new data.

An easily checked but fundamental observation is that $M(\pi, \lambda, \tau)$ and $M(\widetilde{\pi}, \widetilde{\lambda}, \widetilde{\tau})$ are **canonically isomorphic**. This is best seen by contemplating the picture, Figure 4.

The canonical bases of the homology groups $H_1(M, \Sigma, \mathbb{Z}), H_1(M - \Sigma, \mathbb{Z})$ are related as follows : If α_0 is the winner and α_1 is the loser of the step of the algorithm, one has, with the notations of Section 4.5,

$$\begin{aligned}
[\widetilde{\zeta}_\alpha] &= [\zeta_\alpha] && \text{if } \alpha \neq \alpha_0 \,, \\
[\widetilde{\zeta}_{\alpha_0}] &= [\zeta_{\alpha_0}] - [\zeta_{\alpha_1}] \,, \\
[\widetilde{\theta}_\alpha] &= [\theta_\alpha] && \text{if } \alpha \neq \alpha_1 \,, \\
[\widetilde{\theta}_{\alpha_1}] &= [\theta_{\alpha_1}] + [\theta_{\alpha_0}] \,.
\end{aligned}$$

7.5. Formalism for the iteration of the algorithm. Given an i.e.m. T_0 on an interval $I^{(0)}$ with no connection and irreducible combinatorial data $(\mathcal{A}, \pi^{(0)})$, the iteration of the Rauzy-Veech algorithm will produce a sequence of i.e.m. T_n on shorter and shorter intervals $I^{(n)}$ with combinatorial data $\pi^{(n)}$ (on the same alphabet $\mathcal{A}$). The sequence $(\pi^{(n)})_{n \geqslant 0}$ represents an infinite path in the Rauzy diagram $\mathcal{D}$ containing $\pi^{(0)}$ which is determined by its starting vertex $\pi^{(0)}$ and the types of the successive arrows.

To relate the length vectors and the translation vectors, as well as the suspension data that we could associate to the i.e.m., we introduce the following matrices in $SL(\mathbb{Z}^{\mathcal{A}})$.

Let γ be an arrow of $\mathcal{D}$, with winner α and loser β. We define

$$B_\gamma = \mathbb{I} + E_{\beta\alpha}$$

where $\mathbb{I}$ is the identity matrix and $E_{\beta\alpha}$ is the elementary matrix with only one non-zero coefficient, equal to 1, in position $\beta\,\alpha$. We extend the definition to a path

$\underline{\gamma} = (\gamma_1, \cdots, \gamma_n)$ defining

$$B_{\underline{\gamma}} = B_{\gamma_n} \cdots B_{\gamma_1}.$$

The matrices $B_{\underline{\gamma}}$ belong to $SL(\mathbb{Z}^{\mathcal{A}})$ and have nonnegative coefficients. For $n \geqslant 0$, let $\lambda^{(n)}$ be the length vector for T_n (considered as a row vector), let $\delta^{(n)}$ be the translation vector (considered as a column vector) ; for $m \leqslant n$, let $\gamma(m,n)$ the finite path in $\mathcal{D}$ from $\pi^{(m)}$ to $\pi^{(n)}$ determined by the algorithm. The following formulas are trivially checked when $n = m+1$ and then extended by functoriality :

$$\lambda^{(m)} = \lambda^{(n)} \, B_{\gamma(m,n)} \, ,$$

$$\delta^{(n)} = B_{\gamma(m,n)} \, \delta_{(m)} \, .$$

The following interpretation of the coefficients of the matrices $B_{\gamma(m,n)}$ is also immediately checked by induction on $n-m$: for $\alpha, \beta \in \mathcal{A}$, the coefficient of $B_{\gamma(m,n)}$ in position $\alpha\beta$ is the time spent in $I^{(m)}_{\beta}$ by a point in $I^{(n)}_{\alpha}$ under iteration by T_m before coming back to $I^{(n)}$. In particular, the sum over β of the row of the matrix of index α gives the return time under T_m of $I^{(n)}_{\alpha}$ in $I^{(n)}$.

7.6. Symplecticity of B_{γ}. Let $\underline{\gamma}$ be a finite path in a Rauzy diagram $\mathcal{D}$, starting at a vertex π and ending at a vertex π'. Let $\Omega_{\pi}, \Omega_{\pi'}$ be the matrices associated to π, π' as in subsection 3.4. We have seen in subsection 4.5 that $rk\ \Omega_{\pi} = rk\ \Omega_{\pi'} = 2g$, where g is the genus of the translation surface obtained by the zippered rectangle construction from any vertex in $\mathcal{D}$ (and any choice of length and suspension data).

From the relation between length and translation vectors given in subsection 3.4 and in the last section, we obtain

$$\Omega_{\pi'} = B_{\underline{\gamma}} \, \Omega_{\pi} \, {}^{t}B_{\underline{\gamma}} \, .$$

From this we see that :

- $B^{-1}_{\underline{\gamma}}$, acting on row vectors, sends the kernel of Ω_{π} onto the kernel of $\Omega_{\pi'}$;
- $B_{\underline{\gamma}}$, acting on column vectors, sends the image of Ω_{π} onto the image of $\Omega_{\pi'}$;
- if we equip the quotients $\mathbb{R}^{\mathcal{A}}/\mathrm{Ker}\,\Omega_{\pi} \simeq \mathrm{Im}\Omega_{\pi}$, $\mathbb{R}^{\mathcal{A}}/\mathrm{Ker}\ \Omega_{\pi'} \simeq \mathrm{Im}\Omega'_{\pi}$ of the symplectic structures determined by $\Omega_{\pi}, \Omega_{\pi'}$ respectively, then $B_{\underline{\gamma}}$ (acting on column vectors) is symplectic w.r.t these structures.

PROPOSITION 7.7. *One can choose, for each vertex π of $\mathcal{D}$, a basis of row vectors for $Ker\Omega_{\pi}$ such that, for all $\underline{\gamma} : \pi \to \pi'$, the matrix of the restriction of $B^{-1}_{\underline{\gamma}}$ w.r.t. the selected bases of $Ker\Omega_{\pi}$, $Ker\Omega_{\pi'}$ is the identity. In particular, if $\underline{\gamma}$ is a loop at π, the restriction of $B^{-1}_{\underline{\gamma}}$ to the kernel of Ω_{π} is the identity.*

PROOF. We construct, for each vertex π of $\mathcal{D}$, an isomorphism i_{π} from $\mathrm{Ker}\,\Omega_{\pi}$ onto the same subspace K of $\mathbb{R}^{\mathcal{A}}$, such that $i_{\pi'} \circ {}^{t}B^{-1}_{\gamma} = i_{\pi}$ for any arrow $\gamma : \pi \to \pi'$. Choosing a basis for K and transferring it to each $\mathrm{Ker}\,\Omega_{\pi}$ via i_{π} then achieves the required property.

For $0 \leqslant k < d$, let u^t_k, u^b_k be the linear forms on the space of row vectors defined by

$$u^t_k(\lambda) = \sum_{\pi_t \alpha \leqslant k} \lambda_{\alpha} \, ,$$

$$u^b_k(\lambda) = \sum_{\pi_b \alpha \leqslant k} \lambda_\alpha \, .$$

For each vertex π, define linear maps i^t_π, i^b_π of $\mathbb{R}^{\mathcal{A}}$ into itself by

$$i^t_\pi(\lambda) = (u^t_{\pi^t(\alpha)-1})_{\alpha\in\mathcal{A}}, \qquad i^b_\pi(\lambda) = (u^b_{\pi^b(\alpha)-1})_{\alpha\in\mathcal{A}}.$$

Then the map $(i^t_\pi, i^b_\pi) : \mathbb{R}^{\mathcal{A}} \to \mathbb{R}^{\mathcal{A}} \times \mathbb{R}^{\mathcal{A}}$ is injective and $\operatorname{Ker}\Omega_\pi$ is the inverse image by this map of the diagonal of $\mathbb{R}^{\mathcal{A}} \times \mathbb{R}^{\mathcal{A}}$. Let K_π be the image of $\operatorname{Ker}\Omega_\pi$ by i^t_π; it is also the image by i^b_π. Let i_π be the common restriction of i^t_π, i^b_π to $\operatorname{Ker}\Omega_\pi$.

When we perform a single step of the algorithm, corresponding to an arrow $\gamma : \pi \to \pi'$, of top type for instance, the λ_α with $\pi_t(\alpha) < d$ and π_t itself do not change, hence the u^t_k for $0 \leqslant k < d$ stay the same. This means that $K_\pi = K_{\pi'}$ and $i_{\pi'} \circ {}^tB^{-1}_\gamma = i_\pi$. □

7.7. Complete paths.

Definition 7.8. A (finite) path in a Rauzy diagram is **complete** if every letter in $\mathcal{A}$ is the winner of at least one arrow in the path. An infinite path in a Rauzy diagram is **∞-complete** if every letter in $\mathcal{A}$ is the winner of infinitely many arrows in the path. Equivalently, an **∞-complete** path is one can be written as the concatenation of infinitely many complete paths.

This is a relevant notion because of the following characterization of paths associated to an i.e.m.

Proposition 7.9. *An infinite path in a Rauzy diagram is associated to some i.e.m. iff it is ∞-complete.*

We prove first that a path associated to an i.e.m. is ∞-complete, then an important auxiliary result, and then that an ∞-complete path is associated to some i.e.m .

Proof. Let $\mathcal{A}'$ be the set of letters which are the winners of at most finitely many arrows in the path γ_T associated to an i.e.m. $T = T_0$.

Let $(T_n)_{n\geqslant 0}$ be the sequence of i.e.m. obtained from T by iterating the Rauzy-Veech algorithm, $\lambda^{(n)}, \pi^{(n)}$ the length and combinatorial data of T_n.

There exists n_0 such that no letter in $\mathcal{A}'$ is a winner for $n \geqslant n_0$. Then the lengths $\lambda^{(n)}_\alpha$ for $\alpha \in \mathcal{A}'$, $n \geqslant n_0$, are independent of n.

At each step, the length of the loser is subtracted from the length of the winner. As lengths are always positive, there must exist $n_1 \geqslant n_0$ such that no letter in $\mathcal{A}'$ is a loser for $n \geqslant n_1$. This means that, for $\alpha \in \mathcal{A}'$, both $\pi^{(n)}_t(\alpha)$ and $\pi^{(n)}_b(\alpha)$ are non-decreasing with n for $n \geqslant n_1$, hence there exists $n_2 \geqslant n_1$ such that these quantities are independent of n for $n \geqslant n_2$.

Let $\alpha \in \mathcal{A}', \beta \in \mathcal{A} - \mathcal{A}'$. We claim that $\pi^{(n_2)}_t(\alpha) < \pi^{(n_2)}_t(\beta)$ and $\pi^{(n_2)}_b(\alpha) < \pi^{(n_2)}_b(\beta)$. As $\mathcal{A} - \mathcal{A}'$ is not empty and $\pi^{(n_2)}$ is irreducible, this implies that $\mathcal{A}'$ is empty, and therefore γ_T is ∞-complete.

Assume by contradiction, for instance, that $\pi^{(n_2)}_t(\beta) < \pi^{(n_2)}_t(\alpha)$. We have $\pi^{(n)}_t(\alpha) = \pi^{(n_2)}_t(\alpha)$ for $n \geqslant n_2$, hence also $\pi^{(n)}_t(\beta) = \pi^{(n_2)}_t(\beta)$ for $n \geqslant n_2$. Thus β is not the winner of an arrow of top type for $n \geqslant n_2$. As $\beta \in \mathcal{A} - \mathcal{A}'$, β is the winner of an arrow of bottom type for some $n \geqslant n_2$, which gives

$$\pi^{(n+1)}_t(\alpha) = \pi^{(n)}_t(\alpha) + 1,$$

a contradiction. The claim is proved; this completes the proof of the first part of the proposition. Before proving the second half of Proposition 7.9, we give some Corollaries of the first half.

COROLLARY 7.10. *The length of the interval $I^{(n)}$ on which T_n acts goes to zero as n goes to $+\infty$.*

PROOF. Each length $\lambda_\alpha^{(n)}$ is a non-increasing function of n hence has a limit $\lambda_\alpha^{(\infty)}$. Let $\varepsilon > 0$, n_0 such that $\lambda_\alpha^{(n)} \leqslant \lambda_\alpha^{(\infty)} + \varepsilon$ for all $n \geqslant n_0$, $\alpha \in \mathcal{A}$.

Let $\beta \in \mathcal{A}$. There exists $n_1 > n_0$ such that β is the winner of the arrow of index $n_1 - 1$ but not of the next arrow of index n_1. Then β is the loser of the arrow of index n_1. Let α be the winner of this arrow. We have

$$\lambda_\alpha^{(n_1)} = \lambda_\alpha^{(n_1-1)} - \lambda_\beta^{(n_1-1)},$$

hence $\lambda_\beta^{(\infty)} \leqslant \lambda_\beta^{(n_1-1)} \leqslant \varepsilon$. As ε is arbitrary, we have $\lambda_\beta^{(\infty)} = 0$ for all $\beta \in \mathcal{A}$. □

COROLLARY 7.11. *The Rauzy-Veech algorithm stops iff the i.e.m. has a connection.*

PROOF. We already know that the algorithm does not stop if the i.e.m. has no connection. Assume that T has a connection (m, u^t, u^b) ; here u^t is a singularity of T, u^b a singularity of T^{-1} and m is a nonnegative integer such that $T^m(u^b) = u^t$. Assume that one can apply the algorithm once to get an i.e.m. $\widetilde{T}$ on an interval $\widetilde{I}$; the intersection $\{u^b, T(u^b), \dots T^m(u^b) = u^t\} \cap \widetilde{I}$ will produce a connection $(\widetilde{m}, \widetilde{u}^t, \widetilde{u}^b)$ for $\widetilde{T}$ with $\widetilde{m} \leqslant m$, and $\widetilde{m} = m$ iff $\{u^b, T(u^b), \dots, T^m(u^b)\} \subset \widetilde{I}$. When we iterate the algorithm, the length of the interval goes to zero unless the algorithm stops ; this must therefore happen at some point. □

PROPOSITION 7.12. [**MmMsY, Y1**] *Let $\underline{\gamma}$ be a finite path in a Rauzy diagram that can be written as the concatenation of $2d-3$ complete paths (where $d = \#\mathcal{A}$). Then all coefficients of $B_{\underline{\gamma}}$ are positive.*

PROOF. Write $\underline{\gamma} = \underline{\gamma}_1 * \cdots * \underline{\gamma}_{2d-3}$, with each $\underline{\gamma}_i$ complete, and let $\underline{\gamma}(i) = \underline{\gamma}_1 * \cdots * \underline{\gamma}_i$. Recall that the diagonal coefficients of $B_{\underline{\gamma}}$ (for any path $\underline{\gamma}$) are always positive. It is therefore sufficient to show that, for any distinct letters α_1, α_0 in $\mathcal{A}$, we have $(B_{\underline{\gamma}(i)})_{\alpha_0\alpha_1} > 0$ for some i.

As α_1 is the winner of an arrow in $\underline{\gamma}_1$, the loser of which we call α_2, we have

$$(B_{\underline{\gamma}(1)})_{\alpha_2\alpha_1} > 0.$$

When $d = 2$, we must have $\alpha_2 = \alpha_0$ and the result is achieved. Assume $d > 2$ and $\alpha_2 \neq \alpha_0$. Because $\underline{\gamma}_2$ and $\underline{\gamma}_3$ are complete, there exists in $\underline{\gamma}_2 * \underline{\gamma}_3$ an arrow with winner $\alpha_3 \neq \alpha_1, \alpha_2$ immediately followed by an arrow with winner α_1 or α_2. This leads to

$$(B_{\underline{\gamma}(3)})_{\alpha_3\alpha_1} > 0\,.$$

If $d = 3$, we must have $\alpha_0 = \alpha_3$ and we are done. If $d > 3$ and $\alpha_3 \neq \alpha_0$, we go on in the same way : there exists in $\underline{\gamma}_4 * \underline{\gamma}_5$ an arrow with winner $\alpha_4 \neq \alpha_1, \alpha_2, \alpha_3$ immediately followed by an arrow with winner α_1, α_2 or α_3. This leads to

$$(B_{\underline{\gamma}(5)})_{\alpha_4\alpha_1} > 0\,.$$

and we go on ... □

End of proof of Proposition 7.9 : We want to show that if an infinite path γ can be written as the concatenation $\underline{\gamma}_1 * \underline{\gamma}_2 * \underline{\gamma}_3 \cdots$ of complete paths, then γ is associated to some i.e.m. with no connection by the Rauzy-Veech algorithm.

Define the convex open cone

$$\mathcal{C}_n = (\mathbb{R}_+^*)^{\mathcal{A}} B_{\underline{\gamma}_n} B_{\underline{\gamma}_{n-1}} \dots B_{\underline{\gamma}_1}$$

This is the set of length data (for i.e.m having the starting point of γ as combinatorial data) which lead to a path starting with $\underline{\gamma}_1 * \cdots * \underline{\gamma}_n$. The set of length data corresponding to γ is therefore

$$\mathcal{C}(\gamma) = \bigcap_{n \geqslant 0} \mathcal{C}_n \ .$$

By Proposition 2, the closure of $\mathcal{C}_{n+2d-3}$ is contained in $\mathcal{C}_n \cup \{0\}$. Therefore

$$\{0\} \cup \mathcal{C}(\gamma) = \bigcap_{n \geqslant 0} \overline{\mathcal{C}}_n \ .$$

which shows that $\mathcal{C}(\gamma)$ is not empty. □

We will describe more precisely $\mathcal{C}(\gamma)$ in the next section.

8. Invariant measures

8.1. Invariant measures and topological conjugacy. Let T be an i.e.m on an interval I, with combinatorial data $\pi = (\pi_t, \pi_b)$ on an alphabet $\mathcal{A}$. We assume that T has no connection and denote by $\gamma = \gamma_T$ the infinite path associated to T in the Rauzy diagram $\mathcal{D}$ of π.

Let $\mathcal{C}(\gamma)$ be the convex cone considered above ; its elements are the length data of the i.e.m with combinatorial data π which have no connection and γ as associated path. Let $\mathcal{M}(T)$ be the set of finite measures on I invariant under T.

The sets $\mathcal{C}(\gamma)$ and $\mathcal{M}(T)$ are in one-to-one correspondence as follows. Let $\lambda \in \mathcal{C}(\gamma)$ and let T_λ be an i.e.m with these length data (and combinatorial data π) on an interval I_λ. Let u (resp. u_λ) be the largest singularity of T^{-1} (resp. T_λ^{-1}). The sets $(T^n(u))_{n \geqslant 0}$ and $(T_\lambda^n(u_\lambda))_{n \geqslant 0}$ are dense in I and I_λ respectively because T and T_λ are minimal, having no connection. The bijection

$$H : T_\lambda^n(u_\lambda) \longmapsto T^n(u)$$

is increasing because T and T_λ have the same path for the Rauzy-Veech algorithm. Therefore H extends uniquely to an homeomorphism from I_λ onto I, which obviously satisfies

$$H \circ T_\lambda = T \circ H$$

Thus, T_λ and T are topologically conjugated. The image under H_* of the Lebesgue measure on I_λ is a measure on I (of total mass $|I_\lambda|$) which is invariant under T.

Conversely, let μ be a finite measure invariant under T. We set, for $\alpha \in \mathcal{A}$

$$\lambda_\alpha = \mu(I_\alpha^t) = \mu(I_\alpha^b) \ .$$

We also define, for $x \in I$

$$K(x) = \mu(\{y \in I \ ; \ y < x\}) \ .$$

As T is minimal, μ has no atom and the support of μ is I ; therefore, K is an homeomorphism from I onto $(0, \mu(I)) =: I_\mu$.

Define then

$$T_\mu = K \circ T \circ K^{-1}\ .$$

Then T_μ preserves the Lebesgue measure and it is easy to check that T_μ is an i.e.m on I_μ with combinatorial data π and length data λ.

It is immediate to check that the two maps $\mathcal{C}(\gamma) \to \mathcal{M}(T)$, $\mathcal{M}(T) \to \mathcal{C}(\gamma)$ just defined are inverse to each other.

8.2. Number of invariant ergodic probability measures. Let T be an i.e.m on an interval I. Let g be the genus of the translation surfaces that can be constructed from T by the zippered rectangle construction. Let $\mathcal{M}(T)$ be the cone of finite invariant measures for T, which can be identified with the cone $\mathcal{C}(\gamma)$ determined by the infinite path γ associated to T by the Rauzy-Veech algorithm.

PROPOSITION 8.1. *The cone $\mathcal{C}(\gamma) \cup \{0\}$ is a closed simplicial cone of dimension $\leqslant g$. The number of invariant ergodic probability measures is therefore $\leqslant g$.*

PROOF. We have seen in the second part of the proof of Proposition 7.9 in Subsection 7.7 that $\mathcal{C}(\gamma) \cup \{0\}$ is a closed cone. That this closed cone is simplicial follows from the identification of $\mathcal{C}(\gamma)$ with $\mathcal{M}(T)$: extremal rays of $\mathcal{C}(\gamma)$ correspond to ergodic invariant probability measures and invariant probability measures can be written in a unique way as convex combination of ergodic ones.

It remains to be seen that the subspace E of $\mathbb{R}^{\mathcal{A}}$ generated by $\mathcal{C}(\gamma)$ has dimension $\leqslant g$. Let $(\mathcal{A}, \pi)$ be combinatorial data for T, let Ω be the corresponding antisymmetric matrix.

We first claim that $E \cap \mathrm{Ker}\ \Omega = \{0\}$. Indeed, let $v, v' \in \mathcal{C}(\gamma)$ such that $v - v' \in \mathrm{Ker}\ \Omega$. Write $\gamma(n)$ for the initial part of γ of length n. According to the Proposition in Section 7.6, the vector $(v - v')B^{-1}_{\gamma(n)}$ depends only on the endpoint of $\gamma(n)$. On the other hand, from Corollary 7.10 in Subsection 7.7, we have that $v\, B^{-1}_{\gamma(n)}$ and $v'\, B^{-1}_{\gamma(n)}$ go to zero. Hence $v = v'$, proving the claim.

We now show that the image of E in $\mathbb{R}^{\mathcal{A}}/\,\mathrm{Ker}\ \Omega$ is isotropic for the symplectic form determined by Ω. Otherwise, there would exist $v, v' \in \mathcal{C}(\gamma)$ with

$$v\, \Omega\, {}^t v' > 0\ .$$

Again, $v\, B^{-1}_{\gamma(n)}, v'\, B^{-1}_{\gamma(n)}$ go to zero. But according to Section 7.6 we have

$$v\, B^{-1}_{\gamma(n)}\ \Omega_n\ {}^t B^{-1}_{\gamma(n)}\ {}^t v' = v\, \Omega\, {}^t v'\ ,$$

where Ω_n is the matrix associated to the endpoint of $\gamma(n)$. This gives a contradiction; as $\mathrm{rk}\ \Omega = 2g$, we conclude that $\dim\ E \leqslant g$. □

In the next Section, we see that the bound in the proposition is optimal. However, as mentioned in Subsection 6.11, a theorem of Masur and Veech guarantees that $\mathcal{C}(\gamma)$ is a ray for almost all i.e.m.

8.3. Examples of non uniquely ergodic i.e.m. [Kea1, KeyNew]
We will construct in a Rauzy diagram of genus g an infinite path γ which is an infinite concatenation of complete paths but has the property that the subspace generated by $\mathcal{C}(\gamma)$ has dimension g.

Let $d \geqslant 2$. Define $\mathcal{A}^{(d)} = \{1, \ldots, d\}$ and

$$\pi_t^{(d)}(k) = k\ ,\ \pi_b^{(d)}(k) = d + 1 - k\ ,$$

for $1 \leqslant k \leqslant d$. Let $\mathcal{R}(d)$ be the Rauzy class for $\pi^{(d)} = (\pi_t^{(d)}, \pi_b^{(d)})$, $\mathcal{D}(d)$ the associated Rauzy diagram. From Section 4.4, we check that the translation surfaces constructed from these combinatorial data through the zippered rectangle construction satisfy :

- if d is even, $d = 2g, s = 1, k_1 = 2g - 1$;
- if d is odd, $d = 2g + 1$, $s = 2$, $k_1 = k_2 = g$.

The diagrams $\mathcal{D}(d)$ for $d = 2, 3, 4$ have been pictured in Subsection 7.3. Their structure can be described as follows.

There is a canonical involution i of $\mathcal{D}(d)$ defined on vertices by $i(\pi) = \widehat{\pi}$ with

$$\widehat{\pi}_t(k) = \pi_b(d + 1 - k) ,$$

$$\widehat{\pi}_b(k) = \pi_t(d + 1 - k) .$$

The unique fixed point of i is $\pi^{(d)}$, and i changes the type of arrows from top to bottom and back. If one defines

$$\mathcal{D}_t(d) = \{\pi \in \mathcal{R}(d) , \pi_t(2) = 2\}$$

$$\mathcal{D}_b(d) = \{\pi \in \mathcal{R}(d) , \pi_b(d - 1) = 2\}$$

then $i(\mathcal{D}_t(d)) = \mathcal{D}_b(d)$, $i(\mathcal{D}_b(d)) = \mathcal{D}_t(d)$, $\mathcal{D}_t(d) \cap \mathcal{D}_b(d) = \{\pi^{(d)}\}$ and any arrow has both endpoints in $\mathcal{D}_t(d)$ or both endpoints in $\mathcal{D}_b(d)$. Moreover, if one defines, for $3 \leqslant k \leqslant d$

$$\mathcal{D}_{b,k}(d) = \{\pi \in \mathcal{R}(d) ; \pi_b(d - 1) = 2 , \pi_t(k) = 2\} ,$$

then $\mathcal{D}_{b,k}(d)$ is isomorphic to $\mathcal{D}_t(k - 1)$ through an isomorphism which respects type, winner and loser.

A cycle of length $d - 1$ of arrows of bottom type starting at $\pi^{(d)}$ connects together the vertex in $\mathcal{D}_{b,k}(d)$ corresponding to $\pi^{(k-1)}$ in $\mathcal{D}_t(k - 1)$.

Let us now assume that $d = 2g$ is even. Consider, for positive integers $m_1, \dots, m_g$, the loop $\gamma(m_1, \dots, m_g)$ at $\pi^{(d)}$ in $\mathcal{D}(d)$ whose successful winners are (in exponential notation for repetition)

$$(1^{d-2}2^{m_1}1)d^2 1(d^{d-4}4^{m_2}3)d^2 \dots ((d-3)^2(d-2)^{m_{g-1}}(d-3))d^2(d-1)^{m_g} .$$

This is a complete loop in $\mathcal{D}(d)$.

Assume that $0 \ll m_1 \ll m_2 \cdots \ll m_g$ and let $e_1, \dots, e_d$ be the canonical basis of $\mathbb{R}^d$. One checks that

- $e_1 B_\gamma$ and $e_2 B_\gamma$ have size $\sim m_1$ in the approximate direction of $f_1 := e_2$;
- $e_3 B_\gamma$ and $e_4 B_\gamma$ have size $\sim m_2$ in the approximate direction of $f_2 := e_4 + e_1$;
- $e_5 B_\gamma$ and $e_6 B_\gamma$ have size $\sim m_3$ in the approximate direction of $f_3 := e_6 + e_3 + 2e_1$;
- $\vdots$
- $e_{d-3} B_\gamma$ and $e_{d-2} B_\gamma$ have size $\sim m_{g-1}$ in the approximate direction of $f_{g-1} := e_{d-2} + e_{d-5} + \cdots + 2^{g-3} e_1$;
- $e_{d-1} B_\gamma$ and $e_d B_\gamma$ have size $\sim m_g$ in the approximate direction of $f_g := e_{d-1} + e_{d-3} + \cdots + 2^{g-2} e_1$.

Observe that $f_1, \dots, f_g$ are linearly independent.

Now take a sequence $(m_\ell)_{\ell > 0}$ increasing very fast, define

$$\gamma_i = \gamma(m_{ig+1}, m_{ig+2}, \dots, m_{ig+g-1}),$$

$$\gamma(i) = \gamma_0 * \gamma_1 \cdots * \gamma_{i-1},$$

$$\gamma = \gamma_0 * \gamma_1 * \ldots$$

One checks that for all $i > 0$ and $1 \leqslant k \leqslant g$, the vectors $e_{2k-1}\, B_{\gamma(i)}$ and $e_{2k}\, B_{\gamma(i)}$ have approximate directions f_k ; more precisely, as $i \to \infty$, their directions converge to the same limit $f_k(\infty)$ which can be chosen arbitrarily close to f_k. In particular, if the sequence $(m_\ell)_{\ell>0}$ increases fast enough, the limit directions $f_k(\infty), 1 \leqslant k \leqslant g$, are linearly independent, which implies that the vector space spanned by $\mathcal{C}(\gamma)$ has dimension g.

9. Rauzy-Veech dynamics and Teichmüller flow

We establish in this section a relation between the Rauzy-Veech continued fraction algorithm and the Teichmüller flow on the moduli space $\mathcal{M}(M, \Sigma, \kappa)$ that generalizes the classical case of the usual continued fraction and the geodesic flow on the modular surface.

This will also exhibit the moduli space in a form which allows to check that its volume is finite. Throughout this section, we fix an alphabet $\mathcal{A}$, a Rauzy class $\mathcal{R}$ and denote by $\mathcal{D}$ the associated Rauzy diagram.

9.1. Rauzy-Veech dynamics. With

$$\Delta = \{\lambda \in \mathbb{R}^{\mathcal{A}};\ \lambda_\alpha > 0,\ \ \forall\, \alpha \in \mathcal{A}\,\},$$

we set

$$\Delta(\mathcal{D}) = \mathcal{R} \times \mathbb{P}(\Delta).$$

We denote by $V_+ : \Delta(\mathcal{D}) \to \Delta(\mathcal{D})$ the map induced by one step of the Rauzy-Veech algorithm. More precisely, let $\gamma : \pi \to \pi'$ be an arrow of $\mathcal{D}$. Let α_0 be the winner of γ and let α_1 be the loser of γ. Define

$$\Delta_\gamma = \{\lambda \in \Delta;\ \lambda_{\alpha_0} > \lambda_{\alpha_1}\,\}.$$

Then the domain of V_+ is the union, over all arrows γ, of the $\{\pi\} \times \mathbb{P}(\Delta_\gamma)$ and the restriction of V_+ to this set is induced by

$$(\pi, \lambda) \mapsto (\pi', \lambda B_\gamma^{-1}).$$

Each simplex in $\Delta(\mathcal{D})$ (identified by a vertex π of $\mathcal{D}$) contains two components of the domain of V_+ (associated to the two arrows starting at π), each being sent to a full simplex of $\Delta(\mathcal{D})$ (corresponding to the endpoint of the arrow). The map V_+ is therefore essentially 2-to-1.

Introducing the suspension variables τ leads to a map V which is essentially the natural extension of V_+. Let

$$S(\mathcal{D}) = \bigsqcup_{\mathcal{R}} (\{\pi\} \times \mathbb{P}(\Delta) \times \mathbb{P}(\Theta_\pi)),$$

where we recall from Subsection 7.4 that

$$\Theta_\pi = \{\tau \in \mathbb{R}^{\mathcal{A}};\ \sum_{\pi_t(\alpha)<k} \tau_\alpha > 0,\quad \sum_{\pi_b(\alpha)<k} \tau_\alpha < 0,\ \forall\, 1 < k \leqslant d\}.$$

For an arrow $\gamma : \pi \to \pi'$ of $\mathcal{D}$, we set

$$\Theta_\gamma = \{\tau \in \Theta_{\pi'};\ \epsilon \sum_\alpha \tau_\alpha > 0\},$$

where $\epsilon = -1$ (resp.$\epsilon = +1$) if γ is of top type (resp. bottom type). Define then

$$S_\gamma(\mathcal{D}) = \{\pi\} \times \mathbb{P}(\Delta_\gamma) \times \mathbb{P}(\Theta_\pi),$$

$$S^\gamma(\mathcal{D}) = \{\pi'\} \times \mathbb{P}(\Delta) \times \mathbb{P}(\Theta_\gamma).$$

The domain of $V : S(\mathcal{D}) \to S(\mathcal{D})$ is the (disjoint) union, over all arrows of $\mathcal{D}$, of the $S_\gamma(\mathcal{D})$; the image of V is the (disjoint) union of the $S^\gamma(\mathcal{D})$, the restriction of V to $S_\gamma(\mathcal{D})$ sends this set in a one-to-one way onto $S^\gamma(\mathcal{D})$ through the map induced by

$$(\pi, \lambda, \tau) \mapsto (\pi', \lambda B_\gamma^{-1}, \tau B_\gamma^{-1}).$$

The map V is therefore, up to codimension one sets, invertible.

9.2. Rauzy diagrams and Teichmüller spaces. Let π be an element of $\mathcal{R}$. Recall that the canonical length and suspension data are given by

$$\lambda_\alpha^{can} = 1, \qquad \tau_\alpha^{can} = \pi_b(\alpha) - \pi_t(\alpha), \qquad \forall\, \alpha \in \mathcal{A}.$$

With these data, we construct (using the zippered rectangle construction of Subsection 4.3, or the simplified version of Subsection 4.2) a translation surface $(M_\pi, \Sigma_\pi, \kappa_\pi, \zeta_\pi)$.

On the other hand, starting from data $(\lambda, \tau) \in \Delta \times \Theta_\pi$, the zippered rectangle construction produces a translation surface which is a deformation of $(M_\pi, \Sigma_\pi, \kappa_\pi, \zeta_\pi)$, i.e homeomorphic to $(M_\pi, \Sigma_\pi, \kappa_\pi)$ through an homeomorphism whose isotopy class is canonically defined. We therefore obtain a canonical embedding

$$i_\pi\ : \Delta \times \Theta_\pi \longrightarrow \widetilde{Q}(M_\pi, \Sigma_\pi, \kappa_\pi)$$

in the marked Teichmüller space. This is an embedding because it is a local section of the period map.

Let now $\gamma : \pi \to \pi'$ be an arrow of $\mathcal{D}$. The data $\lambda = \lambda^{can} B_\gamma, \tau = \tau^{can}$ produce a translation surface $(M_\pi, \Sigma_\pi, \kappa_\pi, \zeta_\pi^0)$; the data $\lambda = \lambda^{can}, \tau = \tau^{can} B_\gamma^{-1}$ with the combinatorial data π' produce a translation surface $(M_{\pi'}, \Sigma_{\pi'}, \kappa_{\pi'}, \zeta_{\pi'}^1)$. As observed in Subsection 7.4, these two translation surfaces are canonically isomorphic. This means that there exists an homeomorphism between the topological surfaces $(M_\pi, \Sigma_\pi, \kappa_\pi)$ and $(M_{\pi'}, \Sigma_{\pi'}, \kappa_{\pi'})$ whose isotopy class is canonically defined by γ. This leads to a canonical homeomorphism

$$j_\gamma : \widetilde{Q}(M_\pi, \Sigma_\pi, \kappa_\pi) \longrightarrow \widetilde{Q}(M_{\pi'}, \Sigma_{\pi'}, \kappa_{\pi'})$$

between marked Teichmüller spaces.

Let us observe that the isomorphic translation surfaces $(M_\pi, \Sigma_\pi, \kappa_\pi, \zeta_\pi^0)$, $(M_{\pi'}, \Sigma_{\pi'}, \kappa_{\pi'}, \zeta_{\pi'}^1)$ above define a point in

$$i_\pi(\Delta_\pi \times \Theta_\pi) \cap j_\gamma^{-1}(i_{\pi'}(\Delta_{\pi'} \times \Theta_{\pi'})).$$

As a consequence the union

$$i_\pi(\Delta_\pi \times \Theta_\pi) \cup j_\gamma^{-1}(i_{\pi'}(\Delta_{\pi'} \times \Theta_{\pi'}))$$

is a connected subset of $\widetilde{Q}(M_\pi, \Sigma_\pi, \kappa_\pi)$.

We introduce the groupoid $\Gamma(\mathcal{D})$ of paths in the **non-oriented** Rauzy diagram $\widetilde{\mathcal{D}}$: the vertices of $\widetilde{\mathcal{D}}$ are those of $\mathcal{D}$ (i.e the elements of the Rauzy class $\mathcal{R}$) but for each arrow $\gamma : \pi \to \pi'$ in $\mathcal{D}$ we have two arrows $\gamma^+ : \pi \to \pi'$ and $\gamma^- : \pi' \to \pi$ in $\widetilde{\mathcal{D}}$.

The groupoid $\Gamma(\mathcal{D})$ is the groupoid of oriented paths in $\widetilde{\mathcal{D}}$, quotiented out by the cancellation rules $\gamma^+ * \gamma^- = \gamma^- * \gamma^+ = 1$. We denote by $\Gamma_\pi(\mathcal{D})$ the subset of reduced paths starting at π and by $\pi_1(\widetilde{\mathcal{D}}, \pi)$ the group of reduced loops at π.

For each arrow γ of $\mathcal{D}$, we have defined above an isomorphism j_γ between marked Teichmüller spaces. There is a unique way to extend functorially this definition to $\Gamma(\mathcal{D})$: for each $\gamma \in \Gamma(\mathcal{D})$ starting at π and ending at π', we have an isomorphism

$$j_\gamma : \widetilde{Q}(M_\pi, \Sigma_\pi, \kappa_\pi) \longrightarrow \widetilde{Q}(M_{\pi'}, \Sigma_{\pi'}, \kappa_{\pi'}),$$

and $j_{\gamma_1 * \gamma_2} = j_{\gamma_2} \circ j_{\gamma_1}$ whenever $\gamma_1 * \gamma_2$ is defined. In particular, when $\gamma \in \pi_1(\widetilde{\mathcal{D}}, \pi)$, j_γ is an automorphism of $\widetilde{Q}(M_\pi, \Sigma_\pi, \kappa_\pi)$. We obtain in this way a group homomorphism

$$\gamma \longmapsto j_\gamma \,,$$
$$\pi_1(\widetilde{\mathcal{D}}, \pi) \longrightarrow \mathrm{Mod}^+(M_\pi, \Sigma_\pi)\,.$$

We now define

$$\mathcal{U}_\pi = \bigcup_{\Gamma_\pi(\mathcal{D})} j_\gamma^{-1}(i_{\pi'}(\Delta_{\pi'} \times \Theta_{\pi'}))\,,$$

where π' is the endpoint of $\gamma \in \Gamma_\pi(\mathcal{D})$. It follows immediately from the observation at the end of subsection 9.1 that $\mathcal{U}_\pi$ is an open connected subset of $\widetilde{Q}(M_\pi, \Sigma_\pi, \kappa_\pi)$. We will denote by $\mathcal{C}_\pi$ the component of $\widetilde{Q}(M_\pi, \Sigma_\pi, \kappa_\pi)$ which contains $\mathcal{U}_\pi$.

9.3. The following result shows that, when considering some component $\mathcal{C}$ of a (marked) Teichmüller space $\widetilde{Q}(M, \Sigma, \kappa)$, there is no loss of generality if we assume that $(M, \Sigma, \kappa) = (M_\pi, \Sigma_\pi, \kappa_\pi)$ (for some appropriate combinatorial data $(\mathcal{A}, \pi)$) and $\mathcal{C} = \mathcal{C}_\pi$.

PROPOSITION 9.1. *Let (M, Σ, κ) be combinatorial data for a translation surface, let $\mathcal{C}$ be a connected component of the marked Teichmüller space $\widetilde{Q}(M, \Sigma, \kappa)$, and let $\mathcal{U}$ be the open subset of $\mathcal{C}$ formed by the translation surface structures in $\mathcal{C}$ that can be obtained through the zippered rectangle construction.*

(1) *The set $\mathcal{C} - \mathcal{U}$ has real codimension $\geqslant 2$ in $\mathcal{C}$.*
(2) *There exist combinatorial data $(\mathcal{A}, \pi)$ and a homeomorphism $g : (M_\pi, \Sigma_\pi, \kappa_\pi) \to (M, \Sigma, \kappa)$ such that the corresponding isomorphism g_* of marked Teichmüller spaces satisfy*

$$g_*(\mathcal{U}_\pi) = \mathcal{U}.$$

(3) *Assume that $(\mathcal{A}', \pi')$ are combinatorial data and $g' : (M_{\pi'}, \Sigma_{\pi'}, \kappa_{\pi'}) \to (M, \Sigma, \kappa)$ is an homeomorphism such that*

$$g'_*(\mathcal{U}_{\pi'}) \subset \mathcal{C}.$$

Then, the Rauzy diagrams $\mathcal{D}$, $\mathcal{D}'$ spanned by π, π' are isomorphic. Moreover, assuming that $\mathcal{D} = \mathcal{D}'$, $\pi = \pi'$, the element of $\mathrm{Mod}^+(M_\pi, \Sigma_\pi)$ determined by $g^{-1} \circ g'$ belongs to the image of the group homomorphism

$$\pi_1(\mathcal{D}, \pi) \longrightarrow \mathrm{Mod}^+(M_\pi, \Sigma_\pi)$$

defined in the last subsection.

REMARK 9.2. It is quite possible that this homomorphism is always onto. This has been checked for $g = 1$, with any number of marked points, by Wang Zhiren.

PROOF. Part 1. of the proposition is a consequence of the Proposition 5.7 and the Corollary 5.6: if a translation surface structure on (M, Σ, κ) has no vertical connection or no horizontal connection, it can be represented with the appropriate marking as a suspension through the zippered rectangle construction. Having both horizontal and vertical connections is indeed a codimension 2 property: this can already be seen on each orbit of the $SL(2, \mathbb{R})$ action (for instance).

By definition of $\mathcal{U}$, this open set is the union, over all combinatorial data $(\mathcal{A}, \pi)$, and all homeomorphisms $g : (M_\pi, \Sigma_\pi, \kappa_\pi) \to (M, \Sigma, \kappa)$ such that $g_*(i_\pi(\Delta_\pi \times \Theta_\pi)) \subset \mathcal{C}$, of the sets $g_*(i_\pi(\Delta_\pi \times \Theta_\pi))$. As its complement in $\mathcal{C}$ has codimension $\geqslant 2$, the open set $\mathcal{U}$ is **connected**.

CLAIM 9.3. *If $(\mathcal{A}, \pi, g)$, $(\mathcal{A}', \pi', g')$ satisfy*

$$g_*(i_\pi(\Delta_\pi \times \Theta_\pi)) \bigcap g'_*(i_{\pi'}(\Delta_{\pi'} \times \Theta_{\pi'})) \neq \emptyset,$$

then the Rauzy diagrams $\mathcal{D}$, $\mathcal{D}'$ spanned by π, π' are isomorphic and (assuming $\mathcal{A} = \mathcal{A}'$, $\mathcal{D} = \mathcal{D}'$) either ${g_*}^{-1} \circ g'_*$ or ${g'_*}^{-1} \circ g_*$ is equal to j_γ for a finite **oriented** *path γ in $\mathcal{D}$.*

PROOF. By hypothesis, there are two isomorphic translation surface structures ζ, ζ' on (M, Σ, κ) such that:

- ζ is obtained by the zippered rectangle construction from an i.e.m T acting on an interval I with combinatorial data $(\mathcal{A}, \pi)$, length data λ, suspension data τ;
- ζ' is obtained by the zippered rectangle construction from an i.e.m T' acting on an interval I' with combinatorial data $(\mathcal{A}', \pi')$, length data λ', suspension data τ'.

Let $G : (M, \Sigma, \kappa, \zeta) \to (M, \Sigma, \kappa, \zeta')$ be an isomorphism. It sends the marked outgoing horizontal separatrix for ζ isometrically onto the marked outgoing separatrix for ζ'.

If $|I| = |I'|$, we can already conclude that $T = T'$ and $\tau = \tau'$. Assume for instance that $|I| > |I'|$. Then T' must be the first return map of T on the interval of length $|I'|$ with the same left endpoint than I. That T' is obtained from T by a finite number of steps of the Rauzy-Veech algorithm now follows from Corollary 7.10 in Subsection 7.7 (applying if necessary the same small rotation to both ζ and ζ', we may assume that ζ has no vertical connection) and the last exercise in Subsection 7.2 □

End of proof of proposition: A first consequence of the claim is that the combinatorial data $(\mathcal{A}, \pi)$ such that $g_*(i_\pi(\Delta_\pi \times \Theta_\pi)) \subset \mathcal{C}$ all belong to the same Rauzy class (up to isomorphism): otherwise, the set $\mathcal{U}$ would not be connected. Once we know that, both the second and the third part of the proposition are immediate consequences of the claim. □

9.4. Rauzy diagrams and moduli spaces. Let $\mathcal{A}$, $\mathcal{R}$, $\mathcal{D}$ as above. We fix a vertex π^* of $\mathcal{D}$ and denote simply $(M_{\pi^*}, \Sigma_{\pi^*}, \kappa_{\pi^*})$, $\mathcal{U}_{\pi^*}$, $\mathcal{C}_{\pi^*}$ by (M, Σ, κ), $\mathcal{U}$, $\mathcal{C}$.

It follows from the third part of the proposition that the stabilizer of $\mathcal{C}$ (for the action of $\mathrm{Mod}^+(M, \Sigma)$ on $\widetilde{Q}(M, \Sigma, \kappa)$) is the subgroup image of $\pi_1(\mathcal{D}, \pi^*)$, which will be denoted by $\mathrm{Mod}_0(M, \Sigma)$.

We now define what amounts to a fundamental domain for the action of $\mathrm{Mod}_0(M, \Sigma)$ on $\mathcal{C}$. For each vertex π of $\mathcal{D}$, define

$$\Delta_\pi^0 = \{\lambda \in \Delta;\ 1 \leqslant \sum_\alpha \lambda_\alpha \leqslant 1 + \min(\lambda_{\alpha_t}, \lambda_{\alpha_b})\ \}$$

where $\pi_t(\alpha_t) = \pi_b(\alpha_b) = d$.

Consider then the disjoint union, over elements of $\mathcal{R}$, of the $\Delta_\pi^0 \times \Theta_\pi$ and perform the following identifications on the boundaries of these sets.

The part of the boundary of $\Delta_\pi^0 \times \Theta_\pi$ where $\sum_\alpha \lambda_\alpha = 1$ is called the *lower* boundary of $\Delta_\pi^0 \times \Theta_\pi$; it is divided into a *top half* where $\sum_\alpha \tau_\alpha < 0$ and a *bottom half* where $\sum_\alpha \tau_\alpha > 0$.

The part of the boundary of $\Delta_\pi^0 \times \Theta_\pi$ where $\sum_\alpha \lambda_\alpha = 1 + \min(\lambda_{\alpha_t}, \lambda_{\alpha_b})$ is called the *upper* boundary of $\Delta_\pi^0 \times \Theta_\pi$; it is divided into a *top half* where $\lambda_{\alpha_t} > \lambda_{\alpha_b}$ and a *bottom half* where $\lambda_{\alpha_t} < \lambda_{\alpha_b}$.

For each arrow $\gamma : \pi \to \pi'$ in $\mathcal{D}$, of top type, we identify the top half of the upper boundary of $\Delta_\pi^0 \times \Theta_\pi$ with the top half of the lower boundary of $\Delta_{\pi'}^0 \times \Theta_{\pi'}$ through $(\lambda, \tau) \mapsto (\lambda B_\gamma^{-1}, \tau B_\gamma^{-1})$; when γ is of bottom type, we identify similarly bottom halves.

We denote by $\mathcal{M}(D)$ the space obtained from $\bigsqcup_\pi \Delta_\pi^0 \times \Theta_\pi$ by these identifications. From its definition in Subsection 9.2, it is clear that the set $\mathcal{U}$ is invariant under $\mathrm{Mod}_0(M, \Sigma)$. The same is true for the smaller set

$$\mathcal{V} := \bigcup_{\gamma \in \Gamma_{\pi^*}(\mathcal{D})} j_\gamma^{-1}(i_\pi(\Delta_\pi^0 \times \Theta_\pi)) ,$$

where π is the endpoint of a path $\gamma \in \Gamma_{\pi^*}(\mathcal{D})$.

PROPOSITION 9.4. *There exists a unique continuous map*

$$p : \mathcal{V} \longrightarrow \mathcal{M}(D)$$

such that for every $\gamma \in \Gamma_{\pi^*}(\mathcal{D})$ (with endpoint π), the composition $p \circ j_\gamma^{-1} \circ i_\pi$ is the canonical map from $\Delta_\pi^0 \times \Theta_\pi$ to $\mathcal{M}(D)$. Moreover, p is a covering map which identifies $\mathcal{M}(D)$ with the quotient of $\mathcal{V}$ by the action of $\mathrm{Mod}_0(M, \Sigma)$. *The set $\mathcal{U} - \mathcal{V}$ has codimension* 1.

PROOF. Let γ be a path in $\Gamma_{\pi^*}(\mathcal{D})$ with endpoint π, and let γ_0 be an arrow from π to some vertex π'. The intersection

$$j_\gamma^{-1}(i_\pi(\Delta_\pi^0 \times \Theta_\pi)) \bigcap j_{\gamma * \gamma_0}^{-1}(i_{\pi'}(\Delta_{\pi'}^0 \times \Theta_{\pi'}))$$

is non empty; if γ_0 is for instance of top type, it is equal to the image $j_\gamma^{-1} \circ i_\pi$ of the top half of the upper boundary of $\Delta_\pi^0 \times \Theta_\pi$ and also to the image by $j_{\gamma * \gamma_0}^{-1} \circ i_{\pi'}$ of the top half of the lower boundary of $\Delta_{\pi'}^0 \times \Theta_{\pi'}$, the identification between these halves being exactly as in $\mathcal{M}(D)$. Moreover, it follows from the claim in the proof of Proposition 9.1 that this is the only case where a non empty intersection occurs. As a consequence, a map p with the property required in the statement of the proposition exists, is continuous, and is uniquely defined by this property.

From the property defining p, two points in $\mathcal{V}$ have the same image under p iff they belong to the same $\mathrm{Mod}_0(M, \Sigma)$ orbit. This implies that p is a covering map. Finally, let $[\zeta] = j_\gamma^{-1} \circ i_\pi(\lambda, \tau)$ be a point of $\mathcal{U}$ (with $\gamma \in \Gamma_{\pi^*}(\mathcal{D})$, π the endpoint of γ, $\lambda \in \Delta_\pi$, $\tau \in \Theta_\pi$). If $\lambda \in \Delta_\pi^0$, then $[\zeta]$ belongs to $\mathcal{V}$. Otherwise, $\sum_\alpha \lambda_\alpha$ is either too large or too small. If it is too large, we apply one step of the Rauzy-Veech algorithm unless $\lambda_{\alpha_t} = \lambda_{\alpha_b}$. If it is too small, we apply one step backwards unless $\sum_\alpha \tau_\alpha = 0$. Iterating this process, we will end up in $\mathcal{V}$ unless we run into one of the codimension one conditions that stops the algorithm (forwards or backwards). This proves that $\mathcal{U} - \mathcal{V}$ has codimension 1. □

9.5. Canonical volumes. Zorich acceleration. The last proposition allows us to identify $\mathcal{M}(D)$ with a subset of the marked moduli space $\widetilde{\mathcal{M}}(M, \Sigma, \kappa)$ whose complement has codimension 1. In particular this subset has full measure for the canonical volume of the moduli space. Observe that, in view of its relation with the period map, the canonical volume in $\mathcal{M}(D)$ is nothing else than the standard Lebesgue measure $d\lambda\, d\tau$ restricted to each $\Delta_\pi^0 \times \Theta_\pi$.

The model $\mathcal{M}(D)$ for (part of) the moduli space provides us also with a natural transversal section for the Teichmüller flow, namely the union over the vertices π of $\mathcal{D}$ of the lower boundaries of the $\Delta_\pi^0 \times \Theta_\pi$. Indeed, in each $\Delta_\pi^0 \times \Theta_\pi$, the Teichmüller flow reads as

$$(\lambda, \tau) \mapsto (e^t\, \lambda, e^{-t}\, \tau)$$

and flows from the lower boundary of $\Delta_\pi^0 \times \Theta_\pi$ to its upper boundary, being then glued as prescribed by the Rauzy-Veech algorithm to the lower boundary of some $\Delta_{\pi'}^0 \times \Theta_{\pi'}$.

When computing volumes, we have to normalize the area $A = \tau\, \Omega_\pi\, {}^t\lambda$. Let $\mathcal{M}^{(1)}(\mathcal{D})$ be the subset of $\mathcal{M}(\mathcal{D})$ defined by $\{A = 1\}$. We can identify the set $S(\mathcal{D})$ of Subsection 9.1 with the transverse section to the Teichmüller flow in $\mathcal{M}^{(1)}(\mathcal{D})$

$$\{(\pi, \lambda, \tau) \in \bigsqcup_\pi \{\pi\} \times \Delta \times \Theta_\pi;\ \sum_\alpha \lambda_\alpha = 1,\ \ \tau\, \Omega_\pi\, {}^t\lambda = 1\}.$$

With this identification, the return map of the Teichmüller flow on $S(\mathcal{D})$ is precisely given by the Rauzy-Veech dynamics V defined in Subsection 9.1. The return time is equal to

$$\log \frac{||\lambda||_1}{||\lambda\, B_\gamma^{-1}||_1},$$

where $||\ .\ ||_1$ is the ℓ^1-norm.

Observe that the return time is bounded from above, but not bounded away from 0. The unfortunate consequence, as we see below, is that the measure of $S(\mathcal{D})$ is infinite; this already happens in the elementary case $d = 2$.

In order to get nicer dynamical properties, Zorich [**Zo2**] considered instead a smaller transversal section $S^*(\mathcal{D}) \subset S(\mathcal{D})$ which still gives an easily understood return map but has finite measure. For an arrow $\gamma : \pi \to \pi'$ of top type (resp. of bottom type), let $S_\gamma^*(\mathcal{D})$ be the set of $(\pi, \lambda, \tau) \in S_\gamma(\mathcal{D})$ such that $\sum \tau_\alpha > 0$ (resp. $\sum \tau_\alpha < 0$). Let $S^*(\mathcal{D})$ be the union of the $S_\gamma^*(\mathcal{D})$ over all arrows of $\mathcal{D}$.

The return map of the Teichmüller flow to $S^*(\mathcal{D})$, which is also the return map of V to $S^*(\mathcal{D})$, will be denoted by V^*. It is obtained as follows: one iterates V as long as the type of the corresponding arrow does not change. It is easy to check that it is the same than to ask that the winner does not change.

This property of V^* shows the return time for V to $S^*(\mathcal{D})$ does not depends on the τ-coordinate. One can therefore define a map $V_+^* : \Delta(\mathcal{D}) \to \Delta(\mathcal{D})$ such that V^* is fibered over V_+^*.

EXAMPLE 9.5. For $d = 2$, we have

$$\begin{aligned} S(\mathcal{D}) &= \{(\lambda_A, \lambda_B, \tau_A, \tau_B);\ \lambda_A > 0,\ \lambda_B > 0,\ \tau_A > 0, \tau_B < 0, \\ &\qquad \lambda_A + \lambda_B = 1, \lambda_A \tau_B - \lambda_B \tau_A = 1\}, \\ S_t(\mathcal{D}) &= \{(\lambda_A, \lambda_B, \tau_A, \tau_B) \in S(\mathcal{D});\ \lambda_A > \lambda_B\}, \\ S_b(\mathcal{D}) &= \{(\lambda_A, \lambda_B, \tau_A, \tau_B) \in S(\mathcal{D});\ \lambda_A < \lambda_B\}, \\ S^t(\mathcal{D}) &= \{(\lambda_A, \lambda_B, \tau_A, \tau_B) \in S(\mathcal{D});\ \tau_A + \tau_B < 0\}, \\ S^b(\mathcal{D}) &= \{(\lambda_A, \lambda_B, \tau_A, \tau_B) \in S(\mathcal{D});\ \tau_A + \tau_B > 0\}, \\ S_t^*(\mathcal{D}) &= \{(\lambda_A, \lambda_B, \tau_A, \tau_B) \in S(\mathcal{D});\ \lambda_A > \lambda_B,\ \tau_A + \tau_B > 0\}, \\ S_b^*(\mathcal{D}) &= \{(\lambda_A, \lambda_B, \tau_A, \tau_B) \in S(\mathcal{D});\ \lambda_A < \lambda_B,\ \tau_A + \tau_B > 0\}. \end{aligned}$$

For $(\lambda_A, \lambda_B, \tau_A, \tau_B) \in S_t(\mathcal{D})$, we have

$$V(\lambda_A, \lambda_B, \tau_A, \tau_B) = (\lambda_A \lambda_B^{-1}, 1 - \lambda_A \lambda_B^{-1}, \lambda_B \tau_A, \lambda_B(\tau_B - \tau_A)).$$

For $(\lambda_A, \lambda_B, \tau_A, \tau_B) \in S_t^*(\mathcal{D})$, we have

$$V^*(\lambda_A, \lambda_B, \tau_A, \tau_B) = (\lambda_A \Lambda^{-1}, 1 - \lambda_A \Lambda^{-1}, \Lambda \tau_A, \Lambda(\tau_B - n\tau_A)),$$

where $\Lambda = \lambda_B - (n-1)\lambda_A,\ n\lambda_A < \lambda_B < (n+1)\lambda_A,\ n \geqslant 1$.

On the λ-coordinate, V_+^* is essentially given by the Gauss map.

9.6. Volume estimates: the key combinatorial lemmas. We will present three volume estimates: two for the measures of $S(\mathcal{D})$ and $S^*(\mathcal{D})$ and one for the measure of $\mathcal{M}^{(1)}(\mathcal{D})$, i.e the integral over $S(\mathcal{D})$ of the return time for the Teichmüller flow.

Before doing that, we consider the case $d = 2$ as an example of what happens in general. We first integrate over the τ variables. For a point (λ_A, λ_B) with $\lambda_B > \lambda_A > 0,\ \lambda_A + \lambda_B = 1$,

- the integral over $\{\tau_A > 0, \tau_B < 0, \lambda_B \tau_A - \lambda_A \tau_B = 1\}$ gives $\lambda_A^{-1} \lambda_B^{-1}$;
- the integral over $\{\tau_A + \tau_B > 0, \tau_B < 0, \lambda_B \tau_A - \lambda_A \tau_B = 1\}$ gives $(\lambda_A + \lambda_B)^{-1} \lambda_B^{-1} = \lambda_B^{-1}$.

Formulas for $\lambda_A > \lambda_B > 0,\ \lambda_A + \lambda_B = 1$ are symmetric.

For the measure of $S(\mathcal{D})$, we have therefore to integrate

$$\int_0^{\frac{1}{2}} \frac{d\lambda}{\lambda(1-\lambda)}$$

with the pole at 0 making the integral divergent.

For the measure of $S^*(\mathcal{D})$, we have to integrate

$$\int_0^{\frac{1}{2}} \frac{d\lambda}{1-\lambda}$$

on a domain away from the pole; the integral is equal to $\log 2$.

For the measure of $\mathcal{M}^{(1)}(\mathcal{D})$, the return time is $-\log(1-\lambda)$; the zero at 0 cancels the pole and we obtain

$$-\int_0^{\frac{1}{2}} \frac{d\lambda}{\lambda(1-\lambda)} \log(1-\lambda) = \frac{\pi^2}{12}.$$

The measure of $\mathcal{M}^{(1)}(\mathcal{D})$ is twice this.

We come back to the general case. Again, we want first to perform the integration over the τ variables. These variables run over the convex cone Θ_π but are restricted by the area condition. Define as usual α_t, α_b by $\pi_t(\alpha_t) = \pi_b(\alpha_b) = d$. Set

$$\begin{aligned} h^t_\alpha &= \sum_{\pi_t(\beta)\leqslant\pi_t(\alpha)} \tau_\beta \,, \\ h^b_\alpha &= -\sum_{\pi_b(\beta)\leqslant\pi_b(\alpha)} \tau_\beta \,, \\ \check{h}^t_\alpha &= \sum_{\pi_t(\beta)<\pi_t(\alpha)} \tau_\beta \,, \\ \check{h}^b_\alpha &= -\sum_{\pi_b(\beta)<\pi_b(\alpha)} \tau_\beta \,. \end{aligned}$$

With $h = -\Omega\, {}^t\tau$ as in Subsection 4.3, we have

$$h_\alpha = h^t_\alpha + h^b_\alpha = \check{h}^t_\alpha + \check{h}^b_\alpha,$$

for all $\alpha \in \mathcal{A}$ and $h^t_{\alpha_t} + h^b_{\alpha_b} = 0$. The suspension conditions are

$$h^t_\alpha > 0 \quad \text{for } \alpha \neq \alpha_t, \qquad\qquad h^b_\alpha < 0 \quad \text{for } \alpha \neq \alpha_b.$$

Consider for instance the top half of $\Delta = \Delta_\pi$ where $\lambda_{\alpha_t} > \lambda_{\alpha_b}$ (the other case is symmetric); we write

$$\begin{aligned} \widehat{\lambda}_\alpha &= \lambda_\alpha & &\text{for } \alpha \neq \alpha_t, \\ \widehat{\lambda}_{\alpha_t} &= \lambda_{\alpha_t} - \lambda_{\alpha_b} \,, & & \\ \widehat{h}_\alpha &= h_\alpha & &\text{for } \alpha \neq \alpha_b, \\ \widehat{h}_{\alpha_b} &= h_{\alpha_b} + h_{\alpha_t}. & & \end{aligned}$$

The area is given by

$$A = \sum_\alpha \lambda_\alpha h_\alpha = \sum_\alpha \widehat{\lambda}_\alpha \widehat{h}_\alpha.$$

We decompose Θ_π into a finite family $\mathcal{G}(\pi)$ of simplicial disjoint cones. Let Γ be a cone in this family, and let $\tau^{(1)}, \cdots, \tau^{(d)}$ be a base of $\mathbb{R}^\mathcal{A}$ of volume 1 such that

$$\Gamma \;=\; \{ \sum_1^d t_i \tau_i \;;\; t_i > 0 \;\}\,.$$

Writing $h^{(i)} = -\Omega_\pi\, {}^t\tau^{(i)}$ for $1 \leqslant i \leqslant d$, the area condition becomes

$$\sum_1^d t_i \,(\sum_\alpha \widehat{\lambda}_\alpha \, \widehat{h}^{(i)}_\alpha \,) \;=\; 1,$$

and therefore the integral over Γ gives

$$\frac{1}{(d-1)!} \; [\prod_1^d (\sum_\alpha \widehat{\lambda}_\alpha \, \widehat{h}^{(i)}_\alpha \,) \,]^{-1}.$$

To get the measure of $S(\mathcal{D})$, we should then integrate this quantity over the top half of Δ_π (normalized by $\sum_\alpha \lambda_\alpha = 1$), sum over $\Gamma \in \mathcal{G}(\pi)$, sum over π and finally add the symmetric contribution of the bottom halves.

Only the first step presents a finiteness problem. To deal with it, given a proper subset $\mathcal{B}$ of $\mathcal{A}$, we introduce the subspace $E_{\mathcal{B}}$ of $\mathbb{R}^{\mathcal{A}}$ generated by the τ in the closure of Θ_π such that $\widehat{h}_\alpha = 0$ for all $\alpha \in \mathcal{B}$ (with again $h = -\Omega\, {}^t\tau$).

LEMMA 9.6. *We have* $\operatorname{codim} E_{\mathcal{B}} \geqslant \#\mathcal{B}$, *and even* $\operatorname{codim} E_{\mathcal{B}} > \#\mathcal{B}$ *when* $\alpha_b \in \mathcal{B}$.

PROOF. We will find sufficiently many independent linear forms vanishing on $E_{\mathcal{B}}$. Assume first that $\alpha_b \notin \mathcal{B}$. Let τ be a vector in the closure of Θ_π, such that $\widehat{h}_\alpha = 0$ for all $\alpha \in \mathcal{B}$. For $\alpha \in \mathcal{B}$, we have

$$\widehat{h}_\alpha = h_\alpha = h^t_\alpha + h^b_\alpha = \check{h}^t_\alpha + \check{h}^b_\alpha,$$

with $h^b_\alpha \geqslant 0$, $h^t_\alpha \geqslant 0$ (if $\alpha \neq \alpha_t$), $\check{h}^t_\alpha \geqslant 0$, $\check{h}^b_\alpha \geqslant 0$.

We have therefore $h^t_\alpha = 0$ for $\alpha \in \mathcal{B}$, $\alpha \neq \alpha_t$, and also $\check{h}^t_\alpha = 0$ for $\alpha \in \mathcal{B}$, $\pi_t(\alpha) > 1$. This gives at least $\#\mathcal{B}$ independent linear forms vanishing on such vectors τ, and thus also on $E_{\mathcal{B}}$ (the independence of the forms come from the triangular form of the h^t_α, $\check{h}^t_\alpha$).

Assume now that $\alpha_b \in \mathcal{B}$. Let τ be a vector in the closure of Θ_π, such that $\widehat{h}_\alpha = 0$ for all $\alpha \in \mathcal{B}$. The relation $\widehat{h}_{\alpha_b} = 0$ implies $h_{\alpha_t} = h_{\alpha_b} = 0$, hence

$$h^t_{\alpha_t} + h^b_{\alpha_t} = h^b_{\alpha_b} + h^t_{\alpha_b} = 0.$$

As we have $h^b_{\alpha_t} \geqslant 0$, $h^t_{\alpha_b} \geqslant 0$, $h^t_{\alpha_t} + h^b_{\alpha_b} = 0$, we conclude that

$$h^t_{\alpha_t} = h^b_{\alpha_t} = h^b_{\alpha_b} = h^t_{\alpha_b} = 0.$$

We have therefore

- $h^b_\alpha = \check{h}^b_\alpha = 0$ for all $\alpha \in \mathcal{B}$;
- $h^t_\alpha = \check{h}^t_\alpha = 0$ for all $\alpha \in \mathcal{B}$.

The first set of relations gives at least $\#\mathcal{B} + 1$ independent linear forms vanishing on $E_{\mathcal{B}}$ unless $\pi_b(\mathcal{B}) = \{1, \cdots, \#\mathcal{B}\}$. The same is true for the second set of relations unless $\pi_t(\mathcal{B}) = \{1, \cdots, \#\mathcal{B}\}$. By irreducibility, the two exceptional cases are mutually exclusive and the proof of the lemma is complete. □

When we deal with $S^*(\mathcal{D})$, we should replace Θ_π by

$$\Theta^t_\pi = \{\tau \in \Theta_\pi;\ \sum_\alpha \tau_\alpha > 0\}$$

when we deal with the top half of Δ_π. We proceed in the same way, decomposing Θ_π into a finite family of simplicial cones Γ^*. We now define $E^*_{\mathcal{B}}$ as the subspace of $\mathbb{R}^{\mathcal{A}}$ generated by the vectors τ in the closure of Θ^t_π, such that $\widehat{h}_\alpha = 0$ for all $\alpha \in \mathcal{B}$.

LEMMA 9.7. *We have* $\operatorname{codim} E^*_{\mathcal{B}} > \#\mathcal{B}$ *for all proper subsets* $\mathcal{B}$ *of* $\mathcal{A}$.

PROOF. Obviously we have $E^*_{\mathcal{B}} \subset E_{\mathcal{B}}$, therefore the case where $\alpha_b \in \mathcal{B}$ is given by Lemma 9.6. We therefore assume that $\alpha_b \notin \mathcal{B}$.

Let τ be a vector in the closure of Θ^t_π, such that $\widehat{h}_\alpha = 0$ for all $\alpha \in \mathcal{B}$. For $\alpha \in \mathcal{B}$, we have

$$0 = \widehat{h}_\alpha = h_\alpha = h^t_\alpha + h^b_\alpha = \check{h}^t_\alpha + \check{h}^b_\alpha,$$

with $\check{h}^t_\alpha \geqslant 0$, $\check{h}^b_\alpha \geqslant 0$, $h^b_\alpha \geqslant 0$ (because $\alpha \neq \alpha_b$), $h^t_\alpha \geqslant 0$ (even for $\alpha = \alpha_t$). We therefore have

- $h^b_\alpha = \check{h}^b_\alpha = 0$ for all $\alpha \in \mathcal{B}$,
- $h^t_\alpha = \check{h}^t_\alpha = 0$ for all $\alpha \in \mathcal{B}$,

and conclude as in Lemma 9.6. □

9.7. Finiteness of volume for $\mathcal{M}^{(1)}(\mathcal{D})$ and $S^*(\mathcal{D})$. The combinatorial facts proven in the last subsection will be combined with the following simple analytic lemma. Let

$$\Delta^{(1)} = \{\lambda \in \mathbb{R}^{\mathcal{A}};\ \lambda_\alpha > 0,\ \sum_\alpha \lambda_\alpha = 1\,\}.$$

For $\mathcal{B} \subset \mathcal{A}$, define also

$$\Delta^{(1)}_{\mathcal{B}} = \{\lambda \in \Delta^{(1)};\ \lambda_\alpha = 0 \ \text{ for } \alpha \notin \mathcal{B}\}.$$

Consider linear forms $L_1, \cdots, L_p, M_1, \cdots, M_q$ on $\mathbb{R}^{\mathcal{A}}$ which are positive on $\Delta^{(1)}$, and the rational map

$$R := \frac{L_1 \cdots L_p}{M_1 \cdots M_q}.$$

For $\mathcal{B} \subset \mathcal{A}$, let

$$\begin{aligned} m_+(\mathcal{B}) &= \#\{i;\ L_i(\lambda) = 0 \text{ for all } \lambda \in \Delta^{(1)}_{\mathcal{B}}\,\}, \\ m_-(\mathcal{B}) &= \#\{j;\ M_j(\lambda) = 0 \text{ for all } \lambda \in \Delta^{(1)}_{\mathcal{B}}\,\}, \\ m(\mathcal{B}) &= m_+(\mathcal{B}) - m_-(\mathcal{B}). \end{aligned}$$

LEMMA 9.8. *Assume that $d + m(\mathcal{B}) > \#\mathcal{B}$ holds for all proper subsets of $\mathcal{A}$. Then R is integrable on $\Delta^{(1)}$.*

REMARK 9.9. The converse is also true but will not be used.

PROOF. We decompose $\Delta^{(1)}$ as follows: let

$$\mathcal{N} := \{n \in \mathbb{N}^{\mathcal{A}};\ \min_\alpha n_\alpha = 0\}.$$

For $n \in \mathcal{N}$, let $\Delta^{(1)}(n)$ be the set of $\lambda \in \Delta^{(1)}$ such that $\lambda_\alpha \geqslant \frac{1}{2d}$ if $n_\alpha = 0$ and

$$\frac{1}{2d} 2^{1-n_\alpha} > \lambda_\alpha \geqslant \frac{1}{2d} 2^{-n_\alpha}$$

if $n_\alpha > 0$. We have indeed

$$\Delta^{(1)} = \bigsqcup_{\mathcal{N}} \Delta^{(1)}(n)$$

and also

$$C^{-1}\, 2^{-\sum n_\alpha} \leqslant \operatorname{vol} \Delta^{(1)}(n) \leqslant C\, 2^{-\sum n_\alpha}.$$

Fix $n \in \mathcal{N}$. Let $0 = n^0 < n^1 < \cdots$ be the distinct values, in increasing order, taken by the n_α, and let

$$\mathcal{B}_i := \{\alpha \in \mathcal{A};\ n_\alpha \geqslant n^i\,\}.$$

Let L be a linear form on $\mathbb{R}^{\mathcal{A}}$, positive on $\Delta^{(1)}$. There is a maximal subset $\mathcal{B}(L) \subset \mathcal{A}$ such that $L(\lambda) = 0$ for all $\lambda \in \Delta^{(1)}_{\mathcal{B}(L)}$. We have then, for $n \in \mathcal{N}$, $\lambda \in \Delta^{(1)}(n)$

$$C_L^{-1} 2^{-m} \leqslant L(\lambda) \leqslant C_L 2^{-m}, \qquad \text{with } m = \min_{\mathcal{A} - \mathcal{B}(L)} n_\alpha.$$

The definition of m shows that $m \geqslant n^i$ iff $\mathcal{A} - \mathcal{B}(L) \subset \mathcal{B}_i$ and $m = n^i$ iff $\mathcal{B}_i^c \subset \mathcal{B}(L)$ but $\mathcal{B}_{i+1}^c \not\subset \mathcal{B}(L)$. From this, we see that for $n \in \mathcal{N}$, $\lambda \in \Delta^{(1)}(n)$, we have

$$C_R^{-1} 2^{-N} \leqslant R(\lambda) \leqslant C_R 2^{-N},$$

with

$$N = \sum_{i \geqslant 0} n^i\, (m(\mathcal{B}_i^c) - m(\mathcal{B}_{i+1}^c)) = \sum_{i > 0} (n^i - n^{i-1})\, m(\mathcal{B}_i^c).$$

Using the hypothesis of the lemma, we have, for $i > 0$

$$m(\mathcal{B}_i^c) \geqslant \#\mathcal{B}_i^c - d + 1 = 1 - \#\mathcal{B}_i ,$$

and therefore

$$N \geqslant \sum_{i>0}(n^i - n^{i-1}) - \sum_{i\geqslant 0} n^i(\#\mathcal{B}_i - \#\mathcal{B}_{i+1}) = \max_\alpha n_\alpha - \sum_\alpha n_\alpha .$$

We conclude that the integral of R on $\Delta^{(1)}(n)$ is at most of the order of $2^{-\max_\alpha n_\alpha}$. Summing over $\mathcal{N}$ gives the required result. □

We can now prove the finiteness of the measures of $\mathcal{M}^{(1)}(\mathcal{D})$ and $S^*(\mathcal{D})$. As explained in Subsection 9.6, the total masses of these measures are expressed as finite sums of certain integrals over top or bottom halves of the $\Delta_\pi^{(1)}$. We will consider the case of top halves, the other case being symmetric. Observe that the top half of $\Delta_\pi^{(1)}$ is characterized by the inequalities $\widehat{\lambda}_\alpha > 0, \forall \alpha \in \mathcal{A}$. We will therefore in both cases apply the lemma above **in the $\widehat{\lambda}$ variables**. We don't have $\sum \widehat{\lambda}_\alpha = 1$, but observe that $\sum \lambda_\alpha = 1$ implies $\frac{1}{2} \leqslant \sum \widehat{\lambda}_\alpha \leqslant 1$, which is good enough.

- We start with $\mathcal{M}^{(1)}(\mathcal{D})$. The return time of the Teichmüller flow to $S(\mathcal{D})$ is equal to $-\log \sum \widehat{\lambda}_\alpha = -\log(1 - \lambda_{\alpha_b})$ on the top half of $\Delta_\pi^{(1)}$.

  According to Subsection 9.6, we have to integrate

  $$\frac{-\log(1-\lambda_{\alpha_b})}{(d-1)!} \; [\prod_1^d (\, \Sigma_\alpha \, \widehat{\lambda}_\alpha \, \widehat{h}_\alpha^{(i)} \,)\,]^{-1}$$

  over the top half of $\Delta_\pi^{(1)}$. The vectors $h^{(i)} = -\Omega_\pi \, {}^t\tau^{(i)}$ are obtained here from vectors $\tau^{(i)}$ generating a simplicial cone $\Gamma \subset \Theta_\pi$.

  We apply the lemma above with $p = 1$, $q = d$. We take $L(\lambda) = \lambda_{\alpha_b} = \widehat{\lambda}_{\alpha_b}$, a linear form of the same order than the return time $-\log(1 - \lambda_{\alpha_b})$. The linear forms M_i are the $\Sigma_\alpha \, \widehat{\lambda}_\alpha \, \widehat{h}_\alpha^{(i)}$.

  We check the hypothesis of the lemma. Let $\mathcal{B} \subset \mathcal{A}$ be a proper subset. First, we have $m_+(\mathcal{B}) = 0$ if $\alpha_b \in \mathcal{B}$, $m_+(\mathcal{B}) = 1$ if $\alpha_b \notin \mathcal{B}$. Next we have

  $$\Sigma_\alpha \, \widehat{\lambda}_\alpha \, \widehat{h}_\alpha^{(i)} = 0, \quad \text{for all } \widehat{\lambda} \in \Delta_{\mathcal{B}}^{(1)}$$

  iff $\widehat{h}_\alpha^{(i)} = 0$ for all $\alpha \in \mathcal{B}$. By definition of $E_{\mathcal{B}}$, this happens iff $\tau^{(i)} \in E_{\mathcal{B}}$. As the $\tau^{(i)}$ are independent, Lemma 1 in the last subsection gives $m_-(\mathcal{B}) \leqslant d - \#\mathcal{B}$ if $\alpha_b \notin \mathcal{B}$, $m_-(\mathcal{B}) < d - \#\mathcal{B}$ if $\alpha_b \in \mathcal{B}$. The hypothesis of the lemma above is thus satisfied, and its conclusion gives the finiteness of the measure of $\mathcal{M}^{(1)}(\mathcal{D})$.
- We now deal with $S^*(\mathcal{D})$. According to subsection 9.6, we have to integrate

  $$\frac{1}{(d-1)!} \; [\prod_1^d (\, \Sigma_\alpha \, \widehat{\lambda}_\alpha \, \widehat{h}_\alpha^{(i)} \,)\,]^{-1}$$

  over the top half of $\Delta_\pi^{(1)}$. The vectors $h^{(i)} = -\Omega_\pi \, {}^t\tau^{(i)}$ are obtained here from vectors $\tau^{(i)}$ generating a simplicial cone $\Gamma^* \subset \Theta_\pi^t$.

  We will apply the lemma above with $p = 0$, $q = d$. The linear forms M_i are the $\Sigma_\alpha \, \widehat{\lambda}_\alpha \, \widehat{h}_\alpha^{(i)}$.

We check the hypothesis of the lemma . For a proper subset $\mathcal{B} \subset \mathcal{A}$, we have

$$\Sigma_\alpha \, \widehat{\lambda}_\alpha \, \widehat{h}_\alpha^{(i)} = 0, \qquad \text{for all } \widehat{\lambda} \in \Delta_{\mathcal{B}}^{(1)}$$

iff $\widehat{h}_\alpha^{(i)} = 0$ for all $\alpha \in \mathcal{B}$. By definition of $E^*_{\mathcal{B}}$, this happens iff $\tau^{(i)} \in E^*_{\mathcal{B}}$. As the $\tau^{(i)}$ are independent, Lemma 2 in the last subsection guarantees that there are less than $d - \#\mathcal{B}$ such indices i. The hypothesis of the lemma above is thus satisfied, and its conclusion gives the finiteness of the measure of $S^*(\mathcal{D})$.

We have thus proved a first statement in the theorems of Masur and Veech presented in Subsection 6.11, the finiteness of the volume of the moduli space of translation surfaces. Except in the simplest cases, it seems difficult to get the exact value of this volume through this method. Exact formulas for the volumes of the moduli spaces have been obtained by Eskin and Okounkov [**EOk**] using a different approach.

We end this section with the following statement, which is an easy consequence of the lemma above.

PROPOSITION 9.10. *The canonical measure on $S^*(\mathcal{D})$ satisfies, for all $\varepsilon > 0$*

$$m(\{(\pi, \lambda, \tau) \in S^*(\mathcal{D});\ min_\alpha \lambda_\alpha < \varepsilon\} \leqslant C\varepsilon(\log \varepsilon)^{d-2},$$

where the constant C depends only on d.

PROOF. In the context of the proof of the lemma, it is sufficient to observe that the number of $n \in \mathcal{N}$ such that $\max_\alpha n_\alpha = N$ is of the order of N^{d-2}. □

10. Ergodicity and unique ergodicity

In this section, we complete the proofs of the theorems of Masur and Veech presented in Subsection 6.11.

10.1. Hilbert metric. Let C be an open set in the projective space $\mathbb{P}^N$ which is the image of an open convex cone in $\mathbb{R}^{N+1}$ whose closure intersects some hyperplane only at the origin.

Given two distinct points $x, y \in C$, the intersection of the line through x, y with C is a segment (a, b). The crossratio of the points a, b, x, y gives rise to a distance on C called the Hilbert metric on C:

$$d_C(x, y) := |\log \frac{x-a}{y-a} \frac{x-b}{y-b}|.$$

EXERCISE 10.1. Check the triangle inequality.

The following properties are easily verified.

- Let X be a subset of C; then the closure $\overline{X}$ of X in $\mathbb{P}^N$ is contained in C iff X has finite diameter for d_C.
- If $\varphi : \mathbb{P}^N \to \mathbb{P}^N$ is a projective isomorphism, then, for all $x, y \in C$
$$d_{\varphi(C)}(\varphi(x), \varphi(y)) = d_C(x, y).$$
- If $C' \subset C$ is a smaller set satisfying the same hypothesis than C, then, for all $x, y \in C'$
$$d_C(x, y) \leqslant d_{C'}(x, y).$$

- If C' is a set satisfying the same hypothesis than C and $\overline{C}' \subset C$, there exists $k \in (0,1)$ such that, for all $x, y \in C'$

$$d_C(x,y) \leqslant k\, d_{C'}(x,y).$$

Thus, if $\varphi : \mathbb{P}^N \to \mathbb{P}^N$ is a projective isomorphism satisfying $\overline{\varphi(C)} \subset C$, there exists $k \in (0,1)$ such that, for all $x, y \in C'$ we have

$$d_C(\varphi(x), \varphi(y)) \leqslant k\, d_C(x,y).$$

10.2. Almost sure unique ergodicity. We prove that, for every combinatorial data $(\mathcal{A}, \pi)$, and almost every length vector $\lambda \in \mathbb{R}^{\mathcal{A}}$, the corresponding i.e.m is uniquely ergodic.

The set of i.e.m having a connection has codimension 1. Therefore, almost surely the Rauzy-Veech algorithm does not stop and associates to the i.e.m T an infinite path γ_T starting at π in the Rauzy diagram $\mathcal{D}$ constructed from $(\mathcal{A}, \pi)$. According to Subsection 8.1, we have to prove that the closed convex cone $\mathcal{C}(\gamma_T)$ determined by γ_T is almost surely a ray.

By Poincaré recurrence of the Teichmüller flow and Subsection 7.7, for almost every length vector λ, there exists an initial segment γ_s of γ_T which occurs infinitely many times in γ_T and such that all coefficients of the matrix B_{γ_s} are positive. We write γ_T as a concatenation

$$\gamma_T = \gamma_s * \gamma_1 * \gamma_s * \gamma_2 * \cdots .$$

Let C be the open set in $\mathbb{P}(\mathbb{R}^{\mathcal{A}})$ image of the positive cone in $\mathbb{R}^{\mathcal{A}}$. From the last property in the last subsection, there exists $k \in (0,1)$ such that B_{γ_s} decreases the Hilbert metric d_C at least by a factor k, while the B_{γ_i}, $i = 1, 2, \cdots$ do not increase d_C. The first image CB_{γ_s} has closure contained in C hence has finite diameter K for d_C. We then have

$$\operatorname{diam}(CB_{\gamma_s * \cdots \gamma_i}) \leqslant K k^{i-1}.$$

It follows that the image in $\mathbb{P}(\mathbb{R}^{\mathcal{A}})$ of $\mathcal{C}(\gamma_T)$ is a point. The result is proved.

10.3. Ergodicity of the Teichmüller flow. We will prove in this subsection that the Teichmüller flow on $\mathcal{M}^{(1)}(\mathcal{D})$ and its return maps V on $S(\mathcal{D})$ and V^* on $S^*(\mathcal{D})$ are ergodic. In view of the relation between these three dynamical systems, the three statements are equivalent. We will prove that V^* is ergodic.

From the ergodicity of V and V^*, it follows that the maps V_+ and V_+^* on $\Delta(\mathcal{D})$ are also ergodic.

By Birkhoff's ergodic theorem, for every continuous function φ on $S^*(\mathcal{D})$, there exists an almost everywhere defined fuction $\overline{\varphi}$ such that, for almost every $(\pi, \lambda, \tau) \in S^*(\mathcal{D})$, one has

$$\lim_{n \to +\infty} \frac{1}{n} \sum_{0}^{n-1} \varphi((V^*)^m(\pi, \lambda, \tau)) = \overline{\varphi}(\pi, \lambda, \tau),$$

and also

$$\lim_{n \to +\infty} \frac{1}{n} \sum_{0}^{n-1} \varphi((V^*)^{-m}(\pi, \lambda, \tau)) = \overline{\varphi}(\pi, \lambda, \tau).$$

To prove ergodicity, it is sufficient to show that $\overline{\varphi}$ is almost everywhere constant, for any continuous function φ.

Starting from almost every (π, λ, τ), one can iterate the Rauzy-Veech algorithm both forward and backward. This leads to a biinfinite path $\gamma = \gamma^+ * \gamma^-$ in the Rauzy diagram $\mathcal{D}$, where γ^+ depends only on (π, λ) and γ^- depends only on (π, τ).

By Poincaré recurrence, for almost every (π, τ), there is a finite path γ_e at the end of γ^- such that all coefficients of B_{γ_e} are positive and which appears infinitely many times in γ^-. Let again C be the open set in $\mathbb{P}(\mathbb{R}^{\mathcal{A}})$ image of the positive cone in $\mathbb{R}^{\mathcal{A}}$, d_C the associated Hilbert metric. Let λ, $\lambda' \in \Delta_\pi$; for $m \geqslant 0$, let $\lambda_{-m}, \lambda'_{-m}$ be the respective λ-components of $(V^*)^{-m}(\pi, \lambda, \tau), (V^*)^{-m}(\pi, \lambda', \tau)$. By the same argument that in the last subsection, we have

$$\lim_{m \to +\infty} d_C(\lambda_{-m}, \lambda'_{-m}) = 0.$$

This implies that, for almost every (π, τ), $\overline{\varphi}(\pi, \lambda, \tau)$ does not depend on λ.

We claim that the same argument works exchanging λ and τ, future and past. For almost every (π, λ), we want to find a finite path γ_s at the beginning of γ^+ which appears infinitely many times in γ^+ (this is guaranteed by Poincaré recurrence) and satisfies

$$\overline{\Theta}_\pi B_{\gamma_s}^{-1} \subset \Theta_{\pi'} \cup \{0\}$$

where π' is the endpoint of γ_s. Then, using the Hilbert metrics relative to the open sets images in $\mathbf{P}(\mathbb{R}^{\mathcal{A}})$ of the Θ_π, we conclude in the same way as above that, for almost every (π, λ, τ), $\overline{\varphi}(\pi, \lambda, \tau)$ does not depend on τ. Thus, almost surely, $\overline{\varphi}(\pi, \lambda, \tau)$ does not depend on λ and τ. But $\overline{\varphi}(\pi, \lambda, \tau)$ is also V^*-invariant, therefore it must be almost everywhere constant.

It remains to prove that, almost surely, some initial path γ_s of γ^+ satisfies $\overline{\Theta}_\pi B_{\gamma_s}^{-1} \subset \Theta_{\pi'} \cup \{0\}$. This is a consequence of the following result.

LEMMA 10.2. *If a finite path $\underline{\gamma}$ in $\mathcal{D}$, from a vertex π to a vertex π', is the concatenation of $3d - 4$ complete paths, then we have*

$$\overline{\Theta}_\pi B_{\underline{\gamma}}^{-1} \subset \Theta_{\pi'} \cup \{0\}.$$

PROOF. For combinatorial data π and $\tau \in \mathbb{R}^{\mathcal{A}}$, we write as before

$$h_\alpha^t = \sum_{\pi_t(\beta) \leqslant \pi_t(\alpha)} \tau_\beta, \qquad h_\alpha^b = -\sum_{\pi_b(\beta) \leqslant \pi_b(\alpha)} \tau_\beta, \qquad h_\alpha = h_\alpha^t + h_\alpha^b.$$

We write $\gamma_1, \gamma_2, \cdots, \gamma_m$ for the successive arrows of $\underline{\gamma}$.
Starting from $\pi =: \pi^0$ with a nonzero vector $\tau^0 \in \mathbb{R}^{\mathcal{A}}$ satisfying

$$(10.1) \qquad h_\alpha^{0,t} \geqslant 0 \quad \text{for } \pi_t^0(\alpha) < d, \qquad\qquad h_\alpha^{0,b} \geqslant 0 \quad \text{for } \pi_b^0(\alpha) < d,$$

we have to show that

$$(10.2) \qquad h_\alpha^{m,t} > 0 \quad \text{for } \pi_t^m(\alpha) < d, \qquad\qquad h_\alpha^{m,b} > 0 \quad \text{for } \pi_b^m(\alpha) < d,$$

where π_j is the endpoint of γ_j and $h^{j,t}$, $h^{j,b}$ are calculated from $\tau^j := \tau^{j-1} B_{\gamma_j}^{-1}$.

The heigth vectors h^j are column vectors related by

$$h^j = B_{\gamma_j} h^{j-1}$$

and their entries are nonnegative. Let $m' < m$ is the smallest integer such that the initial part $\gamma_1 * \cdots * \gamma_{m'}$ of $\underline{\gamma}$ is the concatenation of $2d - 3$ complete paths. By Proposition 7.12 in Subsection 7.7, we have

$$(10.3) \qquad h_\alpha^j > 0, \qquad\qquad \forall\, \alpha \in \mathcal{A}, \forall\, j \geqslant m'.$$

If γ_j is of top type, one has $\pi_t^j = \pi_t^{j-1}$ and

$$h_\alpha^{j,\,t} = h_\alpha^{j-1,\,t}, \qquad \text{if } \pi_t^j(\alpha) < d, \tag{10.4}$$

$$h_\alpha^{j,\,b} = h_\alpha^{j-1,\,b}, \qquad \text{if } \pi_t^{j-1}(\alpha) < d, \text{ and } \pi_b^{j-1}(\alpha) < d, \tag{10.5}$$

$$h_{\alpha_b}^{j,\,b} = h_{\alpha_t}^{j-1,\,b}, \qquad \text{with } \pi_t^{j-1}(\alpha_t) = \pi_b^{j-1}(\alpha_b) = d, \tag{10.6}$$

$$h_{\alpha_t}^{j,\,b} = h_{\alpha_*}^{j-1,\,b} + h_{\alpha_t}^{j-1}, \qquad \text{with } \pi_b^{j-1}(\alpha_*) = d-1. \tag{10.7}$$

Let $\ell^t(j)$ (resp. $\ell^b(j)$) be the largest integer ℓ such that $h_\alpha^{j,\,t} > 0$ for $\pi_t^j(\alpha) < \ell$ (resp. $h_\alpha^{j,\,b} > 0$ for $\pi_b^j(\alpha) < \ell$). We want to show that $\ell^t(m) = \ell^b(m) = d$. This implies the required conclusion.

We always have (trivially) $\ell^t(j) \geqslant 1$, $\ell^b(j) \geqslant 1$. Assume for instance that γ_j is of top type as above. Then relation (10.4) and $\pi_t^j = \pi_t^{j-1}$ imply that $\ell^t(j) \geqslant \ell^t(j-1)$. If $\pi_b^j(\alpha_t) = \pi_b^{j-1}(\alpha_t) > \ell^b(j-1)$, we have $\ell^b(j) \geqslant \ell^b(j-1)$ from (10.5). On the other hand, if $\pi_b^j(\alpha_t) \leqslant \ell^b(j-1)$ and $j > m'$, it follows from relations (10.3),(10.5),(10.6),(10.7) that $\ell^b(j) > \ell^b(j-1)$. We first conclude that ℓ^t, ℓ^b are non-decreasing functions of $j \geqslant m'$.

Let $m' < m_0 < m_1 \leqslant m$ be such that $\gamma_{m_0} * \cdots * \gamma_{m_1-1}$ is complete. Observe that there is a letter ${}_b\alpha$ such that $\pi({}_b\alpha) = 1$ for all vertices π of $\mathcal{D}$. Let $m_0 \leqslant j < m_1$ such that ${}_b\alpha$ is the winner of γ_j. Then γ_j is of top type so, in the notations above, we have ${}_b\alpha = \alpha_t$, $1 = \pi_b^j(\alpha_t) \leqslant \ell^b(j-1)$ and $\ell^b(j) > \ell^b(j-1)$. As we can find $d-1$ disjoint such complete subpaths between m' and m, this shows that $\ell^b(m) = d$. The proof that $\ell^t(m) = d$ is symmetric. $\square$

The proof of ergodicity is now complete. We recall the full statement.

THEOREM 10.3. *The maps V (on $S(\mathcal{D})$), V^* (on $S^*(\mathcal{D})$),V_+ and V_+^* (on $\Delta(\mathcal{D})$) are ergodic. The restriction of the Teichmüller flow to any component of the marked moduli space $\widetilde{\mathcal{M}}^{(1)}(M, \Sigma, \kappa)$ is ergodic. The action of $SL(2,\mathbb{R})$ on any such component is therefore also ergodic.*

11. Lyapunov exponents

The remaining sections are planned as introductions to further reading. The results are presented mostly without proofs. In this section, we introduce the Kontsevich-Zorich cocycle [**Kon**] and present the results of Forni [**For2**] and Avila-Viana [**AvVi1**].

11.1. Oseledets multiplicative ergodic theorem. Let $(X, \mathcal{B}, \mu)$ be a probability space, and let $T : X \to X$ be a measure-preserving **ergodic** transformation. Let also

$$A : X \longrightarrow GL(d, \mathbb{R})$$

be a measurable function. We assume that both $\log ||A||$ and $\log ||A^{-1}||$ are integrable. These data allow to define a linear cocycle

$$X \times \mathbb{R}^d \longrightarrow X \times \mathbb{R}^d$$

$$(x, v) \longmapsto (Tx, A(x)v).$$

Iterating this map leads to consider, for $n \geqslant 0$, the matrices

$$A^{(n)}(x) := A(T^{n-1}x) \cdots A(x).$$

When T is invertible, one can also consider, for $n < 0$

$$A^{(n)}(x) := (A^{(-n)}(T^n x))^{-1} = (A(T^n x)^{-1} \cdots (A(T^{-1}x))^{-1}.$$

To state Oseledets multiplicative theorem, we distinguish the case where T is invertible, which allows a stronger conclusion, from the general case.

THEOREM 11.1. *(Oseledets* [**Os**]*)*

1. **The invertible case:** *There exist numbers $\lambda_1 > \cdots > \lambda_r$ (the* **Lyapunov exponents***) and, at almost every point $x \in X$, a decomposition*
$$\mathbb{R}^d = F_1(x) \oplus \cdots \oplus F_r(x)$$
depending measurably on x, which is invariant under the action of the cocycle
$$A(x)F_i(x) = F_i(Tx)$$
and such that, for $1 \leqslant i \leqslant r$, $v \in F_i(x)$, $v \neq 0$, one has
$$\lim_{n\to\pm\infty} \frac{1}{n} \log ||A^{(n)}(x)v|| \; = \; \lambda_i.$$
2. **The general case:** *There exist numbers $\lambda_1 > \cdots > \lambda_r$ and, at almost every point $x \in X$, a filtration*
$$\mathbb{R}^d = E_0(x) \supset E_1(x) \supset \cdots \supset E_r(x) = \{0\}$$
depending measurably on x, which is invariant under the action of the cocycle
$$A(x)E_i(x) = E_i(Tx)$$
and such that, for $v \in E_{i-1}(x) - E_i(x)$, one has
$$\lim_{n\to+\infty} \frac{1}{n} \log ||A^{(n)}(x)v|| \; = \; \lambda_i.$$

Remarks

(1) In the invertible case, one obtains the second statement from the first by setting
$$E_i(x) = \oplus_{i+1}^r F_j(x).$$
(2) When A is independent of x, the Lyapunov exponents are the logarithms of the moduli of the eigenvalues of A and the F_i are the sums of the corresponding generalized eigenspaces.
(3) The statements above require obvious modifications for continuous time, i.e for flows and semiflows.

11.2. The Kontsevich-Zorich cocycle (discrete version). Let $\mathcal{R}$ be a Rauzy class, $\mathcal{D}$ the associated Rauzy diagram.

We have defined in subsection 9.1 the map V_+ on the space $\Delta(\mathcal{D})$ which is the dynamics in parameter space defined by the Rauzy-Veech algorithm. There is a partition mod.0

$$\Delta(\mathcal{D}) = \bigcup_\gamma \{\pi\} \times \mathbb{P}(\Delta_\gamma)$$

over arrows $\gamma : \pi \to \pi'$ of $\mathcal{D}$, such that on $\{\pi\} \times \mathbb{P}(\Delta_\gamma)$, V_+ is given by

$$V_+(\pi, \lambda) = (\pi', \lambda\, B_\gamma^{-1})\,.$$

The **(extended) Kontsevich-Zorich cocycle** is the linear cocycle $V_{+,KZ}$: $\Delta(\mathcal{D}) \times \mathbb{R}^{\mathcal{A}} \to \Delta(\mathcal{D}) \times \mathbb{R}^{\mathcal{A}}$ over V_+ defined on $\{\pi\} \times \mathbb{P}(\Delta_\gamma) \times \mathbb{R}^{\mathcal{A}}$ by

$$V_{+,KZ}(\pi, \lambda, w) = (V_+(\pi, \lambda), B_\gamma w).$$

Over the accelerated Zorich dynamics V_+^* on $\Delta(\mathcal{D})$, we similarly define

$$V_{+,KZ}^*(\pi, \lambda, w) = (V_+^*(\pi, \lambda), B_{\underline{\gamma}} w),$$

where $\underline{\gamma}$ is the path in $\mathcal{D}$ (formed of arrows of the same type, having the same winner) associated to a single iteration of V^* at the point (π, λ) under consideration.

The extended Kontsevich-Zorich cocycle has a natural interpretation in terms of Birkhoff sums. Let T be an i.e.m with combinatorial data π, length data λ, acting on an interval I. Assume that T has no connection. Let T_n (with combinatorial data $\pi^{(n)}$, length data $\lambda^{(n)}$, acting on an interval $I^{(n)} \subset I$) be the i.e.m obtained from T after n steps of the Rauzy-Veech algorithm.

For any function φ on I, one can associate a new function $S^{(n)}\varphi$ on $I^{(n)}$ by

$$S^{(n)}\varphi(x) = \sum_{0 \leqslant i < r(x)} \varphi(T^i(x)),$$

where $r(x)$ is the return time in $I^{(n)}$ of $x \in I^{(n)}$.

Let $w \in \mathbb{R}^{\mathcal{A}}$. Consider w as the function on I which takes on I_α^t the constant value w_α. Then it is easy to see that the function $S^{(n)}w$ is constant on each interval $I_\alpha^{(n),t} \subset I^{(n)}$ and thus can also be considered as a vector in $\mathbb{R}^{\mathcal{A}}$. It follows from the properties of the matrices $B_{\gamma(m,n)}$ mentioned at the end of section 7.5 that one has

$$V_{+,KZ}^n(\pi, \lambda, w) = (\pi^{(n)}, \lambda^{(n)}, S^{(n)}w).$$

As was mentioned in subsection 7.6, for any arrow $\gamma : \pi \to \pi'$, the image of $\operatorname{Im} \Omega_\pi$ under B_γ is equal to $\operatorname{Im} \Omega_{\pi'}$. One obtains the **restricted** Kontsevich-Zorich cocycle by allowing only, in the definition of $V_{+,KZ}$ or $V_{+,KZ}^*$, the vector w to vary in $\operatorname{Im} \Omega_\pi$.

When necessary, the Kontsevich-Zorich cocycle (in its extended or restricted version) can also be viewed as a linear cocycle over V or V^*. This is important when one wants to use the Oseledets theorem for invertible maps.

11.3. The Kontsevich-Zorich cocycle (continuous version). The continuous version of the Kontsevich-Zorich cocycle is defined over the Teichmüller flow $(\mathcal{T}_t)_{t\in\mathbb{R}}$ (on the moduli space $\mathcal{M}(M, \Sigma, \kappa)$, or the marked moduli space $\widetilde{\mathcal{M}}(M, \Sigma, \kappa)$) in the following way.

Consider for instance the case of the marked moduli space. Recall that we denote by $\widetilde{\mathcal{Q}}(M, \Sigma, \kappa)$ the associated marked Teichmüller space. On the product $\widetilde{\mathcal{Q}}(M, \Sigma, \kappa) \times H^1(M - \Sigma, \mathbb{R})$, we define a linear cocycle over the Teichmüller flow on $\widetilde{\mathcal{Q}}(M, \Sigma, \kappa)$ by

$$\mathcal{T}_t^{KZ}(\zeta, \theta) = (\mathcal{T}_t(\zeta), \theta).$$

The modular group $\operatorname{Mod}(M, \Sigma)$ acts in a non trivial canonical way on both factors of the product $\widetilde{\mathcal{Q}}(M, \Sigma, \kappa) \times H^1(M - \Sigma, \mathbb{R})$. The quotient is a vector bundle over the marked moduli space $\widetilde{\mathcal{M}}(M, \Sigma, \kappa)$, equipped with a flow fibered over the Teichmüller flow: this flow is the continuous version of the extended Kontsevich-Zorich cocycle. One gets the restricted version by restricting the fiber to the subspace $H^1(M, \mathbb{R}) \subset H^1(M - \Sigma, \mathbb{R})$.

Let us explicit the relation between the discrete and continuous version of the KZ-cocycle.

Let (π, λ, τ) be an element of $S(\mathcal{D})$, viewed both as (cf. Subsection 9.1) the domain of the natural extension of the Rauzy-Veech dynamics and as (cf. Subsection 9.5) a transverse section to the Teichmüller flow in $\mathcal{M}^{(1)}(\mathcal{D})$. Let $w \in \mathbb{R}^{\mathcal{A}}$. Let $(M, \Sigma, \kappa, \zeta)$ be the translation surface obtained from (π, λ, τ) by the zippered rectangle construction. As seen in Subsection 4.5, this construction provides us with a canonical basis $(\zeta_\alpha)_{\alpha \in \mathcal{A}}$ of the homology group $H_1(M, \Sigma, \mathbb{Z})$. We associate to w the homology class $\zeta_w = \sum_\alpha w_\alpha \zeta_\alpha \in H_1(M, \Sigma, \mathbb{R})$, which can also be viewed as a cohomology class in $H^1(M - \Sigma, \mathbb{R})$ from the duality provided by the intersection form.

We assume that $(M, \Sigma, \kappa, \zeta)$ has no vertical connection. From (π, λ, τ) viewed as a point in $\mathcal{M}^{(1)}(\mathcal{D}) \subset \widetilde{\mathcal{M}}(M, \Sigma, \kappa)$, we flow with the Teichmüller flow during a time t to a point $(\pi', \lambda', \tau') \in \mathcal{M}^{(1)}(\mathcal{D})$. The continuous Teichmüller trajectory corresponds to a path $\underline{\gamma}$ from π to π' in $\mathcal{D}$. As seen in Subsection 7.4, the translation surface $(M, \Sigma, \kappa, \zeta)$ is canonically isomorphic to the translation surface constructed from the data $(\pi', e^{-t}\lambda', e^t\tau')$. This isomorphism and the combinatorial data π' provides another basis $(\zeta'_\alpha)_{\alpha \in \mathcal{A}}$ for $H_1(M, \Sigma, \mathbb{Z})$ (or $H^1(M - \Sigma, \mathbb{Z})$). We express ζ_w as $\zeta_w = \sum_\alpha w'_\alpha \zeta'_\alpha$. Then, we have

$$w' = B_{\underline{\gamma}} w.$$

The two versions of the KZ-cocycle are thus seen to be equivalent.

11.4. Lyapunov spectrum of the Kontsevich-Zorich cocycle. We start with some simple observations which follow from Subsections 7.6, 9.7 and 10.3.

It follows from the proposition in Subsection 7.6 that one can choose, for each vertex π of $\mathcal{D}$, a basis for the quotient space $\mathbb{R}^{\mathcal{A}}/\operatorname{Im}\Omega_\pi$, in such a way that, for every arrow $\gamma : \pi \to \pi'$, the homomorphism from $\mathbb{R}^{\mathcal{A}}/\operatorname{Im}\Omega_\pi$ to $\mathbb{R}^{\mathcal{A}}/\operatorname{Im}\Omega_{\pi'}$ induced by B_γ corresponds to the identity matrix in the selected bases.

As a consequence, vectors in these quotient spaces stay bounded under the action of the KZ-cocycle. It follows that 0 is the unique Lyapunov exponent associated with this part of the KZ-cocycle. The multiplicity of this exponent is $s-1 = d-2g$.

By the Masur-Veech theorem stated in Subsection 6.11 and proved in Subsections 9.7 10.3, the canonical measures on $\mathcal{M}(M, \Sigma, \kappa)$ and $\widetilde{\mathcal{M}}(M, \Sigma, \kappa)$ have finite total masses, and the Teichmüller flow is ergodic with respect to these invariant measures. As seen in Subsection 9.7 and first proved by Zorich, the canonical invariant measure on $\widetilde{\mathcal{M}}(M, \Sigma, \kappa)$ induces on $S^*(\mathcal{D})$ a finite measure which is equivalent to Lebesgue measure and invariant under V^*. This measure can be projected to $\Delta(\mathcal{D})$ to obtain a finite measure, equivalent to Lebesgue measure, which is invariant under V_+^*.

We can thus apply the Oseledets theorem to the restricted KZ-cocycle, either in the continuous version over the Teichmüller flow or in the discrete version over V^* or V_+^*.

However, one has first to check the integrability condition of Subsection 10.1 We do that for the discrete version of the cocycle. From the definition of the Zorich acceleration V_+^* of the Rauzy-Veech dynamics, the norm of the matrix $B_{\underline{\gamma}}$ defining the KZ-cocycle at a point (π, λ) is bounded by

$$||B_{\underline{\gamma}}|| \leqslant C \, \frac{\sum_\alpha \lambda_\alpha}{\min_\alpha \lambda_\alpha}.$$

The same estimate holds for the inverse of this matrix. But the proposition at the end of Subsection 9.7 states that the majorant in the inequality above is larger than A on a set of measure at most $A^{-1}(\log A)^{d-2}$, which easily implies the required integrability.

Observe that the same computation shows that the return time for the Teichmüller flow on $S^*(\mathcal{D})$ is integrable. By Birkhoff's ergodic theorem, the mean value θ_1^* over $S^*(\mathcal{D})$ of this return time has the following property: for almost any point in $\zeta \in \widetilde{\mathcal{M}}(M, \Sigma, \kappa)$, we have

$$\lim_{T\to+\infty} \frac{1}{T} \, \#\{t \in [0,T];\ \mathcal{T}_t(\zeta) \in S^*(\mathcal{D})\} = \frac{1}{\theta_1^*}.$$

As a consequence, the Lyapunov exponents for the discrete KZ-cocycle over V^* or V_+^* are proportional by a factor θ_1^* to those of the continuous KZ-cocycle over $\mathcal{T}$.

EXERCISE 11.2. Show that the largest Lyapunov exponent of the continuous KZ-cocycle over $\mathcal{T}$ is equal to 1, and that the largest Lyapunov exponent of the discrete KZ-cocycle over V^* or V_+^* is equal to θ_1^*.

EXERCISE 11.3. Use the ergodicity of V_+^* to show that the largest Lyapunov exponent of the KZ-cocycle is simple.

Let $\gamma : \pi \to \pi'$ be an arrow of $\mathcal{D}$. We have also seen in Subsection 7.6 that, when we equip $\operatorname{Im}\Omega_\pi$ and $\operatorname{Im}\Omega_{\pi'}$ with the symplectic structures defined by Ω_π, $\Omega_{\pi'}$ respectively, the restriction of B_γ to $\operatorname{Im}\Omega_\pi$ is symplectic. This implies that the Lyapunov spectrum (i.e the Lyapunov exponents, counted with multiplicities) of the restricted KZ-cocycle is symmetric with respect to 0: counted with multiplicities the Lyapunov exponents of the continuous restricted KZ-cocycle have the form

$$1 = \theta_1 > \theta_2 \geqslant \dots \theta_g \geqslant \theta_{g+1} = -\theta_g \geqslant \dots \geqslant \theta_{2g-1} = -\theta_2 > \theta_{2g} = -1,$$

the Lyapunov exponents for the discrete restricted KZ-cocycle over V^* or V_+^* being the $\theta_i^* := \theta_1^* \, \theta_i$.

Kontsevich and Zorich conjectured that all Lyapunov exponents of the restricted KZ-cocycle are simple. In particular, this stipulates that $\theta_g > \theta_{g+1} = -\theta_g$, hence that the restricted KZ-cocycle is *hyperbolic* in the sense that it does not have 0 as Lyapunov exponent. Forni then proved the hyperbolicity of the restricted KZ-cocycle before Avila and Viana proved the full conjecture of Kontsevich and Zorich.

THEOREM 11.4. *(Forni* [**For2, Kri**]*) The restricted Kontsevich-Zorich cocycle is hyperbolic.*

The (Lyapunov) hyperbolicity of the KZ-cocycle holds w.r.t the invariant measure equivalent to Lebesgue measure, but not to any invariant measure.

EXERCISE 11.5. In the Rauzy diagram with $g = 2, d = 4$, find a **complete** loop γ such that B_γ has two eigenvalues of modulus 1.

Observe that when $g = 2$, Forni's theorem already implies that the Lyapunov spectrum of the KZ-cocycle is simple. For higher genus, we have

THEOREM 11.6. *(Avila-Viana* [**AvVi1, AvVi2**]*) The Lyapunov spectrum of the restricted Kontsevich-Zorich cocycle is simple.*

The proofs of both theorems (Avila-Viana's approach is quite different from Forni's) are beyond the scope of these notes.

The Lyapunov exponents of the restricted discrete KZ-cocycle over V^* and V_+^* are the same. The conclusions of the Oseledets theorem are however slightly different.

- For almost every $(\pi, \lambda, \tau) \in S^*(\mathcal{D})$, there exists a direct sum decomposition into 1-dimensional subspaces
$$\operatorname{Im}\Omega_\pi = \oplus_1^{2g} F_i(\pi, \lambda, \tau),$$
such that, for $w \in F_i(\pi, \lambda, \tau), w \neq 0$, we have, writing $(V_{KZ}^*)^n(\pi, \lambda, \tau, w) = ((V^*)^n(\pi, \lambda, \tau), w_n)$
$$\lim_{n \to \pm\infty} \frac{1}{n} \log \frac{||w_n||}{||w||} = \theta_i^*.$$
- For almost every $(\pi, \lambda) \in \Delta(\mathcal{D})$, there exists a filtration
$$\operatorname{Im}\Omega_\pi = E_0(\pi, \lambda) \supset E_1(\pi, \lambda) \supset \ldots \supset E_{2g}(\pi, \lambda) = \{0\},$$
with $\operatorname{codim} E_i(\pi, \lambda) = i$, such that, for $w \in E_{i-1}(\pi, \lambda) - E_i(\pi, \lambda)$, writing $(V_{+,KZ}^*)^n(\pi, \lambda, w) = ((V_+^*)^n(\pi, \lambda), w_n)$, we have
$$\lim_{n \to +\infty} \frac{1}{n} \log \frac{||w_n||}{||w||} = \theta_i^*.$$

For almost every $(\pi, \lambda, \tau) \in S^*(\mathcal{D})$, and every $0 \leqslant i < 2g$, the direct sum $\oplus_{i+1}^{2g} F_i(\pi, \lambda, \tau)$ is independent of τ and equal to $E_i(\pi, \lambda)$. Symmetrically, for almost every $(\pi, \lambda, \tau) \in S^*(\mathcal{D})$, and every $0 < i \leqslant 2g$, the direct sum $\oplus_1^i F_i(\pi, \lambda, \tau)$ is independent of λ.

When one considers (assuming $s > 1$) the **extended** KZ-cocycle over V^* or V_+^*, one obtains moreover

- In the invertible case, a subspace $F_*(\pi, \lambda, \tau)$, which complements $\operatorname{Im}\Omega_\pi$ and has dimension $s - 1$, associated to the exponent 0;
- In the non invertible case, the subspaces associated to the positive exponents are the
$$E_i^*(\pi, \lambda) := E_i(\pi, \lambda) \oplus F_*(\pi, \lambda, \tau), \qquad \forall\, 0 \leqslant i \leqslant g,$$
which satisfy $\operatorname{codim} E_i^*(\pi, \lambda) = i$. The subspace associated to the exponent 0 is
$$E^*(\pi, \lambda) := F_*(\pi, \lambda, \tau) \oplus E_{g+1}(\pi, \lambda),$$
and those associated with the negative exponents θ_i^*, $g < i \leqslant 2g$ are the $E_i(\pi, \lambda)$.

11.5. Lyapunov exponents of the Teichmüller flow. Recall that $S(\mathcal{D})$ was identified in Subsection 9.5 with the transverse section to the Teichmüller flow in $\mathcal{M}^{(1)}(\mathcal{D})$
$$\{(\pi, \lambda, \tau) \in \bigsqcup_\pi \{\pi\} \times \Delta \times \Theta_\pi;\ \sum_\alpha \lambda_\alpha = 1,\ \tau\, \Omega_\pi\, {}^t\lambda = 1\},$$

the return map being given by the Rauzy-Veech invertible dynamics V. Thus, a number of iterations of V, associated to a path $\underline{\gamma} : \pi \to \pi'$ in $\mathcal{D}$, correspond to the Teichmüller time

$$\log \frac{||\lambda||_1}{||\lambda \, B_{\underline{\gamma}}^{-1}||_1} = -\log ||\lambda \, B_{\underline{\gamma}}^{-1}||_1$$

and to the return map

$$(\pi, \lambda, \tau) \mapsto (\pi', \frac{\lambda \, B_{\underline{\gamma}}^{-1}}{||\lambda \, B_{\underline{\gamma}}^{-1}||_1}, ||\lambda \, B_{\underline{\gamma}}^{-1}||_1 \, \tau \, B_{\underline{\gamma}}^{-1}).$$

From Subsection 7.6, we know that the action of $B_{\underline{\gamma}}^{-1}$ on row vectors in $\operatorname{Ker}\Omega_\pi$ is neutral and the action on the quotient $\mathbb{R}^{\mathcal{A}}/\operatorname{Ker}\Omega_\pi \simeq \operatorname{Im}\Omega_\pi$ is given by $B_{\underline{\gamma}}$. From this we deduce immediately the Lyapunov exponents of the Teichmüller flow on $\mathcal{M}^{(1)}(\mathcal{D})$ (with respect to the canonical invariant measure)

- There are, counted with multiplicities, $d-1 = (2g-1)+(s-1)$ positive Lyapunov exponents which are the simple exponents
$$2 = 1+\theta_1 > 1+\theta_2 > \ldots > 1+\theta_{2g-1}$$
and, when $s > 1$, the exponent 1 (between $1+\theta_g$ and $1+\theta_{g+1}$) with multiplicity $s-1$.
- There are symmetrically $d-1 = (2g-1)+(s-1)$ negative Lyapunov exponents which are the simple exponents
$$-1+\theta_2 > \ldots > -1+\theta_{2g-1} > -1+\theta_{2g} = -2$$
and, when $s > 1$, the exponent -1 with multiplicity $s-1$.
- Finally, the exponent $0 = 1+\theta_{2g} = -1+\theta_1$ was killed by the normalization conditions on λ and τ, but is still present with multiplicity 1 in the direction of the flow.
- When considering the flow in $\mathcal{M}(\mathcal{D})$, the exponent 0 has multiplicity 2 because the foliation by the levels of the area map A is invariant.
- The strong local stable manifold of a point $(\pi, \lambda_0, \tau_0) \in \mathcal{M}(\mathcal{D})$ has equation $\{\lambda = \lambda_0, \ (\tau - \tau_0)\,\Omega_\pi\, {}^t\lambda_0 = 0\}$. Similarly, the strong local unstable manifold has equation $\{\tau = \tau_0, \ \tau_0\,\Omega_\pi\, {}^t(\lambda - \lambda_0) = 0\}$.

11.6. Deviation of ergodic averages. Let T be an i.e.m with combinatorial data $(\mathcal{A}, \pi)$ and domain $\sqcup I_\alpha^t$. Given a point x_0, a letter $\alpha \in \mathcal{A}$ and an integer k, denote the number of visits to I_α^t of the orbit of x_0 up to time k by

$$\chi_\alpha(k) := \#\{i \in [0,k)\,;\ T^i(x_0) \in I_\alpha^t\}\,.$$

How do these numbers behave as k goes to $+\infty$? This was one of the questions that led Kontsevich and Zorich to introduce their cocycle.

A first answer is provided by Birkhoff's theorem: by the theorem of Masur and Veech, for almost all length data λ, T is ergodic w.r.t Lebesgue measure. Therefore, for such a T, one has , for all $\alpha \in \mathcal{A}$ and almost all x_0

$$\lim_{k \to +\infty} \frac{1}{k}\chi_\alpha(k) = |I_\alpha^t| = \lambda_\alpha\,.$$

A slightly better answer is obtained by using that, by the same theorem of Masur and Veech, almost all T are actually uniquely ergodic. Indeed, if f is a uniquely ergodic minimal homeomorphism of a compact metric space X and φ is a continuous function on X, the convergence of the Birkhoff sums of φ holds for **any** initial value x_0 and is uniform in x_0. Here, T is not a homeomorphism and the characteristic function of I^t_α is not continuous but this is not a problem in reason of the following trick.

Split any point in the forward orbit of the singularities of T^{-1} and the backward orbit of the singularities of T into its left and right limit. One obtains, equipped with the order topology, a compact metric space $\widehat{I}$. The i.e.m T induces on $\widehat{I}$ a homeomorphism $\widehat{T}$ which is easily seen to be uniquely ergodic when T is. Also, the interval I^t_α corresponds to a clopen set in $\widehat{I}$ so its characteristic function is continuous.

A much more precise answer on the speed of convergence of the $\frac{1}{k}\chi_\alpha(k)$ is obtained using the KZ-cocycle.

Assume that T has no connection. Let $(I^{(n)})_{n\geqslant 0}$ be the intervals of induction for the Rauzy-Veech algorithm, $(T_n)_{n\geqslant 0}$ the corresponding i.e.m, $w\in\mathbb{R}^{\mathcal{A}}$. Viewing w as the function on $\sqcup I^t_\alpha$ with constant value w_α on I^t_α, the Birkhoff sums of w are given by

$$S_k w(x_0) = \sum_\alpha w_\alpha \chi_\alpha(k)\ .$$

On the other hand, we have seen in Subsection 11.2 that the KZ-cocycle is directly related to the Birkhoff sums $S^{(n)}w$ of w corresponding to the return to $I^{(n)}$.

In order to relate $S_k w(x_0)$ to the $S^{(n)}w$, we introduce the point x^* of the orbit $\{T^j(x_0);\ 0\leqslant j\leqslant k\}$ which is closest to the left endpoint of I. We consider separately in $S_k w(x_0)$ the part of the sum which is before x^* and the part which is after x^*. Thus, we have just to consider Birkhoff sums $S_j w(x^*)$ (with $j\in\mathbb{Z}$).

Consider such a sum $S_j w(x^*)$, with for instance $j\geqslant 0$ (the case $j\leqslant 0$ is similar). In the orbit $\{T^\ell(x^*);\ 0\leqslant \ell\leqslant j\}$, there exists a unique subsequence $(x^*_s)_{0\leqslant s\leqslant r} = (T^{j_s}(x^*))_{0\leqslant s\leqslant r}$, and a sequence $(n_s)_{0\leqslant s\leqslant r}$ with the following properties:

- $0 = j_0 < j_1 < \ldots < j_r = j$;
- $0 \leqslant n_r \leqslant \ldots \leqslant n_0$;
- the point x^*_s belongs to $I^{(n_s)}$ for $0\leqslant s\leqslant r$;
- the point x^*_s does not belong to $I^{(n_s+1)}$ for $1\leqslant s\leqslant r$;
- the point $T^\ell(x^*)$ does not belong to $I^{(n_s)}$ for $1\leqslant s\leqslant r$, $j_{s-1}<\ell<j_s$;

This means that the sum $\sum_{j_{s-1}}^{j_s-1} w(T^\ell(x^*))$ corresponds to a first return in $I^{(n_s)}$. Writing α_s for the letter such that $x^*_{s-1}\in I^{t,(n_s)}_{\alpha_s}$, we have

$$S_j w(x^*) = \sum_1^r (S^{(n_s)}w)_{\alpha_s}\ .$$

As the return time of T_n in $I^{(n+1)}$ is 1 or 2, we have actually $n_0 > n_1 > \ldots > n_r$. On the other hand, assume that the data (π,λ) of T are typical for Oseledets theorem applied to the KZ-cocycle; when $w\in E^*_i(\pi,\lambda)$ for some $0\leqslant i<g$ (resp. $w\in E^*_g(\pi,\lambda)$), one has

$$\limsup \frac{\log ||S^{(n)}w||}{\log ||S^{(n)}1||} = \theta_{i+1},$$

$$(\text{resp.} \limsup \frac{\log ||S^{(n)}w||}{\log ||S^{(n)}1||} = 0.\ )$$

From this, one obtains the following result

THEOREM 11.7. [**Zo3**] *For almost every i.e.m. $T = T_{\pi,\lambda}$, and all $x \in I$, one has*

$$\limsup \frac{\log |S_k w(x)|}{\log k} \leqslant \theta_{i+1}$$

if $w \in E_i^*(\pi, \lambda)$ for some $0 \leqslant i < g$ and

$$\limsup \frac{\log |S_k w(x)|}{\log k} = 0$$

if $w \in E_g^*(\pi, \lambda)$.

There is a similar interpretation of the KZ-cocycle in terms of the way that the orbits of the vertical flow of a typical translation surface wind around the surface: see [**Zo1, Zo4**].

12. The cohomological equation

We present in this section the main result of [**MmMsY**]. Let $f : X \to X$ be a map. The *cohomological equation* associated to this dynamical system is

$$\psi \circ f - \psi = \varphi,$$

where φ is a given function on X (usually assumed to have some degree of smoothness), and ψ is an unknown function on X (generally required to have another degree of smoothness).

12.1. Irrational numbers of Roth type.

DEFINITION 12.1. An irrational number α is of *Roth type* if, for every $\varepsilon > 0$, there exists $C = C_\varepsilon > 0$ such that, for every rational $\frac{p}{q}$, one has

$$|\alpha - \frac{p}{q}| \geqslant \frac{C}{q^{2+\varepsilon}}.$$

The reason for the terminology is the celebrated result

THEOREM 12.2. *(Roth) Every irrational algebraic number is of Roth type.*

Let $\alpha = [a_0; a_1, \ldots]$ be the continuous fraction decomposition of the irrational number α, and let $(\frac{p_n}{q_n})$ be the associated convergents of α. Then α is of Roth type iff $q_{n+1} = \mathcal{O}(q_n^{1+\varepsilon})$ for all $\varepsilon > 0$; this can be reformulated as $a_{n+1} = \mathcal{O}(q_n^\varepsilon)$ for all $\varepsilon > 0$.

The set of irrational numbers of Roth type has full Lebesgue measure: indeed, for every $q \geqslant 1$, $C > 0$, the set of $\alpha \in (0,1)$ such that

$$|\alpha - \frac{p}{q}| < \frac{C}{q^{2+\varepsilon}}$$

for some $p \in \mathbb{Z}$ has measure $\leqslant 2Cq^{-1-\varepsilon}$ and the series $\sum_{q \geqslant 1} q^{-1-\varepsilon}$ is convergent.

Standard methods of harmonic analysis allow to prove the following fundamental result, where R_α denotes the rotation $x \mapsto x + \alpha$ on $\mathbb{T}$.

THEOREM 12.3. *Let α be an irrational number of Roth type and let r, s be nonnegative real numbers with $r - s > 1$. For every function $\varphi \in C^r(\mathbb{T})$ of mean value 0, there exists a unique function $\psi \in C^s(\mathbb{T})$ of mean value 0 such that*

$$\psi \circ R_\alpha - \psi = \varphi.$$

12.2. Interval exchange maps of Roth type. Let T be an interval exchange map, $(\mathcal{A}, \pi)$ its combinatorial data; denote by $\mathcal{R}$ the Rauzy class of π and by $\mathcal{D}$ the associated Rauzy diagram.

We assume that T has no connection. The Rauzy-Veech algorithm applied to T produces an infinite path γ in $\mathcal{D}$ starting from π. From Proposition 1 in Subsection 7.7, the path γ is ∞-complete. We can therefore write in a unique way γ as a concatenation

$$\gamma = \gamma_1 * \gamma_2 * \ldots * \gamma_n * \ldots$$

where each γ_i is complete but no strict initial subpath of γ_i is complete. We write $\gamma(n)$ for the initial part

$$\gamma(n) = \gamma_1 * \gamma_2 * \ldots * \gamma_n$$

of γ.

We say that T is an i.e.m *of Roth type* if it satisfies the three conditions (a), (b), (c) below.

(a): *For every $\varepsilon > 0$, there exists $C = C_\varepsilon$ such that, for all $n > 0$, one has*

$$||B_{\gamma_n}|| \leqslant C\,||B_{\gamma(n-1)}||^{\varepsilon}.$$

EXERCISE 12.4. Let $x \mapsto x + \alpha$ be an irrational rotation on $\mathbb{T}$ and let T be the associated i.e.m with two intervals. Show that α is of Roth type iff T satisfies condition (a).

Let $\lambda \in \mathbb{R}^{\mathcal{A}}$ be the length data of T and let $E_1 = \{\sum_\alpha \lambda_\alpha w_\alpha = 0\}$; this hyperplane of $\mathbb{R}^{\mathcal{A}}$ should be viewed as the space of functions w, constant on each I^t_α, of mean value 0.

(b): *There exists $\theta > 0$, $C > 0$, such that, for all $n > 0$, one has*

$$||B_{\gamma(n)}|_{E_1}|| \leqslant C\,||B_{\gamma(n)}||^{1-\theta}.$$

EXERCISE 12.5. Show that condition (b) is always satisfied when $d = 2$.

EXERCISE 12.6. Show that condition (b) implies that T is uniquely ergodic.

EXERCISE 12.7. Assume that T satisfies the following reinforcement of condition (a): there exists $C > 0$ such that $||B_{\gamma_n}|| < C$ for all $n > 0$. Show that this imply that T satisfies condition (b).

EXERCISE 12.8. Show that the condition of the last exercise is satisfied iff the orbit of (π, λ) under V is relatively compact in $\Delta(\mathcal{D})$.

In order to state part (c) of the definition of Roth type i.e.m, we define, for $\ell \geqslant k$

$$\gamma(k, \ell) = \gamma_{k+1} * \ldots * \gamma_\ell$$

and introduce, for $k \geqslant 0$

$$E^s(k) := \{w \in \mathbb{R}^{\mathcal{A}}\,;\ \limsup_{\ell \to +\infty} \frac{\log ||B_{\gamma(k,\ell)} w||}{\log ||B_{\gamma(k,\ell)}||} < 0\}.$$

Observe that $E^s(k)$ is a vector subspace of $\mathbb{R}^{\mathcal{A}}$ which is sent by $B_{\gamma(k,\ell)}$ onto $E^s(\ell)$. Denote by $B^\flat_{k,\ell}$ the restriction of $B_{\gamma(k,\ell)}$ to $E^s(k)$ and by $B^\sharp_{k,\ell}$ the map from $\mathbb{R}^{\mathcal{A}}/E^s(k)$ to $\mathbb{R}^{\mathcal{A}}/E^s(\ell)$ induced by $B_{\gamma(k,\ell)}$.

(c): *For every* $\varepsilon > 0$, *there exists* $C = C_\varepsilon$ *such that, for all* $\ell \geqslant k$, *we have*

$$\begin{aligned} ||B^\flat_{k,\ell}|| &\leqslant C\,||B_{\gamma(\ell)}||^\varepsilon, \\ ||(B^\sharp_{k,\ell})^{-1}|| &\leqslant C\,||B_{\gamma(\ell)}||^\varepsilon. \end{aligned}$$

Assume that μ is a probability measure which is invariant under the dynamics V generated by the Rauzy-Veech algorithm or the accelerated version V^*. Assume also that the integrability condition of Oseledets's theorem is satisfied by the Kontsevich-Zorich cocycle w.r.t μ. For instance, μ could be the canonical V^*-invariant measure absolutely continuous w.r.t Lebesgue, or could be supported by a periodic orbit of V (or more generally a compact V-invariant subset of $\Delta(\mathcal{D})$).

Then, property (c) is satisfied by μ-almost all T. The spaces $E^s(k)$ are the stable subspaces associated to the negative Lyapunov exponents (relative to μ) and the estimates in (c) follow from the conclusions of Oseledets's theorem.

Property (b) is also satisfied by μ-almost all T. Indeed, the largest Lyapunov exponent for μ is simple, with associated hyperplane equal to E_1 (the simplicity of the largest exponent for μ is proven from the positivity of the matrices B as in Subsection 10.2).

Regarding property (a), no general statement w.r.t any invariant probability μ as above is known. On the other hand, with respect to the canonical V^*-invariant measure absolutely continuous w.r.t Lebesgue, (or equivalently w.r.t Lebesgue measure), almost all T satisfy property (a): this follows from a stronger statement that will be presented in Section 14 We thus obtain

PROPOSITION 12.9. *For any combinatorial data* $(\mathcal{A}, \pi)$, *and Lebesgue almost any length vector* λ, *the i.e.m* T *constructed from these data is of Roth type.*

12.3. The cohomological equation for interval exchange maps. The first and decisive breakthrough concerning the cohomological equation for i.e.m of higher genus was obtained by Forni [**For1**]. He actually works with the (nonzero) constant vectorfields X on a translation surface $(M, \Sigma, \kappa, \zeta)$ for which the cohomological equation takes the form

$$X.\Psi = \Phi.$$

He defines from the flat metrics associated to the structure of translation surface a family $H^s(M)$ of Sobolev spaces and obtains the following result

THEOREM 12.10. *(Forni* [**For1, For3**]*) Let* $k \geqslant 0$ *be an integer and* r, s *be real numbers satisfying* $s - 3 > k > r$. *For almost all constant unit vectorfields* X *on* $(M, \Sigma, \kappa, \zeta)$, *and all functions* $\Phi \in H^s(M)$ *satisfying* $D.\Phi = 0$ *for all* $D \in \mathcal{I}^s_X$, *there exists* $\Psi \in H^r(M)$ *such that* $X.\Psi = \Phi$. *Here,* $\mathcal{I}^s_X$ *is the finite-dimensional space of* X*-invariant distributions in* $H^{-s}(M)$.

A slight drawback of Forni's theorem is that no explicit description of the set of "good" directions for which it is possible to solve the cohomological equation is given. This is addressed by the next result.

Let T be an interval exchange map, $(\mathcal{A}, \pi)$ its combinatorial data, $\sqcup I^t_\alpha$ the domain of T. We denote by $\mathrm{BV}^1_*(\sqcup I^t_\alpha)$ the Banach space of functions φ on $\sqcup I^t_\alpha$ with the following properties

- the restriction of φ to each I^t_α is absolutely continuous and its derivative is a function of bounded variation;

- the mean value of the derivative $D\varphi$ over $\sqcup I_\alpha^t$ is 0.

REMARK 12.11. The first property implies that the limits $\varphi((u_i^t)^+)$ (for $0 \leqslant i < d$) and $\varphi((u_i^t)^-)$ (for $0 < i \leqslant d$) exist, where $u_0 = u_0^t$, $u_d = u_d^t$ are the endpoints of the domain of t and $u_1^t < \ldots < u_{d-1}^t$ are the singularities of T. Then the second condition is

$$\sum_1^{d-1}(\varphi((u_i^t)^+) - \varphi((u_i^t)^-)) + \varphi(u_0^+) - \varphi(u_d^-) = 0.$$

THEOREM 12.12. [**MmMsY**] *Assume that T has no connection and is of Roth type. Then, for every function $\varphi \in \mathrm{BV}_*^1(\sqcup I_\alpha^t)$, there exists a bounded function ψ on $\sqcup I_\alpha^t$ and a function χ which is constant on each I_α^t such that*

$$\psi \circ T - \psi = \varphi - \chi \quad .$$

REMARK 12.13. The solution (ψ, χ) of the equation is unique if one restricts ψ, χ to smaller subspaces. More precisely, let E_T be the subspace of $\mathbb{R}^{\mathcal{A}}$ formed of the functions χ, constant on each I_α^t, which can be written as $\psi \circ T - \psi$ for some bounded function ψ; let E_T^* be a complementary subspace of E_T in $\mathbb{R}^{\mathcal{A}}$. Then, under the hypotheses of the theorem, one can find a unique pair (ψ, χ) satisfying moreover that ψ has mean value 0 and that $\chi \in E_T^*$. The quotient space $\mathbb{R}^{\mathcal{A}}/E_T$ can thus be seen as the obstruction to solve the cohomological equation for the smoothness data under consideration.

As the derivative of T is 1 on each I_α^t, differentiating the cohomological equation leads to the same equation for derivatives of φ, ψ, with only constants of integration to keep under control. A result on the cohomological equation in higher smoothness is therefore easily deduced from the basic result above.

For $r \geqslant 1$, let $\mathrm{BV}_*^r(\sqcup I_\alpha^t)$ be the space of functions φ on $\sqcup I_\alpha^t$ such that

- the restriction of φ to each I_α^t is of class C^{r-1}, $D^{r-1}\varphi$ is absolutely continuous on I_α^t and $D^r\varphi$ is a function of bounded variation;
- the mean value of the derivative $D^j\varphi$ over $\sqcup I_\alpha^t$ is 0 for every integer $0 < j \leqslant r$.

On the other hand, let I be the interval supporting the action of T. Denote for $r \geqslant 2$ by $C^{r-2+\mathrm{Lip}}(I)$ the space of functions ψ on I which are of class C^{r-2} **on all of** I and such that $D^{r-2}\psi$ is Lipschitz on I.

Finally, for $r \geqslant 1$, let $E(r)$ be the space of functions χ on $\sqcup I_\alpha^t$ such that

- the restriction of χ to each I_α^t is a polynomial of degree $< r$;
- the mean value of the derivative $D^j\chi$ over $\sqcup I_\alpha^t$ is 0 for every integer $0 < j < r$.

One has then

THEOREM 12.14. *Assume that T has no connection and is of Roth type. Let r be an integer $\geqslant 2$. Then, for every function $\varphi \in \mathrm{BV}_*^r(\sqcup I_\alpha^t)$, there exists a function $\psi \in C^{r-2+\mathrm{Lip}}(I)$ and a function $\chi \in E(r)$ such that*

$$\psi \circ T - \psi = \varphi - \chi \quad .$$

12.4. Sketch of the proof. We give some indications about the steps of the proof of the theorem.

We want to use the following classical result.

THEOREM 12.15. *(Gottschak-Hedlund) Let f be a minimal homeomorphism of a compact metric space X, let x_0 be a point of X, and let φ be a continuous function on X. The following are equivalent:*

(1) *The Birkhoff sums $\sum_0^{n-1} \varphi \circ f^i(x_0)$ are bounded.*
(2) *There exists a continuous function ψ on X such that*

$$\psi \circ f - \psi = \varphi.$$

By splitting each point in the orbits of the singularities of T and T^{-1} into its left and right limit, one obtain a compact metric space $\widehat{I}$ on which T induces a minimal homeomorphism. Moreover, every continuous function $\widehat{\psi}$ on $\widehat{I}$ induces a bounded function on I. Therefore, in view of the theorem of Gottschalk-Hedlund, it is sufficient to find, for every $\varphi \in \mathrm{BV}^1_*(\sqcup I^t_\alpha)$, a function χ, constant on each I^t_α, such that the Birkhoff sums of $\varphi - \chi$ are bounded.

Let $\mathrm{BV}(\sqcup I^t_\alpha)$ be the Banach space of functions φ_1 of bounded variation on $\sqcup I^t_\alpha$, equipped with the norm

$$\begin{aligned} ||\varphi_1||_{BV} &:= \sup_{\sqcup I^t_\alpha} |\varphi_1(x)| + |\varphi_1|_{BV}, \\ |\varphi_1|_{BV} &:= \sum_\alpha \mathrm{Var}_{I^t_\alpha} \varphi_1 . \end{aligned}$$

Let $I^{(n)} = \sqcup I^{t,(n)}_\alpha \subset I$ be the interval of induction for the step of the Rauzy-Veech algorithm associated to the initial path $\gamma(n)$ of γ (notations of Subsection 12.2). A simple but crucial observation, in the spirit of the Denjoy estimates for circle diffeomorphisms, is that, for $\varphi_1 \in \mathrm{BV}(\sqcup I^t_\alpha)$, the Birkhoff sum $S^{(n)}\varphi_1$ corresponding to returns in $I^{(n)}$ (see Subsection 11.2) satisfy $S^{(n)}\varphi_1 \in \mathrm{BV}(\sqcup I^{t,(n)}_\alpha)$ with

$$|S^{(n)}\varphi_1|_{BV} \leqslant |\varphi_1|_{BV} .$$

This estimate is the basic ingredient in the proof of the

PROPOSITION 12.16. *Assume that T has no connection and satisfy conditions (a) and (b) of Subsection 12.2. For every function $\varphi_1 \in \mathrm{BV}(\sqcup I^t_\alpha)$ of mean value 0, and every $n \geqslant 0$, we have*

$$\sup_{\sqcup I^{t,(n)}_\alpha} |S^{(n)}\varphi_1(x)| \leqslant C\, ||B_{\gamma(n)}||^{1-\frac{\theta}{2d}} ||\varphi_1||_{BV},$$

where C depends only on the constants in condition (a) and (b).

From condition (a), the lengths $|I^{t,(n)}_\alpha|$ satisfy

$$\lim_{n\to+\infty} \frac{\log |I^{t,(n)}_\alpha|}{\log ||B_{\gamma(n)}||} = -1 .$$

Therefore, for every $\varphi_1 \in \mathrm{BV}(\sqcup I^t_\alpha)$ of mean value 0, and every $n \geqslant 0$, there exists a primitive $\varphi_0 \in \mathrm{BV}^1_*(\sqcup I^t_\alpha)$ of φ_1 (one constant of integration being chosen for each I^t_α) such that

$$\sup_{\sqcup I^{t,(n)}_\alpha} |S^{(n)}\varphi_0(x)| \leqslant C\, ||B_{\gamma(n)}||^{-\frac{\theta}{3d}} ||\varphi_1||_{BV}.$$

Using condition (c) of Subsection 12.2, one can change the order of the quantifiers to make the primitive φ_0 independent of n and still satisfy

$$\sup_{\sqcup I_\alpha^{t,(n)}} |S^{(n)}\varphi_0(x)| \leqslant C\,||B_{\gamma(n)}||^{-\omega}||\varphi_1||_{BV},$$

for some $\omega > 0$. But the last estimate, together with condition (a) of 12.2, easily imply that the Birkhoff sums of φ_0 are bounded. This proves the required result: starting from any $\varphi \in \mathrm{BV}^1_*(\sqcup I^t_\alpha)$, we take $\varphi_1 := D\varphi \in \mathrm{BV}(\sqcup I^t_\alpha)$; it has mean value 0 and therefore has a primitive φ_0 such that the Birkhoff sums of φ_0 are bounded. The difference $\varphi - \varphi_0$ is constant on every I^t_α.

13. Connected components of the moduli spaces

We present in this section the classification of the connected components of the moduli space $\mathcal{M}(M,\Sigma,\kappa)$ by Kontsevich and Zorich [**KonZo**]. The classification of the connected components of the marked moduli space is the same: it is easy to see that the canonical covering map from $\widetilde{\mathcal{M}}(M,\Sigma,\kappa)$ to $\mathcal{M}(M,\Sigma,\kappa)$ induces a bijection at the π_0 level. Observe also that for classification purposes, we can and will assume that all ramification indices κ_i are > 1.

13.1. Hyperelliptic components. Let $d \geqslant 4$ be an integer, and let $P \in \mathbb{C}[z]$ be a polynomial of degree $d+1$ with simple roots. Adding 1 or 2 points at infinity (depending on whether d is even or odd) to the complex curve $\{w^2 = P(z)\}$, one obtains an hyperelliptic compact Riemann surface M of genus $g = [\frac{d}{2}]$. The holomorphic 1-form $\omega := \frac{dz}{w}$ has no zero at finite distance. When d is even, it has a zero of order $d-2 = 2g-2$ at the single point A_1 at infinity. When d is odd, it has a zero of the same order $g-1 = \frac{d-3}{2}$ at each of the two points A_1, A_2 at infinity.

The translation surface defined by (M,ω) has therefore the following data:

- $s = 1$, $\kappa_1 = 2g-1$ if d is even;
- $s = 2$, $\kappa_1 = \kappa_2 = g$ if d is odd.

Moreover we have $d = 2g + s - 1$ in all cases so d is the complex dimension of the corresponding moduli space.

Observe that, for $a \in \mathbb{C}^*, b \in \mathbb{C}$, the polynomials P and $a^{-2}P(az+b)$ produce isomorphic translation surfaces. One has therefore exactly d independent complex parameters to deform the translation surface through a change of polynomial P. It is not difficult to see that one gets in this way, for each integer $d \geqslant 4$, a whole connected component of the corresponding moduli space. Such connected components are called *hyperelliptic*.

Hyperelliptic components correspond to the simplest Rauzy classes. Let $\#\mathcal{A} = d$. A Rauzy class containing some combinatorial data $\pi = (\pi_t, \pi_b)$ such that $\pi_t(\alpha) + \pi_b(\alpha) = d+1$ for all $\alpha \in \mathcal{A}$ is associated to the hyperelliptic component of dimension d.

When $g = 2$, the values $d = 4$ and $d = 5$ correspond to a double zero or two simple zeros for ω respectively. It is immediate to check that the hyperelliptic Rauzy classes described above are the only ones giving these values of (g, s, κ). Therefore, the two strata of the moduli space in genus 2 are connected and hyperelliptic.

Kontsevich and Zorich discovered that the situation is quite different in genus $\geqslant 3$.

13.2. Parity of spin structure. Let $(M, \Sigma, \kappa, \zeta)$ be a translation surface such that all κ_i are **odd**. We denote as usual $\Sigma = (A_1, \ldots, A_s)$. The divisor $D = \sum \frac{(\kappa_i - 1)}{2} A_i$ defines a *spin structure* on the Riemann surface M (equipped with the complex structure defined by the structure of translation surface). The *parity* of this spin structure is the parity of the dimension of the space of meromorphic functions f on M such that $(f) + D \leqslant 0$.

The reader should consult [At], [Mil] for some fundamental facts and results about spin structures and their parity. A fundamental result is that the parity of the spin structure is invariant under deformation, and is therefore the same for all translation surfaces in a same connected component of the moduli space.

The parity of the spin structure can be computed in the following way. For a smooth loop $c : \mathbb{S}^1 \to M - \Sigma$, define the index $ind(c)$ to be the degree mod 2 of the map which associates to $t \in \mathbb{S}^1$ the angle between the tangent vector $\dot{c}(t)$ and the horizontal direction at $c(t)$. As all ramification indices κ_i are odd, the index depends only on the class of c in $H_1(M, \mathbb{Z})$. Now let a_i, b_i, $1 \leqslant i \leqslant g$ be smooth loops in $M - \Sigma$ such that their homology classes form a standard symplectic basis of $H_1(M, \mathbb{Z})$. The parity of the spin structure for the translation surface $(M, \Sigma, \kappa, \zeta)$ is then given by

$$\sum_1^g (ind(a_i) + 1)(ind(b_i) + 1) \quad \text{mod. } 2.$$

13.3. Classification. Kontsevich and Zorich have shown that hyperellipticity and parity of spin structure are sufficient to classify components. More precisely

THEOREM 13.1. [**KonZo**] *Let (g, s, κ) be combinatorial data (with all $\kappa_i > 1$) determining a moduli space for translation surfaces.*

(1) *If at least one of the κ_i is even, the moduli space is connected, except when $s = 2$, $\kappa_1 = \kappa_2 = g \geqslant 4$. In this case, the moduli space has two components, one hyperelliptic and the other not hyperelliptic.*
(2) *If all κ_i are odd and either $s \geqslant 3$ or $s = 2$ and $\kappa_1 \neq \kappa_2$, then the moduli space has two connected components, one with even spin structure and the other with odd spin structure.*
(3) *If either $s = 1, g \geqslant 4$ or $s = 2, \kappa_1 = \kappa_2 = g$ odd $\geqslant 5$, the moduli space has three connected components: one hyperelliptic and two non hyperelliptic distinguished by the parity of the spin structure.*
(4) *If $g = 3$, $s = 1$ or $s = 2$, $\kappa_1 = \kappa_2 = 3$, the moduli space has two components, one hyperelliptic and the other not hyperelliptic. If $g = 2$,$s = 1$, the moduli space is connected.*

We just say a few words of the scheme of the proof. The confluence of the zeros of the 1-form associated to the structure of translation surface organizes the various moduli spaces as the strata of a stratification. The minimal stratum S_{min} corresponds to a single zero of maximal multiplicity $2g - 2$.

Kontsevich and Zorich establish the following fact, which allows to rely any stratum to S_{min}: for any stratum S, and any connected component C of S_{min}, there exists exactly one component of S which contains C in its closure.

The determination of the connected components of the minimal stratum S_{min} is by induction on the genus g. First, using a local construction first described in [**EMaZo**], they show that there are at least as many components as stated in the theorem: given a translation surface with a single zero A_1 of multiplicity $2g-2$, they split A_1 into two zeros A_1', A_1'' of respective multiplicities k_1', k_1'', slit the surface along a segment joining A_1' and A_1'', and glue the two sides to the two boundary components of a cylinder. The resulting translation surface has genus $g+1$, a single zero of maximal multiplicity $2g$ and the parity of its spin structure changes when the parity of k_1' change.

That there are no more components of S_{min} that as stated in the theorem is also proved by induction. The idea is to present any generic translation surface in S_{min} as the suspension, via the zippered rectangle construction, of an i.e.m and then take off a handle by an appropriate reduction operation.

14. Exponential mixing of the Teichmüller flow

We present in this section the main results from [**AvGoYo**].

14.1. Exponential mixing. Let $(X, \mathcal{B}, m)$ be a probability space, and let (T^t) be a measure-preserving dynamical system. We allow here for discrete time $(t \in \mathbb{Z})$ as well as continuous time $(t \in \mathbb{R})$. We denote by $L_0^2(X)$ the Hilbert space of square-integrable functions of mean value 0, by U^t the unitary operator $\varphi \mapsto \varphi \circ T^t$ of $L_0^2(X)$. For φ, $\psi \in L_0^2(X)$, we define the **correlation coefficient** of φ, ψ by

$$c_{\varphi,\psi}(t) := < \varphi, U^t \psi > .$$

We recall that

- T^t is ergodic iff, for all φ, $\psi \in L_0^2(X)$, $c_{\varphi,\psi}(t)$ converges to 0 in the sense of Cesaro as $t \to +\infty$;
- T^t is mixing iff, for all φ, $\psi \in L_0^2(X)$, $c_{\varphi,\psi}(t)$ converges to 0 as $t \to +\infty$.

Exponential mixing requires that this convergence is exponentially fast. However, simple examples (for instance, the shift map) show that this cannot happen, even in the most chaotic dynamical systems, for **all** functions φ, $\psi \in L_0^2(X)$. One generally requires that φ, ψ belong to some Banach space E of "regular" functions on X, dense in $L_0^2(X)$. Then the correlation coefficients should satisfy

$$c_{\varphi,\psi}(t) \leqslant C||\varphi||_E \, ||\psi||_E \exp(-\delta\, t),$$

where $\delta > 0$ is independent of φ, $\psi \in E$. Observe that this indeed imply mixing.

Exponential mixing, unlike ergodicity or mixing, is **not** a spectral notion (one which depends only on the properties of the unitary operators U^t).

THEOREM 14.1. [**AvGoYo**] *The Teichmuller flow is exponentially mixing on any connected component of any marked moduli space* $\widetilde{\mathcal{M}}^{(1)}(M, \Sigma, \kappa)$.

The subspace E of "regular" functions will be explicited below; for any $1 \geqslant \beta > 0$, it can be chosen to contain all β-Hölder functions with compact support.

14.2. Exponential mixing and irreducible unitary representations of $SL(2,\mathbb{R})$. The theorem has an interesting consequence with respect to the representation of $SL(2,\mathbb{R})$ determined by the action of this group on the marked moduli spaces.

Ler $\mathcal{C}$ be a connected component of some marked moduli space $\widetilde{\mathcal{M}}^{(1)}(M, \Sigma, \kappa)$. Denote by H the Hilbert space of zero mean L^2 functions on $\mathcal{C}$. The action of

$SL(2,\mathbb{R})$ induces an unitary representation of $SL(2,\mathbb{R})$ in H. As any unitary representation of $SL(2,\mathbb{R})$, it decomposes as an hilbertian sum

$$H = \int H_\xi d\mu(\xi),$$

where, for each ξ, the representation of $SL(2,\mathbb{R})$ in H_ξ is irreducible.

According to Bargmann, the nontrivial irreducible unitary representations of $SL(2,\mathbb{R})$ are divided into three families, the *discrete, principal and complementary series.* This corresponds to an orthogonal decomposition into invariant subspaces

$$H = H_{tr} \oplus H_d \oplus H_{pr} \oplus H_c\ .$$

The ergodicity of the action of $SL(2,\mathbb{R})$ (Masur-Veech) means that $H_{tr} = \{0\}$.

Write g_t for the diagonal element $diag(e^t, e^{-t})$ of $SL(2,\mathbb{R})$ corresponding to the Teichmüller flow. In general, for vectors v, v' belonging both to the discrete component H_d or the principal component H_{pr} of the representation, one has, for $t \leqslant 1$

$$< g_t(v), v' > \ \leqslant\ C\, t\, e^{-t}\, ||v||\, ||v'||\ .$$

On the other hand, the complementary series is parametrized by a parameter $s \in (0,1)$, with

$$\mathcal{H}_s = \{f : \mathbb{S}^1 \to \mathbb{C}\,,\ ||f||^2 := \int\int \frac{f(z)\overline{f}(z')}{|z-z'|^{1-s}}\, dz\, dz' < +\infty\},$$

the representation of $SL(2,\mathbb{R})$ in $\mathcal{H}_s$ being given by

$$\begin{pmatrix} a & b \\ c & d \end{pmatrix} . f(z) = |\overline{\beta} z + \overline{\alpha}|^{-1-s}\, f\left(\frac{\alpha z + \beta}{\overline{\beta} z + \overline{\alpha}}\right),$$

with

$$\begin{pmatrix} i & i \\ -1 & 1 \end{pmatrix}^{-1} \begin{pmatrix} a & b \\ c & d \end{pmatrix} \begin{pmatrix} i & i \\ -1 & 1 \end{pmatrix} = \begin{pmatrix} \alpha & \beta \\ \overline{\beta} & \overline{\alpha} \end{pmatrix}.$$

We observe that the norm in $\mathcal{H}_s$ is equivalent to the norm

$$||f||' = \left(\sum (1+|n|)^{-s} |\widehat{f}(n)|^2\right)^{\frac{1}{2}}.$$

The integral powers $e_n(z) := z^n$, $n \in \mathbb{Z}$, are eigenfunctions for the action of $SO(2,\mathbb{R})$:

$$\begin{pmatrix} \cos\theta & \sin\theta \\ -\sin\theta & \cos\theta \end{pmatrix} e_n \ = \ \exp(2i\pi n\theta)\, e_n\ .$$

An easy calculation show that, for $m, n \in \mathbb{Z}$, $t \geqslant 1$

$$|< g_t\, e_m\,,\, e_n >| \quad \leqslant \quad < g_t\, e_0\,,\, e_0 >,$$

$$C_s^{-1}\, e^{t(s-1)} \leqslant \ < g_t\, e_0\,,\, e_0 > \ \leqslant C_s\, e^{t(s-1)},$$

with $C_s > 0$ depending on s but not on t.

DEFINITION 14.2. A unitary representation $\mathcal{H}$ of $SL(2,\mathbb{R})$ has an *almost invariant vector* if, given any compact subset K of $SL(2,\mathbb{R})$ and $\varepsilon > 0$, there exists a unit vector $v \in \mathcal{H}$ such that

$$||g.v - v|| < \varepsilon$$

for all $g \in K$.

A unitary representation $\mathcal{H}$ of $SL(2,\mathbb{R})$ with no almost invariant vector is said to have a *spectral gap*.

Let $H = \int H_\xi d\mu(\xi)$ be the decomposition of a unitary representation $\mathcal{H}$ of $SL(2,\mathbb{R})$ into irreducible representations. Then $\mathcal{H}$ has a spectral gap iff there exists $s_0 \in (0,1)$ such that, for almost every ξ, H_ξ is neither the trivial representation nor isomorphic to a representation in the complementary series with parameter $s \in (s_0, 1)$.

DEFINITION 14.3. Let $\mathcal{H}$ be a unitary representation of $SL(2,\mathbb{R})$. A vector $v \in \mathcal{H}$ is C^r-$SO(2,\mathbb{R})$-smooth if the function

$$\theta \to \begin{pmatrix} \cos\theta & \sin\theta \\ -\sin\theta & \cos\theta \end{pmatrix} .v$$

is of class C^r.

PROPOSITION 14.4. *(Ratner [**Rat**]) If the unitary representation $\mathcal{H}$ has a spectral gap, then it is exponential mixing for C^2-$SO(2,\mathbb{R})$-smooth vectors: there exists $\delta > 0$ and $C > 0$ such that, for any C^2-$SO(2,\mathbb{R})$-smooth $v, v' \in \mathcal{H}$, $t \geqslant 1$*

$$| < g_t.v, v' > | \leqslant C \exp(-\delta t) \, ||v||_2 \, ||v'||_2 \, ,$$

where $||v||_2$ is the sum of the norm of v and the norm of the second derivative at 0 of $\theta \to \begin{pmatrix} \cos\theta & \sin\theta \\ -\sin\theta & \cos\theta \end{pmatrix} .v$.

SKETCH OF PROOF. It is sufficient to consider unit vectors v, v' in the complementary component of the representation. There exists $s_0 \in (0,1)$ such that $s(\xi) \notin [s_0, 1)$ for almost every ξ, hence we have

$$| < g_t.e_m, e_n >_\xi | \leqslant C \exp(t(s_0 - 1))$$

for all $t \geqslant 1$, $m, n \in \mathbb{Z}$, and almost every ξ. Let v, v' be C^2-$SO(2,\mathbb{R})$-smooth vectors in the complementary component of $\mathcal{H}$. Write

$$v = \int v(\xi) d\mu(\xi) = \int \sum v_m(\xi) e_m d\mu(\xi),$$

and similarly for v'. Then $v(\xi)$ is C^2-$SO(2,\mathbb{R})$-smooth for almost all ξ. From the remark on the norm in $\mathcal{H}_s$ above, this gives, for all $m \in \mathbb{Z}$

$$|v_m(\xi)| \leqslant C ||v(\xi)||_2 (1 + |m|)^{\frac{s(\xi)}{2} - 2}.$$

We conclude that

$$\begin{aligned} | < g_t.v, \, v' > | &\leqslant \int | < g_t.v(\xi), \, v'(\xi) > | d\mu(\xi) \\ &\leqslant \int | \sum_m \sum_n v_m(\xi) \overline{v}'_n(\xi) < g_t.e_m, e_n >_\xi | d\mu(\xi) \\ &\leqslant C \, \exp(t(s_0 - 1)) \int ||v(\xi)||_2 ||v'(\xi)||_2 d\mu(\xi) \\ &\leqslant C \, \exp(t(s_0 - 1)) \, ||v||_2 ||v'||_2. \end{aligned}$$

□

REMARK 14.5. The absence of a trivial component, i.e the ergodicity of the action of $SL(2,\mathbb{R})$, already imply that the action of the diagonal subgroup is mixing: for vectors of the form

$$v = \int_{0 < s(\xi) < s_0} \sum_{|m| \leqslant M} v_m(\xi) e_m d\mu(\xi), \quad v' = \int_{0 < s(\xi) < s_0} \sum_{|m| \leqslant M} v'_m(\xi) e_m d\mu(\xi)$$

, for some $M > 0$, $s_0 \in (0,1)$, we have that $| < g_t.v, v' > |$ converges to 0 by the calculation above. These vectors are dense in the complementary component of $\mathcal{H}$, and the mixing property follows.

Conversely

PROPOSITION 14.6. *Assume that there exists $\delta > 0$ and a dense subset E of vectors v in the space of $SO(2,\mathbb{R})$-invariant vectors in $\mathcal{H}$ for which the correlation coefficients $< g_t.v, v >$ are $\mathcal{O}(\exp(-\delta t))$. Then $\mathcal{H}$ has a spectral gap.*

PROOF. We may assume that $0 < \delta < 1$. Assume by contradiction that $\mathcal{H}$ has no spectral gap. The complementary component v_c of any $SO(2,\mathbb{R})$-invariant vector takes the form $v_c = \int v_0(\xi) e_0 d\mu(\xi)$, with $v_0 \in L^2(\mu)$. As E is dense in the space of $SO(2,\mathbb{R})$-invariant vectors in $\mathcal{H}$, we can find $v \in E$ such that

$$\mu\{\xi,\ s(\xi) > 1 - \delta \text{ and } v_0(\xi) \neq 0\} > 0\ .$$

Then we have

$$\begin{aligned} < g_t.v_c, v_c > &= \int |\phi(\xi)|^2 < g_t.e_0, e_0 > d\mu(\xi) \\ &\geqslant \int |\phi(\xi)|^2 C_{s(\xi)}^{-1} \exp(t(s(\xi) - 1)) d\mu(\xi); \end{aligned}$$

here $s(\xi)$ is the parameter in the complementary series associated to H_ξ. Thus $< g_t.v_c, v_c >$ is not $\mathcal{O}(\exp(-\delta t))$. But v does not have a discrete component, and the principal component v_p satisfies $< g_t.v_p, v_p >= \mathcal{O}(t \exp(-t))$. This contradicts the property of E. □

Coming back to the setting of the theorem in Subsection 14.1, let $\mathcal{C}$ be a component of some marked moduli space $\widetilde{\mathcal{M}}^{(1)}(M, \Sigma, \kappa)$. The space of compactly supported mean zero smooth $SO(2,\mathbb{R})$-invariant functions on $\mathcal{C}$ is dense in the subspace of $SO(2,\mathbb{R})$-invariant functions in $L_0^2(\mathcal{C})$. Therefore the representation of $SL(2,\mathbb{R})$ in $L_0^2(\mathcal{C})$ has a spectral gap.

14.3. Diophantine estimates. Exponential mixing is a classical property of uniformly hyperbolic transformations preserving a smooth volume form.

EXERCISE 14.7. Let $A \in SL(d,\mathbb{R})$ be a hyperbolic matrix. The induced diffeomorphism of $\mathbb{T}^d$ preserves Lebesgue measure. Prove that , if φ, ψ are Hölder functions on $\mathbb{T}^d$ with zero mean-value, the correlation coefficient $c_{\varphi,\psi}(n) := \int_{\mathbb{T}^d} \varphi\ \psi \circ A^n$ satisfy

$$|c_{\varphi,\psi}(n)| \leqslant C ||\varphi||\ ||\psi||\ \exp(-\delta n),$$

where δ depends only on A and the Hölder exponent of φ, ψ.

With respect to this very basic case, the Teichmüller flow presents three difficulties:

- the time is continuous rather than discrete;
- hyperbolicity is non uniform;
- distortion for large time is not controlled as simply than in the uniformly hyperbolic setting on a compact manifold.

As the constant time suspension of an Anosov diffeomorphism is obviously not mixing, the first difficulty is quite serious. The ideas which allow to deal with it were first introduced by Dolgopyat [**Do**] and later developped by Baladi-Vallée [**BaVa**].

The other two difficulties are related to a lack of compactness of the moduli spaces of translation surfaces. To get uniform hyperbolicity and bounded distortion, one is led to introduce the return map of the Teichmüller flow to a suitably small transversal section (smaller than the ones considered in Section 9). The problem is then to control the return time to this transversal section. This is done through diophantine estimates which we will now present.

Let $\mathcal{R}$ be a Rauzy class on an alphabet $\mathcal{A}$, let $\mathcal{D}$ be the associated Rauzy diagram. The estimates depend on a parameter $q \in \mathbb{R}_+^{\mathcal{A}}$. For such q, we define a probability measure P_q on $\mathbb{P}(\mathbb{R}_+^{\mathcal{A}})$ by

$$P_q(A) := \frac{\mathrm{Leb}(\mathbb{R}_+ A \bigcap \Lambda_q)}{\mathrm{Leb}(\Lambda_q)},$$

where $\Lambda_q = \{\lambda \in \mathbb{R}_+^{\mathcal{A}};\ < \lambda, q > \ < 1\ \}$. Define also, for $q \in \mathbb{R}_+^{\mathcal{A}}$, $M(q) := \max_{\alpha\in\mathcal{A}} q_\alpha$, $m(q) := \min_{\alpha\in\mathcal{A}} q_\alpha$. For a finite path γ in $\mathcal{D}$, starting from a vertex π, we denote by Δ_γ the set of $\lambda \in \Delta_\pi$ whose Rauzy-Veech path starts with γ.

Let now $0 \leqslant m \leqslant M$ be integers, $q \in \mathbb{R}_+^{\mathcal{A}}$, $\pi \in \mathcal{R}$. Define $\Gamma_0 = \Gamma_0(m, M, q, \pi)$ to be the set of finite paths $\gamma \in \mathcal{D}$ starting from π such that

$$M(B_\gamma q) > 2^M\, M(q), \qquad\qquad m(B_\gamma q) < 2^{M-m}\, M(q).$$

THEOREM 14.8. [**AvGoYo**] *There exist constants θ, C depending only on $\#\mathcal{A}$ such that*

$$P_q\Big(\bigcup_{\gamma\in\Gamma_0} \mathbb{P}(\Delta_\gamma)\Big) \leqslant C(m+1)^\theta 2^{-m}.$$

A closely connected estimate is the following. Let M be an integer and $q \in \mathbb{R}_+^{\mathcal{A}}$, $\pi \in \mathcal{R}$. Define $\Gamma_1 = \Gamma_1(M, q, \pi)$ to be the set of finite paths $\gamma \in \mathcal{D}$ starting from π such that γ is not complete and $M(B_\gamma q) > 2^M\, M(q)$.

THEOREM 14.9. [**AvGoYo**] *There exist constants θ, C depending only on $\#\mathcal{A}$ such that*

$$P_q\Big(\bigcup_{\gamma\in\Gamma_1} \mathbb{P}(\Delta_\gamma)\Big) \leqslant C(M+1)^\theta 2^{-M}.$$

EXERCISE 14.10. Use these estimates to show that almost all i.e.m are of Roth type.

References

[At] M. ATIYAH – *Riemann surfaces and spin structures.* Ann. Sci. École Norm. Sup. (4), 4 (1971), 47-62.

[AvVi1] A. AVILA, M. VIANA – *Simplicity of Lyapunov spectra : proof of the Zorich-Kontsevich conjecture.* Acta Math. 198 (2007), n^0 1, 1-56.

[AvVi2] A. AVILA, M. VIANA – *Simplicity of Lyapunov spectra: a sufficient criterion.*Portugaliae Mathematica 64 (2007), 311-376.

[AvGoYo] A. AVILA, S. GOUEZEL, J-C. YOCCOZ – *Exponential mixing for the Teichmüller flow.* Publ. Math. IHÉS, N^0 104(2006), 143-211.

[BaVa] V . BALADI, B. VALLÉE – *Exponential decay of correlations for surface semi-flows without finite Markov partitions.* Proc. Amer. Math. Soc. 133 (2005), n^{0}3, 865-874.

[Do] D. DOLGOPYAT – *On decay of correlations in Anosov flows.* Ann. of Math. (2) 147 (1988), n^{0}2, 357-390.

[EMaZo] A. ESKIN, H. MASUR, A. ZORICH – *Moduli spaces of Abelian differentials : the principal boundary, counting problems, and the Siegel-Veech constants.* Publ. Math. IHÉS, N^{0}97 (2003), 61-179.

[EOk] A. ESKIN, A. OKOUNKOV – *Asymptotics of number of branched coverings of a torus and volumes of moduli spaces of holomorphic differentials.* Invent. Math. 145 (2001), 59-104.

[For1] G. FORNI – *Solutions of the cohomological equation for area-preserving flows on compact surfaces of higer genus.* Ann. of Math. 146 (1997), 295-344.

[For2] G. FORNI – *Deviation of ergodic averages for area-preserving flows on surfaces of higher genus.* Ann. of Math. 155 n^{0}1, (2002), 1-103.

[For3] G. FORNI –*Sobolev regularity of solutions of the cohomological equation.* ArXiv:0707.0940

[HuMaScZo] P. HUBERT, H. MASUR, T. SCHMIDT, A. ZORICH – *Problems on billiards, flot surfaces and translation surfaces in problems on mapping class groups and related topics.* Proc. Symp. Pure Math. 74, Am. Math. Soc., Providence, RI (2006), 233-243.

[Ka] A. KATOK – *Interval exchange transformations and some special flows are not mixing.* Israel Journal of Math. 35 (1980), n^0 4, 301-310.

[Kea1] M . KEANE – *Interval exchange transformations.* Math. Z. 141 (1975), 25-31.

[Kea2] M . KEANE – *Non-ergodic interval exchange transformations.* Israel Journal of Math. 26, (1977), n^0 2, 188-196.

[KeyNew] H. B. KEYNES, D. NEWTON – *A "Minimal", Non-Uniquely Ergodic Interval Exchange Transformaation.* Math. Z. 148 (1976) 101-105.

[Kon] M . KONTSEVICH – *Lyapunov exponents and Hodge theory. "The mathematical beauty of physics".* (Saclay, 1996), (in honor of C. Itzykson) 318-332, Adv. Ser. Math. Phys. 24. World Sci. Publishing, River Edge, NJ (1997).

[KonZo] M . KONTSEVICH, A. ZORICH – *Connected components of the moduli spaces of Abelian differentials.* Invent. Math. 153, (2003), 631-678.

[Kri] R . KRIKORIAN – *Déviations de moyennes ergodiques, d'après Forni, Kontsevich, Zorich.* Séminaire Bourbaki 2003-2004, 56ème année, exposé n^0 927, novembre 2003.

[Ma] H. MASUR – *Interval exchange transformations and measured foliations.* Ann. of Math, 115, (1982), 169-200.

[Mil] J. MILNOR – *Remarks concerning spin manifolds.* In: Differential and Combinatorial Topology (in Honor of Marston Morse), Princeton (1995).

[MmMsY] S. MARMI, P. MOUSSA, J-C. YOCCOZ – *The cohomological equation for Roth-type interval exchange maps.* J. Ann. Math. Soc. 18, (2005), n^0 4, 823-872.

[Os] V.I . OSELEDETS – *A Multiplicative Ergodic Theorem. Lyapunov characteristic numbers for dynamical systems.* Trans. Moscow Math. Soc. 19, (1968) 197-231.

[Rat] M. RATNER – *The rate of mixing for geodesic and horocycle flows.* Erg. Th. Dyn. Sys. 7 (1987), 267-288.

[Rau] G . RAUZY – *Echanges d'intervalles et transformations induites.* Acta Arith. 34, (1979) 315-328.

[Ve1] W.A . VEECH – *Interval exchange transformations.* Journal Anal. Math. 33, (1978) 222-278.

[Ve2] W.A . VEECH – *Gauss measures for transformations on the space of interval exchange maps.* Annals of Math. 115, (1982) 201-242.

[Ve3] W.A . VEECH – *The metric theory of interval exchange transformations I. Generic spectral properties.* Amer. Journal of Math. 106 (1984), 1331-1359.

[Ve4] W.A . VEECH – *The metric theory of interval exchange transformations II. Approximation by primitive interval exchanges.* Amer. Journal of Math. 106 (1984), 1361-1387.

[Ve5] W.A . VEECH – *Teichmüller geodesic flow.* Annals of Math. 124 (1986), 441-530.

[Y1] J-C. YOCCOZ – *Continuous fraction algorithms for interval exchange maps : an introduction.* In: Frontiers in Number Theory, Physics and Geometry Vol. 1, P. Cartier, B. Julia, P. Moussa, P. Vanhove (Editors), Springer Verlag, (2006), 403-437.

[Y2] J-C. YOCCOZ – *Cours 2005: Échange d'intervalles.* http://www.college-de-france.fr/default/EN/all/equ_dif/

[Zo1] A. ZORICH – *Flat surfaces.* In: Frontiers in Number Theory, Physics and Geometry Vol. 1, P. Cartier, B. Julia, P. Moussa, P. Vanhove (Editors), Springer Verlag, (2006), 439-586.

[Zo2] A. ZORICH – *Finite Gauss measure on the space of interval exchange transformations. Lyapunov exponents.* Annales de l'Institut Fourier, 46:2 (1996), 325-370.

[Zo3] A. ZORICH – *Deviation for interval exchange transformations.* Ergodic Theory and Dynamical Systems, 17 (1997), 1477-1499.

[Zo4] A. ZORICH – *How do the leaves of a closed 1-form wind around a surface.* In the collection : Pseudoperiodic Topology, AMS Translations, Ser. 2, vol. 197, AMS, Providence, RI, (1999), 135-178.

Collège de France, 3, rue d'Ulm, 75005, Paris, France

Clay Mathematics Proceedings
Volume 10, 2010

Unipotent Flows and Applications

Alex Eskin

1. General introduction

1.1. Values of indefinite quadratic forms at integral points. The Oppenheim Conjecture. Let

$$Q(x_1, \dots, x_n) = \sum_{1 \le i \le j \le n} a_{ij} x_i x_j$$

be a quadratic form in n variables. We always assume that Q is *indefinite* so that (so that there exists p with $1 \le p < n$ so that after a linear change of variables, Q can be expresses as:

$$Q_p^*(y_1, \dots, y_n) = \sum_{i=1}^{p} y_i^2 - \sum_{i=p+1}^{n} y_i^2$$

We should think of the coefficients a_{ij} of Q as real numbers (not necessarily rational or integer). One can still ask what will happen if one substitutes *integers* for the x_i. It is easy to see that if Q is a multiple of a form with rational coefficients, then the set of values $Q(\mathbb{Z}^n)$ is a discrete subset of $\mathbb{R}$. Much deeper is the following conjecture:

CONJECTURE 1.1 (Oppenheim, 1929). *Suppose Q is not proportional to a rational form and $n \ge 5$. Then $Q(\mathbb{Z}^n)$ is dense in the real line.*

This conjecture was extended by Davenport to $n \ge 3$.

THEOREM 1.2 (Margulis, 1986). *The Oppenheim Conjecture is true as long as $n \ge 3$. Thus, if $n \ge 3$ and Q is not proportional to a rational form, then $Q(\mathbb{Z}^n)$ is dense in $\mathbb{R}$.*

This theorem is a triumph of ergodic theory. Before Margulis, the Oppenheim Conjecture was attacked by analytic number theory methods. (In particular it was known for $n \ge 21$, and for diagonal forms with $n \ge 5$).

Failure of the Oppenheim Conjecture in dimension 2. Let $\alpha > 0$ be a quadratic irrational such that $\alpha^2 \notin \mathbb{Q}$ (e.g. $\alpha = (1+\sqrt{5})/2$), and let

$$Q(x_1, x_2) = x_1^2 - \alpha^2 x_2^2.$$

PROPOSITION 1.3. *There exists $\epsilon > 0$ such that for all $x_1, x_2 \in \mathbb{Z}$, $|Q(x_1, x_2)| > \epsilon$.*

Proof. Suppose not. Then for any $1 > \epsilon > 0$ there exist $x_1, x_2 \in \mathbb{Z}$ such that

$$|Q(x_1, x_2)| = |x_1 - \alpha x_2||x_1 + \alpha x_2| \leq \epsilon. \tag{1}$$

We may assume $x_2 \neq 0$. If $\epsilon < \alpha^2$, one of the factors must be smaller then α. Without loss of generality, we may assume $|x_1 - \alpha x_2| < \alpha$, so $|x_1 - \alpha x_2| < \alpha |x_2|$. Then,

$$|x_1 + \alpha x_2| = |2\alpha x_2 + (x_1 - \alpha x_2)| \geq 2\alpha |x_2| - |x_1 - \alpha x_2| \geq \alpha |x_2|.$$

Substituting into (1) we get

$$\left| \frac{x_1}{x_2} - \alpha \right| \leq \frac{\epsilon}{|x_2||x_1 + \alpha x_2|} \leq \frac{\epsilon}{\alpha} \frac{1}{|x_2|^2}. \tag{2}$$

But since α is a quadratic irrational, there exists $c_0 > 0$ such that for all $p, q \in \mathbb{Z}$, $|\frac{p}{q} - \alpha| \geq \frac{c_0}{q^2}$. This is a contradiction to (2) if $\epsilon < c_0 \alpha$. □

A relation to flows on homogeneous spaces. This was noticed by Raghunathan, and previously in implicit form by Cassels and Swinnerton-Dyer. However the Cassels-Swinnerton-Dyer paper was mostly forgotten. Raghunathan made clear the connection to unipotent flows, and explained from the point of view of dynamics what is different in dimension 2. See §5.1.

1.2. Some basic Ergodic Theory. Transformations, flows and Ergodic Measures. Let X be a locally compact separable topological space, and $T : X \to X$ a map. We assume that there is a finite measure μ on X which is preserved by T. One usually normalizes μ so that $\mu(X) = 1$, in which case μ is called a probability measure.

Sometimes, instead of a transformation T one considers a flow ϕ_t, $t \in \mathbb{R}$. For a fixed t, ϕ_t is a map from X to X. In this section we state definitions and theorems for transformations only, even though we will use them for flows later.

DEFINITION 1.4 (Ergodic Measure). An T-invariant probability measure μ is called *ergodic* for T if for every measurable T-invariant subset E of X one has $\mu(E) = 0$ or $\mu(E) = 1$.

Every measure can be written as a linear combination (possibly uncountable, dealt with via integration) of ergodic measures. This is called the "ergodic decomposition".

Ergodic measures always exist. In fact the probability measures form a convex set, and the ergodic probability measures are the extreme points of this set (cf. the Krein-Milman theorem).

Birkhoff's Ergodic Theorem.

THEOREM 1.5 (Birkhoff Ergodic Theorem). *Suppose μ is ergodic for T, and suppose $f \in L^1(X, \mu)$. Then for μ-almost all $x \in X$, we have*

$$\lim_{n \to \infty} \frac{1}{n} \sum_{k=0}^{n-1} f(T^n x) = \int_X f \, d\mu. \tag{3}$$

The sum on the left-hand side is called the "time average", and the integral on the right is the "space average". Thus the theorem says that for almost all base points x, the time average along the orbit of x converges to the space average.

This theorem is amazing in its generality: the only assumption is ergodicity of the measure μ. (This is a some sort of irreducibility assumption).

The set of $x \in X$ for which (3) holds is called the *generic* set for μ.

Mutually singular measures. Recall that two probability measures μ_1 and μ_2 are called *mutually singular* (written as $\mu_1 \perp \mu_2$ if there exists a set E such that $\mu_1(E) = 1$, $\mu_2(E) = 0$ (so $\mu_2(E^c) = 1$).

In our proofs we will use repeatedly the following:

LEMMA 1.6. *Suppose μ_1 and μ_2 are distinct ergodic measures for the map $T : X \to X$. Then $\mu_1 \perp \mu_2$.*

Proof. This is an immediate consequence of the Birkhoff ergodic theorem. Since $\mu_1 \neq \mu_2$ we can find an f such that $\int_X f\, d\mu_1 \neq \int_X f\, d\mu_2$. Now let E denote the set where (3) holds with $\mu = \mu_1$. □

Remark. It is not difficult to give another proof of Lemma 1.6 using the Radon-Nikodym theorem.

Given an invariant measure μ for T, we want to find conditions under which it is ivariant under the action of a larger group. Now if H commutes with T, then for each $h_0 \in H$ the measure $h_0\mu$ is T-invariant. So if μ is ergodic, so is $h_0\mu$, and Lemma 1.6 applies. More can be said, ([cf. [**Ra4**, Thm. 2.2], [**Mor**, Lem. 5.8.6]]):

LEMMA 1.7. *Suppose $T : X \to X$ is preserving an ergodic measure μ. Suppose H is a group with acts continuously on X and commutes with T. Also suppose that there exists $h_0 \in H$ such that $h_0\mu \neq \mu$. Then there exists a neighborhood B of $h_0 \in H$ and a conull T-invariant subset Ω of X such that*

$$h\Omega \cap \Omega = \emptyset \quad \textit{for all } h \in B.$$

Proof. Since h_0 commutes with T, the measure $h_0\mu$ is T-invariant and ergodic. Thus by Lemma 1.6, $h_0\mu \perp \mu$. This implies there is a compact subset K_0 of X, such that $\mu(K_0) > 0.99$ and $K_0 \cap h_0 K_0 = \emptyset$. By continuity and compactness, there are open neighborhoods $\mathcal{U}$ and $\mathcal{U}^+$ of K_0, and a symmetric neighborhood B_e of e in H, such that $\mathcal{U}^+ \cap h_0\mathcal{U}^+ = \emptyset$ and $B_e\mathcal{U} \subset \mathcal{U}^+$. From applying (3) with f the characteristic function of $\mathcal{U}$, we know there is a conull T-invariant subset Ω_{h_0} of X, such that the T-orbit of every point in Ω_{h_0} spends 99% of its life in $\mathcal{U}$. Now suppose there exists $h \in B_e h_0$, such that $\Omega_{h_0} \cap h\Omega_{h_0} \neq \emptyset$. Then there exists $x \in \Omega_{h_0}$, $n \in \mathbb{N}$, and $c \in B_e$, such that $T^n x$ and $ch_0T^n x$ both belong to $\mathcal{U}$. This implies that $T^n x$ and $h_0T^n x$ both belong to $\mathcal{U}^+$. This contradicts the fact that $\mathcal{U}^+ \cap h_0\mathcal{U}^+ = \emptyset$. □

Uniquely ergodic systems. In some applications (in particular to number theory) we need some analogue of (3) for *all* points x (and not almost all). For example, we want to know if $Q(\mathbb{Z}^n)$ is dense for a specific quadratic form Q (and not for almost all forms). Then the Birkhoff ergodic theorem is not helpful. However, there is one situation where we can show that (3) holds for all x.

DEFINITION 1.8. A map $T : X \to X$ is called *uniquely ergodic* if there exists a unique invariant probability measure μ.

PROPOSITION 1.9. *Suppose X is compact, $T : X \to X$ is uniquely ergodic, and let μ be the invariant probability measure. Suppose $f : X \to \mathbb{R}$ is continuous. Then for all $x \in X$, (3) holds.*

Proof. This is quite easy (as opposed to the Birkhoff ergodic theorem which is hard). Let δ_n be the probability measure on X defined by

$$\delta_n(f) = \frac{1}{n}\sum_{k=0}^{n-1} f(T^n x)$$

(we are now thinking of measures as elements of the dual space to the space $C(X)$ of continuous functions on X). Note that

$$\delta_n(f \circ T) = \frac{1}{n}\sum_{k=0}^{n-1}(f \circ T)(T^n x) = \frac{1}{n}\sum_{k=1}^{n} f(T^n x),$$

so

$$\delta_n(f \circ T) - \delta_n(f) = \frac{1}{n}(f(x) - f(T^n x)), \tag{4}$$

(since the sum telescopes). Suppose some subsequence δ_{n_j} converges to some limit δ_∞ (in the weak-* topology). Then, by (4), $\delta_\infty(f \circ T) = \delta_\infty(f)$, i.e. δ_∞ is T-invariant.

Since X is compact, δ_∞ is a probability measure, and thus by the assumption of unique ergodicity, we have $\delta_\infty = \mu$. Thus all possible limit points of the sequence δ_n are μ. Also the space of probability measures on X is compact (in the weak-* topology), so there exists a convergent subsequence. Hence $\delta_n \to \mu$, which is the same as (3). □

Remarks.

- The main point of the above proof is the construction of an invariant measure (namely δ_∞) supported on the closure of the orbit of x. The same construction works with flows, or more generally with actions of amenable groups.
- We have used the compactness of X to argue that δ_∞ is a probability measure: this might fail if X is not compact. This phenomenon is called "loss of mass".
- Of course the problem with Proposition 1.9 is that most of the dynamical systems we are interested in are not uniquely ergodic. For example any system which has a closed orbit which is not the entire space is not uniquely ergodic.
- However, the proof of Proposition 1.9 suggests that (at least in the amenable case) the classification of the invariant measures is one of the most powerful statements one can make about a dynamical system, in the sense that it allows one to try to understand every orbit (and not just almost every orbit).

Exercise 1. (To be used in §3.)

(a) Show that if α is irrational then the map $T_\alpha : [0,1] \to [0,1]$ given by $T_\alpha(x) = x + \alpha \pmod 1$ is uniquely ergodic. *Hint:* Use Fourier analysis.
(b) Use part (a) to show that the flow on $\mathbb{R}^2/\mathbb{Z}^2$ given by $\phi_t(x,y) = (x + t\alpha, y + t)$ is uniquely ergodic.

1.3. Unipotent Flows. Let G be a semisimple Lie group (I will usually assume the center of G is finite), and let Γ be a lattice in G (this means that $\Gamma \subset G$ is a discrete subgroup, and the quotient G/Γ has finite Haar measure). A lattice Γ is *uniform* if G/Γ is compact.

Let $U = \{u_t\}_{t\in\mathbb{R}}$ be a unipotent one-parameter subgroup of G. Then U acts on G/Γ by left multiplication. (Recall that in $SL(n,\mathbb{R})$ a matrix is unipotent if all its eigenvalues are 1. In a general Lie group an element is unipotent if its Adjoint (acting on the Lie algebra) is a unipotent matrix.) Examples of unipotent one parameter subgroups:

$$\left\{\begin{pmatrix} 1 & t \\ 0 & 1 \end{pmatrix}, \qquad t \in \mathbb{R}\right\},$$

and

$$\left\{\begin{pmatrix} 1 & t & t^2/2 \\ 0 & 1 & t \\ 0 & 0 & 1 \end{pmatrix}, \qquad t \in \mathbb{R}\right\},$$

Ratner's measure classification theorem.

DEFINITION 1.10. A probability measure μ on G/Γ is called *algebraic* if there exists $\bar{x} \in G/\Gamma$ and a subgroup F of G such that $F\bar{x}$ is closed, and μ is the F-invariant probability measure supported on $F\bar{x}$.

THEOREM 1.11 (Ratner's measure classification theorem). *Let G be a Lie group, $\Gamma \subset G$ a lattice. Let U be a one-parameter unipotent subgroup of G. Then, any ergodic U-invariant measure is algebraic. (Also the group F in the definition of algebraic is generated by unipotent elements, and contains U).*

Loosely speaking, this theorem says that all U-invariant ergodic measures are very nice. The assumption that U is unipotent is crucial: if we consider instead arbitrary one-parameter subgroups, then there are ergodic invariant measures supported on Cantor sets (and worse). This phenomenon is responsible in particular for the failure of the Oppenheim conjecture in dimension 2.

Theorem 1.11 has many applications, some of which we will explore in this course. I will give some indication of the ideas which go into the proof of this theorem in the next two lectures.

Remark on algebraic measures. Let $\pi : G \to G/\Gamma$ be the projection map. Suppose $\bar{x} \in G/\Gamma$, and $F \subset G$ is a subgroup. Let $\mathrm{Stab}_F(\bar{x})$ denote the stabilizer in F of $\bar{x}$, i.e. the set of elements $g \in F$ such that $g\bar{x} = \bar{x}$. Then $\mathrm{Stab}_F(\bar{x}) = F \cap x\Gamma x^{-1}$, where $x \in G$ is any element such that $\pi(x) = \bar{x}$. Thus there is a continuous map from $F\bar{x}$ to $F/(F \cap x\Gamma x^{-1})$, which is a bijection, but is in general not a homeomorphism.

However, in the case of algebraic measures, we are making the additional assumption that $F\bar{x}$ is closed. In this case, the above map is a homeomorphism, and thus μ is the image under this map of the Haar measure on $F/(F \cap x\Gamma x^{-1})$. The assumption that μ is a probability measure thus implies that $F \cap x\Gamma x^{-1}$ is a lattice in F. (The last condition is usually taken to be part of the definition of an algebraic measure).

Uniform Distribution and the classification of orbit closures.

THEOREM 1.12 (Ratner's uniform distribution theorem). *Let G be a Lie group, Γ a lattice in G, and $U = \{u_t\}_{t\in\mathbb{R}}$ a one-parameter unipotent subgroup. Then for any $\bar{x} \in G/\Gamma$ there exists a subgroup $F \supset U$ (generated by unipotents) with $F\bar{x}$ closed, and an F-invariant algebraic measure μ supported on $F\bar{x}$, such that for any $f \in C(G/\Gamma)$,*

$$\lim_{T\to\infty} \frac{1}{T}\int_0^T f(u_t\bar{x})\,dt = \int_{F\bar{x}} f\,d\mu \tag{5}$$

Remarks.

- It follows from (5) that the closure of the orbit $U\bar{x}$ is $F\bar{x}$. Thus Theorem 1.12 can be rephrased as "any orbit is uniformly distributed in its closure".
- Theorem 1.12 is derived from Theorem 1.11 by an argument morally similar to the proof of Proposition 1.9. There is one more ingredient: one has to show that the set of subgroups F which appear in Theorem 1.11 is countable up to conjugation (Proposition 4.1 below). For proofs of this fact see [**Ra6**, Theorem 1.1] and [**Ra7**, Cor. A(2)]), or alternatively [**DM4**, Proposition 2.1].

An immediate consequence of Theorem 1.12 is the following:

THEOREM 1.13 (Raghunathan's topological conjecture). *Let G be a Lie group, $\Gamma \subset G$ a lattice, and $U \subset G$ a one-parameter unipotent subgroup. Suppose $\bar{x} \in G/\Gamma$. Then there exists a subgroup F of G (generated by unipotents) such that the closure $\overline{U\bar{x}}$ of the orbit $U\bar{x}$ is $F\bar{x}$.*

This theorem is due to Ratner in the general case, but several cases were known previously. See §5.1 for a discussion and the relation to the Oppenheim Conjecture.

Uniformity of convergence. In many applications it is important to somehow ensure that the time averages converge to the space average uniformly in the base point $\bar{x}$ (for example we may have an additional integral over $\bar{x}$). In the context of Birkhoff's ergodic theorem, we have the following:

LEMMA 1.14. *Suppose $\phi_t : X \to X$ is a flow preserving an ergodic probability measure μ. Suppose $f \in L^1(X,\mu)$. Then for any $\epsilon > 0$ and $\delta > 0$, there exists $T_0 > 0$ and a set $E \subset X$ with $\mu(E) < \epsilon$, such that for any $x \in E^c$ and any $T > T_0$ we have*

$$\left|\frac{1}{T}\int_0^T f(\phi_t(x))\,dt - \int_X f\,d\mu\right| < \delta$$

(In other words, one has uniform convergence outside of a set of small measure.)

Proof. Let E_n denote the set of $x \in X$ such that for some $T > n$,

$$\left|\frac{1}{T}\int_0^T f(\phi_t(x))\,dt - \int_X f\,d\mu\right| \geq \delta.$$

Then by the Birkhoff ergodic theorem, $\mu(\bigcap_{n=1}^{\infty} E_n) = 0$. Hence there exists $n \in \mathbb{N}$ such that $\mu(E_n) < \epsilon$. Now let $T_0 = n$, and $E = E_n$. □

The uniform distribution theorem of Dani-Margulis. One problem with Lemma 1.14 is that it does not provide us with any information about the exceptional set E (other then the fact that it has small measure). In the setting of unipotent flows, Dani and Margulis proved a theorem (see §4.2 below for the precise statement) which is the analogue of Lemma 1.14, but with an explicit geometric description of the set E. This theorem is crucial for many applications. Its proof is based on the Ratner measure classification theorem (Theorem 1.11) and the "linearization" technique of Dani and Margulis (see §4).

2. The case of $SL(2,\mathbb{R})/SL(2,\mathbb{Z})$

In this lecture I will be loosely following Ratner's paper [**Ra8**].

2.1. Basic Preliminaries. The space of lattices. Let $G = SL(n,\mathbb{R})$, and let $\mathcal{L}_n$ denote the space of unimodular lattices in $\mathbb{R}^n$. (By definition, a lattice Δ is unimodular if an only if the volume of $\mathbb{R}^n/\Delta = 1$.) G acts on $\mathcal{L}_n$ as follows: if $g \in G$ and $\Delta \in \mathcal{L}_n$ is the $\mathbb{Z}$-span of the vectors $v_1, \dots v_n$, then gv is the $\mathbb{Z}$-span of $gv_1, \dots, gv_n$. This action is clearly transitive. The stabilizer of the standard lattice $\mathbb{Z}^n$ is $\Gamma = SL(n,\mathbb{Z})$. This gives an identification of $\mathcal{L}_n$ with G/Γ. We choose a *right-invariant* metric $d(\cdot,\cdot)$ on G; then this metric descends to G/Γ.

The set $\mathcal{L}_n(\epsilon)$. For $\epsilon > 0$ let $\mathcal{L}_n(\epsilon) \subset \mathcal{L}_n$ denote the set of lattices whose shortest non-zero vector has length at least ϵ.

THEOREM 2.1 (Mahler Compactness). *For any $\epsilon > 0$ the set $\mathcal{L}_n(\epsilon)$ is compact.*

The upper half plane. In the rest of this section, we set $n = 2$. Let $K = SO(2) \subset G$. Given a pair of vectors v_1, v_2 we can find a unique rotation matrix $k \in K$ so that kv_1 is pointing along the positive x-axis and kv_2 is in the upper half plane. The map $g = \begin{pmatrix} v_1 & v_2 \end{pmatrix} \to kv_2$ gives an identification of $K\backslash G$ with the hyperbolic upper half plane $\mathbb{H}^2$. Now G (and in particular $\Gamma \subset G$) acts on $K\backslash G$ by multiplication on the right. Using the identification of $K\backslash G$ with $\mathbb{H}^2$ this becomes (a variant of) the usual action by fractional linear transformations.

The horocycle and geodesic flows. We use the following notation:

$$u_t = \begin{pmatrix} 1 & t \\ 0 & 1 \end{pmatrix} \qquad a_t = \begin{pmatrix} e^t & 0 \\ 0 & e^{-t} \end{pmatrix} \qquad v_t = \begin{pmatrix} 1 & 0 \\ t & 1 \end{pmatrix}.$$

Let $U = \{u_t \;:\; t \in \mathbb{R}\}$, $A = \{a_t \;:\; t \in \mathbb{R}\}$, $V = \{v_t \;:\; t \in \mathbb{R}\}$. The action of U is called the *horocycle flow* and the action of A is called the *geodesic flow*. Some basic commutation relations are the following:

$$a_t u_s a_t^{-1} = u_{e^{2t}s} \qquad a_t v_s a_t^{-1} = v_{e^{-2t}s} \tag{6}$$

Thus conjugation by a_t for $t > 0$ contracts V and expands U.

Orbits of the geodesic and horocycle flow in the upper half plane. Let $p : G \to K\backslash G$ denote the natural projection. Then for $x \in G$, $p(Ux)$ is either a horizontal line or a circle tangent to the x-axis. Also $p(Ax)$ is either a vertical line or a semicircular arc orthogonal to the x-axis.

Flowboxes. Let $W_+ \subset U$, $W_- \subset V$, $W_0 \subset A$ be intervals containing the identity (we have identified all three subgroups with $\mathbb{R}$). By a *flowbox* we mean a subset of G of the form $W_- W_0 W_+$, or one of its right translates by $g \in G$. Clearly, $W_- W_0 W_+ g$ is an open set containing g. (Recall that in our conventions, right multiplication by g is an isometry).

2.2. An elementary non-divergence result. Much more is proved in [**Kl1**].

LEMMA 2.2. *There exists an absolute constant $\epsilon_0 > 0$ such that the following holds: Suppose $\Delta \in \mathcal{L}_2$ is a unimodular lattice. Then Δ cannot contain two linearly independent vectors each of length less than ϵ_0.*

Proof. Let v_1 be the shortest vector in Δ, and let v_2 be the shortest vector in Δ linearly independent from v_1. Then v_1 and v_2 span a sublattice Δ' of Δ. (In fact $\Delta' = \Delta$ but this is not important for us right now). Since Δ is unimodular, this implies that $\mathrm{Vol}(\mathbb{R}^2/\Delta') \geq 1$. But $\mathrm{Vol}(\mathbb{R}^2/\Delta') = \|v_1 \times v_2\| \leq \|v_1\|\|v_2\|$. Hence $\|v_1\|\|v_2\| \geq 1$, so the lemma holds with $\epsilon_0 = 1$. □

Remark. In general ϵ_0 depends on the choice of norm on $\mathbb{R}^2$.

The following lemma is a simple "nondivergence" result for unipotent orbits:

LEMMA 2.3. *Suppose $\Delta \in \mathcal{L}_2$ is a unimodular lattice. Then at least one of the following holds:*

(a) *Δ contains a horizontal vector.*
(b) *There exists $t \geq 0$ such that $a_t^{-1}\Delta \in \mathcal{L}_2(\epsilon_0)$.*

Proof. Suppose Δ does not contain a horizontal vector, and $\Delta \notin \mathcal{L}_2(\epsilon_0)$. Then Δ contains a vector v with $\|v\| < \epsilon_0$. Since v is not horizontal, there exists a smallest $t_0 > 0$ such that $\|a_t^{-1}v\| = \epsilon_0$. Then by Lemma 2.2 for $t \in [0, t_0]$, $a_t^{-1}\Delta$ contains no vectors shorter then ϵ_0 (other then $a_t^{-1}v$ and possibly its multiples). In particular $a_{t_0}^{-1}\Delta$, contains no vectors shorter then ϵ_0. This means $a_{t_0}^{-1}\Delta \in \mathcal{L}_2(\epsilon_0)$. □

Remark. We note that Lemma 2.2 and thus Lemma 2.3 are specific to dimension 2.

2.3. The classification of U-invariant measures. Note that for $\Delta \in \mathcal{L}_2$, the U-orbit of Δ is closed if and only if Δ contains a horizontal vector. (The horizontal vector is fixed by the action of U). Any closed U-orbit supports a U-invariant probability measure. All such measures are ergodic.

Let ν denote the Haar measure on $\mathcal{L}_2 = G/\Gamma$. The measure ν is normalized so that $\nu(\mathcal{L}_2) = 1$. Recall that ν is ergodic for both the horocycle and the geodesic flows (this follows from the Moore ergodicity theorem, see e.g. [**BM**]).

Our main goal in this lecture is the following:

THEOREM 2.4. *Suppose μ is an ergodic U-invariant probability measure on $\mathcal{L}_2$. Then either μ is supported on a closed orbit, or μ is the Haar measure ν.*

Proof. Let $\mathcal{L}_2' \subset \mathcal{L}_2$ denote the set of lattices which contain a horizontal vector. Note that the set $\mathcal{L}_2'$ is U-invariant.

Suppose μ is an ergodic U-invariant probability measure on $\mathcal{L}_2$. By ergodicity of μ, $\mu(\mathcal{L}_2') = 0$ or $\mu(\mathcal{L}_2') = 1$. If the latter holds, it is easy to show that μ is supported on a closed orbit. Thus we assume $\mu(\mathcal{L}_2') = 0$ and we must show that $\mu = \nu$.

Suppose not. Then there exists a compactly supported continuous function $f : \mathcal{L}_2 \to \mathbb{R}$ and $\epsilon > 0$ such that

$$\left| \int_{\mathcal{L}_2} f\, d\mu - \int_{\mathcal{L}_2} f\, d\nu \right| > \epsilon. \tag{7}$$

Since f is uniformly continuous, there exists a neighborhoods of the identity $W_0' \subset A$ and $W_-' \subset V$ such that for $a \in W_0'$, $v \in W_-'$ and $\Delta'' \in \mathcal{L}_2$,

$$|f(va\Delta'') - f(\Delta'')| < \epsilon/3. \tag{8}$$

Recall that $\pi : G \to G/\Gamma \cong \mathcal{L}_2$ denotes the natural projection. Since $\mathcal{L}_2(\epsilon_0)$ is compact the injectivity radius on $\mathcal{L}_2(\epsilon_0)$ is bounded from below, hence there exist $W_+ \subset U$, $W_0 \subset A$, $W_- \subset V$ so that for any $g \in G$ with $\pi(g) \in \mathcal{L}_2$, the restriction of π to the flowbox $W_-W_0W_+g$ is injective. We may also assume that $W_- \subset W_-'$ and $W_0 \subset W_0'$. Let $\delta = \nu(W_-W_0W_+)$ denote the Lebesque measure of the flowbox.

By Lemma 1.14 applied to the Lebesque measure ν, there exists a set $E \subset \mathcal{L}_2$ with $\nu(E) < \delta$ and $T_1 > 0$ such that for any interval I with $|I| \geq T_1$ and any $\Delta' \notin E$,

$$\left| \frac{1}{|I|} \int_I f(u_t\Delta')\, dt - \int_{\mathcal{L}_2} f\, d\nu \right| < \frac{\epsilon}{3}. \tag{9}$$

Now let Δ be a generic point for U (in the sense of the Birkhoff ergodic theorem). This implies that there exists $T_2 > 0$ such that for any interval I containing the origin of length greater then T_2,

$$\left| \frac{1}{|I|} \int_I f(u_t\Delta)\, dt - \int_{\mathcal{L}_2} f\, d\mu \right| < \frac{\epsilon}{3}. \tag{10}$$

Since $\mu(\mathcal{L}_2') = 0$, we may assume that Δ does not contain any horizontal vectors. Then by repeatedly applying Lemma 2.3 we can construct arbitrarily large $t > 0$ such that

$$a_t^{-1}\Delta \in \mathcal{L}_2(\epsilon). \tag{11}$$

Now suppose t is such that (11) holds, and consider the set $Q = a_tW_-W_0W_+a_t^{-1}\Delta$. Then Q can be rewritten as

$$Q = (a_tW_-a_t^{-1})W_0(a_tW_+a_t^{-1})\Delta$$

(so when t is large, Q is long in the U direction and short in A and V directions.) The set Q is an embedded copy of a flowbox in $\mathcal{L}_2$, and $\nu(Q) = \delta$.

If t is sufficiently large and W_-, W_0 and W_+ are sufficiently small, it is possible to find for each $\Delta' \in Q$ intervals $I(\Delta') \subset \mathbb{R}$ and $I(\Delta) \subset \mathbb{R}$ with the following properties: $|I(\Delta')| \geq \max(T_1, T_2)$, $|I(\Delta)| \geq \max(T_1, T_2)$ and

$$\left| \frac{1}{|I(\Delta')|} \int_{I(\Delta')} f(u_t\Delta')\, dt - \frac{1}{|I(\Delta)|} \int_{I(\Delta)} f(u_t\Delta)\, dt \right| < \frac{\epsilon}{3}. \tag{12}$$

(this says that the integral of f over a suitably chosen interval of each U-orbit is nearly the same).

Since $\nu(E) < \delta$ and $\nu(Q) = \delta$, there exists $\Delta' \in Q \cap E^c$. Now (9) holds with $I = I(\Delta')$, and (10) holds with $I = I(\Delta)$. These estimates together with (12) contradict (7). □

Remarks.

- The above proof works with minor modifications if Γ is an arbitrary lattice in $SL(2,\mathbb{R})$ (not just $SL(2,\mathbb{Z})$).
- If Γ is a uniform lattice in $SL(2,\mathbb{R})$ then the horocycle flow on G/Γ is uniquely ergodic. This is a theorem of Furstenberg [**F**].

- The proof of Theorem 2.4 does not generalize to classification of measures invariant under a one-parameter unipotent subgroup on e.g. $\mathcal{L}_n$, $n \geq 3$. Completely different ideas are needed. (I will introduce some of them in the next lecture).

Horospherical subgroups and a theorem of Dani. The key property of U in dimension 2 which is used in the proof is that U is *horospherical*, i.e. that it is equal to the set contracted by a one-parameter diagonal subgroup. (One-parameter unipotent subgroups are horospherical only in $SL(2,\mathbb{R})$). An argument similar in spirit to the proof of Theorem 2.4 can be used to classify the measures invariant under the action of a horospherical subgroup. This is a theorem of Dani [**Dan2**] (which was proved before Ratner's measure classification theorem). However, the details, and in particular the non-divergence results needed are much more complicated.

The horospherical case also allows for an analytic approach, see e.g. [**Bu**].

3. The case of $SL(2,\mathbb{R}) \ltimes \mathbb{R}^2$.

In this section we will outline a proof of Ratner's measure classification theorem Theorem 1.11 in the special case $G = SL(2,\mathbb{R}) \ltimes \mathbb{R}^2$, $\Gamma = SL(2,\mathbb{Z}) \ltimes \mathbb{Z}^2$. We will be following the argument of Ratner [**Ra1, Ra2, Ra3, Ra4, Ra5, Ra6**] and Margulis-Tomanov [**MT**]. An introduction to these ideas can be found in the books [**Mor**], and also [**BM**]. Another exposition of a closely related case is in [**EMaMo**].

Let $X = G/\Gamma$. Then X can be viewed as a space of pairs (Δ, v), where Δ is a unimodular lattice in $\mathbb{R}^2$ and v is a marked point on the torus $\mathbb{R}^2/\Delta$. (We remove the translation invariance on the torus $\mathbb{R}^2/\Delta$ since we consider the origin as a special point. Alternatively we consider a pair of marked points, and use the translation invariance of the torus to place one of the points at the origin). X is thus naturally a fiber bundle where the base is $\mathcal{L}_2$ and the fiber above the point $\Delta \in \mathcal{L}_2$ is the torus $\mathbb{R}^2/\Delta$. (X is also sometimes called the universal elliptic curve).

The action of $SL(2,\mathbb{R}) \subset G$ on X is by left multiplication. It amounts to

$$g \cdot (\Delta, v) = (g\Delta, gv).$$

The action of the $\mathbb{R}^2$ part of G on X is by translating the marked point, i.e for $w \in \mathbb{R}^2$, $w \cdot (\Delta, v) = (\Delta, w + v)$. Let U be the subgroup of $SL(2,\mathbb{R})$ defined in §2.1. In this lecture our goal is the following special case of Theorem 1.11:

THEOREM 3.1. *Let μ be an ergodic U-invariant measure on X. Then μ is algebraic.*

Let μ be an ergodic U-invariant measure on X. Let $\pi_1 : X \to \mathcal{L}_2$ denote the natural projection (i.e. $\pi_1(\Delta, v) = \Delta$). Then $\pi_1^*(\mu)$ is an ergodic U-invariant measure on $\mathcal{L}_2$. Thus by Theorem 2.4, either $\pi_1^*(\mu)$ is supported on a closed orbit of U, or $\pi_1^*(\mu)$ is the Haar measure ν on $\mathcal{L}_2$. The first case is easy to handle, so in the rest of this section we assume that $\pi_1^*(\mu) = \nu$. Then we can disintegrate

$$d\mu(\Delta, v) = d\nu(\Delta) d\lambda_\Delta(v)$$

where $\lambda_\Delta(v)$ is some probability measure on the torus $\mathbb{R}^2/\Delta$.

3.1. Finiteness of the fiber measures. Many of the ideas behind the proof of Ratner's measure classification theorem Theorem 1.11 can be illustrated in the proof of the following:

PROPOSITION 3.2. *Either μ is Haar measure on X, or for almost all $\Delta \in \mathcal{L}_2$, the measure λ_Δ is supported on a finite set of points.*

We will give an almost complete proof of Proposition 3.2 in this subsection, and then indicate how to complete the proof of Theorem 3.1 in the next subsection.

The subgroups U,V,A,H, and W. Let U, V, A be the subgroups of $SL(2,\mathbb{R})$ defined in §2.1. We also give names to certain subgroups of the $\mathbb{R}^2$ part of G. In particular, let $H = \{h_s, s \in \mathbb{R}\}$ be the subgroup of G whose action on X is given by $h_s(\Delta, v) = (\Delta, v + s\begin{pmatrix}1\\0\end{pmatrix})$, and $W = \{w_r, r \in \mathbb{R}\}$ be the subgroup of G whose action on X is given by $w_r(\Delta, v) = (\Delta, v + r\begin{pmatrix}0\\1\end{pmatrix})$. The action of H is called the *horizontal flow* and the action of W the *vertical flow.*

Action of the centralizer. A key observation is that H commutes with U (and so the action of H commutes with the action of U). This implies that if μ is an ergodic U-invariant measure, so is $h_s\mu$ for any $h_s \in H$. (See the discussion preceeding Lemma 1.7).

Thus, either μ is invariant under H or there exists $s \in \mathbb{R}$ such that $h_s\mu$ is distinct from μ. Suppose μ is invariant under H. Then so are the fiber measures λ_Δ for all $\Delta \in \mathcal{L}_2$. Then by Exercise 1 (b), for ν-almost all $\Delta \in \mathcal{L}_2$, λ_Δ is the Lebesque measure on $\mathbb{R}^2/\Delta$. Thus μ coincides with Haar measure on X for almost all fibers. Then by the ergodicity of μ we can conclude that μ is the Haar measure on X.

Thus, Proposition 3.2 follows from the following:

PROPOSITION 3.3. *Suppose μ is not H-invariant. Then for almost all $\Delta \in \mathcal{L}_2$, the measure λ_Δ is supported on a finite set of points.*

The element h and the compact set K. From now on, we assume that μ is not H-invariant. Then there exists $h_{s_0} \in H$ such that $h_{s_0}\mu \neq \mu$. (We may assume that h_{s_0} is fairly close to the identity). Since $h_{s_0}\mu$ and μ are both ergodic U-invariant measures, by Lemma 1.6 we have $h_{s_0}\mu \perp \mu$. Thus the sets of generic points of μ and $h_{s_0}\mu$ are disjoint. It follows from Lemma 1.7 that there exists $\delta > 0$ and a subset $\Omega \subset X$ with $\mu(\Omega) = 1$ such that $h_s\Omega \cap \Omega = \emptyset$ for all $s \in (s_0 - \delta s_0, s_0]$. It follows that there exists a compact set K with $\mu(K) > 0.999$ such that for all $s \in [(1-\delta_0)s_0, s_0]$, $h_s K \cap K = \emptyset$. Since K is compact and the action of H is continuous, there exist $\epsilon > 0$ and $\delta > 0$ such that

$$d(h_s K, K) > \epsilon \qquad \text{for all } s \in [(1-\delta)s_0, s_0]. \tag{13}$$

The set Ω_ρ. In view of Lemma 1.14 (with f the characteristic function of K), for any $\rho > 0$ we can find a set Ω_ρ with $\mu(\Omega_\rho) > 1 - \rho$ and $T_0 > 0$ such that for all $T > T_0$ and all $p \in \Omega_\rho$ we have

$$\frac{1}{T}|\{t \in [0,T] \ : \ u_t x \in K\}| \geq 1 - (0.01)\delta \tag{14}$$

Shearing. Suppose $p = (\Delta, v)$ and $p' = (\Delta, v')$ are two nearby points in the same fiber. We want to study how they diverge under the action of U. Note that $u_t p$ and $u_t p'$ are always in the same fiber (i.e. $\pi_1(u_t p) = \pi_1(u_t p') = u_t \Delta$), but within the fiber $\pi_1^{-1}(u_t\Delta)$ they will slowly diverge. More precisely, if we let $v = (x, y)$ and $v' = (x', y')$ we have

$$u_t v' - u_t v = (x' - x + t(y' - y), y' - y).$$

Note that if $y = y'$ (i.e. p and p' are in the same orbit of H) then $u_t p$ and $u_t p'$ will not diverge at all.

Now suppose $y \neq y'$. We are considering the regime where $|x' - x|$, $|y' - y|$ are very small, but t is so large that $d(p, p')$ is comparable to 1 (this amounts to $|t(y' - y)|$ comparable to 1). Under these assumptions, *the leading divergence is along* H, i.e.

$$u_t p' = h_s u_t p + \text{ small error} \tag{15}$$

where $s = t(y' - y)$.

LEMMA 3.4. *Suppose that for some positive measure set of $\Delta \in \mathcal{L}_2$, the support of λ_Δ is infinite. Then for any $\rho > 0$ we can find $\Delta \in \mathcal{L}_2$ and a sequence of points $p_n = (\Delta, (x_n, y_n)) \in \Omega_\rho$ which converge to $p = (\Delta, (x, y)) \in \Omega_\rho$ so that $y_n \neq y$ for all n.*

We postpone the proof of this lemma (which is intuitively reasonable anyway).

Proof of Proposition 3.3. Suppose the conclusion of Proposition 3.3 is false, so that for some positive measure set of $\Delta \in \mathcal{L}_2$, the support of λ_Δ is infinite. Then Lemma 3.4 applies.

Let $T_n = s_0/(y_n - y)$. Then by (15) we have for $t \in [(1-\delta)T_n, T_n]$,

$$d(u_t p_n, h_s u_t p) < \epsilon_n, \qquad \text{where } s = t/(y' - y). \tag{16}$$

and $\epsilon_n \to 0$ as $n \to \infty$. If n is sufficiently large, then $T_n > T_0$ where T_0 is as in the definition of Ω_ρ. Then (14) applies to both p and p_n, and we can thus find $t \in [(1-\delta)T_n, T_n]$ such that $u_t p_n \in K$ and also $u_t p \in K$. Then $s = t/(y' - y) \in [(1 - \delta_0)s_0, s_0]$, and so (16) contradicts (13). □

Proof of Lemma 3.4. Suppose that for some positive measure set of $\Delta \in \mathcal{L}_2$, the support of λ_Δ is infinite. Then (by the ergodicity of the action of U on $\mathcal{L}_2$), the support of λ_Δ is infinite for almost all fibers Δ.

Suppose for the moment that the support of λ_Δ is countable for almost all Δ, so λ_Δ is supported on a sequence of points p_n with weights λ_n. But then the collection of points with the same weight is a U-invariant set, so by ergodicity of μ all the points must have the same weight. Thus, since λ_Δ is a probability measure if the support of λ_Δ is countable it must be finite.

Hence we may assume that the support of λ_Δ is uncountable. Then so is $\Omega_\rho \cap \lambda_\Delta$ for almost all Δ. Since any uncountable set contains one of its accumulation points, we may construct a sequence $p_n \in \Omega_\rho$ with $p_n \to p$, where $p \in \Omega_\rho$. It only remains to verify that if we write $p_n = (\Delta, (x_n, y_n))$ and $p = (\Delta, (x, y))$ then we can ensure $y_n \neq y$.

If it is not possible to do so, then it is easy to see that the support of λ_Δ is contained in a finite union of H-orbits. Thus given $a < b$ we can define a function $u((\Delta, v)) = \lambda_\Delta(\{h_s v \ : \ s \in [a, b]\})$. This function is U-invariant hence constant for each choice of $[a, b]$. It is easy to conclude from this that the support of λ_Δ must be finite. □

3.2. Outline of the Proof of Theorem 3.1. The following general lemma is a stronger version of Lemma 1.14:

LEMMA 3.5 (cf. [**MT**, Lem. 7.3]). *Suppose $\phi_t : X \to X$ is a flow preserving an ergodic probability measure μ. For any $\rho > 0$, there is a "uniformly generic set" Ω_ρ in X, such that*

(1) $\mu(\Omega_\rho) > 1 - \rho$,
(2) *for every $\epsilon > 0$ and every compact subset K of X, with $\mu(K) > 1 - \epsilon$, there exists $L_0 \in \mathbb{R}^+$, such that, for all $x \in \Omega_\rho$ and all $L > L_0$, we have*
$$|\{\, t \in [-L, L] \mid d(\phi_t(x), K) < \epsilon \,\} > (1 - \epsilon)(2L).$$

Outline of proof. This is similar to that of Lemma 1.14, except that one also chooses a countable basis of functions and approximates K by elements of the basis. □

We now return to the setting of §3. Let μ be an ergodic invariant measure for the action of U on $X = G/\Gamma = (SL(2, R) \ltimes \mathbb{R}^2)/(SL(2, \mathbb{Z}) \ltimes \mathbb{Z}^2)$. For any $\rho > 0$ we chose a "uniformly generic" set Ω_ρ for μ as in Lemma 3.5.

The argument of §3.1 is the basis of the following more general proposition (which we state somewhat imprecisely):

PROPOSITION 3.6. *Suppose Q is a subgroup of G normalizing U, and suppose that for any $\rho > 0$ we can find sequences p_n and p'_n in Ω_ρ such that $d(p_n, p'_n) \to 0$, and under the action of U the leading transverse divergence of the trajectories $u_t p_n$ and $u_t p'_n$ is in the direction of Q (i.e the analogue of (15) holds with $q \in Q$ instead of $h \in H$).*

Then the measure μ is Q-invariant.

Remark. The analogous statement for unipotent flows is a cornerstone of the proof of Ratner's Measure Classification Theorem [**Ra5**, Lem. 3.3], [**MT**, Lem. 7.5], [**Mor**, Prop. 5.2.4′].

Remark. For two points in the same fiber, the leading divergence is always along H (if the points diverge at all). For an arbitrary pair of nearby points in X this is not the case.

Remark. It is possible that the leading direction of divergence is along U. In that case we want to consider the leading "transverse" divergence. In other words we compare $u_t p_n$ and $u_{t'} p'_n$ where t' is chosen to cancel the divergence along U (i.e. one trajectory waits for the other). In that case we say that the leading transverse divergence is along Q if for some $q \in Q$,
$$u_t p_n = q u_{t'} p'_n + \text{ small error}$$

Remark. To prove Proposition 3.6 we must use Lemma 3.5 instead of Lemma 1.14 as in §3.1 because we must choose Ω_ρ before we know what subgroup Q (and thus what compact set K) we will be dealing with.

We now continue the proof of Theorem 3.1. We assume that μ projects to Haar measure on $\mathcal{L}_2$, but that μ is not Haar measure.

PROPOSITION 3.7. *The measure μ is invariant under some subgroup of AH other then H.*

Proof. Choose Ω_ρ as in Lemma 3.5, with $\rho = 0.01$. By Proposition 3.2, the measure on each fiber is supported on a finite set. Also we are assuming that μ projects to Haar measure on $\mathcal{L}_2$. Then it is easy to see that there exist $p \in \Omega_\rho$, $\{v_n\} \subset V \setminus \{e\}$, and $\{w_n\} \subset HW$, such that $p_n = v_n w_n p \in \Omega_\rho$, $v_n \to e$, and $w_n \to e$.

It is not difficult to compute that (after passing to a subsequence), the leading direction of divergence of $u_t p_n$ and $u_t p$ is a one-parameter subgroup Q which is contained in AH. Then by Proposition 3.6, μ is invariant under Q. By §3.1, we have $Q \neq H$. □

Invariance under A. Any one-parameter subgroup Q of AH other then H is conjugate to A (via an element of H). Thus, by replacing μ with a translate under H, we may (and will) assume μ is A-invariant.

Note. At this point we do not know that μ is A-ergodic.

PROPOSITION 3.8 (cf. [**MT**, Cor. 8.4], [**Mor**, Cor. 5.5.2]). *There is a conull subset Ω of X, such that*

$$\Omega \cap VWp = \Omega \cap Vp,$$

for all $p \in \Omega$.

Proof. Let Ω be a generic set for for the action of A on X; thus, Ω is conull and, for each $p \in \Omega$,

$$a_t p \in \Omega_\rho \text{ for most } t \in \mathbb{R}^+.$$

(The existence of such a set follows e.g. from the full version of the Birkhoff ergodic theorem, in which one does not assume ergodicity). Given $p, p' \in \Omega$, such that $p' = vwp$ with $v \in V$ and $w \in W$, we wish to show $w = e$.

Choose a sequence $t_n \to \infty$, such that $a_{t_n} p$ and $a_{t_n} p'$ each belong to Ω_ρ. Because $t_n \to \infty$ and VW is the foliation that is contracted by $a_{\mathbb{R}^+}$, we know that $a_{-t_n}(vw)a_{t_n} \to e$. Furthermore, because A acts on the Lie algebra of V with twice the weight that it acts on the Lie algebra of W, we see that

$$\|a_{-t_n} v a_{t_n}\| / |a_{-t_n} w a_{t_n}\| \to 0.$$

Thus $p'_n = a_{-t_n} p' a_{t_n}$ approaches $p_n = a_{-t_n} p a_{t_n}$ from the direction of W.

If two points p'_n and p_n approach each other along W, then an easy computation shows that $u_t p_n$ and $u_t p'_n$ diverge along H. (This observation motivates Proposition 3.8). Thus by Proposition 3.6 μ must be invariant under H. But this impossible by §3.1 (since we are assuming that μ is not Haar measure). □

We require the following entropy estimate, (see [**EL**] for a proof).

LEMMA 3.9 (cf. [**MT**, Thm. 9.7], [**Mor**, Prop. 2.5.11]). *Suppose $\mathcal{W}$ is a closed connected subgroup of VW that is normalized by $a \in A^+$, and let*

$$J(a^{-1}, \mathcal{W}) = \det\big((\operatorname{Ad} a^{-1})|_{\operatorname{Lie} \mathcal{W}}\big)$$

be the Jacobian of a^{-1} on $\mathcal{W}$.

(1) *If μ is $\mathcal{W}$-invariant, then $h_\mu(a) \geq \log J(a^{-1}, \mathcal{W})$.*
(2) *If there is a conull, Borel subset Ω of X, such that $\Omega \cap VWp \subset \mathcal{W}p$, for every $p \in \Omega$, then $h_\mu(a) \leq \log J(a^{-1}, \mathcal{W})$.*

(3) *If the hypotheses of 2 are satisfied, and equality holds in its conclusion, then μ is $\mathcal{W}$-invariant.*

PROPOSITION 3.10 (cf. [**MT**, Step 1 of 10.5], [**Mor**, Prop. 5.6.1]). *μ is V-invariant.*

Proof. From Lemma 3.9(1), with a^{-1} in the role of a, we have

$$\log J(a, UX) \leq h_\mu(a^{-1}).$$

From Proposition 3.8 and Lemma 3.9(2), we have

$$h_\mu(a) \leq \log J(a^{-1}, VY).$$

Combining these two inequalities with the facts that

- $h_\mu(a) = h_\mu(a^{-1})$ and
- $J(a, UX) = J(a^{-1}, VY)$,

we have

$$\log J(a, UX) \leq h_\mu(a^{-1}) = h_\mu(a) \leq \log J(a^{-1}, VY) = \log J(a, UX).$$

Thus, we must have equality throughout, so the desired conclusion follows from Lemma 3.9(3). □

PROPOSITION 3.11. *μ is the Lebesgue measure on a single orbit of $SL(2,\mathbb{R})$ on X.*

Proof We know:

- U preserves μ (by assumption),
- A preserves μ (by Proposition 3.7) and
- V preserves μ (by Proposition 3.10).

Since $SL(2,\mathbb{R})$ is generated by U, A and V, μ is $SL(2,\mathbb{R})$ invariant. Because $SL(2,\mathbb{R})$ is transitive on the quotient $\mathcal{L}_2$ and the support of μ on each fiber is finite (see Proposition 3.2), this implies that some orbit of $SL(2,\mathbb{R})$ has positive measure. By ergodicity of U, then this orbit is conull. □

This completes the proof of Theorem 3.1.

4. Linearization and ergodicity

4.1. Non-ergodic measures invariant under a unipotent. The collection $\mathcal{H}$.

(Up to conjugation, this should be the collection of groups which appear in the definition of algebraic measure).

Let G be a Lie group, Γ a discrete subgroup of G, and $\pi : G \to G/\Gamma$ the natural quotient map. Let $\mathcal{H}$ be the collection of all closed subgroups F of G such that $F \cap \Gamma$ is a lattice in F and the subgroup generated by unipotent one-parameter subgroups of G contained in F acts ergodically on $\pi(F) \cong F/(F \cap \Gamma)$ with respect to the F-invariant probability measure.

PROPOSITION 4.1. *The collection $\mathcal{H}$ is countable.*

Proof. See [**Ra6**, Theorem 1.1] or [**DM4**, Proposition 2.1] for different proofs of this result. □

Let U be a unipotent one-parameter subgroup of G and $F \in \mathcal{H}$. Define

$$\begin{aligned} N(F,U) &= \{g \in G : U \subset gFg^{-1}\} \\ S(F,U) &= \bigcup \{N(F',U) : F' \in \mathcal{H},\ F' \subset F,\ \dim F' < \dim F\}. \end{aligned}$$

LEMMA 4.2. ([**MS**, Lemma 2.4]) *Let $g \in G$ and $F \in \mathcal{H}$. Then $g \in N(F,U) \setminus S(F,U)$ if and only if the group gFg^{-1} is the smallest closed subgroup of G which contains U and whose orbit through $\pi(g)$ is closed in G/Γ. Moreover in this case the action of U on $g\pi(F)$ is ergodic with respect to a finite gFg^{-1}-invariant measure.*

As a consequence of this lemma,

$$\pi(N(F,U) \setminus S(F,U)) = \pi(N(F,U)) \setminus \pi(S(F,U)), \qquad \forall F \in \mathcal{H}. \tag{17}$$

Ratner's theorem [**Ra6**] states that given any U-ergodic invariant probability measure on G/Γ, there exists $F \in \mathcal{H}$ and $g \in G$ such that μ is $g^{-1}Fg$-invariant and $\mu(\pi(F)g) = 1$. Now decomposing any finite invariant measure into its ergodic component, and using Lemma 4.2, we obtain the following description for any U-invariant probability measure on G/Γ (see [**MS**, Theorem 2.2]).

THEOREM 4.3 (Ratner). *Let U be a unipotent one-parameter subgroup of G and μ be a finite U-invariant measure on G/Γ. For every $F \in \mathcal{H}$, let μ_F denote the restriction of μ on $\pi(N(F,U) \setminus S(F,U))$. Then μ_F is U-invariant and any U-ergodic component of μ_F is a gFg^{-1}-invariant measure on the closed orbit $g\pi(F)$ for some $g \in N(F,U) \setminus S(F,U)$.*

In particular, for all Borel measurable subsets A of G/Γ,

$$\mu(A) = \sum_{F \in \mathcal{H}^*} \mu_F(A),$$

where $\mathcal{H}^* \subset \mathcal{H}$ is a countable set consisting of one representative from each Γ-conjugacy class of elements in $\mathcal{H}$.

Remark. We will often use Theorem 4.3 in the following form: suppose μ is any U-invariant measure on G/Γ which is not Lebesque measure. Then there exists $F \in \mathcal{H}$ such that μ gives positive measure to some compact subset of $N(F,U) \setminus S(F,U)$.

4.2. The theorem of Dani-Margulis on uniform convergence. The "linearization" technique of Dani and Margulis was devised to understand which measures give positive weight to compact subsets subsets of $N(F,U) \setminus S(F,U)$. Using this technique Dani and Margulis proved the following theorem (which is important for many applications, in particular §5):

THEOREM 4.4 ([**DM4**], Theorem 3). *Let G be a connected Lie group and let Γ be a lattice in G. Let μ be the G-invariant probability measure on G/Γ. Let $U = \{u_t\}$ be an Ad-unipotent one-parameter subgroup of G and let f be a bounded continuous function on G/Γ. Let $\mathcal{D}$ be a compact subset of G/Γ and let $\epsilon > 0$ be given. Then there exist finitely many proper closed subgroups $F_1 = F_1(f,\mathcal{D},\epsilon), \cdots, F_k = F_k(f,\mathcal{D},\epsilon)$ such that $F_i \cap \Gamma$ is a lattice in F_i for all i, and compact subsets $C_1 = C_1(f,\mathcal{D},\epsilon), \cdots, C_k = C_k(f,\mathcal{D},\epsilon)$ of $N(F_1,U), \cdots, N(F_k,U)$ respectively, for which* *the following holds: For any compact subset K of $\mathcal{D} - \bigcup_{1\le i\le k}\pi(C_i)$ there exists a $T_0 \geq 0$ such that for all $x \in K$ and $T > T_0$*

$$\left|\frac{1}{T}\int_0^T f(u_t x)\,dt - \int_{G/\Gamma} f\,d\mu\right| < \epsilon. \tag{18}$$

Remarks.

- This theorem can be informally stated as follows: Fix f and $\epsilon > 0$. Then (18) holds (i.e. the space average of f is within ϵ of the time average of f) uniformly in the base point x, as long as x is restricted to compact sets away from a finite union of "tubes" $N(F, U)$. (The $N(F, U)$ are associated with orbits which do not become equidistributed in G/Γ, because their closure is strictly smaller.)
- It is a key point that only finitely many F_k are needed in Theorem 4.4. This has the remarkable implication that if $F \in \mathcal{H}$ but not one of the F_k, then (18) holds for $x \in N(F, U)$ even though Ux is not dense in G/Γ (the closure of Ux is Fx). Informally, this means the non-dense orbits of U are themselves becoming equidistributed as they get longer.

A full proof of Theorem 4.4 is beyond the scope of this course. However, we will describe the "linearization" technique used in its proof in §4.3.

4.3. Ergodicity of limits of ergodic measures. In this subsection we are following [**MS**], which refers many times to [**DM4**].

Let $\mathcal{P}(G/\Gamma)$ be the space of all probability measures on G/Γ.

THEOREM 4.5 (Mozes-Shah). *Let U_i be a sequence of unipotent one-parameter subgroups of G, and for each i, let μ_i be an ergodic U_i-invariant probability measure on G/Γ. Suppose $\mu_i \to \mu$ in $\mathcal{P}(G/\Gamma)$. Then there exists a unipotent one-parameter subgroup U such that μ is an ergodic U-invariant measure on G/Γ. In particular, μ is algebraic.*

Remarks.

- Let $\mathcal{Q}(G/\Gamma) \subset \mathcal{P}(G/\Gamma)$ denote the set of measures ergodic for the action of a unipotent one-parameter subgroup of G, and let $\mathcal{Q}_0(G/\Gamma)$ denote $\mathcal{Q}(G/\Gamma)$ union the zero measure. If combined with the results of [**Kl1**, §3], Theorem 4.5 shows that $\mathcal{Q}_0(G/\Gamma)$ is *compact.*
- The theorem actually proved by Mozes and Shah in [**MS**] gives more information about what kind of limits of ergodic U-invariant measures are possible. Here is an easily stated consequence:

  Suppose $x_i \in G/\Gamma$ converge to $x_\infty \in G/\Gamma$, and also $x_i \in \overline{Ux_\infty}$. For $i \in \mathbb{N} \cup \{\infty\}$ let μ_i be the algebraic measures supported on $\overline{Ux_i}$, so that the trajectories Ux_i are equidistributed with respect to the measures μ_i. Then $\mu_i \to \mu_\infty$.

We now give some indication of the proof of Theorem 4.5. Let U_i, μ_i, μ be as in Theorem 4.5. Write $U_i = \{u_i(t)\}_{t\in\mathbb{R}}$.

Invariance of μ under a unipotent.

LEMMA 4.6. *Suppose $U_i \neq \{e\}$ for all large $i \in \mathbb{N}$. Then μ is invariant under a one-parameter unipotent subgroup of G.*

Proof. For each $i \in \mathbb{N}$ there exists w_i in the Lie algebra $\mathfrak{g}$ of G, such that $\|w_i\| = 1$ and $U_i = \{\exp(tw_i), t \in \mathbb{R}\}$. (Here $\|\cdot\|$ is some Euclidean norm on $\mathfrak{g}$). By passing to a subsequence we may assume that $w_i \to w$ for some $w \in \mathfrak{g}$, $\|w\| = 1$. For any $t \in \mathbb{R}$ we have $Ad(\exp(tw_i)) \to Ad(\exp(tw))$ as $i \to \infty$. Note that $Ad(\exp(tw))$ is unipotent, since the set of unipotent matrices is closed (consider e.g. the characteristic polynomial). Therefore $U = \{\exp(tw) \ : \ t \in \mathbb{R}\}$ is a nontrivial unipotent subgroup of G. Since $\exp tw_i \to \exp tw$ for all t and $\mu_i \to \mu$, it follows that μ is invariant under the action of U on G/Γ. □

Application of Ratner's measure classification theorem. We want to analyze the case when the limit measure μ is not the G-invariant measure. By Ratner's description of μ as in Theorem 4.3, there exists a proper subgroup $F \in \mathcal{H}$, $\epsilon_0 > 0$, and a compact set $C_1 \subset N(F,U) \setminus S(F,U)$ such that $\mu(\pi(C_1)) > \epsilon_0$. Thus for any neighborhood Φ of $\pi(C_1)$, we have $\mu_i(\Phi) > \epsilon_0$ for all large $i \in \mathbb{N}$. Thus the unipotent trajectories which are equidistributed with respect to the measures μ_i spend a fixed proportion of time in Φ.

Linearization of neighborhoods of singular subsets. Let $F \in \mathcal{H}$. Let $\mathfrak{g}$ denote the Lie algebra of G and let $\mathfrak{f}$ denote its Lie subalgebra associated to F. For $d = \dim \mathfrak{f}$, put $V_F = \wedge^d \mathfrak{f}$, the d-th exterior power, and consider the linear G-action on V_F via the representation $\wedge^d \operatorname{Ad}$, the d-th exterior power of the Adjoint representation of G on $\mathfrak{g}$. Fix $p_F \in \wedge^d \mathfrak{f} \setminus \{0\}$, and let $\eta_F : G \to V_F$ be the map defined by $\eta_F(g) = g \cdot p_F = (\wedge^d \operatorname{Ad} g) \cdot p_F$ for all $g \in G$. Note that

$$\eta_F{}^{-1}(p_F) = \{g \in N_G(F) : \det(\operatorname{Ad} g|_{\mathfrak{f}}) = 1\}.$$

Remark. The idea of Dani and Margulis is to work in the representation space V_F (or more precisely $\bar{V}_F$, which is the quotient of V_F by the involution $v \to -v$) instead of G/Γ. In fact, for most of the argument one works only with the oribit $G \cdot p_F \subset V_F$. The advantage is that F is collapsed to a point (since it stabilizes p_F). The difficulty is that the map $\eta_F : G \to \bar{V}_F$ is not Γ-equivariant, and so becomes multivalued if considered as a map from G/Γ to V_F.

PROPOSITION 4.7 ([**DM4**, Theorem 3.4]). *The orbit $\Gamma \cdot p_F$ is discrete in V_F.*

Remark. In the arithmetic case the above proposition is immediate.

PROPOSITION 4.8. *(*[**DM4**, Prop. 3.2]*) Let A_F be the linear span of $\eta_F(N(F,U))$ in V_F. Then*

$$\eta_F{}^{-1}(A_F) = N(F,U).$$

Let $N_G(F)$ denote the normalizer in G of F. Put $\Gamma_F = N_G(F) \cap \Gamma$. Then for any $\gamma \in \Gamma_F$, we have $\gamma\pi(F) = \pi(F)$, and hence γ preserves the volume of $\pi(F)$. Therefore $|\det(\operatorname{Ad}\gamma|_{\mathfrak{f}})| = 1$. Hence $\gamma \cdot p_F = \pm p_F$. Now define

$$\bar{V}_F = \begin{cases} V_F/\{\text{Id,-Id}\} & \text{if } \Gamma_F \cdot p_F = \{p_F, -p_F\} \\ V_F & \text{if } \Gamma_F \cdot p_F = p_F \end{cases}$$

The action of G factors through the quotient map of V_F onto $\bar{V}_F$. Let $\bar{p}_F$ denote the image of p_F in $\bar{V}_F$, and define $\bar{\eta}_F : G \to \bar{V}_F$ as $\bar{\eta}_F(g) = g \cdot \bar{p}_F$ for all $g \in G$. Then $\Gamma_F = \bar{\eta}_F{}^{-1}(\bar{p}_F) \cap \Gamma$. Let $\bar{A}_F$ denote the image of A_F in $\bar{V}_F$. Note that the inverse image of $\bar{A}_F$ in V_F is A_F.

For every $x \in G/\Gamma$, define the set of representatives of x in $\bar{V}_F$ to be

$$\mathrm{Rep}(x) = \bar{\eta}_F(\pi^{-1}(x)) = \bar{\eta}_F(x\Gamma) \subset \bar{V}_F.$$

Remark. If one attempts to consider the map $\bar{\eta}_F : G \to \bar{V}_F$ as a map from G/Γ to $\bar{V}_F$, one obtains the multivalued map which takes $x \in G/\Gamma$ to the set $\mathrm{Rep}(x) \subset \bar{V}_F$.

The following lemma allows us to understand the map Rep in a special case:

LEMMA 4.9. *If $x = \pi(g)$ and $g \in N(F,U) \setminus S(F,U)$*

$$\mathrm{Rep}(x) \cap \bar{A}_F = \{g \cdot p_F\}.$$

Thus x has a single representative in $\bar{A}_F \subset V_F$.

Proof. Indeed, using Proposition 4.8,

$$\mathrm{Rep}(\pi(g)) \cap \bar{A}_F = (g\Gamma \cap N(F,U)) \cdot \bar{p}_F$$

Now suppose $\gamma \in \Gamma$ is such that $g\gamma \in N(F,U)$. Then g belongs to $N(\gamma F \gamma^{-1}, U)$ as well as $N(F,U)$. Since $g \notin S(F,U)$, we must have $\gamma F \gamma^{-1} = F$, so $\gamma \in \Gamma_F$. Then $\gamma \bar{p}_F = \bar{p}_F$, so $(g\Gamma \cap N(F,U)) \cdot \bar{p}_F = \{g \cdot \bar{p}_F\}$ as required. □

We extend this observation in the following result (cf. [**Sha1**, Prop. 6.5]).

PROPOSITION 4.10 ([**DM4**, Corollary 3.5]). *Let D be a compact subset of $\bar{A}_F$. Then for any compact set $K \subset G/\Gamma \setminus \pi(S(F,U))$, there exists a neighborhood Φ of D in $\bar{V}_F$ such that any $x \in K$ has at most one representative in Φ.*

Remark. This proposition constructs a "fundamental domain" Φ around any compact subset D of $\bar{A}_F$, so that for any x in a compact subset of G/Γ away from $\pi(S(F,U))$, $\mathrm{Rep}(x)$ has at most one element in Φ. Using this proposition, one can uniquely represent in Φ the parts of the unipotent trajectories in G/Γ lying in K.

PROPOSITION 4.11 ([**DM4**, Proposition 4.2]). *Let a compact set $C \subset \bar{A}_F$ and an $\epsilon > 0$ be given. Then there exists a (larger) compact set $D \subset \bar{A}_F$ with the following property: For any neighborhood Φ of D in $\bar{V}_F$ there exists a neighborhood Ψ of C in $\bar{V}_F$ with $\Psi \subset \Phi$ such that the following holds: For any unipotent one parameter subgroup $\{u(t)\}$ of G, an element $w \in \bar{V}_H$ and and interval $I \subset \mathbb{R}$, if $u(t_0)w \notin \Phi$ for some $t_0 \in I$ then,*

$$|\{t \in I \ : \ u(t)w \in \Psi\}| \leq \epsilon \cdot |\{t \in I \ : \ u(t)w \in \Phi\}|. \tag{19}$$

Proof. This is a "polynomial divergence" estimate similar to these in [**Kl1**, §2] and [**Kl1**, §3] □

PROPOSITION 4.12. *Let $\epsilon > 0$, a compact set $K \subset G/\Gamma \setminus \pi(S(F,U))$, and a compact set $C \subset \bar{A}_F$ be given. Then there exists a neighborhood Ψ of C in $\bar{V}_F$ such that for any unipotent one-parameter subgroup $\{u(t)\}$ of G and any $x \in G/\Gamma$, at least one of the following conditions is satisfied:*

1. *There exists $w \in \mathrm{Rep}(x) \cap \overline{\Psi}$ such that $\{u(t)\} \subset G_w$, where $G_w = \{g \in G \ : \ gw = w\}$.*
2. *For all large $T > 0$,*

$$|\{t \in [0,T] : u(t)x \in K \cap \pi(\bar{\eta}_F^{-1}(\Psi))\}| \leq \epsilon T.$$

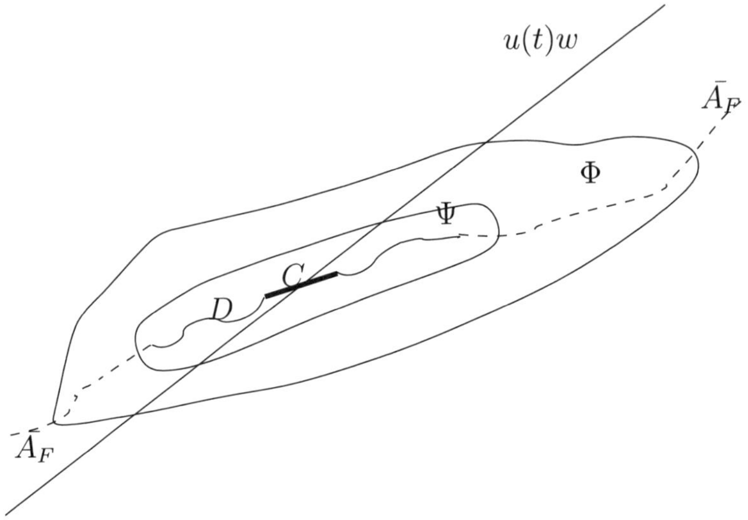

Figure 1. Proposition 4.11.

Proof. Let a compact set $D \subset \bar{A}_F$ be as in Proposition 4.11. Let Φ be a given neighborhood of D in $\bar{V}_F$. Replacing Φ by a smaller neighborhood of D, by Proposition 4.10 the set $\operatorname{Rep}(x) \cap \Phi$ contains at most one element for all $x \in K$. By the choice of D there exists a neighborhood Ψ of C contained in Φ such that equation (19) holds.

Now put $\Omega = \pi(\bar{\eta}_F^{-1}(\Psi)) \cap K$, and define

$$E = \{t \geq 0 \ : \ u(t)x \in \overline{\Omega}\}. \tag{20}$$

Let $t \in E$. By the choice of Φ, there exists a unique $w \in \bar{V}_F$ such that $\operatorname{Rep}(u(t)x) \cap \Phi = \{u(t)w\}$.

Since $s \to u(s)w$ is a polynomial function, either it is constant or it is unbounded as $s \to \pm\infty$. In the first case condition 1) is satisfied and we are done. Now suppose that condition 1 does not hold. Then for every $t \in E$, there exists a largest open interval $I(t) \subset (0,T)$ containing t such that

$$u(s)w \in \Phi \qquad \text{for all } s \in I(t). \tag{21}$$

Put $\mathcal{I} = \{I(t) : t \in E\}$, Then for any $I_1 \in \mathcal{I}$ and $s \in I_1 \cap E$, we have $I(s) = I_1$. Therefore for any $t_1, t_2 \in E$, if $t_1 < t_2$ then either $I(t_1) = I(t_2)$ or $I(t_1) \cap I(t_2) \subset (t_1, t_2)$. Hence any $t \in [0,T]$ is contained in at most two distinct elements of $\mathcal{I}$. Thus

$$\sum_{I \in \mathcal{I}} |I| \leq 2T. \tag{22}$$

Now by equations (19) and (21), for any $t \in E$,

$$|\{s \in I(t) : u(s)w \in \Psi\}| < \epsilon \cdot |I(t)|. \tag{23}$$

Therefore by equations (22) and (23), we get

$$|E| \leq \epsilon \cdot \sum_{I \in \mathcal{I}} |I| \leq (2\epsilon)T,$$

which is condition 2 for 2ϵ in place of ϵ. □

Outline of Proof of Theorem 4.5. Suppose μ is not Haar measure on G/Γ. By Lemma 4.6 μ is invariant under some one-parameter unipotent subgroup μ. Then by Theorem 4.3 there exists $F \in \mathcal{H}$ such that $\mu(N(F,U)) > 0$ and $\mu(S(F,U)) = 0$. Thus there exists a compact subset C_1 of $N(F,U) \setminus S(F,U)$ and $\alpha > 0$ such that

$$\mu(\pi(C_1)) > \alpha. \tag{24}$$

Take any $y \in \pi(C_1)$. It is easy to see that for each $i \in \mathbb{N}$ there exists $y_i \in \operatorname{supp}(\mu_i)$ such that $\{u_i(t)y_i\}$ is uniformly distributed with respect to μ_i, and also $y_i \to y$ as $i \to \infty$. Let $h_i \to e$ be a sequence in G such that $h_i y_i = y$ for all $i \in \mathbb{N}$.

We now replace μ_i by $\mu_i' = h_i \mu_i$. We still have $\mu_i' \to \mu$, but now we also have $y \in \operatorname{supp}(\mu_i')$ for all i. Let $u_i'(t) = h_i u_i(t) h_i^{-1}$. Then the trajectory $\{u_i'(t)y\}$ is uniformly distributed with respect to μ_i'.

We now apply Proposition 4.12 for $C = \bar{\eta}_F(C_1)$ and $\epsilon = \alpha/2$. We can choose a compact neighborhood K of $\pi(C_1)$ such that $K \cap S(F,U) = \emptyset$. Put $\Omega = \pi(\bar{\eta}_F^{-1}(\Psi)) \cap K$. Since $\mu_i' \to \mu$, due to (24) there exists $k_0 \in \mathbb{N}$ such that $\mu_i'(\Omega) > \epsilon$ for all $i \geq k_0$. This means that Condition 2) of Proposition 4.12 is violated for all $i \geq k_0$. Therefore according to condition 1) of Proposition 4.12, for each $i \geq k_0$,

$$\{u_i'(t)y\}_{t \in \mathbb{R}} \subset G_w y,$$

where G_w is as in Proposition 4.12. By Proposition 4.7, $G_w y$ is closed in G/Γ.

The rest of the proof is by induction on $\dim G$. If $\dim G_w < \dim G$ then everything is taking place in the homogeneous space $G_w y$, and therefore μ is ergodic by the induction hypothesis. If $\dim G_w = \dim G$ then $G_w = G$ and hence F is a normal subgroup of G. In this case one can project the measures to the homogeneous space $G/(F\Gamma)$ and apply induction. □

5. Oppenheim and Quantitative Oppenheim

5.1. The Oppenheim Conjecture. Let Q be an indefinite nondegenerate quadratic form in n variables. Let $Q(\mathbb{Z}^n)$ denote the set of values of Q at integral points. The Oppenheim conjecture, proved by Margulis (cf. [**Mar3**]) states that if $n \geq 3$, and Q is not proportional to a form with rational coefficients, then $Q(\mathbb{Z}^n)$ is dense. The Oppenheim conjecture enjoyed attention and many studies since it was conjectured in 1929 mostly using analytic number theory methods.

In the mid seventies Raghunathan observed a remarkable connection between the Oppenheim Conjecture and unipotent flows on the space of lattices $\mathcal{L}_n = SL(n,\mathbb{R})/SL(n,\mathbb{Z})$. It can be summarized as the following:

OBSERVATION 5.1 (Raghunathan). *Let Q be an indefinite quadratic form Q and let $H = SO(Q)$ denote its orthogonal group. Consider the orbit of the standard lattice $\mathbb{Z}^n \in \mathcal{L}_n$ under H. Then the following are equivalent:*

(a) *The orbit $H\mathbb{Z}^n$ is not relatively compact in $\mathcal{L}_n$.*
(b) *For all $\epsilon > 0$ there exists $u \in \mathbb{Z}^n$ such that $0 < |Q(u)| < \epsilon$.*
(c) *The set $Q(\mathbb{Z}^n)$ is dense in $\mathbb{R}$.*

Proof. Suppose (a) holds, so some sequence $h_k\mathbb{Z}^n$ leaves all compact sets. Then in view of the Mahler compactness criterion there exist $v_k \in h_k\mathbb{Z}^n$ such that $\|v_k\| \to 0$. Then also by continuity, $Q(v_k) \to 0$. But then $h_k^{-1}v_k \in \mathbb{Z}^n$, and $Q(h_k^{-1}v_k) = Q(v_k) \to 0$. Thus (b) holds.

It is easy to see that (b) implies (a). It is also possible to show that (b) implies (c). □

The Oppenheim Conjecture, the Raghunathan Conjecture and Unipotent Flows. Raghunathan also explained why the case $n = 2$ is different: in that case $H = SO(Q)$ is not generated by unipotent elements. Margulis's proof of the Oppenheim conjecture, given in [Mar 2-4] uses Raghunathan's observation. In fact Margulis showed that that any relatively compact orbit of $SO(2,1)$ in $SL(3,\mathbb{R})/SL(3,\mathbb{Z})$ is compact; this implies the Oppenheim Conjecture.

Raghunathan also conjectured Theorem 1.13. In the literature it was first stated in the paper [**Dan2**] and in a more general form in [**Mar3**] (when the subgroup U is not necessarily unipotent but generated by unipotent elements). Raghunathan's conjecture was eventually proved in full generality by M. Ratner (see [**Ra7**]). Earlier it was known in the following cases: (a) G is reductive and U is horospherical (see [**Dan2**]); (b) $G = SL(3,\mathbb{R})$ and $U = \{u(t)\}$ is a one-parameter unipotent subgroup of G such that $u(t) - I$ has rank 2 for all $t \neq 0$, where I is the identity matrix (see [**DM2**]); (c) G is solvable (see [**Sta1**] and [**Sta2**]). We remark that the proof given in [**Dan2**] is restricted to horospherical U and the proof given in [**Sta1**] and [**Sta2**] cannot be applied for nonsolvable G.

However the proof in [**DM2**] together with the methods developed in [Mar 2-4] and [**DM1**] suggest an approach for proving the Raghunathan conjecture in general by studying the minimal invariant sets, and the limits of orbits of sequences of points tending to a minimal invariant set. This strategy can be outlined as follows: Let x be a point in G/Γ, and U a connected unipotent subgroup of G. Denote by X the closure of Ux and consider a minimal closed U-invariant subset Y of X. Suppose that Ux is not closed (equivalently X is not equal to Ux). Then X should contain "many" translations of Y by elements from the normalizer $N(U)$ of U not belonging to U. After that one can try to prove that X contains orbits of bigger and bigger unipotent subgroups until one reaches horospherical subgroups. The basic tool in this strategy is the following fact. Let y be a point in X, and let g_n be a sequence of elements in G such that g_n converges to 1, g_n does not belong to $N(U)$, and $y_n = g_n y$ belongs to X. Then X contains AY where A is a nontrivial connected subset in $N(U)$ containing 1 and "transversal" to U. To prove this one has to observe that the orbits Uy_n and Uy are "almost parallel" in the direction of $N(U)$ most of the time in "the intermediate range". (cf. Proposition 3.6).

In fact the set AU as a subset of $N(U)/U$ is the image of a nontrivial rational map from U into $N(U)/U$. Moreover this rational map sends 1 to 1 and also comes from a polynomial map from U into the closure of G/U in the affine space V containing G/U. This affine space V is the space of the rational representation of G such that V contains a vector the stabilizer of which is U (Chevalley theorem).

This program was being actively pursued at the time Ratner's results were announced (cf. [**Sha3**]).

5.2. A quantitative version of the Oppenheim Conjecture. References for this subsection are [**EMM1**] and [**EMM2**].

In this section we study some finer questions related to the distribution of the values of Q at integral points.

Let ν be a continuous positive function on the sphere $\{v \in \mathbb{R}^n \mid \|v\| = 1\}$, and let $\Omega = \{v \in \mathbb{R}^n \mid \|v\| < \nu(v/\|v\|)\}$. We denote by $T\Omega$ the dilate of Ω by T. Define the following set:

$$V^Q_{(a,b)}(\mathbb{R}) = \{x \in \mathbb{R}^n \mid a < Q(x) < b\}$$

We shall use $V_{(a,b)} = V^Q_{(a,b)}$ when there is no confusion about the form Q. Also let $V_{(a,b)}(\mathbb{Z}) = V^Q_{(a,b)}(\mathbb{Z}) = \{x \in \mathbb{Z}^n \mid a < Q(x) < b\}$. The set $T\Omega \cap \mathbb{Z}^n$ consists of $O(T^n)$ points, $Q(T\Omega \cap \mathbb{Z}^n)$ is contained in an interval of the form $[-\mu T^2, \mu T^2]$, where $\mu > 0$ is a constant depending on Q and Ω. Thus one might expect that for any interval $[a, b]$, as $T \to \infty$,

$$|V_{(a,b)}(\mathbb{Z}) \cap T\Omega| \sim c_{Q,\Omega}(b-a)T^{n-2} \tag{25}$$

where $c_{Q,\Omega}$ is a constant depending on Q and Ω. This may be interpreted as "uniform distribution" of the sets $Q(\mathbb{Z}^n \cap T\Omega)$ in the real line. The main result of this section is that (25) holds if Q is not proportional to a rational form, and has signature (p, q) with $p \geq 3$, $q \geq 1$. We also determine the constant $c_{Q,\Omega}$.

If Q is an indefinite quadratic form in n variables, Ω is as above and (a, b) is an interval, we show that there exists a constant $\lambda = \lambda_{Q,\Omega}$ so that as $T \to \infty$,

$$\mathrm{Vol}(V_{(a,b)}(\mathbb{R}) \cap T\Omega) \sim \lambda_{Q,\Omega}(b-a)T^{n-2} \tag{26}$$

The main result is the following:

THEOREM 5.2. *Let Q be an indefinite quadratic form of signature (p, q), with $p \geq 3$ and $q \geq 1$. Suppose Q is not proportional to a rational form. Then for any interval (a, b), as $T \to \infty$,*

$$|V_{(a,b)}(\mathbb{Z}) \cap T\Omega| \sim \lambda_{Q,\Omega}(b-a)T^{n-2} \tag{27}$$

where $n = p + q$, and $\lambda_{Q,\Omega}$ is as in (26).

The asymptotically exact lower bound was proved in [**DM4**]. Also a lower bound with a smaller constant was obtained independently by M. Ratner, and by S. G. Dani jointly with S. Mozes (both unpublished). The upper bound was proved in [**EMM1**].

If the signature of Q is $(2, 1)$ or $(2, 2)$ then no universal formula like (25) holds. In fact, we have the following theorem:

THEOREM 5.3. *Let Ω_0 be the unit ball, and let $q = 1$ or 2. Then for every $\epsilon > 0$ and every interval (a, b) there exists a quadratic form Q of signature $(2, q)$ not proportional to a rational form, and a constant $c > 0$ such that for an infinite sequence $T_j \to \infty$,*

$$|V_{(a,b)}(\mathbb{Z}) \cap T\Omega_0| > cT_j^q(\log T_j)^{1-\epsilon}.$$

The case $q = 1$, $b \leq 0$ of Theorem 5.3 was noticed by P. Sarnak and worked out in detail in [**Bre**]. The quadratic forms constructed are of the form $x_1^2 + x_2^2 - \alpha x_3^2$, or $x_1^2 + x_2^2 - \alpha(x_3^2 + x_4^2)$, where α is extremely well approximated by squares of rational numbers.

However in the $(2, 1)$ and $(2, 2)$ cases, one can still establish an upper bound of the form $cT^q \log T$. This upper bound is effective, and is uniform over compact sets in the set of quadratic forms. We also give an effective uniform upper bound for the case $p \geq 3$.

THEOREM 5.4 ([**EMM1**]). *Let $\mathcal{O}(p, q)$ denote the space of quadratic forms of signature (p, q) and discriminant ± 1, let $n = p + q$, (a, b) be an interval, and let $\mathcal{D}$ be a compact subset of $\mathcal{O}(p, q)$. Let ν be a continuous positive function on the* *unit sphere and let* $\Omega = \{v \in \mathbb{R}^n \mid \|v\| < \nu(v/\|v\|)\}$. *Then, if* $p \geq 3$ *there exists a constant* c *depending only on* $\mathcal{D}$, (a,b) *and* Ω *such that for any* $Q \in \mathcal{D}$ *and all* $T > 1$,

$$|V_{(a,b)}(\mathbb{Z}) \cap T\Omega| < cT^{n-2}$$

If $p = 2$ *and* $q = 1$ *or* $q = 2$, *then there exists a constant* $c > 0$ *depending only on* $\mathcal{D}$, (a,b) *and* Ω *such that for any* $Q \in \mathcal{D}$ *and all* $T > 2$,

$$|V_{(a,b)} \cap T\Omega \cap \mathbb{Z}^n| < cT^{n-2}\log T$$

Also, for the $(2,1)$ and $(2,2)$ cases, we have the following "almost everywhere" result:

THEOREM 5.5. *For almost all quadratic forms* Q *of signature* $(p,q) = (2,1)$ *or* $(2,2)$

$$|V_{(a,b)}(\mathbb{Z}) \cap T\Omega| \sim \lambda_{Q,\Omega}(b-a)T^{n-2}$$

where $n = p + q$, *and* $\lambda_{Q,\Omega}$ *is as in* (26).

Theorem 5.5 may be proved using a recent general result of Nevo and Stein [**NS**]; see also [**EMM1**].

It is also possible to give a "uniform" version of Theorem 5.2, following [**DM4**]:

THEOREM 5.6. *Let* $\mathcal{D}$ *be a compact subset of* $\mathcal{O}(p,q)$, *with* $p \geq 3$. *Let* $n = p+q$, *and let* Ω *be as in Theorem 5.4. Then for every interval* $[a,b]$ *and every* $\theta > 0$, *there exists a finite subset* $\mathcal{P}$ *of* $\mathcal{D}$ *such that each* $Q \in \mathcal{P}$ *is a scalar multiple of a rational form and for any compact subset* $\mathcal{F}$ *of* $\mathcal{D} - \mathcal{P}$ *there exists* T_0 *such that for all* Q *in* $\mathcal{F}$ *and* $T \geq T_0$,

$$(1-\theta)\lambda_{Q,\Omega}(b-a)T^{n-2} \leq |V_{(a,b)}(\mathbb{Z}) \cap T\Omega| \leq (1+\theta)\lambda_{Q,\Omega}(b-a)T^{n-2}$$

where $\lambda_{Q,\Omega}$ *is as in* (26).

As in Theorem 5.2 the upper bound is from [**EMM1**]; the asymptotically exact lower bound, which holds even for $SO(2,1)$ and $SO(2,2)$, was proved in [**DM4**].

REMARK 5.7. If we consider $|V_{(a,b)}(\mathbb{R}) \cap T\Omega \cap \mathcal{P}(\mathbb{Z}^n)|$ instead of $|V_{(a,b)}(\mathbb{Z}) \cap T\Omega|$ (where $\mathcal{P}(\mathbb{Z}^n)$ denotes the set of primitive lattice points, then Theorem 5.2 and Theorem 5.6 hold provided one replaces $\lambda_{Q,\Omega}$ by $\lambda'_{Q,\Omega} = \lambda_{Q,\Omega}/\zeta(n)$, where ζ is the Riemann zeta function.

More on signature (2,2). Recall that a subspace is called isotropic if the restriction of the quadratic form to the subspace is identically zero. Observe also that whenever a form of signature $(2,2)$ has a rational isotropic subspace L then $L \cap T\Omega$ contains on the order of T^2 integral points x for which $Q(x) = 0$, hence $N_{Q,\Omega}(-\epsilon,\epsilon,T) \geq cT^2$, independently of the choice of ϵ. Thus to obtain an asymptotic formula similar to (27) in the signature $(2,2)$ case, we must exclude the contribution of the rational isotropic subspaces. We remark that an irrational quadratic form of signature $(2,2)$ may have at most 4 rational isotropic subspaces (see [**EMM2**, Lemma 10.3]).

The space of quadratic forms in 4 variables is a linear space of dimension 10. Fix a norm $\|\cdot\|$ on this space.

DEFINITION 5.8. **(EWAS)** A quadratic form Q is called *extremely well approximable by split forms (EWAS)* if for any $N > 0$ there exists a split integral form Q' and $2 \le k \in \mathbb{R}$ such that

$$\left\| Q - \frac{1}{k} Q' \right\| \le \frac{1}{k^N}.$$

The main result of [**EMM2**] is:

THEOREM 5.9. *Suppose Ω is as above. Let Q be an indefinite quadratic form of signature $(2,2)$ which is not EWAS. Then for any interval (a,b), as $T \to \infty$,*

$$\tilde{N}_{Q,\Omega}(a,b,T) \sim \lambda_{Q,\Omega}(b-a)T^2, \tag{28}$$

where the constant $\lambda_{Q,\Omega}$ is as in (26), and $\tilde{N}_{Q,\Omega}$ counts the points not contained in isotropic subspaces.

Open Problem. State and prove a result similar to Theorem 5.9 for the signature $(2,1)$ case.

Eigenvalue spacings on flat 2-tori. It has been suggested by Berry and Tabor that the eigenvalues of the quantization of a completely integrable Hamiltonian follow the statistics of a Poisson point-process, which means their consecutive spacings should be i.i.d. exponentially distributed. For the Hamiltonian which is the geodesic flow on the flat 2-torus, it was noted by P. Sarnak [**Sar**] that this problem translates to one of the spacing between the values at integers of a binary quadratic form, and is related to the quantitative Oppenheim problem in the signature $(2,2)$ case. We briefly recall the connection following [**Sar**].

Let $\Delta \subset \mathbb{R}^2$ be a lattice and let $M = \mathbb{R}^2/\Delta$ denote the associated flat torus. The eigenfunctions of the Laplacian on M are of the form $f_v(\cdot) = e^{2\pi i \langle v, \cdot \rangle}$, where v belongs to the dual lattice Δ^*. The corresponding eigenvalues are $4\pi^2\|v\|^2$, $v \in \Delta^*$. These are the values at integral points of the binary quadratic $B(m,n) = 4\pi^2\|mv_1 + nv_2\|^2$, where $\{v_1, v_2\}$ is a $\mathbb{Z}$-basis for Δ^*. We will identify Δ^* with $\mathbb{Z}^2$ using this basis.

We label the eigenvalues (with multiplicity) by

$$0 = \lambda_0(M) < \lambda_1(M) \le \lambda_2(M) \dots$$

It is easy to see that Weyl's law holds, i.e.

$$|\{j \;:\; \lambda_j(M) \le T\}| \sim c_M T,$$

where $c_M = (\operatorname{area} M)/(4\pi)$. We are interested in the distribution of the local spacings $\lambda_j(M) - \lambda_k(M)$. In particular, for $0 \notin (a,b)$, set

$$R_M(a,b,T) = \frac{|\{(j,k) \;:\; \lambda_j(M) \le T, \lambda_k(M) \le T, a \le \lambda_j(M) - \lambda_k(M) \le b\}|}{T}.$$

The statistic R_M is called the pair correlation. The Poisson-random model predicts, in particular, that

$$\lim_{T\to\infty} R_M(a,b,T) = c_M^2(b-a). \tag{29}$$

Note that the differences $\lambda_j(M) - \lambda_k(M)$ are precisely the integral values of the quadratic form $Q_M(x_1,x_2,x_3,x_4) = B(x_1,x_2) - B(x_3,x_4)$.

P. Sarnak showed in [**Sar**] that (29) holds on a set of full measure in the space of tori. Some remarkable related results for forms of higher degree and higher dimensional tori were proved in [**V1**], [**V2**] and [**V3**]. These methods, however, cannot be used to explicitly construct a specific torus for which (29) holds. A corollary of Theorem 5.9 is the following:

THEOREM 5.10. *Let M be a 2 dimensional flat torus rescaled so that one of the coefficients in the associated binary quadratic form B is 1. Let A_1, A_2 denote the two other coefficients of B. Suppose that there exists $N > 0$ such that for all triples of integers (p_1, p_2, q) with $q \geq 2$,*

$$\max_{i=1,2} \left| A_i - \frac{p_i}{q} \right| > \frac{1}{q^N}.$$

Then, for any interval (a, b) not containing 0, (29) holds, i.e.

$$\lim_{T \to \infty} R_M(a, b, T) = c_M^2 (b - a).$$

In particular, the set of $(A_1, A_2) \subset \mathbb{R}^2$ for which (29) does not hold has zero Hausdorff dimension.

Thus, if one of the A_i is Diophantine's (e.g. algebraic), then M has a spectrum whose pair correlation satisfies the Berry-Tabor conjecture.

This establishes the pair correlation for the flat torus or "boxed oscillator" considered numerically by Berry and Tabor. We note that without some diophantine condition, (29) may fail.

5.3. Passage to the space of lattices. We now relate the counting problem of Theorem 5.2 to a certain integral expression involving the orthogonal group of the quadratic form and the space of lattices $SL(n, \mathbb{R})/SL(n, \mathbb{Z})$. Roughly this is done as follows. Let f be a bounded function on $\mathbb{R}^n - \{0\}$ vanishing outside a compact subset. For a lattice $\Delta \in \mathcal{L}_n$ let

$$\tilde{f}(\Delta) = \sum_{v \in \Delta \setminus \{0\}} f(\Delta) \tag{30}$$

(the function $\tilde{f}$ is called the "Siegel Transform" of f). The proof is based on the identity of the form

$$\int_K \tilde{f}(a_t k \Delta)\, dk = \sum_{v \in \Delta \setminus \{0\}} \int_K f(a_t k v)\, dk \tag{31}$$

obtained by integrating (30). In (31) $\{a_t\}$ is a certain diagonal subgroup of the orthogonal group of Q, and K is a maximal compact subgroup of the orthogonal group of Q. Then for an appropriate function f, the right hand side is then related to the number of lattice points $v \in [e^t/2, e^t]\partial\Omega$ with $a < Q(v) < b$. The asymptotics of the left-hand side is then established using the ergodic theory of unipotent flows and some other techniques.

Quadratic Forms, and the lattice Δ_Q. Let $n \geq 3$, and let $p \geq 2$. We denote $n - p$ by q, and assume $q > 0$. Let $\{e_1, e_2, \dots e_n\}$ be the standard basis of $\mathbb{R}^n$. Let Q_0 be the quadratic form defined by

$$Q_0\left(\sum_{i=1}^{n} v_i e_i\right) = 2v_1 v_n + \sum_{i=2}^{p} v_i^2 - \sum_{i=p+1}^{n-1} v_i^2 \quad \text{for all } v_1, \dots, v_n \in \mathbb{R}. \tag{32}$$

It is straightforward to verify that Q_0 has signature (p, q). Let $G = SL(n, \mathbb{R})$, the group of $n \times n$ matrices of determinant 1. For each quadratic form Q and $g \in G$, let Q^g denote the quadratic form defined by $Q^g(v) = Q(gv)$ for all $v \in \mathbb{R}^n$. By the well known classification of quadratic forms over $\mathbb{R}$, for each $Q \in \mathcal{O}(p,q)$ there exists $g \in G$ such that $Q = Q_0^g$. Then let Δ_Q denote the lattice $g\mathbb{Z}^n$, so that $Q_0(\Delta_Q) = Q(\mathbb{Z}^n)$.

For any quadratic form Q let $SO(Q)$ denote the special orthogonal group corresponding to Q; namely $\{g \in G \mid Q^g = Q\}$. Let $H = SO(Q_0)$. Then the map $H\backslash G \to \mathcal{O}(p,q)$ given by $Hg \to Q_0^g$ is a homeomorphism.

The map a_t and the group K. For $t \in \mathbb{R}$, let a_t be the linear map so that $a_t e_1 = e^{-t}e_1$, $a_t e_n = e^t e_n$, and $a_t e_i = e_i$, $2 \le i \le n-1$. Then the one-parameter group $\{a_t\}$ is contained in H. Let $\hat{K}$ be the subgroup of G consisting of orthogonal matrices, and let $K = H \cap \hat{K}$. It is easy to check that K is a maximal compact subgroup of H, and consists of all $h \in H$ leaving invariant the subspace spanned by $\{e_1 + e_n, e_2, \dots, e_p\}$. We denote by m the normalized Haar measure on K.

A Lemma about vectors in $\mathbb{R}^n$. In this section we will be somewhat informal. For a completely rigorous argument see [**EMM1**, §§3.4-3.5]. Also for simplicity we let $\nu = 1$ in this section.

Let $W \subset \mathbb{R}^n$ be the characteristic function of the region defined by the inequalities on $x = (x_1, \dots, x_n)$:

$$a \le Q_0(x) \le b, \quad (1/2) \le \|x\| \le 2,$$
$$x_1 > 0, \quad (1/2)x_1 \le |x_i| \le (1/2)x_1 \text{ for } 2 \le i \le n-1.$$

Let f be the characteristic function of W.

LEMMA 5.11. *There exists $T_0 > 0$ such that for every t with $e^t > T_0$, and every $v \in \mathbb{R}^n$ with $\|v\| > T_0$,*

$$(33) \quad c_{p,q} e^{(n-2)t} \int_K f(a_t k v)\, dm(k) \approx \begin{cases} 1 & \text{if } a \le Q_0(x) \le b \text{ and } \frac{e^t}{2} \le \|v\| \le e^t, \\ 0 & \text{otherwise} \end{cases}$$

where $c_{p,q}$ is a constant depending only on p and q.

Proof. This is a direct calculation. □

Remark. The $\approx$ in (33) is essentially equality up to "edge effects". These edge effects can be overcome if one approximated f from above and below by continuous functions f_+ and f_- in such a way that the L^1 norm of $f_+ - f_-$ is small. We choose not to do this here in order to not clutter the notation.

In (33), we let $T = e^t$ and sum over $v \in \Delta_Q$. We obtain:

PROPOSITION 5.12. *As $T \to \infty$,*

$$c_{p,q} T^{n-2} \int_K \tilde{f}(a_t k \Delta_Q) \approx |\{v \in \Delta_Q \ : \ a < Q_0(v) < b \text{ and } \tfrac{1}{2}T \le \|v\| \le T\}|,$$

where $t = \log T$. Note that the right-hand side is by definition $|V^Q_{(a,b)}(\mathbb{Z}) \cap [T/2, T]\Omega_0|$, where Ω_0 is the unit ball.

We also note without proof the following lemma:

LEMMA 5.13. *Let ρ be a continuous positive function on the sphere, and let $\Omega = \{v \in \mathbb{R}^n \ \|\|v\| < \rho(v/\|v\|)\}$. Then there exists a constant $\lambda = \lambda_{Q,\Omega}$ so that as $T \to \infty$,*

$$\mathrm{Vol}(V^Q_{(a,b)}(\mathbb{R}) \cap T\Omega) \sim \lambda_{Q,\Omega}(b-a)T^{n-2}.$$

Also (using Siegel's formula), $c_{p,q}\int_{\mathcal{L}_n} \tilde{f} = c_{p,q}\int_{\mathbb{R}^n} f = (1-2^{2-n})\lambda_{Q,\Omega}$.

Remark. One can verify that:

$$\lambda_{Q,\Omega} = \int\limits_{L\cap\Omega} \frac{dA}{\|\nabla Q\|},$$

where L is the lightcone $Q = 0$ and dA is the area element on L.

The main theorems. In view of Proposition 5.12 and Lemma 5.13, to prove Theorem 5.2 one may use the following theorem:

THEOREM 5.14. *Suppose $p \geq 3$, $q \geq 1$. Let $\Lambda \in \mathcal{L}_n$ be a unimodular lattice such that $H\Lambda$ is not closed. Let ν be any continuous function on K. Then*

$$\lim_{t\to+\infty} \int_K \tilde{f}(a_t k\Lambda)\nu(k)\, dm(k) = \int_K \nu\, dm \int_{\mathcal{L}_n} \tilde{f}(\Delta)\, d\mu(\Delta). \tag{34}$$

To prove Theorem 5.6 we use the following generalization:

THEOREM 5.15. *Suppose $p \geq 3$, $q \geq 1$. Let ν be as in Theorem 5.14, and let $\mathcal{C}$ be any compact set in $\mathcal{L}_n$. Then for any $\epsilon > 0$ there exist finitely many points $\Lambda_1, \dots, \Lambda_\ell \in \mathcal{L}_n$ such that*

(i) *The orbits $H\Lambda_1, \dots, H\Lambda_\ell$ are closed and have finite H-invariant measure.*

(ii) *For any compact subset F of $\mathcal{C} \setminus \bigcup_{1\leq i\leq \ell} H\Lambda_i$, there exists $t_0 > 0$, so that for all $\Lambda \in F$ and $t > t_0$,*

$$\left| \int_K \tilde{f}(a_t k\Lambda)\nu(k)\, dm(k) - \int_{\mathcal{L}_n} \tilde{f}\, d\mu \int_K \nu\, dm \right| \leq \epsilon \tag{35}$$

Theorem 5.14 and Theorem 5.15 if $\tilde{f}$ is replaced by a bounded function ϕ. If we replace $\tilde{f}$ by a bounded continuous function ϕ then (34) and (35) follow easily from Theorem 4.4. (This was the original motivation for Theorem 4.4). The fact that Theorem 4.4 deals with unipotents and Theorem 5.15 deals with large spheres is not a serious obstacle, since large spheres can be approximated by unipotents. In fact, the integral in (34) can be rewritten as

$$\int_B \left(\frac{1}{T(x)} \int_0^{T(x)} \phi(u_t x)\, dm(k) \right) dx,$$

where B is a suitable subset of G and U is a suitable unipotent. Now by Theorem 4.4, the inner integral tends to $\int_{G/\Gamma} \phi$ uniformly as long as x is in a compact set away from an explicitly described set E, where E is a finite union of neighborhoods of sets of the form $\pi(C)$ where C is a compact subset of some $N(F,U)$. By direct calculation one can show that only a small part of B is near E, hence Theorem 5.14 and Theorem 5.15 both hold.

Remark. Both Theorem 4.4 and Ratner's uniform distribution theorem Theorem 1.12 hold for bounded continuous functions, but not for arbitrary continuous functions from $L^1(G/\Gamma)$. However, for a non-negative bounded continuous function

f on $\mathbb{R}^n$, the function $\tilde{f}$ defined in (30) is non-negative, continuous, and L^1 but unbounded (it is in $L^s(G/\Gamma)$ for $1 \le s < n$, where $G = SL(n, \mathbb{R})$, and $\Gamma = SL(n, \mathbb{Z})$).

The lower bounds. As it was done in [**DM4**] it is possible to obtain asymptotically exact lower bounds by considering bounded continuous functions $\phi \le \tilde{f}$. However, to prove the upper bounds in the theorems stated above we need to examine carefully the situation at the "cusp" of G/Γ, i.e outside of compact sets. This will be done in §6.

6. Quantitative Oppenheim (upper bounds)

The references for this section are [**EMM1**] and [**EMM2**].

Lattices. Let Δ be a lattice in $\mathbb{R}^n$. We say that a subspace L of $\mathbb{R}^n$ is Δ-*rational* if $L \cap \Delta$ is a lattice in L. For any Δ-rational subspace L, we denote by $d_\Delta(L)$ or simply by $d(L)$ the volume of $L/(L \cap \Delta)$. In the notation of [**Kl1**, §3], $d_\Delta(L) = \|L \cap \Delta\|$.

Let us note that $d(L)$ is equal to the norm of $e_1 \wedge \cdots \wedge e_\ell$ in the exterior power $\bigwedge^\ell(\mathbb{R}^n)$ where $\ell = \dim L$ and $(e_1, \cdots, e_\ell)$ is a basis over $\mathbb{Z}$ of $L \cap \Delta$. If $L = \{0\}$ we write $d(L) = 1$.

Let us introduce the following notation:

$$\alpha_i(\Delta) = \sup \left\{ \frac{1}{d(L)} \,\middle|\, L \text{ is a } \Delta\text{-rational subspace of dimension } i \right\}, \quad 0 \le i \le n,$$

(36)

$$\alpha(\Delta) = \max_{0 \le i \le n} \alpha_i(\Delta).$$

The following lemma is known as the "Lipshitz Principle":

LEMMA 6.1 ([**Sch**, Lemma 2]). *Let f be a bounded function on $\mathbb{R}^n$ vanishing outside a compact subset. Then there exists a positive constant $c = c(f)$ such that*

$$\tilde{f}(\Delta) < c\alpha(\Delta)$$

for any lattice Δ in $\mathbb{R}^n$. Here $\tilde{f}$ is the function on the space of lattices defined in (30).

Replacing $\tilde{f}$ by α. By Lemma 6.1, the function $\tilde{f}(g)$ on the space of unimodular lattices $\mathcal{L}_n$ is majorized by the function $\alpha(g)$. The function α is more convenient since it is invariant under the left action of the maximal compact subgroup $\hat{K}$ of G, and its growth rate at infinity is known explicitly. Theorems 5.2 and 5.6 are proved by combining Theorem 4.4 with the following integrability estimate:

THEOREM 6.2 ([**EMM1**]). *If $p \ge 3$, $q \ge 1$ and $0 < s < 2$, or if $p = 2$, $q \ge 1$ and $0 < s < 1$, then for any lattice Δ in $\mathbb{R}^n$*

$$\sup_{t>0} \int_K \alpha(a_t k \Delta)^s \, dm(k) < \infty.$$

The upper bound is uniform as Δ varies over compact sets in the space of lattices.

This result can be interpreted as follows. For a lattice Δ in $\mathcal{L}_n$ and for $h \in H$, let $f(h) = \alpha(h\Delta)$. Since α is left-$\hat{K}$ invariant, f is a function on the symmetric space $X = K \backslash H$. Theorem 6.2 is the statement that if if $p \ge 3$, then the averages of f^s, $0 < s < 2$ over the sets $K a_t K$ in X remain bounded as $t \to \infty$, and the bound is uniform as one varies the base point Δ over compact sets. We remark that in the case $q = 1$, the rank of X is 1, and the sets Ka_tK are metric spheres of radius t, centered at the origin.

If $(p,q) = (2,1)$ or $(2,2)$, Theorem 6.2 does not hold even for $s = 1$. The following result is, in general, best possible:

THEOREM 6.3 ([**EMM1**]). *If $p = 2$ and $q = 2$, or if $p = 2$ and $q = 1$, then for any lattice Δ in $\mathbb{R}^n$,*

$$\sup_{t>1} \frac{1}{t} \int_K \alpha(a_t k\Delta)\, dm(k) < \infty, \tag{37}$$

The upper bound is uniform as Δ varies over compact sets in the space of lattices.

Proof of Theorem 5.15 assuming Theorem 6.2. We can assume that $\tilde{f}$ is nonnegative. Let $A(r) = \{x \in G/\Gamma \;:\; \alpha(x) > r\}$. Choose a continuous nonnegative function g_r on G/Γ such that $g_r(x) = 1$ if $x \in A(r+1)$, $g_r(x) = 0$ if $x \notin A(r)$ and $0 \le g_r(x) \le 1$ if $x \in A(r) - A(r+1)$. Then

$$\begin{aligned}&\int_K \tilde{f}(a_t k x)\nu(k)\, dm(k) = \\ &= \int_K (\tilde{f} g_r)(a_t k x)\nu(k)\, dm(k) + \int_K (\tilde{f} - \tilde{f} g_r)(a_t k x)\nu(k)\, dm(k).\end{aligned} \tag{38}$$

But (letting $\beta = 2 - s$), $(\tilde{f} g_r)(y) \le B_1 \alpha(y)^{2-\beta} g_r(y) = B_1 \alpha(y)^{2-\frac{\beta}{2}} g_r(y) \alpha(y)^{-\frac{\beta}{2}} \le B_1 r^{-\frac{\beta}{2}} \alpha(y)^{2-\frac{\beta}{2}}$ (the last inequality is true because $g_r(y) = 0$ if $\alpha(y) \le r$). Therefore

$$\int_K (\tilde{f} g_r)(a_t k x)\nu(k)\, dm(k) \le B_1 r^{-\frac{\beta}{2}} \int_K \alpha(a_t k x)^{2-\frac{\beta}{2}} \nu(k)\, dm(k). \tag{39}$$

According to Theorem 6.2 there exists B such that

$$\int_K \alpha(a_t k x)^{2-\frac{\beta}{2}}\, dm(k) < B$$

for any $t \ge 0$ and uniformly over $x \in \mathcal{C}$. Then (39) implies that

$$\int_K (\tilde{f} g_r)(a_t k x)\nu(k)\, dm(k) \le B B_1 (\sup \nu) r^{-\frac{\beta}{2}}. \tag{40}$$

Since the function $\tilde{f} - \tilde{f} g_r$ is continuous and has a compact support, the "bounded function" case of Theorem 5.15 implies that for every $\epsilon > 0$ there exists a finite set of points $x_1, \dots, x_\ell$ with Hx_i closed for each i so that for every compact subset F of $\mathcal{C} \setminus \bigcup_{i=1}^{\ell} Hx_i$ there exists $t_0 > 0$ such that for every $t > t_0$ and every $x \in F$,

$$\left| \int_K (\tilde{f} - \tilde{f} g_r)(a_t k x)\nu(k)\, dm(k) - \int_{G/\Gamma} (\tilde{f} - \tilde{f} g_r)(y)\, d\mu(y) \int_K \nu(k)\, dm(k) \right| < \frac{\epsilon}{2}. \tag{41}$$

It is easy to see that (38), (40) and (41) imply (35) if r is sufficiently large. This implies Theorem 5.15. □

In the rest of this section, we prove Theorem 6.2 and Theorem 6.3. We recall the notation from §5: G is $SL(n,\mathbb{R})$, $\Gamma = SL(n,\mathbb{Z})$, $\hat{K} \cong SO(n)$ is a maximal compact subgroup of G, $H \cong SO(p,q) \subset G$, $K = H \cap \hat{K}$ is a maximal compact subgroup of H, and X is the symmetric space $K \backslash H$. From its definition (36), the function $\alpha(\Delta)$ is the maximum over $1 \le i \le n$ of $\hat{K}$ invariant functions $\alpha_i(\Delta)$. The main idea of the proof is to show that the α_i^s satisfy a certain system of integral inequalities which imply the desired bounds.

If $p \geq 3$ and $0 < s < 2$, or if $(p,q) = (2,1)$ or $(2,2)$ and $0 < s < 1$, we show that for any $c > 0$ there exist $t > 0$, and $\omega > 1$ so that the the functions α_i^s satisfy the following system of integral inequalities in the space of lattices:

$$A_t\alpha_i^s \leq c_i\alpha_i^s + \omega^2 \max_{0<j\leq\min(n-i,i)} \sqrt{\alpha_{i+j}^s\alpha_{i-j}^s}, \tag{42}$$

where A_t is the averaging operator $(A_t f)(\Delta) = \int_K f(a_t k\Delta)$, and $c_i \leq c$ (Lemma 6.7). If $(p,q) = (2,1)$ or $(2,2)$ and $s = 1$, then (42) also holds (for suitably modified functions α_i), but some of the constants c_i cannot be made smaller than 1.

Let $f_i(h) = \alpha_i(h\Delta)$, so that each f_i is a function on the symmetric space X. When one restricts to an orbit of H, (42) becomes:

$$A_t f_i^s \leq c_i f_i^s + \omega^2 \max_{0<j\leq\min(n-i,i)} \sqrt{f_{i+j}^s f_{i-j}^s}. \tag{43}$$

If $\operatorname{rank} X = 1$, then $(A_t f)(h)$ can be interpreted as the average of f over the sphere of radius $2t$ in X, centered at h. In §6.4 we show that if the f_i satisfy (43) then for any $\epsilon > 0$, the function $f = f_{\epsilon,s} = \sum_{0\leq i\leq n} \epsilon^{i(n-i)} f_i^s$ satisfies the scalar inequality:

$$A_t f \leq cf + b, \tag{44}$$

where t, c and b are constants. This inequality is studied in §6.3. We show that if c is sufficiently small, then (44) for a fixed t together with the uniform continuity of $\log f$ imply that $(A_r f)(1)$ is bounded as a function of r, which is the conclusion of Theorem 6.2. If $c = 1$, which will occur in the $SO(2,1)$ and $SO(2,2)$ cases, then (44) implies that $(A_r f)(1)$ is growing at most linearly with the radius. In §6.4, we complete the proof of Theorem 6.2, and also prove Theorems 6.3 and 5.15.

Throughout the proof we consider the functions $\alpha(g)^s$ for $0 < s < 2$ even though for the application to quadratic forms we only need $s = 1 + \delta$. This yields a better integrability result, and is also necessary for the proof of Theorem 5.14 and Theorem 5.15.

6.1. Averages of the functions $1/d_i^s$ over spheres. Recall that the function d_i is the norm of a certain vector in the exterior power $\bigwedge^i(\mathbb{R}^n)$. We have the following:

PROPOSITION 6.4. *Let $\{a_t \mid t \in \mathbb{R}\}$ be a self-adjoint one-parameter subgroup of $SO(2,1)$. Let p and q be positive integers and let $0 < i < p+q$. Let*

$$F(i) = \{x_1 \wedge x_2 \wedge \cdots \wedge x_i \mid x_1, x_2, \cdots, x_i \in \mathbb{R}^{p+q}\} \subset \bigwedge^i(\mathbb{R}^{p+q}).$$

Then, if $p \geq 3$, or if $p = 2$, $q = 2$ and $i \neq 2$, then for any s, $0 < s < 2$,

$$\lim_{t\to\infty} \sup_{v\in F(i),\ \|v\|=1} \int_K \frac{dm(k)}{\|a_t kv\|^s} = 0. \tag{45}$$

where $K = SO(p) \times SO(q)$ and $SO(2,1)$ is embedded into $SO(p,q)$. If $p = 2$ and $q = 1$, or if $p = 2$, $q = 2$ and $i = 2$, then (45) holds for any s, $0 < s < 1$.

Proof. This is a direct calculation. □

The next lemma we obtain an analogous result for the case $(p,q) = (2,1)$, $s = 1$.

LEMMA 6.5. *Let $H \cong SO(2,1)$ be the orthogonal group of the quadratic form $x^2 + y^2 - z^2$. Let $\{a_t \mid t \in \mathbb{R}\}$ be a self-adjoint one-parameter subgroup of H, and let $K = H \cap O(3)$ denote the maximal compact of H. We define another norm $\|\cdot\|^*$ on $\mathbb{R}^3$ by*

$$\|(x,y,z)\|^* = \max(\sqrt{x^2+y^2}, |z|). \tag{46}$$

Then, for any $v \in \mathbb{R}^3$, $v \neq 0$, and any $t > 0$,

$$\int_K \frac{dm(k)}{\|a_t k v\|^*} \leq \frac{1}{\|v\|^*}. \tag{47}$$

6.2. A system of inequalities.

LEMMA 6.6. *For any two Δ-rational subspaces L and M*

$$d(L)d(M) \geq d(L \cap M)d(L+M). \tag{48}$$

Proof. Let $\pi : \mathbb{R}^n \to \mathbb{R}^n/(L \cap M)$ denote the natural projection. Then $d(L) = d(\pi(L))d(L\cap M)$, $d(M) = d(\pi(M))d(L\cap M)$ and $d(L+M) = d(\pi(L+M))d(L\cap M)$. On the other hand the inequality (48) is equivalent to the inequality

$$\frac{d(L)}{d(L\cap M)} \frac{d(M)}{d(L \cap M)} \geq \frac{d(L+M)}{d(L\cap M)}.$$

Therefore replacing L, M and $L+M$ by $\pi(L)$, $\pi(M)$ and $\pi(L+M)$ we can assume that $L \cap M = \{0\}$. Let $(e_1, \cdots, e_\ell)$, $\ell = \dim L$, and $(e_{\ell+1}, \cdots, e_{\ell+m})$, $m = \dim M$, be bases in L and M respectively. Then

$$\begin{aligned} d(L)d(M) &= \|e_1 \wedge \cdots \wedge e_\ell\| \, \|e_{\ell+1} \wedge \cdots \wedge e_{\ell+m}\| \\ &\geq \|e_1 \wedge \cdots \wedge e_\ell \wedge e_{\ell+1} \wedge \cdots \wedge e_{\ell+m}\| \geq d(L+M) \end{aligned} \tag{49}$$

that proves (48) (the second inequality in (49) is true because $(L\cap\Delta)+(M\cap\Delta) \subset (L+M)\cap\Delta$. □

LEMMA 6.7. *Let $\{a_t \mid t \in \mathbb{R}\}$ be a self-adjoint one-parameter subgroup of $SO(2,1)$. Let p and q be positive integers, and denote $p+q$ by n. Denote $SO(p) \times SO(q)$ by K. Suppose $p \geq 3$, $q \geq 1$ and $0 < i < n$, or $p = 2$, $q = 2$ and $i = 1$ or 3. Then for any s, $0 < s < 2$, and any $c > 0$ there exist $t > 0$ and $\omega > 1$ such that for any lattice Λ in $\mathbb{R}^n$*

$$\int_K \alpha_i(a_t k\Lambda)^s \, dm(k) < \frac{c}{2}\alpha_i(\Lambda)^s + \omega^2 \max_{0<j\leq \min\{n-i,i\}} \left(\sqrt{\alpha_{i+j}(\Lambda)\alpha_{i-j}(\Lambda)}\right)^s. \tag{50}$$

If $p = 2$, $q = 1$ and $i = 1, 2$, or if $p = 2$, $q = 2$ and $i = 2$, then for any s, $0 < s < 1$, and any $c > 0$ there exist $t > 0$ and $\omega > 1$ such that (50) holds.

Proof. Fix $c > 0$. In view of Proposition 6.4 one can find $t > 0$ such that

$$\int_K \frac{dm(k)}{\|a_t k v\|^s} < \frac{c}{2},$$

whenever $v \in F(i)$, $\|v\| = 1$. It follows that

$$\int_K \frac{dm(k)}{\|a_t k v\|^s} < \frac{c}{2} \cdot \frac{1}{\|v\|^s}, \tag{51}$$

for any $v \in F(i), v \neq 0$. Let Λ be a lattice in $\mathbb{R}^n$. There exists a Λ-rational subspace L_i of dimension i such that

$$\frac{1}{d_\Lambda(L_i)} = \alpha_i(\Lambda). \tag{52}$$

The inequality (51) implies

$$\int_K \frac{dm(k)}{d_{a_tk\Lambda}(a_tkL_i)^s} < \frac{c}{2}\frac{1}{d_\Lambda(L_i)^s}. \tag{53}$$

Let $\omega = \max_{0<j<n} \| \bigwedge^j (a_t)\|$. (In fact $\omega = e^t$). We have that

$$\omega^{-1} \leq \frac{\|a_t v\|}{\|v\|} \leq \omega,\ 0 < j < n,\ v \in F(j). \tag{54}$$

Let us denote the set of Λ-rational subspaces L of dimension i with $d_\Lambda(L) < \omega^2 d_\Lambda(L_i)$ by Ψ_i. We get from (54) that for a Λ-rational i-dimensional subspace $L \notin \Psi_i$

$$d_{a_tk\Lambda}(a_tkL) > d_{a_tk\Lambda}(a_tkL_i), \qquad k \in K. \tag{55}$$

It follows from (53), (55) and the definition of α_i that

$$\int_K \alpha_i(a_tk\Lambda)^s\, dm(k) < \frac{c}{2}\alpha_i(\Lambda)^s \text{ if } \Psi_i = \{L_i\}. \tag{56}$$

Assume now that $\Psi_i \neq \{L_i\}$. Let $M \in \Psi_i,\ M \neq L_i$. Then $\dim(M+L_i) = i+j,\ j > 0$. Now using (52), (54) and Lemma 6.6 we get that for any $k \in K$

$$\begin{aligned} \alpha_i(a_tk\Lambda) < \omega\alpha_i(\Lambda) = \frac{\omega}{d_\Lambda(L_i)} &< \frac{\omega^2}{\sqrt{d_\Lambda(L_i)d_\Lambda(M)}} \\ &\leq \frac{\omega^2}{\sqrt{d_\Lambda(L_i\cap M)d_\Lambda(L_i+M)}} \\ &\leq \omega^2\sqrt{\alpha_{i+j}(\Lambda)\alpha_{i-j}(\Lambda)}. \end{aligned} \tag{57}$$

Hence if $\Psi_i \neq \{L_i\}$

$$\int_K \alpha_i(a_tk\Lambda)^s\, dm(k) \leq \omega^2 \max_{0<j\leq \min\{n-i,i\}} \left(\sqrt{\alpha_{i+j}(\Lambda)\alpha_{i-j}(\Lambda)}\,\right)^s. \tag{58}$$

Combining (56) and (58) we get that for any lattice $\Lambda \subset \mathbb{R}^n$, (50) holds. $\square$

In the rest of this subsection we obtain similar systems of inequalities for the $SO(2,1)$ and $SO(2,2)$ cases, with $s = 1$. For $H = SO(2,1)$, Δ a lattice in $\mathbb{R}^3$, and L a Δ-rational subspace of $\mathbb{R}^3$, let $d^*_\Delta(L) = \|e_1 \wedge \dots e_\ell\|^*$ where $(e_1, \dots e_l)$ is a basis for $\Delta \cap L$. (The norm $\|\cdot\|^*$ defined in (46) on $\mathbb{R}^3 = \bigwedge^1(\mathbb{R}^3)$ can be extended to $\bigwedge^2(\mathbb{R}^3)$ by duality.) For $1 \leq i \leq 2$, let

$$\alpha^*_i(\Delta) = \sup\left\{\frac{1}{d^*_\Delta(L)}\middle| L \text{ is a } \Delta\text{-rational subspace of dimension } i\right\}. \tag{59}$$

Clearly for any Δ,

$$(1/2)\alpha_i(\Delta) < \alpha^*_i(\Delta) < 2\alpha_i(\Delta). \tag{60}$$

LEMMA 6.8. *Let $\{a_t \mid t \in \mathbb{R}\}$ be a self-adjoint one-parameter subgroup of $H = SO(2,1)$, and denote $SO(2)$ by K. Then there exist $t_0 > 0$ and $\omega > 1$, such that for any $t < t_0$, for any unimodular lattice Λ in $\mathbb{R}^3$, and $1 \le i \le 2$,*

$$\int_K \alpha_i^*(a_t k \Lambda)\, dm(k) < \alpha_i^*(\Lambda) + \omega^2 \sqrt{\alpha_{3-i}(\Lambda)}. \tag{61}$$

Proof. The argument is identical to the proof of Lemma 6.7 except that one uses Lemma 6.5 instead of Proposition 6.4. □

Now let $H = SO(2,2)$. The space $V = \bigwedge^2(\mathbb{R}^4)$ splits as a direct sum $V_1 \oplus V_2$ of two invariant subspaces, where on each V_i, H preserves a quadratic form Q_i of signature $(2,1)$. We define on each V_i a Euclidean norm $\|\cdot\|_i^*$ by (46) (adapted to Q_i). Let π_i denote the orthogonal projections from V to V_i. Now let Δ be a lattice in $\mathbb{R}^4$, and let L be a two-dimensional Δ-rational subspace of $\mathbb{R}^4$. For $1 \le i \le 2$, let

$$d_\Delta^{i,\#}(L) = \|\pi_i(e_1 \wedge e_2)\|_i^*, \tag{62}$$

where $\{e_1, e_2\}$ is a basis over $\mathbb{Z}$ for $\Delta \cap L$. Then let

$$\alpha_2^\#(\Delta) = \sup_L \left\{ \min\left(\frac{1}{d_\Delta^{1,\#}(L)}, \frac{1}{d_\Delta^{2,\#}(L)} \right) \right\}. \tag{63}$$

The supremum is taken over Δ-rational two dimensional subspaces L. By construction, for any Δ,

$$C^{-1}\alpha_2^\#(\Delta) < \alpha_2(\Delta) < C\alpha_2^\#(\Delta), \tag{64}$$

where C is an absolute constant.

LEMMA 6.9. *Let $\{a_t \mid t \in \mathbb{R}\}$ be a self-adjoint one-parameter subgroup of $SO(2,1)$, where $SO(2,1)$ is diagonally embedded in $H = SO(2,2)$, under its local identification with $SL(2,\mathbb{R}) \times SL(2,\mathbb{R})$. Denote $SO(2) \times SO(2)$ by K, and the maximal compact of $SO(2,1)$ by $\tilde{K}$. Then there exist $t_0 > 0$ and $\omega > 1$, such that for any $t < t_0$ and for any unimodular lattice Λ in $\mathbb{R}^4$,*

$$\int_{\tilde{K}} \alpha_2^\#(a_t \tilde{k} \Lambda)\, dm(\tilde{k}) < \alpha_2^\#(\Lambda) + \omega^2 \sqrt{\alpha_1(\Lambda)\alpha_3(\Lambda)}. \tag{65}$$

Proof. The group $\tilde{K}$ is diagonally embedded in K. Recall that each $SO(2,2)$ invariant subspace V_i of $\bigwedge^2(\mathbb{R}^4)$ is fixed pointwise by one of the $SL(2,\mathbb{R})$ factors, while the other fixes a quadratic form of signature $(2,1)$. Thus, for $1 \le i \le 2$, the inequalities:

$$\int_{\tilde{K}} \frac{dm(\tilde{k})}{\|\pi_i(a_t \tilde{k} v)\|_i^*} \le \frac{1}{\|\pi_i(v)\|_i^*} \tag{66}$$

follow immediately from Lemma 6.5. Hence,

$$\int_{\tilde{K}} \min\left(\frac{1}{\|\pi_1(a_t\tilde{k}v)\|_1^*}, \frac{1}{\|\pi_2(a_t\tilde{k}v)\|_2^*}\right) dm(k)$$
$$\leq \min\left(\int_{\tilde{K}} \frac{dm(\tilde{k})}{\|\pi_1(a_t\tilde{k}v)\|_1^*}, \int_{\tilde{K}} \frac{dm(\tilde{k})}{\|\pi_2(a_t\tilde{k}v)\|_2^*}\right)$$
$$\leq \min\left(\frac{1}{\|\pi_1(v)\|_1^*}, \frac{1}{\|\pi_2(v)\|_2^*}\right). \tag{67}$$

The rest of the proof is identical to that of Lemma 6.7 except that (67) is used in place of Proposition 6.4. □

6.3. Coarsely Superharmonic Functions. Let $n \in \mathbb{N}^+$ and let D_n^+ denote the set of diagonal matrices $d(\lambda_1, \cdots, \lambda_n) \in GL(n, \mathbb{R})$ with $\lambda_1 \geq \lambda_2 \geq \cdots \geq \lambda_n > 0$. For any $g \in GL(n, \mathbb{R})$, consider the Cartan decomposition $g = k_1(g)d(g)k_2(g), k_1(g), k_2(g) \in K = O(n, \mathbb{R})$, $d(g) \in D_n^+$ and denote by $\lambda_1(g) \geq \lambda_2(g) \geq \cdots \geq \lambda_n(g)$ the eigenvalues of $d(g)$.

LEMMA 6.10. *For every $\epsilon > 0$ there exists a neighborhood U of e in $O(n, \mathbb{R})$ such that*

$$\left|\frac{\lambda_i(d_1kd_2)}{\lambda_i(d_1)\lambda_i(d_2)} - 1\right| < \epsilon \tag{68}$$

for any $d_1, d_2 \in D_n^+$, $k \in U$ and $1 \leq i \leq n$.

Proof. Let $(e_1, \cdots, e_n)$ be the standard orthonormal basis in $\mathbb{R}^n$. If $k \in O(n, \mathbb{R})$ and $\langle ke_1, e_1\rangle > 1 - \epsilon$ then

$$\|d_1kd_2e_1\| > (1 - \epsilon)\lambda_1(d_1)\lambda_1(d_2). \tag{69}$$

On the other hand, for any $g \in GL(n, \mathbb{R})$.

$$\lambda_1(g) = \|g\| \geq \|ge_1\|. \tag{70}$$

Since $\|d_1kd_2\| \leq \|d_1\|\ \|d_2\|$ it follows from (69) and (70) that

$$1 \geq \frac{\lambda_1(d_1kd_2)}{\lambda_1(d_1)\lambda_1(d_2)} > 1 - \epsilon, \tag{71}$$

if $\langle ke_1, e_1\rangle > 1 - \epsilon$. Analogously considering the representation of $GL(n, \mathbb{R})$ in the i-th exterior product $\bigwedge^i(\mathbb{R}^n)$ of $\mathbb{R}^n$ we get that

$$1 \geq \frac{(\lambda_1\lambda_2\cdots\lambda_i)(d_1kd_2)}{(\lambda_1\lambda_2\cdots\lambda_i)(d_1d_2)} > 1 - \epsilon, \tag{72}$$

if $k \in O(n, \mathbb{R})$ and $\langle\bigwedge^i(k)(e_1\wedge\cdots\wedge e_i), e_1\wedge\cdots\wedge e_i\rangle > 1-\epsilon$. It is clear that there exists a neighborhood U of identity in $O(n, \mathbb{R})$ such that $\langle\bigwedge^i(k)(e_1\wedge\cdots\wedge e_i),\ e_1\wedge\cdots\wedge e_i\rangle > \sqrt{1-\epsilon}$ for every $k \in U$ and $1 \leq i \leq n$. But

$$\lambda_i(g) = \frac{(\lambda_1\lambda_2\cdots\lambda_i)(g)}{(\lambda_1\lambda_2\cdots\lambda_{i-1})(g)}.$$

Therefore (68) follows from (72). □

LEMMA 6.11. *(cf. the "wavefront lemma" [**EMc**, Theorem 3.1]) Let H be a self-adjoint connected reductive subgroup of $GL(n,\mathbb{R})$, let $K = O(n,\mathbb{R}) \cap H$ be a maximal compact subgroup of H and let $\{a_t \mid t \in \mathbb{R}\}$ be a self-adjoint one-parameter subgroup of H. Then for every neighborhood V of e in H there exists a neighborhood U of e in K such that*

$$a_t U a_s \subset K V a_t a_s K \tag{73}$$

for any $t \geq 0$ and $s \geq 0$.

Proof. Conjugating a_t by an element of K we can assume that $\{a_t \mid t \geq 0\} \subset D_n^+$. It is easy to see that there exists $\epsilon > 0$ such that $h_1 \in V h_2$ whenever $h_1, h_2 \in D_n^+$ and $\left|\frac{\lambda_i(h_1)}{\lambda_i(h_2)} - 1\right| < \epsilon$ for every $1 \leq i \leq n$. Take a neighborhood U such that (68) is satisfied. Then (73) is true for this U. □

PROPOSITION 6.12. *Let H be a self-adjoint reductive subgroup of $GL(n,\mathbb{R})$, let $K = O(n,\mathbb{R}) \cap H$, let m denote the normalized measure on K, and let $A = \{a_t \mid t \in \mathbb{R}\}$ be a self-adjoint one-parameter subgroup of H. Let $\mathcal{F}$ be a family of strictly positive functions on H having the following properties:*

(a) *The logarithms $\log f$ for $f \in \mathcal{F}$ are equicontinuous with respect to a left-invariant uniform structure on H or, equivalently, for any $\epsilon > 0$ there exists a neighborhood $V(\epsilon)$ of 1 in H such that for any $f \in \mathcal{F}$,*

$$(1-\epsilon) f(h) < f(uh) < (1+\epsilon) f(h)$$

for any $h \in H$ and $u \in V(\epsilon)$;

(b) *The functions $f \in \mathcal{F}$ are left K-invariant, that is $f(Kh) = f(h)$, $h \in H$,*

(c) $\sup_{f \in \mathcal{F}} f(1) < \infty$.

Then there exists $0 < c = c(\mathcal{F}) < 1$ such that for any $t > 0$ and $b > 0$ there exists $B = B(t,b) < \infty$ with the following property: If $f \in \mathcal{F}$ and

$$\int_K f(a_t k h)\, dm(k) < c f(h) + b \tag{74}$$

for any $h \in KAK \subset H$, then

$$\int_K f(a_\tau k)\, dm(k) < B$$

for any $\tau > 0$.

Proof. Fix $f \in \mathcal{F}$, and let

$$\tilde{f}(h) = \int_K f(hk)\, dm(k).$$

Properties (a), (b), (c) of the function f imply that $\tilde{f}$ has the same properties. Hence it suffices to show that the conclusion of the proposition holds for $\tilde{f}$. Therefore we can assume that

$$f(KhK) = f(h),\ h \in H, \tag{75}$$

and we have to prove that

$$\sup_{\tau > 0} f(a_\tau) < B < \infty. \tag{76}$$

It follows from property (a) that

$$\frac{1}{2}f(h) < f(uh) < 2f(h),\ h \in H,\ u \in V = V(\frac{1}{2}). \tag{77}$$

According to Lemma 6.11 there exists a neighborhood U of 1 in H such that $a_tUa_\tau \in KVa_ta_\tau K$ for any $t \geq 0$ and $\tau \geq 0$. Then we get from (75) and (77) that

$$\int_K f(a_tka_\tau)\,dm(k) \geq \int_{U\cap K} f(a_tka_\tau)\,dm(k) > \frac{1}{2}m(U\cap K)f(a_ta_\tau). \tag{78}$$

Suppose for some $t > 0$ and $b > 0$

$$\int_K f(a_tkh)\,dm(k) < \frac{1}{4}m(U\cap K)f(h) + b,\ h \in H. \tag{79}$$

It follows from (78) and (79) that for some $b' > 0$,

$$f(a_ta_\tau) < \frac{1}{2}f(a_\tau) + b',\ \text{for all } \tau > 0. \tag{80}$$

Using induction on ℓ we get from (80) that

$$f(a_{\ell t}) < 2\max\{f(1), b'\},\ ;\ell \in \mathbb{N}^+. \tag{81}$$

Since $\{a_r \mid 0 \leq r \leq t\}$ belongs to V^i for some i where $V^1 = V,\ V^i = V\ V^{i-1}$, it follows that $\sup_{h\in H,\ 0\leq r\leq t} \frac{f(a_rh)}{f(h)} < \infty$. Therefore (81) and property (c) imply (76). □

6.4. Averages over large spheres. In this subsection we complete the proofs of Theorem 6.2, Theorem 6.3 and Theorem 5.15.

Proof of Theorem 6.2. Define functions $f_0, f_1, \cdots, f_n$ on $H = SO(p,q)$ by the following equalities

$$f_i(h) = \alpha_i(h\Delta),\ h \in H,\ 0 \leq i \leq n.$$

Since $\alpha(a_tk\Delta)^s = \max_{0\leq i\leq n} f_i(a_tk)^s < \sum_{0\leq i\leq n} f_i(a_tk)^s$ it is enough to show that

$$\sup_{t>0,\ 0\leq i\leq n} \int_K f_i^s(a_tk)\,dm(k) < \infty. \tag{82}$$

Let A_t denote the averaging operator defined by

$$(A_tf)(h) = \int_K f(a_tkh)\,dm(k), \qquad h \in H.$$

As in Proposition 6.4, let

$$F(i) = \{x_1 \wedge x_2 \wedge \cdots \wedge x_i \mid x_1, x_2, \cdots, x_i \in \mathbb{R}^n\} \subset \textstyle\bigwedge^i(\mathbb{R}^n).$$

Since $\|Kv\| = \|v\|$ and $\frac{\|hv\|}{\|v\|} \leq \|\bigwedge^i(h)\|$, for any $v \in F(i)$ and $h \in H$, each f_i has properties (a) and (b) of Proposition 6.12. Applying Lemma 6.7 to $\Lambda = h\Delta$ we see that for any $i, 0 < i < n$, and $h \in H$

$$A_tf_i^s < \frac{c}{2}f_i^s + \omega^2 \max_{0<j\leq\min\{n-i,i\}} \sqrt{f_{i+j}^s f_{i-j}^s}. \tag{83}$$

Let us denote $q(i) = i(n-i)$. Then by direct computations $2q(i)-q(i+j)-q(i-j) = 2j^2$. Therefore we get from (83) that for any $i, 0 < i < n$, and any positive $\epsilon < 1$

(84)

$$A_t(\epsilon^{q(i)} f_i^s) < \frac{c}{2}\epsilon^{q(i)} f_i^s + \omega^2 \max_{0<j\le\min\{n-i,i\}} \epsilon^{q(i)-\frac{q(i+j)+q(i-j)}{2}} \sqrt{\epsilon^{q(i+j)} f_{i+j}^s \epsilon^{q(i-j)} f_{i-j}^s}$$
$$\le \frac{c}{2}\epsilon^{q(i)} f_i^s + \epsilon\omega^2 \max_{0<j\le\min\{n-i,i\}} \sqrt{\epsilon^{q(i+j)} f_{i+j}^s \epsilon^{q(i-j)} f_{i-j}^s}.$$

Consider the linear combination

$$f_{\epsilon,s} = \sum_{0\le i\le n} \epsilon^{q(i)} f_i^s.$$

The function $f_{\epsilon,s}$ then also has properties (a) and (b) of Proposition 6.12. Since $\epsilon^{q(i)} f_i^s < f_{\epsilon,s}, f_0 = 1$ and $f_n = 1/d(\Delta)$, the inequalities (84) imply the following inequality:

$$A_t f_{\epsilon,s} < 1 + d(\Delta)^{-s} + \frac{c}{2} f_{\epsilon,s} + n\epsilon\omega^2 f_{\epsilon,s}. \tag{85}$$

Taking $\epsilon = \frac{c}{2n\omega^2}$ we see that (74) from Proposition 6.12 also holds. Furthermore property (a) and (74) of Proposition 6.12 hold with the same constants for any unimodular lattice $\Delta \in \mathbb{R}^n$. Since $f_{\epsilon,s}(1) \le n\alpha(\Delta)^s$, $f_{\epsilon,s}(1)$ is uniformly bounded as Δ varies over a compact set $\mathcal{C}$ of unimodular lattices. Hence the family $\mathcal{F}$ of functions $f_{\epsilon,s}$ obtained as Δ varies over $\mathcal{C}$ satisfies all the conditions of Proposition 6.12. Since $\alpha_i(h\Delta)^s = f_i(h)^s \le \epsilon^{-q(i)} f_{\epsilon,s}(h)$, Proposition 6.12 implies that there exists a constant $B > 0$ so that for each i, all $t > 0$, and all $\Delta \in \mathcal{C}$,

$$\int_K \alpha_i(a_t k\Delta)^s \, dm(k) < B.$$

From this the theorem follows. □

7. Connections to dynamics of rational billiards

For references to this section see [**E2**].

In this lecture, we describe some counting problems on translation surfaces and outline their connection to the dynamics of the $SL(2,\mathbb{R})$ action on the moduli space of translation surfaces. Much of this is presented in analogy with the quantitative Oppenheim conjecture (see §5 and §6).

Recall that $\mathcal{L}_n = SL(n,\mathbb{R})/SL(n,\mathbb{Z})$ is the space of covolume 1 lattices in $\mathbb{R}^n$. This space is non-compact, since we can have arbitrarily short vectors in a lattice.

The strata and the measure μ. Let $\beta = \beta_1, \dots, \beta_m$ be a partition of $2g-2$. Let $\mathcal{H}(\beta)$ denote the moduli space of translation surfaces with conical singularities of total angles $2\pi(\beta_1+1), \dots, 2\pi(\beta_m+1)$. (I am using the notation from [**Zor**]: Jean-Christophe is using $\mathcal{M}(\cdot)$.) We will sometimes call $\mathcal{H}(\beta)$ a *stratum*. Let $\mathcal{H}_1(\beta) \subset \mathcal{H}(\beta)$ denote the subset consisting of surfaces of area 1. Let μ be the normalized Lebesque measure on $\mathcal{H}_1(\beta)$ (as defined by Jean-Christophe via the period map). We will use the same letter to denote the restriction of μ to $\mathcal{H}_1(\beta)$. A theorem of Masur and Veech (proved in Jean-Christophe's lectures) states that $\mu(\mathcal{H}_1(\beta)) < \infty$. In §7.5 we will describe how to evaluate the numbers $\mu(\mathcal{H}_1(\beta))$.

Note that the case of $n = 2$ in the space of lattices $\mathcal{L}_2$ and the case of stratum $\mathcal{H}_1(\emptyset)$ boil down to the same thing, since we are considering the space of unit volume tori (or more precisely, the space of 1-forms on unit volume tori), which is given by $SL(2,\mathbb{R})/SL(2,\mathbb{Z})$.

Note. I will use the term *saddle connection* to denote what Jean-Christophe is calling a *connection.*

Holonomy and the sets $V_{sc}(S)$ and $V(S)$. Recall that a point $S \in \mathcal{H}(\beta)$ can be viewed as a pair (M,ω) where M is a Riemann surface and ω is a holomorphic 1-form on M. Recall that the holonomy of a curve γ on S is given by

$$hol(\gamma) = \int_\gamma \omega.$$

Let

$$V_{sc}(S) = \{hol(\gamma) \ : \ \gamma \text{ is a saddle connection on } S\},$$

so that $V_{sc}(S) \subset \mathbb{C} \simeq \mathbb{R}^2$. Note that $V_{sc}(S)$ is a discrete subset of $\mathbb{R}^2$, but it is not, in general, a subgroup. We also define the analogous set:

$$V(S) = \{hol(\gamma) : \gamma \text{ is a closed geodesic on } S \text{ not passing through singularities}\}.$$

Note that any such closed geodesic is part of a cylinder and all the closed geodesics in the cylinder have the same holonomy. (If $S = \mathbb{R}^2/\mathbb{Z}^2$ is the standard torus with the standard flat structure, then $V(S) = \mathbb{Z}^2$).

7.1. Counting cylinders and saddle connections. Let $B(R)$ denote a ball of radius R. Then, $|V(S) \cap B(R)|$ is the number of cylinders on S of length at most R, and $|V_{sc}(S) \cap B(R)|$ is the number of saddle connections (not necessarily vertical) of length at most R. Masur proved the following:

THEOREM 7.1. *For all flat surfaces S in a compact set, there are constants c_1 and c_2 so that for $R \gg 1$*

$$c_1 R^2 < |V(S) \cap B(R)| \leq |V_{sc}(S) \cap B(R)| < c_2 R^2.$$

The upper bound is proved in [**Mas2**] and the lower bound is proved in [**Mas3**]. The proof of the lower bound depends on the proof of the upper bound. Another proof of both the upper and lower bounds with explicit constants was given by Vorobets in [**Vo1**] and [**Vo2**]. We will sketch below yet another proof of the upper bound, using the ideas of §6. (See [**EM**] for the details).

We also note that there is a dense set of directions with a closed trajectory and thus a cylinder.

The following theorem, gives asymptotic formulas for the number of saddle connections and cylinders of closed geodesics on a generic surface. It was first proved in this form in [**EM**], but many of the ideas came from [**Ve**], where a slightly weaker version was proved.

THEOREM 7.2. *For a.e. $S \in \mathcal{H}_1(\beta)$, we have*

$$|V_{sc}(S) \cap B(R)| \sim \pi b(\beta) R^2,$$

where $V_{sc}(S)$ is the collection of vectors in $\mathbb{R}^2$ given by holonomy of saddle connections on S, and $b(\beta)$ is the Siegel-Veech constant defined in §7.2 (see also (89)).

Similarly, for closed geodesics, we have that there is a constant $b_1(\beta)$ so that

$$|V(S) \cap B(R)| \sim \pi b_1(\beta) R^2$$

where $V(S)$ is the collection of vectors given by holonomy along (imprimitive) closed geodesics not passing through singularities, and $b_1(\beta)$ is the associated Siegel-Veech constant.

It will turn out that the problem of counting saddle connections or cylinders closed geodesics on a flat surface is analogous to the quantitative Oppenheim problem (§5 and §6).

7.2. The Siegel-Veech formula. The following construction and its analogues play a key role. For any function of compact support $f \in C_c(\mathbb{R}^n)$, let $\hat{f}(\Delta) = \sum_{v \in \Delta \setminus 0} f(v)$. Note that if $f = \chi_{B(1)}$, we get $\hat{f}(\Delta) = |\Delta \cap B(1)|$. We have the *Siegel formula*: For any $f \in C_c(\mathbb{R}^n)$,

$$\frac{1}{\mu(\mathcal{L}_n)} \int_{\mathcal{L}_n} \hat{f}(\Delta) d\mu(\Delta) = \int_{\mathbb{R}^n} f d\lambda, \tag{86}$$

where μ is Haar measure on $\mathcal{L}_n = SL(n, \mathbb{R})/SL(n, \mathbb{Z})$, and λ is Lebesgue measure on $\mathbb{R}^n$.

The generalization of this formula to moduli space was developed, so the legend goes, by Veech while he listened to Margulis lecture on the Oppenheim conjecture. For $f \in C_c(\mathbb{R}^2)$ we define the Siegel-Veech transform $\hat{f}(S) = \sum_{v \in V_{sc}(S)} f(v)$. Just as above, if $f = \chi_{B(1)}$, $\hat{f}$ counts the number of saddle connections of length ≤ 1.

Just as we had the Siegel formula for lattices, here we have the *Siegel-Veech formula*: There is a constant $b(\beta)$, called the *Siegel-Veech constant*, such that for any $f \in C_c(\mathbb{R}^2)$, we have

$$\frac{1}{\mu(\mathcal{H}_1(\beta))} \int_{\mathcal{H}_1(\beta)} \hat{f}(S)\, d\mu(S) = b(\beta) \int_{\mathbb{R}^2} f, \tag{87}$$

where μ is the natural $SL(2, \mathbb{R})$ invariant measue on $\mathcal{H}_1(\beta)$.

Let us sketch the proof of this result (essentially from [**Ve**], also reproduced in [**EM**]). The first step (which is by far the most technical) is to show that $\hat{f} \in L^1(\mathcal{H}_1(\beta))$, so that the left hand side is finite. This can be deduced e.g. from (94) below. Having done this, we denote the quantity on the left hand side of (87) by $\varphi(f)$.

Thus we have a linear functional $\varphi : C_c(\mathbb{R}^2) \to \mathbb{R}$, i.e. a measure. But it also has to be $SL(2, \mathbb{R})$ invariant. Only Lebesgue measue and δ_0, the delta measure at 0 are $SL(2, \mathbb{R})$ invariant. Thus we have $\varphi(f) = af(0) + b\int_{\mathbb{R}^2} f$. It remains to show $a = 0$. Consider the limit of indicator functions $f = \chi_{B(R)}$ as $R \to 0$. Both sides of the equation tend to 0, so we have that $a = 0$, and thus our result.

Returning to lattices, we can apply literally the same arguments to prove the Siegel formula (86). Note that nothing was special about dimension 2 in the above proof sketch. Thus, we have almost proved (86) as well. To be precise, we currently have:

$$\frac{1}{\mu(\mathcal{L}_n)} \int_{\mathcal{L}_n} \hat{f}(\Delta) d\mu(\Delta) = b \int_{\mathbb{R}^n} f d\lambda,$$

for some constant b. We need to show $b = 1$. Here, we once again use $f = \chi_{B(R)}$, but this time consider $R \to \infty$. Recall that $\hat{f}(\Delta) = |\Delta \cap B(R)| \sim \mathrm{Vol}(B(R))$, for $R \to \infty$ and Δ fixed. Thus, we get $b = 1$, and the Siegel formula.

We should remark that for the space of lattices the proof of the Siegel formula indicated above is not the easiest available. In fact, it is possible to avoid proving *apriori* that $\hat{f} \in L^1(\mathcal{L}_n)$. See [**Sie**] or [**Cas**] or [**Ter**] for the details. A well known consequence of the Siegel formula is the following:

$$\mu(\mathcal{L}_n) = \frac{1}{n}\zeta(2)\zeta(3)\dots\zeta(n). \tag{88}$$

For the stata $\mathcal{H}(\beta)$, this method of evaluating $b(\beta)$ (i.e. considering $f = \chi_{B(R)}$ and taking $R \to \infty$) is not avaliable. Essentially the problem is that we do not have an alternative expression for the constant in Theorem 5.5.

Another approach is to let $f = \chi_{B(\epsilon)}$, send $\epsilon \to 0$ and keep track of the leading term in the asymptotics of both sides. This was done in [**EMZ**] where we obtained the following result: For any stratum $\mathcal{H}_1(\beta)$ in the moduli space of translation surfaces the coefficient $b(\beta)$ involved in (87) can be expressed in the following form:

$$b(\beta) = \sum_{\alpha<\beta} c(\alpha,\beta)\frac{\mu(\mathcal{H}_1(\alpha))}{\mu(\mathcal{H}_1(\beta))}, \tag{89}$$

where the sum is over lower dimensional strata α (which lie at the "boundary" of $\mathcal{H}(\beta)$), and $c(\alpha,\beta)$ are explicitly known rational numbers.

We note that (89) fails as a method for calculating the volumes, since (unlike the lattice case) we do not have an independent formula for $b(\beta)$. In §7.5 we will show that the volumes can be computed in a different way; then (89) can be used to evaluate the Siegel-Veech constants $b(\beta)$. These numbers appear in some other contexts as well, in particular in connection with the Lyapunov exponents of the geodesic flow.

7.3. Counting using the $SL(2,\mathbb{R})$ action. This subsection is closely parallel to §5.3. The following exposition will be along the lines of [**EM**], which was heavily influenced by [**Ve**]. To simplify the notation, we only deal with the case of saddle connections. Define $g_t = \begin{pmatrix} e^t & 0 \\ 0 & e^{-t} \end{pmatrix}$ and $r_\theta = \begin{pmatrix} \cos\theta & \sin\theta \\ -\sin\theta & \cos\theta \end{pmatrix}$. Let f be the indicator function of the trapezoid defined by the points

$$(1,1), (-1,1), (-1/2,1/2), (1/2,1/2).$$

LEMMA 7.3. *We have* $\int_0^{2\pi} f(g_t r_\theta v)\, d\theta \approx \begin{cases} 2e^{-2t} & \text{if } e^t/2 \le \|v\| \le e^t, \\ 0 & \text{otherwise.} \end{cases}$

Proof. Let U denote the trapezoid. Note that

$$f(g_t r_\theta v) \neq 0 \Leftrightarrow g_t r_\theta v \in U \Leftrightarrow r_\theta v \in g_t^{-1}U. \tag{90}$$

The set $g_t^{-1}U$ is the shaded region in Figure 2. From (90) it is clear that the integral in Lemma 7.3 is equal to (2π times) the fraction of the circle which lies inside the shaded region $g_t^{-1}U$. If v is too long or too short (not drawn), then the circle would completely miss the shaded region, and the integral would be zero. If it does not miss, then (2π times) the fraction of the circle in the shaded region is approximately $2e^{-2t}$, independent of $\|v\|$. □

We now prove Theorem 7.2. Summing our formula from Lemma 7.3 over all $v \in V_{sc}(S)$ and recalling the definition of the *Siegel-Veech transform* $\hat{f}(S) = \sum_{v\in V_{sc}(S)} f(v)$, we get

$$\tfrac{1}{2}e^{2t}\int_0^{2\pi} \hat{f}(g_t r_\theta S)\, d\theta \approx |V_{sc}(S)\cap B(e^t)| - |V_{sc}(S)\cap B(e^t/2)|.$$

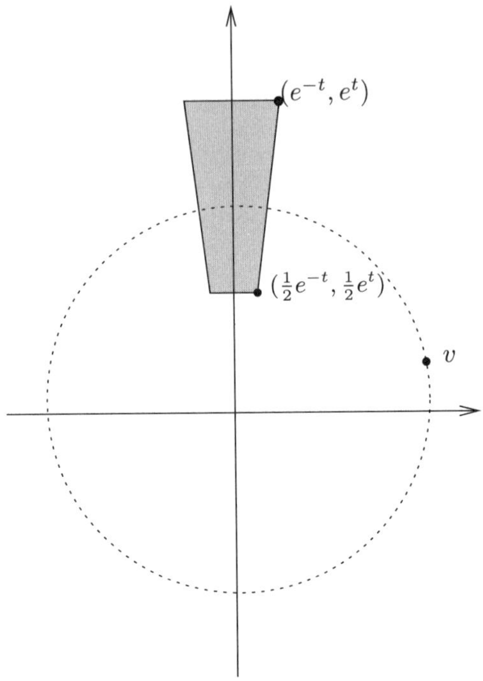

Figure 2. Lemma 7.3.

Writing $R = e^t$, we can rewrite this as

$$\frac{1}{2}R^2 \int_0^{2\pi} \hat{f}(g_t r_\theta S)d\theta \approx |V_{sc}(S) \cap B(R)| - |V_{sc}(S) \cap B(R/2)|. \tag{91}$$

This equation is key to the counting problem, since the right hand side counts saddle connections in an annulus, and the left hand side is an integral over (part of) an $SL(2,\mathbb{R})$ orbit. (The fact that we only have approximate equality does not affect the leading order asymptotics.) Now we are supposed to use some sort of ergodic theory to analyze the behavoir of integral on the left-hand-side of (91) as $t \to \infty$ (or equivalently as $R \to \infty$).

There is an ergodic theorem of Nevo [**Ne**] which implies that[1] for almost all $S \in \mathcal{H}_1(\beta)$, and provided that $\hat{f} \in L^{1+\epsilon}(\mathcal{H}_1(\beta))$, the integral converges to $2\pi \int_{\mathcal{H}_1(\beta)} \hat{f}(S)\, dS = 2\pi b(\beta) \int_{\mathbb{R}^2} f$. The assertion that $\hat{f} \in L^{1+\epsilon}$ can be verified using (94). This immediately implies Theorem 7.2. □

However, this approach is a *failure* if one wants to prove things about billiards: our theorems hold for almost every point S, and the set of translation surfaces arising from rational billiards has measure zero.

One eventual goal is to prove analogues of Ratner's theorems on unipotent flows for the $SL(2,\mathbb{R})$ action on $\mathcal{H}_1(\beta)$. That is, we would like to classify invariant measures, orbit closures, and prove uniform distribution, for both the full $SL(2,\mathbb{R})$ action, and for the horocycle flow. One partial result in this direction is due to

[1]The theorem of Nevo used here is about a general $SL(2,\mathbb{R})$ action, and uses nothing about the geometry of the moduli space.

McMullen [**Mc**]: he has classified the $SL(2,\mathbb{R})$ orbit closures and invariant measures for the moduli space of genus 2 surfaces (i.e., the strata $\mathcal{H}(1,1)$ and $\mathcal{H}(2)$). Note that the integral in (91) is over large circles in $SL(2,\mathbb{R})$, which can be approximated well by horocycles. Thus the action of the horocycle flow (i.e. the action of $\begin{pmatrix} 1 & t \\ 0 & 1 \end{pmatrix}$) is directly relevant. For other very partial results in this direction see [**EMaMo**], [**EMS**] and [**CW**].

7.4. The upper bounds. In this subsection, we will outline a proof of the upper bound in Theorem 7.1, following the scheme of §6.

Let $B(R)$ be the ball of radius R centered at 0 in $\mathbb{R}^n$. For a given lattice $\Delta \in \mathcal{L}_n$. we would like to find out how many lattice points, that is, how many points of Δ are contained in $B(R)$.

It is immediately clear that for a fixed lattice Δ, as $R \to \infty$,

$$|\Delta \cap B(R)| \sim \text{Vol}(B(R)) = \text{Vol}(B(1))R^n. \tag{92}$$

(i.e. the number of lattice points is asymptotic to the volume). However, this is not uniform in Δ. A uniform upper bound has been given in Lemma 6.1, in particular:

$$|\Delta \cap B(1)| < C\alpha(\Delta). \tag{93}$$

The analagous problem in moduli space is as follows: We are interested in $|V_{sc}(S) \cap B(1)|$, i.e. the number of saddle connections of length at most 1 on S.

The result is as follows: Fix $\epsilon > 0$. Then there is a constant $c = c(\beta, \epsilon)$ such that for all $S \in \mathcal{H}(\beta)$ of area 1,

$$|V_{sc}(S) \cap B(1)| \leq \frac{c}{\ell(S)^{1+\epsilon}}, \tag{94}$$

where $\ell(S)$ is the length of the shortest saddle connection on S.

Assuming (94), the proof of the upper bound in Theorem 7.1 can be following the scheme of §6 (with a suitable definition for the functions α_i).

However, it turns out that the proof of (94) is more difficult that that of (93); it itself uses the system of inequalities along the line of §6, as well as induction on the genus.

7.5. Evaluation of the volumes. In this lecture we describe briefly another strategy for calculating volumes of strata, which also has a parallel for the space of lattices. Recall that we are considering the spaces $\mathcal{H}(\beta)$ of flat structures with singularity structure $\beta = (\beta_1, \beta_2, \dots \beta_n)$, where $\beta_i \in \mathbb{N}$, $\sum \beta_i = 2g - 2$. Let the set of singularities be denoted by Σ. We have $|\Sigma| = n$, and we have

$$H_1(S, \Sigma; \mathbb{Z}) = \mathbb{Z}^{2g+n-1}.$$

We can pick a basis by selecting g a-cycles, g b-cycles (from absolute homology), and $n - 1$ relative cycles.

Fix a $\mathbb{Z}$-basis $\gamma_1, \gamma_2, \dots \gamma_k$ of $H_1(S, \Sigma; \mathbb{Z})$, where $k = 2g + n - 1$. We recall the following fact (see [**Ko**]):

THEOREM 7.4. *The map* $(X, \omega) \to (hol(\gamma_1), \dots, hol(\gamma_k))$ *from* $\mathcal{H}(\beta) \to (\mathbb{R}^2)^k$ *is a local coordinate system.*

By pulling back Lebesgue measure on $(\mathbb{R}^2)^k$, we obtain a normalized measure ν on $\mathcal{H}(\beta)$. (For more details on the above constuction, see [**Mas1**, §3].) Now, we would like to define a measure on the hypersurface $\mathcal{H}_1(\beta)$.

This is similar to the lattice setting, where if we pick a basis $v_1, v_2, \dots v_n$ for our lattice $\Delta \subset \mathbb{R}^n$, we get a matrix in $M_n(\mathbb{R})$ by letting v_i be the ith column. Note that since our lattice is unit volume, our matrix has determinant 1. We have a natural (Lebesque) measure ν on $M_n(\mathbb{R})$. Consider the det $= 1$ hypersurface Ω_1 (i.e., $SL(n, \mathbb{R})$). We define a measure μ on this space as follows: let $E \subset \Omega_1$, and let $C_1(E)$ be the cone over E (i.e. the union of all line segments which start at the origin and end at a point of E). We define $\mu(E) = \nu(C_1(E))$. This yields a finite measure since we are considering a fundamental domain under the $SL(n, \mathbb{Z})$-action. This is in fact the measure used in the previous section.

Returning to the setting of surfaces, recall that the area of our surface $S = (X, \omega)$ is given by

$$Area(S) = \frac{1}{2i} \int_X \omega \wedge \bar{\omega} = \frac{1}{2i} \sum_{i=1}^{g} \int_{a_i} \bar{\omega} \int_{b_i} \omega - \int_{b_i} \bar{\omega} \int_{a_i} \omega$$

where a_i and b_i are the a- and b-cycles on X respectively.

This gives that the area is a quadratic form in the coordinate sytem, i.e.,

$$Area(X, \omega) = Q(hol(\gamma_1), \dots, hol(\gamma_k)).$$

However, it is a degenerate form, since it only depends on the absolute cycles a_i and b_i. We can mimic the lattice picture now: we define $\mu(E) = \nu(C_1(E))$ for any subset $E \subset \mathcal{H}_1(\beta)$. Thus,

$$\mu(\mathcal{H}_1(\beta)) = \mu(\mathcal{F}) = \nu(C_1(\mathcal{F})),$$

where $\mathcal{F}$ is a fundamental domain.

We now make a cosmetic step. Let $C_R(\mathcal{F})$ denote the cone of $\mathcal{F}$ extended to the hypersurface of area R-surfaces. Clearly

$$\mu(\mathcal{H}_1(\beta)) = \nu(C_1(\mathcal{F})) = \frac{\nu(C_R(\mathcal{F}))}{R^k}.$$

We have the following fact:

$$|C_R(\mathcal{F}) \cap (\mathbb{Z}^2)^k| \sim \nu(C_R(\mathcal{F}))$$

as $R \to \infty$, i.e. the number of lattice points in a cone is asymptotic to the volume. Ususally this is used to estimate the number of lattice points, but here we use this in reverse and estimate the volume by the number of lattice points. Thus, we get that

$$\mu(\mathcal{H}_1(\beta)) = \frac{\nu(C_R(\mathcal{F}))}{R^k} \sim \frac{|C_R(\mathcal{F}) \cap (\mathbb{Z}^2)^k|}{R^k},$$

or, equivalently,

$$|C_R(\mathcal{F}) \cap (\mathbb{Z}^2)^k| \sim \mu(\mathcal{H}_1(\beta)) R^k. \tag{95}$$

The equation (95) is not useful unless we can find an interpretation of the points of $C_R(\mathcal{F}) \cap (\mathbb{Z}^2)^k$. This is given by the following:

LEMMA 7.5. *$S = (X, \omega) \in C_R(\mathcal{F}) \cap (\mathbb{Z}^2)^k$ if and only if X is a holomorphic branched cover of the standard torus of degree $\leq R$, ω is the pullback of dz under the covering map, and all singularities branch over the same point.*

Proof: Since $S \in C_R(\mathcal{F})$, $area(S) \leq R$. By definition, $S \in (\mathbb{Z}^2)^k$ is equivalent to $hol(\gamma_1), \dots, hol(\gamma_k) \in \mathbb{Z}^2$. Fix a non-singular point z_0 on S, and define $\pi : S \to T$, where T is the standard torus, by $\pi(z) = \int_{z_0}^{z} \omega$. Since $\int_\gamma \omega \in \mathbb{Z} + i\mathbb{Z}$ for any closed curve or saddle connection γ, this is a well defined covering map with all singularities branching over the same point. Since the torus is unit volume, the area of S is equal to the degree of the covering. $\square$

Let $N_\beta(d)$ denote the number[2] of branched covers of T of degree d with branching type β. (Note that $N_\beta(d)$ is defined in purely combinatorial terms).

Combining Lemma 7.5 with (95), we obtain the following: as $R \to \infty$,

$$\sum_{d=1}^{R} N_\beta(d) \sim \mu(\mathcal{H}_1(\beta))R^k. \tag{96}$$

(This relation was discovered by Kontsevich and Zorich, and independently by Masur and the author.) Thus, we can compute $\mu(\mathcal{H}_1(\beta))$ if we can compute the asymptotics of the left-hand-side of (96). This is a purely combinatorial problem.

Suppose we are considering a degree d cover of the torus. Consider the standard basis a and b of curves on the torus (when the torus is viewed as the unit square, the curves correspond to the sides of the square). They give rise to permutations of the sheets, that is, elements of the symmetric group S_d. We will abuse notation by denoting these permutations also by a and b. Singularity types of covers correspond to different conjugacy classes of the commutator $aba^{-1}b^{-1}$. A simple zero is a transposition, a double zero a three cycle, a two simple zeroes is a product of two transpositions, etc. (So for example, if we are considering the stratum $\mathcal{H}(1,1)$, the commutator will be in the same conjugacy class as a product of two transpositions.) The number of pairs $(a, b) \in S_d \times S_d$ satisfying such a commutation relation can be expressed as a sum over the characters of the symmetric group S_d.

However, simply looking at the conjugacy class of the commutator permutation does not guarantee that the resulting surface is connected. We wish to count only the connected covers. However, the disconnected ones dominate the count. If one knows the number of disconnected covers exactly, one can compute the number of connected covers (by using inclusion/exclusion to subtract off all the possible ways a cover can disconnect). Unfortunately, as one does that, the first n terms in the asymptotic formula cancel. Still, it is possible, using the exact formula for the number of disconnected covers in [**BO**], to carry out the computation (see [**EO**]). The result, is a fairly messy but computable formula for $\mu(\mathcal{H}_1(\beta))$.

There are two consequences of the above computations worth mentioning:

THEOREM 7.6. *The generating function* $F_\beta(q) = \sum_{d=0}^{\infty} N_\beta(d)q^d$ *is a quasi-modular form, that is, it is a polynomial in the Eisenstein series* $G_k(q)$, $k = 2, 4, 6$.

THEOREM 7.7. $\pi^{-2g}\mu(\mathcal{H}_1(\beta)) \in \mathbb{Q}$, *where* g *is the genus of any surface in* $\mathcal{H}(\beta)$.

Both of the above theorems were conjectured by Kontsevich. Further work showed that they hold also for the connected components of strata, and that similar results hold for spaces of quadratic differentials. We remark that Theorem 7.7 implies that the Siegel-Veech constants are rational.

For the space of lattices, one can carry out the same construction. The main difference is that one ends up counting *unbranched* covers of the standard torus

---

[2] In order for Theorem 7.6 below to hold, we should, when defining $N_\beta(d)$, weigh each cover by the inverse of its automorphism group. However this does not affect the asymptotics and can be ignored for most purposes.

T^n, or what is equivalent, sublattices of the standard lattice $\mathbb{Z}^n$. By computing the number of sublattices of $\mathbb{Z}^n$ of index at most R, and sending $R \to \infty$, it is not difficult to reproduce (88).

8. Equidistribution of translates and applications to Diophantine equations

We will follow parts of [**EMc**] and [**EMS1**].

In this section, using ergodic properties of subgroup actions on homogeneous spaces of Lie groups, we study asymptotic behavior of number of lattice points on certain affine varieties. Consider for instance the following.

Example 1 Let $p(\lambda)$ be a monic polynomial of degree $n \geq 2$ with integer coefficients and irreducible over $\mathbb{Q}$. Let $\mathrm{M}_n(\mathbb{Z})$ denote the set of $n \times n$ integer matrices, and put

$$V_p(\mathbb{Z}) = \{A \in \mathrm{M}_n(\mathbb{Z}) \ : \ \det(\lambda I - A) = p(\lambda)\}.$$

Hence $V_p(\mathbb{Z})$ is the set of integral matrices with characteristic polynomial $p(\lambda)$. Consider the norm on $n \times n$ real matrices given by $\|(x_{ij})\| = \sqrt{\sum_{ij} x_{ij}^2}$, and let $N(T, V_p)$ denote the number of elements of $V_p(\mathbb{Z})$ with norm less than T.

THEOREM 8.1. *Suppose further that $p(\lambda)$ splits over $\mathbb{R}$, and for a root α of $p(\lambda)$ the ring of algebraic integers in $\mathbb{Q}(\alpha)$ is $\mathbb{Z}[\alpha]$. Then, asymptotically as $T \to \infty$,*

$$N(T, V_p) \sim \frac{2^{n-1} h R \omega_n}{\sqrt{D} \cdot \prod_{k=2}^n \Lambda(k/2)} T^{n(n-1)/2}$$

where h is the class number of $\mathbb{Z}[\alpha]$, R is the regulator of $\mathbb{Q}(\alpha)$, D is the discriminant of $p(\lambda)$, ω_n is the volume of the unit ball in $\mathbb{R}^{n(n-1)/2}$, and $\Lambda(s) = \pi^{-s}\Gamma(s)\zeta(2s)$.

Example 1 is a special case of the following counting problem which was first studied in [**DRS**] and [**EMc**].

The counting problem: Let W be a real finite dimensional vector space with a $\mathbb{Q}$ structure and V a Zariski closed real subvariety of W defined over $\mathbb{Q}$. Let G be a reductive real algebraic group defined over $\mathbb{Q}$, which acts on W via a $\mathbb{Q}$-representation $\rho : G \to \mathrm{GL}(W)$. Suppose that G acts transitively on V. Let $\|\cdot\|$ denote a Euclidean norm on W. Let B_T denote the ball of radius $T > 0$ in W around the origin, and define

$$N(T, V) = |V \cap B_T \cap \mathbb{Z}^n|,$$

the number of integral points on V with norm less than T. We are interested in the asymptotics of $N(T, V)$ as $T \to \infty$.

Let Γ be a subgroup of finite index in $G(\mathbb{Z})$ such that $W(\mathbb{Z})\Gamma \subset W(\mathbb{Z})$. By a theorem of Borel and Harish-Chandra [**BH-C**], $V(\mathbb{Z})$ is a union of finitely many Γ-orbits. Therefore to compute the asymptotics of $N(T, V)$ it is enough to consider each Γ-orbit, say $\mathcal{O}$, separately and compute the asymptotics of

$$N(T, V, \mathcal{O}) = |\mathcal{O} \cap B_T|.$$

Suppose that $\mathcal{O} = \Gamma \cdot v_0$ for some $v_0 \in V(\mathbb{Z})$. Then the stabilizer $H = \{g \in G \ : \ gv_0 = v_0\}$ is a reductive real algebraic $\mathbb{Q}$-subgroup, and $V \cong G/H$. Define

$$R_T = \{gH \in G/H : gv_0 \in B_T\},$$

the pullback of the ball B_T to G/H.

Assume that G^0 and H^0 do not admit nontrivial $\mathbb{Q}$-characters. Then by the theorem of Borel and Harish-Chandra, G/Γ admits a G-invariant (Borel) probability measure, say μ_G, and $H/(\Gamma \cap H)$ admits an H-invariant probability measure, say μ_H. Now the natural inclusion $H/(\Gamma \cap H) \hookrightarrow G/\Gamma$ is an H-equivariant proper map. Let $\pi : G \to G/\Gamma$ be the natural quotient map. Then the orbit $\pi(H)$ is closed, $H/(\Gamma \cap H) \cong \pi(H)$, and μ_H can be treated as a measure on G/Γ supported on $\pi(H)$. Such finite invariant measures supported on closed orbits of subgroups are called *algebraic measures*. Let $\lambda_{G/H}$ denote the (unique) G-invariant measure on G/H induced by the normalization of the Haar measures on G and H.

The following result was proved in [**DRS**]; subsequently a simpler proof appeared in [**EMc**].

THEOREM 8.2. *Suppose that V is affine symmetric and Γ is irreducible (equivalently, H is the set of fixed points of an involution of G, and G is $\mathbb{Q}$-simple). Then asymptotically as $T \to \infty$,*

$$N(T, V, \mathcal{O}) \sim \lambda_{G/H}(R_T).$$

Translates of algebraic measures. For any $g \in G$, let $g\mu_H$ denote the translated measure defined as

$$g\mu_H(E) = \mu_H(g^{-1}E), \qquad \forall \text{ Borel sets } E \subset G/\Gamma.$$

Note that $g\mu_H$ is supported on $g\pi(H)$. A key ingredient in the proofs of Theorem 8.2 in [**DRS**] and [**EMc**] is showing that if H is the set of fixed points of an involution of G then for any sequence $\{g_i\} \subset G$, such that $\{g_iH\}$ has no convergent subsequence in G/H, the translated measures $g_i\mu_H$ get 'equidistributed' on G/Γ as $i \to \infty$; that is, the sequence $\{g_i\mu_H\}$ weakly converges to μ_G. The method of [**DRS**] uses spectral analysis on G/Γ, while the argument of [**EMc**] uses the mixing property of the geodesic flow. However, both methods seem limited essentially to the affine symmetric case. It should be remarked that for the proof of Theorem 8.2 one needs only certain averages of translates of the form $g\mu_H$ to become equidistributed.

One can show that under certain conditions if for some sequence $\{g_i\}$ we have $\lim g_i\mu_H = \nu$ then the measure ν is again algebraic. We give exact algebraic conditions on the sequence $\{g_i\}$ relating it to the limit measure ν. Using this analysis, we show that the counting estimates as in Theorem 8.2 hold for a large class of homogeneous varieties. The following particular cases of homogeneous varieties, which are not affine symmetric, are of interest. We first place Example 1 in this context.

Example 1 continued. Note that $V_p(\mathbb{Z})$ is the set of integral points on the real subvariety $V_p = \{A \in \mathrm{M}_n(\mathbb{R}) : \det(\lambda I - A) = p(\lambda)\}$ contained in the vector space $W = \mathrm{M}_n(\mathbb{R})$. Let $G = \{g \in \mathrm{GL}_n(\mathbb{R}) : \det g = \pm 1\}$. Then G acts on W via conjugations, and V_p is a closed orbit of G (see [**New**, Theorem III.7]). Put $\Gamma = G(\mathbb{Z}) = \mathrm{GL}_n(\mathbb{Z})$. The companion matrix of $p(\lambda)$ is

$$v_0 = \begin{pmatrix} 0 & & 0 & -a_n \\ 1 & & 0 & -a_{n-1} \\ 0 & \cdots & \vdots & \vdots \\ \vdots & & 0 & \\ 0 & & 1 & -a_1 \end{pmatrix} \in V_p(\mathbb{Z}). \tag{97}$$

The centralizer H of v_0 is a maximal $\mathbb{Q}$-torus and H^0 has no nontrivial $\mathbb{Q}$-characters. Note that H is not the set of fixed points of an involution, and the variety $V_p = H\backslash G$ is *not* affine symmetric. Nevertheless, we show that $N(T, V_p, \Gamma v_0) \sim \lambda_{G/H}(R_T)$. By computing the volumes, we obtain the following estimate.

THEOREM 8.3. *Let $N(T, V_p)$ be the number of points on $V_p(\mathbb{Z})$ of norm less than T. Then asymptotically as $T \to \infty$,*

$$N(T, V_p) \sim c_p T^{n(n-1)/2},$$

where $c_p > 0$ is an explicitly computable constant.

We obtain a 'formula' for calculating c_p; for the sake of simplicity we calculate it explicitly only under the additional assumptions on $p(\lambda)$ of Theorem 8.1.

See [**BR**] for some deeper consequences of the above result.

Example 2. Let A be a nondegenerate indefinite integral quadratic form in $n \geq 3$ variables and of signature (p, q), where $p \geq q$, and B a definite integral quadratic form in $m \leq p$ variables. Let $W = \mathrm{M}_{m\times n}(\mathbb{R})$ be the space of $m \times n$ matrices. Consider the norm on W given by $\|(x_{ij})\| = \sqrt{\sum_{i,j} x_{ij}^2}$. Define

$$V_{A,B} = \{X \in \mathrm{M}_{m\times n}(\mathbb{R}) \ : \ XA\,{}^{t}X = B\}.$$

Thus a point on $V_{A,B}(\mathbb{Z})$ corresponds to a way of representing B by A over $\mathbb{Z}$. We assume that $V_{A,B}(\mathbb{Z})$ is not empty.

The group $G = \mathrm{SO}(A)$ acts on W via right multiplication, and the action is transitive on $V_{A,B}$. The stabilizer of a point $\xi \in V_{A,B}$ is an orthogonal group H_ξ in $n - m$ variables. Let $\Gamma = G(\mathbb{Z})$. Then the number of Γ-orbits on $V_{A,B}(\mathbb{Z})$ is finite. Let $\xi_1, \ldots, \xi_h$ be the representatives for the orbits.

THEOREM 8.4. *Let $N(T, V_{A,B})$ denote the number of points on $V_{A,B}(\mathbb{Z})$ with norm less than T. Then asymptotically as $T \to \infty$,*

$$N(T, V_{A,B}) \sim \sum_{i=1}^{h} \frac{\mathrm{vol}(\Gamma \cap H_{\xi_i}\backslash H_{\xi_i})}{\mathrm{vol}(\Gamma\backslash G)} c_{A,B} T^{r(n-r-1)}$$

where $r = \min(m, q)$, and $c_{A,B} > 0$ is an explicitly computable constant.

REMARK 8.5. In some ranges of p, q, m, n this formula may be proved by the Hardy-Littlewood circle method, or by Θ-function techniques. Using our method one also obtains asymptotic formulas for the number of points in the individual orbits $\Gamma\xi_i$.

REMARK 8.6. In the case $m > q$, the asymptotics of the number of integer points does not agree with the heuristic of the Hardy-Littlewood circle method, even if the number of variables mn is very large compared to the number of quadratic equations $m(m+1)/2$. The discrepancy occurs because the null locus $\{X : XA\,{}^{t}X = 0\}$ does not contain a non-singular real point (cf. [**Bir**, Theorem 1]) and so the 'singular integral' vanishes.

8.1. Connection between counting and translates of measures. We recall some observations from [**DRS**, Sect. 2]; see also [**EMc**]. Let the notation be as in the counting problem stated in the introduction. For $T > 0$, define a function F_T on G by

$$F_T(g) = \sum_{\gamma \in \Gamma/(H\cap\Gamma)} \chi_T(g\gamma \cdot v_0),$$

where χ_T is the characteristic function of B_T. By construction F_T is left Γ-invariant, and hence it will be treated as a function on G/Γ. Note that

$$F_T(e) = \sum_{\gamma \in \Gamma/(H\cap\Gamma)} \chi_T(\gamma \cdot v_0) = N(T, V, \mathcal{O}).$$

Since we expect, as in Theorem 8.2, that

$$N(T, V, \mathcal{O}) \sim \lambda_{H\backslash G}(R_T),$$

we define

$$\hat{F}_T(g) = \frac{1}{\lambda_{G/H}(R_T)} F_T(g).$$

Thus the asymptotics in Theorem 8.2 is the assertion

$$\hat{F}_T(e) \to 1 \qquad \text{as } T \to \infty. \tag{98}$$

PROPOSITION 8.7 ([**DRS**, Sect. 2]). *For any compactly supported function ψ on G/Γ,*

$$\langle \hat{F}_T, \psi \rangle = \frac{1}{\lambda_{G/H}(R_T)} \int_{R_T} \overline{\psi^H}\, d\lambda_{G/H},$$

where

$$\psi^H(gH) = \int_{G/\Gamma} \psi \, d(g\mu_H)$$

is a function on G/H.

Proof. Let $\mathcal{F}$ be a fundamental domain for G/Γ. By definition,

$$\begin{aligned}
\langle F_T, \psi \rangle &= \sum_{\gamma \in \Gamma/(H\cap\Gamma)} \int_{\mathcal{F}} \chi_T(g\gamma)\psi(g)\, d\mu_G(g) \\
&= \sum_{\gamma \in \Gamma/(H\cap\Gamma)} \int_{\mathcal{F}\gamma} \chi_T(g)\psi(g)\, d\mu_G(g) \\
&= \int_{G/(H\cap\Gamma)} \chi_T(g)\psi(g)\, d\mu_G(g) \\
&= \int_{G/H} \int_{H/(H\cap\Gamma)} \chi_T(\bar{g})\psi(\bar{g}h)\, d\mu_H(h)\, d\lambda_{G/H}(\bar{g})
\end{aligned}$$

$$\int_{R_T} \left(\int_{G/\Gamma} \psi \, d_{\bar{g}\mu_H} \right) \lambda_{G/H}(\bar{g})$$

□

8.2. Limiting distributions of translates of algebraic measures. The following is the main result of this section which allows us to investigate the counting problems.

THEOREM 8.8. *Let G be a connected real algebraic group defined over $\mathbb{Q}$, $\Gamma \subset G(\mathbb{Q})$ an arithmetic lattice in G with respect to the $\mathbb{Q}$-structure on G, and $\pi : G \to G/\Gamma$ the natural quotient map. Let $H \subset G$ be a connected real algebraic $\mathbb{Q}$-subgroup admitting no nontrivial $\mathbb{Q}$-characters. Let μ_H denote the H-invariant probability measure on the closed orbit $\pi(H)$. For a sequence $\{g_i\} \subset G$, suppose that the translated measures $g_i\mu_H$ converge to a probability measure μ on G/Γ. Then there exists a connected real algebraic $\mathbb{Q}$-subgroup L of G containing H such that the following holds:*

(i) *There exists $c_0 \in G$ such that μ is a $c_0 L c_0{}^{-1}$-invariant measure supported on $c_0\pi(L)$.*

  In particular, μ is a algebraic measure.

(ii) *There exist sequences $\{\gamma_i\} \subset \Gamma$ and $c_i \to c_0$ in G such that $\gamma_i H \gamma_i{}^{-1} \subset L$ and $g_i H = c_i \gamma_i H$ for all but finitely many $i \in \mathbb{N}$.*

The proof of this theorem is based on the following observation.

PROPOSITION 8.9. *Let the notation be as in Theorem 8.8. Then either there exists a sequence $c_i \to c$ in G such that $c_i\mu_i = \mu_H$ for all $i \in \mathbb{N}$ (in which case $\mu = c\mu_H$), or μ is invariant under the action of a nontrivial unipotent one-parameter subgroup of G.*

In order to be able to apply Theorem 8.8 to the problem of counting, we need to know some conditions under which the sequence $\{g_i\mu_H\}$ of probability measures does not escape to infinity. Suppose further that G and H are reductive. Let $Z(H)$ be the centralizer of H in G. By rationality $\pi(Z(H))$ is closed in G/Γ. Now if $\pi(Z(H))$ is noncompact, there exits a sequence $\{z_i\} \subset Z(H)$ such that $\{\pi(z_i)\}$ is divergent; that is, it has no convergent subsequence. Then $z_i\mu_H$ escapes to the infinity; that is $(z_i\mu_H)(K) \to 0$ for any compact set $K \subset G/\Gamma$. The condition that $\pi(Z(H))$ is noncompact is equivalent to the condition that H is contained in a proper parabolic $\mathbb{Q}$-subgroup of G. In the converse direction we have the following (see [**EMS2**]).

THEOREM 8.10. *Let G be a connected real reductive algebraic group defined over $\mathbb{Q}$, and H a connected real reductive $\mathbb{Q}$-subgroup of G, both admitting no nontrivial $\mathbb{Q}$-characters. Suppose that H is not contained in any proper parabolic $\mathbb{Q}$-subgroup of G defined over $\mathbb{Q}$. Let $\Gamma \subset G(\mathbb{Q})$ be an arithmetic lattice in G and $\pi : G \to G/\Gamma$ the natural quotient map. Let μ_H denote the H-invariant probability measure on $\pi(H)$. Then given an $\epsilon > 0$ there exists a compact set $K \subset G/\Gamma$ such that $(g\mu_H)(K) > 1 - \epsilon$, $\forall g \in G$.*

The proof of this result uses generalizations of some results of Dani and Margulis [**DM3**]. Combining this theorem with Theorem 8.8, we deduce the following consequence.

COROLLARY 8.11. *Suppose that H is reductive and a proper maximal connected real algebraic $\mathbb{Q}$-subgroup of G. Then for any sequence $\{g_i\} \subset G$, if the sequence $\{g_iH\}$ is divergent (that is, it has no convergent subsequence) in G/H, then the sequence $\{g_i\mu_H\}$ gets equidistributed with respect to μ_G as $i \to \infty$ (that is, $g_i\mu_H \to \mu_G$ weakly).*

In the general case, one obtains the following analogue of Corollary 8.11. We note that the condition that H is not contained in any proper $\mathbb{Q}$-parabolic subgroup of G, is also equivalent to saying that any real algebraic $\mathbb{Q}$-subgroup L of G containing H is reductive.

COROLLARY 8.12. *Let G be a connected real reductive algebraic group defined over $\mathbb{Q}$, and H a connected real reductive $\mathbb{Q}$-subgroup of G not contained in any proper parabolic $\mathbb{Q}$-subgroup of G. Let $\Gamma \subset G(\mathbb{Q})$ be an arithmetic lattice in G. Suppose that a sequence $\{g_i\} \subset G$ is such that the sequence $\{g_i \mu_H\}$ does not converge to the G-invariant probability measure. Then after passing to a subsequence, there exist a proper connected real reductive $\mathbb{Q}$-subgroup L of G containing H and a compact set $C \subset G$ such that*

$$\{g_i\} \subset CL(Z(H) \cap \Gamma)$$

8.3. Applications to the counting problem. The case where H is maximal. The following is a consequence of Corollary 8.11:

THEOREM 8.13. *Let G and H be as in the counting problem. Suppose that H^0 is reductive and a proper maximal connected real algebraic $\mathbb{Q}$-subgroup of G, where H^0 denotes the connected component of identity in H. Then asymptotically as $T \to \infty$*

$$N(T, V, \mathcal{O}) \sim \lambda_{G/H}(R_T).$$

REMARK 8.14. Suppose that H is the set of fixed point of an involution of G. Let L be a connected real reductive $\mathbb{Q}$-subgroup of G containing H^0. Then there exists a normal $\mathbb{Q}$-subgroup N of G such that $L = H^0 N$. Now if G is $\mathbb{Q}$-simple, then H^0 is a maximal proper connected $\mathbb{Q}$-subgroup of G (see [**Bor**, Lemma 8.0]). Hence Theorem 8.2 follows from Theorem 8.13.

The general case. We now use Corollary 8.12. For applying this result to the counting problem, we need to know that averages of translates of the measure μ_H along the sets R_T become equidistributed as T tends to infinity. I.e., we want the set of 'singular sequences', for which the limit measure is not G-invariant, to have negligible 'measure' in the sets R_T as $T \to \infty$. This does not hold when the sets R_T are 'focused' along $L/H (\subset G/H)$:

DEFINITION 8.15. Let G and H be as in the counting problem. For a sequence $T_n \to \infty$, the sequence $\{R_{T_n}\}$ of open sets in G/H is said to be *focused*, if there exist a proper connected reductive real algebraic $\mathbb{Q}$-subgroup L of G containing H^0 and a compact set $C \subset G$ such that

$$\limsup_{n \to \infty} \frac{\lambda_{G/H}(q_H(CL(Z(H^0) \cap \Gamma)) \cap R_{T_n})}{\lambda_{G/H}(R_{T_n})} > 0,$$

where $q_H : G \to G/H$ is the natural quotient map.

Note that since L is reductive and defined over $\mathbb{Q}$, we have that $\pi(L)$ is closed in G/Γ. In particular, $L(Z(H^0) \cap \Gamma)$ is closed in G. Also $LzH^0 = Lz$ for any $z \in Z(H^0)$. Now since C is compact, the set $q_H(CL(Z(H^0) \cap \Gamma))$ is closed in G/H.

Now if the focusing of $\{R_{T_n}\}$ does not occur, then using Corollary 8.12 we can obtain the following analogue of Corollary 8.11.

COROLLARY 8.16. *Let G and H be as in the counting problem. Suppose that H^0 is not contained in any proper $\mathbb{Q}$-parabolic subgroup of G^0, and for some sequence $T_n \to \infty$, the sequence $\{R_{T_n}\}$ is not focused. Then given $\epsilon > 0$ there exists an open set $\mathcal{A} \subset G/H$ with the following properties:*

$$\liminf_{n\to\infty} \frac{\lambda_{G/H}(\mathcal{A} \cap R_{T_n})}{\lambda_{G/H}(R_{T_n})} > 1 - \epsilon \tag{99}$$

and given any sequence $\{g_i\} \subset q_H{}^{-1}(\mathcal{A})$, if the sequence $\{q_H(g_i)\}$ is divergent in G/H then the sequence $\{g_i\mu_H\}$ converges to μ_G.

This corollary allows us to obtain the counting estimates like in Theorem 8.2 and Theorem 8.13 for a large class of homogeneous varieties.

THEOREM 8.17. *Let G and H be as in the counting problem. Suppose that H^0 is not contained in any proper $\mathbb{Q}$-parabolic subgroup of G^0 (equivalently, $Z(H)/(Z(H)\cap \Gamma)$ is compact), and for some sequence $T_n \to \infty$ with bounded gaps, the sequence $\{R_{T_n}\}$ is not focused. Then asymptotically*

$$N(T, V, \mathcal{O}) \sim \lambda_{G/H}(R_T).$$

Remark. The non-focusing assumption in Theorem 8.17 is not vacuous. In the above setup one is required to verify the condition of nonfocusing in Theorem 8.17 separately for each application of the result.

Outline of the proof of Theorem 8.17, assuming Corollary 8.16.

PROPOSITION 8.18. *Let the notation and conditions be as in Theorem 8.17. Then $\hat{F}_{T_n} \to 1$ in the weak-star topology on $L^\infty(G/\Gamma, \mu_G)$; that is, $\langle \hat{F}_{T_n}, \psi\rangle \to \langle 1, \psi\rangle$ for any compactly supported continuous function ψ on G/Γ.*

Proof. As in Proposition 8.7,

$$\langle \hat{F}_T, \psi\rangle = \frac{1}{\lambda_{G/H}(R_T)} \int_{R_T} \overline{\psi^H}\, d\lambda_{G/H},$$

where

$$\psi^H(gH) = \int_{H\Gamma/\Gamma} \psi(gh\Gamma)\, d\mu_H(h\Gamma) = \int_{G/\Gamma} \psi\, d(g\mu_H)$$

is a function on G/H.

Let $\epsilon > 0$ be given. Since the sequence $\{R_{T_n}\}$ is not focused, we obtain a set $\mathcal{A} \subset G/H$ as in Corollary 8.16. Break up the integral over R_{T_n} into the integrals over $R_{T_n} \cap \mathcal{A}$ and $R_{T_n} \setminus \mathcal{A}$. By equation (99) and the boundedness of ψ, the second integral is $O(\epsilon)$. By Corollary 8.16, for any sequence $\{g_i\} \subset q_H{}^{-1}(\mathcal{A})$, if $\{q_H(g_i)\}$ has no convergent subsequence in G/H, then $g_i \cdot \mu_H \to \mu_G$. Hence

$$\psi^H(g_iH) \to \int_{G/\Gamma} \psi\, d\mu_G = \langle \psi, 1\rangle.$$

We use dominated convergence theorem to justify the interchange of limits. Now

$$\begin{aligned}\lim_{n\to\infty}\langle \hat{F}_{T_n},\psi\rangle &= \lim_{n\to\infty}\frac{1}{\lambda_{G/H}(R_{T_n})}\int_{R_{T_n}\cap\mathcal{A}}\overline{\psi^H}\,d\lambda_{G/H}+O(\epsilon)\\ &= \lim_{n\to\infty}\frac{1}{\lambda_{G/H}(R_{T_n})}\int_{R_{T_n}\cap\mathcal{A}}\overline{\langle\psi,1\rangle}\,d\lambda_{G/H}+O(\epsilon)\\ &= \lim_{n\to\infty}\frac{\lambda_{G/H}(R_{T_n}\cap\mathcal{A})}{\lambda_{G/H}(R_{T_n})}\langle 1,\psi\rangle+O(\epsilon)\\ &= \langle 1,\psi\rangle+O(\epsilon)\end{aligned}$$

Since ϵ is arbitrary, the proof is complete. □

PROPOSITION 8.19 ([**EMS1**]). *There are constants $a(\delta)$ and $b(\delta)$ tending to 1 as $\delta\to 0$ such that*

$$b(\delta)\le\liminf_{T\to\infty}\frac{\lambda_{G/H}(R_{(1-\delta)T})}{\lambda_{G/H}(R_T)}\le\limsup_{T\to\infty}\frac{\lambda_{G/H}(R_{(1+\delta)T})}{\lambda_{G/H}(R_T)}\le a(\delta).$$

Proof of Theorem 8.17. Let ψ in Proposition 8.18 tend to a δ-function at the origin. Then, combining Proposition 8.18 and Proposition 8.19, we obtain that $\hat{F}_{T_i}\to 1$ pointwise on G/Γ as $i\to\infty$. (See [**DRS**, Lemma 2.3] for the details). Thus (98) holds. This completes the proof. □

8.4. Invariance under unipotents.

PROPOSITION 8.20. *Let G be a semisimple Lie group, Γ be a discrete subgroup of G, and $\pi: G\to G/\Gamma$ be the natural quotient map. Let H be a nontrivial reductive subgroup of G and Ω be a relatively compact neighborhood of identity in H. Let μ_Ω be the probability measure on $\pi(\Omega)$ which is the pushforward under π of the restriction to Ω of a Haar measure on H.*

Suppose that for a sequence $\{g_i\}_{i\in\mathbb{N}}\subset G$, the sequence $\{g_i\cdot\mu_\Omega\}_{i\in\mathbb{N}}\subset\mathcal{P}(G/\Gamma)$ converges weakly to a nonzero measure μ on G/Γ. Then one of the following holds:

(1) *There exists a compact set $C\subset G$ such that $\{g_i\}_{i\in\mathbb{N}}\subset CZ_G(H)$.*
(2) *μ is invariant under a nontrivial unipotent one-parameter subgroup of G.*

PROOF. (Cf. [**Moz**, Lemma ??]) Let $\mathfrak{g}$ be the Lie algebra of G and $\mathfrak{h}\subset\mathfrak{g}$ be the Lie subalgebra corresponding to H. Equip $\mathfrak{g}$ with a Euclidean norm, say $\|\cdot\|$.

Claim 1. *If the Condition 1 above does not hold then there exists a sequence $X_i\to 0$ in $\mathfrak{h}$ as $i\to\infty$, such that a subsequence of $\{\operatorname{Ad} g_i\cdot X_i\}_{i\in\mathbb{N}}$ converges to a nonzero element $Y\in\mathfrak{g}$.*

To prove the claim there is no loss of generality if we pass to a subsequence of $\{g_i\}_{i\in\mathbb{N}}$, or replace $\{g_i\}_{i\in\mathbb{N}}$ by $\{g_ic_i\}_{i\in\mathbb{N}}$, where $\{c_i\}_{i\in\mathbb{N}}$ is contained in a compact subset of G.

Since H is reductive, there is a Cartan involution θ of G such that $\theta(H)=H$. Let K be the set of fixed points of θ. Then K is a maximal compact subset of G. There exists a maximal $\mathbb{R}$-split torus A in G such that

(100) $$\theta(a)=a^{-1},\ \forall a\in A.$$

Choose an order on the system of $\mathbb{R}$-roots of A for G and let Δ be the set of simple roots. Let A_+ be the exponential of the closure of the positive Weyl chamber. Then by Cartan decomposition we have

$$G = KA_+K.$$

Hence without loss of generality we can assume that $g_i = a_i k_i$ for all $i \in \mathbb{N}$, where $k_i \to k$ in K as $i \to \infty$ and $\{a_i\}_{i\in\mathbb{N}} \subset A_+$.

Let

$$\Phi = \{\alpha \in \Delta : \sup_{i\in\mathbb{N}} \alpha(a_i) < \infty\}.$$

Then by modifying the sequence $\{a_i\}_{i\in\mathbb{N}}$ from the left by multiplications by elements from a compact set in $A_+ \cap (\cap_{\beta\in\Delta\setminus\Phi} \ker\beta)$, we may assume that

$$\alpha(a_i) = 1, \ \forall \alpha \in \Phi. \tag{101}$$

By passing to a subsequence, we may also assume that

$$\lim_{i\to\infty} \alpha(a_i) = \infty, \ \forall \alpha \in \Delta \setminus \Phi. \tag{102}$$

Let P be the standard parabolic subgroup of G associated to Φ. Let $\mathfrak{p}$ be the Lie algebra of P, and $\mathfrak{n}$ be the Lie algebra of the unipotent radical N of P. Due to (100), we have

$$\mathfrak{g} = \theta(\mathfrak{p}) \oplus \mathfrak{n}.$$

Let $\pi_\mathfrak{n}$ denote the projection onto $\mathfrak{n}$ with $\ker(\pi_\mathfrak{n}) = \sigma(\mathfrak{p})$.

Suppose that the claim fails to hold. Then

$$\sup_{i\in\mathbb{N}} \| \operatorname{Ad} g_i \cdot X\| < \infty, \ \forall X \in \mathfrak{h}. \tag{103}$$

Hence by (102),

$$\lim_{i\to\infty} \pi_\mathfrak{n}(\operatorname{Ad} k_i \cdot X) = 0, \ \forall X \in \mathfrak{h}.$$

Therefore $kHk^{-1} \subset \theta(P)$. Since $\theta(H) = H$ and $\theta(k) = k$, we have that $kHk^{-1} \subset P \cap \theta(P)$. Hence due to (101),

$$\{a_i\}_{i\in\mathbb{N}} \subset Z_G(P \cap \theta(P)) \subset kZ_G(H)k^{-1}.$$

Since $\mathfrak{g} = \theta(\mathfrak{p}) + \mathfrak{n}$ and $k_ik^{-1} \to e$ as $i \to \infty$, by passing to subsequences, there exist sequences $b_i \to e$ in $\theta(P)$ and $n_i \to e$ in N such that

$$k_ik^{-1} = b_in_i, \ \forall i \in \mathbb{N}.$$

Let $\{X_1, \dots, X_m\}$ be a basis of $\mathfrak{h}$ and put $\mathbf{q} = (X_1, \dots, X_m) \in \oplus_{i=1}^m \mathfrak{g}$. Consider the action of G on $\oplus_{i=1}^m \mathfrak{g}$ via the Adjoint action on each of the summands. Then

$$g_i \cdot \mathbf{q} = (g_ik^{-1})(k \cdot \mathbf{q}) = (a_ik_ik^{-1})(k \cdot \mathbf{q}) = (a_ib_ia_i^{-1})(a_in_i{a_i}^{-1})(k \cdot \mathbf{q})$$

By (103), $\{g_i \cdot \mathbf{q}\}_{i\in\mathbb{N}}$ is a bounded sequence. By (100) and (102), $a_ib_i{a_i}^{-1} \to e$ as $i \to \infty$. Therefore $(a_in_i{a_i}^{-1})(k \cdot \mathbf{q}) : i \in \mathbb{N}\}$ is a bounded sequence. Since N is a unipotent group, the orbit $N(k \cdot \mathbf{q})$ is closed. Therefore there exists a compact set $C_1 \subset N$ such that

$${a_i}^{-1}n_ia_i \in C_1(kZ_G(H)k^{-1} \cap N).$$

Therefore, since $\{a_i\} \subset kZ_G(H)k^{-1}$ and $a_ib_i{a_i}^{-1} \to e$ as $i \to \infty$, there exists a compact set $C \subset G$, such that

$$g_ik^{-1} = a_ik_ik^{-1} = (a_ib_i{a_i}^{-1})(a_in_i{a_i}^{-1}) \in CZ_G(H)k^{-1}, \ \forall i \in \mathbb{N}.$$

This contradicts the hypothesis of the claim, and hence the proof of Claim 1 is complete.

Now we can assume that there exists a sequence $X_i \to 0$ in $\mathfrak{h}$ and a nonzero elements $Y \in \mathfrak{g}$ such that

$$\lim_{i\to\infty} \operatorname{Ad} g_i X_i = Y.$$

Consider the one-parameter subgroup $u : \mathbb{R} \to G$ defined as $u(t) = \exp(tY)$ for all $t \in \mathbb{R}$. Since $X_i \to 0$, all the eigenvalues of $\operatorname{Ad} tX_i$ converge to 1 as $i \to \infty$. Since $u(t) = \lim_{i\to\infty} g_i{}^{-1}(\exp tX_i)g_i$ and the eigenvalues are invariant under conjugation, we have that 1 is the only eigenvalue of $\operatorname{Ad} u(t)$ for all $t \in \mathbb{R}$. Therefore u is a unipotent one-parameter subgroup of G.

Claim 2. *The measure μ is invariant under the action of $\{u(t) : t \in \mathbb{R}\}$.*

To prove the claim let $t \in \mathbb{R}$ and put $\delta = \exp(tX_i)$ for all $i \in \mathbb{N}$. Then by the definition of μ_Ω, for any $\psi \in C_c(G/\Gamma)$,

$$\left|\int_{G/\Gamma} \psi(x)\, d\mu_\Omega(x) - \int_{G/\Gamma} \psi(\delta_i x)\, d\mu_\Omega(x)\right| \le \epsilon_i \cdot \sup|\psi|, \tag{104}$$

where ϵ_i depends only on δ_i, and $\epsilon_i \to 0$ as $\delta_i \to 0$. Let $i \in \mathbb{N}$. Applying Eq. 104 for $\psi_i(x) := \psi(g_i x)$ for all $x \in X$, we get

$$\left|\int_{G/\Gamma} \psi(g_i x)\, d\mu_\Omega(x) - \int_{G/\Gamma} \psi((g_i\delta_i g_i{}^{-1})g_i x)\, d\mu_\Omega(x)\right| \le \epsilon_i \cdot \sup|\psi|.$$

We have $g_i \cdot \mu_\Omega \to \mu$ weakly as $i \to \infty$, $g_i{}^{-1}\delta_i g_i \to u(t)$ as $i \to \infty$, and ψ is uniformly continuous. Therefore

$$\int_{G/\Gamma} \psi(x)\, d\mu(x) = \int_{G/\Gamma} \psi(xu(g))\, d\mu(x).$$

This shows that μ is invariant under $\{u(t) : t \in \mathbb{R}\}$. This completes the proof of the theorem. □

8.5. Proving Ergodicity. In view of Proposition 8.20 and the measure classification theorem, Theorem 8.8 would follow immediately if we knew that μ was ergodic. In general the ergodicity of μ does not follow from Theorem 4.5 since we are not assuming that H contains unipotents.

The next part of the proof of Theorem 8.8 parallels §4.3. One applies the measure classification theorem followed by linearization. The analysis is somewhat more complicated then that of §4.3 because of the multi-dimensional situation, and the fact that we have a map only from a compact subset of H. The end result is:

PROPOSITION 8.21. *Let $B \subset H$ be a ball of diameter at most δ_0 in H around e. Let g_i be a sequence of elements in G, and let λ_i be the probability measure on $\pi(g_i(B))$ which is the pushforward under g_i of the normalized Lebesgue measure on B. Suppose that $\lambda_i \to \lambda$ weakly in the space of probability measures on G/Γ. Suppose there exist a unipotent one-parameter subgroup U of G and $F \in \mathcal{H}$ such that $\lambda(\pi(N(F,U))) > 0$ and $\lambda(\pi(S(F,U))) = 0$. Then there exists a compact set $D \subset \mathcal{A}_F$ such that the following holds: For any sequence of neighborhoods $\{\Phi_i\}$ of D in $\bar{V}_F$, there exists a sequence $\{\gamma_i\} \subset \Gamma$ such that for all large $i \in \mathbb{N}$,*

$$g_i(B)\gamma_i \cdot \bar{p}_F \subset \Phi_i. \tag{105}$$

In general the condition (105) is difficult to analyze using linear algebra methods. The idea of the proof of Theorem 8.8 is the following: Since we are assuming that $g_i B$ return to a compact set in G/Γ, we may write $g_i = c_i \gamma_i' h_i$, where c_i is in a compact set, $\gamma_i' \in \Gamma$ and $h_i \in B \subset H$. Without loss of generality, we may then replace g_i by $\gamma_i' h_i$. Consider rational points h_j in BB. The orbit of each rational point under Γ is discrete, so there are only finitely many possibilities for $\gamma_i' h_j \gamma_i \cdot \bar{p}_F$. By passing to a subsequence one can assume that $\gamma_i' h_j \gamma_i \cdot \bar{p}_F$ is constant, which eventually yields the proof of Theorem 8.8.

References

[BM] M. Bekka and M. Mayer, **Ergodic theory and topological dynamics of group actions on homogeneous spaces**, London Math. Soc. Lecture Note Series, Vol. 269, Cambridge University Press, Cambridge, 2000.

[Bir] B.J. Birch, Forms in many variables, *Proc. Roy. Soc. London* **265** (1962), 245–263.

[BO] S. Bloch and A. Okounkov, The Character of the Infinite Wedge Representation, *Adv. Math.* **149** (2000), no. 1, 1–60

[Bor] A. Borel, Values of indefinite quadratic form at integral points and flows on spaces of lattices, *Bull. AMS. (N.S.)* **32** (1995), 184–204.

[BH-C] A. Borel and Harish-Chandra, Arithmetic subgroups of algebraic groups, *Annals of Math* **75** (1962), 485–535.

[BR] M. Borovoi and Z. Rudnick, Hardy–Littlewood varieties and semisimple groups, *Inv. Math.* (1994).

[Bre] T. Brennan, Princeton University undergraduate thesis, 1994.

[Bu] M. Burger, Horocycle flow on geometrically finite surfaces. *Duke Math. J.* **61** (1990), no. 3, 779–803.

[Cas] J.W.S. Cassels, **An introduction to the geometry of numbers**, Springer, 1996.

[CW] K. Calta and K. Wortman, On unipotent flows in $H(1,1)$, arXiv:math/0702238.

[Dan1] S.G. Dani, On invariant measures, minimal sets and a lemma of Margulis, *Invent. Math.* **51** (1979), 239–260.

[Dan2] S.G. Dani, Invariant measures and minimal sets of horoshperical flows, *Invent. Math.* **64** (1981), 357–385.

[Dan3] S.G. Dani, On orbits of unipotent flows on homogeneous spaces, *Ergod. Theor. Dynam. Syst.* **4** (1984), 25–34.

[Dan4] S.G. Dani, On orbits of unipotent flows on homogenous spaces II, *Ergod. Theor. Dynam. Syst.* **6** (1986), 167–182.

[DM1] S.G. Dani and G.A. Margulis, Values of quadratic forms at primitive integral points, *Invent. Math.* **98** (1989), 405–424.

[DM2] S.G. Dani and G.A. Margulis, Orbit closures of generic unipotent flows on homogeneous spaces of $SL(3,\mathbb{R})$, *Math. Ann.* **286** (1990), 101–128.

[DM3] S.G. Dani and G.A. Margulis. Asymptotic behaviour of trajectories of unipotent flows on homogeneous spaces, *Indian. Acad. Sci. J.* **101** (1991), 1–17.

[DM4] S.G. Dani and G.A. Margulis, *Limit distributions of orbits of unipotent flows and values of quadratic forms*, in: **I. M. Gelfand Seminar**, Amer. Math. Soc., Providence, RI, 1993, pp. 91–137.

[DRS] W. Duke, Z. Rudnick and P. Sarnak, Density of integer points on affine homogeneous varieties, *Duke Math. J.* **71** (1993), 181–209.

[E1] A. Eskin, *Counting problems and semisimple groups*, in: **Proceedings of the International Congress of Mathematicians**, Vol. II (Berlin, 1998), Doc. Math., 1998, pp. 539–552.

[E2] A. Eskin, Handbook of Dynamical Systems.

[EL] M. Einsiedler, E. Lindenstrauss, Diagonizable Actions and Arothmetic Applications. Lecture notes this volume.

[EM] A. Eskin and H. Masur, Asymptotic formulas on flat surfaces, *Ergodic Th. Dynam. Syst.*, **21** (2001), 443-478.

[EMaMo] A. Eskin, J. Marklof and D. Morris, Unipotent flows on the space of branched covers of Veech surfaces, *Ergodic Theory Dynam. Systems* **26** (2006), no. 1, 129–162.

[EMM1] A. Eskin, G.A. Margulis and S. Mozes,Upper bounds and asymptotics in a quantitative version of the Oppenheim conjecture, *Ann. Math.* (2), **147** (1998), 93–141.

[EMM2] A. Eskin, G.A. Margulis and S. Mozes, Quadratic forms of signature $(2,2)$ and eigenvalue spacings on rectangular 2-tori, *Ann. Math.* (2) **161** (2005), no. 2, 679–725.

[EMS] A. Eskin, H. Masur and M. Schmoll, Billiards in rectangles with barriers. Duke Math. J. 118 (2003), no. 3, 427–463.

[EMZ] A. Eskin, H. Masur and A. Zorich, Moduli Spaces of Abelian Differentials: The Principal Boundary, Counting Problems and the Siegel–Veech Constants, *Publ. Math. Inst. Hautes Etudes Sci.* **97** (2003), 61–179.

[EMc] A. Eskin and C. McMullen, Mixing, counting, and equidistribution in Lie groups, *Duke Math. J.* **71** (1993), 181–209.

[EMS1] A. Eskin, S. Mozes and N. Shah, Unipotent flows and counting lattice points on homogeneous varieties, *Ann. Math.* **143** (1996), 253–299.

[EMS2] A. Eskin, S. Mozes, and N. Shah, Nondivergence of translates of certain algebraic measures, *Geom. Funct. Anal.* **7** (1997), no. 1, 48–80.

[EO] A. Eskin and A. Okounkov, Asymptotics of numers of branched covers of a torus and volumes of moduli spaces of holomorphic differentials, *Invent. Math.* **145** (2001), 59–104.

[F] H. Furstenberg, *The unique ergodicity of the horocycle flow*, in: **Recent advances in topological dynamics** (Proc. Conf., Yale Univ., New Haven, Conn., 1972), pp. 95–115, Springer, Berlin, 1973.

[Ko] M. Kontsevich, *Lyapunov Exponents and Hodge Theory*, in: **The mathematical beauty of physics (Saclay, 1996)**, 318–332, Adv. Ser. Math. Phys., **24**, World Sci. Publishing, River Edge, NJ, 1997.

[Kl1] D. Kleinbock, *Quantitative Nondivergence and its Diophantine Applications.* lecture notes in this volume.

[Kl2] D. Kleinbock, *Some applications of homogeneous dynamics to number theory*, in: **Smooth ergodic theory and its applications**,Proc. Sympos. Pure Math., 69, Amer. Math. Soc., Providence, RI, 2001, pp. 639–660.

[KSS] D. Kleinbock, N. Shah and A. Starkov, *Dynamics of subgroup actions on homogeneous spaces of Lie groups and applications to number theory*, in: **Handbook on Dynamical Systems**, Volume 1A, Elsevier Science, North Holland, 2002, pp. 813–930.

[Mar1] G.A. Margulis, *On the action of unipotent groups in the space of lattices*, In **Lie Groups and their representations**, Proc. of Summer School in Group Representations, Bolyai Janos Math. Soc., Akademai Kiado, Budapest, 1971, p. 365–370, Halsted, New York, 1975.

[Mar2] G.A. Margulis, Formes quadratiques ind'efinies et flots unipotents sur les spaces homog'enes, *C. R. Acad. Sci. Paris Ser. I* **304** (1987), 247–253.

[Mar3] G.A. Margulis, em Discrete Subgroups and Ergodic Theory, in: **Number theory, trace formulas and discrete subgroups**, a symposium in honor of A Selberg, pp. 377–398. Academic Press, Boston, MA, 1989.

[Mar4] G.A.Margulis, *Indefinite quadratic forms and unipotent flows on homogeneous spaces*, In **Dynamical systems and ergodic theory**, Vol. 23, pp. 399–409, Banach Center Publ., PWN – Polish Scientific Publ., Warsaw, 1989.

[Mar5] G.A. Margulis. *Dynamical and ergodic properties of subgroup actions on homogeneous spaces with applications to number theory*, in: **Proc. of ICM** (Kyoto, 1990), pp. 193–215, Math. Soc. of Japan and Springer, 1991.

[Mar6] G.A. Margulis, *Oppenheim conjecture*, in: **Fields Medalists' lectures**, WorldSci. Publishing, River Edge, NJ, 1997, pp. 272–327.

[Mar7] G.A. Margulis, *Diophantine approximation, lattices and flows on homogeneous spaces*, in: **A panorama of number theory or the view from Baker's garden**, Cambridge Univ. Press, Cambridge, 2002, pp. 280–310.

[MT] G.A. Margulis and G. Tomanov, Invariant measures for actions of unipotent groups over local fields on homogeneous spaces. *Invent. Math.* **116** (1994), 347–392.

[Mas1] H. Masur, *Ergodic theory of translation surfaces*, Handbook of Dynamical Systems.

[Mas2] H. Masur, The growth rate of trajectories of a quadratic differential, *Ergodic Th. Dynam. Syst.*, **10** (1990), 151-176.

[Mas3] H. Masur, *Lower bounds for the number of saddle connections and closed trajectories of a quadratic differential*, in: Holomorphic Functions and Moduli, Vol.1, D.Drasin ed. Springer-Verlag 1988, pp. 215-228

[Mc] C. McMullen, *Dynamics of $SL_2(\mathbb{R})$ actions in genus 2*, Preprint.

[Mor] D. Morris, **Ratner's theorems on unipotent flows**, Chicago Lectures in Mathematics. University of Chicago Press, Chicago, IL, 2005.

[Moz] S. Mozes, Mixing of all orders of Lie group actions, *Invent. Math.* **107** (1992), 235–241.

[MS] S. Mozes, N. Shah, On the space of ergodic invariant measures of unipotent flows, *Ergodic Theory Dynam. Systems* **15** (1995), no. 1, 149–159.

[New] M. Newman, **Integral Matrices**, Academic Press, New York, 1972.

[Ne] A. Nevo, *Equidistribution in measure preserving actions of semisimple groups*, preprint.

[NS] A. Nevo and E. Stein, *A generalization of Wiener's pointwise ergodic theorem*, Preprint, FIXME

[Ra1] M. Ratner, Rigidity of horocycle flows. *Ann. Math.* **115** (1982), 597–614.

[Ra2] M. Ratner, Factors of horocycle flows, *Ergodic Theory Dynam. Systems* **2** (1982), 465–489.

[Ra3] M. Ratner, Horocycle flows, joinings and rigidity of products. *Ann. Math.* **118** (1983), 277–313.

[Ra4] M. Ratner, Strict measure rigidity for unipotent subgroups of solvable groups, *Invent. Math.* **101** (1990), 449–482.

[Ra5] M. Ratner, On measure rigidity of unipotent subgroups of semisimple groups, *Acta Math.* **165** (1990), 229–309.

[Ra6] M. Ratner, On Raghunathan's measure conjecture, *Ann. Math.* **134** (1991), 545–607.

[Ra7] M. Ratner, Raghunathan's topological conjecture and distributions of unipotent flows, *Duke Math. J.* **63** (1991), no. 1, 235–280.

[Ra8] M. Ratner, Raghunathan's conjectures for $\mathrm{SL}(2,\mathbf{R})$, *Israel J. Math.* **80** (1992), no. 1-2, 1–31.

[Sar] P. Sarnak, *Values at integers of binary quadratic forms*, to appear in a volume in memory of C. Herz, S. Drury, editor.

[Sch] W. Schmidt, Asymptotic formulae for point lattices of bounded determinant and subspaces of bounded height, *Duke Math. J.* **35** (1968), 327–339.

[Sie] C.L. Siegel, **Lectures on the geometry of numbers**, Springer, 1989.

[Sha1] N.A. Shah, Uniformly distributed orbits of certain flows on homogeneous spaces, *Math. Ann.* **289** (1991), 315–334.

[Sha2] N.A. Shah, Limit distributions of polynomial trajectories on homogeneous spaces, *Duke Math. J.* **75**(1994), 711–732.

[Sha3] N.A. Shah, PhD thesis, Tata Institute for Fundamental Research.

[Sta1] A.N. Starkov, Solvable homogeneous flows, *Mat. Sbornik* **176** (1987), 242–259, in Russian.

[Sta2] A.N. Starkov, The ergodic decomposition of flows on homogenous spaces of finite volume, *Math. Sbornik* **180** (1989), 1614–1633, in Russian.

[Sta3] A. Starkov, **Dynamical systems on homogeneous spaces**, Translations of Mathematical Monographs, Volume 190, Amer. Math. Soc., Providence, RI, 2000.

[Ter] A. Terras, **Harmonic analysis on Symmetric spaces and Applications II**, Springer, 1988.

[V1] J. Vanderkam, Values at integers of homogeneous polynomials, *Duke Math. J.* **97** (1999), no. 2, 379–412.

[V2] J. Vanderkam, Pair correlation of four-dimensional flat tori, *Duke Math. J.* **97** (1999), no. 2, 413–438.

[V3] J. Vanderkam, Correlations of eigenvalues on multi-dimensional flat tori, *Comm. Math. Phys.* **210** (2000), no. 1, 203–223.

[Ve] W. Veech, Siegel measures, *Ann. Math.* **148** (1998), 895–944

[Vo1] Y. Vorobets, Ergodicity of billiards in polygons, *Mat. Sb.* **188** (1997), 65–112.

[Vo2] Y. Vorobets, *Periodic geodesics on translation surfaces,* Eprint, arXiv:math.DS/0307249.

[Zim] R. J. Zimmer, **Ergodic theory and semisimple groups**, Birkhaüser, 1984.

[Zor] A. Zorich, *Flat surfaces*, in: **Frontiers in Number Theory, Physics and Geometry**, Vol. 1.

Alex Eskin, Department of Mathematics, University of Chicago, 5734 S. University Ave, Chicago, IL 60637
E-mail address: eskin@math.uchicago.edu

Clay Mathematics Proceedings
Volume 10, 2010

Quantitative nondivergence and its Diophantine applications

Dmitry Kleinbock

ABSTRACT. The main goal of these notes is to describe a proof of quantitative nondivergence estimates for quasi-polynomial trajectories on the space of lattices, and show how estimates of this kind are applied to some problems in metric Diophantine approximation.

1. Introduction

These lecture notes constitute part of a course taught together with Alex Eskin at the Clay Mathematics Institute Summer School at Centro de Giorgi, Pisa, in June 2007. The exposition below is a continuation of [**E**]; the reader is referred there, as well as to books [**BM, Mor, St**] and the article [**KSS**] from the Handbook of Dynamical Systems, for background information on homogeneous spaces and unipotent flows.

In what follows, most of the work will be done on the space $\mathcal{L}_n$ of unimodular lattices in $\mathbb{R}^n$. We recall that $G = \mathrm{SL}(n,\mathbb{R})$ acts transitively on $\mathcal{L}_n$ (if $g \in G$ and $\Lambda \in \mathcal{L}_n$ is the $\mathbb{Z}$-span of the vectors $\mathbf{v}_1, \dots \mathbf{v}_n$, then $g\Lambda$ is the $\mathbb{Z}$-span of $\{g\mathbf{v}_1, \dots, g\mathbf{v}_n\}$), and the stabilizer of the standard lattice $\mathbb{Z}^n$ is $\Gamma = \mathrm{SL}(n,\mathbb{Z})$. This gives an identification of $\mathcal{L}_n$ with G/Γ. We choose a right-invariant metric on G; then this metric descends to G/Γ. Equivalently, one can define topology on $\mathcal{L}_n$ by saying that two lattices are close to each other if so are their generating sets.

For $\varepsilon > 0$ we will denote by $\mathcal{L}_n(\varepsilon) \subset \mathcal{L}_n$ the set of lattices whose shortest non-zero vector has norm at least ε. It is clear from the above description of the topology on $\mathcal{L}_n$ that any compact subset of $\mathcal{L}_n$ is contained in $\mathcal{L}_n(\varepsilon)$ for some positive ε. Conversely, one has

THEOREM 1.1 (Mahler Compactness Criterion). *For any $\varepsilon > 0$ the set $\mathcal{L}_n(\varepsilon)$ is compact.*

See [**Cas**] or [**BM**] for a proof. We note that the set $\mathcal{L}_n(\varepsilon)$ depends on the choice of the norm on $\mathbb{R}^n$, but in a rather mild way: change of one norm for another would result in multiplication/division of ε by at most a fixed positive constant.

Recall that an element g of G is unipotent if all its eigenvalues are equal to 1. If $n = 2$, every one-parameter unipotent subgroup of $G = \mathrm{SL}(2,\mathbb{R})$ is conjugate to

$$U = \{u_x : x \in \mathbb{R}\} \text{ where } u_x = \begin{pmatrix} 1 & x \\ 0 & 1 \end{pmatrix}. \tag{1.1}$$

In general, a crucial property of an arbitrary unipotent subgroup $\{u_x\}$ of $\mathrm{SL}(n,\mathbb{R})$ is that the map $x \mapsto u_x$ is polynomial of degree depending only on n. This observation was instrumental in the proof due to Margulis that one-parameter unipotent trajectories on $\mathcal{L}_n$ are never divergent. Namely the following theorem was conjectured by Piatetski-Shapiro in the late 1960s and showed in 1971 by Margulis [**Mar**] as part of the program aimed at proving arithmeticity of lattices in higher rank algebraic groups:

THEOREM 1.2. *Let $\{u_x\}$ be a one-parameter unipotent subgroup of* $\mathrm{SL}(n,\mathbb{R})$. *Then for any $\Lambda \in \mathcal{L}_n$, $u_x\Lambda$ does not tend to ∞ as $x \to \infty$. Equivalently, there exists $\varepsilon > 0$ such that the set $\{x \in \mathbb{R}_+ : u_x\Lambda \in \mathcal{L}_n(\varepsilon)\}$ is unbounded.*

In fact, ε in the above theorem can be chosen independent on the choice of $\{u_x\}$, although it does depend on Λ, see §3 for more detail. The above statement is very easy to prove when $n = 2$, but much more difficult for bigger n. In this exposition we first discuss the easy special case, then the general strategy of Margulis in various modifications, and then some applications and further extensions of the general result.

Acknowledgements: The author is grateful to the Clay Mathematics Institute for a wonderful opportunity to participate in the 2007 Summer School, to the co-organizers of the event for their help and encouragement, and to the staff at Centro di Georgi for being so very helpful and attentive. Special thanks are due to Alex Eskin, the co-lecturer of this course, to Elon Lindenstrauss for careful reading of a preliminary version of these notes, and to many participants of the summer school for their patience and valuable comments during and after the lectures. The work on the manuscript was partially supported by NSF grants DMS-0239463 and DMS-0801064.

2. Non-divergence of unipotent flows: the case of $\mathrm{SL}(2,\mathbb{R})$.

2.1. Geometry of lattices in $\mathbb{R}^2$. Recall the following lemma from [**E**]:

LEMMA 2.1. *There exists $\varepsilon_0 > 0$ (depending on the choice of the norm on $\mathbb{R}^2$) such that no $\Lambda \in \mathcal{L}_2$ contains two linearly independent vectors each of norm less than ε_0.*

Let us now use this lemma to prove a nondivergence result for the U-action on $\mathcal{L}_2$, where U is as in (1.1):

PROPOSITION 2.2. *For any $\Lambda \in \mathcal{L}_2$, $u_x\Lambda$ does not tend to ∞ as $x \to \infty$.*

In other words, for any $\Lambda \in \mathcal{L}_2$ there exists a compact subset K of $\mathcal{L}_2$ such that the set $\{x > 0 : u_x\Lambda \in K\}$ is unbounded.

PROOF. Assume the contrary; in view of Theorem 1.1, this would amount to assuming that the norm of the shortest nonzero vector of $u_x\Lambda$ tends to zero as $x \to \infty$. Note that an obvious example of a divergent orbit would be constructed if one could find a vector $\mathbf{v} \in \Lambda \smallsetminus \{0\}$ such that $u_x\mathbf{v} \to 0$. But this is impossible:

either $\mathbf{v}$ is horizontal and thus fixed by U, or its y-component is nonzero and does not change under the action. Thus the only allowed scenario for a divergent U-trajectory would be the following: for some $\mathbf{v} \in \Lambda \smallsetminus \{0\}$, $u_x\mathbf{v}$ gets very small, say shorter than ε, then starts growing but before it grows too big (longer than ε), another vector in $\Lambda \smallsetminus \{0\}$ not proportional to $\mathbf{v}$ gets shrunk by u_x to the length less than ε. This however is prohibited by Lemma 2.1. □

REMARK 2.3. Note that the analogue of this proposition is false if U is replaced by

$$A = \{a_t : t \in \mathbb{R}\} \text{ where } a_t = \begin{pmatrix} e^t & 0 \\ 0 & e^{-t} \end{pmatrix}, \tag{2.1}$$

since a_t can contract nonzero vectors. However the same argument as above shows that for any continuous function $h : \mathbb{R}_+ \to \mathrm{SL}(2, \mathbb{R})$ and any $\Lambda \in \mathcal{L}_2$ such that $h(x)\Lambda$ diverges, it must do so in a degenerate way (a terminology suggested by Dani, see [**D3**]), that is, shrinking some nonzero vector $\mathbf{v} \in \Lambda$. This phenomenon is specific to dimension 2: if $n > 2$, as shown in [**D3**], one can construct divergent trajectories $\{a_t\Lambda\} \subset \mathcal{L}_n$ of diagonal one-parameter semigroups $\{a_t\} \subset \mathrm{SL}(n, \mathbb{R})$ in $\mathcal{L}_n$ which diverge in a non-degenerate way (without shrinking any subpace of $\mathbb{R}^n$).

Despite the above remark, Theorem 1.2, which is an analogue of Proposition 2.2, holds for $n > 2$ as well. An attempt to replicate the proof of Proposition 2.2 verbatim fails miserably: there are no obstructions to having many short linear independent vectors. We will prove Theorem 1.2 in the next section in a much stronger (quantitative) form, which also happens to have important applications to problems arising in Diophantine approximation theory. But first, following the methodology of [**E**] where the exposition of Ratner's theorem begins with an extensive discussion of the U-action on $\mathcal{L}_2$, we explain how one can easily establish a stronger form of Proposition 2.2, just for $n = 2$.

2.2. Quantitative nondivergence in $\mathcal{L}_2$. We are going to fix an interval $B \subset \mathbb{R}$ and $\Lambda \in \mathcal{L}_2$, and will look at the piece of trajectory $\{u_x\Lambda : x \in B\}$. Applying the philosophy of the proof of Proposition 2.2, one can see that one of the following two alternatives can take place:

Case 1. There exists a vector $\mathbf{v} \in \Lambda \smallsetminus \{0\}$ such that $\|u_x\mathbf{v}\|$ is small, say less than ε_0, for all $x \in B$. (For example this $\mathbf{v}$ may be fixed by U.) This case is not so interesting: again by Lemma 2.1, we know that this vector $\mathbf{v}$ is "the only source of trouble", namely no other vector can get small at the same time.

Case 2. The contrary, i.e.

$$\forall \mathbf{v} \in \Lambda \smallsetminus \{0\} \quad \sup_{x \in B} \|u_x\mathbf{v}\| \geq \rho\,. \tag{2.2}$$

In other words, every nonzero vector grows big enough at least at some point $x \in B$. This assumption turns out to be enough to conclude that for small ε the trajectory $\{u_x\Lambda : x \in B\}$ spends relatively small proportion of time, in terms of Lebesgue measure λ on $\mathbb{R}$, outside of $\mathcal{L}_2(\varepsilon)$.

THEOREM 2.4. *Suppose an interval $B \subset \mathbb{R}$, $\Lambda \in \mathcal{L}_2$ and $0 < \rho < \varepsilon_0$ are such that* (2.2) *holds. Then for any $\varepsilon > 0$,*

$$\lambda(\{x \in B : u_x\Lambda \notin \mathcal{L}_2(\varepsilon)\}) \leq 2\frac{\varepsilon}{\rho}\lambda(B)\,.$$

Thus, if one studies the curve $\{u_x\Lambda\}$ where x ranges from 0 to T, it suffices to look at the starting point Λ of the trajectory, find its shortest vector $\mathbf{v}$, choose $\rho < \min(\varepsilon_0, \|\mathbf{v}\|)$, and apply the theorem to get a quantitative statement concerning the behavior of $\{u_x\Lambda : 0 \le x \le T\}$ for any T. Note that it is meaningful, and requires proof, only when ε is small enough (not greater than $\rho/2$).

Proof. Denote by $P(\Lambda)$ the set of primitive vectors in Λ ($\mathbf{v}$ is said to be *primitive* in Λ if $\mathbb{R}\mathbf{v}\cap\Lambda$ is generated by $\mathbf{v}$ as a $\mathbb{Z}$-module). Clearly in all the argument it will suffice to work with primitive vectors.

Now for each $\mathbf{v} \in P(\Lambda)$ consider

$$B_{\mathbf{v}}(\varepsilon) \stackrel{\text{def}}{=} \{x \in B : \|u_x\mathbf{v}\| < \varepsilon\} \quad \text{and} \quad B_{\mathbf{v}}(\rho) \stackrel{\text{def}}{=} \{x \in B : \|u_x\mathbf{v}\| < \rho\},$$

where $\|\cdot\|$ is the supremum norm. Let $\mathbf{v} = \begin{pmatrix} a \\ b \end{pmatrix} \in P(\Lambda)$ be such that $B_{\mathbf{v}}(\varepsilon) \neq \varnothing$. Then, since $u_x\mathbf{v} = \begin{pmatrix} a+bx \\ b \end{pmatrix}$, it follows that $|b| < \varepsilon$, and (2.2) implies that b is nonzero. Therefore, if we denote $f(x) = a + bx$, we have

$$B_{\mathbf{v}}(\varepsilon) = \{x \in B : |f(x)| < \varepsilon\} \quad \text{and} \quad B_{\mathbf{v}}(\rho) = \{x \in B : |f(x)| < \rho\}.$$

Clearly the ratio of lengths of intervals $B_{\mathbf{v}}(\varepsilon)$ and $B_{\mathbf{v}}(\rho)$ is bounded from above by $2\varepsilon/\rho$ (by looking at the worst case when $B_{\mathbf{v}}(\varepsilon)$ is close to one of the endpoints of B). Lemma 2.1 guarantees that the sets $B_{\mathbf{v}}(\rho)$ are disjoint for different $\mathbf{v} \in P(\Lambda)$, and also that $u_x\Lambda \notin \mathcal{L}_2(\varepsilon)$ whenever $x \in B_{\mathbf{v}}(\rho) \smallsetminus B_{\mathbf{v}}(\varepsilon)$ for some $\mathbf{v} \in P(\Lambda)$. Thus we conclude that

$$\lambda(\{x \in B : u_x\Lambda \notin \mathcal{L}_2(\varepsilon)\}) \le \sum_{\mathbf{v}} \lambda\big(B_{\mathbf{v}}(\varepsilon)\big) \le 2\frac{\varepsilon}{\rho}\sum_{\mathbf{v}} \lambda\big(B_{\mathbf{v}}(\rho)\big) \le 2\frac{\varepsilon}{\rho}\lambda(B). \quad \square$$

REMARK 2.5. Before proceeding to the more general case, let us summarize the main features of the argument. Each primitive vector $\mathbf{v}$ came with a function, $x \mapsto \|u_x\mathbf{v}\|$, which

[2.5-i] allowed to compare measure of the subsets of B where this function is less than ε and ρ respectively, and

[2.5-ii] attained value at least ρ on B.

Let us say that a point $x \in B$ is *(ε/ρ)-protected* if $x \in \overline{B_{\mathbf{v}}(\rho)} \smallsetminus B_{\mathbf{v}}(\varepsilon)$ for some $\mathbf{v} \in P(\Lambda)$. [2.5-i] and [2.5-ii] imply that for each $\mathbf{v}$, the relative measure of protected points inside $B_v(\rho)$ is big. Then Lemma 2.1 shows that protected points are safe (no other vector can cause trouble), i.e. brings us to the realm of Case 1 when restricted to $B_{\mathbf{v}}(\rho)$.

In the analog of the argument for $n > 2$, properties [2.5-i] and [2.5-ii] of certain functions will play an important role. However it will be more difficult to protect points from small vectors, and the final step, that is, an application of Lemma 2.1, will be replaced by an inductive procedure, described in the next section.

3. Quantitative non-divergence in $\mathcal{L}_n$.

3.1. The main concepts needed for the proof. The crucial idea that serves as a substitute for the absence of Lemma 2.1 in dimensions 3 and up is an observation that whenever a lattice Λ in $\mathbb{R}^n$ contains two linearly independent short vectors, one can consider a subgroup of rank two generated by them, and this subgroup will be "small", which should eventually contribute to preventing other small vectors from showing up. (Here and hereafter by the *rank* $\text{rk}(\Delta)$ of a discrete subgroup Δ of $\mathbb{R}^n$ we mean its rank as a free $\mathbb{Z}$-module, or, equivalently, the dimension of the real vector space spanned by its elements.) Thus we are led to consider all subgroups of Λ, not just of rank one. In fact, similarly to the $n = 2$ case, it suffices to work with *primitive* subgroups. Namely, a subgroup Δ of Λ is called *primitive* in Λ if $\Delta = \mathbb{R}\Delta \cap \Lambda$; equivalently, if Δ admits a generating set which can be completed to a generating set of Λ. The inclusion relation makes the set $P(\Lambda)$ of all nonzero primitive subgroups of Λ a partially ordered set of length equal to $\text{rk}(\Lambda)$ (any two primitive subgroups properly included in one another must have different ranks). This partial order turns out to be instrumental in creating a substitute for Lemma 2.1.

We also need a way to measure the size of a discrete subgroup Δ of $\mathbb{R}^n$. The best solution seems to be to use Euclidean norm $\|\cdot\|$ and extend it by letting $\|\Delta\|$ to be the volume of the quotient space $\mathbb{R}\Delta/\Delta$. This is clearly consistent with the one-dimensional picture, since $\|\mathbb{Z}\mathbf{v}\| = \|\mathbf{v}\|$. This is also consistent with the induced Euclidean structure on the exterior algebra of $\mathbb{R}^n$: if Δ is generated by $\mathbf{v}_1, \dots, \mathbf{v}_k$, then $\|\Delta\| = \|\mathbf{v}_1 \wedge \cdots \wedge \mathbf{v}_k\|$.

Our goal is to understand the trajectories $u_x\Lambda$ as in Theorem 1.2. However, observe that the group structure of U was not used at all in the proof in the previous section. Thus we are going to consider "trajectories" of a more general type. Namely, we will work with continuous functions h from an interval $B \subset \mathbb{R}$ into $\text{SL}(n, \mathbb{R})$, and replace the map $x \mapsto u_x\Lambda$ with $x \mapsto h(x)\mathbb{Z}^n$ (then in the case of Theorem 1.2 we are going to have $h(x) = u_x g$ where $\Lambda = g\mathbb{Z}^n$).

Among the assumptions to be imposed on h, the central role is played by an analogue of [2.5-i]. This is taken care of by introducing a certain class of functions and then demanding that all functions of the form $x \mapsto \|h(x)\Delta\|$ where $\Delta \in P(\mathbb{Z}^n)$, belong to this class.

If C and α are positive numbers and B a subset of $\mathbb{R}$, let us say that a function $f : B \mapsto \mathbb{R}$ is *(C, α)-good on B* if for any open interval $J \subset B$ and any $\varepsilon > 0$ one has

$$\lambda\big(\{x \in J : |f(x)| < \varepsilon\}\big) \le C \left(\frac{\varepsilon}{\sup_{x \in J} |f(x)|}\right)^{\alpha} \lambda(J)\,. \tag{3.1}$$

Informally speaking, graphs of good functions are not allowed to spend a big proportion of "time" near the x-axis and then suddenly jump up. Several elementary facts about (C, α)-good functions are listed below:

LEMMA 3.1. (a) *f is (C, α)-good on B $\Leftrightarrow$ so is $|f|$ $\Rightarrow$ so is cf $\forall\, c \in \mathbb{R}$;*
(b) *f_i, $i = 1, \dots, k$, are (C, α)-good on B $\Rightarrow$ so is $\sup_i |f_i|$;*
(c) *If f is (C, α)-good on B and $c_1 \le \left|\frac{f(x)}{g(x)}\right| \le c_2$ for all $x \in B$, then g is $\big(C(c_2/c_1)^\alpha, \alpha\big)$-good on B;*

The proofs are left as exercises. Another exercise is to construct a C^∞ function which is not good on (a) some interval (b) any interval.

The notion of (C, α)-good functions was introduced in [**KM1**] in 1998, but the importance of (3.1) for measure estimates on the space of lattices was observed earlier. For instance, the next proposition, which describes what can be called a model example of good functions, can be traced to [**DM2**, Lemma 4.1]. We will prove a slightly stronger version paying more atention to the constant C (which will not really matter for the main results).

PROPOSITION 3.2. *For any $k \in \mathbb{N}$, any polynomial of degree not greater than k is $\big(k(k+1)^{1/k}, 1/k\big)$-good on $\mathbb{R}$.*

PROOF. Fix an open interval $J \subset \mathbb{R}$, a polynomial f of degree not exceeding k, and a positive ε. We need to show that

$$\lambda\big(\{x \in J : |f(x)| < \varepsilon\}\big) \le k(k+1)^{1/k} \left(\frac{\varepsilon}{\sup_{x\in J}|f(x)|}\right)^{1/k} \lambda(J)\,. \tag{3.2}$$

Suppose that the left hand side of (3.2) is strictly bigger than some number m. Then it is possible to choose $x_1, \dots, x_{k+1} \in \{x \in J : |f(x)| < \varepsilon\}$ with $|x_i - x_j| \ge m/k$ for each $1 \le i \ne j \le k+1$. (Exercise.) Using Lagrange's interpolation formula one can write down the exact expression for f:

$$f(x) = \sum_{i=1}^{k+1} f(x_i)\frac{\prod_{j=1,\, j\ne i}^{k+1}(x - x_j)}{\prod_{j=1,\, j\ne i}^{k+1}(x_i - x_j)}\,. \tag{3.3}$$

Note that $|f(x_i)| < \varepsilon$ for each i, $|x - x_j| < \lambda(J)$ for each j and $x \in J$, and also $|x_i - x_j| \ge m/k$. Therefore

$$\sup_{x\in J}|f(x)| < (k+1)\varepsilon\frac{\lambda(J)^k}{(m/k)^k}\,.$$

which can be rewritten as

$$m < k(k+1)^{1/k}\left(\frac{\varepsilon}{\sup_{x\in J}|f(x)|}\right)^{1/k}\lambda(J)\,,$$

proving (3.2). □

Observe that in the course of the proof of Theorem 2.4 it was basically shown that linear functions are $(2,1)$-good on $\mathbb{R}$. The relevance of the above proposition for the nondivergence of unipotent flows on $\mathcal{L}_n$ is highlighted by

COROLLARY 3.3. *For any $n \in \mathbb{N}$ there exist (explicitly computable) $C = C(n)$, $\alpha = \alpha(n)$ such that for any one-parameter unipotent subgroup $\{u_x\}$ of $\mathrm{SL}(n,\mathbb{R})$, any $\Lambda \in \mathcal{L}_n$ and any subgroup Δ of Λ, the function $x \mapsto \|u_x\Delta\|$ is (C,α)-good.*

PROOF. Represent Δ by a vector $w \in \bigwedge^k(\mathbb{R}^n)$ where k is the rank of Δ; the action of u_x on $\bigwedge^k(\mathbb{R}^n)$ is also unipotent, therefore every component of $u_x w$ (with respect to some basis) is a polynomial in x of degree uniformly bounded in terms of n. Thus the claim follows from Proposition 3.2, Lemma 3.1(b) for the supremum norm, and then Lemma 3.1(c) for the Euclidean norm. □

3.2. The main nondivergence result and its history. Let us now state a generalization of Theorem 2.4 to the case of arbitrary n.

THEOREM 3.4. *Suppose an interval $B \subset \mathbb{R}$, $C, \alpha > 0$, $0 < \rho < 1$ and a continuous map $h : B \to \mathrm{SL}(n,\mathbb{R})$ are given. Assume that for any $\Delta \in P(\mathbb{Z}^n)$,*

[3.4-i] *the function $x \mapsto \|h(x)\Delta\|$ is (C,α)-good on B, and*
[3.4-ii] $\sup_{x\in B}\|h(x)\Delta\| \ge \rho^{\mathrm{rk}(\Delta)}$.

Then for any $\varepsilon < \rho$,

$$\lambda(\{x \in B : h(x)\mathbb{Z}^n \notin \mathcal{L}_n(\varepsilon)\}) \leq n2^n C \left(\frac{\varepsilon}{\rho}\right)^\alpha \lambda(B)\,. \tag{3.4}$$

This is a simplified version of a theorem from [**Kl5**], which sharpens the one proved in [**KM1**]. The latter had a slightly stronger assumptions, with ρ in place of $\rho^{\mathrm{rk}(\Delta)}$ in [3.4-ii]. In most of the applications this improvement is not needed – but there are some situations in metric Diophantine approximation, described later in the notes, where it becomes important. Anyway, the scheme of the proof, see §3.3, is the same for both original and new versions, and also there are some reasons why the sharpening appears to be more natural, as will be seen below. See [**KLW**] for another exposition of the proof.

It is straightforward to verify that Theorem 1.2 follows from Theorem 3.4: take $B = [0, T]$ and $h(x) = u_x g$ where $\Lambda = g\mathbb{Z}^n$. Condition [3.4-i] has already been established in Corollary 3.3, and [3.4-ii] clearly holds with some ρ dependent of Λ: just put $x = 0$ and

$$\rho = \rho(\Lambda) = \inf_{\Delta \in P(\Lambda)} \|\Delta\|^{1/\operatorname{rk}(\Delta)}\,, \tag{3.5}$$

positive since Λ is discrete. Furthermore, Theorem 3.4 implies the following

COROLLARY 3.5. *For any* $\Lambda \in \mathcal{L}_n$ *and any positive* δ *there exists a compact subset* K *of* $\mathcal{L}_n$ *such that for any unipotent one-parameter* $\{u_x\} \subset \mathrm{SL}(n, \mathbb{R})$ *and any positive* T *one has*

$$\frac{1}{T}\lambda(\{0 \leq x \leq T : u_x\Lambda \notin K\}) \leq \delta\,. \tag{3.6}$$

This was proved by Dani in 1979 [**D1**]. For the proof using Theorem 3.4, just take $K = \mathcal{L}_n(\varepsilon)$ where ε is such that

$$n2^n C(n)\,(\varepsilon/\rho)^{\alpha(n)} < \delta\,, \tag{3.7}$$

$C(n), \alpha(n)$ are as in Corollary 3.3 and $\rho(\Lambda)$ as defined in (3.5). Thus, on top of Dani's result, one can recover an expression for the "size" of K in terms of δ.

But this is not the end of the story – one can conclude much more. It immediately follows from Minkowski's Lemma that if $\mathrm{rk}(\Delta)$ is, say, k, then the intersection of Δ with any compact convex subset of $\mathbb{R}\Delta$ of volume $2^k\|\Delta\|$ contains a nonzero vector. Thus such a Δ must contain a nonzero vector of length $\leq 2\|\Delta\|/\nu_k^{1/k}$, where ν_k is the volume of the unit ball in $\mathbb{R}^k$. Consequently, if we know that $\Lambda \in \mathcal{L}_n(\rho')$ for some positive ρ', then $\rho(\Lambda)$ as defined in (3.5) is at least $c'\rho'$ where $c' = c'(n)$ depends only on n. Thus we have derived (modulo elementary computations left as an exercise) the following statement:

COROLLARY 3.6. *For any* $\delta > 0$ *there exists (explicitly computable)* $c = c(n, \delta)$ *such that whenever* $\{u_x\Lambda : 0 \leq x \leq T\} \subset \mathcal{L}_n$ *is a unipotent trajectory nontrivially intersecting* $\mathcal{L}_n(\rho)$ *for some* $\rho > 0$, (3.6) *holds with* $K = \mathcal{L}_n(c\rho)$.

In order to appreciate a geometric meaning of the above corollary and other related results, it will be convenient to choose a right-invariant Riemannian metric on $\mathrm{SL}_n(\mathbb{R})$ and use it to induce a Riemannian metric on $\mathcal{L}_n$. Then it is not hard to see that the distance between $\mathcal{L}_n(\rho)$ and the complement of $\mathcal{L}_n(c\rho)$ is uniformly bounded from above by a constant depending only on c, not on ρ. Thus Corollary

3.6 guarantees that, regardless of the size of the compact set where a unipotent trajectory begins, one only needs to increase the set by a bounded distance to make sure that the trajectory spends, say, at least half the time in the bigger set. Note that for the last conclusion it is important to have $\rho^{\mathrm{rk}(\Delta)}$ and not ρ in the right hand side of [3.4-ii]; previously available non-divergence estimates forced a much more significant expansion of $\mathcal{L}_n(\rho)$.

Let us now turn our attention to another non-divergence theorem, proved by Dani in 1986 [**D4**], and later generalized by Eskin, Mozes and Shah [**EMS**]:

COROLLARY 3.7. *For any $\delta > 0$ there exists a compact subset $K \subset \mathcal{L}_n$ such that for any unipotent one-parameter subgroup $\{u_x\} \subset \mathrm{SL}(n,\mathbb{R})$ and any $\Lambda = g\mathbb{Z}^n \in \mathcal{L}_n$, either* (3.6) *holds for all large T, or there exists a $(g^{-1}u_x g)$-invariant proper subspace of $\mathbb{R}^n$ defined over $\mathbb{Q}$.*

PROOF. Apply Theorem 3.4 with an arbitrary $\rho < 1$ and ε as in (3.7), as before choosing K to be equal to $\mathcal{L}_n(\varepsilon)$. Assume that the first alternative in the statement of the corollary is not satisfied for some $\{u_x\}$, Λ and this K. This means that there exists an unbounded sequence T_k such that for each k, the conclusion of Theorem 3.4 with $\rho = 1$, ε chosen as above and $h(x) = u_x g$, does not hold for $B = [0, T_k]$. Since assumption [3.4-i] is always true, [3.4-ii] must go wrong, i.e. for each k there must exist $\Delta_k \in P(\mathbb{Z}^n)$ such that $\|u_x g\Delta_k\| < 1$ for all $0 \le x \le T_k$. However, by the discreteness of $\bigwedge(g\mathbb{Z}^n)$ in $\bigwedge(\mathbb{R}^n)$, there are only finitely many choices for such subgroups; hence one of them, Δ, works for infinitely many k. But $\|u_x g\Delta\|^2$ is a polynomial, therefore it must be constant, which implies that u_x fixes $g(\mathbb{R}\Delta) \Leftrightarrow g^{-1}u_x g$ fixes the proper rational subspace $\mathbb{R}\Delta$. □

3.3. The proof. In order to prove Theorem 3.4, we are going to create a substitute for the procedure of marking points by vectors (and thereby declaring them safe from any other small vectors) used in the proof of Theorem 2.4. However now vectors will not be sufficient for our purposes, we will need to replace it with *flags*, that is, linearly ordered subsets of the partially ordered set (poset) $P(\Lambda)$, $\Lambda \in \mathcal{L}_n$. Furthermore, to set up the induction we will need to prove a version of the theorem with $P(\mathbb{Z}^n)$ repalced by its subsets (more precisely, sub-posets) P. The induction will be on the *length* of P, i.e. the number of elements in its maximal flag. In this more general theorem we will also get rid of the expressions $\rho^{\mathrm{rk}(\Delta)}$ in the right hand side of [3.4-ii], replacing them with $\eta(\Delta)$, where η is an arbitrary function $P \to (0,1]$ (to be called the *weight function*).

Now let us fix an interval $B \subset \mathbb{R}$, a sub-poset $P \subset P(\mathbb{Z}^n)$, a weight function η and a map $h : B \to \mathrm{SL}(n,\mathbb{R})$. Then say that, given $\varepsilon > 0$, a point $x \in B$ is *ε-protected relative to P* if there exists a flag $F \subset P$ with the following properties:

(M1) $\varepsilon\eta(\Delta) \le \|h(x)\Delta\| \le \eta(\Delta) \quad \forall\, \Delta \in F$;
(M2) $\|h(x)\Delta\| \ge \eta(\Delta) \quad \forall\, \Delta \in P \setminus F$ comparable with every element of F.

We are going to show that with the choice $\eta(\Delta) = \rho^{\mathrm{rk}(\Delta)}$ and $P = P(\mathbb{Z}^n)$, any (ε/ρ)-protected point $x \in B$ is indeed protected from vectors in $h(x)\mathbb{Z}^n$ of length less than ε. But first let us check that the above definition reduces to the one used for the proof of Theorem 2.4 when $P = P(\mathbb{Z}^2)$. Indeed, for $h(x) = u_x g$, $\Delta = \mathbb{Z}\mathbf{v}$ of rank 1, $\eta(\Delta) = \rho$ and ε substituted with ε/ρ, (M1) reduces to $\varepsilon \le \|u_x g\mathbf{v}\| \le \rho$, which was exactly the condition satisfied by some vector $\mathbf{v} \in \mathbb{Z}^2$ for $x \in \overline{B_{g\mathbf{v}}(\rho)} \setminus B_{g\mathbf{v}}(\varepsilon)$. Further, (M2) in that case holds trivially, since the only element of $P(\mathbb{Z}^2) \setminus \{\Delta\}$

comparable with Δ is $\mathbb{Z}^2$ itself, and $\|g\mathbb{Z}^2\| = 1 > \rho^2$. And the conclusion was that the existence of such $\mathbf{v}$ forces $u_x g\mathbb{Z}^2$ to belong to $\mathcal{L}_2(\varepsilon)$.

Here is a generalization:

PROPOSITION 3.8. *Let η be given by $\eta(\Delta) = \rho^{\mathrm{rk}(\Delta)}$ for some $0 < \rho < 1$. Then for any $\varepsilon < \rho$ and any $x \in B$ which is (ε/ρ)-protected relative to $P(\mathbb{Z}^n)$, one has $h(x)\mathbb{Z}^n \in \mathcal{L}_n(\varepsilon)$.*

PROOF. For x as above, let $\{0\} = \Delta_0 \subsetneq \Delta_1 \subsetneq \cdots \subsetneq \Delta_\ell = \mathbb{Z}^n$ be all the elements of $F \cup \{\{0\}, \mathbb{Z}^n\}$. Properties (M1) and (M2) translate into:

(M1) $\frac{\varepsilon}{\rho} \cdot \rho^{\mathrm{rk}(\Delta_i)} \le \|h(x)\Delta_i\| \le \rho^{\mathrm{rk}(\Delta_i)} \quad \forall\, i = 0, \dots, \ell - 1$;

(M2) $\|h(x)\Delta\| \ge \rho^{\mathrm{rk}(\Delta)} \quad \forall\, \Delta \in P(\mathbb{Z}^n) \smallsetminus F$ comparable with every Δ_i.

(Even though $\Delta_0 = \{0\}$ is not in $P(\mathbb{Z}^n)$, it would also satisfy (M1) with the convention $\|\{0\}\| = 1$.)

Take any $\mathbf{v} \in \mathbb{Z}^n \smallsetminus \{0\}$. Then there exists j, $1 \le j \le \ell$, such that $\mathbf{v} \in \Delta_j \smallsetminus \Delta_{j-1}$. Denote $\mathbb{R}(\Delta_{j-1} + \mathbb{Z}\mathbf{v}) \cap \Lambda$ by Δ. Clearly it is a primitive subgroup of Λ satisfying $\Delta_{j-1} \subset \Delta \subset \Delta_j$, therefore Δ is comparable with Δ_i for every i (and may or may not coincide with one of the Δ_is). Now one can use properties (M1) and (M2) to deduce that

$$\|h(x)\Delta\| \ge \min\left(\frac{\varepsilon}{\rho} \cdot \rho^{\mathrm{rk}(\Delta)}, \rho^{\mathrm{rk}(\Delta)}\right) = \varepsilon\rho^{\mathrm{rk}(\Delta)-1} = \varepsilon\rho^{\mathrm{rk}(\Delta_{i-1})}. \tag{3.8}$$

On the other hand, from the submultiplicativity of the covolume it follows that $\|h(x)\Delta\|$ is not greater than $\|h(x)\Delta_{i-1}\| \cdot \|\mathbf{v}\|$ (recall a similar step in the proof of Lemma 2.1). Thus

$$\|h(x)\mathbf{v}\| \ge \frac{\|h(x)\Delta\|}{\|h(x)\Delta_{i-1}\|} \underset{\text{by (M1) and (3.8)}}{\ge} \frac{\varepsilon\rho^{\mathrm{rk}(\Delta_{i-1})}}{\rho^{\mathrm{rk}(\Delta_{i-1})}} = \varepsilon\,.$$

Hence $\Lambda \in \mathcal{L}_n(\varepsilon)$ and the proof is finished. □

This is perhaps the crucial point in the proof: we showed that a flag with certain properties does exactly what a single vector was doing in the case of $\mathrm{SL}(2,\mathbb{R})$; namely, it guarantees that in the lattices corresponding to protected points, no vector can be shorter than ε.

Now that the above proposition is established, we will forget about the specific form of the weight function and work with an arbitrary η. Here is a more general theorem:

THEOREM 3.9. *Fix $0 \le k \le n$, and suppose an interval $B \subset \mathbb{R}$, $C, \alpha > 0$, a continuous map $h : B \to \mathrm{SL}(n, \mathbb{R})$, a poset $P \subset P(\mathbb{Z}^n)$ of length k and a weight function $\eta : P \to (0, 1]$ are given. Assume that for any $\Delta \in P$*

[3.9-i] *the function $x \mapsto \|h(x)\Delta\|$ is (C, α)-good on B, and*

[3.9-ii] $\sup_{x \in B} \|h(x)\Delta\| \ge \eta(\Delta)$.

Then for any $0 < \varepsilon < 1$,

$$\lambda(\{x \in B : x \text{ is not } \varepsilon\text{-protected relative to } P\}) \le k2^k C\varepsilon^\alpha \lambda(B)\,.$$

We remark that the use of an arbitrary P in place of $P(\mathbb{Z}^n)$ is justified not only by a possibility to prove the theorem by induction, but also by some applications to Diophantine approximation, see e.g. [**BKM, Kl3, G1**], where proper sub-posets of $P(\mathbb{Z}^n)$ arise naturally.

PROOF. We will break the argument into several steps.

Step 0. First let us see what happens when $k = 0$, the base case of the induction. In this case P is empty, and the flag $F = \varnothing$ will satisfy both (M1) and (M2). Thus all points of B are ε-protected relative to P for any ε, which means that in the case $k = 0$ the claim is trivial. So we can take $k \geq 1$ and suppose that the theorem is proved for all the smaller lengths of P.

Step 1. For any $y \in B$ let us define

$$S(y) \stackrel{\text{def}}{=} \{\Delta \in P : \|h(y)\Delta\| < \eta(\Delta)\}\,.$$

Roughly speaking, $S(y)$ is the set of Δs which gets small enough at y, i.e. potentially could bring trouble. By the discreteness of $h(y)\mathbb{Z}^n$ in $\mathbb{R}^n$, this is a finite subset of P. Note that if this set happens to be empty, then $\|h(y)\Delta\| \geq \eta(\Delta)$ for all $\Delta \in P$, which means that $F = \varnothing$ can be used to ε-protect y for any ε. So let us define

$$E \stackrel{\text{def}}{=} \{y \in B : S(y) \neq \varnothing\} = \{y \in B : \exists\, \Delta \in P \text{ with } \|h(y)\Delta\| < \eta(\Delta)\}\,;$$

then to prove the theorem it suffices to estimate the measure of the set of points $x \in E$ which are not ε-protected relative to P. A flashback to the proof for $n = 2$: there $S(y)$ consisted of primitive vectors $\mathbf{v}$ for which $\|u_y\mathbf{v}\|$ was less than ρ, not more than one such vector was allowed, and nonexistence of such vectors automatically placed the lattice in $\mathcal{L}_n(\varepsilon)$.

Step 2. Take $y \in E$ and $\Delta \in S(y)$, and define $B_{\Delta,y}$ to be the maximal interval of the form $B \cap (y-r, y+r)$ on which the absolute value of $\|h(\cdot)\Delta\|$ is not greater than $\eta(\Delta)$. From the definition of $S(y)$ and the continuity of functions $\|h(\cdot)\Delta\|$ it follows that $B_{\Delta,y}$ contains some neighborhood of y. Further, the maximality property of $B_{\Delta,y}$ implies that

$$\sup_{x \in B_{\Delta,y}} \|h(x)\Delta\| = \eta(\Delta)\,. \tag{3.9}$$

Indeed, either $B_{s,y} = B$, in which case the claim follows from [3.9-ii], or at one of the endpoints of $B_{\Delta,y}$, the function $\|h(\cdot)\Delta\|$ must attain the value $\eta(\Delta)$ – otherwise one can enlarge the interval and still have $\|h(\cdot)\Delta\|$ not greater than $\eta(\Delta)$ for all its points. (Another flashback: intervals $B_{\Delta,y}$ are analogues of $B_{\mathbf{v}}(\rho)$ from the proof of Theorem 2.4 – but this time there is no disjointness, since many Δs can get small simultaneously.)

Step 3. For any $y \in E$ let us choose an element Δ_y of $S(y)$ such that $B_{\Delta_y,y} = \bigcup_{\Delta \in S(y)} B_{\Delta,y}$ (this can be done since $S(y)$ is finite). In other words, $B_{\Delta_y,y}$ is maximal among all $B_{\Delta,y}$. For brevity we will denote $B_{\Delta_y,y}$ by B_y. We now claim that

$$\sup_{x \in B_y} \|h(x)\Delta\| \geq \eta(\Delta) \text{ for any } y \in E \text{ and } \Delta \in P\,. \tag{3.10}$$

Indeed, if not, then $\|h(x)\Delta\| < \eta(\Delta)$ for all $x \in B_y$, in particular one necessarily has $\|h(y)\Delta\| < \eta(\Delta)$, hence $\Delta \in S(y)$ and $B_{\Delta,y}$ is defined. But $B_{\Delta,y}$ is contained in B_y, so (3.10) follows from (3.9). This step allows one to replace the covering $\{B_{\Delta,y} : \Delta \in S(y),\ y \in E\}$ of E by a more efficient covering $\{B_y : y \in E\}$; informally speaking, this is achieved by selecting $\Delta = \Delta_y$ which works best for every given y.

Step 4. Now we are ready to perform the induction step. For any $y \in E$ define

$$P_y \stackrel{\text{def}}{=} \{\Delta \in P \setminus \{\Delta_y\} : \Delta \text{ is comparable with } \Delta_y\}.$$

We claim that P_y (a poset of length $k-1$) in place of P and B_y in place of B satisfy all the conditions of the theorem. Indeed, [3.9-i] is clear since B_y is a subset of B, and [3.9-ii] follows from (3.10). Therefore, by induction,

$$\lambda(\{x \in B_y : x \text{ is not } \varepsilon\text{-protected relative to } P_y\}) \leq (k-1)2^{k-1}C\varepsilon^\alpha \lambda(B_y). \tag{3.11}$$

Step 5. Does the previous step help us, and how? let us take x outside of this set of relatively small measure, that is, assume that x is ε-protected relative to P_y, and try to use this protection. By definition, there exists a flag F' inside P_y such that

$$\varepsilon\eta(\Delta) \leq \|h(x)\Delta\| \leq \eta(\Delta) \quad \forall\, \Delta \in F' \tag{3.12}$$

and

$$\|h(x)\Delta\| \geq \eta(\Delta) \quad \forall\, \Delta \in P_y \setminus F' \text{ comparable with every element of } F'. \tag{3.13}$$

However this F' will NOT protect x relative to the bigger poset P, because Δ_y, comparable with every element of F', would not satisfy (M2) – on the contrary, recall that it was chosen so that the reverse inequality, $\|h(x)\Delta_y\| \leq \eta(\Delta_y)$, holds for all $x \in B_y$, see (3.10)! Thus our only choice seems to be to add Δ_y to F', for extra protection, and put $F \stackrel{\text{def}}{=} F' \cup \{\Delta_y\}$. Then $\Delta \in P \setminus F$ is comparable with every element of F if and only if Δ is in $P_y \setminus F'$, and is comparable with every element of F'. Because of that, (M2) immediately follows from (3.13). As for (M1), we already know it for for $\Delta \neq \Delta_y$ by (3.12), so it remains to put $\Delta = \Delta_y$. The upper estimate in (M1) is immediate from (3.10). The lower estimate, on the other hand, can fail – but only on a set of relatively small measure, because of assumption [3.9-i] which, by the way, has not been used so far at all:

$$\begin{aligned} \lambda(\{x \in B_y : \|h(x)\Delta_y\| < \varepsilon\eta(\Delta_y)\}) &\leq C\left(\frac{\varepsilon\eta(\Delta_y)}{\sup_{x\in B_y}\|h(x)\Delta_y\|}\right)^\alpha \lambda(B_y) \\ &\underset{(3.9)}{\leq} C(\varepsilon)^\alpha \lambda(B_y). \end{aligned} \tag{3.14}$$

The union of the two sets above, in the left hand sides of (3.11) and (3.3), has measure at most $k2^{k-1}C\varepsilon^\alpha\lambda(B_y)$. We have just shown that this union exhausts all the unprotected points as long as we are restricted to B_y. Thus we have achieved an analogue of what was extremely easy for $n = 2$: bounded the measure of the set of points where things can go wrong on each of the intervals $B_{\mathbf{v}}(\rho)$.

Step 6. It remains to produce a substitute for the disjointness of the intervals, that is, put together all the B_ys. For that, consider the covering $\{B_y : y \in E\}$ of E and choose a subcovering $\{B_i\}$ of multiplicity at most 2. (Exercise: this is always possible.) Then the measure of $\{x \in E : x \text{ is not } \varepsilon\text{-protected relative to } P\}$ is not greater than

$$\begin{aligned} \sum_i \lambda(\{x \in B_i : x \text{ is not } \varepsilon\text{-protected relative to } P\}) &\leq k2^{k-1}C\varepsilon^\alpha \sum_i \lambda(B_i) \\ &\leq k2^k Ce^\alpha \lambda(B), \end{aligned}$$

and the theorem is proven. □

4. Applications of non-divergence to metric Diophantine approximation

Here we present applications of Theorem 3.4 to number theory which reach beyond the unipotent, or even polynomial, case.

4.1. Inheritance of sublinear growth. Our main object of studying will be a fixed parametrized curve $B \to \mathcal{L}_n$, where $B \subset \mathbb{R}$ is an interval. Given such a curve, we will consider a family of curves which are translations of the initial one by some group elements a_t. That is, put

$$h(x) = h_t(x) = a_t h_0(x) \tag{4.1}$$

in Theorem 3.4, where h_0 is a fixed map from B to $\mathrm{SL}(n, \mathbb{R})$. We would like to investigate the following two questions:

(1) What are interesting examples of h_0 and a_t for which one can establish conditions [3.4-i] and [3.4-ii] uniformly for all $t > 0$?
(2) What would be consequences of that for the initial curve $h_0(x)\mathbb{Z}^n$?

Let us start with the second question, since it is easier. That is, suppose we are given an interval $B \subset \mathbb{R}$, $C, \alpha > 0$, $0 < \rho < 1$, a continuous map $h_0 : B \to \mathrm{SL}(n, \mathbb{R})$ and $h = h_t$ as in (4.1). Also let us assume that for any $\Delta \in P(\mathbb{Z}^n)$ and any $t > 0$, conditions [3.4-i] and [3.4-ii] are satisfied. The trick is now to choose $\varepsilon = e^{-\gamma t}$ for some positive γ. From Theorem 3.4 it follows that there exists a constant $\tilde{C}$ (depending on n, C, ρ, B) such that for any t,

$$\lambda\big(\{x \in B : a_t h_0(x)\mathbb{Z}^n \notin \mathcal{L}_n(e^{-\gamma t})\}\big) \le \tilde{C} e^{-\alpha\gamma t}\,.$$

The sum of the right hand sides of the above equation will converge if added up say for $t \in \mathbb{N}$. This immediately calls for an application of the following standard principle from elementary probability theory (the proof is left as an exercise):

LEMMA 4.1 (Borel-Cantelli Lemma). *If μ is a measure on a space X and $\{A_i\}$ is a countable collection of measurable subsets of X with $\sum_i \mu(A_i) < \infty$, then μ-almost every $x \in X$ is contained in at most finitely many sets A_i.*

The conclusion from this is: given an arbitrary $\gamma > 0$, for λ-almost every $x \in B$ we have $a_t h_0(x)\mathbb{Z}^n \in \mathcal{L}_n(e^{-\gamma t})$ if $t \in \mathbb{N}$ is sufficiently large. In fact, by changing γ just a little bit it is easily seen that $t \in \mathbb{N}$ in the last statement can be replaced by $t > 0$. (Exercise.) That is, for all $\gamma > 0$ we have

$$\{a_t\Lambda \;:\; t > 0\} \text{ eventually grows slower than the family } \mathcal{L}_n(e^{-\gamma t}) \tag{4.2}$$

for (Lebesgue) almost every Λ of the form $h_0(x)\mathbb{Z}^n$.

To put this conclusion in an appropriate context, we need to describe the family of sets $\mathcal{L}_n(\varepsilon)$ in a more detailed way. It is not hard to see, using reduction theory for $\mathrm{SL}(n, \mathbb{R})/\mathrm{SL}(n, \mathbb{Z})$, that minus logarithm of the biggest ε such that $\Lambda \in \mathcal{L}_n(\varepsilon)$ is (asymptotically for far away Λ) roughly the same as the distance[1] from Λ to $\mathbb{Z}^n$ or some other base point. Thus the validity of (4.2) for any $\gamma > 0$ can, and will, be referred to as the *sublinear growth* of $\{a_t\Lambda\}$. More generally, we will say, for fixed $\gamma_0 \ge 0$, that $\{a_t\Lambda\}$ has *growth rate* $\le \gamma_0$ if (4.2) holds for any $\gamma > \gamma_0$.

Now denote by ν the Haar probablilty measure on $\mathcal{L}_n$. One can show using Siegel's Formula (see [**E**] for more detail) that $\nu\big(\mathcal{L}_n \smallsetminus \mathcal{L}_n(\varepsilon)\big) \le \mathrm{const}_n\, \varepsilon^n$. (Exercise:

[1] See a remark after Corollary 3.6 for a description of a metric on $\mathcal{L}_n$; note also that $\mathrm{dist}(\Lambda, \mathbb{Z}^n)$ is also roughly the same as minus logarithm of $\rho(\Lambda)$ defined in (3.5)

compute this constant; a more difficult exercise: prove that the right hand side captures the asymptotics of $\nu\big(\mathcal{L}_n \smallsetminus \mathcal{L}_n(\varepsilon)\big)$ as $\varepsilon \to 0$; this is done in [**KM2**].) Since a_t preserves ν, for any $\gamma > 0$ and any t we have

$$\nu\big(\{\Lambda \in \mathcal{L}_n : a_t\Lambda \notin \mathcal{L}_n(e^{-\gamma t})\}\big) \le \mathrm{const}_n\, e^{-n\gamma t};$$

therefore for the same (Borel-Cantelli) reason as above, for any positive γ (4.2) is satisfied by ν-a.e. $\Lambda \in \mathcal{L}_n$. Thus we have proved that, assuming all the functions of the form (4.1) satisfy [3.4-i] and [3.4-ii], certain dynamical behavior (sublinear growth of trajectories) of generic points of the phase space is inherited by generic points on the curve $\{h_0(x)\mathbb{Z}^n\}$.

We note that problems of this type, i.e. studying rates of growth of trajectories, or rates with which dense trajectories approximate points, are sometimes referred to as *shrinking target problems*. Indeed, the family of complements of the sets $\mathcal{L}_n(e^{-\gamma t})$ can be thought of as a shrinking target zooming at the cusp of $\mathcal{L}_n$, and to hit this target means to get into those "neighborhoods of infinity" infinitely many times. See [**KM2**] for a detailed discussion.

Also, observe that we haven't really used the full strength of Theorem 3.4, with $\rho^{\mathrm{rk}(\Delta)}$ in place of ρ, and it was promised that it is supposed to be important for applications. The next theorem summarizes the above discussion and strengthens its conclusions:

THEOREM 4.2 ([**Kl5**]). *Suppose an interval $B \subset \mathbb{R}$, $C, \alpha, \gamma_0 > 0$, a continuous map $h_0 : B \to \mathrm{SL}(n, \mathbb{R})$ and a subgroup $\{a_t\} \subset \mathrm{SL}(n, \mathbb{R})$ are given.*

(a) *Assume that:*

[4.2-i] *for all $\Delta \in P(\mathbb{Z}^n)$ and $t > 0$, functions $x \mapsto \|a_t h_0(x)\Delta\|$ are (C, α)-good on B, and*

[4.2-ii] *for any $\beta > \gamma_0$ there exists T such that $\sup_{x\in B} \|a_t h_0(x)\Delta\| \ge (e^{-\beta t})^{\mathrm{rk}(\Delta)}$ for all $\Delta \in P(\mathbb{Z}^n)$ and $t > T$.*

Then for λ-a.e. $x \in B$, $\{a_t h_0(x)\mathbb{Z}^n\}$ has growth rate $\le \gamma_0$.

(b) *Suppose that* [4.2-ii] *does not hold; then $\{a_t h_0(x)\mathbb{Z}^n\}$ has growth rate $> \gamma_0$ for all $x \in B$.*

PROOF. Part (a) follows from a minor modification of the argument preceding the theorem: for any $\gamma > \gamma_0$ choose β between γ and γ_0, and apply Theorem 3.4 with $\rho = e^{-\beta t}$, and then the Borel-Cantelli Lemma. For part (b), if for some $\beta > \gamma_0$ there exist $t_k \to \infty$ and $\Delta_k \in P(\mathbb{Z}^n)$ such that $\|a_{t_k} h_0(x)\Delta_k\| < (e^{-\beta t_k})^{\mathrm{rk}(\Delta_k)}$ for all $x \in B$, then for each x, using Minkowski Lemma, one can choose a nonzero vector $v_k \in \Delta_k$ such that $\|a_{t_k} h_0(x) v_k\| < e^{-\beta t_k}$, which implies that $a_{t_k} h_0(x)\mathbb{Z}^n \notin \mathcal{L}_n(e^{-\beta t_k})$. □

We have therefore established a remarkable dichotomy: for curves satisfying [4.2-i], either almost all trajectories grow slowly, or all trajectories grow fast. See [**Kl6**] for a further exploration of this theme.

4.2. Checking [3.4-i] and [3.4-ii]. Of course there would be no point in the argument of the previous section if we didn't know that there exist examples, and moreover very naturally arising in number theory, of functions h_t as in (4.1) satisfying the assumptions of Theorem 3.4 uniformly in t. We are going to describe a special case which is very useful for applications.

For this, it will be convenient to upgrade the dimension of the space where all the lattices live from n to $n+1$. Then choose

$$a_t = \mathrm{diag}(e^{nt}, e^{-t}, \dots, e^{-t}),$$

that is, consider a generalization of $\{a_t\} \subset \mathrm{SL}(2,\mathbb{R})$ as in (2.1). One can easily see that the unstable leaves of the action of a_t, $t > 0$, on $\mathcal{L}_{n+1}$ are given by the orbits of the group

$$\left\{ u_{\mathbf{y}} \stackrel{\text{def}}{=} \begin{pmatrix} 1 & \mathbf{y}^T \\ 0 & I_n \end{pmatrix} : \mathbf{y} \in \mathbb{R}^n \right\},$$

a higher-dimensional analogue of $U \subset \mathrm{SL}(2,\mathbb{R})$ (This group is denoted by $G^+_{a_1}$ in the notation of [**EL**] and is also known as the *expanding horospherical subgroup* corresponding to a_1). We are going to put our initial curve $\{h_0(x)\}$ inside this group; that is, consider

$$h_0(x) = \begin{pmatrix} 1 & \mathbf{f}(x)^T \\ 0 & I_n \end{pmatrix},$$

where $\mathbf{f}$ is a map $B \to \mathbb{R}^n$. The question now becomes: under what conditions on $\mathbf{f}$ can we verify the assumptions of Theorem 3.4 with $h_t(x) = a_t u_{\mathbf{f}(x)}$ uniformly in t.

In order to do that, we need to understand the action of the elements $u_{\mathbf{y}}$ on the exterior powers of $\mathbb{R}^{n+1}$. Choose the standard basis $\mathbf{e}_0, \mathbf{e}_1, \dots, \mathbf{e}_n$ of $\mathbb{R}^{n+1}$, and denote by V the space spanned by $\mathbf{e}_1, \dots, \mathbf{e}_n$. It will be convenient to identify $\mathbf{y} \in \mathbb{R}^n$ with $y_1\mathbf{e}_1 + \cdots + y_n\mathbf{e}_n$. Note that $\mathbf{e}_0$ is expanded by a_t (eigenvalue e^{nt}) and V is the contracting subspace (eigenvalue e^{-t}). Similarly for any $k \le n$, the k-th exterior power of $\mathbb{R}^{n+1}$ splits into the expanding (subspaces containing $\mathbf{e}_0$) and contracting (contained in V) parts.

Observe that $u_{\mathbf{y}}$ leaves $\mathbf{e}_0$ fixed and sends vectors $\mathbf{v} \in V$ to $\mathbf{v} + (\mathbf{y}\cdot\mathbf{v})\mathbf{e}_0$. From this is is easy to conclude how $u_{\mathbf{y}}$ acts on $\bigwedge^k(\mathbb{R}^{n+1})$: elements of the form $\mathbf{e}_0 \wedge \mathbf{w}$ are fixed, and

$$\begin{aligned} \mathbf{v}_1 \wedge \cdots \wedge \mathbf{v}_k &\stackrel{u_{\mathbf{y}}}{\to} (\mathbf{v}_1 + (\mathbf{y}\cdot\mathbf{v}_1)\mathbf{e}_0) \wedge \cdots \wedge (\mathbf{v}_k + (\mathbf{y}\cdot\mathbf{v}_k)\mathbf{e}_0) \\ &= \mathbf{v}_1 \wedge \cdots \wedge \mathbf{v}_k + \mathbf{e}_0 \wedge \left(\sum_{i=1}^{k} \pm(\mathbf{y}\cdot\mathbf{v}_i) \bigwedge_{j\ne i} \mathbf{v}_j \right). \end{aligned} \tag{4.3}$$

Now let us see what conditions on $\mathbf{f}$ are sufficient to establish [3.4-i] and [3.4-ii]. Take $\Delta \in P(\mathbb{Z}^{n+1})$ and represent it (up to $\pm$) by the exterior product of generators of Δ, let us call it $\mathbf{w}$. First of all it follows from the above formula that for any $\mathbf{w} \in \bigwedge(\mathbb{R}^{n+1})$, all the coordinates of $u_{\mathbf{y}}\mathbf{w}$, and hence of $a_t u_{\mathbf{y}}\mathbf{w}$ for any t, are linear combinations of $1, y_1, \dots, y_n$ (coefficients in these linear combinations depend on t). Thus property [3.4-i] uniformly over all t would follow if we could find C, α such that all the linear combinations of $1, f_1, \dots, f_n$ are (C,α)-good on B.

It turns out that condition [3.4-ii], that is, $\sup_{x\in B} \|a_t h_0(x)\Delta\| \ge \rho^{\mathrm{rk}(\Delta)}$ for all Δ and all (large enough) t is also easy to check:

LEMMA 4.3. *Suppose that* $\mathbf{f}(B)$ *is not contained in any affine hyperplane (equivalently, the restrictions of* $1, f_1, \dots, f_n$ *to* B *are linearly independent over* $\mathbb{R}$*). Then:*

(a) *there exists* $\rho > 0$ *such that* [3.4-ii] *holds for any* $\Delta \in P(\mathbb{Z}^{n+1})$ *and any* $t > 0$*;*

(b) $\exists\, t_0 > 0$ *such that* [3.4-ii] *holds with* $\rho = 1$ *for all* $\Delta \in P(\mathbb{Z}^{n+1})$ *and* $t > t_0$.

PROOF. Both claims are trivial if $\mathbb{R}\Delta$ contains $\mathbf{e}_0$: indeed, from the previous discussion it follows that $\mathbf{w}$ representing Δ is then fixed by $u_{\mathbf{f}(x)}$ and expanded by a_t. If not, suppose that $\text{rk}(\Delta) = k$; then $\dim(\mathbb{R}\Delta \oplus \mathbb{R}\mathbf{e}_0) = k+1$. One can choose an orthonormal set $\{\mathbf{v}_1, \dots, \mathbf{v}_{k-1}\} \subset \mathbb{R}\Delta \cap V$ and complete it to an orthonormal basis $\{\mathbf{v}_1, \dots, \mathbf{v}_k, \mathbf{e}_0\}$ of $\mathbb{R}\Delta \oplus \mathbb{R}\mathbf{e}_0$. Then

$$\mathbf{w} = a\mathbf{e}_0 \wedge \mathbf{v}_1 \cdots \wedge \mathbf{v}_{k-1} + b\mathbf{v}_1 \wedge \cdots \wedge \mathbf{v}_k\,,$$

where $a^2 + b^2 \geq 1$. (Note: a, b do not have to be integers, and vectors $\mathbf{v}_i$ are not necessarily with integer coordinates; however orthonormality is important.) Now we can, using (4.3), simply look at the projection of $u_{\mathbf{f}(x)}\mathbf{w}$ onto $\mathbf{e}_0 \wedge \mathbf{v}_1 \cdots \wedge \mathbf{v}_{k-1}$:

$$u_{\mathbf{f}(x)}\mathbf{w} = \big(a + b(\mathbf{f}(x)\cdot \mathbf{v}_k)\big)\mathbf{e}_0 \wedge \mathbf{v}_1 \cdots \wedge \mathbf{v}_{k-1} + \dots .$$

Regardless of the choice of a, b, the coefficient in front of $\mathbf{e}_0 \wedge \mathbf{v}_1 \cdots \wedge \mathbf{v}_{k-1}$ is of the form $c_0 + c_1 f_1 + \cdots + c_n f_n$ with $\sum |c_i|^2 \geq 1$. In view of the linear independence assumption and the compactness of the unit sphere in $\mathbb{R}^{n+1}$, there exists $\rho = \rho(B) > 0$ such that the supremum of the absolute value of every such function, and hence $\sup_{x\in B} \|u_{\mathbf{f}(x)}\Delta\|$ is at least ρ. But $\mathbf{e}_0 \wedge \mathbf{v}_1 \cdots \wedge \mathbf{v}_{k-1}$ is expanded by a_t with a rate at least e^t, and both conclusions follow. □

Now, abusing terminology for some more, let us introduce the following definitions. Say that a map $\mathbf{f}$ from a subset U of $\mathbb{R}$ to $\mathbb{R}^n$ is *good* if for λ-a.e. $x \in U$ there exists a neighborhood $B \subset U$ of x and $C, \alpha > 0$ such that any linear combination of $1, f_1, \dots, f_n$ is (C, α)-good on B. We will also say that $\mathbf{f}$ is (C, α)-good if C and α can be chosen uniformly for all x as above. Polynomial maps form a basic example. Later we will explain how one can prove that real analytic maps also have this property.

Also, say that $\mathbf{f}$ is *nonplanar* if for any nonempty interval $B \subset U$, the restrictions of $1, f_1, \dots, f_n$ to B are linearly independent over $\mathbb{R}$; in other words, no nonempty relatively open piece of $\mathbf{f}(U)$ is contained in a proper affine subspace of $\mathbb{R}^n$. The above discussion can be thus summarized in the following way:

THEOREM 4.4. *Let U be a subset of $\mathbb{R}$ and let $\mathbf{f} : U \to \mathbb{R}^n$ be a continuous good nonplanar map. Then for λ-a.e. $x \in U$, the a_t-trajectory of $u_{\mathbf{f}(x)}\mathbb{Z}^{n+1}$ has sublinear growth.*

Note: it follows from remarks made at the end of the previous section and a "flowbox" argument (see [**E**]) that the a_t-trajectory of $u_{\mathbf{y}}\mathbb{Z}^n$ has sublinear growth for λ-a.e. $\mathbf{y} \in \mathbb{R}^n$. Thus the above theorem describes examples of curves in $\mathbb{R}^n$ whose generic points inherit certain property of generic points of $\mathbb{R}^n$.

4.3. Inheritance of Diophantine properties. Of course a reasonable question concerning all the argument above would be – why would anybody at all care about orbit growth properties of typical poitns on some curves. The answer is – that all along, like monsieur Jourdain speaking in prose, we were actively involved in proving theorems in Diophantine approximation without knowing it.

Indeed, let us see how $\mathbf{y} \in \mathbb{R}^n$ is characterized by the fact that $\{a_t u_{\mathbf{y}}\mathbb{Z}^{n+1}\}$ has growth $\leq \gamma_0$. Suppose that for any $\gamma > \gamma_0$ there exists $T > 0$ such that for any $t > T$ and any nonzero $(p, \mathbf{q}) \in \mathbb{Z} \times \mathbb{Z}^n$ one has

$$\left\| a_t u_{\mathbf{y}} \begin{pmatrix} p \\ \mathbf{q} \end{pmatrix} \right\| = \max\left(e^{nt}|p + \mathbf{y}\cdot\mathbf{q}|, e^{-t}\|\mathbf{q}\|\right) \geq e^{-\gamma t}\,. \tag{4.4}$$

For such γ, given $\mathbf{q} \in \mathbb{Z}^n$, choose t such that $e^{-t}\|\mathbf{q}\| = e^{-\gamma t} \Leftrightarrow e^t = \|\mathbf{q}\|^{\frac{1}{1-\gamma}}$. For large enough $\|\mathbf{q}\|$ this t will be greater than T. In view of (4.4), $e^{nt}|p + \mathbf{y}\cdot\mathbf{q}|$ must be at least $e^{-\gamma t}$, which translates into

$$|p + \mathbf{y}\cdot\mathbf{q}| \geq e^{-(n+\gamma)t} = \|\mathbf{q}\|^{-\frac{n+\gamma}{1-\gamma}}.$$

We proved that $\{a_t u_{\mathbf{y}}\mathbb{Z}^{n+1}\}$ having growth rate $\leq \gamma_0$ implies that $\mathbf{y}$ is Diophantine[2] of order v for any $v > \frac{n+\gamma_0}{1-\gamma_0}$. In fact, converse implication is also true, and is left as an exercise; see [**KM2, Kl2**]. Consequently, sublinear growth of $\{a_t u_{\mathbf{y}}\mathbb{Z}^{n+1}\}$ is equivalent to $\mathbf{y}$ being Diophantine of all orders $> n$; those $\mathbf{y}$ are called *not very well approximable*, to be abbreviated as not VWA. It is an elementary fact, immediately implied by the Borel-Cantelli Lemma, that λ-a.e. $\mathbf{y} \in \mathbb{R}^n$ is not VWA. Thus we can reformulate the theorem proved in the previous section as follows:

THEOREM 4.5. *Let U be an open subset of $\mathbb{R}$ and let $\mathbf{f} : U \to \mathbb{R}^n$ be a continuous good nonplanar map. Then for λ-a.e. $x \in U$, $\mathbf{f}(x)$ is not VWA.*

Results of this type have a long history, see [**BD**] and surveys [**Sp3, Kl1, Kl4**]. The above statement was conjectured by Mahler [**Mah**] in 1932 for

$$\mathbf{f}(x) = (x, x^2, \ldots, x^n)\,. \tag{4.5}$$

This curve is indeed somewhat special: for any $x \in \mathbb{R}$, $(x, x^2, \ldots, x^n)$ is VWA if and only if for some $v > n$ there are infinitely many integer polynomials P of degree $\leq n$ such that $|P(x)| <$ (height of $P)^{-v}$. Thus Mahler's Conjecture asserts, roughly speaking, that almost all transcendental numbers are "not very algebraic". Mahler himself proved a bound with a weaker exponent, and the full strength of the conjecture was established in 1964 by Sprindžuk. [**Sp1, Sp2**]. Then Sprindžuk in 1980 [**Sp3**] made the following

CONJECTURE 4.6 (now a theorem). *For open $U \subset \mathbb{R}$, let $\mathbf{f} : U \to \mathbb{R}^n$ be nonplanar and real analytic. Then for λ-a.e. $x \in U$, $\mathbf{f}(x)$ is not VWA.*

This was proved in [**KM1**] via deducing it from a more general Theorem 4.5. Even for general polynomial maps, not of the form (4.5), this was new.

At this point the only missing part for us is to understand why real analytic implies good. The explanation involves passing from C^∞ to C^k class. The next lemma produces a wide variety of examples of good functions:

LEMMA 4.7. *For any $k \in \mathbb{N}$ there exists $C_k > 0$ such that whenever an interval $B \subset \mathbb{R}$ and $f \in C^k(B)$, $k \in \mathbb{N}$, are such that for some $0 < a \leq A$ one has*

$$a \leq |f^{(k)}(x)| \leq A \quad \forall x \in B\,, \tag{4.6}$$

then f is $\big(C_k(A/a)^{1/k}, 1/k\big)$-good on B.

This can be seen as a generalization of Proposition 3.2: indeed, polynomials of degree k satisfy the above assumptions with $A = a$.

---

[2]Definition: $\mathbf{y}$ is *Diophantine of order v* if $|\mathbf{q}\cdot\mathbf{y}+p| \geq \text{const}\,\|\mathbf{q}\|^{-v}$ for all large enough $\mathbf{q}$ and all p. Note that here we interpret $\mathbf{y}$ as a linear form, but the method is equally well applicable to treating $\mathbf{y}$ as a vector, that is, looking at inequalities of type $\|q\mathbf{y}+\mathbf{p}\| \geq \text{const}\,|q|^{-v}$, where $q \in \mathbb{Z}$ and $\mathbf{p} \in \mathbb{Z}^n$. The book [**Sch**] by Schmidt is an excellent reference.

PROOF. We outline the argument in the case $k = 2$, with $C_2 = 2\sqrt{22}$; an extension to arbitrary degree of smoothness is straightforward and is left as an exercise, see [**KM1**] for hints. (However it is interesting that letting $A = a$ in the above lemma produces a constant which is not as good as the one for polynomials.)

Fix a subinterval J of B and denote by d the length of J and by s the supremum of $|f|$ on J. Take $\varepsilon > 0$; since, by the lower estimate in (4.6), the second derivative of f does not vanish on J, the set $\{x \in J : |f(x)| < \varepsilon\}$ consists of at most 2 intervals. Let I be the maximal of those, and denote its length by r. Then

$$\lambda(\{x \in J : |f(x)| < \varepsilon\}) \leq 2r\,, \tag{4.7}$$

so it suffices to estimate r from above.

SUBLEMMA 4.8. $r \leq 2\sqrt{6\varepsilon/a}$.

PROOF. Let x_1, x_2, x_3 be the left endpoint, midpoint and right endpoint of I respectively, and let P be the Lagrange polynomial of degree 2 formed by using values of f at these points, i.e. given by the expression in the right hand side of (3.3) with $k = 2$. Then there exists $x \in I$ such that $P''(x) = f''(x)$. Hence, by the lower estimate in (4.6), $|P''(x)| \geq a$. On the other hand, one can differentiate the right hand side of (3.3) twice to get $|P''(x)| \leq 3\varepsilon\dfrac{2}{(r/2)^2} = 24\varepsilon/r^2$. Combining the last two inequalities yields the desired estimate. □

Now recall that, since we are after the (C, α)-good property, we would like to have an upper estimate for r in the form $r \leq C(\varepsilon/s)^\alpha d$. Thus let us rewrite the conclusion of the lemma as

$$r \leq 2\sqrt{\frac{6s}{ad^2}}\left(\frac{\varepsilon}{s}\right)^{1/2} d = 2\sqrt{\frac{6t}{a}}\left(\frac{\varepsilon}{s}\right)^{1/2} d\,, \tag{4.8}$$

where we introduced a parameter $t \stackrel{\text{def}}{=} s/d^2$. We see that the above estimate is useful when t is small, and to finish the proof it suffices to produce an estimate improving (4.8) for large values of t. Here it goes:

SUBLEMMA 4.9. $r \leq \sqrt{\frac{10A/a}{1-A/2t}} \cdot \left(\frac{\varepsilon}{s}\right)^{1/2} d.$

PROOF. Let Q be the Taylor polynomial of f of degree 1 at x_1. By Taylor's formula,

$$|f(x_2) - Q(x_2)| \leq \sup_{x\in I} |f''(x)| \frac{(r/2)^2}{2} \underset{(4.6)}{\leq} \frac{Ar^2}{8} \underset{\text{Lemma 4.8}}{\leq} \frac{A}{8}\frac{24\varepsilon}{a} = 3\frac{A}{a}\varepsilon\,.$$

But also $|f(x_2)| \leq \varepsilon$, therefore

$$|Q(x_2)| \leq \left(3\frac{A}{a} + 1\right)\varepsilon \underset{\text{to simplify computations}}{\leq} 4\frac{A}{a}\varepsilon\,.$$

We now apply Lagrange's formula to reconstruct Q on B by its values at x_1, x_2. As in the proof of Proposition 3.2, we get

$$\|Q\|_B \leq \left(4\frac{A}{a}\varepsilon + \varepsilon\right)\frac{d}{r/2} \underset{\text{to simplify computations}}{\leq} 10\frac{A}{a} \cdot \varepsilon\frac{d}{r}\,. \tag{4.9}$$

Finally, the difference between f and Q on B is, again by the upper estimate in (4.6), bounded from above by $Ad^2/2$, so from (4.9) one deduces that

$$s \le 10\frac{A}{a}\cdot\varepsilon\frac{d}{r} + Ad^2/2 \underset{\text{to simplify computations}}{\le} 10\frac{A}{a}\cdot\varepsilon\frac{d^2}{r^2} + Ad^2/2\,,$$

or, equivalently,

$$r \le \sqrt{\frac{10\frac{A}{a}\cdot\varepsilon d^2}{s - Ad^2/2}} = \sqrt{\frac{10\frac{A}{a}}{1 - Ad^2/2s}}\cdot\left(\frac{\varepsilon}{s}\right)^{1/2} d\,.$$

which is what we wanted to prove. □

It remains to observe (Exercise) that the right hand sides of the two inequalities in Sublemmas 4.8 and 4.9 are equal to each other when $t = 11A/12$, and substitute $t = 11A/12$ in (4.8) to obtain $r \le 2\sqrt{\frac{11A}{2a}}\left(\frac{\varepsilon}{s}\right)^{1/2} d$, which, in view of (4.7), gives the conclusion of Lemma 4.7. □

Here is another important definition. Say that $\mathbf{f}$ is *ℓ-nondegenerate at x* if $\mathbb{R}^n$ is spanned by $\mathbf{f}'(x), \mathbf{f}''(x), \dots, \mathbf{f}^{(\ell)}(x)$. We will say that $\mathbf{f} : U \to \mathbb{R}^n$ is *ℓ-nondegenerate* if it is ℓ-nondegenerate at almost every point. It is clear that nondegeneracy implies nonplanarity (if $f(B)$ belongs to a proper affine hyperplane for some interval B, derivatives of all orders at any point of B won't generate anything more than the tangent space to this hyperplane). On top of this, we also have

PROPOSITION 4.10. *Nondegenerate maps are good. More precisely, if $\mathbf{f}$ is ℓ-nondegenerate at x_0, then there exists a neighborhood B of x_0 and positive C such that any linear combination of $1, f_1, \dots, f_n$ is $(C, 1/\ell)$-good on B; and with a little more work this C can be chosen uniformly if we are given an ℓ-nondegenerate $\mathbf{f} : U \to \mathbb{R}^n$.*

PROOF. The ℓ-nondegeneracy of $\mathbf{f}$ at x_0 implies that for any $\mathbf{c} = (c_1, \dots, c_n) \ne 0$ there exists $1 \le k \le \ell$ such that $\mathbf{c}\cdot\mathbf{f}^{(k)}(x_0) \ne 0$, and in fact for this k depending on $\mathbf{c}$ there is a uniform lower bound on $|\mathbf{c}\cdot\mathbf{f}^{(k)}(x)|$ over all $\mathbf{c}$ on (or outside) the unit sphere and x in some neighborhood B of x_0. But $\mathbf{c}\cdot\mathbf{f}^{(k)} = f^{(k)}$ where $f = c_0 + \sum_{i=1}^n c_i f_i$; this produces an a as in Lemma 4.7, and using it with some upper bound A (which can be made closer to a by making B smaller) one concludes that f is $(C, 1/k)$-good (and therefore $(C, 1/\ell)$-good) on B. □

PROOF OF CONJECTURE 4.6. It remains to take a nonplanar analytic $\mathbf{f} : U \to \mathbb{R}$, $U \subset \mathbb{R}$ a bounded interval, and verify that it must be ℓ-nondegenerate with some uniform ℓ. This is an easy exercise. (Hint: if derivatives of $\mathbf{f}$ at x_k of order up to k are contained in a hyperplane L_k, then all derivatives of $\mathbf{f}$ at $\lim x_k$ will be contained in $\lim L_k$.) □

We remark that as long as the nonplanarity of $\mathbf{f}$ is assumed, we are guaranteed to have condition [3.4-ii] with some ρ uniform in t, and do not really care to distinguish between ρ and $\rho^{\mathrm{rk}(\Delta)}$. However this distinction becomes important when $\mathbf{f}(B)$ belongs to a proper affine subspace $L \subset \mathbb{R}^n$. Then it matters how fast L can be approximated by rational subspaces. It is possible to use Theorem 4.2 to write a condition on L equivalent to almost all ($\Leftrightarrow$ at least one of) its points being not VWA, or more generally, Diophantine of order v for all $v > v_0$, and also to prove that this condition is inherited by $\mathbf{f}(B)$ whenever $\mathbf{f}$ is good and "nonplanar in L", in particular, smooth and "nondegenerate in L" [**Kl5**]. (Exercises: give definitions of the terms in quotation marks, show that nondegenerate in $L \Rightarrow$ good and nonplanar in L, and real analytic $\Rightarrow$ nondegenerate in some L.)

4.4. More about the correspondence between approximation and dynamics. The principle that was used to connect growth rate of $\{a_t u_{\mathbf{y}} \mathbb{Z}^{n+1}\}$ and approximation properties of $\mathbf{y}$ has various manifestations and has been extensively used to relate Diophantine approximation to dynamics. In a nutshell, a very good approximation for $\mathbf{y}$ amounts to a very small value of the function $(p, \mathbf{q}) \mapsto \|\mathbf{q}\|^n |p + \mathbf{y} \cdot \mathbf{q}|$ at a nonzero integer point $\Leftrightarrow$ a small value of the function $\mathbf{v} = (v_0, v_1, \dots, v_n) \mapsto |v_0| \cdot \max(|v_1|, \dots, |v_n|)^n$ at a nonzero vector $\mathbf{v}$ from the latice $u_{\mathbf{y}} \mathbb{Z}^{n+1}$. And the reason for the n-th power is precisely to make the latter function invariant by $a_t \in \mathrm{SL}(n+1, \mathbb{R})$ and use the action to produce a very small nonzero vector in the lattice $a_t u_{\mathbf{y}} \mathbb{Z}^{n+1}$ ($\Leftrightarrow$ a very deep excursion into the cusp).

The same principle was involved in the reduction of the Oppenheim conjecture to dynamics of the stabilizer of a quadratic form on the space of lattices, described in detail in [**E**]. For the same reasons, the trajectory $a_t u_{\mathbf{y}} \mathbb{Z}^{n+1}$ is bounded in $\mathcal{L}_{n+1}$ if and only if $\mathbf{y}$ is *badly approximable*[3]; this is a theorem of Dani [**D3**].

As another application of this principle, consider the product of $n+1$ linear forms, $\mathbf{v} \mapsto \prod_{i=0}^{n} |v_i|$; its stabilizer is the full diagonal subgroup of $\mathrm{SL}(n+1, \mathbb{R})$. It is not hard to show that the orbit of the lattice $u_{\mathbf{y}} \mathbb{Z}^{n+1}$ under the action of the semigroup

$$a_{\mathbf{t}} = \mathrm{diag}(e^{t_1 + \cdots + t_n}, e^{-t_1}, \dots, e^{-t_n}), \quad t_i > 0$$

is bounded in $\mathcal{L}_{n+1}$ if and only if $\mathbf{y}$ is an exception to the (n-dimensional version of the) Littlewood's Conjecture. More about it can be found in [**EL**]

In fact, it is worthwhile to mention that all the results discussed in this section have their multi-parameter analogues; if $h_{\mathbf{t}}(x) = a_{\mathbf{t}} h_0(x)$ is used instead of h_t, it is often possible to establish conditions [3.4-ii] and [3.4-ii] uniformly over $\mathbf{t} \in \mathbb{R}^n_+$. This yields the proof of a *multiplicative* version of Sprindžuk's Conjecture 4.6 (a special case for the curve (4.5) was conjectured earlier by A. Baker) and many of its generalizations.

Also note that the correspondence described above already made an appearance in two more lecture courses in this volume: by Svetlana Katok in the case $n = 1$ [**K**], where it was shown that the diagonal action on $\mathcal{L}_2$ is a suspension of the Gauss map, and by Jean-Christophe Yoccoz [**Y**], who treated the "$n = 1$" case as a platform for a generalization of the aforementioned suspension to the moduli space of translation surface structures in higher genus. Here we are talking about a generalization of a different kind. In fact one can also treat the a_t-action on $\mathcal{L}_{n+1}$ as a suspension of certain first return map, thus obtaining a higher-dimensional version of continued fractions. However in order to obtain results in Diophantine approximation it is often efficient to simply work with the suspension itself, as was demonstrated during these lectures.

Let us describe one more example of the use of Theorem 3.4 in a slightly different context. Suppose we fix $\varepsilon > 0$ and want to look at the a_t-trajectories which, starting from some t_0, decide to leave the compact set $\mathcal{L}_{n+1}(\varepsilon)$ and never come back again. This is possible, for example the a_t-trajectory can be divergent; but of course this can only happen on a null set by ergodicity, i.e. for any ε, the

[3]that is, if and only if $\inf_{p \in \mathbb{Z},\, \mathbf{q} \in \mathbb{Z}^n \smallsetminus \{0\}} \|\mathbf{q}\|^n |p + \mathbf{y} \cdot \mathbf{q}| > 0$

trajectory $\{a_t u_{\mathbf{y}} \mathbb{Z}^{n+1}\}$ will do it for almost no $\mathbf{y}$. Now what about $\mathbf{y}$ of the specific form $\mathbf{f}(x)$, for $\mathbf{f}$ as above, for Lebesgue-generic x?

Before answering this question, let me restate it in Diophantine language: such behavior amounts to existence of t_0 such that for any $t > t_0$ the system

$$|\mathbf{y} \cdot \mathbf{q} + p| < \varepsilon e^{-tn} \quad \text{and} \quad \|\mathbf{q}\| < \varepsilon e^t$$

has a nontrivial integer solution $(p, \mathbf{q})$. Number theorists say in this situation that *Dirichlet's Theorem can be ε-improved* for $\mathbf{y}$, see [**DS1, DS2**]. Partial results, such as for $\mathbf{f}$ of the form (4.5), has been known due to Davenport–Schmidt [**DS1**], Baker [**Ba1, Ba2**] and Bugeaud [**Bu**].

Let us now see what one can do with the dynamical method. Assume that $\mathbf{f} : U \to \mathbb{R}$ is (C, α)-good; then condition [3.4-i] will hold with uniform C, α for all $B \subset U$ except for those touching a null set. If $\mathbf{f}$ is also nonplanar, we can use Lemnma 4.3(b) and for any B find t_0 such that [3.4-ii] holds with $\rho = 1$. Then take ε such that $n 2^n C \varepsilon^\alpha < 1$. It follows that for almost every $x_0 \in U$, any interval $B \subset U$ centered at x_0, the intersection of B with

$$\{x \in U : a_t u_{\mathbf{y}} \mathbb{Z}^{n+1} \notin \mathcal{L}_{n+1}(\varepsilon) \text{ for large enough } t\} \tag{4.10}$$

has relative measure in B strictly less than one. By the Lebesgue Density Theorem, the set (4.10) must have measure zero. This conclusion can be phrased as follows:

THEOREM 4.11. *For any n, C, α there exists $\varepsilon_0 > 0$ with the following property: let U be an open subset of $\mathbb{R}$ and let $\mathbf{f} : U \to \mathbb{R}^n$ be a continuous (C, α)-good nonplanar map. Then for any $\varepsilon < \varepsilon_0$ and for λ-a.e. $x \in U$, Dirichlet's Theorem cannot be ε-improved for $\mathbf{f}(x)$.*

This was done several years ago in [**KW2**] in a much more general, multiplicative, context. Then, in the special case when $\mathbf{f}$ is real analytic and nonplanar, the above result was sharpened by Shah [**Sh1, Sh2**], who established that the conclusion of Theorem 4.11 holds for arbitrary $\varepsilon_0 < 1$ using approximation by unipotent trajectories and the linearization technique described in [**E**].

5. Concluding remarks

5.1. Generalizations. Quantitative nondivergence results can be proved, and have applications, in a much more general situations then described above. In particular, intervals $B \subset \mathbb{R}$ can be replaced by balls in a metric space satisfying certain (Besicovitch) covering property; other measures, including those supported on fractals, can be used instead of Lebesgue measure λ. These generalizations are based on the work done by [**KLW, KT**], but most of the main ideas are contained in the proof presented in §3.3. Among applications to number theory was the proof of so-called Khintchine-type theorems on nondegenerate manifolds, both convergence and divergence cases (with Beresnevich, Bernik and Margulis, see [**BKM, BBKM**]). More recent developments include studying Diophantine properties of points on fractal subsets of $\mathbb{R}^n$. For example, a repeated application of Theorem 3.4 to measures supported on certain self-similar fractals (or, more generally, satisfying certain decay conditions) allows to construct many bounded orbits (read: badly approximable vectors) in the supports of those measures. This was recently done in [**KW1**]. Other applications involve analogues of Diophantine approximation results in the S-arithmetic [**KT**] and positive characteristic [**G2**] setting.

5.2. What else is in the proof. It is fair to say that the approach to metric Diophantine approximation described above is definitely not the only one available. Methods originally developed by Sprindžuk have produced many important results, including an independent proof of Conjecture 4.6 by Beresnevich [**Ber**], and also including many theorems which do not seem to be attainable by dynamics on the space of lattices. However in some cases one can see how the main constructions used in the the proof of Theorem 3.4 (primitive subgroups of various ranks) show up in the other proofs in various disguises, and the passage to the space of latices seems to work as a tool to somehow change variables and compress all the induction argument into one scheme based on the partial order of subgroups of $\mathbb{Z}^n$.

Another classical argument somewhat similar to that proof is the reduction algorithm of Minkowski (mentioned in [**K**]) and Siegel (its generalization to $\mathcal{L}_n$). In fact, flags naturally appear in the construction of Siegel sets (fundamental sets for $\mathrm{SL}(n,\mathbb{R})/\mathrm{SL}(n,\mathbb{Z})$), hence in the proof of the finiteness of Haar measure on $\mathcal{L}_n$. Another feature of the proof of Theorem 3.4 is the fact that it does not use the fact that $\mathcal{L}_n$ has finite volume. And indeed, it can be perhaps thought of as an alternative approach to reduction theory. Using Corollary 3.5 (high frequency of visits of unipotent trajectories to compact sets), one can construct an everywhere positive u_x-invariant integrable function on $\mathcal{L}_n$, which would contradict to Moore's Ergodicity Theorem unless $\nu(\mathcal{L}_n)$ is finite, thus providing an alternative proof of a theorem of Borel and Harish-Chandra. Details are left as an exercise. A more general version of this exercise is a theorem of Dani [**D1**], which he derived from Corollary 3.5, that any locally finite ergodic u_x-invariant measure on $\mathcal{L}_n$ is finite.

5.3. From $\mathcal{L}_n$ to G/Γ and beyond. The reader is invited to look at Dani's papers from late 1970s and early 1980s [**D1, D2, D4**], see also [**DM1**], where it is shown how Corollary 3.5 and other quantitative nondivergence results can be extended to an arbitrary homogeneous space, using the Margulis Arithmeticity Theorem and some standard facts from the theory of algebraic groups. See also [**KT**] and [**GO**] where a similar reduction is carried out in the S-arithmetic case. Finally, it is worthwhile to mention another important theme of the summer school, the analogy between homogeneous space of Lie groups and moduli spaces of translation surface structures. The scheme of proof presented in §3 was used in [**MW**] to establish quantitative nondivergence of horocyclic flows on the moduli space of quadratic differentials, see also [**LM**, Appendix].

References

[Ba1] R.C. Baker, *Metric diophantine approximation on manifolds*, J. Lond. Math. Soc. (2) **14** (1976), 43–48.

[Ba2] ———, *Dirichlet's theorem on diophantine approximation*, Math. Proc. Cambridge Phil. Soc. **83** (1978), 37–59.

[BM] M. Bekka and M. Mayer, *Ergodic theory and topological dynamics of group actions on homogeneous spaces*, London Math. Soc. Lecture Note Series, Vol. 269, Cambridge University Press, Cambridge, 2000.

[Ber] V. Beresnevich, *A Groshev type theorem for convergence on manifolds*, Acta Mathematica Hungarica **94** (2002), 99–130.

[BBKM] V. Beresnevich, V.I. Bernik, D. Kleinbock and G.A. Margulis, *Metric Diophantine approximation: the Khintchine-Groshev theorem for nondegenerate manifolds*, Moscow Math. J. **2** (2002), 203–225.

[BD] V.I. Bernik and M.M. Dodson, *Metric Diophantine approximation on manifolds*, Cambridge University Press, Cambridge, 1999.

[BKM] V.I. Bernik, D. Kleinbock and G.A. Margulis, *Khintchine-type theorems on manifolds: convergence case for standard and multiplicative versions*, Internat. Math. Res. Notices **2001**, 453–486.

[Bu] Y. Bugeaud, *Approximation by algebraic integers and Hausdorff dimension*, J. London Math. Soc. (2) **65** (2002), no. 3, 547–559.

[Cas] J.W.S. Cassels, *An introduction to the geometry of numbers*, Springer, Berlin, 1997.

[D1] S.G. Dani, *On invariant measures, minimal sets and a lemma of Margulis*, Invent. Math. **51** (1979), 239–260.

[D2] ———, *On orbits of unipotent flows on homogeneous spaces*, Ergod. Theor. Dynam. Syst. **4** (1984), 25–34.

[D3] ———, *Divergent trajectories of flows on homogeneous spaces and diophantine approximation*, J. Reine Angew. Math. **359** (1985), 55–89.

[D4] ———, *On orbits of unipotent flows on homogenous spaces II*, Ergod. Theor. Dynam. Syst. **6** (1986), 167–182.

[DM1] S.G. Dani and G.A. Margulis. *Asymptotic behaviour of trajectories of unipotent flows on homogeneous spaces*, Indian. Acad. Sci. J. **101** (1991), 1–17.

[DM2] ———, *Limit distributions of orbits of unipotent flows and values of quadratic forms*, in: *I.M. Gelfand Seminar*, Amer. Math. Soc., Providence, RI, 1993, pp. 91–137.

[DS1] H. Davenport and W. M. Schmidt, *Dirichlet's theorem on diophantine approximation*, in: Symposia Mathematica, Vol. IV (INDAM, Rome, 1968/69), pp. 113–132, 1970.

[DS2] ———, *Dirichlet's theorem on diophantine approximation. II*, Acta Arith. **16** (1969/1970) 413–424.

[E] A. Eskin, *Unipotent Flows and Applications*, lecture notes in this volume.

[EL] M. Einsiedler and E. Lindenstrauss, *Diagonalizable actions and arithmetic applications*, lecture notes in this volume.

[EMS] A. Eskin, S. Mozes and N. Shah, *Unipotent flows and counting lattice points on homogeneous varieties*, Ann. Math. **143** (1996), 253–299.

[G1] A. Ghosh, *A Khintchine-type theorem for hyperplanes*, J. London Math. Soc. (2) **72** (2005), no. 2, 293–304.

[G2] ———, *Metric Diophantine approximation over a local field of positive characteristic*, J. Number Theory **124** (2007), 454–469.

[GO] A. Gorodnik and H. Oh, *Rational points on homogeneous varieties and equidistribution of adelic periods* (with an appendix by M. Borovoi), Preprint, `arXiv:0803.1996v2`.

[K] S. Katok, *Fuchsian groups, geodesic flows on surfaces of constant negative curvature and symbolic coding of geodesics*, lecture notes in this volume.

[Kl1] D. Kleinbock, *Some applications of homogeneous dynamics to number theory*, in: *Smooth ergodic theory and its applications*, Proc. Sympos. Pure Math., Vol. 69, Amer. Math. Soc., Providence, RI, 2001, pp. 639–660.

[Kl2] ———, *Extremal subspaces and their submanifolds*, Geom. Funct. Anal. **13** (2003), 437–466.

[Kl3] ———, *Baker-Sprindžuk conjectures for complex analytic manifolds*, in: *Algebraic groups and Arithmetic*, Tata Inst. Fund. Res., Mumbai, 2004, 539–553.

[Kl4] ———, *Diophantine properties of measures and homogeneous dynamics*, Pure Appl. Math. Quarterly **4** (2008), 81–97.

[Kl5] ———, *An extension of quantitative nondivergence and applications to Diophantine exponents*, Trans. Amer. Math. Soc. **360** (2008), 6497–6523.

[Kl6] ———, *An 'almost all versus no' dichotomy in homogeneous dynamics and Diophantine approximation*, `arXiv:0904.1614`, Geom. Dedicata, to appear.

[KLW] D. Kleinbock, E. Lindenstrauss and B. Weiss, *On fractal measures and diophantine approximation*, Selecta Math. **10** (2004), 479–523.

[KM1] D. Kleinbock and G.A. Margulis, *Flows on homogeneous spaces and Diophantine approximation on manifolds*, Ann. Math. **148** (1998), 339–360.

[KM2] ———, *Logarithm laws for flows on homogeneous spaces*, Invent. Math. **138** (1999), 451–494.

[KSS] D. Kleinbock, N. Shah and A. Starkov, *Dynamics of subgroup actions on homogeneous spaces of Lie groups and applications to number theory*, in: *Handbook on Dynamical Systems*, Volume 1A, Elsevier Science, North Holland, 2002, pp. 813–930.

[KT] D. Kleinbock and G. Tomanov, *Flows on S-arithmetic homogeneous spaces and applications to metric Diophantine approximation*, Comm. Math. Helv. **82** (2007), 519–581.

[KW1] D. Kleinbock and B. Weiss, *Badly approximable vectors on fractals*, Israel J. Math. **149** (2005), 137–170.

[KW2] ———, *Dirichlet's theorem on diophantine approximation and homogeneous flows*, J. Mod. Dyn. **2** (2008), 43–62.

[LM] E. Lindenstrauss and M. Mirzakhani, *Ergodic theory of the space of measured laminations*, Internat. Math. Res. Notices **2008**, no. 4, doi:10.1093/imrn/rnm126.

[Mah] K. Mahler, *Über das Mass der Menge aller S-Zahlen*, Math. Ann. **106** (1932), 131–139.

[Mar] G.A. Margulis, *On the action of unipotent groups in the space of lattices*, In *Lie Groups and their representations*, Proc. of Summer School in Group Representations, Bolyai Janos Math. Soc., Akademai Kiado, Budapest, 1971, pp. 365–370, Halsted, New York, 1975.

[MW] Y. Minsky and B. Weiss, *Nondivergence of horocyclic flows on moduli space*, J. Reine Angew. Math. **552** (2002), 131–177.

[Mor] D. Morris, *Ratner's theorems on unipotent flows*, Chicago Lectures in Mathematics. University of Chicago Press, Chicago, IL, 2005.

[Sch] W. Schmidt, *Diophantine approximation*, Springer-Verlag, Berlin, 1980.

[Sh1] N. Shah, *Equidistribution of expanding translates of curves and Dirichlet's theorem on Diophantine approximation*, Invent. Math. **177** (2009), no. 3, 509–532.

[Sh2] ———, *Expanding translates of curves and Dirichlet-Minkowski theorem on linear forms*, Preprint, arXiv:0804.1424.

[Sp1] V.G. Sprindžuk, *More on Mahler's conjecture*, Dokl. Akad. Nauk SSSR, **155** (1964), 54–56.

[Sp2] ———, *Mahler's problem in metric number theory*, Translations of Mathematical Monographs, Amer. Math. Soc., Providence, R.I., 1969.

[Sp3] ———, *Achievements and problems in Diophantine approximation theory*, Russian Math. Surveys **35** (1980), 1–80.

[St] A. Starkov, *Dynamical systems on homogeneous spaces*, Translations of Mathematical Monographs, Amer. Math. Soc., Providence, RI, 2000.

[Y] J-C. Yoccoz, *Interval exchange maps and translation surfaces*, lecture notes in this volume.

Brandeis University, Waltham MA 02454-9110
E-mail address: kleinboc@brandeis.edu

Clay Mathematics Proceedings
Volume 10, 2010

Diagonal actions on locally homogeneous spaces

M. Einsiedler and E. Lindenstrauss

Contents

1. Introduction

1.1. In these notes we present some aspects of work we have conducted, in parts jointly with Anatole Katok, regarding dynamics of higher rank diagonalizable groups on (locally) homogeneous spaces[(1)] $\Gamma\backslash G$. A prototypical example of such an action is the action of the group of determinant one diagonal matrices A on the space of lattices in $\mathbb{R}^n$ with covolume one for $n \geq 3$ which can be identified with the quotient space $\mathrm{SL}(n,\mathbb{Z})\backslash \mathrm{SL}(n,\mathbb{R})$. More specifically, we consider the problem of classifying measures invariant under such an action, and present two of the applications of this measure classification.

[(1)]The space $X = \Gamma\backslash G$ we define is in fact a homogeneous space for the group G in the abstract sense of algebra but if we also consider the metric structure, see §7.1, the phrase "locally homogeneous" seems more appropriate.

There have been several surveys on this topic, including some that we have written (specifically, [**Lin05**] and [**EL06**]). For this reason we will be brief in our historical discussions and the discussion of the important work of the pioneers of the subject.

1.2. For the more general setup let $G = \mathbb{G}(\mathbb{R})$ be the group of $\mathbb{R}$-points of a linear algebraic group over $\mathbb{R}$, and let $\Gamma < G$ be a lattice (i.e., a discrete, finite covolume subgroup). In this setup it is natural to consider for any subgroup $H < G$, in particular for any algebraic subgroup, the action of H on the symmetric space $\Gamma\backslash G$. Ratner's landmark measure classification theorem (which is somewhat more general as it considers the case of G a general Lie group) states the following:

1.3. Theorem (M. Ratner [**Rt91**]). *Let G, Γ be as above, and let $H < G$ be an algebraic subgroup generated by one parameter unipotent subgroups. Then any H-invariant and ergodic probability measure μ is the natural (i.e., L-invariant) probability measure on a single orbit of some closed subgroup $L < G$ ($L = G$ is allowed).*

We shall call a probability measure of the type above (i.e., supported on a single orbit of its stabilizer group) *homogeneous.*

1.4. For one parameter diagonalizable flows the (partial) hyperbolicity of the flow guarantees the existence of many invariant measures. It is, however, not unreasonable to hope that for multiparameter diagonalizable flows the situation is better. For example one has the following conjecture attributed to Furstenberg, Katok-Spatzier and Margulis:

1.5. Conjecture. *Let A be the group of diagonal matrices in $\mathrm{SL}(n,\mathbb{R})$, $n \geq 3$. Then any A-invariant and ergodic probability measure on $\mathrm{SL}(n,\mathbb{Z})\backslash \mathrm{SL}(n,\mathbb{R})$ is homogeneous.*

The reader may note that we have phrased Conjecture 1.5 in a much more specialized way than Theorem 1.3. While the basic phenomena behind the conjecture is expected to be quite general, care must be exercised when stating it more generally (even for the groups A and G given above).

1.6. Conjecture 1.5 is open, but some progress has been made. Specifically, in our joint paper with Katok [**EKL06**], Conjecture 1.5 is proved under the condition that there is some $a \in A$ with positive entropy (see Theorem 11.5 below for a more formal statement).

1.7. These lecture notes are based on our joint course given in the CMI Pisa summer school as well as a graduate course given by the second named author in Princeton the previous semester. Notes for both were carefully taken by Shimon Brooks and thoroughly edited by us. The material presented here has almost entirely been published in several research papers, in particular [**EK03, Lin06, EK05, EKL06, EL08**].

1.8. The treatment here differs from the original treatment in places, hopefully for the better. In particular, we use this opportunity to give an alternative simplified treatment of the high entropy method developed by M.E. and Katok in [**EK03, EK05**]. For this reason our treatment of the high entropy method in §9 is much more careful and thorough than our treatment of the low entropy method in the following section (the reader who wishes to learn this technique in greater detail is advised to look at our recent paper [**EL08**]).

It is interesting to note that what we call the low entropy method for studying measures invariant under diagonalizable groups uses heavily unipotent dynamics, and, in particular, ideas of Ratner developed in her study of isomorphism and joining rigidity in [**Rt82b, Rt82a, Rt83**] which was a precursor to her more general results on unipotent flows in [**Rt90, Rt91**].

1.9. More generally, the amount of detail given on the various topics is not uniform. Our treatment of the basic machinery of leafwise measures as well as entropy in §3-7 is very thorough as are the next two sections §8-9. This has some correlation to the material given in the Princeton graduate course, though the presentation of the high entropy method given here is more elaborate.

The last two sections of these notes give a sample of some of the applications of the measure classification results given in earlier chapters. We have chosen to present only two: our result with Katok on the set of exceptions to Littlewood's Conjecture from [**EKL06**] and the result of E.L. on Arithmetic Quantum Unique Ergodicity from [**Lin06**]. The measure classification results presented here also have other applications; in particular we mention our joint work with P. Michel and A. Venkatesh on the distribution properties of periodic torus orbits [**ELMV09, ELMV07**].

1.10. One day a more definitive and complete treatment of these measure rigidity results would be written, perhaps by us. Until that day we hope that these notes, despite their obvious shortcomings, might be useful.

Acknowledgements. This work owes a debt to the Clay Mathematical Institute in more than one way. We thank CMI for its support of both of us (E.L. was supported by CMI during the years 2003-2005, and M.E. was supported by CMI in the second half of 2005). Many of the ideas we present here were developed during this period. We also thank CMI for the opportunity it provided us to present our work to a wide and stimulating audience in the Pisa summer school. We also thank Shimon Brooks for his careful notetaking. Finally we thank Shirali Kadyrov, Beverly Lytle, Fabrizio Polo, Alex Ustian, and in particular Uri Shapira for comments on the manuscript. The work presented here has been obtained over several years and supported by several NSF grants, in particular DMS grants DMS-0554373, 0622397 (ME), 0500205 and 0554345 (EL).

2. Ergodic theory: some background

We start by summarizing a few basic notions of ergodic theory, and refer the reader with the desire to see more details to any book on ergodic theory, e.g. [**Wal82**], [**Gla03**], or [**EW09**].

2.1. Definition. *Let X be a locally compact space, equipped with an action of a noncompact (but locally compact) group*[(2)] *H which we denote by $(h, x) \mapsto h.x$ for $h \in H$ and $x \in X$. An H-invariant probability measure μ on X is said to be* ergodic *if one of the following equivalent conditions holds:*

[(2)] All groups will be assumed to be second countable locally compact, all measures Borel probability measures unless otherwise specified.

(i) *Suppose $Y \subset X$ is a measurable H-invariant set, i.e., $h.Y = Y$ for every $h \in H$. Then $\mu(Y) = 0$ or $\mu(X \setminus Y) = 0$.*
(ii) *Suppose f is a measurable function on X with the property that for every $h \in H$, for μ-a.e. x, $f(h.x) = f(x)$. Then f is equal to a constant a.e.*
(iii) *μ is an extreme point of the convex set of all H-invariant Borel probability measures on X.*

2.2. A stronger condition which implies ergodicity is mixing:

2.3. Definition. *Let X, H and μ be as in Definition 2.1. The action of H is said to be* mixing *if for any sequence $h_i \to \infty$ in H*[(3)] *and any measurable subsets $B, C \subset X$,*

$$\mu(B \cap h_i.C) \to \mu(B)\mu(C) \qquad \text{as } i \to \infty.$$

Recall that two sets B, C in a probability space are called *independent* if

$$\mu(B \cap C) = \mu(B)\mu(C).$$

So mixing is asking for two sets to be asymptotically independent (when one of the sets is moved by bigger and bigger elements of H).

2.4. A basic fact about H-invariant measures is that any H-invariant measure is an average of ergodic measures, i.e., there is some auxiliary probability space (Ξ, ν) and a (measurable) map attaching to each $\xi \in \Xi$ an H-invariant and ergodic probability measure μ_ξ on X so that

$$\mu = \int_\Xi \mu_\xi d\nu(\xi).$$

This is a special case of Choquet's theorem on representing points in a compact convex set as generalized convex combinations of extremal points.

2.5. Definition. *An action of a group H on a locally compact topological space X is said to be* uniquely ergodic *if there is only one H-invariant probability measure on X.*

2.6. The simplest example of a uniquely ergodic transformation is the map $T_\alpha : x \mapsto x + \alpha$ on the one dimensional torus $\mathbb{T} = \mathbb{R}/\mathbb{Z}$ where α is irrational. Clearly Lebesgue measure m on $\mathbb{T}$ is T_α-invariant; we need to show it is the only such probability measure.

To prove this, let μ be an arbitrary T_α-invariant probability measure. Since μ is T_α-invariant,

$$\hat{\mu}(n) = \int_\mathbb{T} e(nx)d\mu(x) = \int_\mathbb{T} e(n(x+\alpha))d\mu(x) = e(n\alpha)\hat{\mu}(n),$$

where as usual $e(x) = \exp(2\pi i x)$. Since α is irrational, $e(n\alpha) \neq 1$ for all $n \neq 0$, hence $\hat{\mu}(n) = 0$ for all $n \neq 0$ and clearly $\hat{\mu}(0) = 1$. Since the functions $e(nx)$ span a dense subalgebra of the space of continuous functions on $\mathbb{T}$ we have $\mu = m$.

2.7. Definition. *Let X be a locally compact space, and suppose that $H = \{h_t\} \cong \mathbb{R}$ acts continuously on X. Let μ be an H-invariant measure on X. We say that $x \in X$*

---

(3) I.e., a sequence so that for any compact $K \subset H$ only finitely many of the h_i are in K.

is generic for μ *if for every* $f \in C_0(X)$ *we have*(4):

$$\frac{1}{T}\int_0^T f(h_t.x)\,dt \to \int_X f(y)\,d\mu(y) \qquad \text{as } T \to \infty.$$

Equidistribution is another closely related notion:

2.8. Definition. *A sequence of probability measures* μ_n *on a locally compact space* X *is said to be* equidistributed *with respect to a (usually implicit) measure* m *if they converge to* m *in the weak* topology, i.e., if* $\int f\,d\mu_n \to \int f\,dm$ *for every* $f \in C_0(X)$.

A sequence of points $\{x_n\}$ *in* X *is said to be equidistributed if the sequence of probability measures* $\mu_N = N^{-1}\sum_{n=1}^N \delta_{x_n}$ *is equidistributed, i.e., if for every* $f \in C_0(X)$

$$\frac{1}{N}\sum_{n=1}^N f(x_n) \to \int_X f(y)\,dm(y) \qquad \text{as } N \to \infty.$$

Clearly there is a lot of overlap between the two definitions, and in many situations "equidistributed" and "generic" can be used interchangeably.

2.9. For an arbitrary invariant measure μ on X with respect to an action of $H \cong \mathbb{R}$, the Birkhoff pointwise ergodic theorem shows that μ-almost every point $x \in X$ is generic with respect to an invariant and ergodic probability measure on X (which leads to a construction of the ergodic decomposition). If μ is ergodic, μ-a.e. $x \in X$ is generic for μ.

If X is compact, and if the action of $H \cong \mathbb{R}$ on X is uniquely ergodic with μ being the unique H-invariant measure, then something much stronger is true: *every* $x \in X$ is generic for μ!

Indeed, let μ_T denote the probability measure

$$\mu_T = \frac{1}{T}\int_0^T \delta_{h_t.x}\,dt$$

for any $T > 0$. Then any weak* limit of μ_T as $T \to \infty$ will be H-invariant. However, there is only one H-invariant probability measure(5) on X, namely μ, so $\mu_T \to \mu$, i.e., x is generic for μ.

For the irrational rotation considered in §2.6 it follows that all orbits are equidistributed. A more interesting example is provided by the horocycle flow on *compact* quotients $\Gamma\backslash \mathrm{SL}(2,\mathbb{R})$. The unique ergodicity of this system is a theorem due to Furstenberg [**Fur73**] and is covered in the lecture notes [**Esk**] by Eskin.

3. Entropy of dynamical systems: some more background

3.1. A very basic and important invariant in ergodic theory is *entropy*. It can be defined for any action of a (not too pathological) unimodular amenable group H preserving a probability measure [**OW87**], but for our purposes we will only need (and only consider) the case $H \cong \mathbb{R}$ or $H \cong \mathbb{Z}$. For more details we again refer to [**Wal82**], [**Gla03**], or [**ELW09**].

(4)Where $C_0(X)$ denotes the space of continuous functions on X which decay at infinity, i.e., so that for any $\epsilon > 0$ the set $\{x : |f(x)| \geq \epsilon\}$ is compact.

(5)This uses the assumption that X is compact. If X is non-compact, one would have to address the possibility of the limit not being a probability measure. This possibility is often described as *escape of mass*.

Entropy is an important tool also in the study of unipotent flows[(6)], but plays a much more prominent role in the study of diagonalizable actions which we will consider in these notes.

3.2. Let (X,μ) be a probability space. The static entropy $H_\mu(\mathcal{P})$ of a finite or countable partition $\mathcal{P}$ of X is defined to be

$$H_\mu(\mathcal{P}) = -\sum_{P\in\mathcal{P}} \mu(P)\log\mu(P),$$

which in the case where $\mathcal{P}$ is countable may be finite or infinite.

One basic property of entropy is sub-additivity; the entropy of the refinement $\mathcal{P}\vee\mathcal{Q} = \{P\cap Q : P\in\mathcal{P}, Q\in\mathcal{Q}\}$ satisfies

$$H_\mu(\mathcal{P}\vee\mathcal{Q}) \leq H_\mu(\mathcal{P}) + H_\mu(\mathcal{Q}). \tag{3.2a}$$

However, this is just a starting point for many more natural identities and properties of entropy, e.g. equality holds in (3.2a) if and only if $\mathcal{P}$ and $\mathcal{Q}$ are independent, the latter means that any element of $\mathcal{P}$ is independent of any element of $\mathcal{Q}$ in the sense of probability theory. All these natural properties find good explanations if one interprets $H_\mu(\mathcal{P})$ as the average of the information function

$$I_\mu(\mathcal{P})(x) = -\log\mu(P) \text{ for } x\in P\in\mathcal{P}$$

which measures the amount of information revealed about x if one is told the partition element $P\in\mathcal{P}$ that contains $x\in P$.

3.3. The ergodic theoretic entropy $h_\mu(T)$ associated to a measure-preserving map $T: X\to X$ can be defined using the entropy function H_μ as follows:

3.4. Definition. *Let μ be a probability measure on X and $T: X\to X$ a measurable map preserving μ. Let $\mathcal{P}$ be either a finite or a countable[(7)] partition of X with $H_\mu(\mathcal{P})<\infty$. The entropy of the four-tuple $(X,\mu,T,\mathcal{P})$ is defined to be[(8)]*

$$h_\mu(T,\mathcal{P}) = \lim_{N\to\infty}\frac{1}{N}H_\mu\left(\bigvee_{n=0}^{N-1} T^{-n}\mathcal{P}\right). \tag{3.4a}$$

The ergodic theoretic entropy of (X,μ,T) is defined to be

$$h_\mu(T) = \sup_{\mathcal{P}:H_\mu(\mathcal{P})<\infty} h_\mu(T,\mathcal{P}).$$

The ergodic theoretic entropy was introduced by A. Kolmogorov and Ya. Sinai and is often called the Kolmogorov-Sinai entropy.[(9)] We may interpret the entropy $h_\mu(T)$ as a measure of the complexity of the transformation with respect to the measure μ. We will discuss this in greater detail later, but the geodesic flow has positive entropy with respect to the Haar measure on $\Gamma\backslash\mathrm{SL}(2,\mathbb{R})$ while the horocycle flow has zero entropy. However, vanishing entropy does not mean that the

(6)In particular, in [**MT94**] Margulis and Tomanov give a shorter proof of Ratner's measure classification theorem using entropy theory.

(7)One may also restrict oneself to finite partitions without changing the outcome, but we will see situations where it will be convenient to allow countable partitions.

(8)Note that by the subadditivity of the entropy function H_μ the limit in (3.4a) exists and is equal to $\inf_N \frac{1}{N}H_\mu(\bigvee_{n=0}^{N-1}T^{-n}\mathcal{P})$.

(9)Ergodic theoretic entropy is also somewhat confusingly called the metric entropy (even though it has nothing to do with any metric that might be defined on X!).

dynamics of the transformation or the flow is simple, e.g. the horocycle flow is mixing with respect to the Haar measure on $\Gamma\backslash \mathrm{SL}(2,\mathbb{R})$. Also, one can find quite complicated measures μ on $\Gamma\backslash \mathrm{SL}(2,\mathbb{R})$ that are invariant under the geodesic flow and with respect to which the geodesic flow has zero entropy.

3.5. If μ is a T-invariant but not necessarily ergodic measure, it can be shown that the entropy of μ is the average of the entropies of its ergodic components: i.e., if μ has the ergodic decomposition $\mu = \int \mu_\xi d\nu(\xi)$, then

$$h_\mu(T) = \int h_{\mu_\xi}(T) d\nu(\xi). \tag{3.5a}$$

Therefore, it follows that an invariant measure with positive entropy has in its ergodic decomposition a *positive fraction* of ergodic measures with positive entropy.

3.6. We will see in §7 concrete formulas and estimates for the entropy of flows on locally homogeneous spaces $\Gamma\backslash G$. To obtain these the main tool is the following notion: A partition $\mathcal{P}$ is said to be a *generating partition* for T and μ if the σ-algebra $\bigvee_{n=-\infty}^{\infty} T^{-n}\mathcal{P}$ (i.e., the σ-algebra generated by the sets $\{T^n P : n \in \mathbb{Z}, P \in \mathcal{P}\}$) separates points; that is, for μ-almost every x, the atom of x with respect to this σ-algebra is $\{x\}$.[10] The Kolmogorov-Sinai theorem asserts the non-obvious fact that $h_\mu(T) = h_\mu(T,\mathcal{P})$ whenever $\mathcal{P}$ is a generating partition.

3.7. We have already indicated that we will be interested in the entropy of flows. So we need to define the ergodic theoretic entropy for flows (i.e., for actions of groups $H \cong \mathbb{R}$). Suppose $H = \{a_t\}$ is a one parameter group acting on X. Then it can be shown that for $s \neq 0$, $\frac{1}{|s|} h_\mu(x \mapsto a_s.x)$ is independent of s. We define the entropy of μ with respect to $\{a_t\}$, denoted $h_\mu(a_\bullet)$, to be this common value of $\frac{1}{|s|} h_\mu(x \mapsto a_s.x)$.[11]

3.8. Suppose now that (X,d) is a compact metric space, and that $T : X \to X$ is a homeomorphism (the pair (X,T) is often implicitly identified with the generated $\mathbb{Z}$-action and is called a topological dynamical system).

3.9. Definition. *The $\mathbb{Z}$-action on X generated by T is said to be* expansive *if there is some $\delta > 0$ so that for every $x \neq y \in X$ there is some $n \in \mathbb{Z}$ so that $d(T^n x, T^n y) > \delta$.*

If X is expansive then any measurable partition $\mathcal{P}$ of X for which the diameter of every element of the partition is $< \delta$ is generating (with respect to any measure μ) in the sense of §3.6.

3.10. Problem. *Let A be a $d \times d$ integer matrix with determinant 1 or -1. Then A defines a dynamical system on $X = \mathbb{R}^n/\mathbb{Z}^n$. Characterize expansiveness of A with respect to the metric derived from the Euclidean metric on $\mathbb{R}^n$. Also determine whether an element of the geodesic flow on a compact quotient $\Gamma\backslash \mathrm{SL}(2,\mathbb{R})$ is expansive.*

---

[10] Recall that the atom of x with respect to a countably generated σ-algebra $\mathcal{A}$ is the intersection of all $B \in \mathcal{A}$ containing x and is denoted by $[x]_{\mathcal{A}}$. We will discuss that and related notions in greater detail in §5.

[11] Note that $h_\mu(a_\bullet)$ depends not only on H as a group but on the particular parametrization a_t.

3.11. For some applications presented later, an important fact is that for many dynamical systems (X, T) the map $\mu \mapsto h_\mu(T)$ defined on the space of T-invariant probability measures on X is semicontinuous. This phenomenon is easiest to see when (X, T) is expansive.

3.12. Proposition. *Suppose (X, T) is expansive, and that μ_i, μ are T-invariant probability measures on X with $\mu_i \to \mu$ in the weak* topology. Then*

$$h_\mu(T) \geq \overline{\lim_{i\to\infty}}\, h_{\mu_i}(T).$$

In less technical terms, for expansive dynamical systems, a "complicated" invariant measure might be approximated by a sequence of "simple" ones, but not vice versa.

3.13. Proof. Let $\mathcal{P}$ be a partition of X such that for each $P \in \mathcal{P}$

(i) $\mu(\partial P) = 0$

(ii) P has diameter $< \delta$ (δ as in the definition of expansiveness).

As X is compact, such a partition can easily obtained from a (finite sub-cover of a) cover of X consisting of small enough balls satisfying (i).

Since $\mu(\partial P) = 0$ and $\mu_i \to \mu$ weak*, for every $P \in \mathcal{P}$ we have that $\mu_i(P) \to \mu(P)$. Then for a fixed N we have (using footnote (8) for the measure μ_i) that

$$\frac{1}{N} H_\mu\left(\bigvee_{n=0}^{N-1} T^{-n}\mathcal{P}\right) = \lim_{i\to\infty} \frac{1}{N} H_{\mu_i}\left(\bigvee_{n=0}^{N-1} T^{-n}\mathcal{P}\right)$$
$$\geq \overline{\lim_{i\to\infty}}\, h_{\mu_i}(T, \mathcal{P}) \overset{\text{(by (ii))}}{=} \overline{\lim_{i\to\infty}}\, h_{\mu_i}(T).$$

Taking the limit as $N \to \infty$ we get

$$h_\mu(T) = h_\mu(T, \mathcal{P}) = \lim_{N\to\infty} \frac{1}{N} H_\mu\left(\bigvee_{n=0}^{N-1} T^{-n}\mathcal{P}\right) \geq \overline{\lim_{i\to\infty}}\, h_{\mu_i}(T)$$

as required. □

Note that we have used both (ii) and expansiveness only to establish

(ii′) $h_\nu(T) = h_\nu(T, \mathcal{P})$ for $\nu = \mu_1, \mu_2, \ldots$.

We could have used the following weaker condition: for every ϵ, there is a partition $\mathcal{P}$ satisfying (i) and

(ii″) $h_\nu(T) \leq h_\nu(T, \mathcal{P}) + \epsilon$ for $\nu = \mu_1, \mu_2, \ldots$.

3.14. We are interested in dynamical systems of the form $X = \Gamma\backslash G$ (G a connected Lie group and $\Gamma < G$ a lattice) and

$$T : x \mapsto g.x = xg^{-1}.$$

Many such systems[(12)] will not be expansive, and furthermore in the most interesting case of $X_n = \mathrm{SL}(n, \mathbb{Z})\backslash \mathrm{SL}(n, \mathbb{R})$ the quotient is not compact (which we assumed throughout the above discussion of expansiveness).

Even worse, on $X_2 = \mathrm{SL}(2, \mathbb{Z})\backslash \mathrm{SL}(2, \mathbb{R})$ one may have a sequence of probability measures μ_i ergodic and invariant under the one parameter group

$$\left\{ a_t = \begin{pmatrix} e^{t/2} & 0 \\ 0 & e^{-t/2} \end{pmatrix} \right\}$$

(12) For example, the geodesic flow defined on quotients of $G = \mathrm{SL}(2, \mathbb{R})$.

with $\underline{\lim}_{i\to\infty} h_{\mu_i}(a_\bullet) > 0$ converging weak* to a measure μ which is not a probability measure — there is escape of mass — and furthermore has zero entropy(13).

However, one has the following "folklore theorem"(14) :

3.15. Proposition. *Let G be a connected Lie group, $\Gamma < G$ a lattice, and $H = \{a_t\}$ a one parameter subgroup of G. Suppose that μ_i, μ are H-invariant* probability(15) *measures on X with $\mu_i \to \mu$ in the weak* topology. Then*

$$h_\mu(a_\bullet) \geq \overline{\lim_{i\to\infty}}\, h_{\mu_i}(a_\bullet).$$

For X compact (and possibly by some clever compactification also for general X), this follows from deep (and complicated) work of Yomdin, Newhouse and Buzzi (see e.g. [**Buz97**] for more details); however Proposition 3.15 can be established quite elementarily. In order to prove this proposition, one shows that any sufficiently fine finite partition of X satisfies §3.11(ii$''$).

3.16. The following example shows that this semicontinuity does not hold for a general dynamical system:

3.17. Example. *Let $S = \left\{1, \frac{1}{2}, \frac{1}{3}, \dots, 0\right\}$, and $X = S^{\mathbb{Z}}$ (equipped with the usual Tychonoff topology). Let $\sigma : X \to X$ be the shift map defined by $\sigma(x)_n = x_{n+1}$ for $x = (x_n)_{n\in\mathbb{Z}} \in X$.*

Let μ_n be the probability measure on X obtained by taking the product of the probability measures on S giving equal probability to 0 and $\frac{1}{n}$, and $\delta_{\mathbf{0}}$ the probability measure supported on the fixed point $\mathbf{0} = (\dots, 0, 0, \dots)$ of σ. Then $\mu_n \to \delta_{\mathbf{0}}$ weak, $h_{\mu_n}(\sigma) = \log 2$ but $h_{\delta_{\mathbf{0}}}(\sigma) = 0$.*

3.18. Let (X, d) be a compact metric space and let $T : X \to X$ be continuous. Two points $x, x' \in X$ are said to be *k, ϵ-separated* if for some $0 \leq \ell < k$ we have that $d(T^\ell x, T^\ell x') \geq \epsilon$. Let $N(X, T, k, \epsilon)$ denote the maximal cardinality of a k, ϵ-separated subset of X.

3.19. Definition. *The* topological entropy(16) *of (X, T) is defined by*

$$h_{\text{top}}(X, T) = \lim_{\epsilon\to 0} H(X, T, \epsilon),$$

where

$$H(X, T, \epsilon) = \underline{\lim}_{k\to\infty} \frac{\log N(X, T, k, \epsilon)}{k}$$

The topological entropy of a flow $\{a_t\}$ is defined as in §3.7 and denoted by $h_{\text{top}}(X, a_\bullet)$.

---

(13) Strictly speaking, we define entropy only for probability measures, so one needs to rescale μ first.

(14) Which means in particular that there seems to be no good reference for it. A special case of this proposition is proved in [**EKL06**, Section 9]. The proof of this proposition is left as an exercise to the energetic reader.

(15) Here we assume that the weak* limit is a probability measure as, unlike the case of unipotent flows, there is no general fact that rules out various weird situations. E.g., for the geodesic flow on a noncompact quotient X of $\mathrm{SL}(2, \mathbb{R})$ it is possible to construct a sequence of invariant probability measures whose limit μ satisfies $\mu(X) = 1/2$.

(16) For X which is only locally compact, one can extend T to a map $\tilde{T}$ on its one-point compactification $\tilde{X} = X \cup \{\infty\}$ fixing ∞ and define $h_{\text{top}}(X, T) = h_{\text{top}}(\tilde{X}, \tilde{T})$.

3.20. Topological entropy and the ergodic theoretic entropy are related by the *variational principle* (see e.g. [**Gla03**, Theorem 17.6] or [**KH95**, Theorem 4.5.3]).

3.21. Proposition. *Let X be a compact metric space and $T : X \to X$ a homeomorphism.*[(17)] *Then*

$$h_{\mathrm{top}}(X,T) = \sup_{\mu} h_\mu(T)$$

where the sup *runs over all T-invariant probability measures on X.*

Note that when $\mu \mapsto h_\mu(T)$ is upper semicontinuous (see §3.11) the supremum is actually attained by some T-invariant measure on X. These *measures of maximal entropy* are often quite natural measures, e.g. in many cases they are Haar measures on $\Gamma\backslash G$.

3.22. To further develop the theory of entropy we need to recall in the next few sections some more notions from measure theory.

4. Conditional Expectation and Martingale theorems

The material of this and the following section can be found in greater detail e.g. in [**EW09**].

4.1. Proposition. *Let $(X, \mathcal{B}, \mu)$ be a probability space, and $\mathcal{A} \subset \mathcal{B}$ a sub-σ-algebra. Then there exists a continuous linear functional*

$$E_\mu(\cdot|\mathcal{A}) : L^1(X,\mathcal{B},\mu) \to L^1(X,\mathcal{A},\mu)$$

called the conditional expectation of f given $\mathcal{A}$, *such that*

$$E_\mu(f|\mathcal{A}) \text{ is } \mathcal{A}\text{-measurable} \tag{4.1a}$$

for any $f \in L^1(X,\mathcal{B},\mu)$, and we have

$$\int_A E_\mu(f|\mathcal{A})d\mu = \int_A f d\mu \text{ for all } A \in \mathcal{A}. \tag{4.1b}$$

Moreover, together equations (4.1a)–(4.1b) *characterizes the function $E_\mu(f|\mathcal{A}) \in L^1(X,\mathcal{B},\mu)$.*

On $L^2(X,\mathcal{B},\mu)$ the operator $E_\mu(\cdot|\mathcal{A})$ is simply the orthogonal projection to the closed subspace $L^2(X,\mathcal{A},\mu)$. From there one can extend the definition by continuity to $L^1(X,\mathcal{B},\mu)$. Often, when we only consider one measure we will drop the measure in the subscript.

Below we will base our arguments on the dynamical behavior of points. Because of that we prefer to work with functions instead of equivalence classes of functions and hence the above uniqueness has be understood accordingly. We will need the following useful properties of the conditional expectation $E(f|\mathcal{A})$, which we already phrase in terms of functions rather than equivalence classes of functions:

4.2. Proposition.
- (i) *$E(\cdot|\mathcal{A})$ is a positive operator of norm 1, and moreover, $|E(f|\mathcal{A})| \le E(|f||\mathcal{A})$ almost everywhere.*
- (ii) *For $f \in L^1(X,\mathcal{B},\mu)$ and $g \in L^\infty(X,\mathcal{A},\mu)$, we have $E(gf|\mathcal{A}) = gE(f|\mathcal{A})$ almost everywhere.*

---

(17) This proposition also easily implies the analogous statement for flows $\{a_t\}$.

(iii) *If $\mathcal{A}' \subset \mathcal{A}$ is a sub-σ-algebra, then*

$$E(E(f|\mathcal{A})|\mathcal{A}') = E(f|\mathcal{A}')$$

almost everywhere. Moreover, if $f \in L^1(X, \mathcal{A}, \mu)$, then $E(f|\mathcal{A}) = f$ almost everywhere.

(iv) *If $T : X \to Y$ sends the probability measure μ on X to $T_*\mu = \mu \circ T^{-1} = \nu$ on Y, and if $\mathcal{C}$ is a sub-σ-algebra of the σ-algebra $\mathcal{B}_Y$ of measurable sets on Y, then $E_\mu(f \circ T|T^{-1}\mathcal{C}) = E_\nu(f|\mathcal{C}) \circ T$ for any $f \in L^1(Y, \mathcal{B}_Y, \nu)$.*

We only prove the last two claims. Take any $A \in \mathcal{A}' \subset \mathcal{A}$. By the characterizing property of conditional expectation, we have

$$\int_A E(E(f|\mathcal{A})|\mathcal{A}') = \int_A E(f|\mathcal{A}) = \int_A f.$$

Therefore by uniqueness, we have $E(E(f|\mathcal{A})|\mathcal{A}') = E(f|\mathcal{A}')$ almost everywhere. If $f \in L^1(X, \mathcal{A}, \mu)$, then f satisfies the first characterizing property of $E(f|\mathcal{A})$, while trivially satisfying the second. Again invoking uniqueness, we have $E(f|\mathcal{A}) = f$ almost everywhere.

We consider now the situation of the pushforward $T_*\mu = \nu$ of the measure and the pullback $T^{-1}\mathcal{C}$ of the σ-algebra. By the definitions we have

$$\int_{T^{-1}C} E_\nu(f|\mathcal{C}) \circ T d\mu = \int_C E_\nu(f|\mathcal{C}) d\nu = \int_C f d\nu = \int_{T^{-1}C} f \circ T d\mu$$

for any $C \in \mathcal{C}$, which implies the claim by the uniqueness properties of conditional expectation.

4.3. The next two theorems describes how the conditional expectation behaves with respect to a sequence of sub-σ-algebras, and can be thought of as continuity properties.

4.4. Theorem (Increasing Martingale Convergence Theorem). *Let $\mathcal{A}_1, \mathcal{A}_2, \dots$ be a sequence of σ-algebras, such that $\mathcal{A}_i \subset \mathcal{A}_j$ for all $i < j$. Let $\mathcal{A}$ be the smallest σ-algebras containing all of the $\mathcal{A}_n$ (in this case, we write $\mathcal{A}_n \nearrow \mathcal{A}$). Then*

$$E(f|\mathcal{A}_n) \to E(f|\mathcal{A})$$

almost everywhere and in L^1.

4.5. Theorem (Decreasing Martingale Convergence Theorem). *Suppose that we have a sequence of σ-algebras $\mathcal{A}_i \searrow \mathcal{A}$, i.e., such that $\mathcal{A}_i \supset \mathcal{A}_j$ for $i < j$, and $\mathcal{A} = \bigcap \mathcal{A}_i$. Then $E(f|\mathcal{A}_n)(x) \to E(f|\mathcal{A})(x)$ almost everywhere and in L^1.*

4.6. Remark. In many ways, the Decreasing Martingale Convergence Theorem is similar to the pointwise ergodic theorem. Both theorems have many similarities in their proof with the pointwise ergodic theorem and other theorems; the proofs consists of two steps, convergence in L^1, and a maximum inequality to deduce pointwise convergence.

5. Countably generated σ-algebras and conditional measures

Note that the algebra generated by a countable set of subsets of X is countable, but that in general the same is not true for the σ-algebra generated by a countable set of subsets of X. E.g. the Borel σ-algebra of any space we consider is countably generated in the following sense.

5.1. Definition. *A σ-algebra $\mathcal{A}$ in a space X is* countably generated *if there is a countable set (or equivalently algebra) $\mathcal{A}_0$ of subsets of X such that the smallest σ-algebra $\sigma(\mathcal{A}_0)$ that contains $\mathcal{A}_0$ is precisely $\mathcal{A}$.*

5.2. One nice feature of countably generated σ-algebras is that we can study the atoms of the algebra. If $\mathcal{A}$ is generated by a countable algebra $\mathcal{A}_0$, then we define the $\mathcal{A}$-*atom* of a point x to be

$$[x]_{\mathcal{A}} := \bigcap_{x\in A\in\mathcal{A}_0} A = \bigcap_{x\in A\in\mathcal{A}} A.$$

The equality follows since $\mathcal{A}_0$ is a generating algebra for the σ-algebra $\mathcal{A}$. In particular, it shows that the atom $[x]_{\mathcal{A}}$ does not depend on a choice of the generating algebra. Notice that by countability of $\mathcal{A}_0$ we have $[x]_{\mathcal{A}} \in \mathcal{A}$. In other words, $[x]_{\mathcal{A}}$ is the smallest set of $\mathcal{A}$ containing x. Hence the terminology — the atom of x cannot be broken up into smaller sets within the σ-algebra $\mathcal{A}$.

Note, in particular, that $[x]_{\mathcal{A}}$ could consist of the singleton x; in fact, this is the case for all atoms of the Borel σ-algebra on, say, $\mathbb{R}$. The notion of atoms is convenient when we want to consider conditional measures for smaller σ-algebras.

5.3. CAUTION. A sub σ-algebra of a countably generated σ-algebra need not be countably generated!

5.4. Lemma. *Let $(X, \mathcal{B}, \mu, T)$ be an invertible ergodic probability preserving system such that individual points have zero measure. Then the σ-algebra $\mathcal{E}$ of T–invariant sets (i.e., sets $B \in \mathcal{B}$ such that $B = T^{-1}B = TB$) is* not *countably generated.*

5.5. PROOF. Since T is ergodic, any set in $\mathcal{E}$ has measure 0 or 1, and in particular, this holds for any generating set. Suppose that $\mathcal{E}$ is generated by a countable collection $\{E_1, E_2, \ldots\}$, each E_i having measure 0 or 1. Taking the intersection of all generators E_i of measure one and the complement $X \setminus E_i$ of those of measure zero, we obtain an $\mathcal{E}$-atom $[x]_{\mathcal{E}}$ of measure 1. Since the orbit of x is invariant under T, we have that $[x]_{\mathcal{E}}$ must be the orbit of x. Since the orbit is at most countable, this is a contradiction since $\mu_x^{\mathcal{E}}([x]_{\mathcal{E}}) = 1$. □

5.6. We will now restrict ourselves to the case of X a locally compact, second-countable metric space; $\mathcal{B}$ will be the Borel σ-algebra on X. A space and σ-algebra of this form will be referred as a *standard Borel space*, and we will always take μ to be a Borel measure. We note that for such X, the Borel σ-algebra is countably generated by open balls with rational radius and center belonging to a countable dense subset of X. When working with a Borel measure on X, we may replace X by the one-point-compactification of X, extend the measure trivially to the compactification, and assume without loss of generality that X is compact.

5.7. Definition. *Let $\mathcal{A}$, $\mathcal{A}'$ be sub-σ-algebras of the σ-algebra $\mathcal{B}$ of a probability space $(X, \mathcal{B}, \mu)$. We say that $\mathcal{A}$ is* equivalent *to $\mathcal{A}'$* modulo μ *(denoted $\mathcal{A} \doteq_\mu \mathcal{A}'$) if for every $A \in \mathcal{A}$ there exists $A' \in \mathcal{A}'$ such that $\mu(A \triangle A') = 0$, and vice versa.*

5.8. Proposition. *Let $(X, \mathcal{B})$ be a standard Borel space, and let μ be a Borel probability measure on X. Then for every sub-σ-algebra $\mathcal{A} \subset \mathcal{B}$, there exists $\tilde{\mathcal{A}} \subset \mathcal{A}$ such that $\tilde{\mathcal{A}}$ is countably generated, and $\tilde{\mathcal{A}} \doteq_\mu \mathcal{A}$.*

Roughly speaking the proposition follows since the space $L^1(X, \mathcal{A}, \mu)$ is separable, which in turn is true because it is a subspace of $L^1(X, \mathcal{B}, \mu)$. One can define

$\tilde{\mathcal{A}}$ by a countable collection of sets $A_i \in \mathcal{A}$ for which the characteristic functions χ_{A_i} are dense in the set of all characteristic functions χ_A with $A \in \mathcal{A}$.

This Proposition conveniently allows us to ignore issues of countable generation, as long as we do so with respect to a measure (i.e., up to null sets) on a nice space.

We now wish to prove the existence and fundamental properties of conditional measures:

5.9. Theorem. *Let $(X, \mathcal{B}, \mu)$ be a probability space with $(X, \mathcal{B})$ being a standard Borel space, and let $\mathcal{A} \subset \mathcal{B}$ a sub-σ-algebra. Then there exists a subset $X' \subset X$ of full measure (i.e., $\mu(X \backslash X') = 0$), belonging to $\mathcal{A}$, and Borel probability measures $\mu_x^{\mathcal{A}}$ for $x \in X'$ such that:*

(i) *For every $f \in L^1(X, \mathcal{B}, \mu)$ we have $E(f|\mathcal{A})(x) = \int f(y) d\mu_x^{\mathcal{A}}(y)$ for almost every x. In particular, the right-hand side is $\mathcal{A}$-measurable as a function of x.*

(ii) *If $\mathcal{A} \doteq_\mu \mathcal{A}'$ are equivalent σ-algebras modulo μ, then we have $\mu_x^{\mathcal{A}} = \mu_x^{\mathcal{A}'}$ for almost every x.*

(iii) *If $\mathcal{A}$ is countably generated, then $\mu_x^{\mathcal{A}}([x]_{\mathcal{A}}) = 1$ for every $x \in X'$, and for $x, y \in X'$ we have that $[x]_{\mathcal{A}} = [y]_{\mathcal{A}}$ implies $\mu_x^{\mathcal{A}} = \mu_y^{\mathcal{A}}$.*

(iv) *The set X' and the map $x \overset{\tau}{\mapsto} \mu_x^{\mathcal{A}}$ are $\mathcal{A}$-measurable on X'; i.e., if U is open in $\mathcal{P}(X)$, the space of probability measures on X equipped with the weak* topology, then $\tau^{-1}(U) \in \mathcal{A}|_{X'}$.*

Moreover, the family of conditional measures $\mu_x^{\mathcal{A}}$ is almost everywhere uniquely determined by its relationship to the conditional expectation described above.

If $\mathcal{A}$ is countably generated, then $x, y \in X$ are called equivalent w.r.t. $\mathcal{A}$ if $[x]_{\mathcal{A}} = [y]_{\mathcal{A}}$. Hence (iii) also says that equivalent points have identical conditional measures.

5.10. Caution. In general we will only prove facts concerning the conditional measures $\mu_x^{\mathcal{A}}$ for almost every $x \in X$. In fact, we even restricted ourselves to a set X' of full measure in the existence of $\mu_x^{\mathcal{A}}$. However, even the set X' is by no means canonical. We also must understand the last claim regarding the uniqueness in that way; if we have two families of conditional measure defined on sets of full measure X' and X'', then one can find a *subset* of $X' \cap X''$ of full measure where they agree.

5.11. Comments. If $N \subset X$ is a null set, it is clear that $\mu_x^{\mathcal{A}}(N) = 0$ for a.e. x. (Use Theorem 5.9.(i) and Proposition 4.1 to check this.) However, we cannot expect more as, for a given x, the set $[x]_{\mathcal{A}}$ is often a null set.

If $B \subset X$ is measurable, then

$$\mu(\{x \in B : \mu_x^{\mathcal{A}}(B) = 0\}) = 0. \tag{5.11a}$$

To see this define $A = \{x : \mu_x^{\mathcal{A}}(B) = 0\} \in \mathcal{A}$ and use again Theorem 5.9.(i) and Proposition 4.1 to get

$$\mu(A \cap B) = \int_A \chi_B d\mu = \int_A \mu_x^{\mathcal{A}}(B) d\mu(x) = 0.$$

5.12. PROOF. Since we are working in a standard Borel space, we may assume that X is a compact, metric space. Hence, we may choose a countable set of continuous functions which give a dense $\mathbb{Q}$-vector space $\{f_0 \equiv 1, f_1, \ldots\} \subset C(X)$. Set $g_0 = f_0 \equiv 1$, and for each f_i with $i \geq 1$, pick[18] $g_i = E(f_i|\mathcal{A}) \in L^1(X, \mathcal{A}, \mu)$. Taking the union of countably many null sets there exists a null set N for the measure μ such that for all $\alpha, \beta \in \mathbb{Q}$ and all i, j, k:

- If $\alpha \leq f_i \leq \beta$ (on all of X), then $\alpha \leq g_i(x) \leq \beta$ for all $x \notin N$.
- If $\alpha f_i + \beta f_j = f_k$, then $\alpha g_i(x) + \beta g_j(x) = g_k(x)$ for $x \notin N$.

Now for all $x \notin N$, we have a continuous linear functional $\mathcal{L}_x : f_i \mapsto g_i(x)$ from $C(X) \to \mathbb{R}$ of norm $||\mathcal{L}_x|| \leq 1$. By the Riesz Representation Theorem, this yields a measure $\mu_x^{\mathcal{A}}$ on $C(X)$. This measure is characterized by $E(f|\mathcal{A})(x) = \mathcal{L}_x(f) = \int f(y)d\mu_x^{\mathcal{A}}(y)$ for all $f \in C(X)$. Using monotone convergence this can be extended to other class of functions: first to characteristic functions of compact and of open sets, then to characteristic functions of all Borel sets and finally to integrable functions, i.e., we have part (i) of the Theorem. As already remarked, this implies that $x \mapsto \int f(y)d\mu_x^{\mathcal{A}}(y)$ is an $\mathcal{A}$-measurable function for $x \notin N$. This implies part (iv).

Now suppose we have two equivalent σ-algebras $\mathcal{A}$ and $\mathcal{A}'$ modulo μ, and take their common refinement $\tilde{\mathcal{A}}$. Then for any $f \in C(X)$, we see that both $g = E(f|\mathcal{A})$ and $g' = E(f|\mathcal{A}')$ satisfy the characterizing properties of $E(f|\tilde{\mathcal{A}})$, and so they are equal almost everywhere. Again taking a countable union of null sets, corresponding to a countable dense subset of $C(X)$, we see that $\mu_x^{\mathcal{A}} = \mu_x^{\mathcal{A}'}$ almost everywhere, giving part (ii).

For part (iii), suppose that $\mathcal{A} = \sigma(\{A_1, \ldots\})$ is countably generated. For every i, we have that $\chi_{A_i}(x) = E(\chi_{A_i}|\mathcal{A})(x) = \mu_x^{\mathcal{A}}(A_i)$ almost everywhere. Hence there exists a set N of μ-measure 0, given by the union of the these null sets for each i, such that $\mu_x^{\mathcal{A}}(A_i) = 1$ for all i and every $x \in A_i \backslash N$. Therefore, since $[x]_{\mathcal{A}}$ is the countable intersection of A_i's containing x, we have $\mu_x^{\mathcal{A}}([x]_{\mathcal{A}}) = 1$ for all $x \notin N$. Finally, since $x \to \mu_x^{\mathcal{A}}$ is $\mathcal{A}$-measurable, we have that $[x]_{\mathcal{A}} = [y]_{\mathcal{A}} \Rightarrow \mu_x^{\mathcal{A}} = \mu_y^{\mathcal{A}}$ whenever both are defined (i.e., $x, y \in X'$). □

5.13. ANOTHER CONSTRUCTION. An alternate construction for the conditional measure for a countably generated σ-algebra is to start by finding a sequence of finite partitions $\mathcal{A}_n \nearrow \mathcal{A}$. For finite partitions, the conditional measures are particularly simple; we have

$$\mu_x^{\mathcal{A}_n} = \frac{\mu|_{[x]_{\mathcal{A}_n}}}{\mu([x]_{\mathcal{A}_n})}.$$

Now, for any $f \in C(X)$, the Increasing Martingale Convergence Theorem tells us that for any continuous f and for almost every x, we have $\int f d\mu_x^{\mathcal{A}_n} = E(f|\mathcal{A}_n)(x) \to E(f|\mathcal{A})(x)$. Again by choosing a countable dense subset of $C(X)$ we show a.e. that $\mu_x^{\mathcal{A}_n}$ converge in the weak* topology to a measure $\mu_x^{\mathcal{A}}$ as in (i) of the theorem.

5.14. THE ERGODIC DECOMPOSITION REVISITED. One application for the notion of conditional measures is that it can be used to prove the existence of the ergodic decomposition. In fact, for any H-invariant measure μ, we have the ergodic

(18) Here the word "pick" refers to the choice of a representative of the equivalence class of integrable measurable functions.

decomposition

$$\mu = \int \mu_x^{\mathcal{E}} d\mu(x),$$

where $\mathcal{E}$ is (alternatively a countably generated σ-algebra equivalent to) the σ-algebra of all H-invariant sets, and $\mu_x^{\mathcal{E}}$ is the conditional measure (on the $\mathcal{E}$-atom of x). This is a somewhat more intrinsic way to write the ergodic decomposition as one does not have to introduce an auxiliary probability space.

5.15. Definition. *Two countably generated σ-algebras $\mathcal{A}$ and $\mathcal{C}$ on a space X are* countably equivalent *if any atom of $\mathcal{A}$ can be covered by at most countably many atoms of $\mathcal{C}$, and vice versa.*

5.16. REMARK. This *is* an equivalence relation. Symmetry is part of the definition, reflexivity is obvious, and transitivity can be readily checked.

5.17. Proposition. *Suppose $\mathcal{A}$ and $\mathcal{A}'$ are countably equivalent sub-σ-algebras. Then for μ-a.e. x, we have*

$$\mu_x^{\mathcal{A}}|_{[x]_{\mathcal{A}\vee\mathcal{A}'}} \propto \mu_x^{\mathcal{A}'}|_{[x]_{\mathcal{A}\vee\mathcal{A}'}}.$$

Or, put another way,

$$\mu_x^{\mathcal{A}\vee\mathcal{A}'} = \frac{\mu_x^{\mathcal{A}}|_{[x]_{\mathcal{A}\vee\mathcal{A}'}}}{\mu_x^{\mathcal{A}}([x]_{\mathcal{A}\vee\mathcal{A}'})} = \frac{\mu_x^{\mathcal{A}'}|_{[x]_{\mathcal{A}\vee\mathcal{A}'}}}{\mu_x^{\mathcal{A}'}([x]_{\mathcal{A}\vee\mathcal{A}'})}.$$

Here and in the following the notation $\mu \propto \nu$ for two measures on a space X denotes proportionality, i.e. that there exists some $c > 0$ with $\mu = c\nu$.

5.18. PROOF. As a first step, we observe that $\mathcal{A}$ is countably equivalent to $\mathcal{A}'$ if and only if $\mathcal{A}$ is countably equivalent to the σ-algebra generated by $\mathcal{A}$ and $\mathcal{A}'$. Hence we may assume that $\mathcal{A} \subset \mathcal{A}'$, and the statement of the Proposition reduces to

$$\mu_x^{\mathcal{A}'} = \frac{\mu_x^{\mathcal{A}}|_{[x]_{\mathcal{A}'}}}{\mu_x^{\mathcal{A}}([x]_{\mathcal{A}'})}.$$

The next step is to verify that the denominator on the right-hand side is actually $\mathcal{A}'$-measurable (as a function of x). As $\mathcal{A}'$ is countably generated, we may take a sequence $\mathcal{A}'_n \nearrow \mathcal{A}'$ of finite algebras, and consider the decreasing chain of sets $[x]_{\mathcal{A}'_n}$. Notice that $E(1_{[x]_{\mathcal{A}'_n}}|\mathcal{A})(x) = \mu_x^{\mathcal{A}}([x]_{\mathcal{A}'_n})$ is a perfectly good $\mathcal{A} \vee \mathcal{A}'_n$-measurable function. In the limit as $n \to \infty$, the set $[x]_{\mathcal{A}'_n} \searrow [x]_{\mathcal{A}'} = \bigcap_n [x]_{\mathcal{A}'_n}$ as $(\mathcal{A}'_n \vee \mathcal{A}) \nearrow \mathcal{A}'$, and so $x \mapsto \mu_x^{\mathcal{A}}([x]_{\mathcal{A}'})$ is $\mathcal{A}'$-measurable.

We still also have to verify that this denominator is non-zero (almost everywhere). Consider the set $Y = \{x : \mu_x^{\mathcal{A}}([x]_{\mathcal{A}'}) = 0\}$. We must show that $\mu(Y) = 0$ when $\mathcal{A}$ and $\mathcal{A}'$ are countably equivalent. The previous step guarantees that Y is measurable, and we can integrate fibre by fibre: $\mu(Y) = \int \mu_x^{\mathcal{A}}(Y) d\mu(x)$. But $[x]_{\mathcal{A}}$ is a finite or countable union $\bigcup_{i \in I} [x_i]_{\mathcal{A}'}$ of $\mathcal{A}'$-atoms, and so

$$\mu_x^{\mathcal{A}}(Y) = \sum_{i \in I} \mu_x^{\mathcal{A}}([x_i]_{\mathcal{A}'} \cap Y)$$

and so it suffices to show that each term on the right-hand side is 0. If $[x_i]_{\mathcal{A}'} \cap Y = \emptyset$, then there is nothing to show. On the other hand, if there exists some $y \in [x_i]_{\mathcal{A}'} \cap Y$, then by definition of Y we have $\mu_y^{\mathcal{A}}([x_i]_{\mathcal{A}'}) = 0$. But $[x_i]_{\mathcal{A}'} \subset [x]_{\mathcal{A}}$, and so $y \in [x]_{\mathcal{A}}$, which by Theorem 5.9 (and the subsequent Remark) implies that $\mu_x^{\mathcal{A}}([x_i]_{\mathcal{A}'}) = \mu_y^{\mathcal{A}}([x_i]_{\mathcal{A}'}) = 0$.

We now know that $\frac{\mu_x^{\mathcal{A}}|_{[x]_{\mathcal{A}'}}}{\mu_x^{\mathcal{A}}([x]_{\mathcal{A}'})}$ makes sense. We easily verify that it satisfies the characterizing properties of $\mu_x^{\mathcal{A}'}$, and we are done. □

6. Leaf-wise Measures, the construction

We will need later (e.g. in the discussion of entropy) another generalization of conditional measures that allows us to discuss "the restrictions of the measure" to the orbits of a group action just like the conditional measures describe "the restriction of the measure" to the atoms. However, as we have seen in Lemma 5.4, one cannot expect to have a σ-algebra whose atoms are precisely the orbits.

As we will see these restricted measures for orbits, which we will call *leaf-wise measures*, can be constructed by patching together conditional measures for various σ-algebras whose atoms are pieces of orbits. Such a construction (with little detail provided) is used by Katok and Spatzier in [**KS96**]; we follow here the general framework outlined in [**Lin06**], with some simplifications and improvements (e.g. Theorem 6.30 which in this generality seems to be new).

6.1. A FEW ASSUMPTIONS. Let T be a locally compact, second countable group. We assume that T is equipped with a right-invariant metric such that any ball of finite radius has compact closure. We write $B_r^T(t_0) = \{t \in T : d(t, t_0) < r\}$ for the open ball of radius r around $t_0 \in T$, and write $B_r^T = B_r^T(e)$ for the ball around the identity $e \in T$. Also let X be a locally compact, second countable metric space. We assume that T acts continuously on X, i.e., that there is a continuous map $(t, x) \mapsto t.x \in X$ defined on $T \times X \to X$ satisfying $s.(t.x) = (st).x$ and $e.x = x$ for all $s, t \in T$ and $x \in X$. We also assume the T-action to be *locally free* in the following uniform way: for every compact $K \subset X$ there is some $\eta > 0$ such that $t \in B_\eta^T$, $x \in K$, and $t.x = x$ imply $t = e$. In particular, the identity element $e \in T$ is isolated in $\mathrm{Stab}_T(x) = \{t \in T : t.x = x\}$, so that the latter becomes a discrete group, for every $x \in X$— this property allows a nice foliation of X into T-orbits. Finally we assume that μ is a *Radon* (or *locally finite*) measure on X, meaning that $\mu(K) < \infty$ for any compact $K \subset X$.

6.2. Definition. *Let*[19] $x \in X$. *A set* $A \subset T.x$ *is an* open T-plaque *if for every* $a \in A$ *the set* $\{t : t.a \in A\}$ *is open and bounded.*

Note that by the above assumptions on T a set is bounded only if its closure is compact. We recall that $\mu \propto \nu$ for two measures on a space X denotes proportionality, i.e., that there exists some $c > 0$ with $\mu = c\nu$.

6.3. Theorem (Provisional[20]!). *In addition to the above assume also that* $\mathrm{Stab}_T(x) = \{e\}$ *for* μ*-a.e.* $x \in X$, *i.e.,* $t \mapsto t.x$ *is injective for a.e.* x. *Then there is a system* $\{\mu_x^T\}_{x \in X'}$ *of Radon measures on* T *which we will call the* leaf-wise measures *which are determined uniquely, up to proportionality and outside a set of measure zero, by the following properties:*

(i) *The domain* $X' \subset X$ *of the function* $x \mapsto \mu_x^T$ *is a full measure subset in the sense that* $\mu(X \backslash X') = 0$.
(ii) *For every* $f \in C_c(T)$, *the map* $x \mapsto \int f d\mu_x^T$ *is Borel measurable.*

(19) Below we will work mostly with points x for which $t \in T \mapsto t.x$ is injective.

(20) Ideally, we would like to "normalize" by looking at equivalence classes of proportional Radon measures, but this will require further work. See Theorem 6.30.

(iii) *For every* $x \in X'$ *and* $s \in T$ *with* $s.x \in X'$, *we have* $\mu_x^T \propto (\mu_{s.x}^T)s$, *where the right-hand side is the push-forward of* $\mu_{s.x}^T$ *by the right translation (on* T*)* $t \mapsto ts$, *see Figure 1.*

(iv) *Suppose* $Z \subset X$ *and that there exists a countably generated* σ*-algebra* $\mathcal{A}$ *of subsets of* Z *such that for any* $x \in Z$, *the set* $[x]_{\mathcal{A}}$ *is an open* T*-plaque; i.e.,* $U_{x,\mathcal{A}} := \{t : t.x \in [x]_{\mathcal{A}}\}$ *is open and bounded satisfying* $[x]_{\mathcal{A}} = U_{x,\mathcal{A}}.x$. *Then for* μ*-a.e.* $x \in Z$,

$$(\mu|_Z)_x^{\mathcal{A}} \propto \left(\mu_x^T|_{U_{x,\mathcal{A}}}\right).x$$

where the latter is the push-forward under the map $t \in U_{x,\mathcal{A}} \mapsto t.x \in [x]_{\mathcal{A}}$.

(v) *The identity element* $e \in T$ *is in the support of* μ_x^T *for* μ*-a.e.* x.

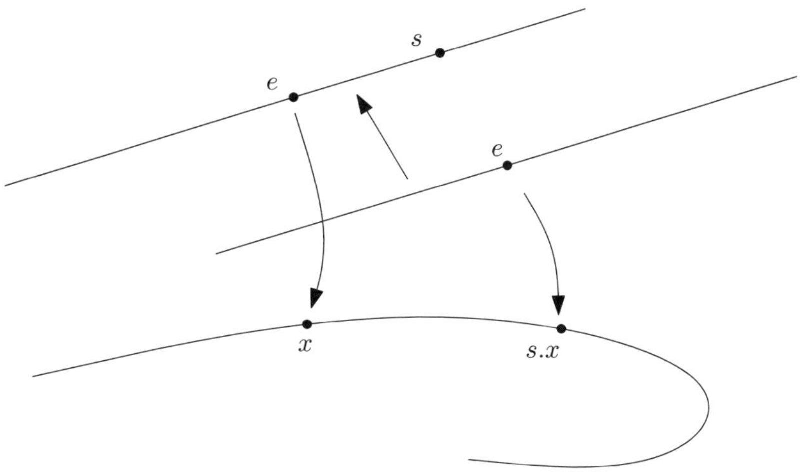

FIGURE 1. The two straight lines represent two copies of the group T and the curved line represents the orbit $T.x = T.(s.x)$. The arrows from the groups to the orbit represent the orbit maps $t \to t.x$ and $t \to t.(s.x)$. Right translation by s from T to T makes the diagram commutative. In other words, Thm. 6.3(iii) only says that the infinite measures $\mu_x^T.x$ and $\mu_{s.x}^T.(s.x)$ on X are proportional.

6.4. REMARKS.

(i) The properties of leaf-wise measures are analogous to those of the conditional measures described in Theorem 5.9. With leaf-wise measures, we demand that the "atoms" correspond to entire (non-compact!) T-orbits, and herein lie most of the complications. On the other hand, these orbits inherit the group structure from T, and so the conditional measures μ_x^T are actually measures on the *group* T, which has structure that we can exploit.

(ii) Property 6.3.(iii) is the analogue of Property 5.9.(iii). Ideally, we would like to say that, since x and $g.x$ are in the same T-orbit, their leaf-wise measures should be the same. However, we prefer to work with measures on T so we move the measures from $T.x$ to T via $t.x \mapsto t$

(which implicitly makes use of the initial point x). Therefore, *points* on the orbit correspond to different *group* elements depending on the base point; hence we need to employ the right translation in order to have our measures (defined as measures on the group) agree at points of the orbit. Another difficulty is that the μ_x^T need not be probability measures, or even finite measures. There being no good way to "normalize" them, we must make do with proportionality instead of equality.

(iii) Property 6.3.(iv) is the most restrictive; this is the heart of the definition. It essentially says that one can restrict μ_x^T to $U_{x,\mathcal{A}}$ and get a finite measure, which looks just like (up to normalization) a good old conditional measure $\mu_x^{\mathcal{A}}$ derived from $\mathcal{A}$. So μ_x^T is in essence a global "patching" together of local conditional measures (up to proportionality issues).

6.5. EXAMPLES.

6.5.1. Let $X = \mathbb{T}^2$, on which $T = \mathbb{R}$ acts by $t.x = x + t\vec{v} \mod \mathbb{Z}^2$, for some irrational vector $\vec{v}$. If $\mu = \lambda$ is the Lebesgue measure on $\mathbb{T}^2$, then we can take $\mu_x^T = \lambda_{\mathbb{R}}$ to be Lebesgue measure on $\mathbb{R}$. Note that, even though the space X is quite nice (eg., compact), none of the leaf-wise measures are finite. Also, notice that the naive approach to constructing these measures would be to look at conditional measures for the sub-σ-algebra $\mathcal{A}$ of T-invariant Borel sets. Unfortunately, this σ-algebra is not countably generated, and is equivalent (see Lemma 5.4 and Proposition 5.8) to the trivial σ-algebra! This is a situation where passing to an equivalent σ-algebra to avoid uncountable generation actually destroys the information we want (T-orbits have measure 0). Instead, we define the leaf-wise measures on small pieces of T-orbits and then glue them together.

6.5.2. We now give an example of a p-adic group action. Let $X = (\mathbb{Q}_p \times \mathbb{R})/\mathbb{Z}[\frac{1}{p}] \cong (\mathbb{Z}_p \times \mathbb{R})/\mathbb{Z}$ where both $\mathbb{Z}[\frac{1}{p}]$ and $\mathbb{Z}$ are considered as subgroups via the canonical diagonal embedding. We let $T = \mathbb{Q}_p$ act on X by translations (where our group law is given by addition). To describe an interesting example of leaf-wise measures, we (measurably) identify X with the space of 2-sided sequences $\{x(i)\}_{i=-\infty}^{\infty}$ in base p (up to countably many nuisances) as follows: Note that $\mathbb{T} = \mathbb{R}/\mathbb{Z}$ is the quotient of X by the subgroup $\mathbb{Z}_p$ and that we may use p-nary digit expansion in $[0, 1) \cong \mathbb{T}$. This way $x \in X$ determines a one-sided sequence of digits $x(i)$ for $i = 1, 2, \ldots$. Since multiplication by p is invertible on X, we may recover all digits $x(i)$ for $i = \ldots, -1, 0, 1, \ldots$ by applying the above to the points $p^{-n}x$. (The reader should verify that this procedure is well-defined at all but countably many points and that the assigned sequence of digits uniquely defines the initial point $x \in X$.)

Under this isomorphism of X with the space of sequences the action of translation by $\mathbb{Z}_p$ corresponds to changing (in a particular manner) the coordinates of the sequence corresponding to $i \leq 0$ such that the orbit under $\mathbb{Z}_p$ consists of all sequences that agree with the original sequence on all positive coordinates. For this recall that $\mathbb{Z}_p$ is isomorphic to $\{0, \ldots, p-1\}^{\mathbb{N}_0}$. More generally, the orbit of a point under $p^{-n}\mathbb{Z}_p$ corresponds to all sequences that have the same coordinates as the original sequence for $i > n$. Hence the $\mathbb{Q}_p$-orbit corresponds to all sequences that have the same digits as the original sequence for all $i > n$ for some n.

We now define a measure and discuss the leaf-wise measures for the action by $\mathbb{Q}_p$. Let μ be an identically independently distributed but biased Bernoulli measure – in other words we identify X again with the space of all 2-sided sequences, i.e., with $\{0, 1, \ldots, p-1\}^{\mathbb{Z}}$, and define μ as the infinite product measure using some fixed probability vector $v = (v_0, \ldots, v_{p-1}) \neq (\frac{1}{p}, \ldots, \frac{1}{p})$. We note that the map $\alpha : x \mapsto px$ defined by multiplication with p (which corresponds to shifting the sequences) preserves the measure μ and acts ergodically w.r.t. μ (in fact as one can check directly it is mixing w.r.t. μ which as mentioned before implies ergodicity). Note also that α preserves the foliation into $\mathbb{Q}_p$-orbits and in fact contracts them, i.e., $\alpha(x + \mathbb{Q}_p) = \alpha(x) + \mathbb{Q}_p$ and $\alpha(x+t) = \alpha(x) + pt$ for $t \in \mathbb{Q}_p$ and pt is p-adically smaller than t. Finally note that the $\mathbb{Q}_p$-action does not preserve the measure μ unless $v = (\frac{1}{p}, \ldots, \frac{1}{p})$. In this case there is very little difference to the above example on $\mathbb{T}^2$ – the leaf-wise measures end up being Haar measures on $\mathbb{Q}_p$. So let us assume the almost opposite extreme: suppose $v_0, \ldots, v_{p-1} \in (0,1)$ and no two components of v are equal.

Let $\mathcal{A}$ be the countably generated σ-algebra (contained in the Borel σ-algebra of X) whose atoms are the $\mathbb{Z}_p$-orbits; it is generated by the *cylinder sets* of the form $\{x : x(i) = \epsilon_i \text{ for } 1 \le i \le N\}$ for any $N > 0$ and all possible finite sequences $(\epsilon_1, \ldots, \epsilon_N) \in \{0, \ldots, p-1\}^N$. Equivalently, the $\mathcal{A}$-atoms are all sequences that agree with a given one on all coordinates for $i \ge 1$ so that the atom has the structure of a one-sided shift space. By independence of the coordinates (w.r.t. μ) the conditional measures $\mu_x^{\mathcal{A}}$ are all Bernoulli i.i.d. measures according to the original probability vector v of μ; in other words, a random element of $[x]_{\mathcal{A}}$ according to $\mu_x^{\mathcal{A}}$ is a sequence $\{y(i)\}$ such that $y(i) = x(i)$ for $i \ge 1$, and the digits $y(i)$ for $-\infty < i \le 0$ are picked independently at random according to the probability vector defining μ.

What does μ_x^T look like (where $T = \mathbb{Q}_p$)? For this notice that $\mathbb{Z}_p$ is open in $\mathbb{Q}_p$, so that the atoms for $\mathcal{A}$ are open T-plaques. Therefore, if we restrict μ_x^T to the subgroup $U = \mathbb{Z}_p$ of $T = \mathbb{Q}_p$, we should get by Theorem 6.3 (iv) that

$$x + \mu_x^T|_U \propto \mu_x^{\mathcal{A}}.$$

To understand this better, let's examine what a random point of $\frac{1}{\mu_x^T(U)}\mu_x^T|_U$ looks like. Of course, an element belonging to $\mathbb{Z}_p$ corresponds to a sequence $\{t(i)\}_{i=-\infty}^0$; how are the digits $t(i)$ distributed? Recall that if we translate by x, the resulting digits $(t+x)(i)$ (with addition formed in $\mathbb{Z}_p$ where the carry goes to the left) should be randomly selected according to the original probability vector. Hence the probability of $t(0) = \epsilon$ with respect to the normalized $\mu_x^T|_U$ becomes the original probability $v_{\epsilon+x(0)}$ of selecting the digit $\epsilon + x(0)$. By our assumption on the vector v this shift in the distribution determines $x(0)$. However, by using σ-algebras whose atoms are orbits of $p^n\mathbb{Z}^p$ for all $n \in \mathbb{Z}$ we conclude that μ_x^T determines all coordinates of x and hence x! (Of course had we used the theorem to construct the leaf-wise measures instead of directly finding it by using the structure of the given measure then the leaf-wise measure would only be defined on a set of full measure and the above conclusion would only hold on a set of full measure.)

This example shows that the seemingly mild assumption (which we will see satisfied frequently later) that there are different points with the same leaf-wise measures (after moving the measures to T as we did) is a rather special property of the underlying measure μ.

6.5.3. The final example is really more than an example – it is the reason we are developing the theory of leaf-wise measures and we will return to it in great detail (and greater generality) in the following sections. Let G be a Lie group, let T

be a closed subgroup, and let Γ be a discrete subgroup of G. Then T acts by right translation on $X = \Gamma\backslash G$, i.e., for $t \in T$ and $x = \Gamma g \in X$ we may define $t.x = xt^{-1}$. For a probability measure μ on X we have therefore a system of leaf-wise measures μ_x^T defined for a.e. $x \in X$ (provided the injectivity requirement is satisfied a.e.) which as we will see describes the properties of the measure along the direction of T. Moreover, if right translation by some $a \in G$ preserves μ, then with the correctly chosen subgroup T (namely the horospherical subgroups defined later) the leaf-wise measures for T will allow us to describe entropy of a w.r.t. μ.

The following definition and the existence established in Proposition 6.7 established afterwards will be a crucial tool for proving Theorem 6.3.

6.6. Definition. *Let $E \subset X$ be measurable and let $r > 0$. We say $C \subset X$ is an r-cross-section for E if*

(i) *C is Borel measurable,*
(ii) $|B_{r+1}^T.x \cap C| = |B_1^T.x \cap C| = 1$ *for all $x \in E \cup C$,*
(iii) *$t \in B_{r+1}^T \mapsto t.x$ is injective for all $x \in C$,*
(iv) *$B_{r+1}^T.x \cap B_{r+1}^T.x' = \emptyset$ if $x \neq x' \in C$, and*
(v) *the restriction of the action map $(t,x) \mapsto t.x$ to $B_{r+1}^T \times C \to B_{r+1}^T.C \supseteq B_r^T.E$ is a Borel isomorphism.*

The second property describes the heart of the definition; the piece $B_{r+1}^T.x$ of the T-orbit through $x \in E$ intersects C exactly once which justifies the term cross-section, see Figure 2. Also note that by the second property there is for

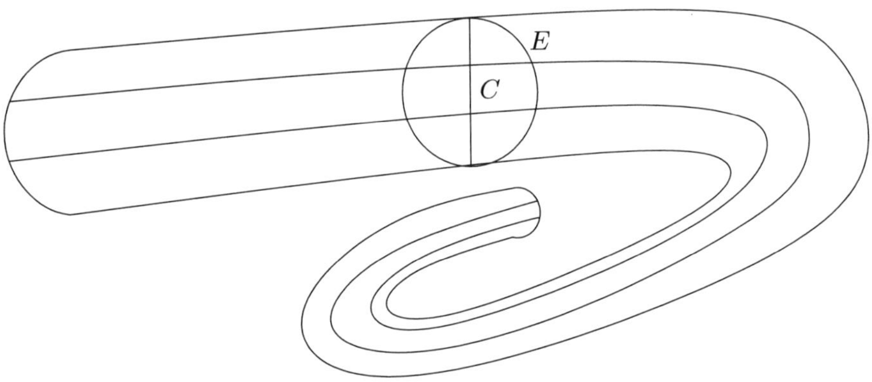

FIGURE 2. E (the circle) needs to be "small enough" in order for an r-cross section C (the vertical line through the circle) to exist. Otherwise, there may be large returns of points in E to E (in the picture if the circle is just a bit bigger) along the action of T (indicated by the curved lines).

every $x \in E$ some $t \in B_1^T$ with $t.x = x' \in C$. Hence, by right invariance of the metric on T we have $B_r^T t^{-1} \subset B_{r+1}^T$ and so the inclusion $B_{r+1}^T.C \supseteq B_r^T.E$ stated in the final property follows from the second property. Moreover, it is clear that the restriction of the continuous action is measurable, so the only requirement in the final property is injectivity of the map and the Borel measurability of the inverse. However, injectivity of this map is precisely the assertion in property (iii) and (iv). Finally, the measurability of the image and the inverse map are guaranteed by a general fact, see [**Sri98**, §4.5], saying that the image and the inverse of an injective Borel map are again Borel measurable. The reader who is unfamiliar with this theorem may construct (replacing the following general proposition) concrete cross-sections of sufficiently small balls in the important example in §6.5.3 using a transverse subspace to the Lie algebra of T inside the Lie algebra of G. This way one may obtain a compact cross-section and this implies measurability of the inverse map rather directly as the restriction of a continuous map to a compact set has compact image and a continuous inverse.

6.7. Proposition. *Let T act continuously on X satisfying the assumptions discussed in the beginning of this section. Assume $x_0 \in X$ is such that $t \in \overline{B^T_{r+1}} \mapsto t.x_0$ is injective for some $r > 1$. Then there exists some $\delta > 0$ such that for all $x \in E = \overline{B_\delta(x_0)}$ the map $t \in B^T_{r+1} \mapsto t.x$ is also injective and such that $t.x = t'.x'$ for some $x, x' \in E$ and $t, t' \in B^T_{r+1}$ implies $t't^{-1}, t^{-1}t' \in B^T_1$ and so $x' \in B^T_1.x$. Moreover, there exists some $C \subset E$ which is an r-cross-section for E.*

6.8. Problem. Prove the proposition in the case where $X = \Gamma\backslash G$ for a Lie group G (or a p-adic Lie group) and a closed subgroup $T < G$ by using a transverse to the Lie algebra of T as suggested above. The reader interested in only these cases may continue with §6.14.

6.9. Proof, Construction of E. If for every δ there exists some $x_\delta \in B_\delta(x_0)$ for which the restricted action $t \in B^T_{r+1} \mapsto t.x_\delta$ fails to be injective then there are $t_\delta \neq t'_\delta \in B^T_{r+1}$ with $t_\delta.x_\delta = t'_\delta.x_\delta$. Choosing converging subsequences of t_δ, t'_δ we get $t, t' \in \overline{B^T_{r+1}}$ with $t.x_0 = t'.x_0$. Moreover, we would have $t \neq t'$ as otherwise we would get a contradiction to the uniform local freeness of the action in §6.1 for the compact set $\overline{B^T_{r+1}}.\overline{B_\epsilon(x_0)}$ (where ϵ is small enough so that $\overline{B_\epsilon(x_0)}$ is compact).

Similarly, if for every $\delta > 0$ there are $x_\delta, x'_\delta \in B_\delta(x_0)$ and $t_\delta, t'_\delta \in B^T_{r+1}$ so that $t_\delta.x_\delta = t'_\delta.x'_\delta$ then in the limit we would have $t, t' \in \overline{B^T_{r+1}}$ with $t.x_0 = t'.x_0$. By assumption this implies $t = t'$, which shows that for sufficiently small δ, we must have $t'_\delta t_\delta^{-1}, t_\delta^{-1}t'_\delta \in B^T_1$ as claimed. Also notice that $\left(B^T_1\right)^{-1} = B^T_1$ by right invariance of the metric.

We now fix some $\delta > 0$ with the above properties and let $E = \overline{B_{\delta/2}(x_0)}$. Below we will construct a Borel subset $C \subset E$ such that $|B^T_1.x \cap C| = 1$ for all $x \in E$. This implies that C is an r-cross-section by the above properties: $t \in B^T_{r+1}$ and $x \in E$ with $t.x \in C \subset E$ implies $t \in B^T_1$ and so property (ii) of the definition holds. Injectivity of $t \in B^T_{r+1} \mapsto t.x$ for all $x \in E$ we have already checked. For the property (iv), note that $x, x' \in C$ and $t, t' \in B^T_{r+1}$ with $t.x = t'.x$ implies $x = t^{-1}t'.x' \in B^T_1.x'$ by the construction of E and so $x = x'$ by the assumed property of C. As explained after the definition the last property follows from the first four. Hence it remains to find a Borel subset $C \subset E$ with $|B^T_1.x \cap C| = 1$ for all $x \in E$.

6.10. Outline of construction of C. We will construct C by an inductive procedure where at every stage we define a set $C_{n+1} \subset C_n$ such that for every $x \in E$ the set $\{t \in B^T_1 : t.x \in C_n\}$ is nonempty, compact, and the diameter of this set decreases to 0 as $n \to \infty$.

6.11. Construction of P_w. For the construction of C_n we first define for every n a partition of E which refines all prior partitions: For $n = 1$ we choose a finite cover of E by closed balls of radius[21] 1, choose some order of these balls, and define P_1 to be the first ball in this cover intersected with E, P_2 the second ball intersected with E minus P_1, and more generally if $P_1, \dots, P_i$ have been already defined then P_{i+1} is the $(i+1)$-th ball intersected with E and with $P_1 \cup \dots \cup P_i$ removed from it.

For $n = 2$ we cover P_1 by finitely many closed balls of radius $1/2$ and construct with the same algorithm as above a finite partition of P_1 into sets $P_{1,1}, \dots, P_{1,i_1}$ of diameter less than $1/2$. We repeat this also for $P_2, \dots$

Continuing the construction we assume that we already defined the sets P_w where w is a word of length $|w| \le n$ (i.e., w is a list of m natural numbers and m is called the length $|w|$) with the obvious compatibilities arising from the construction: for any w of length $|w| = m \le n-1$ the sets $P_{w,1}, P_{w,2}, \dots$ (there are only finitely many) all have diameter less than $1/m$ and form a partition of P_w.

Roughly speaking, we will use these partitions to make decisions in a selection process: Given some $x \in E$ we want to make sure that there is one and only one element of the desired set C that belongs to $B_1^T.x$. Assuming this is not the case for $C = E$ (which can only happen for discrete groups T) we wish to remove, in some inductive manner obtaining the sets C_n along the way, some parts of E so as to make this true for the limiting object $C = \bigcap_n C_n$. Removing too much at once may be fatal as we may come to the situation where $B_1^T.x \cap C_n$ is empty for some $x \in E$. The partition elements P_w give us a way of ordering the elements of the space which we will use below.

6.12. Definition of Q_w and C_n. From the sequence of partitions defined by $\{P_w : w \text{ is a word of length } n\}$ we now define subsets $Q_w \subset P_w$ to define the C_n: We let $Q_1 = P_1$, and let $Q_2 = P_2 \setminus B_1^T.Q_1$, i.e. we remove from P_2 all points that already have on their B_1^T-orbit a point in Q_1. More generally, we define $Q_i = P_i \setminus \big(B_1^T.(Q_1 \cup \dots \cup Q_{i-1})\big)$ for all i and define $C_1 = \bigcup_i Q_i$ (which as before is just a finite union). We now prove the claim from §6.10 for $n = 1$ that for every $x \in E$ the set $\{t \in B_1^T : t.x \in C_1\}$ is nonempty and compact. Here we will use without explicitly mentioning, as we will also do below, the already established fact that $t \in B_2^T$ and $x, t.x \in E$ implies $t \in B_1^T$ (note that by assumption $r > 1$). If i is chosen minimally with $B_1^T.x \cap P_i$ nonempty, then

$$\{t \in B_1^T : t.x \in C_1\} = \{t \in B_1^T : t.x \in Q_i\} = \{t \in B_1^T : t.x \in P_i\} = \{t \in \overline{B_1^T} : t.x \in P_1 \cup \dots \cup P_i\}.$$

Now note that $P_1 \cup \dots \cup P_i$ is closed by the above construction (we used closed balls to cover E and $P_1 \cup \dots \cup P_i$ equals the union of the first i balls intersected with E, a closed ball itself), and so the claim follows for $n = 1$ and any $x \in E$.

Proceeding to the general case for n, we assume $Q_w \subset P_w$ has been defined for $|w| = m < n$ with the following properties: we have $Q_{w,i} \subset Q_w$ for $i = 1, 2, \dots$ and for all $|w| < n-1$, for $|w| = |w'| < n$ and $w \ne w'$ the sets $B_1^T.Q_w$ and $B_1^T.Q_{w'}$ are disjoint, and the claim holds for $C_m = \bigcup\{Q_w : |w| = m\}$ and all $m < n$. Now fix some word w of length $n-1$, we define $Q_{w,1} = Q_w \cap P_{w,1}$,

(21) We ignore, for simplicity of notation, the likely possibility that $\delta < 1$.

$Q_{w,2} = Q_w \cap P_{w,2} \setminus (B_1^T.Q_{w,1})$, and for a general i we define inductively

$$Q_{w,i} = Q_w \cap P_{w,i} \setminus \big(B_1^T.(Q_{w,1} \cup \cdots \cup Q_{w,i-1})\big).$$

By the inductive assumption we know that for a given $x \in E$ there is some w of length $n-1$ such that the set

$$\{t \in B_1^T : t.x \in C_{n-1}\} = \{t \in B_1^T : t.x \in Q_w\} \tag{6.12a}$$

is closed and nonempty. Choose i minimally such that $B_1^T.x \cap Q_{w,i}$ (or equivalently $B_1^T.x \cap Q_w \cap P_{w,i}$) is nonempty, then as before

$$\{t \in B_1^T : t.x \in C_n\} = \{t \in B_1^T : t.x \in Q_{w,i}\} = \\ \big\{t \in B_1^T : t.x \in Q_w \cap (P_{w,1} \cup \cdots \cup P_{w,i})\big\} \tag{6.12b}$$

is nonempty. Now recall that by construction $P_{w,1} \cup \cdots \cup P_{w,i}$ is relatively closed in P_w, so that the set in (6.12b) is relatively closed in the set in (6.12a). The latter is closed by assumption which concludes the induction that indeed for every n the set $\{t \in B_1^T : t.x \in C_n\}$ is closed and nonempty.

6.13. Conclusion. The above shows that $C_n = \bigcup_w Q_w$ (where the union is over all words w of length n) satisfies the claim that $\{t \in B_1^T : t.x \in C_n\}$ is compact and non-empty for every $x \in E$. Therefore, $C = \bigcap_n C_n \subset E$ satisfies that $C \cap B_1^T.x \neq \emptyset$ for every $x \in E$. Suppose now $t_1.x, t_2.x \in C$ for some $x \in E$ and $t_1, t_2 \in B_1^T$. Fix some $n \geq 1$. Recall that $\{t \in B_1^T : t.x \in C_n\} = \{t \in B_1^T : t.x \in Q_w\}$ for some Q_w corresponding to a word w of length n. As the diameter of $Q_w \subset P_w$ is less than $1/n$ we have $d(t_1.x, t_2.x) < 1/n$. This holds for every n, so that $t_1.x = t_2.x$ and so $t_1 = t_2$ as required. □

6.14. σ-algebras. Proposition 6.7 allows us to construct σ-algebras as they appear in Theorem 6.3(iv) in abundance. In fact we have found closed balls E and r-cross-sections $C \subset E$ such that $B_{r+1}^T \times C$ is measurably isomorphic to $Y = B_{r+1}^T.C$ (with respect to the natural map) so that we may take the countably generated σ-algebra on $B_{r+1}^T \times C$ whose atoms are of the form $B_{r+1}^T \times \{z\}$ for $z \in C$ and transport it to Y via the isomorphism. As we will work very frequently with σ-algebras of that type we introduce a name for them.

6.15. Definition. *Let $r > 1$. Given two measurable subsets $E \subset Y$ of X and a countably generated σ-algebra $\mathcal{A}$ of subsets of Y, we say that $(Y, \mathcal{A})$ is an (r,T)-*flower with base E, *if and only if:*

(i) *For every $x \in E$, we have that $[x]_{\mathcal{A}} = U_x.x$ is an open T-plaque such that $B_r^T \subset U_x \subset B_{r+2}^T$.*
(ii) *Every $y \in Y$ is equivalent to some $x \in E$, i.e., the atom $[y]_{\mathcal{A}} = [x]_{\mathcal{A}}$ is always an open T-plaque intersecting E nontrivially.*

We note that often the cross-section C will be a nullset (for the measure μ on X), but that the base E will not be a null set, hence it is important to introduce it — it may be thought of as a slightly thickened version of the cross-section so that we still know the rough shape of the atoms as required in (i). We may visualize the flower and the base using Figure 2. The base is the circle and the flower is the σ-algebra on the tube-like set whose atoms are the curved lines.

6.16. Corollary. *Assume as in Theorem 6.3 that $t \mapsto t.x$ for $t \in T$ is injective for μ-a.e. $x \in X$. Then for every n there exists a countable list of (n, T)-flowers such that the union of their bases is a set of full measure. In other words, there exists a countable collection of σ-algebras $\mathcal{A}_k$ of Borel subsets of Borel sets Y_k for $k = 1, 2, \ldots$ such that all of the $\mathcal{A}_k$-atoms are open T-plaques for all k, and such that for a.e. $x \in X$ and all $n \geq 1$ there exists k such that the $\mathcal{A}_k$-atom $[x]_{\mathcal{A}_k}$ contains $B_n^T.x$.*

6.17. PROOF. By our assumption there exists a set X_0 of full measure such that $t \in T \mapsto t.x_0$ is injective for $x \in X_0$. Fix some n. By Prop. 6.7 applied to $r = n$ there exists an uncountable collection of closures E_x of balls for $x \in X_0$ such that x is contained in the interior E_x° and there is an n-cross-section $C_x \subset E_x$ for $x \in X_0$. Since X is second countable, there is a countable collection of these sets $C_m \subset E_m$ for which the union of the interiors is the same as the union of interiors of all of them.

As C_m is an n-cross-section for E_m, we have that $B_{n+1}^T.C_m \supset B_n^T.E_m$ and that $B_{n+1}^T \times C_m$ is measurably isomorphic to $Y_m = B_{n+1}^T.C_m$. We now define $\mathcal{A}_m$ to be the σ-algebra of subsets of Y_m which corresponds under the isomorphism to $\{B_{n+1}^T, \emptyset\} \otimes \mathcal{B}_{C_m}$ — here $\mathcal{B}_{C_m}$ is the Borel σ-algebra of the set C_m. It is clear that $\mathcal{A}_m$ is an (n, T)-flower with base E_m. Using this construction for all n, we get the countable list of (n, T)-flowers as required. □

It is natural to ask how the various σ-algebras in the above corollary fit together, where the next lemma gives the crucial property.

6.18. Lemma. *Let Y_1, Y_2 be Borel subsets of X, and $\mathcal{A}_1, \mathcal{A}_2$ be countably generated σ-algebras of Y_1, Y_2 respectively, such that atoms of each $\mathcal{A}_i$ are open T-plaques. Then the σ-algebras $\mathcal{C}_1 := \mathcal{A}_1|_{Y_1 \cap Y_2}$ and $\mathcal{C}_2 := \mathcal{A}_2|_{Y_1 \cap Y_2}$ are countably equivalent.*

6.19. PROOF. Let $x \in Y_1 \cap Y_2$, and consider $[x]_{\mathcal{C}_1} = [x]_{\mathcal{A}_1} \cap Y_2$. By this and the assumption on $\mathcal{A}_1$ there exists a bounded set $U \subset T$ such that $[x]_{\mathcal{C}_1} = U.x$. Now, for each $t \in U$, we have the open T-plaque $[t.x]_{\mathcal{A}_2}$, which must be of the form $U_t.x$ for some open, bounded $U_t \subset T$. Now the collection $\{U_t\}_{t \in U}$ covers U, and since T is locally compact second countable, there exists a countable subcollection of the $\{U_t\}$ covering U. But this means that a countable collection of atoms of $\mathcal{A}_2$ covers $[x]_{\mathcal{C}_1}$; we then intersect each atom with Y_1 to get atoms of $\mathcal{C}_2$. Switch $\mathcal{C}_1$ and $\mathcal{C}_2$ and repeat the argument to get the converse. □

6.20. PROOF OF THEOREM 6.3, BEGINNING. We now combine Corollary 6.16, Lemma 6.18, and Proposition 5.17: Let $\mathcal{A}_k$ be the sequence of σ-algebras of subsets of Y_k as in Corollary 6.16. We define $Y_{k,\ell} = Y_k \cap Y_\ell$ and get that $(\mathcal{A}_k)|_{Y_{k,\ell}}$ and $(\mathcal{A}_\ell)|_{Y_{k,\ell}}$ are countably equivalent by Lemma 6.18. By Proposition 5.17 we get that

$$\mu_x^{\mathcal{A}_k}|_{[x]_{\mathcal{A}_\ell}} \text{ and } \mu_x^{\mathcal{A}_\ell}|_{[x]_{\mathcal{A}_k}} \tag{6.20a}$$

are proportional for a.e. $x \in Y_{k,\ell}$ (where we used additionally that the conditional measure for $\mu|_{Y_{k,\ell}}$ with respect to the σ-algebra $\mathcal{A}_k|_{Y_{k,\ell}}$ is just the normalized restriction of $\mu_x^{\mathcal{A}_k}$ to $Y_{k,\ell}$). Also recall that by Theorem 5.9(iii) for every k there is a null set in Y_k such that for $x, y \in Y_k$ not belonging to this null set and $[x]_{\mathcal{A}_k} = [y]_{\mathcal{A}_k}$ we have $\mu_x^{\mathcal{A}_k} = \mu_y^{\mathcal{A}_k}$. We collect all of these null sets to one null set $N \subset X$ and let X'' be the set of all points $x \in X \setminus N$ for which $t \mapsto t.x$ is injective. By construction of $\mathcal{A}_k$ we have $[x]_{\mathcal{A}_k} = U_{x,k}.x$ for some open and bounded $U_{x,k} \subset T$. For a bounded measurable set $D \subset T$ and $x \in X''$ we define

$$\mu_x^T(D) = \frac{1}{\mu_x^{\mathcal{A}_k}(B_1^T.x)} \mu_x^{\mathcal{A}_k}(D.x) \tag{6.20b}$$

where we choose k such that $D.x \subset [x]_{\mathcal{A}_k}$ which by the construction of the sequence of σ-algebras, i.e., by Corollary 6.16, is possible. Notice this definition is independent of k by the proportionality of the conditional measures in (6.20a).

However, we need to justify this definition by showing that the denominator does not vanish, at least for a.e. $x \in X''$. We prove this in the following lemma which will also prove Theorem 6.3(v).

6.21. Lemma. *Suppose $\mathcal{A}$ is a countably generated sub-σ-algebra of Borel subsets of a Borel set $Y \subset X$. Suppose further that the $\mathcal{A}$-atoms are open T-plaques. Let $U \subset T$ be an open neighborhood of the identity. Then for μ-a.e. $x \in Y$, we have $\mu_x^{\mathcal{A}}(U.x) > 0$.*

6.22. PROOF. Set $B = \{x \in Y' : \mu_x^{\mathcal{A}}(U.x) = 0\}$, where $Y' \subset Y$ is a subset of full measure on which the conclusion of Theorem 5.9(iii) holds. We wish to show that $\mu(B) = 0$, and since we can integrate first over the atoms and then over the space (Theorem 5.9(i) and Proposition 4.1), it is sufficient to show for each $x \in Y'$ that $\mu_x^{\mathcal{A}}(B) = \mu_x^{\mathcal{A}}([x]_{\mathcal{A}} \cap B) = 0$. Now since atoms of $\mathcal{A}$ are open T-plaques, we can write $[x]_{\mathcal{A}} = (U_x).x$. Set $V_x \subset U_x$ to be the set of those t such that $t.x \in [x]_{\mathcal{A}} \cap B$.

Now clearly the collection $\{Ut\}_{t \in V_x}$ covers V_x, and we can find a countable subcollection $\{Ut_i\}_{i=1}^{\infty}$ that also covers V_x. This implies that $\{(Ut_i).x\}_{i=1}^{\infty}$ covers $[x]_{\mathcal{A}} \cap B$ by definition of V_x, so we have

$$\mu_x^{\mathcal{A}}([x]_{\mathcal{A}} \cap B) \leq \mu_x^{\mathcal{A}}(\bigcup_{i=1}^{\infty}(Ut_i).x) \leq \sum_{i=1}^{\infty} \mu_x^{\mathcal{A}}((Ut_i).x)$$

On the other hand, $t_i.x \in B$, so by definition of B we have that each term $\mu_x^{\mathcal{A}}((Ut_i).x) = \mu_x^{\mathcal{A}}(U.(t_i.x))$ on the right-hand side is 0. □

6.23. PROOF OF THEOREM 6.3, SUMMARY. We let $X' \subset X''$ be a subset of full measure such that the conclusion of Lemma 6.21 holds for the σ-algebra $\mathcal{A}_k$, all $x \in Y_k \cap X'$, all k, and every ball $U = B_{1/n}^T$ for all n. This shows that for $x \in X'$ the expression on the right of (6.20b) is well defined. By the earlier established property it is also independent of k (as long as $D.x \subset [x]_{\mathcal{A}_k}$ as required before). Therefore, (6.20b) defines a Radon measure on T satisfying Theorem 6.3 (v). Property (iii) follows directly from the definition and the requirement that for $x, g.x \in X'' \cap Y_k$ with $[x]_{\mathcal{A}_k} = [g.x]_{\mathcal{A}_k}$ (which will be the case for many k) we have $\mu_x^{\mathcal{A}_k} = \mu_{g.x}^{\mathcal{A}_k}$, where we may have a proportionality factor appearing as μ_x^T is normalized via the set $B_1^T.x$ and $\mu_{g.x}^T$ is normalized via the set $B_1^T g.x$. Property (iv) follows from Lemma 6.18 and Proposition 5.17 similar to the discussion in 6.20. We leave property (ii) to the reader. □

We claimed before that the leaf-wise measure describes properties of the measure μ along the direction of the T-leaves, we now give three examples of this.

6.24. PROBLEM. The most basic question one can ask is the following: What does it mean to have $\mu_x^T \propto \delta_e$ a.e.? Here δ_e is the Dirac measure at the identity of T, and this case is often described as *the leaf-wise measures are trivial a.e.* Show this happens if and only if there is a global cross-section of full measure, i.e., if there is a measurable set $B \subset X$ with $\mu(X \setminus B) = 0$ such that $x, t.x \in B$ for some $t \in T$ implies $t = e$.

6.25. Definition. *Suppose we have a measure space X, a group T acting on X, and μ a locally finite measure on X. Then μ is T-*recurrent *if for every measurable $B \subset X$ of positive measure, and for a.e. $x \in B$, the set $\{t : t.x \in B\}$ is unbounded (i.e., does not have compact closure in T).*

6.26. Theorem. *Let X, T, μ be as before, and suppose additionally that μ is a probability measure. Then μ is T-recurrent if and only if μ_x^T is infinite for almost every x.*

6.27. PROOF. Assume T-recurrence. Let $Y = \{x : \mu_x^T(T) < \infty\}$, and suppose that $\mu(Y) > 0$. We may find a sufficiently large n such that the set $Y' = \{x \in Y : \mu_x^T(B_n^T) > 0.9\mu_x^T(T)\}$ also has positive measure. We will show that, for any $y \in Y'$, the set of return times $\{t : t.y \in Y'\}$ is bounded; in fact, that $\{t : t.y \in Y'\} \subset B_{2n}^T$ for any $y \in Y'$. Since $\mu(Y') > 0$, this then shows that μ is not T-recurrent.

Pick any return time t. By definition of Y', we know that $\mu_y^T(B_n^T) > 0.9\mu_y^T(T)$ and $\mu_{t.y}^T(B_n^T) > 0.9\mu_{t.y}^T(T)$. On the other hand, from Theorem 6.3.(iii) we know that $\mu_{t.y}^T \propto (\mu_y^T)t$, so that we have $\mu_y^T(B_n^T t) > 0.9\mu_y^T(Tt) = 0.9\mu_y^T(T)$. But now we have two sets B_n^T and $B_n^T t$ of very large μ_y^T measure, and so we must have $B_n^T \cap B_n^T t \neq \emptyset$. This means $t \in (B_n^T)^{-1} B_n^T$, as required.

Assume now that the leaf-wise measures satisfy $\mu_x^T(T) = \infty$ for a.e. x, but μ is not T-recurrent. This means there exists a set B of positive measure, and some compact $K \subset T$ such that $\{t : t.x \in B\} \subset K$ for every $x \in B$.

We may replace B by a subset of B of positive measure and assume that $B \subset E$ for a measurable $E \subset X$ for which there is an r-cross-section $C \subset E$ as in Proposition 6.7, where we chose r sufficiently big so that $B_r^T \supset B_1^T K B_1^T$. Let $(B_{r+1}^T.C, \mathcal{A})$ be the (r,T)-flower for which the atoms are of the form $B_{r+1}^T.z$ for $z \in C$. As C is a cross-section, the atoms of $\mathcal{A}$ are in one-to-one correspondence with elements of C. We define $D = \{z \in C : \mu_z^{\mathcal{A}}(B) > 0\}$, where we may require that $\mu_x^{\mathcal{A}}$ is defined on a set $X' \in \mathcal{A}$ and is strictly $\mathcal{A}$-measurable by removing possibly a null set from B. Therefore, the definition of D as a subset of the likely nullset C makes sense. Note that $B \setminus (B_{r+1}^T.D)$ is a null set, and so we may furthermore assume $B \subset B_1^T.D$ by the properties of C and E in Proposition 6.7.

Suppose now $t.z = t'.z'$ for some $t, t' \in T$ and $z, z' \in D$. By construction of D and by Proposition 6.7 we may write $z = t_x.x$ and $z' = t_{x'}.x'$ for some $t_x, t_{x'} \in B_1^T$ and $x, x' \in B$. Therefore, $tt_x.x = t't_{x'}.x'$ which implies that $t_{x'}^{-1}(t')^{-1}tt_x \in K$ by the assumed property of B. Thus $(t')^{-1}t \in B_1^T K B_1^T \subset B_r^T$, which implies $t = t'$ and $z = z'$ since $C \supset D$ is an r-cross-section. This shows that for every n we have that $B_{n+1}^T \times D \to B_{n+1}^T.D$ is injective and just as in Corollary 6.16 this gives rise to the (n,T)-flower $(B_{n+1}^T.D, \mathcal{A}_n)$ with center $B_1^T.D$ such that the atoms are of the form $B_{n+1}^T.z$ for $z \in D$.

By Theorem 6.3.(iv), we know that

$$\mu_x^{\mathcal{A}_n}(B) = \frac{\mu_x^T(\{t \in U_{x,n} : t.x \in B\})}{\mu_x^T(U_{x,n})}$$

for a.e. $x \in B_n^T.D$. Here $U_{x,n} \subset T$ is the shape of the atom, i.e., is such that $[x]_{\mathcal{A}_n} = U_{x,n}.x$. Clearly, for $z \in D$ we have $U_{z,n} = B_{n+1}^T$ by construction. Therefore, we have for $y \in B \subset E \subset B_1^T.C$ that $U_{y,n} \supset B_n^T$. Also recall that by assumption $y \in B$, $t \in T$, and $t.y \in B$ implies $t \in K$. Together we get for a.e. $y \in B$ that

$$\mu_y^{\mathcal{A}_n}(B) \leq \frac{\mu_y^T(K)}{\mu_y^T(B_n^T)},$$

which approaches zero for a.e. $y \in B$ as $n \to \infty$ by assumption on the leaf-wise measures.

We define

$$B' = \{y \in B : \mu_y^{\mathcal{A}_n}(B) \to 0\},$$

which by the above is a subset of B of full measure. We also define the function f_n by the rule $f_n(x) = 0$ if $x \notin B_n^T.D$ and $f_n(x) = \mu_x^{\mathcal{A}_n}(B')$ if $x \in B_n^T.D$. Clearly, if $y \notin T.D$ then $f_n(y) = 0$ for all n. While if $y \in B_{n_0}^T.D$ and $f_{n_0}(y) = \mu_x^{\mathcal{A}_{n_0}}(B') > 0$ for some n_0 then we may find some $x \in B'$ equivalent to y with respect to all $\mathcal{A}_n$ for $n \geq n_0$, so that $f_n(y) = f_n(x)$ for $n \geq n_0$ by the properties of conditional measures. Therefore, $f_n(y) \to 0$ for a.e. $y \in X$. By dominated convergence (μ is a finite measure by assumption and $f_n \leq 1$) we have

$$\mu(B) = \int_{B_n^T.D} \mu_x^{\mathcal{A}_n}(B')d\mu = \int f_n d\mu \to 0,$$

i.e., $\mu(B) = 0$ contrary to the assumptions.□

6.28. PROBLEM. With triviality of leaf-wise measures as one possible extreme for the behavior of μ along the T-leaves and T-recurrence in between, on the opposite extreme we have the following fact: μ is T-invariant if and only if the leaf-wise measures μ_x^T are a.e. left Haar measures on T. Show this using the flowers constructed in Corollary 6.16.

6.29. NORMALIZATION. One possible normalization of the leaf-wise measure μ_x^T, which is uniquely characterized by its properties up to a proportionality factor, is to normalize by a scalar (depending on x measurably) so that $\mu_x^T(B_1^T) = 1$. However, under this normalization we have no idea how big $\mu_x^T(B_n^T)$ can be for $n > 1$.

It would be convenient if the leaf-wise measures μ_x^T would belong to a fixed compact metric space in a natural way — then we could ask (and answer in a positive manner) the question whether the leaf-wise measures depend measurably on x where we consider the natural Borel σ-algebra on the compact metric space. Compare this with the case of conditional measures $\mu_x^{\mathcal{A}}$ for a σ-algebra $\mathcal{A}$ and a finite measure μ on a compact metric space X, here the conditional measures belong to the compact metric space of probability measures on X (where we use the weak* topology on the space of measures). Unfortunately, the lack of a bound of $\mu_x^T(B_2^T)$ shows, with μ_x^T normalized using the unit ball, that the leaf-wise measures do not belong to a compact subset in the space of Radon measures (using the weak* topology induced by compactly supported continuous functions on T). For that reason we are interested(22) in the possibly growth rate of $\mu_x^T(B_n^T)$, so that we can introduce a different normalization with respect to which we get values in a compact metric space.

(22)While convenient, this theorem is not completely necessary for the material presented in the following sections. The reader who is interested in those could skip the proof of this theorem and return to it later.

6.30. Theorem. *Assume in addition to the assumptions of Theorem 6.3 that μ is a probability measure on X and that T is unimodular. Denote the bi-invariant Haar measure on T by λ. Fix weights b_n such that $\sum_{n=1}^{\infty} b_n^{-1} < \infty$ (eg., think of $b_n = n^2$) and a sequence $r_n \nearrow \infty$. Then for μ-a.e. x we have*

$$\lim_{n\to\infty} \frac{\mu_x^T(B_{r_n}^T)}{b_n \lambda(B_{r_n+5}^T)} = 0$$

where B_r^T is the ball of radius r around $e \in T$.

In other words, the leaf-wise measure of big balls $B_{r_n}^T$ can't grow much faster than the Haar measure of a slightly bigger ball $B_{r_n+5}^T$. This is useful as it gives us a function $f : T \to \mathbb{R}^+$ which is integrable w.r.t. μ_x^T for a.e. $x \in X$, e.g. $f(x) = \frac{1}{b_n^2 \lambda(B_{r_n+5}^T)}$ for $x \in B_{r_n}^T \setminus B_{r_{n-1}}^T$. Hence we may normalize μ_x^T such that $\int_T f d\mu_x^T = 1$ and we get that μ_x^T belongs to the compact metric space of measures ν on T for which $\int_T f d\nu \leq 1$, where the latter space is equipped with the weak* topology induced by continuous functions with compact support. Hence it makes sense, and this is essentially Theorem 6.3.(ii), to ask for measurable dependence of μ_x^T as a function of x.

Before proving this theorem, we will need the following refinement regarding the existence of (r, T)-flowers.

6.31. Lemma. *For any measurable set $B \subset X$, $R > 0$, we can find a countable collection of (R, T)-flowers $(Y_k, \mathcal{A}_k)$ with base E_k so that*

(i) *any $x \in X$ is contained in only finitely many bases E_k, in fact the multiplicity is bounded with the bound depending only on T,*
(ii) $\mu(B \setminus \bigcup_k E_k) = 0$,
(iii) *for every $x \in E_k$ there is some $y \in [x]_{\mathcal{A}_k} \cap E_k \cap B$ so that*

$$B_1^T.y \subset [x]_{\mathcal{A}_k} \cap E_k,$$

for any two equivalent[23] *$x, y \in E_k$ we have $[x]_{\mathcal{A}_k} \cap E_k \subset B_4^T.y$, and*
(iv) *for every $x \in Y_k$ there is some $y \in [x]_{\mathcal{A}_k} \cap E_k \cap B$.*

The third property may, loosely speaking, be described as saying that for points x in the base E_k we require that there is some $y \in B \cap E_k$ equivalent to x such that y is deep inside the base E_k (has distance one to the complement) in the direction of T.

6.32. Proof. By Corollary 6.16 we already know that we can cover a subset of full measure by a countable collection of bases $\tilde{E}_k$ of $(R+1, T)$-flowers $(\tilde{Y}_k, \tilde{\mathcal{A}}_k)$ such that additionally there is some $(R+2)$-cross-section $\tilde{C}_k \subset \tilde{E}_k$, $\tilde{Y}_k = B_{R+2}^T.\tilde{C}_k$, and $\tilde{E}_k \subset B_1^T.\tilde{C}_k$. We will construct Y_k by an inductive procedure as subsets of $\tilde{Y}_k$ and will use the restriction $\mathcal{A}_k$ of $\tilde{\mathcal{A}}_k$ to Y_k as the σ-algebra.

For $k = 1$ we define

$$Y_1 = \{x \in \tilde{Y}_1 : \mu_x^{\tilde{\mathcal{A}}_1}(B \cap \tilde{E}_1) > 0\}, \tag{6.32a}$$

and $\mathcal{A}_1 = \tilde{\mathcal{A}}_1|_{Y_1}$. By definition we remove from $\tilde{Y}_1$ complete atoms to obtain Y_1, so that the shape of the remaining atoms is unchanged. From this it follows that

---

[23] Recall that x and y are equivalent w.r.t. $\mathcal{A}_k$ if $[x]_{\mathcal{A}_k} = [y]_{\mathcal{A}_k}$.

$(Y_1, \mathcal{A})$ is an $(R+1,T)$-flower with base $\tilde{E}_1 \cap Y_1$. Also note that $B \cap \tilde{E}_1 \cap Y_1$ is a subset of full measure of $B \cap \tilde{E}_1$ (cf. (5.11a) and (6.32a)). We define

$$E_1 = B_2^T.(\tilde{C}_1 \cap Y_1) \supset \tilde{E}_1 \cap Y_1,$$

where the inclusion follows because $\tilde{E}_1 \subset B_1^T.\tilde{C}_1$ holds by construction of the original flowers. Since we constructed Y_1 by removing whole atoms from $\tilde{Y}_1$, we obtain $E_1 \subset Y_1$.

Finally, by definition of Y_1 we have $\mu_x^{\mathcal{A}_1}(B \cap \tilde{E}_1) > 0$ for every $x \in E_1 \subset Y_1$, so there must indeed be some $y \in B \cap \tilde{E}_1$ which is equivalent to x. Again because Y_1 was obtained from $\tilde{Y}_1$ by removing entire atoms, we have $y \in \tilde{E}_1 \cap Y_1$. Moreover, $y \in B_1^T.C_1$ so that $B_1^T.y \subset (B_2^T.C_1) \cap Y_1 = E_1$. The conclusions in (iii) follow now easily for the case $k = 1$. At last notice that $(Y_1, \mathcal{A}_1)$ is an (R,T)-flower with base E_1.

For a general k we assume that we have already defined for any $\ell < k$ an (R,T)-flower $(Y_\ell, \mathcal{A}_\ell)$ with bases E_ℓ satisfying: $Y_\ell \subset \tilde{Y}_\ell$ is obtained by removing entire $\tilde{\mathcal{A}}_\ell$-atoms, $\mathcal{A}_\ell = \tilde{\mathcal{A}}_\ell|_{Y_\ell}$, properties (iii) and (iv) hold, and that $B \cap \bigcup_{\ell<k} E_\ell$ contains $B \cap \bigcup_{\ell<k} \tilde{E}_\ell$ except possibly for a nullset. The latter is the inductive assumption regarding (ii) as at the end of the construction it will imply (ii) by the assumption that the bases $\tilde{E}_j$ for $j = 1, 2, \ldots$ cover a set of full measure.

We now define

$$Y_k = \{x \in \tilde{Y}_k : \mu_x^{\tilde{\mathcal{A}}_k}(B \cap \tilde{E}_k \setminus \bigcup_{\ell<k} E_\ell) > 0\},$$

which as before is $\tilde{Y}_k$ minus a union of complete $\tilde{\mathcal{A}}_k$-atoms. In particular, we again get that $(Y_k, \mathcal{A}_k)$ (with $\mathcal{A}_k = \tilde{\mathcal{A}}_k|_{Y_k}$) is an $(R+1,T)$-flower with base $\tilde{E}_k \cap Y_k$ and that $B \cap \tilde{E}_k \setminus \bigcup_{\ell<k} E_\ell$ is contained in $\tilde{E}_k \cap Y_k$ except possibly for a null set. The latter ensures the inductive assumption regarding (ii) if we define E_k as a superset of $\tilde{E}_k \cap Y_k$. We define $E_k = B_2^T.(Y_k \cap C_k)$ which implies $\tilde{E}_k \cap Y_k \subset E_k$ and also property (iii) similar to the case $k = 1$. Indeed, if $x \in E_k$, then $x = t.z$ for some $t \in B_2^T$ and $z \in Y_k \cap C_k$ which implies

$$\mu_z^{\mathcal{A}_k}(B \cap \tilde{E}_k \setminus \bigcup_{\ell<k} E_\ell) > 0$$

by definition of Y_k. Hence there is some $y \in B \cap \tilde{E}_k$ equivalent to z (and to x) with $y = t_y.z$ for some $t_y \in B_1^T$ by the properties of $\tilde{E}_k$. This implies $B_1^T.y \subset E_k$ as required.

Suppose now we have completed the above construction defining Y_k and E_k and assume that x belongs to $E_{k_1}, E_{k_2}, \ldots, E_{k_m}$ for some $k_1 < k_2 < \cdots < k_m$. We wish to bound m in order to proof (i). By property (iii) we know for $j = 1, \ldots, m$ that $x = t_j.y_j$ for some $t_j \in B_1^T$ and $y_j \in E_{k_j} \cap B$. In fact, by the construction we know that $y_j \in B \cap \tilde{E}_{k_j} \setminus \bigcup_{\ell<k_j} E_\ell$. Also notice that

$$t_j^{-1}t_i.y_i = t_j^{-1}.x = y_j \text{ for any pair } i, j.$$

However, since $B_1^T.y_i \subset E_{k_i}$ for $i < j$ we must have $t_j^{-1}t_i \notin B_1^T$. As the metric on T is assumed to be right invariant we conclude that the elements $t_1^{-1}, \ldots, t_{k_m}^{-1}$ have all distance ≥ 1, and so m is bounded by the maximal number of 1-separated elements of B_1^T which has compact closure. This proves (i). $\square$

6.33. PROOF OF THEOREM 6.30. We fix some $\delta > 0$, and some integer M. We define

$$B_m = \left\{ y : \frac{\mu_y^T(B_{r_n}^T)}{\mu_y^T(B_4^T)} \geq b_n \delta \frac{\lambda(B_{r_n+5}^T)}{\lambda(B_4^T)} \text{ for at least } m \text{ different } n \leq M \right\}.$$

We want to give a bound on $\mu(B_m)$ which will be independent of M and tends to 0 as $m \to \infty$. Let $R = r_M$, and let E_i and $\mathcal{A}_i$ be as in Lemma 6.31. (Note that by the choice of R the sequence of σ-algebras depends crucially on M.)

Consider the function

$$G = \sum_{n=1}^{M} \sum_{i=1}^{\infty} w_n \chi_{B_{r_n}^T . E_i}$$

with $w_n = \frac{1}{b_n \lambda(B_{r_n+5}^T)}$ and where χ_A denotes the characteristic function of a set A. We claim that G is bounded, with the bound independent of M.

Fixing n and x, let $I = \{ i : x \in B_{r_n}^T . E_i \}$. For each $i \in I$, let $h_i' \in B_{r_n}^T$ be such that $h_i'.x \in E_i$, and by Lemma 6.31.(iii), we can modify h_i' to some $h_i \in B_{r_n+4}^T$ so that $B_1^T h_i.x \subset [x]_{\mathcal{A}_i} \cap E_i$.

As the multiplicity of the sets $E_1, E_2, \ldots$ is bounded by some constant c_1 (that only depends on T) and since $B_1^T h_i.x \subset E_i$ we get that

$$\sum_{i \in I} \chi_{B_1^T h_i} \leq c_1 \chi_{B_{r_n+5}}.$$

This implies that $|I| \, \lambda(B_1^T) \leq c_1 \lambda(B_{r_n+5}^T)$. We conclude that

$$\sum_{i=1}^{\infty} w_n \chi_{B_{r_n}^T . E_i}(x) \leq w_n \, |I| \leq \frac{c_1 \lambda(B_{r_n+5}^T)}{b_n \lambda(B_1^T) \lambda(B_{r_n+5}^T)} \leq \frac{c_2}{b_n},$$

where c_2 again only depends on T. Therefore, $G(x) \leq c_3 = c_2 \sum_{n=1}^{\infty} b_n^{-1}$ for all M as claimed.

On the other hand, consider the (R,T)-flower $(Y_i, \mathcal{A}_i)$ with base E_i. By the properties of leaf-wise measures (Theorem 6.3.(iv)) and Lemma 6.31.(iii), we know that for every $y \in E_i \cap B_m$ and every n,

$$\frac{\mu_y^{\mathcal{A}_i}(E_i)}{\mu_y^{\mathcal{A}_i}(B_{r_n}^T . y)} \leq \frac{\mu_y^T(B_4^T)}{\mu_y^T(B_{r_n}^T)}.$$

So if $z \in Y_i$ and $y \in [z]_{\mathcal{A}_i} \cap B_m \cap E_i$ (the existence of such a y is guaranteed by Lemma 6.31.(iv)), then $\chi_{B_{r_n}^T . E_i} \geq \chi_{B_{r_n}^T . y}$ and so

$$\int_{Y_i} \chi_{B_{r_n}^T . E_i} \, d\mu_z^{\mathcal{A}_i} \geq \mu_y^{\mathcal{A}_i}(B_{r_n}^T . y) \geq \frac{\mu_y^T(B_{r_n}^T)}{\mu_y^T(B_4^T)} \mu_z^{\mathcal{A}_i}(E_i).$$

Multiplying with w_n and summing over $n = 1, \ldots, M$ we get

$$\begin{aligned} \int_{Y_i} \sum_{n=1}^{M} w_n \chi_{B_{r_n}^T . E_i} \, d\mu_z^{\mathcal{A}_i} &\geq \sum_{n=1}^{M} \frac{1}{b_n \lambda(B_{r_n+5}^T)} \frac{\mu_y^T(B_{r_n}^T)}{\mu_y^T(B_4^T)} \mu_z^{\mathcal{A}_i}(E_i) \\ &\geq m\delta \frac{1}{\lambda(B_4^T)} \mu_z^{\mathcal{A}_i}(E_i) \end{aligned}$$

where the latter follows from the definition of B_m. Integrating over $z \in Y_i$ we get

$$\int_{Y_i} \sum_{n=1}^{M} w_n \chi_{B^T_{r_n}.E_i} \, d\mu \geq m\delta c_4 \mu(E_i)$$

for a constant $c_4 > 0$ only depending on T. Summing the latter inequality over i, we get that

$$c_3\mu(X) \geq \int_X G \, d\mu \geq c_4 m\delta \sum_i \mu(E_i) \geq c_4 m\delta\mu(B_m)$$

by Lemma 6.31.(ii). This implies $\mu(B_m) \leq \frac{c_3\mu(X)}{c_4 m\delta}$, independent of M. Hence we may lift the requirement that $n \leq M$ in the definition of B_m without effecting the above estimate and then let $m \to \infty$ and $\delta \to 0$ to obtain the theorem. □

7. Leaf-wise Measures and entropy

We return now to the study of entropy in the context of locally homogeneous spaces.

7.1. General setup, real case. Let $G \subset \mathrm{SL}(n,\mathbb{R})$ be a closed real linear group. (One may also take G to be a connected, simply connected real Lie group if so desired.) Let $\Gamma \subset G$ be a discrete subgroup and define $X = \Gamma\backslash G$. We may endow G with a left-invariant Riemannian metric which then induces a Riemannian metric on X too. With respect to this metric X is locally isometric to G, i.e., for every $x \in X$ there exists some $r > 0$ such that $g \mapsto xg$ is an isometry from the open r-ball B_r^G around the identity in G onto the open r-ball $B_r^X(x)$ around $x \in X$. Within compact subsets of X one may choose r uniformly, and we may refer to r as an *injectivity radius* at x (or on the compact subset).

Clearly any $g \in G$ acts on X simply by right translation $g.x = xg^{-1} = \Gamma(hg^{-1})$ for $x = \Gamma h \in X$, and one may check that this action is by Lipschitz automorphisms of X. For this recall that the metric on X is defined using a left-invariant metric on G, which in general is not right-invariant. By definition of X the G-action is transitive.

Recall that Γ is called a *lattice* if X carries a G-invariant probability measure m_X, which is called the *Haar measure* on X. This is the case if the quotient is compact, and in this case Γ is called a *uniform lattice*. From transitivity of the G-action it follows that the G-action is ergodic with respect to the Haar measure m_X. Although this is not clear a priori it is often true (in the non-commutative setting we are most interested in) that unbounded subgroups of G also act ergodically with respect to m_X.

If Γ is a lattice, then we may fix some $a \in G$ or a one-parameter subgroup $A = \{a_t = \exp(tw) : t \in \mathbb{R}\}$ and obtain a measure-preserving transformation $a.x = xa^{-1}$ or flow $a_t.x = xa_t^{-1}$ with respect to $\mu = m_X$. Our discussion of entropy below may be understood in that context. However, we will not assume that the measure μ on X, which we will be discussing, equals the Haar measure or that Γ is a lattice. Rather we will use the results here to obtain information about an unknown measure μ and in the best possible situations deduce from that μ equals the Haar measure.

7.2. ARITHMETIC SETUP. Fix a prime number p and let G be the group of $\mathbb{Q}_p$-points of an *algebraic subgroup* $\mathbb{G} \subset \mathrm{SL}(n)$, i.e., G would consists of all $\mathbb{Q}_p$-points of a variety $\mathbb{G}$ which is contained in the affine space of all n-by-n-matrices and whose points happen to form a group. Here a $\mathbb{Q}_p$-point of $\mathbb{G}$ is an element of the variety whose matrix entries are elements of $\mathbb{Q}_p$, as a shorthand we will write $G = \mathbb{G}(\mathbb{Q}_p)$ for the group of all $\mathbb{Q}_p$-points. In this setting (more precisely in the zero characteristic case) one may say $\mathbb{G}$ is *defined over* a field F if $\mathbb{G} = \{g \in \mathrm{SL}(n) : \phi(g)v \propto v\}$ where ϕ is an algebraic representation over F, i.e., an action of $\mathrm{SL}(n)$ by linear automorphisms of a finite dimensional vector space with a given basis such that the matrix entries corresponding to $\phi(g)$ are polynomials in the matrix entries of g with coefficients in F, and v equals an F-linear combination of the basis vectors. Again we will let $\Gamma \subset G$ be a discrete subgroup and study dynamics of subgroups of G on $X = \Gamma\backslash G$.

E.g. if G is the group of $\mathbb{Q}_p$-points of $SO(3)$ (defined in the usual way as the group of matrices of determinant one preserving $x_1^2+x_2^2+x_3^2$), which is an algebraic subgroup defined over $\mathbb{Q}$, then one may take Γ to be the group of $\mathbb{Z}[\frac{1}{p}]$-points of $SO(3)$. In this case G is noncompact if $p > 2$ but $X = \Gamma\backslash G$ is compact for any p.

A more general setup would be to allow products

$$G = G_\infty \times G_{p_1} \times \cdots \times G_{p_\ell}$$

over the real and finite places[24] of the group of $\mathbb{R}$-points $G_\infty = \mathbb{G}(\mathbb{R})$, resp., the group of $\mathbb{Q}_p$-points $G_p = \mathbb{G}(\mathbb{Q}_p)$ for some finite list of primes $p \in S_{\mathrm{fin}} = \{p_1, \ldots, p_\ell\}$, of an algebraic group $\mathbb{G}$ defined over $\mathbb{Q}$. In this case one may take $\Gamma = \mathbb{G}(\mathbb{Z}[\frac{1}{p} : p \in S_{\mathrm{fin}}])$ to be the $\mathbb{Z}[\frac{1}{p} : p \in S_{\mathrm{fin}}]$-points of $\mathbb{G}$, which one considers as a subgroup of the product of the real and p-adic groups by sending a matrix γ with coefficients in $\mathbb{Z}[\frac{1}{p} : p \in S_{\mathrm{fin}}]$ to the element $(\gamma, \gamma, \ldots, \gamma) \in G_\infty \times G_{p_1} \times \cdots \times G_{p_\ell}$. This embedding is called the *diagonal embedding*. It can easily be checked that (the image of) Γ forms a discrete subgroup. Often (e.g. when $\mathbb{G}$ is semisimple) Γ defined by this diagonal embedding will form a lattice in G.

A similar construction of arithmetically defined quotients $X = \Gamma\backslash G$ can be used in positive characteristic. Most of what we will discuss in this chapter (and possibly beyond) applies to either of these settings. However, so as to keep the notation at a minimum we will confine ourselves to the situation where $G = \mathbb{G}(k)$ is the group of k-points of an algebraic group $\mathbb{G}$ defined over k, where $k = \mathbb{R}$, $k = \mathbb{Q}_p$, or k equals a local field of positive characteristic. We will refer to this by briefly saying G is an algebraic group over a local field k. Also we only assume that $\Gamma < G$ is a discrete subgroup.

7.3. THE HOROSPHERICAL SUBGROUP DEFINED BY a. For the following fix some $a \in G$. Then we may define the *stable horospherical subgroup for* a by

$$G^- = \{g : a^n g a^{-n} \to e \text{ as } n \to \infty\},$$

[24]The reader who is familiar with adeles may want to consider them instead of finite products.

which in the setting described above is always a closed[25] subgroup of G. Similarly, one can define the *unstable horospherical subgroup* G^+ e.g. as the stable horospherical subgroup for a^{-1}. (We note, that in the theory of algebraic groups G^- and G^+ are also known as the unipotent radicals of the parabolic subgroups defined by a one-parameter subgroup containing a.)

Consider two points $x, xg \in X = \Gamma\backslash G$ for some $g \in G^-$. Then $a^n.x$ and $a^n.xg$ get closer and closer to one another as $n \to \infty$. In fact, $a^n.xg = xa^{-n}(a^n g a^{-n})$ and $a^n.x$ have distance $\leq d(a^n g a^{-n}, e) \to 0$. In that sense we will refer to $G^-.x$ as the *stable manifold through* x. Note that x may not be fixed or even periodic so the statement needs to be understood by the sequence of tuples of points as described. Also note that it is not clear that $G^-.x$ is necessarily the complete set of points y for which $d(a^n.y, a^n.x) \to 0$, but we will show that for all practical purposes it suffices to study $G^-.x$.

7.4. PROBLEM. Suppose X is a compact quotient. Show that in this case $G^-.x \subset X$ is precisely the set of points $y \in X$ with $d(a^n.y, a^n.x) \to 0$ as $n \to \infty$.

7.5. ENTROPY AND THE HOROSPHERICAL SUBGROUP. The following is one of the main results of this section.

7.6. Theorem. *Let μ be an a-invariant probability measure on $\Gamma\backslash G$. Let U be a closed subgroup of G^- normalized by a. Then:*

(i) *The* entropy contribution of U at x

$$D_\mu(a, U)(x) := \lim_{n\to\infty} \frac{\log \mu_x^U(a^{-n} B_1^U a^n)}{n}$$

exists for a.e. x and defines an a-invariant function on X.

(ii) *For a.e. x we have $D_\mu(a, U)(x) \leq h_{\mu_x^{\mathcal{E}}}(a)$, with equality if $U = G^-$. Here $\mathcal{E}$ denotes the σ-algebra of a-invariant sets as in §5.14.*

(iii) *For a.e. x we have $D_\mu(a, U)(x) = 0$ if and only if μ_x^U is finite, which again holds if and only if μ_x^U is trivial*[26].

In particular, the theorem shows that entropy must vanish for all invariant measures if the stable horospherical subgroup G^- is the trivial subgroup. This is the case for the horocycle flow (and all other unipotent flows), hence its entropy vanishes. Therefore, the most interesting case will be the study of the opposite extreme, namely, diagonalizable elements $a \in G$ (and in the proof we will restrict ourselves to this case). For instance, the theorem shows that entropy for the geodesic flow is determined precisely by the leaf-wise measure for the horocyclic subgroup, as for the time-one-map a_1 of the geodesic flow the stable horospherical subgroup is precisely the horocyclic subgroup.

7.7. Corollary. *The measure μ is G^--recurrent if and only if $h_{\mu_x^{\mathcal{E}}}(a) > 0$ a.e. Assume μ is additionally a-ergodic, then μ is G^--recurrent if and only if $h_\mu(a) > 0$.*

---

[25] This is not true for general Lie groups, hence our assumption that G should be a linear group or a simply connected Lie group. In the case of a linear group G^- can easily be defined by linear equations by first bringing a into its Jordan normal form.

[26] Recall that we consider the leaf-wise measures to be trivial if they equal the Dirac measure at the identity.

7.8. Entropy and G^--invariance. To state the second equally important theorem we ask first what is $h_{m_X}(a)$ where Γ is assumed to be a lattice and m_X denotes the Haar measure on X. The answer follows quickly from Theorem 7.6: Since m_X is invariant under G^-, its leaf-wise measures are Haar measures on G^-. Hence the expression in Theorem 7.6.(i) can be calculated and one obtains

$$D_{m_X}(a, G^-) = -\log\left|\det \mathrm{Ad}_a|_{\mathfrak{g}^-}\right|,$$

here Ad_a is the adjoint action of a on the Lie algebra $\mathfrak{g}$ and $\mathfrak{g}^-$ is the Lie algebra of G^- which is, by definition, invariant and being contracted by Ad_a.

The following theorem will characterize when a measure μ is invariant under G^- (or under $U \subset G^-$) in terms of the entropy $h_\mu(a)$ (or the entropy contribution of U). To state it most conveniently, let us define the *entropy contribution* of an a-normalized closed subgroup $U \subset G^-$ by

$$h_\mu(a, U) = \int D_\mu(a, U) d\mu$$

the integral of the entropy contributions at the various x. This way, the entropy contribution of G^- equals the entropy of a (cf. §3.5 and §5.14).

7.9. Theorem. *Let $U < G^-$ be an a-normalized closed subgroup of the horospherical subgroup G^- for some $a \in G$, and let $\mathfrak{u}$ denote the Lie algebra of U. Let μ be an a-invariant probability measure on $X = \Gamma\backslash G$. Then the entropy contribution is bounded by*

$$h_\mu(a, U) \leq -\log\left|\det \mathrm{Ad}_a|_{\mathfrak{u}}\right|$$

and equality holds **if and only if** *μ is U-invariant.*

In many cases this theorem shows that the Haar measure on X is the unique measure of maximal entropy. For example the Haar measure on $\mathrm{SL}(2,\mathbb{Z})\backslash \mathrm{SL}(2,\mathbb{R})$ is the unique measure of maximal entropy as follows from Theorem 7.9: Since the stable horospherical subgroup is the upper unipotent subgroup in $\mathrm{SL}(2,\mathbb{R})$, we have that an a-invariant measure whose entropy equals that of the Haar measure must be invariant under the upper unipotent subgroup. Since $h_\mu(a) = h_\mu(a^{-1})$ we get the same for the lower unipotent subgroup. However, since the upper and the lower unipotent subgroups generate $\mathrm{SL}(2,\mathbb{R})$, we get that $h_\mu(a) = h_{m_X}(a)$ implies $\mu = m_X$. By the same argument one obtains the following more general corollary.

7.10. Corollary. *Suppose Γ is a lattice in G, and let $X = \Gamma\backslash G$. Suppose $a \in G$ is such that G is generated by G^+ and G^-. Then m_X is the unique measure of maximal entropy for the action of a on X, i.e., if μ is an a-invariant probability measure on X with $h_\mu(a) = h_{m_X}(a)$ then $\mu = m_X$.*

7.11. Starting the proofs. Let us start by discussing the technical assumption of the last section that a.e. orbit is embedded.

7.12. Lemma. *Let μ be an a-invariant probability measure on $X = \Gamma\backslash G$. Then for μ-a.e. x the map $u \in G^- \mapsto u.x$ is injective.*

7.13. Proof. Suppose $x = u.x$ for some nontrivial $u \in G^-$. Then $x_n = a^n.x = a^n u a^{-n}.x_n$ for all $n = 1, 2, \ldots$. However, $a^n u a^{-n} \to e$ so that the injectivity radius at x_n goes to 0 as $n \to \infty$. This shows that x does not satisfy Poincaré recurrence. Hence it belongs to a null set. $\square$

7.14. SEMISIMPLE ELEMENTS AND CLASS A ELEMENTS. As before we assume that G is an algebraic group over a local field k (or that G is a simply connected real Lie group), $\Gamma < G$ a discrete subgroup, and $X = \Gamma\backslash G$. We say that $a \in G$ is *k-semisimple* if as an element of $\mathrm{SL}(n,k)$ its eigenvalues belong to k. In particular, this implies that the adjoint action Ad_a of a on the Lie algebra has eigenvalues in k and so is diagonalizable over k. (In the Lie group case the latter would be our assumption with $k = \mathbb{R}$.) We say furthermore that a is *class A* if the following properties hold:

- a is k-semisimple.
- 1 is the only eigenvalue of absolute value 1 for the adjoint action Ad_a.
- No two different eigenvalues of Ad_a have the same absolute value.

For class A elements a we have a decomposition of $\mathfrak{g}$, the Lie algebra of G, into subspaces

$$\mathfrak{g} = \mathfrak{g}_0 \oplus \mathfrak{g}_- \oplus \mathfrak{g}_+$$

where $\mathfrak{g}_0$ is the eigenspace for eigenvalue 1, $\mathfrak{g}_-$ is the direct sum of the eigenspaces with eigenvalues less than 1 in absolute value, and $\mathfrak{g}_+$ is the direct sum of the eigenspaces with eigenvalues greater than 1 in absolute value. These are precisely the Lie algebras of the corresponding subgroups

$$\begin{aligned} G^0 &= \{h : ah = ha\} = C_G(a), \\ G^- &= \{h : a^n h a^{-n} \to e \text{ as } n \to \infty\}, \\ G^+ &= \{h : a^{-n} h a^n \to e \text{ as } n \to \infty\}. \end{aligned}$$

We refer to G^0 as the *centralizer of a*, while G^- and G^+ are the horospherical subgroups of a.

If convenient we will assume[(27)] below that a is of class A as this gives us a convenient description of a neighborhood of $e \in G$ in terms of neighborhoods in the three subgroups G^0, G^-, and G^+. As before we will always assume that $U < G^-$ is a closed a-normalized subgroup of the stable horospherical subgroup.

7.15. PROBLEM. Show that for any Lie group G and any $a \in G$ the Lie algebra generated by $\mathfrak{g}^-$ and $\mathfrak{g}^+$ is a Lie ideal in $\mathfrak{g}$. Deduce that the assumption regarding a in Corollary 7.10 is satisfied whenever G is a simple real Lie group and $\mathfrak{g}^-$ is nontrivial.

7.16. Lemma. *Let $U < G^-$ be a closed a-normalized subgroup for some $a \in G$, denote conjugation by a by $\theta(g) = aga^{-1}$ for $g \in G$. Let μ be an a-invariant probability measure on $X = \Gamma\backslash G$. Then $\mu_{a.x}^U \propto \theta_* \mu_x^U$ for a.e. x.*

7.17. PROOF. As a normalizes U it maps an (r, U)-flower $(Y, \mathcal{A})$ with base E to another σ-algebra $a.\mathcal{A}$ of subsets of $a.Y$ whose atoms are still open U-plaques. More precisely, for $u \in U$ we have $aua^{-1} \in U$ and $a.(u.x) = \theta(u).(a.x)$. As a preserves the measure μ the conditional measures for $\mathcal{A}$ are mapped to those of $a.\mathcal{A}$. Combining this with Theorem 6.3.(iv) gives the lemma. □

---

(27)Replacing G^0 and $\mathfrak{g}^0$ with slightly more complicated versions this assumption can be avoided but in our applications a will always be of class A.

7.18. INDEPENDENCE, a-INVARIANCE. From the definition of $D_\mu(a,U)(x)$ it follows that if the limit defining $D_\mu(a,U)(x)$ exists, then the original set B_1^U can be replaced by any bounded neighborhood O of $e \in U$ without affecting the limit $D_\mu(a,U)(x)$. In fact, if $a^k B_1^U a^{-k} \subset O \subset a^{-k} B_1^U a^k$ (and such a k exists as U is being contracted by a and both B_1^U and O are bounded neighborhoods) then $\mu(a^{-n+k} B_1^U a^{n-k}) \leq \mu(a^{-n} O a^n) \leq \mu(a^{-n-k} B_1^U a^{n+k})$ and this implies the claim (using the sandwich argument for sequences and $\frac{n \pm k}{n} \to 1$).

The a-invariance follows from Lemma 7.16: Replacing x by $a.x$ may be interpreted a.e. as replacing μ_x^U by a measure proportional to $\theta_* \mu_x^U$, and the latter replaces B_1^U by $O = a^{-1} B_1^U a$. Both the proportionality factor and the change to O does not affect the limit $D_\mu(a,U)(x)$ so that $D_\mu(a,U)(x) = D_\mu(a,U)(a.x)$ a.e.

7.19. PREPARING THE REDUCTION TO THE ERGODIC CASE. Recall from §5.14 that for any a-invariant measure μ, we have the ergodic decomposition

$$\mu = \int \mu_x^{\mathcal{E}} d\mu(x)$$

where $\mathcal{E}$ is the σ-algebra of all a-invariant sets, and $\mu_x^{\mathcal{E}}$ is the conditional measure. Also recall from §3.5 that the entropy $h_\mu(a)$ equals the average of the entropies $h_{\mu_x^{\mathcal{E}}}(a)$ of the ergodic components. In what follows we wish to reduce the proof of Theorem 7.6 and 7.9 to the corresponding statements under the assumption of ergodicity. The reader who is willing to assume ergodicity[(28)] of a or to accept this, may continue reading with §7.25.

An important observation (the Hopf argument) is that we can choose the elements of $\mathcal{E}$ to be not only a-invariant, but in fact $\langle U, a\rangle$-invariant. This will allow us to reduce the proof of the main theorems to the case of a-ergodic invariant measures.

7.20. Lemma. *Let C be an a-invariant subset of X. Then there exists a $\langle G^-, a\rangle$-invariant set $\tilde{C}$ such that $\mu(C \triangle \tilde{C}) = 0$.*

7.21. PROOF (USING THE HOPF ARGUMENT). Let $\epsilon > 0$ and choose $f \in C_c(X)$ such that $\|f - 1_C\|_1 < \epsilon$. Set

$$C_\epsilon = \left\{ x : \lim_{n\to\infty} A(f,n)(x) > \frac{1}{2} \right\}$$

where $A(f,n) = \frac{1}{n}\sum_{i=0}^{n-1} f(a^i.x)$. Now

$$\begin{aligned} &(7.21a) \qquad C \triangle C_\epsilon \subset \{x : \lim_{n\to\infty} A(f,n)(x) \text{ does NOT exist }\} \\ &(7.21b) \qquad\qquad \cup \{x \in C : \lim_{n\to\infty} A(f,n)(x) \leq \tfrac{1}{2}\} \\ &(7.21c) \qquad\qquad \cup \{x \notin C : \lim_{n\to\infty} A(f,n)(x) > \tfrac{1}{2}\} \end{aligned}$$

By the pointwise ergodic theorem, the set on the right of (7.21a) has measure 0. We are interested in showing that the measures of (7.21b) and (7.21c) are small. Since C is a-invariant, we have

$$\begin{aligned} x \in C &\Rightarrow A(f - 1_C, n)(x) = A(f,n)(x) - 1 \\ x \notin C &\Rightarrow A(f - 1_C, n)(x) = A(f,n)(x) \end{aligned}$$

(28)This assumption should not be confused with A-ergodicity which we will assume in the later sections but which in general does not imply a-ergodicity.

Let $M(f) = \sup_n |A(f,n)|$ be the maximal function as in the maximal ergodic theorem. Then (7.21b) $\cup$ (7.21c) $\subset \{x : M(f - 1_C)(x) \geq 1/2\}$. Therefore, $\mu\big((7.21\text{b})\cup(7.21\text{c})\big) \leq 2\|f - 1_C\|_1 < 2\epsilon$ by the maximal ergodic theorem.

Furthermore, we claim that C_ϵ is G^--invariant. Notice that for any $h \in G^-$, we have that $a^n.(h.x) = a^n h a^{-n}.(a^n.x)$ and $a^n.x$ are asymptotic to one another. Since f has compact support it is uniformly continuous. Therefore, we have

$$\frac{1}{n}\sum_{i=0}^{n-1}\left[f(a^i.x) - f(a^i.(h.x))\right] \to 0$$

uniformly in x. This shows that C_ϵ is G^--invariant.

To finish the proof we may choose $\epsilon_n = 2^{-n}$ and

$$\tilde{C} = \varlimsup_{n\to\infty} C_{2^{-n}} = \bigcap_n \bigcup_{k\geq n} C_{2^{-n}}$$

to obtain a set as in the lemma. □

7.22. Proposition. *Let $\mu = \int \mu_x^{\mathcal{E}} d\mu(x)$ be an a-invariant probability measure, and $U < G^-$. Then for μ-a.e. x, for $\mu_x^{\mathcal{E}}$-a.e. y, we have $\mu_y^U = (\mu_x^{\mathcal{E}})_y^U$.*

In other words, by changing the leaf-wise measures for $\mu_x^{\mathcal{E}}$ at most on a $\mu_x^{\mathcal{E}}$-nullset, we may define $(\mu_x^{\mathcal{E}})_y^U$ to be equal to μ_y^U. With this definition in place, we also have $(\mu_x^{\mathcal{E}})_x^U = \mu_x^U$. (In the formulation of the proposition we avoided this formula as $\{x\}$ is a null set for $\mu_x^{\mathcal{E}}$ and so making claims for the leaf-wise measure at x would be irrelevant.)

7.23. Proof. Recall that the leaf-wise measures μ_x^U were determined by moving the conditional measures $\mu_x^{\mathcal{A}_i}$ to U and patching them together there. Here $(Y_i, \mathcal{A}_i)$ were U-flowers. By Lemma 7.20 (and Proposition 5.8) we may replace $\mathcal{E}$ by a countably generated σ-algebra consisting of a-invariant and U-invariant sets. In particular, this shows that the atoms of $\mathcal{E}|_{Y_i}$ are unions of the atoms of $\mathcal{A}_i$ (which are open U-plaques). However, using conditional measures for $\mathcal{A}_i$ it is easy to see that a measurable function that is constant on $\mathcal{A}_i$-atoms is in fact $\mathcal{A}_i$-measurable modulo μ. Therefore, we have $\mathcal{E}|_{Y_i} \subset \mathcal{A}_i$ modulo μ. However, this inclusion of σ-algebras implies that

$$E(E(f|\mathcal{A}_i)|\mathcal{E}|_{Y_i}) = E(f|\mathcal{E}|_{Y_i})$$

for any $f \in L^1$. In turn, using the defining properties of conditional measures (in terms of conditional expectations) this gives the following relation between the conditional measures: for μ-a.e. $x \in Y_i$ we have for $\mu_x^{\mathcal{E}|_{Y_i}}$-a.e. y that

$$\left(\mu_x^{\mathcal{E}|_{Y_i}}\right)_y^{\mathcal{A}_i} = \mu_y^{\mathcal{A}_i}.$$

Translating this to a property of leaf-wise measures we see that μ_y^U and $(\mu_x^{\mathcal{E}})_y^U$ agree on the subset of U corresponding to the atom $[x]_{\mathcal{A}_i}$ and the proposition follows by collecting the various null sets of Y_i. □

7.24. PROOF OF REDUCTION TO ERGODIC CASE. Working with double conditional measures as in the above proposition may be confusing, but it is useful for the following purpose: In the proof of Theorem 7.6 and 7.9 we are comparing the entropy of the ergodic components and the entropy contribution arising from the subgroup $U < G^-$. From §5.14 and §3.5 we know that

$$h_\mu(a) = \int h_{\mu_x^{\mathcal{E}}}(a) d\mu.$$

We would like to have a similar relationship between $D_\mu(a,U)(x)$ and $D_{\mu_x^{\mathcal{E}}}(a,U)(x)$. Using $(\mu_x^{\mathcal{E}})_x^U = \mu_x^U$ as in the discussion right after Proposition 7.22 we get

$$D_{\mu_x^{\mathcal{E}}}(a,U)(x) = D_\mu(a,U)(x).$$

Since $\mu_x^{\mathcal{E}}$ is a-invariant and ergodic for μ-a.e. x, and as we assume the statements of Theorem 7.6 and 7.9 in the ergodic case, the general case follows from this. □

7.25. Definition. *We say that a σ-algebra $\mathcal{A}$ is* subordinate to U (mod μ) *if for μ-a.e. x, there exists $\delta > 0$ such that*

$$B_\delta^U.x \subset [x]_{\mathcal{A}} \subset B_{\delta^{-1}}^U.x.$$

We say that $\mathcal{A}$ is subordinate to U on Y *if and only if the above holds for a.e. $x \in Y$.*

We say that $\mathcal{A}$ is a-descending *if $a^{-1}.\mathcal{A} \subset \mathcal{A}$.*

Ignoring null sets to say that $\mathcal{A}$ is subordinate to U is basically equivalent to say that the $\mathcal{A}$-atoms are open U-plaques. Hence we have already established in the last section the existence of σ-algebras which are subordinate to U at least on some sets of positive measure. Also, it is rather easy to find an a-descending σ-algebra as $\bigvee_{n=0}^\infty a^{-n}.\mathcal{P}$ is a-descending for any countable partition (or even σ-algebra) $\mathcal{P}$. We note however, that the existence of an a-descending σ-algebra that is also subordinate is not trivial.

Recall that we may assume that μ is a-ergodic, so that the a-invariant function $D_\mu(a,U)(x)$ (whose existence we still have to show, see Prop. 7.34) must be constant a.e. If we are given an a-descending σ-algebra $\mathcal{A}$ that is subordinate to U, we will show the following properties (which, in particular, gives an independent meaning to the generic value of $D_\mu(a,U)(x)$):

(i) For a.e. x

$$\frac{\log \mu_x^U(a^{-n} B_1^U a^n)}{n} \to H_\mu(\mathcal{A}|a^{-1}.\mathcal{A}) = h_\mu(a,U)$$

as $n \to \infty$.

(ii) $h_\mu(a,U) \leq h_\mu(a)$, with equality if $U = G^-$.

(iii) If $h_\mu(a,U) = 0$ then $a^{-1}.\mathcal{A} = \mathcal{A}$ (mod μ) and $\mu_x^U = \delta_e$ almost surely.

In other words, we will use the σ-algebra $\mathcal{A}$ as a gadget linking the two expressions $D_\mu(a,U)(x)$ and $h_\mu(a)$ appearing in the Theorem 7.6.

Recall that the "empirical entropy" $H_\mu(\mathcal{A}|a^{-1}.\mathcal{A})$ is the average of the "conditional information function"

$$I_\mu(\mathcal{A}|a^{-1}.\mathcal{A})(x) = -\log \mu_x^{a^{-1}.\mathcal{A}}([x]_{\mathcal{A}}).$$

7.26. Hyperbolic torus automorphisms. We first look at a particular example[29] where it is relatively easy to give an a-descending σ-algebra $\mathcal{A}$ that is subordinate to G^- and to see the connection to entropy. Let a be a hyperbolic automorphism of $\mathbb{T}^m = \mathbb{R}^m/\mathbb{Z}^m$. By this we mean that a is defined by a matrix with eigenvalues, real or complex, of absolute value different from one (this is the answer to Problem 3.10). We set $G = \mathbb{R}^m$, $\Gamma = \mathbb{Z}^m$, and write θ for the linear map on $\mathbb{R}^m$ defining a; for consistency we will still write $a.x$ for the action. It is easy to see that G^- is the sum[30] of all generalized eigenspaces for eigenvalues of absolute value less than one. By expansiveness we know that any partition $\mathcal{P}$ whose atoms have sufficiently small diameter will be a generator for a. By the same argument one easily shows that $\mathcal{A} = \bigvee_{n=0}^{\infty} a^{-n}.\mathcal{P}$ satisfies that the $\mathcal{A}$-atoms are of the form $[x]_{\mathcal{A}} = V_x.x$ for bounded subsets $V_x \subset G^-$. Also, $\mathcal{A}$ is a-descending. The remaining property that V_x contains the identity in the interior a.e. is not a general property (as it is likely not true if the boundaries of the partition elements are not null sets) but follows if we are a bit more careful in the choice of the partition $\mathcal{P}$. What we will need is the following quantitative strengthening of $\mu(\partial P) = 0$ for all $P \in \mathcal{P}$.

7.27. Lemma. *Let X be a locally compact metric space and let μ be a Radon measure on X. Then for every $x \in X$ and Lebesgue-a.e. $r > 0$ there exists a constant $c = c_{x,r}$ such that $\mu(\partial_\delta B_r(x)) \leq c\delta$ for all sufficiently small $\delta > 0$. Here we refer to*

$$\partial_\delta B = \{y \in X : \inf_{z \in B} d(y,z) + \inf_{z \notin B} d(y,z) < \delta\}$$

as the δ-neighborhood of the boundary[31] *of a subset $B \subset X$.*

7.28. Problem. *Prove Lemma 7.27 using the function $f(r) = \mu(B_r(x))$. A hint may be found in the footnote*[32] *on the next page.*

We say that a set B has *μ-thin boundary* if there exists some c such that $\mu(\partial_\delta B) \leq c\delta$ for all $\delta > 0$. It is clear that a set obtained from finitely many sets with μ-thin boundary via the set-theoretic operations of intersections, union, or complements also has μ-thin boundary. Hence by Lemma 7.27 any compact space has a partition $\mathcal{P}$ consisting of sets with μ-thin boundary and arbitrarily small diameter.

We also note another property, which is rather easy to verify for the Euclidean metric on $G = \mathbb{R}^m$ and the linear map θ defining the automorphism a.

7.29. Lemma. *There exists some $\alpha > 0$ and $d > 0$ depending on a and G such that for every $r \in (0,1]$ we have*

$$\theta^n(B_r^{G^-}) \subset B_{de^{-n\alpha}r}^{G}$$

for all $n \geq 1$.

7.30. Problem. *Let $a \in G$ be an element of class $\mathcal{A}$. Prove Lemma 7.29 in the context of G being a real Lie group, assuming that G is endowed with a left invariant Riemannian metric.*

---

[29]This example almost fits into the framework under which we work, except that the automorphism we consider is not coming from an element of $G = \mathbb{R}^m$. We could use a bigger subgroup, namely a semidirect product of $\mathbb{Z}$ and $\mathbb{R}^m$, but this is not necessary and may be more confusing.

[30]This is always a real subspace even if some eigenvalues are complex.

[31]We use this phrase even though in general ∂B may be empty with $\partial_\delta B$ nonempty.

Prove Lemma 7.29 in the setting of G being an algebraic group defined over a p-adic field or a finite characteristic local field by first defining a metric on G. (If necessary it would not make a difference to our applications below to replace the upper bound 1 for r by some smaller quantity depending on a and G.)

We now show how the two properties in Lemma 7.27 and Lemma 7.29 can be used in combination.

7.31. Lemma. *Suppose $\mathcal{P}$ is a finite partition of $X = \Gamma\backslash G$ consisting of measurable sets with μ-thin boundary. Then for a.e. $x \in X$ there is some $\delta > 0$ such that*

$$B_\delta^{G^-}.x \subset [x]_{\bigvee_{n\geq 0} a^{-n}.\mathcal{P}}. \tag{7.31a}$$

7.32. PROOF. Let c be the maximal constant as in the definition of μ-thin boundary for the elements of $\mathcal{P}$, and let α and d be as in Lemma 7.29. Also let $r = 1$. We write $\partial_\delta \mathcal{P}$ for the union of the δ-neighborhoods of the boundaries of the elements of $\mathcal{P}$.

Fix some $\delta > 0$ and define for $n \geq 0$ the set

$$E_n = a^{-n}.\partial_{de^{-n\alpha}\delta}\mathcal{P}.$$

By construction we have

$$\mu\left(\bigcup_{n\geq 0} E_n\right) \leq cd\left(\sum_{n\geq 0} e^{-n\alpha}\right)\delta,$$

which shows that for a.e. x there is some δ with $x \notin \bigcup_{n\geq 0} E_n$. Fix such an x and the corresponding δ, we claim that (7.31a) holds. Indeed let $h \in B_\delta^{G^-}$ (which in the case of $X = \mathbb{T}^m$ acts by addition $h.x = x + h$ on $x \in \mathbb{T}^m$) and suppose $h.x \notin [x]_{\bigvee_{n\geq 0} a^{-n}.\mathcal{P}}$. Then there would be some $n \geq 0$ such that $a^n.x$ and $a^n.(h.x)$ belong to different elements of the partition $\mathcal{P}$. However, θ contracts G^- and indeed $d(\theta^n(h), e) < de^{-n\alpha}\delta$ by Lemma 7.29. Therefore, $a^n.(h.x) = \theta^n(h).(a^n.x)$ and $a^n.x$ have distance less than $de^{-n\alpha}\delta$, which shows that both belong to $\partial_{de^{-n\alpha}\delta}\mathcal{P}$. However, this gives a contradiction to the definition of E_n and the choice of x and δ.□

7.33. HYPERBOLIC TORUS AUTOMORPHISM CONCLUDED. The discussion in §7.26 together with Lemma 7.31 shows that it is possible to choose $\mathcal{P}$ such that the σ-algebra $\mathcal{A} = \bigvee_{n=0}^\infty a^{-n}.\mathcal{P}$ is a-decreasing and subordinate to G^-. Recalling that $\mathcal{P}$ was constructed as a generator (c.f. §3.6) we also get

$$h_\mu(a) = h_\mu(a, \mathcal{P}) = H_\mu(\mathcal{A}|a^{-1}.\mathcal{A}).$$

This establishes the link between $H_\mu(\mathcal{A}|a^{-1}.\mathcal{A})$ and the entropy $h_\mu(a)$ in the case at hand; the link between $H_\mu(\mathcal{A}|a^{-1}.\mathcal{A})$ and the entropy contribution we now establish in great generality.

7.34. Proposition. *Suppose $\mathcal{A}$ is a countably generated σ-algebra subordinate to U, such that $\mathcal{A} \supset a^{-1}.\mathcal{A}$. Then*

$$\lim_{n\to\infty} \frac{\log \mu_x^U(a^{-n}B_1^U a^n)}{n} = H_\mu(\mathcal{A}|a^{-1}.\mathcal{A}).$$

In particular, the limit defining the entropy contribution of U at x exists.

---

(32) Notice that $f(r)$ is monotone and hence differentiable a.e.

7.35. PROOF. We start by showing that

$$-\frac{1}{n}\log \mu_x^{a^{-n}.\mathcal{A}}([x]_{\mathcal{A}}) \to H_\mu(\mathcal{A}|a^{-1}.\mathcal{A}).$$

Here notice first that by Proposition 5.17

$$\mu_x^{a^{-1}.\mathcal{A}}\big|_{[x]_{\mathcal{A}}} = \mu_x^{a^{-1}.\mathcal{A}}([x]_{\mathcal{A}})\mu_x^{\mathcal{A}}$$

for a.e. x since $[x]_{a^{-1}.\mathcal{A}}$ is a countable union of $\mathcal{A}$-atoms. More generally we obtain by the same argument that

$$\mu_x^{a^{-n}.\mathcal{A}}([x]_{\mathcal{A}}) = \prod_{i=1}^{n} \mu_x^{a^{-i}.\mathcal{A}}([x]_{a^{-(i-1)}.\mathcal{A}}).$$

Also note that $\mu_{a.x}^{\mathcal{A}} = a_*\mu_x^{a^{-1}.\mathcal{A}}$ (as one may verify from the defining relation of $\mu_x^{\mathcal{A}}$ in terms of the conditional expectation). Combining these one gets by taking logarithms that

$$\begin{aligned} -\frac{1}{n}\log \mu_x^{a^{-n}.\mathcal{A}}([x]_{\mathcal{A}}) &= \sum_{i=1}^{n} \frac{-\log \mu_x^{a^{-i}.\mathcal{A}}([x]_{a^{-(i-1)}.\mathcal{A}})}{n} \\ &= \frac{1}{n}\sum_{i=0}^{n-1} I_\mu(\mathcal{A}|a^{-1}.\mathcal{A})(a^i.x) \\ &\to H_\mu(\mathcal{A}|a^{-1}.\mathcal{A}) \end{aligned}$$

by the pointwise ergodic theorem (since μ is assumed to be a-ergodic).

We may also obtain in a similar manner that

$$\frac{\log \mu_x^U(a^{-n}B_1^U a^n)}{n} \to \int \log \mu_x^U(a^{-1}B_1^U a),$$

where we assume the normalization $\mu_x^U(B_1^U) = 1$. Indeed by Lemma 7.16 we know $\mu_{a.x}^U(a^{-1}B_1^U a) = \frac{\mu_x^U(a^{-2}B_1^U a^2)}{\mu_x^U(a^{-1}B_1^U a)}$, which easily generalizes to higher powers of a and then gives

$$\mu_x^U(a^{-n}B_1^U a^n) = \prod_{i=0}^{n-1} \mu_{a^i.x}^U(a^{-1}B_1^U a).$$

Taking the logarithm and using the pointwise ergodic theorem the above claim follows.

We outline the remainder of the proof of Proposition 7.34: Both of the above limits measure the growth rate of a dynamically expanded set in relation to a fixed set. By Theorem 6.3.(iv) the fact that in one expression we are using the conditional measure $\mu_x^{a^{-n}.\mathcal{A}}$ and in the other the leaf-wise measure μ_x^U seems irrelevant. However, what is unclear is the precise relationship between the shape $V_{n,x} \subset U$ of the atoms $[x]_{a^{-n}.\mathcal{A}} = V_{n,x}.x$ and the set $a^{-n}B_1^U a^n$. We show below that as $n \to \infty$ the influence of the shape is negligible, thus obtaining the proposition.

Fix $\delta > 0$ such that

(7.35a) $$Y := \{x : B_\delta^U.x \subset [x]_{\mathcal{A}} \subset B_{\delta^{-1}}^U.x\}$$

has positive measure. By the argument in §7.18 (which only assumes the existence of the limit for $r = 1$) we know that

(7.35b) $$\lim_{n\to\infty} \frac{\log \mu_x^U(a^{-n}B_r^U a^n)}{n}$$

is independent of r for a.e. x. Moreover, for a.e. x there exists a sequence n_j of integers for which $a^{n_j}.x \in Y$. For those $n = n_j$ we therefore have

$$[x]_{a^{-n}.\mathcal{A}} = a^{-n}.[a^n.x]_{\mathcal{A}} \subset a^{-n}B^U_{\delta^{-1}}a^n.x$$

and similarly,

$$[x]_{a^{-n}.\mathcal{A}} \supset a^{-n}B^U_{\delta}a^n.x.$$

Therefore, $a^{-n}B^U_\delta a^n \subset V_{n,x} \subset a^{-n}B^U_{\delta^{-1}}a^n$. Also recall that $\mu_x^{a^{-n}.\mathcal{A}}$ is proportional to $\mu_x^U|_{V_{n,x}}.x$ by Theorem 6.3.(iv). Hence

$$\mu_x^{a^{-n}\mathcal{A}}([x]_{\mathcal{A}}) = \frac{c(x)}{\mu_x^U(V_{n,x})}$$

where $c(x) = \mu_x^U(V_{0,x})$. With this notation the above inclusions imply

$$\mu_x^U(a^{-n}B^U_\delta a^n) \leq \mu_x^U(V_{n,x}) = c(x)\mu_x^{a^{-n}\mathcal{A}}([x]_{\mathcal{A}})^{-1} \leq \mu_x^U(a^{-n}B^U_{\delta^{-1}}a^n)$$

for a.e. x. Taking the logarithm, letting $n = n_j \to \infty$, and using the independence of the limit in (7.35b) the proposition follows. □

7.36. Returning to the general case. Even though we used in the example of the hyperbolic torus automorphism certain special properties of the system, namely that X is compact and that a is expansive, it does give hope regarding the existence of a subordinate and a-descending σ-algebra in general. In fact, using somewhat similar methods (Lemma 7.29, 7.27, and 7.31 are general) as in the example we now establish the existence of the σ-algebra. However, linking the σ-algebras and the entropy (as we did in §7.33) will need more work.

7.37. Proposition. *Let μ be an a-invariant and ergodic probability measure on $\Gamma\backslash G$, and let $U < G^-$ be a closed subgroup normalized by a. Then there exists a countably generated σ-algebra $\mathcal{A}$ such that:*

(i) *$\mathcal{A}$ is subordinate to U.*
(ii) *$a^{-1}.\mathcal{A} \subset \mathcal{A}$, i.e., $\mathcal{A}$ is a-decreasing.*

We note that this establishes the existence of the limit in Theorem 7.6 (i) by using Proposition 7.34.

7.38. Comment. We note that without the assumption of ergodicity the proof below almost gives the claims of the proposition in the following sense: For every $\epsilon > 0$ there exists a set $Y \subset X$ of measure $\mu(Y) > 1 - \epsilon$ such that $\mathcal{A}$ is subordinate to U on Y and $\mathcal{A}$ is a-decreasing.

7.39. Proof. Applying Lemma 7.27 we can find some open $Y \subset X$ with compact closure such that $\mu(Y) > 1 - \epsilon$ and Y has μ-thin boundary, e.g. by letting $Y = B_r^X(x_0)$ for some large r. Below we will construct the σ-algebra $\mathcal{A}$ which will be subordinate to U on Y and a-decreasing. Note that under the assumption of ergodicity this gives the proposition: For a.e. $x \notin Y$ there exists some positive as well as some negative n with $a^n.x \in Y$ which together with $a^{-1}.\mathcal{A} \subset \mathcal{A}$ gives the correct upper, resp., lower bound for $[x]_{\mathcal{A}}$. More precisely, by correct upper bound we mean that $[x]_{\mathcal{A}} \subset B_x.x$ for some bounded subset $B_x \subset U$ and by correct lower bound we mean that $[x]_{\mathcal{A}} \supset O_x.x$ for some open $O_x \subset U$ containing the identity element.

Again applying Lemma 7.27 we can find a finite partition of Y into sets of small diameter (as specified below) and with μ-thin boundary — here Lemma 7.27

is applied to find for every $x \in \overline{Y}$ a small ball around x with μ-thin boundary and then a finite subcover is chosen using compactness. We add to this partition the set $X \setminus Y$ to obtain the partition $\mathcal{P}$. Since the boundaries of all elements of $\mathcal{P}$ are null sets, we may assume all elements of $\mathcal{P}$ are open (and ignore the remaining null set). By Lemma 7.31 we know that the atoms of $\bigvee_{n\geq 0} a^{-n}.\mathcal{P}$ contain a neighborhood of x in the direction of G^- almost surely, i.e., for a.e. $x \in X$ there is some $\delta > 0$ such that

$$[x]_{\bigvee_{n\geq 0} a^{-n}.\mathcal{P}} \supset B_\delta^{G^-}.x. \tag{7.39a}$$

We will replace $\mathcal{P}$ by a σ-algebra $\mathcal{P}^U$ in such a way that $\mathcal{A} = \bigvee_{n\geq 0} a^{-n}.\mathcal{P}^U$ will be subordinate to U (at least) on Y. Let P denote an element of $\mathcal{P}$ different from $X \setminus Y$. We may assume the diameter of P is smaller than the injectivity radius on Y, we get that P is the injective isometric image of an open subset $\tilde{P}$ of G. By assumption U is closed, so that the Borel σ-algebra $\mathcal{B}_{G/U}$ of the quotient G/U is countably generated. This induces a σ-algebra $\mathcal{C}_P$ first on $\tilde{P}$ and then also on P whose atoms are open U-plaques. We define $\mathcal{P}^U$ to be the countably generated σ-algebra whose elements are unions of elements of $\mathcal{C}_P$ for $P \in \mathcal{P}$ and possibly the set $X \setminus Y$, i.e., the atoms of x for $\mathcal{P}^U$ is either $X \setminus Y$ if $x \notin Y$ or an open U-plaque $V_x.x$ of x if $x \in Y$. We claim that for a.e. $x \in Y$ the atom $[x]_\mathcal{A}$ w.r.t. $\mathcal{A} = \bigvee_{n\geq 0} a^{-n}.\mathcal{P}^U$ is an open U-plaque. Indeed suppose x satisfies (7.39a) for some $\delta > 0$ (which we may assume is smaller than the injectivity radius) and $u \in B_\delta^U$. Then for all $n \geq 0$ we know that $a^n.x$ and $a^n u.x$ belong to the same element $P \in \mathcal{P}$. Fix some $n \geq 0$. If $P = X \setminus Y$, then $a^n.x$ and $a^n u.x$ still belong to the same atom of $\mathcal{P}^U$. If $P \neq X \setminus Y$, then we also claim that $a^n.x$ and $a^n u.x$ belong to the same atom of $\mathcal{P}^U$: The two elements $y = a^n.x, z = a^n u.x \in P$ correspond to two elements $\tilde{y}, \tilde{z} \in \tilde{P}$. Since P and $\tilde{P}$ are isometric and $a^n u a^{-n}$ is being contracted we conclude that these two points are still on the same U-coset $\tilde{y}U = \tilde{z}U$, for otherwise we would get a contradiction to the injectivity property at $z \in Y$. This shows that the atoms $[x]_\mathcal{A}$ for a.e. $x \in Y$ are indeed open U-plaques. □

7.40. Proof of Theorem 7.6.(iii). Clearly if μ_x^U is finite, then the entropy contribution $h_\mu(a, U)$ vanishes (as it measures a growth rate). Assume now on the other hand $h_\mu(a, U) = H_\mu(\mathcal{A}|a^{-1}\mathcal{A}) = 0$, where $\mathcal{A}$ is as in Prop. 7.37 (cf. Prop. 7.34). Then

$$H_\mu(\mathcal{A}|a^{-1}.\mathcal{A}) = \int (-\log \mu_x^{a^{-1}.\mathcal{A}}([x]_\mathcal{A}))d\mu = 0$$

implies $\mu_x^{a^{-1}.\mathcal{A}}([x]_\mathcal{A}) = 1$ a.e. which is equivalent to $\mathcal{A} = a^{-1}.\mathcal{A} \bmod \mu$. Iterating this gives $a^m.\mathcal{A} = a^{-m}.\mathcal{A} \bmod \mu$ and $\mu_x^{a^{-m}\mathcal{A}}([x]_{a^m.\mathcal{A}}) = 1$ a.e. and for all $m \geq 1$. By Theorem 6.3.(iv) this says that $\mu_x^U(V_{-m,x} \setminus V_{m,x}) = 0$ a.e., where $V_{m,x}$ denotes the shape of the $a^m.\mathcal{A}$-atom of x. Using again the set Y in (7.35a) we see that the precise shapes do not matter as $V_{-m,x} \nearrow U$ and $V_{m,x} \searrow \{e\}$ as $m \to \infty$ for a.e. x. It follows that $\mu_x^U \propto \delta_e$. □.

7.41. Proof of the inequality $h_\mu(a, U) \leq h_\mu(a, U')$ for $U \subset U' \subseteq G^-$. Assume both U and U' are closed a-normalized subgroups of G^- such that $U \subset U'$. By the construction of the σ-algebra we see that there exist two σ-algebras $\mathcal{A}$ and $\mathcal{A}'$ which are both a-decreasing and subordinate to U and to U', resp., such that additionally $\mathcal{A} \supset \mathcal{A}'$. In order to obtain these, one may use the same finite partition $\mathcal{P}$ and then carry the construction through with both groups.

We claim that $\mathcal{A}' \vee a^{-1}.\mathcal{A} = \mathcal{A} \pmod{\mu}$. We already know one inclusion, to see the other we describe the atoms for the σ-algebra $\mathcal{C} = \mathcal{A}' \vee a^{-1}.\mathcal{A}$. Suppose y and x are equivalent w.r.t. $\mathcal{C}$, then a.s. there exists some $u \in U$ with $y = u.x$ where u may be rather big because the $a^{-1}.\mathcal{A}$-atoms are in general bigger than the $\mathcal{A}$-atoms. To make this more precise, assume y, x belong to the set Y which was used in the constructions of the σ-algebras. Then we do not know that $d(e, u)$ is smaller than the injectivity radius (of Y). However, we know that $y = u'.x$ for some $u' \in U'$ (as the two points are also $\mathcal{A}'$-equivalent), and that $d(e, u')$ is less than the injectivity radius. Since for a.e. x the G^--leaf is embedded by Lemma 7.12, we must have $u = u'$. This implies that x and $y = u.x$ belong to the same atom of the σ-algebra $\mathcal{C}_P$ (for $x, y \in P \subset Y$) which was used in the construction of $\mathcal{P}^U$. This shows the two points are equivalent w.r.t. $\mathcal{A}$, first under the assumption that $x, y \in Y$ but the general case follows by the same argument and ergodicity by using the minimal n with $a^n.x, a^n.y \in Y$. As the atoms of the σ-algebra determine the σ-algebra at least mod μ the claim follows.

The claim implies the desired inequality since

$$h(a, U) = H_\mu(\mathcal{A}|a^{-1}.\mathcal{A}) = H_\mu(\mathcal{A}'|a^{-1}.\mathcal{A}) \leq H_\mu(\mathcal{A}'|a^{-1}.\mathcal{A}') = h(a, U')$$

by monotonicity of the entropy function with respect to the given (i.e., the second) σ-algebra. □

7.42. First proof of the inequality in Theorem 7.9. Recall that $U \subset G^-$ is a-normalized and that λ denotes the Haar measure on U, which is (unipotent and so) necessarily unimodular. We may normalize λ to have measure one on the unit ball of U. Then it is easy to see that $\lambda(a^{-n}B_1^U a^n) = c^n$, where c is the determinant of the adjoint representation of a^{-1} acting on the Lie algebra $\mathfrak{u}$ of U. Hence for a.e. x, we have by Theorem 6.30[33]

$$\begin{aligned} \overline{\lim_{n\to\infty}} \frac{\mu_x^U(a^{-n}B_1^U a^n)}{n^2 c^n} &= 0 \text{ and so} \\ h_\mu(a, U) = \lim_{n\to\infty} \frac{\log \mu_x^U(a^{-n}B_1^U a^n)}{n} &\leq \log c. \end{aligned}$$

This is the inequality in Theorem 7.9. In §7.55 we will give the proof of Theorem 7.9 in full including an independent proof of the inequality shown here. □

7.43. Where we are. To summarize we have shown Theorem 7.6.(i), the inequality in (ii), (iii), and the inequality in Theorem 7.9 (and also that it suffices to study ergodic measures). However, we still have to show the equality between the entropy contribution $h_\mu(a, G^-)$ and the entropy $h_\mu(a)$ and the relationship between invariance and equality in Theorem 7.9. We now turn to the former problem in general.

7.44. Proposition. *Let μ be an a-invariant and ergodic measure on $X = \Gamma\backslash G$. Then there exists a countable partition $\mathcal{P}$ with finite entropy which is a generator for*

[33] Strictly speaking the sets $a^{-n}B_1^U a^n$ may not be balls, but the proof can be adapted to allow for that and the additional thickening of the balls by the parameter 5 does not change the asymptotical behavior of $\lambda(a^{-n}B_1^U a^n)$.

a mod μ. *Moreover, the* σ*-algebra* $\mathcal{A} = \bigvee_{n\geq 0} a^{-n}.\mathcal{P}$ *is* a*-decreasing and subordinate to* G^-.

Together with Proposition 7.34 this implies the last claim of Theorem 7.6.(ii). We will need a few more elementary lemmas.

7.45. Lemma. *There exists some* $\alpha > 0$ *depending on* a *and* G *such that for every* $r > 0$ *we have*

$$\theta^n(B^G_{e^{-|n|\alpha}r}) \subset B^G_r$$

for all $n \in \mathbb{Z}$. *Here* $\theta(g) = aga^{-1}$ *for* $g \in G$ *again stands for conjugation by* a.

Basically this lemma follows from the fact that conjugation is a Lipschitz map whose Lipschitz constant is the norm of the adjoint representation of a.

7.46. Lemma. *For every* $\Omega \subset \Gamma\backslash G$ *with compact closure, and for every* α *and* $r > 0$, *there exist* $\kappa(G, \alpha), c(\Omega, r)$ *such that for every* n, *the set* Ω *can be covered by* $ce^{\kappa n}$ *balls of radius* $e^{-\alpha n}r$.

For the proof of this lemma notice that the set Ω can be covered by finitely many small balls of fixed radius, and that in each one of these we may argue that the metric is basically flat (e.g. in characteristic zero the logarithm map would be bi-Lipshitz in a neighborhood of the identity and the claim is quite easy for a linear space). In a sense this lemma captures (in some weak way) the finite-dimensionality of the group in question.

7.47. PROOF. Equipped with the lemmas above, we are ready to start the construction of our partition $\mathcal{P}$. Fix an open subset $\Omega \subset X = \Gamma\backslash G$ of compact closure, positive measure, and μ-thin boundary (see Lemma 7.27). We may assume Ω is a ball $B_{r/16}(x_0)$ where r is an injectivity radius at x_0.

7.48. THE PARTITION $\mathcal{Q}$. We define $\mathcal{Q} = \{\Omega, X \setminus \Omega\}$. By Lemma 7.31 we have that for a.e. x there exists some $\delta > 0$ with

$$B^{G^-}_\delta .x \subset [x]_{\bigvee_{n\geq 0} a^{-n}.\mathcal{Q}}. \tag{7.48a}$$

7.49. THE PARTITION $\tilde{\mathcal{Q}}$. Next we define $\tilde{\mathcal{Q}} = \{Q_i : i = 0, 1, 2, \ldots\}$, where we define $Q_0 = X \setminus \Omega$, resp. $Q_1 = \Omega \cap a^{-1}.\Omega$, $Q_2 = (\Omega \setminus a^{-1}.\Omega) \cap a^{-2}\Omega$, ..., in other words we split Ω into countably many sets according to when the points next visit Ω (under forward iterates of a). (Strictly speaking we should also add the set $Q_\infty = \Omega \cap \bigcap_{j=1}^\infty a^{-j}.X \setminus \Omega$ to the partition $\tilde{\mathcal{Q}}$, but by Poincaré Recurrence $\mu(Q_\infty) = 0$, so we may omit it from the discussion.)

We observe that $\tilde{\mathcal{Q}}$ is contained in the σ-algebra $\bigvee_{n=1}^\infty a^{-n}.\mathcal{Q}$. Therefore, $\bigvee_{n=1}^\infty a^{-n}.\mathcal{Q} = \bigvee_{n=1}^\infty a^{-n}.\tilde{\mathcal{Q}}$ and the above claim (7.48a) regarding the atoms remains true for $\tilde{\mathcal{Q}}$.

7.50. FINITE ENTROPY. We will now show that $H_\mu(\tilde{\mathcal{Q}}) < \infty$ (but we will need to refine it further to obtain the desired partition). First, note that $X \setminus \Omega$ can be partitioned according to how much time a point will spend (resp. has already spent) in $X \setminus \Omega$ before returning to (resp. since coming from) Ω, keeping in mind that the set of points which remain in $X \setminus \Omega$ forever (resp. have always been in $X \setminus \Omega$) has measure 0 by ergodicity. Moreover, the set of points that have spent time $t \geq 1$ in $X \setminus \Omega$ (including the current time) and will return to Ω in time $s \geq 1$ iterations of a the first time is exactly $a^t.Q_{t+s}$. This implies that

$X\backslash\Omega = \bigcup_{i=1}^{\infty}\bigcup_{t=1}^{i} a^t Q_{i+1}$ (with the union being disjoint), and since μ is a-invariant, we see that $\mu(X \setminus \Omega) = \sum_{i=1}^{\infty} i\mu(Q_{i+1}) < 1$. As the sets $Q_1, Q_2, \ldots$ partition Ω we also have $\mu(\Omega) = \sum_{i=1}^{\infty} \mu(Q_i)$, and so we conclude that

$$\sum_{i=1}^{\infty} i\mu(Q_i) = 1.$$

We can therefore write

$$H_\mu(\tilde{\mathcal{Q}}) = -\sum_{i=0}^{\infty} \mu(Q_i)\log\mu(Q_i) < \sum_{\mu(Q_i)>e^{-i}} \mu(Q_i)i + \sum_{\mu(Q_i)\le e^{-i}} e^{-i}i + c < \infty$$

by using monotonicity of $-\log t$ in the first case and the monotonicity of $-t\log t$ for small values of t in the second case (the constant c is there to handle the finitely many cases where the latter monotonicity may not apply).

7.51. THE PARTITION $\mathcal{P}$. We now apply Lemma 7.46 to Ω and conclude that for $i \ge 1$ each of the sets $Q_i \subset \Omega$ may be covered with $\le ce^{i\kappa}$ many balls B_j of radius $e^{-\alpha i} r/8$. Here r is the injectivity radius at the center x_0 of the ball Ω and α is chosen as in Lemma 7.45. We will refine the partition $\tilde{\mathcal{Q}}$ by splitting each Q_i into smaller sets. However, so as not to destroy the property (7.48a) we will use instead of the original balls B_j some modified version of them that are "widened" or "smeared out" in the direction of G^-.

Fix some $Q_i \in \tilde{\mathcal{Q}}$ for $i \ge 1$ and write $D = Q_i$ to simplify the notation, also let $B_1, B_2, \ldots, B_N$ with $N = N(i) \le ce^{i\kappa}$ be the cover obtained above. We split D into the sets $D_1, D_2, \ldots$ as follows:

$$\begin{aligned} D_1 &= D \cap (B_{r/4}^{G^-}.B_1),\\ D_2 &= D \cap (B_{r/4}^{G^-}.B_2) \setminus D_1, \ldots \end{aligned}$$

Roughly speaking, since the set $\Omega \supset D$ has small diameter (at most $r/8$) in comparison to its injectivity radius (r) and since the widening by $B_{r/4}^{G^-}$ is by a bigger radius, we should think of the splitting of D into the sets $D_1, \ldots$ as a splitting transversely to the G^--orbits.

This defines a partition of $D = Q_i$ into $\le ce^{i\kappa}$ many sets. Collecting these partitions for the various sets Q_i we obtain one partition $\mathcal{P}$ of X containing all of them and $Q_0 = X \setminus \Omega$.

7.52. FINITE ENTROPY. Now for each n, we define $\mu|_{Q_n}$ to be the restricted measure normalized to be a probability measure. Then the entropy $H_{\mu|_{Q_n}}(\mathcal{P}) \le \log c + \kappa n$ since the partition $\mathcal{P}$ restricted to Q_n has at most $\le ce^{n\kappa}$ many elements by construction. Also

$$H_\mu(\mathcal{P}) = H_\mu(\tilde{\mathcal{Q}}) + H_\mu(\mathcal{P}|\tilde{\mathcal{Q}}),$$

and the latter quantity may be expressed as the weighted average of the entropies $H_{\mu|_{Q_n}}(\mathcal{P})$ so that finally

$$H_\mu(\mathcal{P}) \le H_\mu(\tilde{\mathcal{Q}}) + \log c + \kappa \sum n\mu(Q_n) < \infty.$$

7.53. Upper bound for atom. We claim that the partition $\mathcal{P}$ has the property that for any $x, a^n.x \in \Omega$, we have

$$[x]_{\bigvee_{i=0}^{\infty} a^{-i}\mathcal{P}} \subset \left(\bigcap_{k=0}^{n} a^{-k} B_r^G a^k \right) .x,$$

which is quite similar to what we proved in the case of a hyperbolic torus automorphism. The idea is that, although we do not learn much information about the orbits during the time it spends near the cusp (our partition element $Q_0 = X \setminus \Omega \in \mathcal{P}$ is rather crude there and moreover the injectivity radius is not uniform there), we compensate by learning a great deal about the point at the time at which it leaves Ω.

To prove the claim assume $x, a^n.x \in \Omega$ and $y = g.x \in [x]_{\bigvee_{i=0}^{\infty} a^{-i}\mathcal{P}}$ for some $g \in B_r$. Then $x \in Q_i = D$ for some i — this means that $a^i.x \in \Omega$ — and $x \in D_j$ for some $j \leq N(i)$. We will first show the claim for $n = i$. By equivalence of $y = g.x$ to x and by construction of the set D_j we get that $x = u_x h_x.z_j$ with $u_x \in B_{r/4}^{G^-}$ and $h_x \in B_{e^{-\alpha n} r/8}^G$ and similarly for y, where $z_j \in D$ is the center of the ball B_j used to define D_j. We may remove z_j from the formulas and obtain first that $y = g.x = u_y h_y h_x^{-1} u_x^{-1}.x$ which implies $g = u_y h_y h_x^{-1} u_x^{-1} \in B_r^G$ as r is an injectivity radius. If r is sufficiently small we obtain from this $g = uh$ with $u = u_y u_x^{-1} \in B_{r/2}^{G^-}$ and $h = u_x(h_y h_x^{-1})u_x^{-1} \in B_{e^{-\alpha n} r/2}$ as conjugation by a small element does not change the metric much. This shows that $a^k g a^{-k} = (a^k u a^{-k})(a^k h a^k) \in B_{r/2}^{G^-} B_{r/2}^G$ for $k = 1, \ldots, i$ by Lemma 7.45, which proves the claim in the case of $n = i$.

If $n > i$ we obtain from the above that $a^k g a^{-k} \in B_r^G$ for $k = 1, \ldots, i$ and then we may repeat the argument with $x, y = g.x$, and g replaced by $a^i.x, a^i.y$ and $a^i g a^{-i}$ resp., and with n replaced by $n - i$. Repeating the argument as needed shows the claim.

The claim implies that

$$[x]_{\mathcal{A}} \subset B_r^{G^- G^0}.x$$

for a.e. $x \in \Omega$, where we define $\mathcal{A} = \bigvee_{n=0}^{\infty} a^{-n}\mathcal{P}$ and we recall that $G^0 = C_G(a)$. Indeed for a.e. $x \in \Omega$ we have infinitely many n with $a^n.x \in \Omega$ by Poincarè recurrence.

Moreover, if μ is not compactly supported, then $\mu(Q_n) \neq 0$ for infinitely many n which implies that the above atom is actually contained in $B_r^{G^-}.x$ for a.e. $x \in \Omega$. In fact, suppose $\mu(Q_{n_0}) > 0$ then for a.e. $x \in \Omega$ we know that there are infinitely many n with $a^n.x \in Q_{n_0}$. Take one such x and assume that $g.x$ is equivalent to x and $g = uh$ with $u \in G^-$ and $h \in G^0$. This implies that $a^n g a^{-n} = a^n u a^{-n} h = u'h$ is the displacement between $a^n.x$ and $a^n g.x$ which implies $h \in B_{e^{-\alpha n_0} r/2}^G$. As we know this for infinitely many n_0 we obtain $h = e$.

If however, we have $\mu(Q_n) = 0$ for all but finitely many n, then $\mathcal{P}$ is actually a finite partition mod μ and the last statement may not hold. However, in this case we may artificially split one of the sets of positive measure into countably many sets of positive measures such that for every ϵ we have a partition element of positive measure contained in a set of the form $B_{r/4}^{G^-} B_\epsilon.x_\epsilon$. Making these new partition elements small enough, we may assume that their measure decays rapidly which ensures that the resulting partition still has finite entropy. With this refined partition the above holds also in this case.

Notice that the above statements regarding the upper bound $B_r^{G^-}.x$ of the atom were stated for $x \in \Omega$, but that a slightly weaker form also holds for a.e. $x \in X$. In fact, if $x \in X$ and $n \geq 1$ is such that $a^n.x \in \Omega$ satisfies the inclusion $[a^n.x]_{\mathcal{A}} \subset B_r^{G^-} a^n.x$ then we have that $[x]_{\mathcal{A}} \subset B_s^{G^-}.x$ for some s that depends on n.

7.54. Lower bound for atom. To finish the proof we wish to show that (7.48a) also holds for the partition $\mathcal{P}$. So suppose $x \in \Omega$ and $\delta > 0$ satisfies (7.48a). Here we will use the fact that we "widened" the balls B_j in the direction of G^- to obtain the sets $D_j \subset Q_i$. We may assume $\delta < r/8$, and pick some $u \in B_\delta^{G^-}$. As $\tilde{\mathcal{Q}}$ is contained in the σ-algebra generated by $\bigvee_{n\geq 0} a^{-n}.\mathcal{Q}$ we know that x and $u.x$ belong to the same $Q_i = D \in \tilde{\mathcal{Q}}$. Suppose $x \in D_j$ which shows $x = u_x h_x.z_j$ with $u_x \in B_{r/4}^{G^-}$ and $h_x \in B_{e^{-\alpha i}r/8}^{G}$ where $z_j \in B_j \cap \Omega$ is the center of the ball B_j that was used to construct D_j. Now clearly $u.x = (uu_x)h_x.z_j$ and $uu_xh_x \in B_{r/2}^{G}$. As the diameter of Ω is at most $r/8$ by definition, we obtain $uu_xh_x \in B_{r/8}^{G}$ since r is an injectivity radius on Ω. Together with $h_x \in B_{r/8}^{G}$ this implies $uu_x \in B_{r/4}^{G^-}$. (To see this notice that by left invariance of the metric we have $d(g,e) = d(e,g^{-1}) \leq d(e,h) + d(h,g^{-1}) = d(e,h) + d(gh,e)$ for all $g,h \in G$.) This implies that $u.x$ also belongs to $B_{r/4}^{G^-}.B_j$ and D. In fact this shows $u.x \in D_j$, for if $u.x \notin D_j$ then necessarily $u.x \in D_{j'}$ for some $j' < j$ but then by symmetry of the argument between x and $u.x$ we would have also $x \notin D_j$. Therefore, x and $u.x$ belong to the same element of $\mathcal{P}$. Repeating the argument as needed starting with $a^i.x$ and $a^i u.x$ shows that x and $u.x$ are equivalent with respect to $\bigvee_{n=0}^{\infty} a^{-n}\mathcal{P}$. The points $x \in X \setminus \Omega$ are dealt with in the same manner as before by choosing a minimal n with $a^n.x \in \Omega$. This finishes the proof of Proposition 7.44. □

7.55. Proof of Theorem 7.9. Let $U < G^-$ be a closed a-normalized subgroup. Let μ be an a-invariant and ergodic probability measure on $X = \Gamma\backslash G$. We wish to show that the entropy contribution is bounded by $h_\mu(a,U) \leq J$ where $J = -\log\left|\det \mathrm{Ad}_a|_{\mathfrak{u}}\right|$ is the negative logarithm of the absolute value of the determinant of the adjoint representation of a restricted to the Lie algebra $\mathfrak{u}$ of U. As we will show we only have to use convexity of $\log t$ for $t \in \mathbb{R}$. However, we will have to use it on every atom $[x]_{a^{-1}.\mathcal{A}}$ for an a-decreasing σ-algebra which is subordinate to U.

We fix a Haar measure m_U on U, and note that

$$m_U(a^{-1}Ba) = e^J m_U(B) \text{ for any measurable } B \subset U. \tag{7.55a}$$

For $x \in X$ we write $V_x \subset U$ for the shape of the $\mathcal{A}$-atom so that $V_x.x = [x]_{\mathcal{A}}$ a.e. Recall that $\mu_x^{a^{-1}.\mathcal{A}}$ is a probability measure on $[x]_{a^{-1}.\mathcal{A}} = a^{-1}V_{a.x}a.x$ which is used in the definition of

$$H_\mu(\mathcal{A}|a^{-1}.\mathcal{A}) = -\int \log \mu_x^{a^{-1}.\mathcal{A}}([x]_{\mathcal{A}}).$$

We wish to compare this to a similar expression where we use (in a careful manner) the Haar measure m_U on U as a replacement for the conditional measures. We note however, that we will always work with the given measure μ on X, so our notion of "a.e." is here always meant w.r.t. μ. We define τ_x^{Haar} to be the normalized push forward of $m_U|_{a^{-1}V_{a.x}a}$ under the orbit map, i.e., we define

$$\tau_x^{\mathrm{Haar}} = \frac{1}{m_U(a^{-1}V_{a.x}a)} m_U\big|_{a^{-1}V_{a.x}a}.x,$$

which again is a probability measure on $[x]_{a^{-1}\mathcal{A}}$.

We define

$$p(x) = \mu_x^{a^{-1}.\mathcal{A}}([x]_{\mathcal{A}})$$

which appears in the definition of $H_\mu(\mathcal{A}|a^{-1}.\mathcal{A})$. By analogy we also define

$$p^{\mathrm{Haar}}(x) = \tau_x^{\mathrm{Haar}}([x]_{\mathcal{A}}) = \frac{m_U(V_x)}{m_U(a^{-1}V_{a.x}a)} = \frac{m_U(V_x)}{m_U(V_{a.x})} e^{-J}$$

where we used (7.55a). Taking the logarithm and applying the ergodic theorem (check this) we see that $-\int \log p^{\mathrm{Haar}} d\mu = J$.

Now we recall that both $\mathcal{A}$ and $a^{-1}.\mathcal{A}$ are subordinate to U, which means that after removing a null set they must be countably equivalent. In other words, there exists a null set N such that for $x \notin N$ the $\mathcal{A}$-atom of x contains an open neighborhood of x in the U-orbit. We may also assume that for $x \notin N$ there are infinitely many positive and negative n with $a^n.x \in Y$ where Y is as in (7.35a). Since U is second countable, this implies that

$$[x]_{a^{-1}.\mathcal{A}} \setminus N = \bigcup_{i=1}^{\infty} [x_i]_{\mathcal{A}} \setminus N$$

where the union is disjoint. For a.e. x we wouldn't have to be too careful about the null set N as it is also a null set for the conditional measure, but note that it may not be a null set for τ_x^{Haar}. Therefore, we write

$$[x]_{a^{-1}.\mathcal{A}} = \bigcup_{i=1}^{\infty} [x_i]_{\mathcal{A}} \cup N_x$$

where N_x is a null set for $\mu_x^{a^{-1}.\mathcal{A}}$ but maybe not for τ_x^{Haar}. We may assume $\mu_x^{a^{-1}.\mathcal{A}}([x_i]_{\mathcal{A}}) > 0$, otherwise we just remove this atom from the list and increase N_x accordingly. This shows

$$\sum_{i=1}^{\infty} \mu_x^{a^{-1}.\mathcal{A}}([x_i]_{\mathcal{A}}) = 1$$

but only

$$\sum_{i=1}^{\infty} \tau_x^{\mathrm{Haar}}([x_i]_{\mathcal{A}}) \leq 1.$$

We now integrate $\log p^{\mathrm{Haar}} - \log p$ over the atom $[x]_{a^{-1}.\mathcal{A}}$ to get

$$\int \log p^{\mathrm{Haar}} d\mu_x^{a^{-1}.\mathcal{A}} - \int \log p \, d\mu_x^{a^{-1}.\mathcal{A}}, \tag{7.55b}$$

but as both functions are constant on the $\mathcal{A}$-atoms (and as N_x is a null set w.r.t. the measure w.r.t. which we integrate) this integral is nothing but the countable sum

$$= \sum_{i=1}^{\infty} \left(\log \frac{\tau_x^{\mathrm{Haar}}([x_i]_{\mathcal{A}})}{\mu_x^{a^{-1}.\mathcal{A}}([x_i]_{\mathcal{A}})} \right) \mu_x^{a^{-1}.\mathcal{A}}([x_i]_{\mathcal{A}})$$

Using now convexity of $\log t$ for $t \in \mathbb{R}$ with $\mu_x^{a^{-1}.\mathcal{A}}([x_i]_{\mathcal{A}})$ as the weights at $t_i = \frac{\tau_x^{\mathrm{Haar}}([x_i]_{\mathcal{A}})}{\mu_x^{a^{-1}.\mathcal{A}}([x_i]_{\mathcal{A}})}$ we get

$$(7.55c) \qquad = \sum_{i=1}^{\infty} \log(t_i)\mu_x^{a^{-1}.\mathcal{A}}([x_i]_{\mathcal{A}}) \leq \log\left(\sum_{i=1}^{\infty} t_i \mu_x^{a^{-1}.\mathcal{A}}([x_i]_{\mathcal{A}})\right)$$

$$= \log\left(\sum_{i=1}^{\infty} \tau_x^{\mathrm{Haar}}([x_i]_{\mathcal{A}})\right) = \log \tau_x^{\mathrm{Haar}}\left(\bigcup_{i=1}^{\infty}[x_i]_{\mathcal{A}}\right) \leq 0.$$

Integrating this inequality over all of X and recalling the relation of the function p with the entropy contribution $h_\mu(a,U) = H_\mu(\mathcal{A}|a^{-1}.\mathcal{A})$ and of the function p^{Haar} with J gives the desired inequality.

In case of equality we use strict convexity of $\log t$: If $h_\mu(a,U) = J$, then the integral of the non-positive (due to (7.55c)) expression in (7.55b) vanishes. Therefore, for a.e. atom (7.55b) vanishes, or equivalently we must have 0 on both sides of (7.55c). However, this means that $\tau_x^{\mathrm{Haar}}(N_x) = 0$ and that $t_i = 1$ for all i by strict convexity of $\log t$. Notice that $t_i = 1$ means that the conditional measure $\mu_x^{a^{-1}.\mathcal{A}}$ gives the same weight to the $\mathcal{A}$-atoms $[x_i]_{\mathcal{A}}$ as does the normalized Haar measure τ_x^{Haar} on the $a^{-1}.\mathcal{A}$-atom.

Using that $H_\mu(a^k.\mathcal{A}|a^{-\ell}.\mathcal{A}) = (k+\ell)h_\mu(a,U) = (k+\ell)J$ for any $k,\ell \geq 0$ together with the same argument we obtain that the conditional measure $\mu_x^{a^{-\ell}.\mathcal{A}}$ gives the same weight to the $a^k.\mathcal{A}$-atoms as does the normalized Haar measure on the $a^{-\ell}.\mathcal{A}$-atom. For a.e. x the $a^{-\ell}.\mathcal{A}$-atom can be made arbitrarily large as there is a sequence $\ell_n \to \infty$ with $a^{\ell_n}.x \in Y$. Now fix ℓ, then the various $a^k.\mathcal{A}$-atoms for all $k \geq 0$ generate the Borel σ-algebra on the $a^{-\ell}.\mathcal{A}$-atom, at least on the complement of N which is a null set both for $\mu_x^{a^{-\ell}.\mathcal{A}}$ and for the normalized Haar measure on the atom. This follows as for $\mu_x^{a^{-\ell}.\mathcal{A}}$-a.e. y the $a^k.\mathcal{A}$ atom can be made to have arbitrarily small diameter since for $y \notin N$ there is a sequence $k_n \to \infty$ with $a^{-k_n}.y \in Y$. This shows that $\mu_x^{a^{-\ell}.\mathcal{A}}$ equals the normalized Haar measure on the atom $[x]_{a^{-\ell}.\mathcal{A}}$. Using this for all ℓ we see that the leaf-wise measure μ_x^U is the Haar measure on U, and so that μ is U-invariant (c.f. Problem 6.28). This concludes the proof of Theorem 7.9. □

8. The product structure

8.1. Assumptions. In the previous chapter we considered a measure μ on $\Gamma\backslash G$ invariant under the action of a diagonalizable element $a \in G$, and studied in some detail the leafwise measures induced by μ on orbits of unipotent groups U contracted by a. When considering the action of a multiparameter diagonalizable group $A \subset G$, it is often possible to find some $a \in A$ which contracts some nontrivial unipotent group U but which acts isometrically on orbits of some other group T (which may well be contracted by some other element $a' \in A$). In this case there is a surprisingly simple relation between the leafwise measures of the group generated by U and T (which we assume to be simply the product group) and the leafwise measures for each of these groups: essentially, the leafwise measures for TU will be the product of the leafwise measures for T and U!

Even though a key motivation to looking at these conditional measures is our desire to understand action of multiparameter diagonal groups, we will make use of a single diagonalizable $a \in G$ (more precisely — an element of the class $\mathcal{A}$

defined in §7.14). Let $U \subset G^-$ be a-normalized, and contracted by a. Finally let $T \subset G^0 = C_G(a)$ centralize a and assume that T normalizes U (this is not a very restrictive condition: in particular, the reader can easily verify that the full contracting subgroup for a is normalized by T). We define $H = TU \subset G$, which we can identify with $T \ltimes U$, and will show below that the leaf-wise measure for H is proportional to the product of the leaf-wise measures for T and U.

8.2. This simple relation was discovered by M.E. and A. Katok and is one of the key ingredients in the paper [**EK03**] and extended in [**EK05**] (cf. also [**Lin06**, §6]). A weaker form of this relation can be derived from the work of H. Hu [**Hu93**] on entropy of smooth $\mathbb{Z}^d$-actions, and the relation between entropy and leafwise measures, and was used by Katok and Spatzier in [**KS96**].

8.3. EXAMPLE. The following is an example (for $G = \mathrm{SL}(3, \mathbb{R})$) to have in mind: Let

$$a = \begin{pmatrix} e^{-2} & & \\ & e & \\ & & e \end{pmatrix}, \quad U = \left\{ \begin{pmatrix} 1 & * & * \\ & 1 & 0 \\ & & 1 \end{pmatrix} \right\}, \quad T = \left\{ \begin{pmatrix} 1 & 0 & 0 \\ & 1 & * \\ & & 1 \end{pmatrix} \right\}$$

so that

$$H = \left\{ \begin{pmatrix} 1 & * & * \\ & 1 & * \\ & & 1 \end{pmatrix} \right\}.$$

We note that the Haar measure of H is the three-dimensional Lebesgue measure and so is also the direct product of the Haar measures on T and U.

8.4. SHORT REMINDER. Recall that the equivalence classes by proportionality of the leaf-wise measures live in a compact metric space, because of the growth property from Theorem 6.30. More precisely, recall that we have a function $\rho > 0$ such that $\int \rho d\mu_x^T < \infty$ a.e. Taking a sequence $\{0 \le f_i \le \rho\}_{i=1}^{\infty} \subset C_c(T)$ spanning a dense subset, we may define

$$d([\nu_1], [\nu_2]) := \sum_{i=1}^{\infty} 2^{-i} \left| \frac{\int f_i d\nu_1}{\int \rho d\nu_1} - \frac{\int f_i d\nu_2}{\int \rho d\nu_2} \right|$$

for any two equivalence classes of Radon measures with $\int \rho d\nu_i < \infty$. If we chose a representative of the equivalence class we may assume $\int \rho d\nu_i = 1$. This way, the metric just defined corresponds to the weak* topology in the space of Radon measures

$$\Big\{\nu : \int \rho d\nu = 1\Big\} \subset \Big\{\nu : \int \rho d\nu \le 1\Big\}.$$

This way the leaf-wise measure μ_x^T can be interpreted as a measurable function with values in a compact metric space.

We also recall the property of leaf-wise measures (Theorem 6.3.(iii)):

$$[\mu_x^T] = [(\mu_{t.x}^T).t] \tag{8.4a}$$

whenever $t \in T$, and $x, t.x \in X'$ (a set of full measure). The following proposition extends this by explaining how μ_x^T transforms under the bigger group $H = TU_-$.

8.5. Proposition. *There exists $X' \subset X$ of full measure, such that for every $x \in X'$ and $h \in H$ such that $h.x \in X'$, we have*

$$[\mu_x^T] = [(\mu_{h.x}^T)t]$$

where $h = tu' = u''t$ for some $u'', u' \in U$ and $t \in T$.

The special case of $t = e$ of this proposition implies the following (note, however that Proposition 8.5 is a substantially stronger statement — cf. §8.9):

8.6. Corollary. *Let $u \in U$. Then $x, u.x \in X'$ implies $[\mu_x^T] = [\mu_{u.x}^T]$.*

8.7. Proof of Proposition 8.5. As explained above the map $x \mapsto [\mu_x^T]$ is Borel measurable, from $\Gamma\backslash G$ to a compact metric space. By Luzin's Theorem, for any $\epsilon > 0$, there exists a compact $K_\epsilon \subset \Gamma\backslash G$ such that:

- $\mu(K_\epsilon) > 1 - \epsilon$,
- $x \mapsto [\mu_x^T]$ is continuous on K_ϵ, and
- (8.4a) holds whenever $x, t.x \in K_\epsilon$.

Define

$$X_\epsilon = \{x \in K_\epsilon : \sup_n \frac{1}{n} \sum_{i=0}^{n-1} 1_{X\backslash K_\epsilon}(a^i.x) < 1/2\}$$

Then using the maximal ergodic theorem one easily verifies that $\mu(X_\epsilon) > 1 - 2\epsilon$.

If $x, h.x \in X_\epsilon$, then there is a sequence $n_i \to \infty$ such that $a^{n_i}x, a^{n_i}h.x \in K_\epsilon$. Passing to a subsequence if necessary, we may assume that $a^{n_i}x \to x_0$. We note that a commutes with the elements of T by definition and so Lemma 7.16 implies that $\mu_{a^n.x}^T = \mu_x^T$ for every n and a.e. x. We may assume this holds for any $x \in X_\epsilon$. By continuity on K_ϵ we have

$$[\mu_x^T] = [\mu_{a^{n_i}.x}^T] \to [\mu_{x_0}^T].$$

For $h.x$, we can rewrite and get

$$a^{n_i}h.x = a^{n_i}tu'.x = t(a^{n_i}u'a^{-n_i})a^{n_i}.x \to t.x_0,$$

since the term in parentheses $a^{n_i}u'a^{-n_i} \to e$ as $n_i \to \infty$. So again by continuity we have

$$[\mu_{h.x}^T] = [\mu_{a^{n_i}h.x}^T] \to [\mu_{t.x_0}^T].$$

Together, we have for $x, h.x \in X_\epsilon$

$$[\mu_x^T] = [\mu_{x_0}^T] = [(\mu_{t.x_0}^T)t] = [\mu_{a^{n_i}h.x}^T t] = [\mu_{h.x}^T t]$$

as desired. We conclude the proof by letting $\epsilon = \frac{1}{n} \searrow 0$, choosing K_ϵ increasing, and defining X' to be the union of the $X_{\frac{1}{n}}$. □

8.8. Corollary (Product structure). *Let $H = T \ltimes U$ be as in §8.1. There exists $X' \subset X$ of full measure, such that for every $x \in X'$ we have*

$$\mu_x^H \propto \iota(\mu_x^T \times \mu_x^U),$$

where $\iota : (t, u) \in T \times U \mapsto tu \in H$.

8.9. REMARK. All the essential facts for this corollary have already been proved as we explain now. Suppose for a minute that the product formula as in the last corollary holds, then indeed the leaf-wise measure for T of a point $u.x$ with $u \in U$ should be the same as for x, but as usual we should allow for null sets and just state this for a.e. $u \in U$ (w.r.t. the natural measure μ_x^U there). This is what we proved in Corollary 8.6 for a.e. x. (Recall that $\mu(X \setminus X') = 0$ implies that for μ-a.e. $x \in X$ and μ_x^U-a.e. u we have $u.x \in X'$.)

However, this property does not imply that μ_x^H is a product measure: E.g. if μ_x^H were a measure supported on a measurable graph from T to U then μ_x^U would typically be trivial and this would make the above true while μ_x^H may not be a product measure.

Proposition 8.5, on the other hand, does contradict the prevalence of this type of leafwise measures. We may rephrase Proposition 8.5 as follows: for μ-a.e. x and μ_x^H-a.e. $h = tu$ (with $t \in T$ and $u \in U$) we know that the leaf-wise measure $\mu_{h.x}^T$ is, apart from the shift by t (and possibly a proportionality factor), the same as μ_x^T. This property is incompatible with a graph-like measure we described above unless the graph describes a constant map (which is compatible with the product structure we claim). To convert this heuristic into an argument we need to prove another lemma regarding leaf-wise measures.

8.10. Lemma. *Let H be a locally compact second countable group acting nicely on X (say, locally and measure-theoretically free), and let μ be a Radon (i.e., locally finite) measure on X. Assume $H = LM = \iota(L \times M)$ is topologically isomorphic (under the product map $\iota(\ell, m) = \ell m$ for $\ell \in L, m \in M$) to the product of two closed subgroups $L, M < H$. Then L acts by restriction on X and on H by left translation, and so gives rise to families of leaf-wise measures μ_x^L and $(\mu_x^H)_h^L$ for $x \in X$ and $h \in H$. Then there exists $X' \subset X$ of full measure such that whenever $x \in X'$ we have $[(\mu_x^H)_h^L] = [\mu_{h.x}^L]$ for μ_x^H-a.e. $h \in H$.*

Roughly speaking the above is what we should expect: μ_x^H is the measure on H such that $\mu_x^H.x$ describes μ along on the orbit $H.x$. Similarly, $(\mu_x^H)_h^L$ is the measure on L for which $(\mu_x^H)_h^L h$ describes μ_x^H on the coset Lh, and so we expect that $(\mu_x^H)_h^L$ will be such that $(\mu_x^H)_h^L h.x$ describes μ on the orbit $Lh.x$ which suggests the conclusion.

8.11. PROOF. Let $\Xi \subset X$ be an R-cross-section for the action of H on some set of positive measure (see Definition 6.6). Let $\tilde{\mathcal{A}}_H$ be the σ-algebra $\{B_R^H, \emptyset\} \otimes \mathcal{B}(\Xi)$ on $B_R^H \times \Xi$, where $\mathcal{B}(\Xi)$ is the Borel σ-algebra on Ξ. The map $\iota(h, x) = h.x$ is injective on $B_R^H \times \Xi$ by definition, and so $\mathcal{A}_H = \iota(\tilde{\mathcal{A}}_H)$ is a countably generated σ-algebra of Borel sets. The atom $[x]_{\mathcal{A}_H}$ is an open H-plaque for any $x \in \iota(B_R^H \times \Xi) = B_R^H.\Xi$ (namely equal to $B_R^H.z$ for some $z \in \Xi$).

We further define $\tilde{\mathcal{A}}_L := \{LB \cap B_R^H : B \in \mathcal{B}(M)\}$ where $\mathcal{B}(M)$ is the Borel σ-algebra on M, which is by assumption a global cross-section of L in H. The σ-algebra $\mathcal{A}_L = \iota(\tilde{\mathcal{A}}_L \times \mathcal{B}(\Xi))$ is countably generated, and $[x]_{\mathcal{A}_L}$ is an open L-plaque for all $x \in \iota(B_R^H \times \Xi)$. Note that $\mathcal{A}_L \supset \mathcal{A}_H$.

The measures μ_x^H and μ_x^L can be defined by the values of conditional measures with respect to a countable collection of σ-algebras $\mathcal{A}_H^{(i)}$ and $\mathcal{A}_L^{(i)}$ constructed as above. On each of these σ-algebras a corresponding compatibility condition is satisfied due to the inclusion $\mathcal{A}_L^{(i)} \supset \mathcal{A}_H^{(i)}$; this implies the lemma. $\square$

8.12. Proof of Corollary 8.8. Take $x \in X$ to be typical (i.e., outside of the union of bad null sets from Proposition 8.5, and Lemma 8.10 applied to both $L = T$, $M = U$ and $L = U$, $M = T$). We are going to combine these statements, but for this it will be easier to restrict μ_x^H to the bounded product set $Q = B_r^T B_r^U \subset H$ for some $r > 0$, which we may envision as a rectangle with sides B_r^T and B_r^U.

Using $L = T$ Lemma 8.10 is telling us that the conditional measures for $\mu_x^H|_Q$ with respect to the σ-algebra $\mathcal{A} = \{B_r^T, \emptyset\} \otimes \mathcal{B}(B_r^U)$ can be obtained from the leaf-wise measures $\mu_{h.x}^T$ (for μ_x^H-a.e. $h \in Q$). As usual, we have to shift the leaf-wise measure for T back to the space in question, which after applying the lemma may be taken to be H, and restrict to the atoms of the σ-algebra $\mathcal{A}$ in question. This gives

$$(\mu_x^H)_h^{\mathcal{A}} \propto (\mu_{h.x}^T h)\big|_Q. \tag{8.12a}$$

However, Proposition 8.5 gives

$$\mu_{tu.x}^T t \propto \mu_x^T. \tag{8.12b}$$

for μ_x^H-a.e. $h = tu \in Q$ (which is a form of independence of $\mu_{tu.x}^T$ in terms of $u \in U$). Using (8.12a)-(8.12b) together, we obtain that the conditional measures of $\mu_x^H|_Q$ with respect to $\mathcal{A} = \{B_r^T, \emptyset\} \otimes \mathcal{B}(B_r^U)$ at $h = tu \in Q$ is equal to $\mu_x^T|_{B_r^T} \times \delta_u$ normalized to be a probability measure. However, this just says that $\mu_x^H|_Q$ is a product measure which is proportional to $\iota(\mu_x^T \times \nu_r)$ for some finite measure ν_r on B_r^U.

Varying r it is easy to check that one can patch these measures ν_r together (i.e. that they extend each other up to a proportionality factor) to obtain a Radon measure ν on U and that μ_x^H is in fact proportional to $\iota(\mu_x^T \times \nu)$. We wish to show that $\nu \propto \mu_x^U$.

As $\iota(\mu_x^T \times \nu)$ is a product measure it is clear what the conditional measures for it are with respect to a σ-algebra whose atoms are of the form tV for open subsets $V \subset U$. However, this corresponds really to the right action of U on H while we have to use the left action if we want to apply Lemma 8.10 for $L = U$ and $M = T$. Luckily U is a normal subgroup, so at least the orbits of these two actions are the same even though the way these two actions identify the orbit with the group differs. We now analyze this in more detail.

Restrict again to $Q = B_r^T B_r^U \subset H$ and consider the σ-algebra $\mathcal{A}' = \mathcal{B}(B_r^T) \otimes \{B_r^U, \emptyset\}$, whose atoms are tB_r^U for $t \in B_r^T$. We know that the conditional measure of $\mu_x^H|_Q$ at $h = tu$ equals $\iota(\delta_t \times \nu_r)$. Considering now the action of U by left multiplication on H we see that the atom of $h = tu \in Q$ corresponds to the set $V_h = tB_r^U h^{-1} \subset U$. Using these σ-algebras for all positive integers r we characterize the leaf-wise measures of μ_x^H with respect to the U-action and obtain that $(\mu_x^H)_h^U$ must be proportional to (the push forward) $t\,\nu\,h^{-1}$ for μ_x^H-a.e. h.

To summarize we know

$$\mu_{h.x}^U \propto (\mu_x^H)_h^U \propto t\,\nu\,h^{-1} \text{ where } h = tu. \tag{8.12c}$$

Clearly the above gives the desired statement if we just set $h = e$. However, strictly speaking we are not allowed to use $h = e$ as we only know these two formulas for μ_x^H-a.e. $h \in H$. Instead we may show the corresponding claim not for the x we started with but for $h.x$ for μ_x^H-a.e. $h \in H$. This will show that the corollary holds a.e. In fact, for μ_x^H-a.e. $h = tu$ we know

$$\mu_{h.x}^H \propto \mu_x^H h^{-1} \propto \iota(\mu_x^T \times \nu)h^{-1} \propto \iota(\mu_{h.x}^T t \times \nu)h^{-1} \propto$$
$$\propto \iota(\mu_{h.x}^T \times t\nu h^{-1}) \propto \iota(\mu_{h.x}^T \times \mu_{h.x}^U)$$

by combining Theorem 6.3.(iii) for the action of H on X, with the product structure at μ_x^H already obtained, with Proposition 8.5, with the definition of $\iota(t,u) = tu$, and finally with (8.12c). □

From the product structure just proven we can read off an analogue of Corollary 8.6 for the action of T. We note however that, since T and U may not commute, a full analogue of Proposition 8.5 with the roles of T and U reversed will in general not hold (unless one allows conjugation as in the proof above).

8.13. Corollary. *Let $t \in T$. Then $x, t.x \in X'$ implies $[\mu_x^U] = [\mu_{t.x}^U]$.*

We refer to Example 6.5.2 for a discussion showing that this mild coincidence of leaf-wise measures may indeed be a very special property.

9. Invariant measures and entropy for higher rank subgroups A, the high entropy method

9.1. As before we consider the space $X = \Gamma\backslash G$, where G is an algebraic group over a characteristic zero local field k (say $k = \mathbb{R}$ or $\mathbb{Q}_p$ for simplicity). We fix an algebraic subgroup $A \subset G$ which is diagonalizable over the ground field k. In algebraic terms A is the group of k-points of a k-split torus, but we may simply refer to A as a *torus* and to its action on X as a *torus action.* We will assume that we have a homomorphism $\alpha : (k^\times)^n \hookrightarrow G$ which is defined by polynomials with coefficients in k and whose range equals A. We may often suppress the isomorphism and use A and $(k^\times)^n$ interchangeably. For example, if $G = \mathrm{SL}(4,\mathbb{R})$, we could have

$$\alpha : (t,s) \mapsto \begin{pmatrix} t^2 & & & \\ & ts & & \\ & & s & \\ & & & t^{-3}s^{-2} \end{pmatrix}.$$

Let $\mathfrak{g}$ be the Lie algebra of G. Recall that in zero characteristic, the functions exp and log are homeomorphisms between neighborhoods of $0 \in \mathfrak{g}$ and $e \in G$. Hence, the restriction of the adjoint action of A on the Lie algebra $\mathfrak{g}$ gives a good description of the behavior of conjugation on G which as we have seen is crucial in the study of the action of the elements of A on X.

9.2. A character λ is a homomorphism $\lambda : (k^\times)^n \simeq A \to k^\times$ defined by polynomials with coefficients in k; these polynomials necessarily have the form $\lambda(t_1, \dots, t_n) = t_1^{\ell_1} \cdots t_n^{\ell_n}$ where $\ell_1, \dots, \ell_n \in \mathbb{Z}$.

We say that a character λ is a *weight* (which one also may refer to as eigenvalue, Lyapunov weight, or root) for the action of A if there is some nonzero $\underline{x} \in \mathfrak{g}$ such that for every $a \in A$, we have $Ad_a(\underline{x}) = \lambda(a)\underline{x}$. The set of all such $\underline{x} \in \mathfrak{g}$ is the *weight space* $\mathfrak{g}^\lambda$. By the assumption that A is diagonalizable we get a decomposition $\mathfrak{g} = \bigoplus_{\lambda \in \Phi} \mathfrak{g}^\lambda$ where Φ is the set of all weights.

For $a \in A$ the subspace $\mathfrak{g}_a^- = \bigoplus_{|\lambda(a)|<1} \mathfrak{g}^\lambda$ is a nilpotent subalgebra, and exp gives a global homeomorphism from $\mathfrak{g}_a^-$ to the horospherical group $G_a^- < G$. Here, the absolute value comes from the Archimedean norm on $\mathbb{R}$ resp. the p-adic norms on $\mathbb{Q}_p$. Also, we introduced the subscript in the notation G_a^- to explicate the dependence of the horospherical subgroup on the element $a \in A$ used.

Note that $[\mathfrak{g}^\lambda, \mathfrak{g}^\eta] \subset \mathfrak{g}^{\lambda\eta}$ which follows easily from the formula

$$Ad_a([\underline{x}, \underline{y}]) = [Ad_a(\underline{x}), Ad_a(\underline{y})] \qquad \underline{x}, \underline{y} \in \mathfrak{g}.$$

Note that in general $\mathfrak{g}^\lambda$ is not a sub-Lie-algebra if λ^2 is also a weight.

9.3. Define an equivalence relation on Φ by $\lambda \sim \eta$ if there exist positive integers ℓ, m such that $\lambda^\ell = \eta^m$. This means that λ and η are weights in the same "direction" — characters are in a one-to-one correspondence with $\mathbb{Z}^n$ and under this correspondence $\lambda \sim \eta$ if and only if they are on the same ray from the origin. For a nontrivial weight $\lambda \in \Phi$ we define the *coarse Lyapunov subalgebra*

$$\mathfrak{g}^{[\lambda]} := \bigoplus_{\eta \sim \lambda} \mathfrak{g}^\eta.$$

We note that exp gives a globally defined homeomorphism between $\mathfrak{g}^{[\lambda]}$ and a unipotent subgroup $G^{[\lambda]}$ which we will refer to as the *coarse Lyapunov subgroup.*

Note that λ is nontrivial (i.e., not the constant homomorphism) implies that $\mathfrak{g}^\lambda$ can be made part of some $\mathfrak{g}_a^-$ for some correctly chosen $a \in A$. Moreover, two weights λ and η are equivalent if and only if their corresponding weight spaces are contained in $\mathfrak{g}_a^-$ for the same set of $a \in A$. In this sense, one might say that weights are equivalent if they cannot be distinguished by any elements of a in terms of whether or not the weight space is being contracted.

Similarly, the coarse Lyapunov subgroup $G^{[\lambda]}$ is the intersection of stable horospherical subgroups for various elements of A and is a smallest nontrivial such subgroup. Dynamically speaking, we may say that the orbits of the coarse Lyapunov subgroups are the smallest nontrivial intersections one can obtain by intersecting stable manifolds of various elements of A.

9.4. In this section, we study the structure of the leaf-wise measures on these coarse Lyapunov groups. This study due to Einsiedler and Katok [**EK03, EK05**] by itself gives sufficient information to yield the following measure classification theorem:

9.5. Theorem (Einsiedler and Katok [**EK03**]). *Let Γ be a discrete subgroup in $G = \mathrm{SL}(3, \mathbb{R})$ and define $X = \Gamma \backslash G$. Let A be the full diagonal subgroup of G and suppose μ is an A-invariant and ergodic probability measure on X. Let*

$$a = \begin{pmatrix} t & 0 & 0 \\ 0 & s & 0 \\ 0 & 0 & t^{-1}s^{-1} \end{pmatrix} \in A$$

and suppose that

$$h_\mu(a) > \frac{1}{2}\left(\left|\log|\tfrac{t}{s}|\right| + \left|\log|t^2 s|\right| + \left|\log|ts^2|\right|\right).$$

Then μ is the Haar measure m_X on X and in particular Γ is a lattice.

We note that the expression in the parenthesis is the entropy of the Haar measure m_X. Hence, the theorem (as well as its generalizations below) says that an ergodic measure whose entropy is close to that of the Haar measure must be the Haar measure.

In the next section we present a completely different technique (the low entropy method) that will allow us to sharpen the above theorem, treating all positive entropy measures.

9.6. Fixing some $a \in A$ for which $\mathfrak{g}_a^-$ is nontrivial (equivalently, there is some $\lambda \in \Phi$ so that $|\lambda(a)| < 1$) we obtain a decomposition $\mathfrak{g}_a^- = \bigoplus_{i=1,\dots,\ell} \mathfrak{g}^{[\lambda_i]}$ into finitely many of these "coarse" Lyapunov subalgebras (corresponding to the subgroups $G^{[\lambda_i]}$).

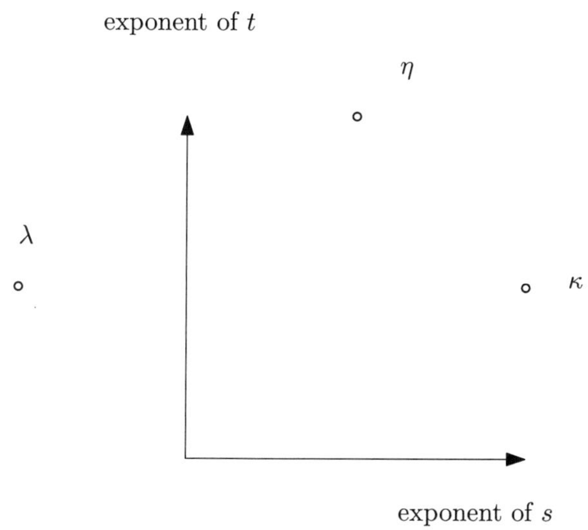

FIGURE 3. Weights for the Heisenberg group

We order these coarse Lyapunov weights $[\lambda_1], [\lambda_2], \dots, [\lambda_\ell]$ so that for each i, the weight λ_i, or more precisely the corresponding point in $\mathbb{Z}^k$ is not in the convex cone generated by the points corresponding to $\lambda_{i+1}, \dots, \lambda_\ell$ — we refer to this by saying that λ_i is *exposed.* This ordering guarantees that for every i there will be an element $a \in A$ so that $\lambda_i(a) = 1$ but $|\lambda_j(a)| < 1$ for $i < j < \ell$.

9.7. EXAMPLE. Take $G = \mathrm{SL}(3, \mathbb{R})$, and A the full diagonal subgroup. Let α be the homomorphism $\alpha : (t, s) \mapsto \begin{pmatrix} t & & \\ & s & \\ & & t^{-1}s^{-1} \end{pmatrix}$. Here $\mathfrak{g}$ is the algebra of traceless matrices. Suppose now $a = \alpha(t, s)$ with $|t| < |s| < |t^{-1}s^{-1}|$. Then $\mathfrak{g}_a^-$ is the algebra of upper triangular nilpotent matrices. Moreover, the coarse Lyapunov subalgebras are the 1-dimensional spaces

$$\mathfrak{g}^{[\lambda]} = \begin{pmatrix} 0 & * & 0 \\ & 0 & 0 \\ & & 0 \end{pmatrix}, \quad \mathfrak{g}^{[\eta]} = \begin{pmatrix} 0 & 0 & * \\ & 0 & 0 \\ & & 0 \end{pmatrix}, \quad \text{and} \quad \mathfrak{g}^{[\kappa]} = \begin{pmatrix} 0 & 0 & 0 \\ & 0 & * \\ & & 0 \end{pmatrix},$$

with the corresponding weights $\lambda = ts^{-1}, \eta = t^2 s$ and $\kappa = ts^2$. As Figure 3 shows, $\lambda_1 = \lambda$, $\lambda_2 = \eta$ and $\lambda_3 = \kappa$ is a legitimate ordering (as would be the reverse ordering and λ, κ, η, but not η, λ, κ). We also have the corresponding subgroups

$$G^{[1]} = \begin{pmatrix} 1 & * & 0 \\ & 1 & 0 \\ & & 1 \end{pmatrix}, \quad G^{[2]} = \begin{pmatrix} 1 & 0 & * \\ & 1 & 0 \\ & & 1 \end{pmatrix}, \quad \text{and} \quad G^{[3]} = \begin{pmatrix} 1 & 0 & 0 \\ & 1 & * \\ & & 1 \end{pmatrix}.$$

where $G^{[1]}$ and $G^{[3]}$ each commute with $G^{[2]}$, and the commutator $[G^{[1]}, G^{[3]}] = G^{[2]}$.

9.8. Theorem. *Let $A = \alpha((k^\times)^n)$, and suppose μ is an A-invariant measure on $X = \Gamma\backslash G$. Fix some $a \in A$, and choose an allowed order of the coarse Lyapunov subalgebras contracted by a (as described above). Then for μ-a.e. $x \in X$, we have*

$$\mu_x^{G^-} \propto \iota(\mu_x^{G^{[1]}} \times \mu_x^{G^{[2]}} \times \cdots \times \mu_x^{G^{[k]}})$$

where $\iota(g_1, g_2, \ldots, g_k) = g_1 g_2 \cdots g_k$ is the product map.

9.9. Proof. By assumption $[\lambda_1]$ is exposed within the set of all Lyapunov weights appearing in G_a^- so that there exists some $a' \in A$ with $G^{[1]} \subset G_{a'}^0$ and $U = G^{[2]} \cdots G^{[k]} = G_a^- \cap G_{a'}^-$. It follows easily that U is a normal subgroup of G_a^- and that $G_a^- \simeq G^{[1]} \ltimes U$ where the isomorphism is just the map ι taking the product. From Corollary 8.8 we deduce that $\mu_x^{G_a^-} \propto \iota(\mu_x^{G^{[1]}} \times \mu_x^U)$. Repeating the argument, starting with $G^{[2]}$ inside U, the theorem follows. □

9.10. Corollary.

$$h_\mu(a, G^-) = \sum_{i=1}^{k} h_\mu(a, G^{[i]})$$

This follows from Theorem 7.6 and Theorem 9.8 since the left hand side is

$$\lim_{n\to\infty} \frac{\log \mu_x^{G^-}(a^n B_1^{G^-} a^{-n})}{n}$$

and we have already shown in §7.18 that the particular shape of the set $B_1^{G^-}$ used in the definition does not matter. Using the product set

$$B_1^{G^{[1]}} B_1^{G^{[2]}} \cdots B_1^{G^{[k]}}$$

instead we obtain with the theorem that the left hand side splits into the corresponding expression for $G^{[i]}$. Hence in this setting our term 'entropy contribution' is quite accurate. We note, however, that in general such a formula does not hold for a finer foliation than the coarse Lyapunov subalgebras.

9.11. Getting invariance. In fact, more is true. Let us for now continue Example 9.7 (which will lead to the proof of Theorem 9.5), and consider $f \in C_c(G_a^-)$, and observe that

$$\begin{aligned}\int f(g) d\mu_x^{G^-} &= \int f(g_1 g_2 g_3) d\mu_x^{[1]}(g_1) d\mu_x^{[2]}(g_2) d\mu_x^{[3]}(g_3) \\ &= \int f(g_3 g_2 g_1) d\mu_x^{[3]}(g_3) d\mu_x^{[2]}(g_2) d\mu_x^{[1]}(g_1)\end{aligned}$$

where $\mu_x^{[i]} := \mu_x^{G^{[i]}}$. This follows from Theorem 9.8 by using the two allowed orders $1, 2, 3$ resp. $3, 2, 1$. Notice that, since both $G^{[1]}$ and $G^{[3]}$ commute with $G^{[2]}$, we can rewrite $g_1 g_2 g_3 = g_2 g_1 g_3$, and $g_3 g_2 g_1 = g_2 g_3 g_1 = (g_2 [g_3, g_1]) g_1 g_3$. Inserting this above, and taking the leaf-wise measure for the $G^{[2]}$-action on G_a^- we find that $\mu_x^{[2]} \propto \mu_x^{[2]}[g_3, g_1]$ for $\mu_x^{[1]}$-a.e. g_1 and $\mu_x^{[3]}$-a.e. g_3 (by using Lemma 8.10 and by recalling that $[g_3, g_1] \in G^{[2]}$).

Now, if $[g_3, g_1]$ has infinite order, in other words if the element $[g_3, g_1]$ is nontrivial, then $\mu_x^{[2]}$ must be $[g_3, g_1]$-invariant; since otherwise, successive translations by $[g_3, g_1]$ would cause $\mu_2(B_r^{G^{[2]}})$ to grow exponentially, contradicting Theorem 6.30. Since the set $\{g_2 : (\mu_x^{[2]}).g_2 = \mu_x^{[2]}\}$ is closed, it follows that if both $\mu_x^{[1]}$ and $\mu_x^{[3]}$ are non-atomic, we must have $\mu_x^{[2]}$ invariant under $[\operatorname{supp} \mu_x^{[1]}, \operatorname{supp} \mu_x^{[3]}]$.

This is a significant restriction. By Poincaré Recurrence, we know that $\mu_x^{[1]} = \delta_e$ or $\operatorname{supp} \mu_x^{[1]}$ contains arbitrarily small (and large) elements; and similarly for $\mu_x^{[2]}$. In the example of the Heisenberg group above, there are only three possible cases: either $\mu_x^{[1]}$ or $\mu_x^{[3]}$ is trivial, or else the closed group generated by $[\operatorname{supp} \mu_x^{[1]}, \operatorname{supp} \mu_x^{[3]}]$ equals $G^{[2]}$, and so $\mu_x^{[2]}$ is a Haar measure on $G^{[2]}$.

9.12. PROOF OF THEOREM 9.5. The above shows (in the notation of Example 9.7) that if $\mu_x^{[1]}$ and $\mu_x^{[3]}$ are both nontrivial at x, then a.s. μ_2 is the Haar measure on $G^{[2]}$. Lemma 7.16 shows that the set of points where $\mu_x^{[i]}$ is trivial is A-invariant, and so has either measure zero or one by ergodicity. Supposing that $\mu_x^{[1]}$ and $\mu_x^{[3]}$ are both nontrivial a.e., we get that $\mu_2 = \mu_x^{G^{[2]}}$ equals the Haar measure on $G^{[2]}$ a.e. and so that μ is invariant under $G^{[2]}$ by Problem 6.28. We now bring in entropy and the assumption to the theorem to justify the assumptions to this 'commutator argument'.

Let now $a \in A$ be as in the theorem. There are essentially two cases for elements of A: An element $a \in A$ is called *regular* if all of its eigenvalues are different, and is called *singular* if two eigenvalues are the same. If a is regular, then we may assume it is as in Example 9.7, for otherwise we get a group isomorphic to the Heisenberg group embedded in some other way into $\mathrm{SL}(3, \mathbb{R})$. If a is singular we may assume (again in the notation of Example 9.7) that $t = s$ with $|t| < 1$.

We define the opposite weight spaces

$$\mathfrak{g}^{[-1]} = \begin{pmatrix} 0 & 0 & 0 \\ * & 0 & 0 \\ 0 & 0 & 0 \end{pmatrix}, \quad \mathfrak{g}^{[-2]} = \begin{pmatrix} 0 & 0 & 0 \\ 0 & 0 & 0 \\ * & 0 & 0 \end{pmatrix}, \quad \text{and} \quad \mathfrak{g}^{[-3]} = \begin{pmatrix} 0 & 0 & 0 \\ 0 & 0 & 0 \\ 0 & * & 0 \end{pmatrix},$$

and similarly the coarse Lyapunov subgroups.

In the singular case $G_a^- = G^{[2]}G^{[3]}$ and

$$h_\mu(a) = h_\mu(a, G_a^-) = h_\mu(a, G^{[2]}) + h_\mu(a, G^{[3]})$$

by Theorem 7.6 and Corollary 9.10. By Theorem 7.9 each summand on the right is bounded by $3|\log|t||$ (which is precisely the entropy contribution for the Haar measure). By assumption on the entropy we have $h_\mu(a) > 3|\log|t||$ (i.e., entropy is more than one half of the maximal entropy), so that both entropy contributions are positive. In turn, this shows that both leaf-wise measures $\mu_x^{[2]}$ and $\mu_x^{[3]}$ are nontrivial a.e. By symmetry of entropy $h_\mu(a) = h_\mu(a^{-1})$ we also get that both $\mu_x^{G^{[-2]}}$ and $\mu_x^{G^{[-3]}}$ are nontrivial a.e. However, the two subgroups $G^{[2]}$ and $G^{[-3]}$ do not commute and have commutator $G^{[1]}$. Moreover, the three groups $G^{[2]}$, $G^{[1]}$, and $G^{[-3]}$ generate a stable horospherical subgroup $G_{a'}^-$ (for some regular $a' \in A$) which is isomorphic to the Heisenberg group studied so far. By the above commutator argument we get that μ is invariant under $G^{[1]}$. Note that we could also have used the triple $G^{[3]}$, $G^{[-1]}$, and $G^{[-2]}$ to obtain invariance under $G^{[-1]}$. So we obtain in the singular case that in fact all leaf wise measures of the coarse Lyapunov subgroups are nontrivial a.e. (and some of them are Haar measures). This is enough to imply that μ is invariant under all coarse Lyapunov subgroups (and so must be the Haar measure m_X) by the commutator argument: Any of the coarse Lyapunov subgroups $G^{[i]}$ is the commutator of two other coarse Lyapunov subgroups $G^{[j]}$ and $G^{[k]}$ such that all three of them generate a stable horospherical subgroup (isomorphic to the Heisenberg group).

In the regular case there are a few more possibilities. We know that $h_\mu(a) = h_\mu(a, G^{[1]}) + h_\mu(a, G^{[2]}) + h_\mu(a, G^{[3]})$. In this case the upper bounds coming from Theorem 7.9 are for the three summands $\log|\frac{s}{t}|$, $-\log|t^2 s|$, resp. $-\log|ts^2|$. Note that the second term equals the sum of the other two, so that our assumption translates to the assumption that at least two out of the three entropy contributions must be positive — any particular entropy contribution coming from one coarse Lyapunov subgroup cannot give more than one half of the maximal entropy. Hence we conclude that at least two of the three leaf-wise measure $\mu_x^{G^{[1]}}$, $\mu_x^{G^{[2]}}$, or $\mu_x^{G^{[3]}}$ must be nontrivial a.e. From the above we know that if $\mu_x^{G^{[1]}}$ and $\mu_x^{G^{[3]}}$ are nontrivial a.e., then $\mu_x^{G^{[2]}}$ is actually the Haar on $G^{[2]}$ a.e. and so again nontrivial a.e. Using again symmetry of entropy and the same commutator argument within various stable horospherical subgroups the theorem follows easily. □

9.13. Problem. Prove the following version of the high entropy theorem for quotients of $G = \mathrm{SL}(n, \mathbb{R})$ (starting with $n = 3$). Suppose μ is an A-invariant and ergodic probability measure on $X = \Gamma\backslash G$ such that all nontrivial elements of A have positive entropy. Deduce that μ is the Haar measure on X.

Generalizing the commutator argument leads to the following theorem.

9.14. Theorem. *(High entropy theorem) Let μ be an A-invariant and ergodic probability measure on $X = \Gamma\backslash G$. Let $[\zeta]$ and $[\eta]$ be coarse Lyapunov weights such that $[\zeta] \neq [\eta] \neq [\zeta^{-1}]$. Then for a.e. x, μ is invariant under the group generated by $[\operatorname{supp}\mu_x^{G^{[\zeta]}}, \operatorname{supp}\mu_x^{G^{[\eta]}}]$. In fact the same holds with $\operatorname{supp}\mu_x^{G^{[\zeta]}}$ and $\operatorname{supp}\mu_x^{G^{[\eta]}}$ replaced by the smallest Zariski closed A-normalized subgroups containing the supports.*

To prove this in general we need a few more preparations.

9.15. Invariance subgroups. Let $a \in A$ and assume $U \subset G_a^-$ is a-normalized. We define for any x the closed subgroup

$$\mathrm{Stab}_x^U = \{u \in U : u\mu_x^U = \mu_x^U\} < U.$$

Over $\mathbb{R}$, there are — in some sense — very few closed subgroups, which restricts the possibilities for Stab_x^U. More precisely, we claim that Stab_x^U equals the connected component $(\mathrm{Stab}_x^U)^0$ of the identity in Stab_x^U, at least for a.e. x.

To see this let $d(x)$ be the distance from $(\mathrm{Stab}_x^U)^0$ to $\mathrm{Stab}_x^U \backslash (\mathrm{Stab}_x^U)^0$ (using a left invariant metric) and define $d(x) = 0$ if the claim holds for x.

Now $a^n \,\mathrm{Stab}_x^U\, a^{-n} = \mathrm{Stab}_{a^n.x}^U$ (see Lemma 7.16), and since a contracts U, we must have $d(a^n x) \to 0$ as $n \to \infty$. Thus, we see that $d(x) = 0$ for a.e. x by Poincaré recurrence. Therefore, Stab_x^U is connected.

9.16. PROBLEM. Over $\mathbb{Q}_p$ it doesn't make sense to speak of the connected component (as it would be the trivial group in any case), but we can speak of the maximal algebraic subgroup contained in Stab_x^U. For this recall that exp and log are polynomial isomorphisms between the Lie algebra of U and U. Also sub-Lie algebras are mapped under this map to Zariski closed subgroups of U. We may define $(\mathrm{Stab}_x^U)^0$ to be the exponential image of the largest subalgebra contained in the logarithmic image of Stab_x^U. (The reader may verify that over $\mathbb{R}$ this defines the connected component.) Show that $(\mathrm{Stab}_x^U)^0 = \mathrm{Stab}_x^U$ a.e. (The situation is in a sense opposite to the real case where one had to apply the contraction by a to obtain small elements — over $\mathbb{Q}_p$ we can simply take a power of an element of Stab_x^U to obtain small elements but one has to apply the expansion a^{-1} and Poincaré recurrence to obtain big elements of Stab_x^U.)

9.17. Stab_x^U IS NORMALIZED BY A. If U is one-dimensional, then this follows simply from $\mathrm{Stab}_x^U = (\mathrm{Stab}_x^U)^0$. However, in general this is a special property which again is a result of Poincaré recurrence.

In fact, as $\mathrm{Stab}_x^U = (\mathrm{Stab}_x^U)^0$ it is uniquely determined by its Lie algebra $\mathfrak{s}_x$. Notice that $\mathfrak{s}_{a^n.x} = \mathrm{Ad}_a^n \mathfrak{s}_x$ a.e. However, A is generated by elements $a \in A$ whose eigenvalues are all powers of a single number t. For these it follows that either $\mathfrak{s}_{a.x} = \mathfrak{s}_x$ or that $\mathfrak{s}_{a^n.x}$ approaches a sub Lie algebra $\mathfrak{h}$ for which $\mathrm{Ad}_a \mathfrak{h} = \mathfrak{h}$. In fact, this follows from considering the alternating tensor product of the Lie algebra of U of degree equal to the dimension of $\mathfrak{s}_x$ (which is independent of x for a.e. x by ergodicity of A): The action of the class A element a on that space still has all eigenvalues equal to a power of t and either the point corresponding to $\mathfrak{s}_x$ is an eigenvector for that action or it approaches projectively one when the iterates of a are applied to it. By Poincaré recurrence the latter is not possible for a.e. x, hence the conclusion.

In particular, the above shows that $\mathrm{Stab}^U = \mathrm{Stab}_x^U$ is independent of x for a.e. x as μ is A-ergodic. This makes the following lemma useful for $H = U$ and $L = \mathrm{Stab}^U$.

9.18. Lemma. *Let H act on X, and $L < H$ be a subgroup. Suppose that for every $l \in L$, we have $l\mu_x^H = \mu_x^H$ for a.e. $x \in X$. Then μ is L-invariant.*

This follows literally from Problem 6.28 and Lemma 8.10, but also purely from the argument behind Problem 6.28.

9.19. PROOF OF THEOREM 9.14. We take two coarse Lyapunov weights $[\zeta]$ and $[\eta]$ satisfying $[\zeta] \neq [\eta] \neq [\zeta^{-1}]$ as in the theorem. Then there exists an A-normalized subgroup $H \subset G_a^-$ (for some $a \in A$) which is a product of coarse Lyapunov subgroups for which $[\zeta], [\eta]$ are both exposed. This implies as in Theorem 9.8 that

$$\mu_x^H \propto \iota(\mu_x^{G^{[\zeta]}} \times \mu_x^{G^{[\eta]}} \times \mu_x^U) \propto \iota(\mu_x^{G^{[\eta]}} \times \mu_x^{G^{[\zeta]}} \times \mu_x^U)$$

where U is the product of all coarse Lyapunov subgroups that are contained in H except for $G^{[\zeta]}$ and $G^{[\eta]}$. The argument in 9.11 now shows that μ_x^U must be invariant under $[\operatorname{supp}\mu_x^{G^{[\zeta]}}, \operatorname{supp}\mu_x^{G^{[\eta]}}] \in \mathrm{Stab}^U$. Together with the above discussion, this implies that μ is invariant under $[\operatorname{supp}\mu_x^{G^{[\zeta]}}, \operatorname{supp}\mu_x^{G^{[\eta]}}]$ for a.e. x as claimed.

We now wish to prove the additional claim that μ is, for a.e. x, also invariant under the commutators $[h^{[\zeta]}, h^{[\eta]}]$ of elements of the smallest Zariski closed

a-normalized subgroups $P^{[\zeta]} \ni h^{[\zeta]}$ and $P^{[\eta]} \ni h^{[\eta]}$ containing $\operatorname{supp} \mu_x^{G^{[\zeta]}}$ resp. $\operatorname{supp} \mu_x^{G^{[\eta]}}$. We may assume ζ (and similarly η) is an indivisible weight, i.e., all other weights which are coarsely equivalent to ζ are powers of ζ. Now notice that Zariski closed subgroups of the unipotent group $G^{[\zeta]}$ are precisely the exponential images of subalgebras of the Lie algebra of $G^{[\zeta]}$. To prove the above we are first claiming that if $g^{[\zeta]} \in \operatorname{supp} \mu_x^{G^{[\zeta]}}$ and $h^{[\eta]} \in \operatorname{supp} \mu_x^{G^{[\eta]}}$ and we write

$$\begin{aligned} \log g^{[\zeta]} &= u^{[\zeta]} = u_\zeta + u_{\zeta^2} + \cdots \\ \log h^{[\eta]} &= v^{[\eta]} = v_\eta + v_{\eta^2} + \cdots \end{aligned}$$

with $u_\xi, v_\xi \in \mathfrak{g}^\xi$, then a.s. $\exp[u_{\zeta^m}, v_{\eta^n}]$ preserves the measure μ (or equivalently $\mu_x^{G^{[\zeta^m\eta^n]}}$). For this we have to proceed by induction on the complexity of the subgroup U. If U is the trivial subgroup, there is nothing to prove as in this case $G^{[\zeta]}$ and $G^{[\eta]}$ commute. For the general case we have to compare the group theoretic commutator

$$[g^{[\zeta]}, h^{[\eta]}] = (g^{[\zeta]})^{-1}(h^{[\eta]})^{-1} g^{[\zeta]} h^{[\eta]}$$

with the Lie theoretic commutator in the Lie algebra $\mathfrak{g}$. By the Campbell-Baker-Hausdorff formula the former equals

$$[g^{[\zeta]}, h^{[\eta]}] = \exp\big([u^{[\zeta]}, v^{[\eta]}] + \cdots\big), \tag{9.19a}$$

where the dots indicate a finite sum of various iterated commutators of $u^{[\zeta]}$ and $v^{[\eta]}$ with $[u^{[\zeta]}, v^{[\eta]}]$. Let us refer to $[u^{[\zeta]}, v^{[\eta]}]$ as the main term. Note that in $\log g$ the only term of weight $\zeta\eta$ is $[u_\zeta, v_\eta]$ (which is part of the main term), as all terms indicated by the dots only contain terms of weight $\zeta^k\eta^\ell$ with $k + \ell \geq 3$. As $g \in \operatorname{Stab}^U$ and this group is A-normalized and equals the exponential image of its Lie algebra, we see that $[u_\zeta, u_\eta]$ belongs to the Lie algebra of Stab^U. We note that this implies that $\exp[u_\zeta, v_\eta]$ preserves $\mu_x^{G^{[\zeta\eta]}}$ which implies that $\exp[u^\zeta, v^\eta] \in \operatorname{supp} \mu_x^{G^{[\zeta\eta]}}$. If we replace η by $\zeta\eta$ and $h^{[\eta]}$ by $\exp[u_\zeta, v_\eta]$, we obtain a situation as before but with a smaller dimensional subgroup U' replacing U. By the inductive hypothesis we conclude that all terms of the form $\exp[u_{\zeta^m}, [u_\zeta, v_\eta]]$ preserve the measure. However, this now shows that the term inside the exponential in (9.19a) corresponding to weight $\zeta^2\eta$ is the sum of $[u^{\zeta^2}, v^\eta]$ (which is part of the main term) and of a multiple of $[u_\zeta, [u_\zeta, v_\eta]]$. As before we conclude that this sum belongs to the Lie algebra of Stab^U, which in return shows the same for $[u^{\zeta^2}, v^\eta]$ (and similarly for $[u^\zeta, v^{\eta^2}]$). Proceeding inductively one shows in the same manner that all components $[u_{\zeta^m}, u_{\eta^n}]$ of the main term belongs to the Lie algebra of Stab^U.

As the Lie bracket is bilinear, it is clear that we may multiply the various components u_{ζ^m} and u_{η^n} by scalars without affecting the conclusion. It remains to show that if $[u_1, v]$ and $[u_2, v]$ for $u_1, u_2 \in \mathfrak{g}^{[\zeta]}$ and $v \in \mathfrak{g}^{[\eta]}$ belong to the Lie algebra of Stab^U, then the same is true for $[[u_1, u_2], v]$. However, this follows from the Jacobi identity

$$[[u_1, u_2], v] = -[[v, u_1], u_2] - [[u_2, v], u_1]$$

where the terms on the right belong to the Lie algebra of Stab^U by what we already established.

The above together shows that we may take u, v in the Lie algebra generated by $\log \operatorname{supp} \mu_x^{[\zeta]}$ resp. generated by $\log \operatorname{supp} \mu_x^{[\eta]}$ and obtain that $[u, v]$ belongs to the

Lie algebra generated by Stab^U. This together with the Campbell-Baker-Hausdorff formula is the desired result. □

As a corollary, we have the following entropy gap principle.

9.20. Theorem. *Let G be a* simple *algebraic group defined over $\mathbb{R}$ and connected in the Hausdorff topology. Let $\Gamma < G$ be a discrete subgroup. Say $A \subset G$ is a split torus of rank at least 2. Suppose μ is an A-invariant and ergodic probability measure on $X = \Gamma\backslash G$. Then for every a, there exists $h_0 < h_{m_X}(a)$ such that $h_\mu(a) > h_0$ implies that $\mu = m_X$ is the Haar measure on X.*

We note that the entropy $h_{m_X}(a)$ of the Haar measure m_X on X is determined by a concrete formula involving only Ad_a, and so is independent of Γ. If Γ is not assumed to be a lattice, we still write $h_{m_X}(a)$ for this expression. With this in mind, we do not have to assume in the above theorem that Γ is a lattice, rather obtain this as part of the conclusion if only $h_\mu(a) > h_0$.

9.21. Example. We illustrate Theorem 9.20 as well as another formulation of the high entropy theorem in the case of $G = \mathrm{SL}(3,\mathbb{R})$ (as in Problem 9.13).

Say

$$a = \begin{pmatrix} e^{-t} & & \\ & 1 & \\ & & e^{t} \end{pmatrix} \qquad G_a^- = \begin{pmatrix} 1 & G^{[1]} & G^{[2]} \\ & 1 & G^{[3]} \\ & & 1 \end{pmatrix}$$

We have

$$h_{m_X}(a) = \sum_{i=1}^{3} h_\lambda(a, G^{[i]}) = t + 2t + t.$$

If we take $h_0 = 3t$, then $h_\mu(a) > 3t$ implies that there is an entropy contribution from all 3 expanding directions, and so all three leaf-wise measures are non-trivial almost everywhere. Therefore the support of each $\mu_x^{G^{[i]}}$ is all of $G^{[i]}$, and the high-entropy method then implies that μ is invariant under all $G^{[i]}$, and therefore invariant under G, so μ is the Haar measure on X.

Now suppose

$$a = \begin{pmatrix} e^{-2t} & & \\ & e^{t} & \\ & & e^{t} \end{pmatrix} \qquad G_a^- = \begin{pmatrix} 1 & G^{[1]} & G^{[2]} \\ & 1 & \\ & & 1 \end{pmatrix}$$

we have central directions that are neither expanded nor contracted by a. Here, we have

$$h_{m_X}(a) = 3t + 3t$$

and $h_\mu(a) > h_0 = 3t = \frac{1}{2} h_{m_X}(a_t)$ implies that the Zariski closure of $\operatorname{supp} \mu_x^{G^{[i]}}$ is a.s. all of $G^{[i]}$, and so by taking the commutator we get invariance of μ under the central direction as well (eg., since $[G^{[1]}, G^{[-2]}]$ is the lower central direction, μ is invariant under this direction as well.)

Now suppose we know that, for *every* a, we have $h_\mu(a) > 0$. By examining the element $a = \begin{pmatrix} e^{-2t} & & \\ & e^{t} & \\ & & e^{t} \end{pmatrix}$ as above, we find that either $\mu_x^{G^{[1]}}$ or $\mu_x^{G^{[2]}}$ is nontrivial almost everywhere. If we assume that, say, $\mu_x^{G^{[2]}}$ is trivial (and hence that $\operatorname{supp} \mu_x^{G^{[1]}}$

is Zariski dense in $G^{[1]}$ a.s.), then we can use the element $a = \begin{pmatrix} e^{-t} & & \\ & e^{-t} & \\ & & e^{2t} \end{pmatrix}$ to show that $\mu_x^{G^{[3]}}$ has Zariski dense support in $G^{[3]}$ a.s., and we get invariance under $G^{[2]}$ anyway. By similar arguments using other singular elements a, we can get invariance under any $G^{[i]}$, and so μ must be the Haar measure on X.

9.22. Lemma. *Let $V < U$ be a-normalized closed subgroups of the stable horospherical subgroup G_a^-. Suppose that* $\operatorname{supp}\mu_x^U \subset V$ *for a.e. x. Then*

$$h_\mu(a, U) \le h_{m_X}(a, V) \le h_{m_X}(a, U)$$

In fact, the second inequality is strict (and uniformly so) if V is a proper *subgroup of U.*

Note that the assumption on the support of μ_x^U implies that $h_\mu(a, V) = h_\mu(a, U)$. With this in mind the lemma follows from Theorem 7.9.

9.23. Lemma. *Let $a \in A$ and let $[\eta]$ be coarse Lyapunov weight contracted by a. Under the hypotheses of 9.20, for h_0 large enough and μ-a.e. x, we have that $G^{[\eta]}$ is the smallest a-normalized Zariski closed subgroup containing the support of $\mu_x^{G^{[\eta]}}$.*

This follows by combining Lemma 9.22 and Corollary 9.10.

9.24. Proposition. *For any nontrivial $a \in A$, the simple group G is generated by the set of commutators $[G^{[\lambda]}, G^{[\eta]}]$ of all pairs of coarse Lyapunov subgroups which satisfy $[\eta] \ne [\lambda] \ne [\eta^{-1}]$ and $\eta(a) \ne 1 \ne \lambda(a)$.*

We see that Theorem 9.20 follows from Theorem 9.14 together with Proposition 9.24 and Lemma 9.23.

9.25. Proof of Proposition 9.24. Let V be a lower dimensional subgroup of the group of characters of A. Let

$$\mathfrak{w} = \operatorname{span}\{\mathfrak{g}^\lambda, [\mathfrak{g}^\eta, \mathfrak{g}^\lambda] : \lambda, \eta \notin V\}$$

We claim that $\mathfrak{w}$ is a Lie ideal of $\mathfrak{g}$. To check this, we first take $x \in \mathfrak{g}^\delta \subset \mathfrak{w}$ (first type of elements) for some $\delta \notin V$, and some $z \in \mathfrak{g}^\zeta$ and look at $[x, z]$. There are two cases:

(i) If $\zeta \notin V$, then $[x, z] \in \mathfrak{w}$ by definition due to the second type of elements of $\mathfrak{w}$.
(ii) If $\zeta \in V$, then $[x, z] \in \mathfrak{g}^{\delta\zeta} \subset \mathfrak{w}$ due to the first type of elements since $\delta\zeta \notin V$.

Assume now $[x, y] \in \mathfrak{w}$ with $x \in \mathfrak{g}^\lambda$, $y \in \mathfrak{g}^\eta$, and $\lambda, \eta \notin V$ as in the second type of elements of $\mathfrak{w}$. Also let as before $z \in \mathfrak{g}^\zeta$. There are again two cases:

(i) If $\zeta \notin V$ then by the above cases $[[x, y], z] \in \mathfrak{w}$.
(ii) In the remaining case of $\zeta \in V$ we use the Jacobi identity

$$[[x, y], z] = -[[y, z], x] - [[z, x], y]$$

which leads to the expressions $[[y, z], x]$ and $[[z, x], y]$. However, $[y, z] \in \mathfrak{g}^{\eta\zeta}$ with $\eta\zeta \notin V$ and $\lambda \notin V$ shows that $[[y, z], x]$ is an expression of the second type in the definition of $\mathfrak{w}$. The same holds for $[[z, x], y]$ which shows that $[[x, y], z] \in \mathfrak{w}$ as claimed.

As V is assumed to be lower dimensional and the weights span, $\mathfrak{w}$ is nontrivial and hence equal to $\mathfrak{g}$ by the assumption that G is simple.

Now we let V be the kernel of the evaluation map $\lambda \mapsto \lambda(a)$ and let $\zeta \in V$ be a nontrivial weight. Then the above claim shows that all elements of the Lie algebra $\mathfrak{g}^\zeta$ can be written as sums of Lie commutators of elements of $\mathfrak{g}^\eta, \mathfrak{g}^\lambda$ with $\eta, \lambda \notin V$. Here we cannot have $[\eta] = [\lambda]$ or $[\eta] = [\lambda^{-1}]$ as otherwise the commutator would belong to a weight space $\mathfrak{g}^\zeta$ also satisfying that ζ is trivial, $[\eta] = [\zeta]$, or that $[\eta] = [\zeta^{-1}]$ which is impossible as $\eta \notin V$ but $\zeta \in V$.

Similarly we may now set V equal to the subgroup of characters equivalent to a given nontrivial λ or its inverse λ^{-1}. Applying again the above we see that the elements of the weight space $\mathfrak{g}^\lambda$ can be written as sums of Lie commutators of elements of $\mathfrak{g}^\eta, \mathfrak{g}^\lambda$ with $[\eta] \neq [\lambda] \neq [\eta^{-1}]$. (By the argument above we do not have to restrict ourselves any longer to weight spaces that do not commute with a). Therefore, all elements of all nonzero weight spaces can be generated by the Lie brackets that we consider.

Finally, note that the Lie algebra generated (set V equal to the trivial group) by all nonzero weight spaces is the whole of $\mathfrak{g}$, so that $\mathfrak{g}$ is generated indeed by the Lie brackets that we consider. □

10. Invariant measures for higher rank subgroups A, the low entropy method

10.1. In this section we sketch the proof of a theorem regarding A-invariant measures where only positivity of entropy is assumed (instead of entropy close to being maximal). In addition to the ideas we already discussed they use one more method, which we refer to as the low entropy method. This method was first used in [**Lin06**]; one of the main motivations being the Arithmetic Quantum Unique Ergodicity Conjecture which is partially resolved in that paper — see §13.

10.2. A basic feature of this method is that it gives a prominent role to the dynamics of the unipotent groups normalized by A, even though these unipotent groups a-priori do not preserve the measure in any way. Ideas of Ratner, particularly from her work on the horocycle flow [**Rt82a, Rt82b, Rt83**] are used in an essential way.

We first present this method which has been extended to fairly general situations in [**EL08**] in one particular case (the reader who is averse to p-adic numbers is welcome to replace $\mathrm{SL}(2,\mathbb{Q}_p)$ by $\mathrm{SL}(2,\mathbb{R})$ in the theorem and its proof below).

10.3. Theorem. *Let $X = \Gamma\backslash \mathrm{SL}(2,\mathbb{R}) \times \mathrm{SL}(2,\mathbb{Q}_p)$, where Γ is an irreducible lattice in $G = \mathrm{SL}(2,\mathbb{R}) \times \mathrm{SL}(2,\mathbb{Q}_p)$. Let $A = \left(\begin{pmatrix} * & \\ & * \end{pmatrix} \times e\right)$ be the (one parameter) diagonal subgroup in the $\mathrm{SL}(2,\mathbb{R})$ factor. Suppose μ is an A-invariant probability measure such that*

- *μ is $\mathrm{SL}(2,\mathbb{Q}_p)$-recurrent.*
- *Almost all A-ergodic components of μ have positive entropy under the A-flow.*

Then μ is the Haar measure on X.

10.4. We recall that a lattice Γ in $G_1 \times G_2$ is said to be irreducible if the kernel of the projection to each factor is finite. For the case at hand of $G_1 = \mathrm{SL}(2,\mathbb{R})$ and $G_2 = \mathrm{SL}(2,\mathbb{Q}_p)$ this is equivalent to both projections being dense. This assumption of irreducibility is clearly necessary; the assumption that Γ is a lattice (i.e., that it has finite covolume) is not, though it is not clear that classifying *probability* measures in the non-compact case is a very natural question. An example of an irreducible lattice in $\mathrm{SL}(2,\mathbb{R}) \times \mathrm{SL}(2,\mathbb{Q}_p)$ is $\mathrm{SL}(2,\mathbb{Z}[\frac{1}{p}])$ (embedded diagonally).

10.5. We remark that unlike many measure classification theorems it is not possible to reduce Theorem 10.3 to the case of μ being A-ergodic. This is because if one takes an arbitrary measure μ satisfying the condition of the theorem and take its ergodic decomposition with respect to the A action there is no reason to expect the ergodic components to remain $\mathrm{SL}(2,\mathbb{Q}_p)$-recurrent. The fact that we are considering general invariant measures requires us to demand that not only does μ have positive entropy under A, but that each ergodic component has positive entropy.

10.6. The requirement that μ be $\mathrm{SL}(2,\mathbb{Q}_p)$-recurrent is clearly necessary, there are plenty of A invariant and ergodic measures on $\Gamma\backslash \mathrm{SL}(2,\mathbb{R}) \times \mathrm{SL}(2,\mathbb{Q}_p)$ with positive entropy. E.g., when $\Gamma = \mathrm{SL}(2,\mathbb{Z}[\frac{1}{p}])$ as above, $\Gamma\backslash \mathrm{SL}(2,\mathbb{R}) \times \mathrm{SL}(2,\mathbb{Q}_p)$ is a compact extension of $\mathrm{SL}(2,\mathbb{Z})\backslash \mathrm{SL}(2,\mathbb{R})$ (with the action of A respected by the corresponding projection map) and hence any A-invariant and ergodic measure on $\mathrm{SL}(2,\mathbb{Z})\backslash \mathrm{SL}(2,\mathbb{R})$ can be lifted to an invariant and ergodic measure on $\Gamma\backslash \mathrm{SL}(2,\mathbb{R}) \times \mathrm{SL}(2,\mathbb{Q}_p)$ with exactly the same entropy[(34)].

10.7. Outline of Proof. The starting point. Let $T = \mathrm{SL}(2,\mathbb{Q}_p)$ and let $U = \begin{pmatrix} 1 & * \\ & 1 \end{pmatrix}$ be the real upper unipotent subgroup; then $U = G_a^-$ and $T < C_G(a) \cap C_G(U)$ for e.g.[(35)] $a = \left(\begin{pmatrix} e^{-1} & \\ & e \end{pmatrix}, \begin{pmatrix} 1 & \\ & 1 \end{pmatrix}\right) \in A$. In particular, the assumptions to Corollary 8.8 are satisfied and the leaf-wise measures for the subgroup $H = TU$ are product measures a.s. By Corollary 8.13 there is a subset of full measure $X' \subset X$ such that we have $\mu_x^U = \mu_y^U$ whenever $x, y \in X'$ belong to the same $\mathrm{SL}(2,\mathbb{Q}_p)$-orbit. This shows together with the assumed recurrence that we can find many close-by points with the same leaf-wise measures, i.e., $y = (g_1, g_2).x$ with the displacements $g_1 \in \mathrm{SL}(2,\mathbb{R})$ and $g_2 \in \mathrm{SL}(2,\mathbb{Q}_p)$ both close to the identity and $\mu_x^U = \mu_y^U$.

As we have already observed in Example 6.5.2 the coincidence of leaf-wise measures can have strong implications. This is the case here. By replacing both x and y by $u.x$ and $u.y$ for some (in a certain sense) typical $u \in U$, we bring the polynomial shearing properties of the U-flow in the picture.

---

(34) This last claim requires some justification; what is immediate and is sufficient for our purpose is that the lifted measure would have at least the same entropy as the original measure. Also if the lifted measure is not ergodic, then one can take a typical ergodic component of it which will also be a lift of the original measure.

(35) Here e is the constant $2.71828\ldots$; below and above e is also used to denote the identity element of G.

10.8. POLYNOMIAL DIVERGENCE. If, starting with $y = (g_1, g_2).x$, one moves along the U-orbit, the displacement of $x' = u.x$ and $y' = u.y$ is the conjugate (ug_1u^{-1}, g_2) and the U-action by conjugation is shearing depending polynomially on the time parameter in U. More precisely, given x and $y = (g_1, g_2).x$ with $g_1 = \begin{pmatrix} a & b \\ c & d \end{pmatrix} \notin U$ (an assumption which we will need to justify), we apply $u(s) = \begin{pmatrix} 1 & s \\ & 1 \end{pmatrix}$ to both to get

$$x_s = \begin{pmatrix} 1 & s \\ & 1 \end{pmatrix}.x$$

$$y_s = \begin{pmatrix} 1 & s \\ & 1 \end{pmatrix}(g_1, g_2).x = (g_{(s)}, g_2)x_s$$

We compute

$$g_{(s)} = u(s)g_1u(-s) = \begin{pmatrix} a + cs & b + (d-a)s - cs^2 \\ c & d - cs \end{pmatrix}. \tag{10.8a}$$

Hence the terms contributing to the divergence are $|d - a|$ and $\sqrt{|c|}$. (As it turns out using the square root puts the two terms on more equal footing). For $S = \min(\frac{1}{|d-a|}, \frac{1}{\sqrt{|c|}})$, we expect to have $g_{(S)} \approx \begin{pmatrix} 1 & r \\ & 1 \end{pmatrix} = u(r)$ for some nontrivial r. Let us explain this more precisely: starting with the top right corner it is possible that by some coincidence for $s = S$ the terms $(d - a)s$ and cs^2 might cancel each other, but this would be an exception: for most $s \in [-S, S]$ the top right coordinate $b + (d - a)s - cs^2$ in (10.8a) will be of the order 1 (for a linear polynomial this is obvious, for the general quadratic polynomial one only needs that s is sufficiently far from both roots). Also $g_{(s)} = u(s)g_1u(-s)$ is bounded for $s \in [-S, S]$, and except for the top right entry, for s in that range, all other entries will be very close to the corresponding entries in the identity matrix (i.e., to one for the diagonal elements, and to zero for the bottom left corner): indeed, the diagonal entries are $a + cs$ and $d - cs$ which are close to 1 as $|cs| \leq |c|S \leq \sqrt{|c|}$ and the bottom left entry c is close to zero (and unchanged).

Hence $g_{(s)}$ will indeed be approximately $u(r)$ for some nontrivial and bounded r for most $s \in [-S, S]$.

10.9. CHOOSING THE CORRECT $u(s)$, TWO CONDITIONS. We will need to choose some $s \in [-S, S]$ such that $g_{(s)}$ has significant size. By the above discussion this is quite easy and a purely **algebraic condition**. At the same time $x' = x_s$ and $y' = y_s$ should have good properties with respect to the measure μ, which is a **measure theoretic condition**, the verification of which requires more work.

Clearly some condition on the points x' and y' is needed for them to give any meaningful information about the measure μ — the most important for us will be that both x' and y' belong to some compact set K on which the map sending a point z to the leafwise measure μ_z^U is continuous. We can also assume that this set K is contained in the conull set on which Theorem 6.3 (iii) holds so that (by applying that theorem to both x and y) we can deduce from our original assumption $\mu_x^U = \mu_y^U$ that $\mu_{x'}^U = \mu_{y'}^U$.

If we could take a limit, taking the original points x, y from a sequence x_k, y_k ever closer together (which forces $r_k \to \infty$), the limit points x'', y'' of a common subsequence of two points x'_k, y'_k will end up being different points on the same U-orbit. Restricting everything to a large compact set K where $z \mapsto \mu_z^U$ is continuous we would obtain $\mu_{x''}^U = \mu_{y''}^U$ and since $y'' = u.x''$ for some nontrivial $u \in U$ also $\mu_{x''}^U \propto \mu_{y''}^U u$ by Theorem 6.3 (iii). This leads to U-invariance of μ almost[36] as in the last section.

For this to work we need to ensure that given the two close-by points x_k, y_k with the same leaf-wise measures (and other good properties that hold on sets of large measure) we can find some $u(s_k) \in U$ such that $x'_k = u(s_k).x_k, y'_k = u(s_k).y_k \in K$ and the displacement is significant but of bounded size. As explained above it is easy to find some S_k depending on the displacement $g_1^{(k)}$ such that $u(s)g_1^{(k)}u(s)^{-1}$ is significant but bounded for all $s \in [-S_k, S_k]$ except those belonging to two small subintervals of $[-S_k, S_k]$. So basically we have two requirements for s_k, it shouldn't belong to one of two small subintervals which have been found using purely algebraic properties, and we also want both points $x'_k = u(s_k).x_k, y'_k = u(s_k).y_k$ to belong to the compact set K on which everything behaves nicely — which is a measure theoretic property involving μ since all we know about K is that it has large μ-measure.

10.10. A MAXIMAL ERGODIC THEOREM. To prove the latter property we need a kind of ergodic theorem for the U-action with respect to μ, even though we do not know invariance under U. A maximal ergodic theorem for the U-action would imply that for a given set of large measure K, the set of points x, for which there is some scale S for which it is not true that for *most* $s \in [-S, S]$ we have $u(s).x \in K$, has small μ-measure (and so can be avoided in the argument). However, here the correct notion of *most* must come from the measure μ_x^U instead of the Lebesgue measure as μ is not known to be invariant under U.

There are several versions of such maximal ergodic theorems in the literature starting from Hurewicz [**Hur44**]; see also [**Bec83**]. In [**Lin06**] a variant proved in the appendix to that paper jointly with D. Rudolph was used. An alternative approach which we have employed in [**EL08**] is to use the decreasing Martingale theorem by using the sequence of σ-algebras $a^{-n}\mathcal{A}$, where $\mathcal{A}$ is subordinate to U on a set of large measure and a-decreasing as in Definition 7.25. The latter approach has the advantage of working in greater generality, see Comment 7.38.

10.11. COMPATIBILITY ISSUE. Assume now that a sufficient form of such a maximal ergodic theorem holds for the U-action. This then implies, starting with sufficiently well behaved initial points x_k, y_k (with $\mu_{x_k}^U = \mu_{y_k}^U$), that for $\mu_{x_k}^U$-most $s \in [S_k, S_k]$ (say for 90%) we have $u(s).x_k, u(s).y_k \in K$. Even so there is still a gap in the above outline: Can we ensure that the two subintervals of $[-S_k, S_k]$, where $u(s)g_1^{(k)}u(s)^{-1}$ is too little, have also small mass with respect to $\mu_{x_k}^U$? This is desired as it would ensure the compatibility of the algebraic and measure-theoretic properties needed, sincc in this case for $\mu_{x_k}^U$-most $s \in [-S_k, S_k]$ both properties would hold. However, if e.g. $\mu_{x_k}^U$ is trivial, i.e., is supported on the identity only,

[36] The cautious reader may be concerned about the lack of ergodicity assumption here. Indeed one first only obtains that some ergodic component is U-invariant, but one may apply the whole argument to the measure restricted to the subset where μ_x^U is not the Lebesgue measure to obtain a contradiction.

this is not the case. Luckily by assumption entropy is positive for a.e. ergodic component which translates to $\mu_{x_k}^U$ being nontrivial a.e. Even so, $\mu_{x_k}^U$ could give large mass to very small subintervals and the compatibility of the two conditions does not seem automatic.

10.12. SELF-SIMILARITY OF LEAF-WISE MEASURES. What rescues the argument is a kind of self-similarity of the measures μ_x^U. E.g. if one assumes a doubling condition of the form that there exists some $\rho \in (0,1)$ for which

(10.12a) $$\mu_x^U(B_{\rho S}^U) < \frac{1}{2}\mu_x(B_S^U) \text{ for all } S > 0,$$

then sufficiently small symmetric subintervals of a given interval $[-S,S]$ also get small μ_x^U-mass. (Given such a ρ we then would adjust the meaning of 'significant' in the discussion of §10.8 and §10.9.) There is no reason why such a strong regularity property of the conditional measures should hold. However, the A-action on X together with Lemma 7.16 implies some regularity properties: E.g. by Poincaré recurrence there are infinitely many S such that (after rescaling) μ_x^U restricted to B_S^U is very similar to μ_x^U restricted to B_1^U. To obtain something similar to (10.12a) we notice first that there is some $\rho > 0$ such that

(10.12b) $$\mu_x^U(B_\rho^U) < \frac{1}{2}$$

except possibly on a set Z of small μ-measure. Then one can apply the standard maximal ergodic theorem for the action of

$$a^r = \left(\begin{pmatrix} e^{-r} & \\ & e^r \end{pmatrix}, \begin{pmatrix} 1 & \\ & 1 \end{pmatrix}\right) \in A$$

to show that for μ-most x and for any given K of large μ-measure, most $r \in [0,R]$ satisfy that

(10.12c) $$\mu_x^U(B_{\rho e^{2r}}^U) < \frac{1}{2}\mu_x^U(B_{e^{2r}}^U)$$

(which is equivalent by Lemma 7.16 to $a^r.x$ satisfying (10.12b)). As it turns out the weaker (10.12c) is sufficient and one does not need (10.12a).

10.13. THE HEART OF THE ARGUMENT, CHOOSING t. Given the two points x_k, y_k and with them the parameter S_k we would need the regularity (10.12c) for $r = \frac{1}{2}\log S_k$ in order to apply the arguments from above. This may or may not happen but we can increase our chance of succeeding by looking not only at the given points x_k, y_k but also at all the points $a^t x_k, a^t y_k$ for some $t \in [0, T_k]$ for the appropriate choice of T_k (which in this case turns out to be $T_k = \frac{1}{2}\log S_k$).

This is the technical heat of the argument, and we sketch the proof below. To simplify matters, we assume that either $|d-a| \gg \sqrt{|c|}$ or $|d-a| \ll \sqrt{|c|}$ (with $\gg$ used here in a somewhat loose sense that we refrain from making more precise in this sketch).

We will only chose values of t for which the new points $a^t x_k, a^t y_k$ have good properties with respect to μ (i.e., belong to a previously defined set of points with good properties etc.), which in view of the (standard) maximal inequality holds for most $t \in [0, T_k]$ if the original points x_k, y_k were chosen from a suitable set of large measure.

Suppose first $|d-a| \gg \sqrt{|c|}$. In this case, the parameters a, d and with it S_k are unchanged when x_k and y_k are replaced by $a^t.x_k$ and $a^t.y_k$. Therefore, the regularity property (10.12c) is needed for the point $a^t.x_k$ and scale $r = \frac{1}{2}\log S_k$, which is equivalent to (10.12c) holding at the original point x for $r' = t + \frac{1}{2}\log S_k$.

At this stage we still have the freedom to chose t almost arbitrarily in the range $0 \le t \le \frac{1}{2}\log S_k$. As (10.12c) can be assumed to hold at x for most $r' \in [0, \log S_k]$ we can indeed chose t so that at $a^t.x_k$ (10.12c) holds for precisely the value of r we need.

In the second case $|d-a| \ll \sqrt{|c|}$, the important parameter $\sqrt{|c|}$ and with it S_k do change when x_k and y_k are replaced as above. The danger here is that if the parameter S_k changes in a particular way, it may be that one is still interested in the regularity property (10.12c) for x and the very same $r = \frac{1}{2}\log S_k$ even after introducing t. The reader may verify that this is **not** the case, after calculating the parameter $S_k(t)$ for the points $a^t.x_k$ and $a^t.y_k$ as a function of t one sees that $t + S_k(t)$ is affine with a linear component $\frac{1}{2}t$. As before a density argument gives that it is possible to find t as required. In the general case, the function one studies may switch between having linear part t and having linear part $\frac{1}{2}t$, i.e., may be only piecewise linear, but this does not alter the density argument for finding t. Moreover, one easily checks that $a^t.x_k$ and $a^t.y_k$ are still close together.

Having found t, one has the required regularity property to apply the density argument for $s \in [-S_k, S_k]$ and obtains $x_k', y_k' \in K$ which differ mostly by some element of U of bounded but significant size. As mentioned before, taking the limit along some subsequence concludes the argument.

10.14. Justification for $g_1 \notin U$. Let us finish the outline of the proof of Theorem 10.3 by justifying the assertion in §10.8 that one can find $x, y = (g_1, g_2).x$ with $g_1 \notin U$ and the same U leaf-wise measure using the recurrence of the $\mathrm{SL}(2,\mathbb{Q}_p)$-action. By construction, $y = (e,h).x$ for some big $h \in \mathrm{SL}(2,\mathbb{Q}_p)$, and we have already verified that the U leafwise measure at x and y are the same using the product lemma. What remains is to explain why we can guarantee that $g_1 \notin U$.

By Poincaré recurrence we may assume that our initial point x satisfies that there is a sequence $t_n \to \infty$ with $a^{t_n}.x \to x$. If now $g_1 \in U$, then $a^{t_n} g_1 a^{-t_n} \to e$ and applying a^{t_n} to $(e,h).x = y = (g_1, g_2).x$ we would obtain $(e,h).x = (e, g_2).x$. As h is big, but g_2 is small, we obtain the nontrivial identity $x = (e, h^{-1}g_2).x$ which is impossible as the lattice Γ is irreducible.

11. Combining the high and low entropy methods

11.1. Consider now the action of the diagonal group A on the space $X_n = \mathrm{SL}(n,\mathbb{Z})\backslash \mathrm{SL}(n,\mathbb{R})$. The method of proof of Theorem 10.3 can be adapted to study the A-invariant measures also in this case, but there are some extra twists; specifically we will need to combine in the low entropy method we have developed in the previous section with the high entropy method presented in §9. This has been carried out in the paper [**EKL06**] of the authors and A. Katok, and the results of this section are taken from that paper.

11.2. We recall the following conjecture regarding invariant measures on $X_n = \mathrm{SL}(n,\mathbb{Z})\backslash \mathrm{SL}(n,\mathbb{R})$, which is due to Margulis, Katok and Spatzier, and Furstenberg (cf. [**Mar00**]):

11.3. Conjecture. *Let A be the group of diagonal matrices in $\mathrm{SL}(n,\mathbb{R})$, $n \ge 3$. Then any A-invariant and ergodic probability measure μ on X_n is homogeneous.*

It is not hard to classify the possible homogeneous measures (see e.g. [**LW01**]). For n prime, the situation is particularly simple: any A-invariant homogeneous probability measure on X_n is either the natural measure on a periodic A-orbit, or the $\mathrm{SL}(n,\mathbb{R})$ invariant measure m on X_n.

11.4. In [**EKL06**] we give together with A.Katok the following partial result towards Conjecture 11.3:

11.5. Theorem ([**EKL06**, Theorem 1.3]). *Let A be the group of diagonal matrices as above and $n \geq 3$. Let μ be an A-invariant and ergodic probability measure on X_n. Then one of the following holds:*

(i) *μ is an A-invariant homogeneous measure which is not supported on a periodic A-orbit.*
(ii) *for every one-parameter subgroup $\{a_t\} < A$, $h_\mu(a_t) = 0$.*

By the classification of A-invariant homogeneous measures alluded to in §11.3, if (i) holds μ is not compactly supported.

11.6. For $1 \leq i \neq j \leq n$, let U_{ij} denote the one parameter unipotents subgroup of $\mathrm{SL}(n,\mathbb{R})$ which consists of all matrices that have 1 on the diagonal and 0 at all other entries except the (i,j) entry, and let μ be an A-invariant and ergodic probability measure on X_n which has positive entropy with respect to some $a_0 \in A$. By Theorem 7.6 and our assumption regarding positive entropy of μ it follows that the leaf-wise measure $\mu_x^{G^-_{a_0}}$ are nontrivial almost everywhere (this requires a bit of explanation, as μ is A-ergodic but not necessarily a_0-ergodic; however if one takes the ergodic decomposition of μ with respect to a_0 one gets from the ergodicity of μ under A that each ergodic component has the same entropy with respect to a_0 and one can apply Theorem 7.6 to each components separately). Using the product structure of $\mu_x^{G^-}$ given by Corollary 8.8 and the ergodicity under A it follows that there is some $i \neq j$ so that $\mu_x^{U_{ij}} =: \mu_x^{ij}$ is nontrivial almost everywhere. For notational simplicity suppose this happens for $(i,j) = (1,n)$.

11.7. One can now apply the argument described in §10.7– §10.14 to the group U_{1n} and an appropriate $a = \mathrm{diag}(\alpha_1, \dots, \alpha_n) \in A$ (we assume all the $\alpha_i > 0$) . One obvious requirement for a is that it contracts U_{1n}, i.e., that $\alpha_1 < \alpha_n$. It turns out though that in the proof (specifically, in §10.13) additional more subtle conditions on a need to be imposed that are nonetheless easy to satisfy: indeed in this case what one needs is simply that

$$\alpha_1 < \min_{1<i<n} \alpha_i \leq \max_{1<i<n} \alpha_i < \alpha_n.$$

For example, we can take $a = \mathrm{diag}(e^{-1}, 1, \dots, 1, e)$ which together with U_{1n} and U_{n1} form a subgroup of $\mathrm{SL}(n,\mathbb{R})$ isomorphic to $\mathrm{SL}(2,\mathbb{R})$.

11.8. We recall what was the outcome of the argument given in §10.7– §10.14 for $G = \mathrm{SL}(2,\mathbb{R}) \times \mathrm{SL}(2,\mathbb{Q}_p)$. The end result of that long argument was finding two distinct "μ-typical" points x, y with the same leafwise measures (i.e., $\mu_x^U = \mu_y^U$) with $y = u.x$ for some nontrivial $u \in U$.

An appropriate adaptation of this argument to the case at hand (i.e., $G = \mathrm{SL}(n,\mathbb{R})$) will yield at the end two μ-typical points x, y with the same U_{1n}-leafwise measures which differ by some element u obtained in a limiting procedure involving the shearing properties of U_{1n}. It turns out that in this case these limiting directions

u may not belong to U_{1n} but rather one has $u \in C_G(U_{1n}) \cap G^-$; note that this group $C_G(U_{1n}) \cap G^-$ is precisely the group generated by the 1-parameter unipotent groups U_{ij} with either $i = 1$ or $j = n$ (or both).

11.9. Playing around with leaf-wise measures, one can show that the measure μ must satisfy one of the following two possibilities:

(i) One can find a subset $X' \subset X_n$ of full measure such that every two points $x, y \in X'$ on the same $C_G(U_{1n}) \cap G^-$-orbit are in fact on the same U_{1n} orbit.

(ii) There are $(i, j) \neq (1, n)$ with $i = 1$ or $j = n$ so that μ_x^{ij} is nontrivial a.s.

.

If (i) holds, then the points x, y obtained in §11.8 in fact differ along U_{1n} from which one can deduce, exactly as in the proof of Theorem 10.3, that μ is U_{1n} invariant where we are clearly at the endgame; e.g. one can apply Ratner's measure classification theorem, though it is better to first get some more information out of the proof, specifically invariance along U_{n1}. Ratner's measure classification theorem for semisimple groups (such as the group generated by U_{1n} and U_{n1} which is isomorphic to $\mathrm{SL}(2, \mathbb{R})$) is substantially simpler than the general case (for a simple proof see [**Ein06**]). Moreover, also the analysis of all possible cases is much simpler if one first establishes invariance under this bigger group.

If (ii) holds, by using the time-symmetry of entropy for the element $a = \mathrm{diag}(e^{-1}, 1, \ldots, 1, e)$ we obtained that there are some (i', j') with $i' = 1$ or $j' = n$ or both so that $\mu_x^{j'i'}$ is nontrivial a.s. (note the switch in the order of the indices!). If $(i', j') \neq (1, n)$ we can apply Theorem 9.14 to obtain that μ is invariant under the group $[U_{j'i'}, U_{1n}]$ (which is either $U_{j'n}$ or $U_{1i'}$): again arriving at the endgame of the proof. If $(i', j') = (1, n)$ we obtain similarly that μ is invariant under $[U_{n1}, U_{ij}]$.

11.10. The above simplified discussion neglects to mention one crucial point. In Theorem 10.3, an important assumption was that Γ is irreducible, an assumption which only entered in order to show that there are nearby "typical" points x and y which differ in a shearable direction (i.e., not by an element in $U \times \mathrm{SL}(2, \mathbb{Q}_p)$) — c.f. §10.14.

The same issue arises also in the case of $\mathrm{SL}(n, \mathbb{R})$. For the particular lattice we are considering, namely $\mathrm{SL}(n, \mathbb{Z})$, one can show such nearby "shearable" pairs exist; but for a general lattice, even in $\mathrm{SL}(n, \mathbb{R})$, this problem can actually happen, and is precisely the source of an important class of counterexamples discovered by M. Rees to the most optimistic plausible measure classification conjecture for multidimensional diagonalizable groups [**Ree82**] (for a more accessible source, see [**EK03**, Section 9]; the same phenomena has been discovered independently in a somewhat different context by S. Mozes [**Moz95**]).

12. Application towards Littlewood's Conjecture

12.1. In this section we present an application of the measure classification results we have developed in the previous sections towards the following conjecture of Littlewood:

12.2. Conjecture (Littlewood (c. 1930)). *For every $\alpha, \beta \in \mathbb{R}$,*

$$\varliminf_{n \to \infty} n \|n\alpha\| \|n\beta\| = 0, \qquad (12.2a)$$

where $\|w\| = \min_{n\in\mathbb{Z}} |w-n|$ *is the distance of* $w \in \mathbb{R}$ *to the nearest integer.*

12.3. The work we present here toward this conjecture was first presented in the paper [**EKL06**] which is joint paper of A. Katok and us. The presentation of this work is taken essentially verbatim from [**Lin07**, Sec. 6].

12.4. It turns out that Littlewood's conjecture would follow from the Conjecture 11.3. The reduction is not trivial and is essentially due to Cassels and Swinnerton-Dyer [**CSD55**], though there is no discussion of invariant measures in that paper(37). A more recent discussion of the connection highlighting Cassels' and Swinnerton-Dyer's work can be found in [**Mar97**].

We need the following criterion for when α, β satisfy (12.2a):

12.5. Proposition. *(α, β) satisfy (12.2a) if and only if the orbit of*

$$x_{\alpha,\beta} = \mathrm{SL}(3,\mathbb{Z}) \begin{pmatrix} 1 & \alpha & \beta \\ 0 & 1 & 0 \\ 0 & 0 & 1 \end{pmatrix}$$

under the semigroup

$$A^+ = \{a(s,t) : s,t \geq 0\} \qquad a(s,t) = \begin{pmatrix} e^{s+t} & 0 & 0 \\ 0 & e^{-s} & 0 \\ 0 & 0 & e^{-t} \end{pmatrix}$$

is unbounded(38) *in* $X_3 = \mathrm{SL}(3,\mathbb{Z})\backslash \mathrm{SL}(3,\mathbb{R})$. *Moreover, for any* $\delta > 0$ *there is a compact* $C_\delta \subset X_3$, *so that if* $\underline{\lim}_{n\to\infty} n \|n\alpha\| \|n\beta\| \geq \delta$ *then* $A^+.x_{\alpha,\beta} \subset C_\delta$.

12.6. Before we prove Proposition 12.5 we need to understand better what it means for a set $E \subset X_3$ to be bounded. We write $\pi_\Gamma : \mathrm{SL}(n,\mathbb{R}) \to X_n$ for the natural map that sends $g \in \mathrm{SL}(n,\mathbb{R})$ to $\mathrm{SL}(n,\mathbb{Z})g \in X_n$. We have the following important criterion (see e.g. [**Rag72**, Chapter 10]):

12.7. Proposition (Mahler's compactness criterion). *Let* $n \geq 2$. *A set* $E \subset X_n = \mathrm{SL}(n,\mathbb{Z})\backslash \mathrm{SL}(n,\mathbb{R})$ *is bounded if and only if there is some* $\epsilon > 0$ *so that for any* $x = \pi_\Gamma(g) \in E$ *there is no vector* v *in the lattice spanned by the rows of* g *with* $\|v\|_\infty < \epsilon$.

12.8. Proof of Proposition 12.5. We prove only that $A^+.x_{\alpha,\beta}$ unbounded implies that (α,β) satisfies (12.2a); the remaining assertions of this proposition follow similarly and are left as an exercise to the reader.

Let $\epsilon \in (0, 1/2)$ be arbitrary and write

$$g_{\alpha,\beta} = \begin{pmatrix} 1 & \alpha & \beta \\ 0 & 1 & 0 \\ 0 & 0 & 1 \end{pmatrix}.$$

(37)It is worthwhile to note that this remarkable paper appeared in 1955, many years before Conjecture 11.3 was made, and even before 1967 when Furstenberg made his related discoveries about scarcity of invariant sets and measures for the maps $x \mapsto 2x \bmod 1$ and $x \mapsto 3x \bmod 1$ on $\mathbb{R}/\mathbb{Z}$! The same paper also implicitly discusses the connection between Oppenheim's conjecture and the action of $\mathrm{SO}(2,1)$ on X_3.

(38)I.e. $\overline{A^+.x_{\alpha,\beta}}$ is not compact.

By Mahler's compactness criterion (see §12.7), if $A^+.x_{\alpha,\beta}$ is unbounded, there is a $a \in A^+$ such that in the lattice generated by the rows of $g_{\alpha,\beta}a^{-1}$ there is a nonzero vector v with $\|v\|_\infty < \epsilon$. This vector v is of the form

$$v = (ne^{-s-t}, (n\alpha - m)e^s, (n\beta - k)e^t)$$

where n, m, k are integers at least one of which is nonzero, and $s, t \geq 0$. Since $\|v\|_\infty < 1/2$, $n \neq 0$ and $\|n\alpha\| = (n\alpha - m)$, $\|n\beta\| = (n\beta - k)$. Without loss of generality $n > 0$ and

$$n \|n\alpha\| \|n\beta\| \leq \|v\|_\infty^3 < \epsilon^3.$$

As ϵ was arbitrary, (12.2a) follows. □

12.9. We now turn to answering the following question: With the partial information given in Theorem 11.5, what information, if any, do we get regarding Littlewood's conjecture?

12.10. Theorem ([**EKL06**, Theorem 1.5]). *For any $\delta > 0$, the set*

$$\Xi_\delta = \left\{ (\alpha, \beta) \in [0,1]^2 : \varliminf_{n\to\infty} n \|n\alpha\| \|n\beta\| \geq \delta \right\}$$

has zero upper box dimension(39)(40).

12.11. We present a variant of the proof of this theorem given in [**EKL06**]. The first step of the proof, which is where Theorem 11.5 is used, is an explicit sufficient criterion for a single point α, β to satisfy Littlewood's conjecture (§12.2). This criterion is based on the notion of topological entropy; see §3.18 for the definition and basic properties of this entropy.

Let $a_{\sigma,\tau}(t) = a(\sigma t, \tau t)$, with $a(s,t)$ as in §12.5.

12.12. Proposition. *Suppose that $(\alpha, \beta) \in \mathbb{R}^2$ does not satisfy* (12.2a) *or equivalently that $A^+.x_0$ is bounded. Then for any $\sigma, \tau \geq 0$, the topological entropy of $a_{\sigma,\tau}$ acting on the compact set*

$$\overline{\{a_{\sigma,\tau}(t).x_{\alpha,\beta} : t \in \mathbb{R}^+\}}$$

vanishes.

12.13. Proof. Let x_0 be as in the proposition such that $A^+.x_0$ is bounded. If the topological entropy were positive, then by the variational principal in §3.21, there is an $a_{\sigma,\tau}$-invariant measure μ supported on $\overline{\{a_{\sigma,\tau}(t).x_0 : t \in \mathbb{R}^+\}}$ with $h_\mu(a_{\sigma,\tau}) > 0$.

Define for any $S > 0$

$$\mu_S = \frac{1}{S^2} \iint_0^S a(s,t)_* \mu \, ds \, dt,$$

with $a(s,t)_*\mu$ denoting the push forward of μ under the map $x \mapsto a(s,t).x$. Since $a(s,t)$ commutes with the one parameter subgroup $a_{\sigma,\tau}$, for any $a_{\sigma,\tau}$-invariant measure μ' the entropies satisfy

$$h_{\mu'}(a_{\sigma,\tau}) = h_{a(s,t)_*\mu'}(a_{\sigma,\tau}).$$

(39) I.e., for every $\epsilon > 0$, for every $0 < r < 1$, one can cover Ξ_δ by $O_{\delta,\epsilon}(r^{-\epsilon})$ boxes of size $r \times r$.

(40) Since (12.2a) depends only on α, β mod 1 it is sufficient to consider only $(\alpha, \beta) \in [0,1]^2$.

If μ has the ergodic decomposition $\int \mu_\xi \, d\nu(\xi)$, the measure μ_S has ergodic decomposition $S^{-2} \iint_0^S \int a(s,t)_* \mu_\xi \, d\nu(\xi) \, ds \, dt$ and so by §3.5, for every S

$$h_{\mu_S}(a_{\sigma,\tau}) = h_\mu(a_{\sigma,\tau}).$$

All μ_S are supported on the compact set $\overline{A^+.x_0}$, and therefore there is a subsequence converging weak* to some compactly supported probability measure μ_∞, which will be invariant under the full group A. By semicontinuity of entropy (§3.15),

$$h_{\mu_\infty}(a_{\sigma,\tau}) \geq h_\mu(a_{\sigma,\tau}) > 0,$$

hence by Theorem 11.5 the measure μ_∞ is not compactly supported[(41)] — a contradiction. □

12.14. Fix $\sigma, \tau \geq 0$. For $\alpha, \beta \in \mathbb{R}$ we define $X_{\alpha,\beta} = \overline{\{a_{\sigma,\tau}(t).x : t \in \mathbb{R}^+\}}$. Proposition 12.12 naturally leads us to the question of the size of the set of $(\alpha, \beta) \in [0,1]^2$ for which $h_{\text{top}}(X_{\alpha,\beta}, a_{\sigma,\tau}) = 0$. This can be answered using the following general observation:

12.15. Proposition. *Let X' be a metric space equipped with a continuous $\mathbb{R}$-action $(t,x) \mapsto a_t.x$. Let X'_0 be a compact a_t-invariant[(42)] subset of X' such that for any $x \in X'_0$,*

$$h_{\text{top}}(Y_x, a_t) = 0 \qquad Y_x = \overline{\{a_t.x : t \in \mathbb{R}^+\}}.$$

Then $h_{\text{top}}(X'_0, a_t) = 0$.

12.16. Proof. Assume for contradiction that $h_{\text{top}}(X'_0, a_t) > 0$. By the variational principle (§3.21), there is some a_t-invariant and ergodic measure μ on X'_0 with $h_\mu(a_t) > 0$.

By the pointwise ergodic theorem, for μ-almost every $x \in X'_0$ the measure μ is supported on Y_x. Applying the variational principle again (this time in the opposite direction) we get that

$$0 = h_{\text{top}}(Y_x, a_t) \geq h_\mu(a_t) > 0$$

a contradiction. □

12.17. Corollary. *Consider, for any compact $C \subset X_3$ the set*

$$X_C = \left\{x \in X_3 : A^+.x \subset C\right\}.$$

Then for any $\sigma, \tau \geq 0$, it holds that $h_{\text{top}}(X_C, a_{\sigma,\tau}) = 0$.

12.18. Proof. By Proposition 12.12, for any $x \in X_C$ the topological entropy of $a_{\sigma,\tau}$ acting on $\overline{\{a_{\sigma,\tau}(t).x : t \in \mathbb{R}^+\}}$ is zero. The corollary now follows from Proposition 12.15. □

12.19. We are now in position to prove Theorem 12.10, or more precisely to deduce the theorem from Theorem 11.5:

---

(41) Notice that a priori there is no reason to believe μ_∞ will be A-ergodic, while Theorem 11.5 deals with A-ergodic measures. So an implicit exercise to the reader is to understand why we can still deduce from $h_{\mu_\infty}(a_{\sigma,\tau}) > 0$ that μ_∞ is not compactly supported.

(42) Technical point: we only use that $a_t.X' \subset X'$ for $t \geq 0$. The variational principle (§3.21) is still applicable in this case.

12.20. PROOF OF THEOREM 12.10. To show that Ξ_δ has upper box dimension zero, we need to show, for any $\epsilon > 0$ and for any $r \in (0,1)$ that the set Ξ_δ can be covered by $O_\epsilon(r^{-\epsilon})$ boxes of side r, or equivalently that any r-seperated set (i.e., any set S such that for any $x, y \in S$ we have $\|x-y\|_\infty > r$) is of size $O_{\delta,\epsilon}(r^{-\epsilon})$.

Let C_δ be as in Proposition 12.5. Let d denote a left invariant Riemannian metric on $G = \mathrm{SL}(3,\mathbb{R})$. Then d induces a metric, also denote by d on X_3. For $a, b \in \mathbb{R}$ let

$$g_{a,b} = \begin{pmatrix} 1 & a & b \\ 0 & 1 & 0 \\ 0 & 0 & 1 \end{pmatrix}.$$

Since C_δ is compact and d is induced from a left invariant Riemannian metric (so that there is an injectivity radius on C_δ), there will be r_0, c_0 such that for any $x \in C_\delta$ and $|a|, |b| < r_0$

$$d(x, g_{a,b}.x) \geq c_0 \max(|a|, |b|).$$

For any $\alpha, \alpha', \beta, \beta' \in \mathbb{R}$ we have that

$$x_{\alpha,\beta} = g_{\alpha'-\alpha,\beta'-\beta}.x_{\alpha',\beta'}$$

and more generally for any n

$$a_{1,1}^n.x_{\alpha,\beta} = g_{e^{3n}(\alpha'-\alpha),e^{3n}(\beta'-\beta)}.a_{1,1}^n.x_{\alpha',\beta'}.$$

It follows that if $S \subset \Xi_\delta$ is r-separated for $r = e^{-3n} r_0 \in (0, r_0)$ then

$$S' = \{x_{\alpha,\beta} : (\alpha, \beta) \in S\}$$

is $(n, c_0 r_0)$-separated for $a_{1,1}$ in the sense of §3.18. By definition of C_δ and Ξ_δ, we have that (in the notations of §12.17) the set $S' \subset X_{C_\delta}$, a set which has zero topological entropy with respect to the group $a_{1,1}$. It follows that the cardinality of a maximal $(n, c_0 r_0)$-separated set in S' is at most $O_{\delta,\epsilon}(\exp(\epsilon n))$; hence for $r < r_0$ the cardinality of a maximal r-separated subset of Ξ_δ is $O_{\delta,\epsilon}(r^{-\epsilon})$. □

13. Application to Arithmetic Quantum Unique Ergodicity

13.1. We begin by recalling some basic facts about harmonic analysis on $\Gamma\backslash\mathbb{H}$. Here, $\mathbb{H} := \{x + iy : y > 0\}$ is the upper-half plane model of the hyperbolic plane (for more details, see [**Lan75**]). It is isomorphic to $G/K = \mathrm{SL}(2,\mathbb{R})/SO(2,\mathbb{R})$, and carries the Riemannian G-invariant metric $ds^2 = \frac{dx^2+dy^2}{y^2}$, where the action of G is given by fractional linear transformations in the usual way. This metric gives us the invariant area form $d\,\mathrm{area} = \frac{dxdy}{y}$.

We have the Laplacian operator

$$\Delta = y^2[\partial_x^2 + \partial_y^2]$$

which is also invariant under G. We wish to study $L^2(\Gamma\backslash\mathbb{H}, \text{area}) \cong L^2(\Gamma\backslash G, \mu_{\text{Haar}})_K$ for Γ a lattice in $G = \mathrm{SL}(2,\mathbb{R})$. Here $L^2(\Gamma\backslash G, \mu_{\text{Haar}})_K$ denotes the space of K-invariant L^2-functions on $\Gamma\backslash G$.

13.2. For any $\phi \in C_c(G)$, we can write a convolution operator

$$\phi * f := \int_G \phi(g) f(xg^{-1}) dg$$

but we will restrict ourselves to K-bi-invariant functions $\phi \in C_c(K\backslash G/K)$. In this case, we have the nice property that

$$\phi * \psi * f = (\phi * \psi) * f = (\psi * \phi) * f = \psi * \phi * f$$

and so these operators form a large commutative algebra which is called the **Hecke ring**. Since the Laplacian can be written as a limit of such convolution operators, it commutes with this algebra as well.

13.3. We begin with Γ cocompact (harmonic analysis is much easier in this case). The mapping $A_\phi : f \mapsto \phi * f$ is a compact, normal operator (this is false if Γ is not cocompact!). Therefore $L^2(M)$ is spanned by an orthonormal set of eigenfunctions of A_ϕ. Since Δ and A_ϕ commute, we can find an orthonormal basis of joint eigenfunctions.

In fact, if f is an eigenfunction of Δ, say $\Delta f = \lambda f$, it will automatically be an eigenfunction of the convolution operator A_ϕ, and the corresponding eigenvalue is by definition equal to the spherical transform of ϕ.

Weyl's Law (true for general compact surfaces) gives the asymptotic number of eigenvalues of Δ:

$$\#\{\text{eigenvalues of } \Delta \leq T\} \sim \pi \cdot \text{area}(M) \cdot T$$

13.4. In many cases we will be interested in Γ NOT cocompact; eg. $\Gamma_1 = \mathrm{SL}(2,\mathbb{Z})$ or one of the "principle congruence subgroups" $\Gamma_N = \{\gamma \equiv I \mod N\}$. For simplicity, let us assume for the moment that M has one cusp (at ∞).

We have some explicit eigenfunctions of Δ on $\mathbb{H}$. For example,

$$\Delta(y^s) = s(s-1)y^s = -(1/4 + t^2)y^s$$

where we make the convenient substitution $s = 1/2 + it$. These eigenfunctions correspond to planar waves going up; note that they are not Γ-invariant. Closely related are the Eisenstein series, which are Γ-invariant eigenfunctions of the Laplacian, not in $L^2(\Gamma\backslash\mathbb{H})$, satisfying

$$E_{1/2+it}(z) = y^{1/2+it} + \theta(1/2+it)y^{1/2-it} + \text{ (rapidly decaying terms)}$$

We have the following spaces [**CS80**, Ch. 6-7]:

- $L^2_{\text{Eisenstein}}$ spanned by the Eisenstein series, and constituting the continuous part of the spectrum.
- $L^2_{\text{constants}}$ of constant functions.
- L^2_{cusp} of **cusp forms**, the orthogonal complement of the others. This consists of the functions f on $\Gamma\backslash\mathbb{H}$ whose integral along all periodic horocycles vanishes, i.e., (identifying the functions on $\Gamma\backslash\mathbb{H}$ with Γ-invariant functions on $\mathbb{H}$) functions f so that $\int_0^1 f(x+iy)dx = 0$ for all $y > 0$.

13.5. Selberg [**Sel56**] proved that if Γ is a **congruence subgroup**, i.e., $\Gamma_N < \Gamma < \Gamma_1$ for some N, then Weyl's Law holds for *cusp* forms

$$\#\{\text{eigenvalues of cusp forms } \leq T\} \sim \pi \cdot \text{area}(M) \cdot T$$

This is very far from the generic picture, where Phillips and Sarnak have conjectured that L^2_{cusp} is *finite dimensional* for generic Γ. While this remains to present an open question, significant results in this direction have been obtained by them [**PS92**] and Wolpert [**Wol94**].

Why is the case of congruence lattices so special? They carry a lot of extra symmetry, which makes it a lot easier for cusp forms to arise. We will now discuss these symmetries.

13.6. HECKE CORRESPODENCE.

We will define the Hecke correspondence (for a given prime p) in several equivalent ways. For simplicity we work with $\Gamma = \Gamma_1$.

13.7. First, associate to $z \in \mathbb{H}$ the $p+1$ points in $\Gamma\backslash\mathbb{H}$

$$T_p(z) := \Gamma\backslash \left\{ pz, \frac{z}{p}, \frac{z+1}{p}, \dots, \frac{z+p-1}{p} \right\}$$

each of these points is a fractional linear image of z, so each branch of this mapping is an isometry. One needs to check that this passes to the quotient by Γ; i.e., that if $z = \gamma z'$ then $T_p(z) = T_p(z')$. Since Γ is generated by the maps $z \mapsto z+1$ and $z \mapsto -\frac{1}{z}$, and $T_p(z)$ is obviously invariant under the former, it remains to check that $T_p(z) = T_p(-\frac{1}{z})$, which is left to the reader.

13.8. We will now give an alternate way to define T_p. It will be more convenient for us to work in $\text{PGL}(2,\mathbb{R})$ instead of $\text{SL}(2,\mathbb{R})$, and so we take $\Gamma = \text{PGL}(2,\mathbb{Z})$ [this is not quite $\text{SL}(2,\mathbb{Z})$ because matrices with determinant -1 are allowed on the one hand, but on the other hand $\begin{pmatrix} 1 & \\ & 1 \end{pmatrix}$ and $\begin{pmatrix} -1 & \\ & -1 \end{pmatrix}$ which were distinct elements of $\text{SL}(2,\mathbb{R})$ are identified in $\text{PGL}(2,\mathbb{Z})$, but for our purposes this difference is very minor]. The matrix $\gamma_p = \begin{pmatrix} p & \\ & 1 \end{pmatrix} \in \mathit{comm}(\Gamma)$, where $\mathit{comm}(\Gamma)$ denotes the **commensurator** of Γ — the set of $\gamma \in G$ such that $[\Gamma : \gamma\Gamma\gamma^{-1} \cap \Gamma] < \infty$.

Note that

$$\Gamma\gamma_p\Gamma = \Gamma \begin{pmatrix} p & \\ & 1 \end{pmatrix} \sqcup \bigsqcup_{i=0}^{p-1} \Gamma \begin{pmatrix} 1 & i \\ 0 & p \end{pmatrix}$$

The mapping $T_p : \Gamma g \mapsto \Gamma\gamma_p\Gamma g$ gives the same correspondence as above.

Because we defined this correspondence by left multiplication, we can still quotient by K on the right, to get a correspondence on $\Gamma\backslash\mathbb{H}$.

13.9. Here is a third way to define the same correspondence. Since we can identify $X_2 = \text{PGL}(2,\mathbb{Z})\backslash\text{PGL}(2,\mathbb{R})$ with the space of lattices in $\mathbb{R}^2$ (up to homothety), we can define for $x \in X_2$ the set $T_p(x)$ to be the set of all lattices $y \in X_2$ homothetic to a sublattice of x of index p; or equivalently as the set of all $y \in X_2$ which contain a lattice homothetic to x as a sublattice of index p.

One should check that this agrees with the previous definitions (in particular, that $T_p(x)$ consists of $p+1$ points, which is not obvious from this defintion).

13.10. Lastly, we consider $\mathrm{PGL}(2,\mathbb{Z}[\frac{1}{p}])\backslash\mathrm{PGL}(2,\mathbb{R})\times\mathrm{PGL}(2,\mathbb{Q}_p)$, the space of $\mathbb{Z}[\frac{1}{p}]$-modules that are lattices in $\mathbb{R}^2\times\mathbb{Q}_p^2$, again up to homothety. By this we mean that an element of this space looks like $\mathbb{Z}[\frac{1}{p}](v_1,w_1)\oplus\mathbb{Z}[\frac{1}{p}](v_2,w_2)$ where $\{v_1,v_2\}$ is an $\mathbb{R}$-basis for $\mathbb{R}^2$, and (w_1,w_2) is a $\mathbb{Q}_p$-basis for $\mathbb{Q}_p^2$; and the two points $\mathbb{Z}[\frac{1}{p}](v_1,w_1)\oplus\mathbb{Z}[\frac{1}{p}](v_2,w_2)$ and $\mathbb{Z}[\frac{1}{p}](\lambda v_1,\theta w_1)\oplus\mathbb{Z}[\frac{1}{p}](\lambda v_2,\theta w_2)$ are identified for any $\lambda\in\mathbb{R}$ and $\theta\in\mathbb{Q}_p$.

Let $\pi:\mathbb{R}^2\times\mathbb{Q}_p^2\to\mathbb{R}^2$ be the natural projection, and consider the map $\pi_1:x\mapsto\pi(x\cap\mathbb{R}^2\times\mathbb{Z}_p^2)$ for x a lattice as above. Then $\pi_1(x)$ is a lattice in $\mathbb{R}^2$ and it respects equivalence up to homothety. Moreover, for every lattice $y\in X_2$, the inverse image $\pi_1^{-1}(y)=\mathrm{PGL}(2,\mathbb{Z}_p).x$ for some x. We've shown that

$$\mathrm{PGL}(2,\mathbb{Z}[\frac{1}{p}])\backslash\mathrm{PGL}(2,\mathbb{R})\times\mathrm{PGL}(2,\mathbb{Q}_p)/\mathrm{PGL}(2,\mathbb{Z}_p)\cong X_2$$

Using (a p-adic version of) the KAK-decomposition, we can write any $g_p\in\mathrm{PGL}(2,\mathbb{Q}_p)$ as $k_1\begin{pmatrix}p^n & \\ & 1\end{pmatrix}k_2$ for some $k_1,k_2\in K$ and some integer n. Then the map $x\mapsto\pi_1(g_p\cdot\pi_1^{-1}(x))$ yields a finite collection of points: x if $n=0$, the set $T_p(x)$ if $n=1$, and a finite set which we will denote by $T_{p^k}(x)$ if $n=k>1$. This gives our fourth equivalent definition of the Hecke correspondence.

13.11. The Hecke correspondence allows us to define an operator, also denoted by T_p, on $L^2(\Gamma\backslash G)$ (resp. on $L^2(\Gamma\backslash\mathbb{H})$) by

$$T_pf(x)=\frac{1}{\sqrt{p}}\sum_{y\in T_p(x)}f(y)$$

As a side remark, we note that for the Eisenstein series, the eigenvalues of T_p can be computed explicitly, and we have

$$\begin{aligned}T_pE_{1/2+it}&=\cos(t\log p)E_{1/2+it}\\&=(p^{\sqrt{\Delta+1/4}}+p^{-\sqrt{\Delta+1/4}})E_{1/2+it}\end{aligned}$$

The operator $(p^{\sqrt{\Delta+1/4}}+p^{-\sqrt{\Delta+1/4}})$ is essentially the propagating operator of the wave equation.

This property equating two operators which are defined by completely different means (eg., one by global symmetries and one by local differential structure) should be quite rare. This is one indication that $L^2_{\text{Eisenstein}}$ should be very small in the arithmetic situations, and hence L^2_{cusp} should contain the vast majority of the eigenfunctions. This idea can be used to give an alternative elementary proof of the existence of cusp forms [**LV07**].

One should also note that there are compact surfaces $\Gamma\backslash\mathbb{H}$ with Hecke symmetries; one way to construct such lattices Γ is via quaternion algebras (see e.g. [**Mor**, Ch. 7]), for example

$$\Gamma=\left\{\begin{pmatrix}x+\sqrt{2}y & z+\sqrt{2}w\\5(z-\sqrt{2}w) & x-\sqrt{2}y\end{pmatrix}:x,y,z,w\in\mathbb{Z},\right.$$
$$\left.x^2-2y^2-5t^2+10w^2=1\right\}.$$

13.12. We now discuss the quantum unique ergodicity conjecture, in particular in the arithmetic case. We begin with a general compact Riemannian manifold M, on which we have the Laplacian Δ_M, and we wish to understand the distribution properties of eigenfunctions of Δ_M.

According to Schroedinger, the motion of a free (spinless, non-relativistic) quantum particle flowing in the absence of external forces on M is given by the equation

$$i\frac{\partial\psi}{\partial t} = \Delta_M\psi$$

This defines a unitary evolution, i.e., the norm $||\psi(\cdot,t)||_{L^2}$ is independent of t. We will always take $||\psi||_{L^2} = 1$.

The Born interpretation of the "wave function" ψ is that the function $|\psi|^2 d(\mathrm{vol})$ defines a probability measure on M, representing the average position of a particle in the state ψ; i.e., for any (measurable) region $A \subset M$, the probability of finding our particle in A at time t is given by $\int_A |\psi(x,t)|^2 d\,\mathrm{vol}(x)$, where $d\,\mathrm{vol}$ is the Riemannian volume on M. Note that if ψ is an eigenfunction of Δ_M, then the time dependence of ψ only appears as a phase; i.e., $\psi(x,t) = e^{-i\lambda t}\psi(x,0)$. Hence eigenfunctions give rise to steady states, or invariant quantum distributions, $d\tilde{\mu}_\psi = |\psi|^2 d\,\mathrm{vol}$.

Let $\pi : S^*M \to M$ be the canonical projection. One can (see below) lift these $\tilde{\mu}_\psi$ to measures μ_ψ on the unit cotangent bundle S^*M which satisfy:

(i) $\left|\int \tilde{f}\,d\pi_*\mu_\psi - \int \tilde{f}\,d\tilde{\mu}_\psi\right| < \lambda^{-0.1}$ for any $\tilde{f} \in C^\infty(M)$

(ii) $\left|\int Hf\,d\mu_\psi\right| < \lambda^{-0.1}$ for any $f \in C^\infty(S^*M)$, where H is differentiation along the geodesic flow.

Suppose now that $\{\psi_i\}$ is a sequence of (normalized) eigenfunctions whose eigenvalues $\lambda_i \to \infty$, denote by $\tilde{\mu}_i = \tilde{\mu}_{\psi_i}$ the corresponding measures, and let μ_i be the corresponding lifts. The above conditions guarantee that any weak* limit point μ_∞ of the μ_i will satisfy

- $\pi_*\mu_\infty = \tilde{\mu}_\infty$ (the weak* limit of the corresponding $\tilde{\mu}_i$).
- $\int Hf\,d\mu_\infty = 0$, i.e., $\frac{\partial}{\partial t}\int f(g_t.x)\,d\mu_\infty = 0$. This means that μ_∞ is g_t invariant.

We call the μ's "microlocal lifts" of the $\tilde{\mu}$'s.

13.13. Definition. *Any weak* limit μ_∞ of $\{\mu_i\}$ as above is called a* **quantum limit**.

13.14. Here we will be interested in the special case of $M = \Gamma\backslash\mathbb{H}$ for Γ an arithmetic lattice; e.g. Γ a congruence subgroup of $\mathrm{SL}(2,\mathbb{Z})$, or one of the arithmetic compact quotients mentioned earlier. These manifolds carry the extra symmetry of the Hecke operators, and since all of these operators commute, we can find a basis of L^2 (or L^2_{cusp} in the non-compact case) consisting of joint eigenfunctions of Δ_M and all of the T_p, such joint eigenfunctions are called **Hecke-Maass forms**. Any weak* limit of μ_{ψ_i}, where all of the ψ_i are Hecke-Maass forms, is called an **arithmetic quantum limit**.

13.15. For now, we assume that M is compact. Snirlman, Colin de Verdiere, and Zelditch have shown that if $\{\psi_i\}_{i=1}^\infty$ is a full set of (normalized) eigenfunctions ordered by eigenvalue, then the average $\frac{1}{N}\sum_{i=1}^N \mu_{\psi_i}$ converges to the Liouville measure on S^*M. If we assume that the geodesic flow on M is ergodic with respect to Liouville measure (satisfied e.g. if M has negative sectional curvature), then outside a set E of indices of density zero (i.e., $\lim_{N\to\infty} \frac{1}{N}\#\{i \in E : i < N\} = 0$), the sequence $\{\mu_i\}_{i\notin E}$ converges to Liouville measure; this is because an ergodic measure cannot be written as a proper convex combination of other invariant measures.

13.16. Conjecture (Rudnick-Sarnak [**RS94**]). *Let M be a compact, Riemannian manifold of negative sectional curvature. Then the Liouville measure on S^*M is the unique quantum limit.*

13.17. Theorem ([**Lin06**, Theorem 1.4]). *Say $M = \Gamma\backslash\mathbb{H}$, for Γ arithmetic (of finite covolume, but not necessarily co-compact). Then the only arithmetic quantum limits are scalar multiples of Liouville measure (i.e., the measure is a Haar measure).*

13.18. Corollary. *Let $f \in C_c(M)$ be such that $\int_M f = 0$. Then for a sequence $\{\psi_i\}$ of Hecke-Maass forms, we have*

$$\int_M f|\psi_i|^2 d\,\mathrm{area}(x) \to 0$$

as $i \to \infty$.

Note that in Theorem 13.17 we do not know that the limit measure is a probability measure. If Γ is cocompact, then this is immediate; but in the case of Γ a congruence subgroup of $\mathrm{SL}(2,\mathbb{Z})$, there remains the possibility that some (or possibly all) of the mass escapes to the cusp in the limit. We note that since the summer school this problem has been solved: Soundararajan [**Sou09**] proved, by purely number theoretic methods, that the arithmetic quantum limits are probability measures. Together this proves the arithmetic quantum unique ergodicity conjecture.

13.19. For example, we could take f in Corollary 13.18 to also be a Hecke-Maass form (recall these are orthogonal to constants, so the hypothesis is satisfied). In fact, since these span L^2_{cusp}, the statement of Corollary 13.18 will hold for all such f if and only if it holds for all Hecke-Maass forms.

An identity of Watson shows that the quantity $\int \psi_1\psi_2\psi_3 d\,\mathrm{area}$ can be expressed in terms of L-functions, specifically

$$\left|\int \psi_1\psi_2\psi_3 d\,\mathrm{area}\right|^2 = \frac{\pi^4 \Lambda(\frac{1}{2}, \psi_1 \times \psi_2 \times \psi_3)}{\Lambda(1, \mathrm{sym}^2\psi_1)\Lambda(1, \mathrm{sym}^2\psi_2)\Lambda(1, \mathrm{sym}^2\psi_3)}.$$

Hence good estimates on the completed L-function $\Lambda(\frac{1}{2}, \psi_1 \times \psi_2 \times \psi_3)$ would imply Arithmetic QUE. Unfortunately, the best estimates we have for this L-functions gives only a trivial bound, and so Theorem 13.17 does not follow from existing technology in this direction. The Generalized Riemann Hypothesis (GRH) would imply a bound of $\lesssim \lambda_i^{-1/4}$, which would not only imply Theorem 13.17, but would give an optimal rate of convergence. Further discussion of many of these topics can be found in the survey [**Sar03**].

13.20. We also note that Theorem 13.17 has been extended to other $\Gamma\backslash G$ by Silberman and Venkatesh; e.g. for $G = \mathrm{SL}(p,\mathbb{R})$ and Γ a congruence lattice therein [**SV04, SV06**].

13.21. As a first step, we wish to construct the measures μ_i from the $\tilde{\mu}_i$, which satisfy the conditions of 13.12. The "standard" way to do this is via pseudodifferential calculus (see e.g. [**Ana**]), but we wish to give a representation-theoretic construction, which will respect the Hecke symmetries that we wish to exploit.

Given an eigenfunction $\phi \in L^2(\Gamma\backslash\mathbb{H}) = L^2(\Gamma\backslash \mathrm{SL}(2,\mathbb{R}))_K$, we can translate the function via $g.\phi(x) = \phi(xg^{-1})$ for any $g \in \mathrm{SL}(2,\mathbb{R})$. Taking all possible translates, we get a representation

$$V_\phi = \overline{\langle g.\phi : g \in G\rangle} = \overline{\langle \psi * \phi : \psi \in C_c(G)\rangle}$$

where the action of ψ is by convolution as in §13.2. This representation is unitary (since the Riemannian measure is G-invariant), and also irreducible (this is not quite obvious, but follows from the general theory). Moreover, the isomorphism class of this representation is completely determined by the eigenvalue of ϕ.

In fact, we can write down an explicit model $\tilde{V}_t \cong V_\phi$ for this representation (where t is determined by $\Delta_M\phi = (\frac{1}{4}+t^2)\phi)$. The Hilbert space on which the representation acts will be simply $L^2(K)$. To understand the action of $G = \mathrm{SL}(2,\mathbb{R})$, extend any function f on K to a function $\tilde{f}$ on G using the NAK decomposition of $\mathrm{SL}(2,\mathbb{R})$,

$$g = n\hat{a}k = \begin{pmatrix} 1 & s \\ & 1 \end{pmatrix} \begin{pmatrix} a & \\ & a^{-1} \end{pmatrix} \begin{pmatrix} \cos\theta & \sin\theta \\ -\sin\theta & \cos\theta \end{pmatrix}$$

via

$$\tilde{f}(n\hat{a}k) = a^{1+2it} f(k)$$

now define $g.f$ as the restriction of the left translation of $\tilde{f}$ by g to K. It can be shown by an explicit calculation that this representation is unitary as long as $t \in \mathbb{R}$, i.e., as long as the eigenvalue of ϕ under the Laplacian is $\geq 1/4$ (for our purposes, this is all we care about).

For every $n \geq 0$ we may choose the vector $\Phi^{(n)} \in V_\phi$ that corresponds to the (normalized) Dirichlet kernel

$$f(k(\theta)) = \frac{1}{\sqrt{2n+1}} \frac{\sin{(n+\frac{1}{2})\theta}}{\sin\frac{\theta}{2}}$$

which is the n-th order approximation to the δ-function on K. We then set $\mu_i = |\Phi_i^{(n)}|^2 d\mathrm{vol}$ (here $d\,\mathrm{vol}$ is the Riemannian volume on $\Gamma\backslash G$ and n will be chosen as a function of i below), and we will see that the μ_i are close to being invariant under the geodesic flow.

What is the role of n in all of this? There are two competing properties:

- The larger the value of n, the closer f is to a δ function, and the better the invariance properties of μ_i. The problem is that then μ_i loses much of its relation to $\tilde{\mu}_i$ (i.e., $\pi_*\mu_i$ and $\tilde{\mu}_i$ become farther apart).
- The smaller the value of n, the closer f is to a constant function, which means that μ_i agrees well with $\tilde{\mu}_i$, but μ_i loses its invariance properties.

However, as $i \to \infty$ (and simultaneously also $t \to \infty$), both approximations improve; hence if we "split the difference" by choosing an appropriate value of n for each t, we will get both desired properties in the limit.

13.22. We now wish to explain why large n values make μ_i more invariant. We define the following differential operators:

$$\begin{aligned} Hf &= \frac{\partial}{\partial s} f\left(g\begin{pmatrix} e^s & \\ & e^{-s}\end{pmatrix}\right) \\ Vf &= \frac{\partial}{\partial s} f\left(g\begin{pmatrix} \cosh(s) & \sinh(s) \\ \sinh(s) & \cosh(s)\end{pmatrix}\right) \\ Wf &= \frac{\partial}{\partial \theta} f(gk(\theta)) \end{aligned}$$

H is differentiation along the geodesic flow, V is differentiation along the perpendicular direction to the geodesic flow, and W is differentiation in the rotational direction (i.e., fixing the point in M and letting the direction of the tangent vector vary).

We have the **Casimir** element

$$\omega = H^2 + V^2 - W^2$$

which is self-adjoint, commutes with translations by any $g \in G$, and coincides (up to a scalar) with Δ_M on the subspace $\{f : Wf = 0\} = L^2(\Gamma\backslash G)_K$. Every vector $\psi \in V_\phi$ is an eigenfunction of ω, with eigenvalue $1 + 4t^2$.

For t large, consider

$$\begin{aligned} \langle \omega\Phi, \Phi f\rangle &= \langle \Phi, \omega(\Phi f)\rangle \\ &= \langle \Phi, (\omega\Phi) f\rangle + \langle \Phi, \Phi(\omega f)\rangle + \\ &\qquad + 2\langle \Phi, H\Phi H f\rangle + 2\langle \Phi, V\Phi V f\rangle + 2\langle \Phi, W\Phi W f\rangle, \end{aligned}$$

which follows from ω being self-adjoint and from the product rule for differentiation. Note that the first terms in both lines are equal:

$$\langle \omega\Phi, \Phi f\rangle = \langle \Phi, (\omega\Phi) f\rangle = (1 + 4t^2)\langle \Phi, \Phi f\rangle,$$

and also that for fixed f, the quantity $\langle \Phi, \Phi(\omega f)\rangle = O_f(1)$ as the eigenvalue t tends to infinity. On the other hand, if n is large (but much smaller than t), Φ is close to being an eigenfunction of H of large eigenvalue ($\sim it$), and both $||V\Phi||$ and $||W\Phi||$ are much less than $t||\Phi||$. Dividing by the "eigenvalue for H" we must have $\langle \Phi, \Phi H f\rangle = o(1)$.

But what is $\langle \Phi, \Phi H f\rangle$? By definition, it is the integral $\int Hf d\mu_i$ of the derivative of f along the geodesic flow. Since this tends to 0 as t gets large, we get

$$\frac{\partial}{\partial t}\int f(\cdot a_t) d\mu_\infty = 0$$

if μ_∞ is a weak* limit point of the μ_i; i.e., we have that μ_∞ is a_t-invariant.

13.23. We have shown that any weak* limit point of the measures $\tilde{\mu}_i$ is a projection of a measure μ_∞ on $\Gamma\backslash \mathrm{SL}(2,\mathbb{R})$ which is a_t-invariant. But as we know well, there are many a_t-invariant measures here, even with positive entropy!

Thus in order to classify quantum limits, we will have to use additional information about these limits. At this stage, we will abandon the properties coming from the ϕ_i being eigenfunctions of Δ (though we have not harnessed the full power of this assumption), and use properties of Hecke eigenfunctions.

The fact that the ϕ_i are Hecke eigenfunctions implies (since the Hecke operators are defined by translations) that Φ_i (indeed, any vector in V_{ϕ_i}) is a Hecke eigenfunction. Now, one certainly cannot expect $|\Phi_i|^2$ to be a Hecke eigenfunction, but traces of this symmetry do survive in the measures μ_i as well as their limit μ_∞.

13.24. Recall the Hecke correspondence (fourth formulation) given via the projection map

$$\mathrm{PGL}(2,\mathbb{Z}\left[\frac{1}{p}\right])\backslash \mathrm{PGL}(2,\mathbb{R}) \times \mathrm{PGL}(2,\mathbb{Q}_p) \to \mathrm{PGL}(2,\mathbb{Z})\backslash \mathrm{PGL}(2,\mathbb{R}).$$

For each x we have a set of points $T_p(x)$, and its iterates, giving a Hecke tree which is the projection of a full $\mathrm{PGL}(2,\mathbb{Q}_p)$-orbit of x.

13.25. Definition. *A measure μ on $\mathrm{PGL}(2,\mathbb{Z})\backslash \mathrm{PGL}(2,\mathbb{R})$ is p-Hecke recurrent if there is a measure $\tilde{\mu}$ on*

$$\mathrm{PGL}(2,\mathbb{Z}\left[\frac{1}{p}\right])\backslash \mathrm{PGL}(2,\mathbb{R}) \times \mathrm{PGL}(2,\mathbb{Q}_p)$$

such that $\pi_*\tilde{\mu} = \mu$ and $\tilde{\mu}$ is $\mathrm{PGL}(2,\mathbb{Q}_p)$-recurrent.

13.26. PROBLEM. Show that the property of p-Hecke recurrence is independent of the lifting; i.e., μ is p-Hecke recurrent if and only if *any* lifting measure $\tilde{\mu}$ is $\mathrm{PGL}(2,\mathbb{Q}_p)$-recurrent.

13.27. Let $\mathcal{G}$ be an abstract $p+1$-regular tree, with a distinguished base point. For a more direct definition of p-Hecke recurrence, we can define leafwise measures $\mu_x^{\mathcal{G}}$ on these Hecke trees (our space is foliated into Hecke orbits), and then as before Hecke recurrence will hold whenever these leafwise measures are infinite a.e.

Note that unlike the case of group actions, there is no canonical labeling on the p-Hecke tree of a point $x \in X$ in terms of the nodes of $\mathcal{G}$. The only inherent structure on these p-Hecke trees is the (discrete) tree metric; and a construction of leafwise measures in such cases is given in [**Lin06**].

To avoid having to introduce this formalism we can consider instead the corresponding non-locally finite measure $\mu_{x,p} = \mu_x^{\mathcal{G}}.x$ on $\mathrm{PGL}(2,\mathbb{Z})\backslash \mathrm{PGL}(2,\mathbb{R})$.

13.28. These leafwise measures (more precisely, their image under the embedding of the abstract $p+1$-regular tree $\mathcal{G}$ to p-Hecke trees in $\mathrm{PGL}(2,\mathbb{Z})\backslash \mathrm{PGL}(2,\mathbb{R})$ satisfy a.s. that

$$\frac{\mu_{x,p}(y)}{\mu_{x,p}(x)} = \lim_{r\to 0} \frac{\mu(B_r(y))}{\mu(B_r(x))}$$

where $B_r(x) = x.B_r^G(1)$ is a small ball around x in the group G.

Now, since Φ_i are Hecke eigenfunctions, the restriction of Φ_i to each Hecke tree will give an eigenfunction of the tree Laplacian. Hecke recurrence will then follow (after a short argument that can be found in e.g. [**Lin06**, Sec. 8]) from

13.29. Lemma. *Let $\mathcal{G}$ be a $p+1$-regular tree, and $\phi : \mathcal{G} \to \mathbb{C}$ a function such that $\Delta_{\mathcal{G}}\phi = \lambda_p\phi$. Then $\phi \notin L^2(\mathcal{G})$; in fact, there exists a (universal) constant independent of λ_p, such that*

$$\sum_{d(x,y)\leq R} |\phi|^2 \geq cR|\phi(x)|^2$$

13.30. This implies that our quantum limit will be both a_t-invariant and Hecke recurrent. By Theorem 10.3, if a.e. ergodic component of μ has positive entropy (this was shown for arithmetic quantum limits by Bourgain-Lindenstrauss [**BL03**]), then μ is G-invariant; i.e., μ is a multiple of Haar measure.

References

[Ana] Nalini Anantharaman. Eigenfunctions of the Laplacian on negatively curved manifolds: a semiclassical approach. in this volume.

[Bec83] María E. Becker. A ratio ergodic theorem for groups of measure-preserving transformations. *Illinois J. Math.*, 27(4):562–570, 1983.

[BL03] Jean Bourgain and Elon Lindenstrauss. Entropy of quantum limits. *Comm. Math. Phys.*, 233(1):153–171, 2003.

[Buz97] Jérôme Buzzi. Intrinsic ergodicity of smooth interval maps. *Israel J. Math.*, 100:125–161, 1997.

[CS80] Paul Cohen and Peter Sarnak. Notes on the Selberg trace formula. chapters 6 and 7 posted at `http://www.math.princeton.edu/sarnak/`, 1980.

[CSD55] J. W. S. Cassels and H. P. F. Swinnerton-Dyer. On the product of three homogeneous linear forms and the indefinite ternary quadratic forms. *Philos. Trans. Roy. Soc. London. Ser. A.*, 248:73–96, 1955.

[Ein06] Manfred Einsiedler. Ratner's theorem on SL(2,R)-invariant measures. Jahresber. Deutsch. Math.-Verein. 108 (2006), no. 3, 143–164.

[EK03] Manfred Einsiedler and Anatole Katok. Invariant measures on G/Γ for split simple Lie groups G. *Comm. Pure Appl. Math.*, 56(8):1184–1221, 2003. Dedicated to the memory of Jürgen K. Moser.

[EK05] Manfred Einsiedler and Anatole Katok. Rigidity of measures—the high entropy case and non-commuting foliations. *Israel J. Math.*, 148:169–238, 2005. Probability in mathematics.

[EKL06] Manfred Einsiedler, Anatole Katok, and Elon Lindenstrauss. Invariant measures and the set of exceptions to Littlewood's conjecture. *Ann. of Math. (2)*, 164(2):513–560, 2006.

[EL06] Manfred Einsiedler and Elon Lindenstrauss. Diagonalizable flows on locally homogeneous spaces and number theory. In *International Congress of Mathematicians. Vol. II*, pages 1731–1759. Eur. Math. Soc., Zürich, 2006.

[EL08] Manfred Einsiedler and Elon Lindenstrauss. On measures invariant under diagonalizable actions the rank one case and the general low entropy method. *Journal of Modern Dynamics*, 2(1):83–128, 2008.

[ELMV09] Manfred Einsiedler, Elon Lindenstrauss, Philippe Michel, and Akshay Venkatesh. Distribution of periodic torus orbits on homogeneous spaces. *Duke Math. J.* 148 (2009), no. 1, 119–174.

[ELMV07] Manfred Einsiedler, Elon Lindenstrauss, Philippe Michel, and Akshay Venkatesh. Distribution of periodic torus orbits and Duke's theorem for cubic fields, 2007. to appear in Ann. of Math. (58 pages).

[ELW09] Manfred Einsiedler, Elon Lindenstrauss, and Thomas Ward. Entropy in ergodic theory and homogeneous dynamics. In preparation, some chapters available online at `http://www.mth.uea.ac.uk/entropy/`, 2009.

[EW09] Manfred Einsiedler and Thomas Ward. Ergodic theory with a view towards number theory. To appear in Springer GTM, see also `http://www.mth.uea.ac.uk/ergodic/`, 2009.

[Esk] Alex Eskin. Unipotent flows and applications. in this volume.

[Fur73] Harry Furstenberg. The unique ergodicity of the horocycle flow. In *Recent advances in topological dynamics (Proc. Conf., Yale Univ., New Haven, Conn., 1972; in honor of Gustav Arnold Hedlund)*, pages 95–115. Lecture Notes in Math., Vol. 318. Springer, Berlin, 1973.

[Gla03] Eli Glasner. *Ergodic theory via joinings*, volume 101 of *Mathematical Surveys and Monographs*. American Mathematical Society, Providence, RI, 2003.

[Hu93] Hu Yi Hu. Some ergodic properties of commuting diffeomorphisms. *Ergodic Theory Dynam. Systems*, 13(1):73–100, 1993.

[Hur44] Witold Hurewicz. Ergodic theorem without invariant measure. *Ann. of Math. (2)*, 45:192–206, 1944.

[KH95] Anatole Katok and Boris Hasselblatt. *Introduction to the modern theory of dynamical systems*, volume 54 of *Encyclopedia of Mathematics and its Applications*. Cambridge University Press, Cambridge, 1995. With a supplementary chapter by Katok and Leonardo Mendoza.

[KS96] A. Katok and R. J. Spatzier. Invariant measures for higher-rank hyperbolic abelian actions. *Ergodic Theory Dynam. Systems*, 16(4):751–778, 1996.

[Lan75] Serge Lang. $SL_2(\mathbf{R})$. Addison-Wesley Publishing Co., Reading, Mass.-London-Amsterdam, 1975.

[Lin05] Elon Lindenstrauss. Rigidity of multiparameter actions. *Israel J. Math.*, 149:199–226, 2005.

[Lin06] Elon Lindenstrauss. Invariant measures and arithmetic quantum unique ergodicity. *Ann. of Math. (2)*, 163(1):165–219, 2006.

[Lin07] Elon Lindenstrauss. Some examples how to use measure classification in number theory. In *Equidistribution in number theory, an introduction*, volume 237 of *NATO Sci. Ser. II Math. Phys. Chem.*, pages 261–303. Springer, Dordrecht, 2007.

[LV07] Elon Lindenstrauss and Akshay Venkatesh. Existence and Weyl's law for spherical cusp forms. *Geom. Funct. Anal.*, 17(1):220–251, 2007.

[LW01] Elon Lindenstrauss and Barak Weiss. On sets invariant under the action of the diagonal group. *Ergodic Theory Dynam. Systems*, 21(5):1481–1500, 2001.

[Mar97] G. A. Margulis. Oppenheim conjecture. In *Fields Medallists' lectures*, volume 5 of *World Sci. Ser. 20th Century Math.*, pages 272–327. World Sci. Publishing, River Edge, NJ, 1997.

[Mar00] Gregory Margulis. Problems and conjectures in rigidity theory. In *Mathematics: frontiers and perspectives*, pages 161–174. Amer. Math. Soc., Providence, RI, 2000.

[Mor] Dave Witte Morris. Introduction to arithmetic groups. in preparation, `http://people.uleth.ca/%7Edave.morris/LectureNotes.shtml#ArithmeticGroups`.

[Moz95] Shahar Mozes. Actions of Cartan subgroups. *Israel J. Math.*, 90(1-3):253–294, 1995.

[MT94] G. A. Margulis and G. M. Tomanov. Invariant measures for actions of unipotent groups over local fields on homogeneous spaces. *Invent. Math.*, 116(1-3):347–392, 1994.

[OW87] Donald S. Ornstein and Benjamin Weiss. Entropy and isomorphism theorems for actions of amenable groups. *J. Analyse Math.*, 48:1–141, 1987.

[PS92] R. Phillips and P. Sarnak. Automorphic spectrum and Fermi's golden rule. *J. Anal. Math.*, 59:179–187, 1992. Festschrift on the occasion of the 70th birthday of Shmuel Agmon.

[Rag72] M. S. Raghunathan. *Discrete subgroups of Lie groups*. Springer-Verlag, New York, 1972. Ergebnisse der Mathematik und ihrer Grenzgebiete, Band 68.

[Rt82a] Marina Ratner. Factors of horocycle flows. *Ergodic Theory Dynam. Systems*, 2(3-4):465–489, 1982.

[Rt82b] Marina Ratner. Rigidity of horocycle flows. *Ann. of Math. (2)*, 115(3):597–614, 1982.

[Rt83] Marina Ratner. Horocycle flows, joinings and rigidity of products. *Ann. of Math. (2)*, 118(2):277–313, 1983.

[Rt90] Marina Ratner. On measure rigidity of unipotent subgroups of semisimple groups. *Acta Math.*, 165(3-4):229–309, 1990.

[Rt91] Marina Ratner. On Raghunathan's measure conjecture. *Ann. of Math. (2)*, 134(3):545–607, 1991.

[Ree82] M. Rees. Some R^2-anosov flows. unpublished, 1982.

[RS94] Zeév Rudnick and Peter Sarnak. The behaviour of eigenstates of arithmetic hyperbolic manifolds. *Comm. Math. Phys.*, 161(1):195–213, 1994.

[Sar03] Peter Sarnak. Spectra of hyperbolic surfaces. *Bull. Amer. Math. Soc. (N.S.)*, 40(4):441–478 (electronic), 2003.

[Sel56] A. Selberg. Harmonic analysis and discontinuous groups in weakly symmetric Riemannian spaces with applications to Dirichlet series. *J. Indian Math. Soc. (N.S.)*, 20:47–87, 1956.

[Sou09] K. Soundararajan. Quantum unique ergodicity for $SL_2(\mathbb{Z})\backslash\mathbb{H}$. preprint, `arXiv:0901.4060v1[math.NT]`, 2009.

[Sri98] S. M. Srivastava. *A course on Borel sets*, volume 180 of *Graduate Texts in Mathematics*. Springer-Verlag, New York, 1998.

[SV04] Lior Silberman and Akshay Venkatesh. On quantum unique ergodicity for locally symmetric spaces I: a micro local lift. to appear in GAFA (37 pages), 2004.

[SV06] Lior Silberman and Akshay Venkatesh. Entropy bounds for Hecke eigenfunctions on division algebras. preprint (22 pages), 2006.

[Wal82] Peter Walters. *An introduction to ergodic theory*, volume 79 of *Graduate Texts in Mathematics*. Springer-Verlag, New York, 1982.

[Wol94] Scott A. Wolpert. Disappearance of cusp forms in special families. *Ann. of Math. (2)*, 139(2):239–291, 1994.

Department of Mathematics, The Ohio State University, 231 West 18th Avenue, Columbus, OH 43210-1174, USA

Current address: ETH Zürich, Departement Mathematik, Rämistrasse 101, 8092 Zürich, Switzerland

E-mail address: `Manfred.Einsiedler@math.ethz.ch`

Fine Hall, Washington Road, Princeton, NJ 08544-1000, USA

Current address: Einstein Institute of Mathematics, Givat Ram, The Hebrew University of Jerusalem, Jerusalem, 91904, Israel

E-mail address: `elon@math.huji.ac.il`

Clay Mathematics Proceedings
Volume 10, 2010

Fuchsian groups, geodesic flows on surfaces of constant negative curvature and symbolic coding of geodesics

Svetlana Katok

Dedicated to the memory of my father Boris Abramovich Rosenfeld (1917-2008)

Contents

Introduction

These are the notes of an introductory four-lectures course given in the Summer School in Pisa in June, 2007. Lectures I and II cover hyperbolic geometry and the theory of Fuchsian groups; the material of these lectures is mostly an adaptation from the author's book "Fuchsian groups" [**14**]. Lecture III describes the geodesic flow on the surfaces of constant negative curvature and establishes its dynamical properties. Lecture IV is devoted to symbolic coding of the geodesic flow with emphasis on the modular surface. The material of this lecture is based on the survey article [**17**] and two more recent papers [**20, 21**] of the author with I. Ugarcovici.

Lecture I. Hyperbolic geometry

1. Models of hyperbolic geometry

Our first model of hyperbolic geometry is obtained similarly to the model of elliptic geometry on the unit sphere S^2 in $\mathbb{R}^3$,

$$S^2 = \{(x_1, x_2, x_3) \in \mathbb{R}^3 \mid x_1^2 + x_2^2 + x_3^2 = 1\}.$$

The metric (arc length) on S^2 is induced from the Euclidean metric on $\mathbb{R}^3$,

$$ds^2 = dx_1^2 + dx_2^2 + dx_3^2 \tag{1.1}$$

which corresponds to the standard inner product on $\mathbb{R}^3$,

$$(x, y) = x_1y_1 + x_2y_2 + x_3y_3.$$

The geodesics for this metric (i.e. the length minimizing curves) lie on planes through $0 \in \mathbb{R}^3$ and are arcs of great circles. The group of orientation- preserving isometries of S^2 is the group $SO(3)$ that preserves the standard inner product (,) on $\mathbb{R}^3$.

If instead of the metric (1.1) we consider a *pseudo-metric* in $\mathbb{R}^3$:

$$ds_h^2 = dx_1^2 + dx_2^2 - dx_3^2, \tag{1.2}$$

corresponding to the bilinear symmetric form of signature $(2, 1)$:

$$(x, y)_{2,1} = x_1y_1 + x_2y_2 - x_3y_3,$$

then the upper fold of the hyperboloid

$$H^2 = \{(x_1, x_2, x_3) \in \mathbb{R}^3 \mid x_1^2 + x_2^2 - x_3^2 = -1, \quad x_3 > 0\}$$

represents a model of hyperbolic geometry in two dimensions. Notice that outside of H^2 $(x,x)_{2,1} > -1$ and inside H^2 $(x,x)_{2,1} < -1$.

First, we check that the pseudo-metric (1.2) induces Riemannian metric on H^2. Let $x = (x_1, x_2, x_3) \in H^2$. Define $x^\perp = \{y \in \mathbb{R}^3 \mid (y,x)_{2,1} = 0\}$. It is a plane passing though 0, and $x + x^\perp$ is a plane passing through x.

PROPOSITION 1.1. *The tangent plane T_xH^2 to the hyperboloid H^2 at the point x is given by $T_xH^2 = x + x^\perp$, and $(\ ,\ )_{2,1}$ restricted to $x^\perp$ is positive-definite, hence gives a scalar product on T_xH^2, i.e. a Riemannian metric on H^2.*

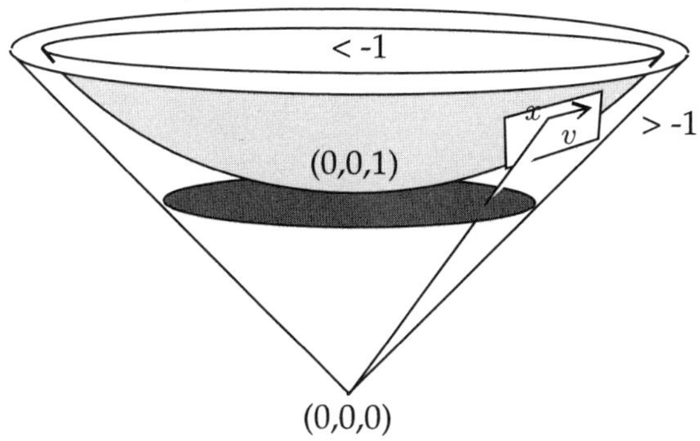

FIGURE 1.1. The hyperboloid model

PROOF. The upper fold of the hyperboloid is given by the equation $x_3 = \sqrt{x_1^2 + x_2^2 + 1}$. Let $x + v \in T_xH^2$. A tangent vector v at x is a linear combination of two basic tangent vectors,

$$v = a\Big(1, 0, \frac{\partial x_3}{\partial x_1}\Big) + b\Big(0, 1, \frac{\partial x_3}{\partial x_2}\Big) = \Big(a, b, \frac{ax_1 + bx_2}{x_3}\Big).$$

We see that $(v,x)_{2,1} = 0$, i.e. $v \in x^\perp$, hence $T_xH^2 \subset x + x^\perp$, and since T_xH^2 is a plane, $T_xH^2 = x + x^\perp$.

Let $v \in x^\perp$. By convexity of the hyperboloid, $x + v$ is outside of H^2, hence

$$-1 < (x+v, x+v)_{2,1} = (x,x)_{2,1} + 2(x,v)_{2,1} + (v,v)_{2,1} = -1 + (v,v)_{2,1}.$$

Therefore for all $v \in x^\perp$, $(v,v)_{2,1} > 0$. □

The geodesics for this metric also lie on planes through $0 \in \mathbb{R}^3$. The group of orientation-preserving isometries of H^2 is $SO(2,1)$, the group preserving the bilinear symmetric form $(x,y)_{2,1}$,

$$SO(2,1) = \{A \in SL(3,\mathbb{R}) \mid {}^TASA = S, \text{ where } S = \begin{pmatrix} 1 & 0 & 0 \\ 0 & 1 & 0 \\ 0 & 0 & -1 \end{pmatrix}\}.$$

Other models of the hyperbolic plane are obtained from the **hyperboloid model** described above.

The Beltrami-Klein model. The group $G = SO(2,1)$ acts transitively on the upper fold of the hyperboloid H^2 by linear transformations. The cone given by the equation $x_1^2 + x_2^2 - x_3^2 = 0$ lies outside of H^2 and is asymptotic to it, as illustrated on Figure 1.1. The intersection of this cone with the plane $x_3 = 1$ tangent to H^2 is the circumference $\partial\mathcal{U} = \{x_1^2 + x_2^2 = 1\}$. Consider the central projection σ of H^2 to the plane $x_3 = 1$ from the origin $0 \in \mathbb{R}^3$. We have

$$\sigma(x_1, x_2, x_3) = (\eta_1, \eta_2),$$

where $\eta_1 = \frac{x_1}{x_3}, \eta_2 = \frac{x_2}{x_3}$. We have $\eta_1^2 + \eta_2^2 = 1 - \frac{1}{x_3^2}$, hence $\sigma(H^2) = \mathcal{U} = \{\eta_1^2 + \eta_2^2 < 1\}$. Equivalently, $\mathcal{U}$ may be viewed as the space of *negative* vectors $x \in \mathbb{R}^3$, i.e. such that $(x, x)_{2,1} < 0$, which will be useful later. The metric on $\mathcal{U}$ is induced by the hyperbolic metric d_h on H^2:

$$d_h^*(\eta_1, \eta_2) = d_h(\sigma^{-1}\eta_1, \sigma^{-1}\eta_2).$$

Geodesics in H^2 are mapped to chords of the unit disc $\mathcal{U}$, which thus become geodesics with respect to the hyperbolic metric d_h^* on $\mathcal{U}$ (in what follows we will omit $*$ in most cases). We define the action of G on $\mathcal{U}$ so that it commutes with σ. It follows that G acts on $\mathcal{U}$ by fractional linear transformations: for (η_1, η_2) and

$$g = \begin{pmatrix} a_{11} & a_{12} & a_{13} \\ a_{21} & a_{22} & a_{23} \\ a_{31} & a_{32} & a_{33} \end{pmatrix}$$

$$g(\eta_1, \eta_2) = \left(\frac{a_{11}\eta_1 + a_{12}\eta_2 + a_{13}}{a_{31}\eta_1 + a_{32}\eta_2 + a_{33}}, \frac{a_{21}\eta_1 + a_{22}\eta_2 + a_{23}}{a_{31}\eta_1 + a_{32}\eta_2 + a_{33}}\right).$$

Notice that this model is not angle-true. Two geodesics which meet at the boundary are in fact asymptotically tangent.

The hemispherical model. Let $\eta_1 = \frac{x_1}{x_3}, \eta_2 = \frac{x_2}{x_3}, \eta_3 = \frac{1}{x_3}$. Then from the equation of H^2 we obtain the equation of the unit hemisphere

$$\eta_1^2 + \eta_2^2 + \eta_3^2 = 1, \ \eta_3 > 0$$

This model is obtained from the Beltrami-Klein model by the orthogonal projection of the unit disc $\mathcal{U}$ to the hemisphere. The geodesics in this model are arcs of circles on the hemisphere orthogonal to the disc—the boundary of the hemispere.

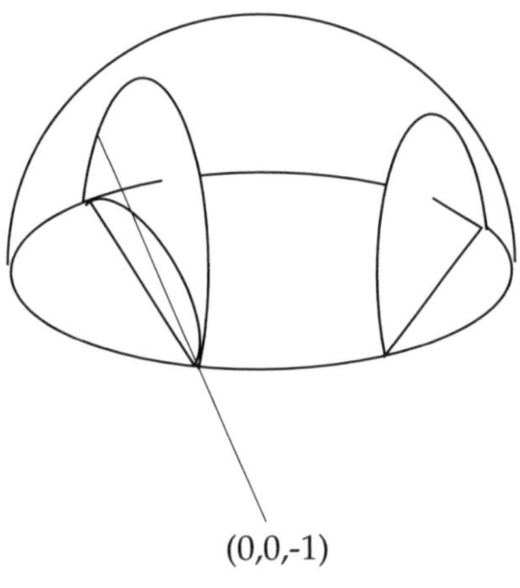

FIGURE 1.2. The hemispherical, Beltrami-Klein, and Poincaré disc models

The Poincaré disc model. The stereographic projection of the hemisphere from the point $(0,0,-1)$ onto the plane $\eta_3 = 0$ (i.e. to the same unit disc $\mathcal{U}$) maps geodesics in the hemisphere to arcs of circles orthogonal to the boundary $\partial\mathcal{U}$. This gives us a new model in the unit disc, the Poincaré disc model. If we go from the Beltrami-Klein model to the Poincaré model (through the hemisphere) we notice that the end points of the geodesics are preserved and each point with polar coordinates (r,φ) is mapped to the point on the same radius (r',φ), where $r' = \frac{r}{\sqrt{1-r^2}+1}$.

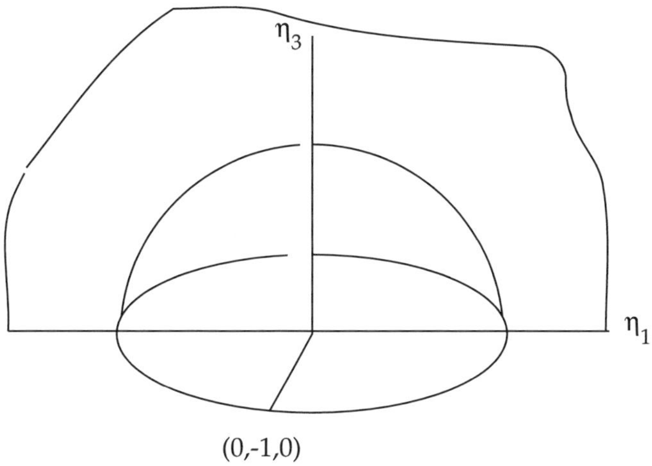

FIGURE 1.3. The Poincaré upper half-plane model

The Poincaré upper half-plane model. The stereographic projection of the upper hemisphere from the point $(0,-1,0)$ onto the plane $\eta_2 = 0$ give the model in the half-plane $\mathcal{H} = \{(\eta_1,\eta_3),\ \eta_3 > 0\}$.

Since the stereographic projection is *conformal* (i.e. preserves angles) the hemispherical model and its derivatives, the Poincaré disc model and the Poincaré upper half-plane model are angle-true.

Models as homogeneous spaces. All three models are obtained algebraically as homogeneous spaces G/K due to the accidental isomorphisms $SL(2,\mathbb{R}) \approx SU(1,1) \approx SO(2,1)$.

In the Beltrami-Klein model

$$G = SO(2,1), \quad K = \left\{ \begin{pmatrix} \cos\varphi & -\sin\varphi & 0 \\ \sin\varphi & \cos\varphi & 0 \\ 0 & 0 & 1 \end{pmatrix}, 0 \le \varphi < 2\pi \right\}.$$

In the Poincaré disc model, $G = SU(1,1)$, the group that preserves the Hermitian form on $\mathbb{C}^2$, $< z, w >= z_1\bar{w}_1 - z_2\bar{w}_2$ (for $z = (z_1,z_2)$ and $w = (w_1,w_2)$), is

$$SU(1,1) = \left\{ g \in SL(2,\mathbb{C}) \mid g = \begin{pmatrix} a & c \\ \bar{c} & \bar{a} \end{pmatrix} \right\},$$

and

$$K = \left\{ \begin{pmatrix} e^{i\varphi} & 0 \\ 0 & e^{-i\varphi} \end{pmatrix}, 0 \le \varphi < 2\pi \right\}.$$

The homogeneous space G/K can be identified with the "projectivized" space of the *negative* vectors in $\mathbb{C}^2$ ($< z, z >< 0$), analogous to that discussed above for $\mathbb{R}^3$, or, in homogeneous coordinates, with the unit disc in $\mathbb{C}$,

$$\mathcal{U} = \{z \in \mathbb{C} \mid |z| < 1\}.$$

In the Poincaré upper half-plane model, $G = SL(2,\mathbb{R})$, and $K = SO(2)$, the stabilizer of the point $i \in \mathcal{H}$. Here the homogeneous space G/K is identified with the upper half-plane

$$\mathcal{H} = \{z \in \mathbb{C} \mid \mathrm{Im}(z) > 0\}.$$

by the following construction. Each matrix in $SL(2,\mathbb{R})$ can be written as a product of upper-triangular and orthogonal (the Iwasawa decomposition):

$$g = \begin{pmatrix} a & b \\ c & d \end{pmatrix} = \begin{pmatrix} \sqrt{y} & \frac{x}{\sqrt{y}} \\ 0 & \frac{1}{\sqrt{y}} \end{pmatrix} \begin{pmatrix} \cos\varphi & -\sin\varphi \\ \sin\varphi & \cos\varphi \end{pmatrix},$$

where $x, y \in \mathbb{R},\ y > 0$ and $0 \le \varphi < 2\pi$. Then $\pi : G/K \to \mathcal{H}$ given by

$$\pi(g) = g(i) = \frac{ai+b}{ci+d} = x + iy = z$$

does the identification.

In the last two conformal models, the corresponding group G acts by fractional-linear transformations: for $g = \begin{pmatrix} a & b \\ c & d \end{pmatrix}$, $g(z) = \frac{az+b}{cz+d}$.

2. The hyperbolic plane

Let $\mathcal{H} = \{z \in \mathbb{C} \mid \mathrm{Im(z)} > 0\}$ be the upper-half plane. Equipped with the metric

$$ds = \frac{\sqrt{dx^2 + dy^2}}{y}, \tag{2.1}$$

it becomes a model of the *hyperbolic* or *Lobachevski plane* (see Exercise 2). We will see that the *geodesics* (i.e., the shortest curves with respect to this metric) will be straight lines and semicircles orthogonal to the real line

$$\mathbb{R} = \{z \in \mathbb{C} \mid \mathrm{Im(z)} = 0\}.$$

Using this fact and elementary geometric considerations, one easily shows that any two points in $\mathcal{H}$ can be joined by a unique geodesic, and that from any point in $\mathcal{H}$ in any direction one can draw a geodesic. We will measure the distance between two points in $\mathcal{H}$ along the geodesic connecting them. It is clear that any geodesic can be continued indefinitely, and that one can draw a circle centered at a given point with any given radius.

The tangent space to $\mathcal{H}$ at a point z is defined as the space of tangent vectors at z. It has the structure of a 2-dimensional real vector space or of a 1-dimensional complex vector space: $T_z\mathcal{H} \approx \mathbb{R}^2 \approx \mathbb{C}$. The Riemannian metric (2.1) is induced by the following inner product on $T_z\mathcal{H}$: for $\zeta_1 = \xi_1 + i\eta_1$ and $\zeta_2 = \xi_2 + i\eta_2$ in $T_z\mathcal{H}$, we put

$$\langle \zeta_1, \zeta_2 \rangle = \frac{(\zeta_1, \zeta_2)}{\mathrm{Im(z)}^2}, \tag{2.2}$$

which is a scalar multiple of the Euclidean inner product $(\zeta_1, \zeta_2) = \xi_1\xi_2 + \eta_1\eta_2$. We define the *angle* between two geodesics in $\mathcal{H}$ at their intersection point z as the angle between their tangent vectors in $T_z\mathcal{H}$. Using the formula

$$\cos\varphi = \frac{\langle \zeta_1, \zeta_2 \rangle}{\|\zeta_1\|\|\zeta_2\|} = \frac{(\zeta_1, \zeta_2)}{|\zeta_1||\zeta_2|},$$

where $\| \;\|$ denotes the norm in $T_z\mathcal{H}$ corresponding to the inner product $\langle\, , \rangle$, and $|\;|$ denotes the norm corresponding to the inner product $(\, ,)$, we see that this notion of angle measure coincides with the Euclidean angle measure.

The first four axioms of Euclid hold for this geometry. However, the fifth postulate of Euclid's *Elements*, the axiom of parallels, does not hold: there is more than one geodesic passing through the point z not lying in the geodesic L that does not intersect L (see Fig. 2.1). Therefore the geometry in $\mathcal{H}$ is *non-Euclidean.* The metric in (2.1) is said to be the *hyperbolic metric.* It can be used to calculate the length of curves in $\mathcal{H}$ the same way the Euclidean metric $\sqrt{dx^2 + dy^2}$ is used to calculate the length of curves on the Euclidean plane. Let $I = [0, 1]$ be the unit

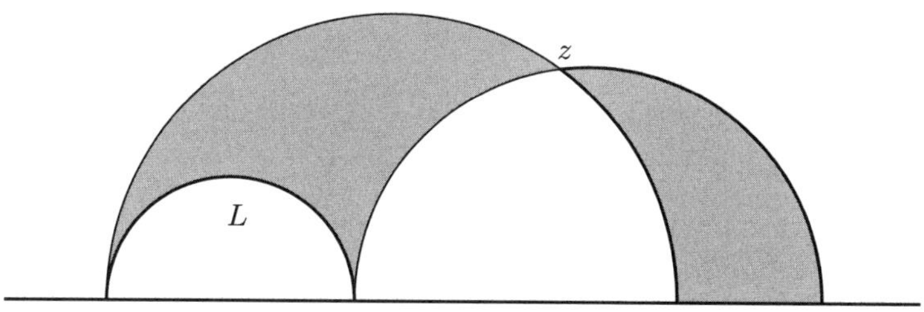

FIGURE 2.1. Geodesics in the upper half-plane

interval, and $\gamma : I \to \mathcal{H}$ be a piecewise differentiable curve in $\mathcal{H}$,

$$\gamma(t) = \{v(t) = x(t) + iy(t) \mid t \in I\}.$$

The length of the curve γ is defined by

$$h(\gamma) = \int_0^1 \frac{\sqrt{(\frac{dx}{dt})^2 + (\frac{dy}{dt})^2}}{y(t)} dt. \tag{2.3}$$

We define the *hyperbolic distance* between two points $z, w \in \mathcal{H}$ by setting

$$\rho(z, w) = \inf h(\gamma),$$

where the infimum is taken over all piecewise differentiable curves connecting z and w.

PROPOSITION 2.1. *The function $\rho : \mathcal{H} \times \mathcal{H} \to \mathbb{R}$ defined above is a distance function, i.e.*

(a) *nonnegative: $\rho(z, z) = 0$; $\rho(z, w) > 0$ if $z \neq w$;*
(b) *symmetric: $\rho(u, v) = \rho(v, u)$;*

(c) *satisfies the triangle inequality:* $\rho(z,u) \le \rho(z,w) + \rho(w,u)$.

PROOF. It is easily seen from the definition that (b), (c), and the first part of property (a) hold. The second part follows from Exercise 3. □

Consider the group $SL(2,\mathbb{R})$ of real 2×2 matrices with determinant one. It acts on $\mathcal{H}$ by *Möbius transformations* if we assign to each $g = \begin{pmatrix} a & b \\ c & d \end{pmatrix} \in SL(2,\mathbb{R})$ the transformation

$$T_g(z) = \frac{az+b}{cz+d}. \tag{2.4}$$

PROPOSITION 2.2. *Any Möbius transformation T_g maps $\mathcal{H}$ into itself.*

PROOF. We can write

$$w = T_g(z) = \frac{(az+b)(c\bar{z}+d)}{|cz+d|^2} = \frac{ac|z|^2 + adz + bc\bar{z} + bd}{|cz+d|^2}.$$

Therefore

$$\mathrm{Im(w)} = \frac{\mathrm{w} - \bar{\mathrm{w}}}{2\mathrm{i}} = \frac{(\mathrm{ad}-\mathrm{bc})(\mathrm{z}-\bar{\mathrm{z}})}{2\mathrm{i}|\mathrm{cz}+\mathrm{d}|^2} = \frac{\mathrm{Im(z)}}{|\mathrm{cz}+\mathrm{d}|^2}. \tag{2.5}$$

Thus Im(z) > 0 implies Im(w) > 0. □

One can check directly that if $g, h \in SL(2,\mathbb{R})$, then $T_g \circ T_h = T_{gh}$ and $T_g^{-1} = T_{g^{-1}}$. It follows that each T_g, $g \in SL(2,\mathbb{R})$, is a bijection, and thus we obtain a *representation* of the group $SL(2,\mathbb{R})$ by Möbius transformations of the upper-half plane $\mathcal{H}$. In fact, the two matrices g and $-g$ give the same Möbius transformation, so formula (2.4) actually gives a representation of the quotient group $SL(2,\mathbb{R})/\{\pm 1_2\}$ (where 1_2 is the 2×2 identity matrix), denoted by $PSL(2,\mathbb{R})$, which we will identify with the group of Möbius transformations of the form (2.4). Notice that $PSL(2,\mathbb{R})$ contains all transformations of the form

$$z \to \frac{az+b}{cz+d} \quad \text{with} \quad ad - bc = \Delta > 0,$$

since by dividing the numerator and the denominator by $\sqrt{\Delta}$, we obtain a matrix for it with determinant equal to 1. In particular, $PSL(2,\mathbb{R})$ contains all transformations of the form $z \to az + b$ $(a, b \in \mathbb{R},\ a > 0)$. Since transformations in $PSL(2,\mathbb{R})$ are continuous, we have the following result.

THEOREM 2.3. *The group $PSL(2,\mathbb{R})$ acts on $\mathcal{H}$ by homeomorphisms.*

DEFINITION 2.4. A transformation of $\mathcal{H}$ onto itself is called an *isometry* if it preserves the hyperbolic distance in $\mathcal{H}$.

Isometries clearly form a group; we will denote it by $\mathrm{Isom}(\mathcal{H})$.

THEOREM 2.5. *Möbius transformations are isometries, i.e., we have the inclusion $PSL(2,\mathbb{R}) \subset \mathrm{Isom}(\mathcal{H})$.*

PROOF. Let $T \in PSL(2,\mathbb{R})$. By Theorem 2.3 T maps $\mathcal{H}$ onto itself. Let $\gamma : I \to \mathcal{H}$ be the piecewise differentiable curve given by $z(t) = x(t) + iy(t)$. Let

$$w = T(z) = \frac{az+b}{cz+d};$$

then we have $w(t) = T(z(t)) = u(t) + iv(t)$ along the curve γ. Differentiating, we obtain

$$\frac{dw}{dz} = \frac{a(cz+d) - c(az+b)}{(cz+d)^2} = \frac{1}{(cz+d)^2}. \tag{2.6}$$

By (2.5) we have

$$v = \frac{y}{|cz+d|^2}, \text{ therefore } \left|\frac{dw}{dz}\right| = \frac{v}{y}.$$

Thus

$$h(T(\gamma)) = \int_0^1 \frac{|\frac{dw}{dt}|dt}{v(t)} = \int_0^1 \frac{|\frac{dw}{dz}||\frac{dz}{dt}|dt}{v(t)} = \int_0^1 \frac{|\frac{dz}{dt}|dt}{y(t)} = h(\gamma).$$

The invariance of the hyperbolic distance follows from this immediately. □

3. Geodesics

THEOREM 3.1. *The geodesics in $\mathcal{H}$ are semicircles and the rays orthogonal to the real axis $\mathbb{R}$.*

PROOF. Let $z_1, z_2 \in \mathcal{H}$. First consider the case in which $z_1 = ia, z_2 = ib$ with $b > a$. For any piecewise differentiable curve $\gamma(t) = x(t) + iy(t)$ connecting ia and ib, we have

$$h(\gamma) = \int_0^1 \frac{\sqrt{(\frac{dx}{dt})^2 + (\frac{dy}{dt})^2}}{y(t)} dt \geq \int_0^1 \frac{|\frac{dy}{dt}|dt}{y(t)} \geq \int_0^1 \frac{\frac{dy}{dt}dt}{y(t)} = \int_a^b \frac{dy}{y} = \ln \frac{b}{a},$$

but this is exactly the hyperbolic length of the segment of the imaginary axis connecting ia and ib. Therefore the geodesic connecting ia and ib is the segment of the imaginary axis connecting them.

Now consider the case of arbitrary points z_1 and z_2. Let L be the unique Euclidean semicircle or a straight line connecting them. Then by Exercise 4, there exists a transformation in $PSL(2, \mathbb{R})$ which maps L into the positive imaginary axis. This reduces the problem to the particular case studied above, and by Theorem 2.5 we conclude that the geodesic between z_1 and z_2 is the segment of L joining them. □

Thus we have proved that any two points z and w in $\mathcal{H}$ can be joined by a unique geodesic, and the hyperbolic distance between them is equal to the hyperbolic length of the geodesic segment joining them; we denote the latter by $[z, w]$. This and the additivity of the integral (2.3) imply the following statement.

COROLLARY 3.2. *If z and w are two distinct points in $\mathcal{H}$, then*

$$\rho(z, w) = \rho(z, \xi) + \rho(\xi, w)$$

if and only if $\xi \in [z, w]$.

THEOREM 3.3. *Any isometry of $\mathcal{H}$, and, in particular, any transformation from $PSL(2, \mathbb{R})$, maps geodesics into geodesics.*

PROOF. The same argument as in the Euclidean case using Corollary 3.2 works here as well. □

The *cross-ratio* of distinct points $z_1, z_2, z_3, z_4 \in \hat{\mathbb{C}} = \mathbb{C} \cup \{\infty\}$ is defined by the following formula:

$$(z_1, z_2; z_3, z_4) = \frac{(z_1 - z_2)(z_3 - z_4)}{(z_2 - z_3)(z_4 - z_1)}.$$

THEOREM 3.4. *Suppose $z, w \in \mathcal{H}$ are two distinct points, the geodesic joining z and w has end points $z^*, w^* \in \mathbb{R} \cup \{\infty\}$, and $z \in [z^*, w]$. Then*

$$\rho(z, w) = \ln(w, z^*; z, w^*).$$

PROOF. Using Exercise 4, let us choose a transformation $T \in PSL(2, \mathbb{R})$ which maps the geodesic joining z and w to the imaginary axis. By applying the transformations $z \mapsto kz$ $(k > 0)$ and $z \mapsto -1/z$ if necessary, we may assume that $T(z^*) = 0$, $T(w^*) = \infty$ and $T(z) = i$. Then $T(w) = ri$ for some $r > 1$, and

$$\rho(T(z), T(w)) = \int_1^r \frac{dy}{y} = \ln r.$$

On the other hand, $(ri, 0; i, \infty) = r$, and the theorem follows from the invariance of the cross-ratio under Möbius transformations, a standard fact from complex analysis (which can be checked by a direct calculation). □

We will derive several explicit formulas for the hyperbolic distance involving the hyperbolic functions

$$\sinh x = \frac{e^x - e^{-x}}{2}, \qquad \cosh x = \frac{e^x + e^{-x}}{2}, \qquad \tanh z = \frac{\sinh x}{\cosh x}.$$

THEOREM 3.5. *For $z, w \in \mathcal{H}$, we have*

(a) $\rho(z, w) = \ln \frac{|z-\overline{w}|+|z-w|}{|z-\overline{w}|-|z-w|}$;

(b) $\cosh \rho(z, w) = 1 + \frac{|z-w|^2}{2\text{Im(z)Im(w)}}$;

(c) $\sinh[\frac{1}{2}\rho(z, w)] = \frac{|z-w|}{2(\text{Im(z)Im(w)})^{1/2}}$;

(d) $\cosh[\frac{1}{2}\rho(z, w)] = \frac{|z-\overline{w}|}{2(\text{Im(z)Im(w)})^{1/2}}$;

(e) $\tanh[\frac{1}{2}\rho(z, w)] = |\frac{z-w}{z-\overline{w}}|$.

PROOF. We will prove that (e) holds. By Theorem 2.5, the left-hand side is invariant under any transformation $T \in PSL(2, \mathbb{R})$. By Exercise 5, the right-hand side is also invariant under any $T \in PSL(2, \mathbb{R})$. Therefore if is sufficient to check the formula for the case when $z = i$, $w = ir$ $(r > 1)$. The right-hand side is equal to $(r-1)/(r+1)$. The left-hand side is equal to $\tanh[\frac{1}{2} \ln r]$. A simple calculation shows that these two expressions are equal. The other formulas are proved similarly. □

4. Isometries

We have seen that transformations in $PSL(2, \mathbb{R})$ are isometries of the hyperbolic plane $\mathcal{H}$ (Theorem 2.5). The next theorem identifies all isometries of $\mathcal{H}$ in terms of Möbius transformations and symmetry in the imaginary axis.

THEOREM 4.1. *The group* $\text{Isom}(\mathcal{H})$ *is generated by the Möbius transformations from $PSL(2, \mathbb{R})$ together with the transformation $z \mapsto -\overline{z}$. The group $PSL(2, \mathbb{R})$ is a subgroup of* $\text{Isom}(\mathcal{H})$ *of index two.*

PROOF. Let φ be any isometry of $\mathcal{H}$. By Theorem 3.3, φ maps geodesics into geodesics. Let I denote the positive imaginary axis. Then $\varphi(I)$ is a geodesic in $\mathcal{H}$, and, according to Exercise 4, there exists an isometry $T \in PSL(2,\mathbb{R})$ that maps $\varphi(I)$ back to I. By applying the transformations $z \mapsto kz$ $(k > 0)$ and $z \mapsto -1/z$, we may assume that $g \circ \varphi$ fixes i and maps the rays (i,∞) and $(i,0)$ onto themselves. Hence, being an isometry, $g\circ\varphi$ fixes each point of I. The same (synthetic) argument as in the Euclidean case shows that

$$g \circ \varphi(z) = z \text{ or } -\overline{z}. \tag{4.1}$$

Let z_1 and z_2 be two fixed points on I. For any point z not on I, draw two hyperbolic circles centered at z_1 and z_2 and passing through z. These circles intersect in two points, z and $z' = -\overline{z}$, since the picture is symmetric with respect to the imaginary axis (note that a hyperbolic circle is a Euclidean circle in $\mathcal{H}$, but with a different center). Since these circles are mapped into themselves under the isometry $g \circ \varphi$, we conclude that $g \circ \varphi(z) = z$ or $g \circ \varphi(z) = -\overline{z}$. Since isometries are continuous (see Exercise 6), only one of the equations (4.1) holds for all $z \in \mathcal{H}$. If $g \circ \varphi(z) = z$, then $\varphi(z)$ is a Möbius transformation of the form (2.4). If $g \circ \varphi(z) = -\overline{z}$, we have

$$\varphi(z) = \frac{a\overline{z}+b}{c\overline{z}+d} \quad \text{with } ad-bc=-1, \tag{4.2}$$

which proves the theorem. □

Thus we have characterized all the isometries of $\mathcal{H}$. The sign of the determinant of the corresponding matrix in (2.4) or (4.2) determines the *orientation* of an isometry. We will refer to transformations in $PSL(2,\mathbb{R})$ as *orientation-preserving* isometries and to transformations of the form (4.2) as *orientation-reversing* isometries.

Now we will study and classify these two types of isometries of the hyperbolic plane $\mathcal{H}$.

Orientation-preserving isometries. The classification of matrices in $SL(2,\mathbb{R})$ into hyperbolic, elliptic, and parabolic depended on the absolute value of their trace, and hence makes sense in $PSL(2,\mathbb{R})$ as well. A matrix $A = \begin{pmatrix} a & b \\ c & d \end{pmatrix} \in SL(2,\mathbb{R})$ with trace $t = a+d$ is called *hyperbolic* if $|t| > 2$, *elliptic* if $|t| < 2$, and *parabolic* if $|t| = 2$. Let

$$T(z) = \frac{az+b}{cz+d} \in PSL(2,\mathbb{R}).$$

The action of the group $PSL(2,\mathbb{R})$ extends from $\mathcal{H}$ to its *Euclidean boundary* $\mathbb{R} \cup \{\infty\}$, hence $PSL(2,\mathbb{R})$ acts on the *Euclidean closure* of $\mathcal{H}$, denoted by $\widetilde{\mathcal{H}}$. The fixed points of T are found by solving the equation

$$z = \frac{az+b}{cz+d}, \quad \text{i.e.,} \quad cz^2 + (d-a)z - b = 0.$$

We obtain

$$w_1 = \frac{a-d+\sqrt{(a+d)^2-4}}{2c}, \quad w_2 = \frac{a-d-\sqrt{(a+d)^2-4}}{2c}.$$

Notice that $\lambda_i = cw_i + d$ $(i = 1, 2)$ are the eigenvalues of the matrix A. A fixed point w_i of T can be expressed in terms of the eigenvector $\begin{pmatrix} x_i \\ y_i \end{pmatrix}$ with eigenvalue λ_i, namely, $w_i = x_i/y_i$. The derivative at the fixed point w_i can be written in terms of the eigenvalue λ_i as

$$T'(w_i) = \frac{1}{(cw_i + d)^2} = \frac{1}{\lambda_i^2}.$$

We see that if T is hyperbolic, then it has two fixed points in $\mathbb{R} \cup \{\infty\}$, if T is parabolic, it has one fixed point in $\mathbb{R} \cup \{\infty\}$, and if T is elliptic, it has two complex conjugate fixed points, hence one fixed point in $\mathcal{H}$. A Möbius transformation T fixes ∞ if and only if $c = 0$, and hence it is in the form $z \mapsto az + b$ $(a, b \in \mathbb{R},\ a > 0)$. If $a = 1$, it is parabolic; if $a \neq 0$, it is hyperbolic and its second fixed point is $b/(1-a)$.

DEFINITION 4.2. A fixed point w of a transformation $f : \widetilde{\mathcal{H}} \to \widetilde{\mathcal{H}}$ is called *attracting* if $|f'(w)| < 1$, and it is called *repelling* if $|f'(w)| > 1$.

Now we are ready to summarize what we know from linear algebra about different kinds of transformations in $PSL(2, \mathbb{R})$ and describe the action of Möbius transformations in $\mathcal{H}$ geometrically.

1. Hyperbolic case. A hyperbolic transformation $T \in PSL(2, \mathbb{R})$ has two fixed points in $\mathbb{R} \cup \{\infty\}$, one attracting, denoted by u, the other repelling, denoted by w. The geodesic in $\mathcal{H}$ connecting them is called the *axis* of T and is denoted by $C(T)$. By Theorem 3.3, T maps $C(T)$ onto itself, and $C(T)$ is the only geodesic with this property. Let λ be the eigenvalue of a matrix corresponding to T with $|\lambda| > 1$. Then the matrix of T is conjugate to the diagonal matrix $\begin{pmatrix} \lambda & 0 \\ 0 & \frac{1}{\lambda} \end{pmatrix}$ that corresponds to the Möbius transformation

$$\Lambda(z) = \lambda^2 z, \tag{4.3}$$

i.e., there exists a transformation $S \in PSL(2, \mathbb{R})$ such that $STS^{-1} = \Lambda$. The conjugating transformation S maps the axis of T, oriented from u to w, to the positive imaginary axis I, oriented from 0 to ∞, which is the axis of Λ (cf. Exercises 4 and 9).

In order to see how a hyperbolic transformation T acts on $\mathcal{H}$, it is useful to look at the all its iterates T^n, $n \in \mathbb{Z}$. If $z \in C(T)$, then $T^n(z) \in C(T)$ and $T^n(z) \to w$ as $n \to \infty$, while $T^n(z) \to u$ as $n \to -\infty$. The curve $C(T)$ is the only geodesic which is mapped onto itself by T, but there are other T-invariant curves, also "connecting" u and w. For the standard hyperbolic transformation (4.3), the Euclidean rays in the upper half-plane issuing from the origin are obviously T-invariant. If we define the distance from a point z to a given geodesic L as $\inf_{v \in L} \rho(z, v)$, we see that the distance is measured over a geodesic passing through z and orthogonal to L (Exercise 7). Such rays have an important property: they are equidistant from the axis $C(\Lambda) = I$ (see Exercise 8), and hence are called *equidistants*. Under S^{-1} they are mapped onto equidistants for the transformation T, which are Euclidean circles passing through the points u and w (see Figure 4.1).

A useful notion in understanding how hyperbolic transformations act is that of an isometric circle. Since $T'(z) = (cz + d)^{-2}$, the Euclidean lengths are multiplied by $|T'(z)| = |cz + d|^{-2}$. They are unaltered in magnitude if and only if $|cz + d| = 1$.

If $c \neq 0$, then the locus of such points z is the circle

$$\left|z + \frac{d}{c}\right| = \frac{1}{|c|}$$

with center at $-d/c$ and radius $1/|c|$. The circle

$$I(T) = \{z \in \mathcal{H} \mid |cz + d| = 1\}$$

is called the *isometric circle* of the transformation T. Since its center $-d/c$ lies in $\mathbb{R}$, we immediately see that isometric circles are geodesics in $\mathcal{H}$. Further, $T(I(T))$ is a circle of the same radius, $T(I(T)) = I(T^{-1})$, and the transformation maps the outside of $I(T)$ onto the inside of $I(T^{-1})$ and vice versa (see Figure 4.1 and Exercise 10).

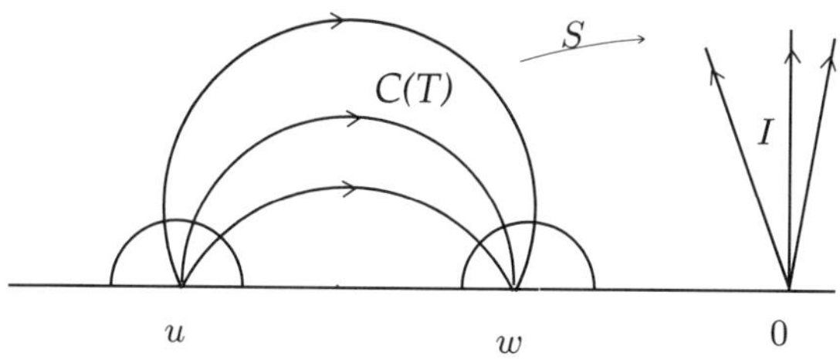

FIGURE 4.1. Hyperbolic transformations

If $c = 0$, then there is no circle with the isometric property: all Euclidean lengths are altered.

2. Parabolic case. A parabolic transformation $T \in PSL(2,\mathbb{R})$ has one fixed point $s \in \mathbb{R} \cup \{\infty\}$, i.e. "at infinity". The transformation T has one eigenvalue $\lambda = \pm 1$ and is conjugate to the transformation $P(z) = z + b$ for some $b \in \mathbb{R}$, i.e., there exists a transformation $S \in PSL(2,\mathbb{R})$ such that $P = STS^{-1}$. The transformation P is an Euclidean translation, and hence it leaves all horizontal lines invariant. Horizontal lines are called *horocycles* for the transformation P. Under the map S^{-1} they are sent to invariant curves—*horocycles*—for the transformation T. Horocycles for T are Euclidean circles tangent to the real line at the parabolic fixed point s (see Figure 4.2 and Exercise 12); we denote a horocycle through a point s at infinity by $\omega(s)$. Figure 4.2 illustrates a family of horocycles through a given point $s \in \mathbb{R}$ and $s = \infty$.

If $c \neq 0$, then the isometric circles for T and T^{-1} are tangent to each other (see Exercise 11). If $c = 0$, then there is no unique circle with the isometric property: in this case T is an Euclidean translation, all Euclidean lengths are unaltered.

3. Elliptic case An elliptic transformation $T \in PSL(2,\mathbb{R})$ has a unique fixed point $e \in \mathcal{H}$. It has the eigenvalues $\lambda = \cos\varphi + i\sin\varphi$ and $\overline{\lambda} = \cos\varphi - i\sin\varphi$, and it is easier to describe its simplest form in the *unit disc model* of hyperbolic geometry: $\mathcal{U} = \{z \in \mathbb{C} \mid |z| < 1\}$. The map

$$f(z) = \frac{zi + 1}{z + i} \tag{4.4}$$

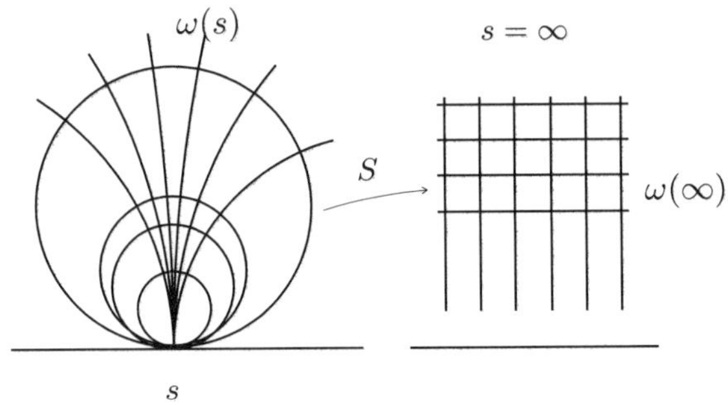

FIGURE 4.2. Parabolic transformations

is a homeomorphism of $\mathcal{H}$ onto $\mathcal{U}$. The distance in $\mathcal{U}$ is induced by means of the hyperbolic distance in $\mathcal{H}$:

$$\rho(z, w) = \rho(f^{-1}z, f^{-1}w) \ (z, w \in \mathcal{U}).$$

The readily verified formula

$$\frac{2|f'(z)|}{1 - |f(z)|^2} = \frac{1}{\mathrm{Im(z)}}$$

implies that this distance in $\mathcal{U}$ is derived from the metric

$$ds = \frac{2|dz|}{1 - |z|^2}.$$

Geodesics in the unit disc model are circular arcs and diameters orthogonal to the *principle circle* $\Sigma = \{z \in \mathbb{C} \mid |z| = 1\}$, the *Euclidean boundary* of $\mathcal{U}$. Isometries of $\mathcal{U}$ are the conjugates of isometries of $\mathcal{H}$, i.e., we can write

$$S = f \circ T \circ f^{-1} \ (T \in PSL(2, \mathbb{R})).$$

Exercise 13 shows that orientation-preserving isometries of $\mathcal{U}$ are of the form

$$z \mapsto \frac{az + \bar{c}}{cz + \bar{a}} \ (a, c \in \mathbb{C}, a\bar{a} - c\bar{c} = 1),$$

and the transformation corresponding to the standard reflection $R(z) = -\bar{z}$ is also the reflection of $\mathcal{U}$ in the vertical diameter.

Let us return to our elliptic transformation $T \in PSL(2, \mathbb{R})$ that fixes $e \in \mathcal{H}$. Conjugating T by f, we obtain an elliptic transformation of the unit disc $\mathcal{U}$. Using an additional conjugation by an orientation-preserving isometry of $\mathcal{U}$ if necessary (see Exercise 14), we bring the fixed point to 0, and hence bring T to the form $z \mapsto e^{2i\varphi}z$. In other words, an elliptic transformation with eigenvalues $e^{i\varphi}$ and $e^{-i\varphi}$ is conjugate to a rotation by 2φ.

EXAMPLE A. Let $z \mapsto -1/z$ be the elliptic transformation given by the matrix $\begin{pmatrix} 0 & -1 \\ 1 & 0 \end{pmatrix}$. Its fixed point in $\mathcal{H}$ is i. It is a transformation of order 2 since the identity in $PSL(2, \mathbb{R})$ is $\{1_2, -1_2\}$, and hence is a half-turn. In the unit disc model, its matrix is conjugate to the matrix $\begin{pmatrix} i & 0 \\ 0 & -i \end{pmatrix}$.

Orientation-reversing isometries. The simplest orientation-reversing isometry of $\mathcal{H}$ is the transformation $R(z) = -\overline{z}$, a reflection in the imaginary axis I, and hence it fixes I pointwise. It is also a hyperbolic reflection in I, i.e., if for each point z we draw a geodesic through z, orthogonally to I that intersects I at the point z_0, then $R(z) = z'$ is on the same geodesic and $\rho(z', z_0) = \rho(z, z_0)$. Let L be any geodesic in $\mathcal{H}$ and $T \in PSL(2,\mathbb{R})$ be any Möbius transformation. Then the transformation

$$TRT^{-1} \tag{4.5}$$

fixes the geodesic $L = T(I)$ pointwise and therefore may be regarded as a "reflection in the geodesic L". In fact, it is the well-known geometrical transformation called *inversion in a circle* (see Exercise 16).

DEFINITION 4.3. Let Q be a circle in $\mathbb{R}^2$ with center K and radius r. Given any point $P \neq K$ in $\mathbb{R}^2$, a point P_1 is called *inverse* to P if

(a) P_1 lies on the ray from K to P,
(b) $|KP_1| \cdot |KP| = r^2$.

The relationship is reciprocal: if P_1 is inverse to P, then P is inverse to P_1. We say that P and P_1 are *inverse with respect to Q*. Obviously, inversion fixes all points of the circle Q. Inversion may be described by a geometric construction (see Exercise 15). We will derive a formula for it. Let P, P_1 and K be the points z, z_1, and k in $\mathbb{C}$. Then the definition can be rewritten as

$$|(z_1 - k)(z - k)| = r^2, \quad \arg(z_1 - k) = \arg(z - k).$$

Since $\arg(z - k) = -\arg(\overline{z} - \overline{k})$, both equations are satisfied if and only if

$$(z_1 - k)(\overline{z} - \overline{k}) = r^2. \tag{4.6}$$

This gives us the following formula for the inversion in a circle:

$$z_1 = \frac{k\overline{z} + r^2 - |k|^2}{\overline{z} - \overline{k}}. \tag{4.7}$$

Now we are able to prove a theorem for isometries of the hyperbolic plane similar to a result in Euclidean geometry.

THEOREM 4.4. *Every isometry of $\mathcal{H}$ is a product of not more than three reflections in geodesics in $\mathcal{H}$.*

PROOF. By Theorem 4.1 it suffices to show that each transformation from the group $PSL(2,\mathbb{R})$ is a product of two reflections. Let

$$T(z) = \frac{az + b}{cz + d}.$$

First consider the case for which $c \neq 0$. Then both T and T^{-1} have well-defined isometric circles (see Exercise 11). They have the same radius $1/|c|$ and their centers are on the real axis at $-d/c$ and a/c, respectively. We will show that $T = R \circ R_{I(T)}$, where $R_{I(T)}$ is the reflection in the isometric circle $I(T)$, or inversion, and R is the reflection in the vertical geodesic passing through the midpoint of the interval $[-d/c, a/c]$. To do this, we use formula (4.6) for inversion:

$$R_{I(T)}(z) = \frac{-\frac{d}{c}\overline{z} + \frac{1}{c^2} - \frac{d^2}{c^2}}{\overline{z} + \frac{d}{c}} = \frac{-d(\overline{z} + \frac{d}{c}) + \frac{1}{c}}{c\overline{z} + d}.$$

The reflection in the line $x = (a-d)/2c$ is given by the formula

$$R(z) = -\overline{z} + 2\frac{a-d}{2c}.$$

Combining the two, we obtain

$$R \circ R_{I(T)} = \frac{az+b}{cz+d}.$$

Now if $c = 0$, the transformation T may be either parabolic $z \mapsto z + b$ or hyperbolic $z \mapsto \lambda^2 z + b$, each fixing ∞. In the first case, the theorem follows from the Euclidean result for translations. For $T(z) = \lambda^2 z + b$, it is easy to see that the reflections should be in circles of radii 1 and λ centered at the second fixed point. □

5. Hyperbolic area and the Gauss-Bonnet formula

Let T be a Möbius transformation. The *differential* of T, denoted by DT, at a point z is the linear map that takes the tangent space $T_z\mathcal{H}$ onto $T_{T(z)}\mathcal{H}$ and is defined by the 2×2 matrix

$$DT = \begin{pmatrix} \frac{\partial u}{\partial x} & \frac{\partial u}{\partial y} \\ \frac{\partial v}{\partial x} & \frac{\partial v}{\partial y} \end{pmatrix}.$$

THEOREM 5.1. *Let $T \in PSL(2,\mathbb{R})$. Then DT preserves the norm in the tangent space at each point.*

PROOF. For $\zeta \in T_z\mathcal{H}$, we have $DT(\zeta) = T'(z)\zeta$ by Exercise 22. Since

$$|T'(z)| = \frac{\mathrm{Im(T(z))}}{\mathrm{Im(z)}} = \frac{1}{|cz+d|^2},$$

we can write

$$\|DT(\zeta)\| = \frac{|DT(\zeta)|}{\mathrm{Im(T(z))}} = \frac{|T'(z)||\zeta|}{\mathrm{Im(T(z))}} = \frac{|\zeta|}{\mathrm{Im(z)}} = \|\zeta\|.$$

□

COROLLARY 5.2. *Any transformation in $PSL(2,\mathbb{R})$ is conformal, i.e., it preserves angles.*

PROOF. It is easy to prove the *polarization identity*, which asserts that for any $\zeta_1, \zeta_2 \in T_z\mathcal{H}$ we have

$$\langle \zeta_1, \zeta_2 \rangle = \frac{1}{2}(\|\zeta_1\|^2 + \|\zeta_2\|^2 - \|\zeta_1 - \zeta_2\|^2);$$

this identity implies that the inner product and hence the absolute value of the angle between tangent vectors is also preserved. Since Möbius transformations preserve orientation, the corollary follows. □

Let $A \subset \mathcal{H}$. We define the *hyperbolic area* of A by the formula

$$\mu(A) = \int_A \frac{dxdy}{y^2}, \tag{5.1}$$

provided this integral exists.

THEOREM 5.3. *Hyperbolic area is invariant under all Möbius transformations $T \in PSL(2,\mathbb{R})$, i.e., if $\mu(A)$ exists, then $\mu(A) = \mu(T(A))$.*

PROOF. It follows immediately from the preservation of Riemannian metric (Theorem 5.1). Here is a direct calculation as well. When we performed the change of variables $w = T(z)$ in the line integral of Theorem 2.5, the coefficient $|T'(z)|$ appeared (it is the coefficient responsible for the change of Euclidean lengths). If we carry out the same change of variables in the plane integral, the Jacobian of this map will appear, since it is responsible for the change of the Euclidean areas. Let $z = x + iy$, and $w = T(z) = u + iv$.

The Jacobian is the determinant of the differential map DT and is customarily denoted by $\partial(u,v)/\partial(x,y)$. Thus

$$\frac{\partial(u,v)}{\partial(x,y)} := \det\begin{pmatrix} \frac{\partial u}{\partial x} & \frac{\partial u}{\partial y} \\ \frac{\partial v}{\partial x} & \frac{\partial v}{\partial y} \end{pmatrix} = \left(\frac{\partial u}{\partial x}\right)^2 + \left(\frac{\partial v}{\partial x}\right)^2 = |T'(z)|^2 = \frac{1}{|cz+d|^4}. \tag{5.2}$$

We use this expression to compute the integral

$$\mu(T(A)) = \int_{T(A)} \frac{dudv}{v^2} = \int_A \frac{\partial(u,v)}{\partial(x,y)} \frac{dxdy}{v^2}$$
$$= \int_A \frac{1}{|cz+d|^4} \frac{|cz+d|^4}{y^2} dxdy = \mu(A),$$

as claimed. □

A *hyperbolic triangle* is a figure bounded by three segments of geodesics. The intersection points of these geodesics are called the *vertices* of the triangle. We allow vertices to belong to $\mathbb{R} \cup \{\infty\}$. There are four types of hyperbolic triangles, depending on whether 0, 1, 2, or 3 vertices belong to $\mathbb{R} \cup \{\infty\}$ (see Figure 5.1).

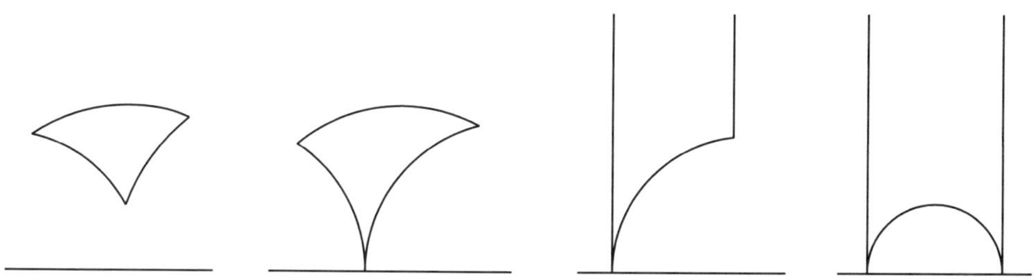

FIGURE 5.1. Hyperbolic triangles

The Gauss-Bonnet formula shows that the hyperbolic area of a hyperbolic triangle depends only on its angles.

THEOREM 5.4 (Gauss-Bonnet). *Let Δ be a hyperbolic triangle with angles α, β, and γ. Then $\mu(\Delta) = \pi - \alpha - \beta - \gamma$.*

PROOF. First we consider the case in which one of the vertices of the triangle belongs to $\mathbb{R} \cup \{\infty\}$. Since transformations from $PSL(2,\mathbb{R})$ do not alter the area and the angles of a triangle, we may apply the transformation from $PSL(2,\mathbb{R})$ which maps this vertex to ∞ and the base to a segment of the unit circle (as in Figure 5.2), and prove the formula in this case. The angle at infinity is equal to 0, and let us assume that the other two angles are equal to α and β.

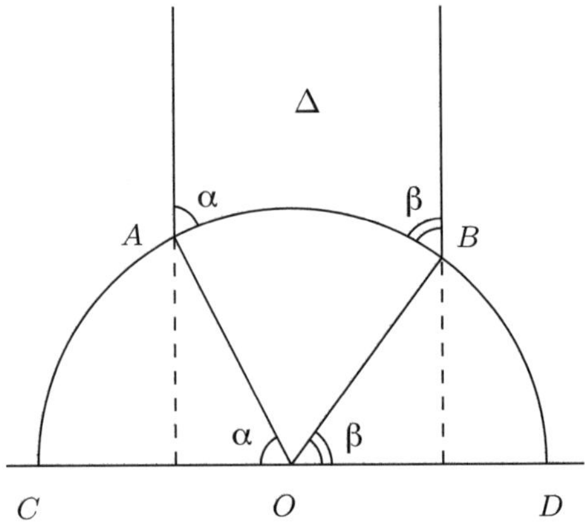

FIGURE 5.2. Proof of the Gauss-Bonnet formula

Since the angle measure in the hyperbolic plane coincides with the Euclidean angle measure, the angles AOC and BOD are equal to α and β, respectively, as angles with mutually perpendicular sides. Assume the vertical geodesics are the lines $x = a$ and $x = b$. Then

$$\mu(\Delta) = \int_\Delta \frac{dxdy}{y^2} = \int_a^b dx \int_{\sqrt{1-x^2}}^\infty \frac{dy}{y^2} = \int_a^b \frac{dx}{\sqrt{1-x^2}}.$$

The substitution $x = \cos\theta$ $(0 \le \theta \le \pi)$ gives

$$\mu(\Delta) = \int_{\pi-\alpha}^{\beta} \frac{-\sin\theta d\theta}{\sin\theta} = \pi - \alpha - \beta.$$

For the case in which Δ has no vertices at infinity, we continue the geodesic connecting the vertices A and B, and suppose that it intersects the real axis at the point D (if one side of Δ is a vertical geodesic, then we label its vertices A and B), and draw a geodesic from C to D. Then we obtain the situation depicted in Figure 5.3.

We denote the triangle ADC by Δ_1 and the triangle CBD by Δ_2. Our formula has already been proved for triangles such as Δ_1 and Δ_2, since the vertex D is at infinity. Now we can write

$$\begin{aligned}\mu(\Delta) = \mu(\Delta_1) - \mu(\Delta_2) &= (\pi - \alpha - \gamma - \theta) - (\pi - \theta - \pi + \beta) \\ &= \pi - \alpha - \beta - \gamma,\end{aligned}$$

as claimed. □

Theorem 5.4 asserts that the area of a triangle depends only on its angles, and is equal to the quantity $\pi - \alpha - \beta - \gamma$, which is called the *angular defect.* Since the area of a nondegenerate triangle is positive, the angular defect is positive, and therefore, in hyperbolic geometry the sum of angles of any triangle is less than π. We will also see that there are no similar triangles in hyperbolic geometry (except isometric ones).

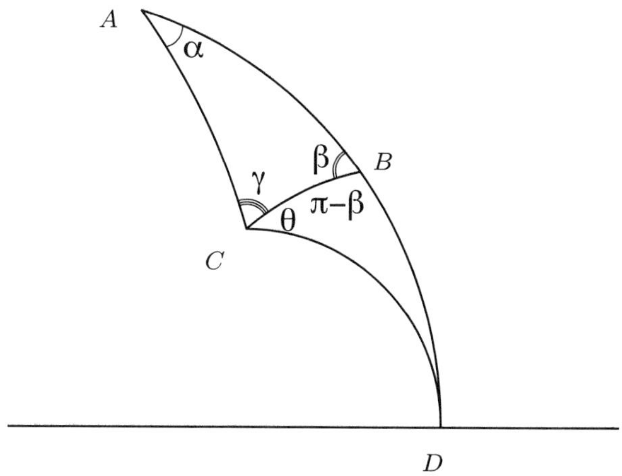

FIGURE 5.3. A general case in the Gauss-Bonnet formula

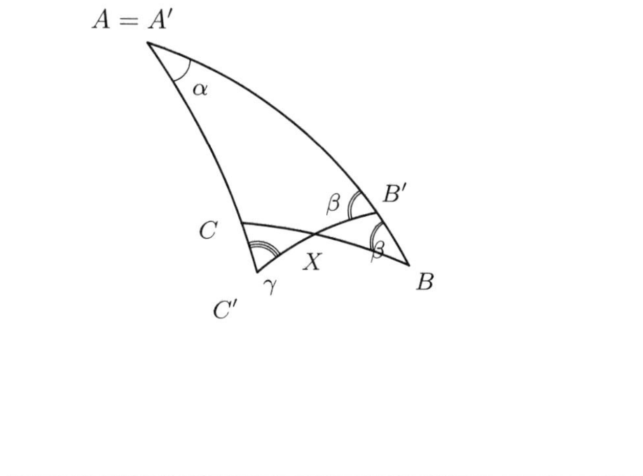

FIGURE 5.4. There are no similar triangles in hyperbolic geometry

THEOREM 5.5. *If two triangles have the same angles, then there is an isometry mapping one triangle into the other.*

PROOF. If necessary, we perform the reflection $z \mapsto -\overline{z}$, so that the respective angles of the triangles ABC and $A'B'C'$ (in the clockwise direction) are equal. Then we apply a hyperbolic transformation mapping A to A' (Exercise 14), and an elliptic transformation mapping the side AB onto the side $A'B'$. Since the angles CAB and $C'A'B'$ are equal, the side AC will be mapped onto the side $A'C'$. We must prove that B is then mapped to B' and C to C'. Assume B' is mapped inside the geodesic segment AB. If we had $C' \in [A, C]$, the areas of triangles ABC and $A'B'C'$ would not be equal, which contradicts Theorem 5.4. Therefore C must belong to the side $A'C'$, and hence the sides BC and $B'C'$ intersect at a point X (see Figure 5.4); thus we obtain the triangle $B'XB$. Its angles are β and $\pi - \beta$

since the angles at the vertices B and B' of our original triangles are equal (to β). Then the sum of the angles of the triangle $B'XB$ is at least π, in contradiction with Theorem 5.4.

□

6. Hyperbolic trigonometry

Let us consider a general hyperbolic triangle with sides of hyperbolic length a, b, c and opposite angles α, β, γ. We assume that α, β, and γ are positive (so a, b, and c are finite) and prove the following results.

THEOREM 6.1. *(i) The Sine Rule:* $\frac{\sinh a}{\sin\alpha} = \frac{\sinh b}{\sin\beta} = \frac{\sinh c}{\sin\gamma}$.

(ii) The Cosine Rule I: $\cosh c = \cosh a \cosh b - \sinh a \sinh b \cos\gamma$.

(iii) The Cosine Rule II: $\cosh c = \frac{\cos\alpha\cos\beta+\cos\gamma}{\sin\alpha\sin\beta}$.

REMARK. Note that Cosine Rule II implies that if two triangles have the same angles, then their sides are also equal, and therefore it has no analogue in Euclidean geometry. It also gives an alternative proof of Theorem 5.5.

PROOF OF (ii). Let us denote the vertices opposite the sides a, b, c by v_a, v_b, v_c respectively. We shall use the model $\mathcal{U}$ and may assume that $v_c = 0$ and $\operatorname{Im} v_a = 0$, $\operatorname{Re} v_a > 0$ (see Figure 6.1).

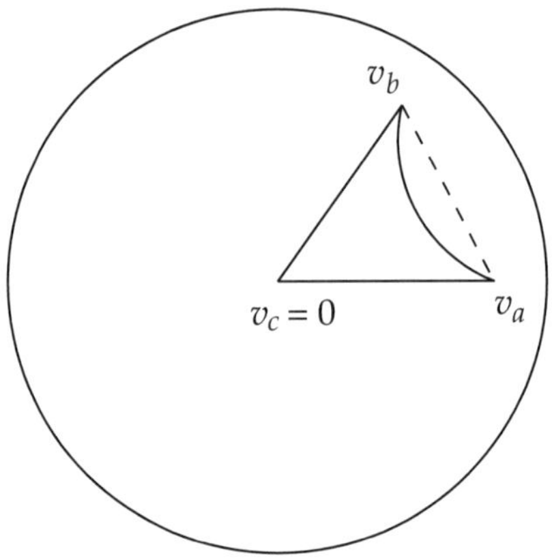

FIGURE 6.1. The Cosine Rule I

By Exercise 20(iv) we have

$$v_a = \tanh\frac{1}{2}\rho(0, v_a) = \tanh(\frac{1}{2}b), \tag{6.1}$$

and similarly,

$$v_b = e^{i\gamma}\tanh(\frac{1}{2}a), \tag{6.2}$$

We have $c = \rho(v_a, v_b)$, and from Exercise 20(iii)

$$\cosh c = \sinh^2[\frac{1}{2}\rho(v_a, v_b)] + 1 = \frac{2|v_a - v_b|^2}{(1-|v_a|^2)(1-|v_b|^2)} + 1. \tag{6.3}$$

The right-hand side of expression (6.3) is equal to $\cosh a \cosh b - \sinh a \sinh b \cos \gamma$ by Exercise 23, and hence (ii) follows. ☐

PROOF OF (i). Using (ii) we obtain

$$\left(\frac{\sinh c}{\sin \gamma}\right)^2 = \frac{\sinh^2 c}{1 - \left(\frac{\cosh a \cosh b - \cosh c}{\sinh a \sinh b}\right)^2}. \tag{6.4}$$

The Sine Rule will be valid if we prove that the expression on the right-hand side of (6.4) is symmetric in a, b, and c. This follows from the symmetry of

$$(\sinh a \sinh b)^2 - (\cosh a \cosh b - \cosh c)^2$$

which is obtained by a direct calculation. ☐

PROOF OF (iii). Let us write A for $\cosh a$, B for $\cosh b$, and C for $\cosh c$. The Cosine Rule I yields

$$\cos \gamma = \frac{(AB - C)}{(A^2 - 1)^{\frac{1}{2}}(B^2 - 1)^{\frac{1}{2}}}$$

and so

$$\sin^2 \gamma = \frac{D}{(A^2 - 1)(B^2 - 1)}$$

where $D = 1 + 2ABC - (A^2 + B^2 + C^2)$ is symmetric in A, B, and C. The expression for $\sin^2 \gamma$ shows that $D \geq 0$. Using analogous expressions for $\cos \alpha$, $\sin \alpha$, $\cos \beta$, and $\sin \beta$ we observe that if we multiply both the numerator and denominator of

$$\frac{\cos \alpha \, \cos \beta + \cos \gamma}{\sin \alpha \, \sin \beta}$$

by the positive value of

$$(A^2 - 1)^{\frac{1}{2}}(B^2 - 1)^{\frac{1}{2}}(C^2 - 1)^{\frac{1}{2}}$$

we obtain

$$\frac{\cos \alpha \cos \beta + \cos \gamma}{\sin \alpha \sin \beta} = \frac{[(BC - A)(CA - B) + (AB - C)(C^2 - 1)]}{D} = C$$

☐

THEOREM 6.2. (Pythagorean Theorem) If $\gamma = \frac{\pi}{2}$ we have $\cosh c = \cosh a \cosh b$.

PROOF. Immediate from the Cosine Rule I. ☐

Exercises

1. Prove that the metric in the Poincaré disc model is given by

$$ds^2 = \frac{4(d\eta_1^2 + d\eta_2^2)}{(1 - (\eta_1^2 + \eta_2^2))^2}.$$

2. Prove that the metric in the upper half-plane model is given by

$$ds^2 = \frac{d\eta_1^2 + d\eta_3^2}{\eta_3^2}.$$

3. Prove that if $z \neq w$, then $\rho(z, w) > 0$.

4. Let L be a semicircle or a straight line orthogonal to the real axis which meets the real axis at a point α. Prove that the transformation

$$T(z) = -(z-\alpha)^{-1} + \beta \in PSL(2,\mathbb{R}),$$

for an appropriate value of β, maps L to the positive imaginary axis.

5. Prove that for $z, w \in \mathcal{H}$ and $T \in PSL(2,\mathbb{R})$, we have

$$|T(z) - T(w)| = |z-w||T'(z)T'(w)|^{1/2}.$$

6. Prove that isometies are continuous maps.

7. (a) Prove that there is a unique geodesic through a point z orthogonal to a given geodesic L.

(b)* Give a geometric construction of this geodesic.

(c) Prove that for $z \notin L$, the greatest lower bound $\inf_{v\in L} \rho(z,v)$ is achieved on the geodesic described in (a).

8. Prove that the rays in $\mathcal{H}$ issuing from the origin are equidistant from the positive imaginary axis I.

9. Let $A \in PSL(2,\mathbb{R})$ be a hyperbolic transformation, and suppose that $B = SAS^{-1}$ ($B \in PSL(2,\mathbb{R})$) is its conjugate. Prove that B is also hyperbolic and find the relation between their axes $C(A)$ and $C(B)$.

10. Prove that isometric circles $I(T)$ and $I(T^{-1})$ have the same radius, and that the image of $I(T)$ under the transformation T is $I(T^{-1})$.

11. Prove that

(a) T is hyperbolic if and only if $I(T)$ and $I(T^{-1})$ do not intersect;
(b) T is elliptic if and only if $I(T)$ and $I(T^{-1})$ intersect;
(c) T is parabolic if and only if $I(T)$ and $I(T^{-1})$ are tangential.

12. Prove that the horocycles for a parabolic transformation with a fixed point $p \in \mathbb{R}$ are Euclidean circles tangent to the real line at p.

13. Show that orientation-preserving isometries of $\mathcal{U}$ are of the form

$$z \mapsto \frac{az+\bar{c}}{cz+\bar{a}} \qquad (a, c \in \mathbb{C}, a\bar{a} - c\bar{c} = 1).$$

14. Prove that for any two distinct points $z_1, z_2 \in \mathcal{H}$ there exists a transformation $T \in PSL(2,\mathbb{R})$ such that $T(z_1) = z_2$.

15. Give a geometric construction of the inversion in a given circle Q in the Euclidean plane $\mathbb{R}^2$.

16. Prove that the transformation (4.5) is an inversion in the circle corresponding to the geodesic L.

17. Prove that any orientation-preserving isometry T of the unit disc $\mathcal{U}$ is an inversion in $I(T)$ followed by a reflection in the straight line L, the Euclidean bisector between the centers of the isometric circles $I(T)$ and $I(T^{-1})$.

18. Prove that two hyperbolic transformations in $PSL(2,\mathbb{R})$ commute if and only if their axes coincide.

19. Let $A \in PSL(2,\mathbb{R})$ be hyperbolic and $B \in PSL(2,\mathbb{R})$ be an elliptic transformation different from the identity. Prove that $AB \neq BA$.

20. Use the map f (4.4) to derive the formulae for the hyperbolic distance in the unit disc model similar to those in Theorem 3.5, for $z, w \in \mathcal{U}$:

(i) $\rho(z,w) \in \ln \frac{|1-z\bar{w}|+|z-w|}{|1-z\bar{w}|-|z-w|}$,

(ii) $\cosh^2[\frac{1}{2}\rho(z,w)] = \frac{|1-z\bar{w}|^2}{(1-|z|^2)(1-|w|^2)}$,

(iii) $\sinh^2[\frac{1}{2}\rho(z,w)] = \frac{|z-w|^2}{(1-|z|^2)(1-|w|^2)}$,

(iv) $\tanh[\frac{1}{2}\rho(z,w)] = |\frac{z-w}{1-z\bar{w}}|$.

21. Justify the calculations in (5.2) by checking that for the Möbius transformation

$$w = T(z) = \frac{az+b}{cz+d} \quad \text{with} \quad z = x+iy,\ w = u+iv$$

we have

$$\frac{\partial u}{\partial x} = \frac{\partial v}{\partial y}, \quad \frac{\partial v}{\partial x} = -\frac{\partial u}{\partial y}$$

(these are the classical *Cauchy-Riemann equations*) and

$$T'(z) = \frac{dw}{dz} = \frac{1}{2}\Big(\frac{\partial w}{\partial x} - i\frac{\partial w}{\partial y}\Big) = \frac{\partial u}{\partial x} + i\frac{\partial v}{\partial x};$$

(*Hint*: express x and y in terms of z and $\overline{z}$ and use the Cauchy-Riemann equations.)

22. If we identify the tangent space $T_z\mathcal{H} \approx \mathbb{R}^2$ with the complex plane $\mathbb{C}$ by means of the map

$$\begin{pmatrix}\xi\\ \eta\end{pmatrix} \mapsto \xi + i\eta = \zeta,$$

then $DT(\zeta) = T'(z)\zeta$, where in the left-hand side we have a linear transformation of $T_z\mathcal{H} \approx \mathbb{R}^2$, and in the right-hand side, the multiplication of two complex numbers.

23. Show that the right-hand side of expression (6.3) is equal to $\cosh a \cosh b - \sinh a \sinh b \cos\gamma$.

Lecture II. Fuchsian Groups and Their Fundamental Regions

7. The group $PSL(2,\mathbb{R})$

Let $S\mathcal{H}$ be the unit tangent bundle of the upper half-plane $\mathcal{H}$. It is homeomorphic to $\mathcal{H} \times S^1$. Let us parametrize it by local coordinates (z,ζ), where $z \in \mathcal{H}$, $\zeta \in \mathbb{C}$ with $|\zeta| = \mathrm{Im}(z)$. (Notice that with this parametrization, $\|\zeta\| = 1$ (see (2.2)), so that ζ is a unit tangent vector.) The group $PSL(2,\mathbb{R})$ acts on $S\mathcal{H}$ by the differentials: for $T : z \to \frac{az+b}{cz+d}$, $T(z,\zeta) = (T(z), DT(\zeta))$, where

$$DT(\zeta) = \frac{1}{(cz+d)^2}\,\zeta. \tag{7.1}$$

As any group, $PSL(2,\mathbb{R})$ acts on itself by left multiplication. The next result connects these two actions.

THEOREM 7.1. *There is a homeomorphism between $PSL(2,\mathbb{R})$ and the unit tangent bundle $S\mathcal{H}$ of the upper half-plane $\mathcal{H}$ such that the action of $PSL(2,\mathbb{R})$ on itself by left multiplication corresponds to the action of $PSL(2,\mathbb{R})$ on $S\mathcal{H}$ induced by its action on $\mathcal{H}$ by fractional linear transformations.*

PROOF. Let (i,ι) be a fixed element of $S\mathcal{H}$, where ι is the unit vector at the point i tangent to the imaginary axis and pointed upwards, and let (z,ζ) be an arbitrary element of $S\mathcal{H}$. There exists a unique $T \in PSL(2,\mathbb{R})$ sending the imaginary axis to the geodesic passing through z and tangent to ζ (Exercise 4) so that $T(i) = z$. By (7.1) we have $DT(\iota) = \zeta$, and hence

$$T(i,\iota) = (z,\zeta). \tag{7.2}$$

It is easy to see that the map $(z,\zeta) \to T$ is a homeomorphism between $S\mathcal{H}$ and $PSL(2,\mathbb{R})$.

For $S \in PSL(2,\mathbb{R})$, suppose that $S(z,\zeta) = (z',\zeta')$. By (7.2) $S(z,\zeta) = ST(i,\iota)$, and hence $S(z,\zeta) \to ST$, and the last assertion follows. □

Let $d\ell = \sqrt{ds^2 + d\theta^2}$ be a Riemannian metric on $S\mathcal{H}$, where ds is the hyperbolic metric on $\mathcal{H}$ (2.1), and $\theta = \frac{1}{2\pi}\arg(\zeta)$; and let $dv = d\mu d\theta$ be the canonical (Liouville) volume on $S\mathcal{H}$, where $d\mu$ is the hyperbolic area on $\mathcal{H}$ (5.1).

PROPOSITION 7.2. *The metric $d\ell$ and the volume dv on $S\mathcal{H}$ are $PSL(2,\mathbb{R})$-invariant.*

PROOF. This can be seen by a direct calculation. Let $f(z) = \frac{az+b}{cz+d} \in PSL(2,\mathbb{R})$. In local coordinates $(z,\zeta) \mapsto (f(z),(Df)(\zeta)) = (z',\zeta')$. The metric $d\ell$ on $S\mathcal{H}$ is a norm in the tangent space to $S\mathcal{H}$:

$$\|(dz,d\zeta)\|^2 = \frac{|dz|^2}{y^2} + (d\phi)^2.$$

Since each summand is invariant:

$$\frac{|dz'|^2}{(\mathrm{Imf(z)})^2} = \frac{|f'(z)|^2|dz|^2}{(\mathrm{Imf(z)})^2} = \frac{|dz|^2}{y^2} \text{ and } (d\phi')^2 = (d\phi)^2,$$

the invariance of the Riemannian metric $d\ell$ follows. The invariance of the volume dv follows from the invariance of the metric. □

Thus, besides being a group, $PSL(2,\mathbb{R})$ is also a topological space. Convergence in $PSL(2,\mathbb{R})$ can be expressed in the matrix language. If $g_n \to g$ in $PSL(2,\mathbb{R})$, this means that there exist matrices $A_n \in SL(2,\mathbb{R})$ representing g_n such that $\lim_{n\to\infty} \|A_n - A\| = 0$, where $\|\cdot\|$ is a norm on $SL(2,\mathbb{R})$ induced from $\mathbb{R}^4$.

DEFINITION 7.3. A subgroup Γ of $\mathrm{Isom}(\mathcal{H})$ is called *discrete* if the induced topology on Γ is a discrete topology, i.e. if Γ is a discrete set in the topological space $\mathrm{Isom}(\mathcal{H})$.

It follows that Γ is discrete if and only if $T_n \to \mathrm{Id}$, $T_n \in \Gamma$ implies $T_n = \mathrm{Id}$ for sufficiently large n.

8. Discrete and properly discontinuous groups

DEFINITION 8.1. A discrete subgroup of Isom $(\mathcal{H})$ is called a *Fuchsian group* if it consists of orientation-preserving transformations, in other words, a Fuchsian group is a discrete subgroup of $PSL(2,\mathbb{R})$.

For any discrete group Γ of Isom $(\mathcal{H})$, its subgroup Γ^+ of index ≤ 2 consisting of orientation-preserving transformations is a Fuchsian group. Thus the main ingredient in the study of discrete subgroups of isometries of $\mathcal{H}$ is the study of Fuchsian groups. The action of $PSL(2,\mathbb{R})$ on $\mathcal{H}$ lifts to the action on its unit tangent bundle $S\mathcal{H}$ by isometries (Proposition 7.2), thus sometimes it is useful to consider Fuchsian groups as discrete groups of isometries of $S\mathcal{H}$. Certain discrete subgroups of Lie groups are called *lattices* by analogy with lattices in $\mathbb{R}^n$ that are discrete groups of isometries of $\mathbb{R}^n$. The latter have the following important property: their action on $\mathbb{R}^n$ is *discontinuous* in the sense that every point of $\mathbb{R}^n$ has a neighborhood which is carried outside itself by all elements of the lattice except for the identity. In general, discrete groups of isometries do not have such discontinuous behavior, for if some elements have fixed points these points cannot have such a neighborhood. However, they satisfy a slightly weaker discontinuity condition. First we need several definitions.

Let X be a locally compact metric space, and let G be a group of isometries of X.

DEFINITION 8.2. A family $\{M_\alpha \mid \alpha \in A\}$ of subsets of X indexed by elements of a set A is called *locally finite* if for any compact subset $K \subset X$, $M_\alpha \bigcap K \neq \emptyset$ for only finitely many $\alpha \in A$.

REMARK. Some of the subsets M_α may coincide but they are still considered different elements of the family.

DEFINITION 8.3. For $x \in X$, a family $Gx = \{g(x) \mid g \in G\}$ is called the G-orbit of the point x. Each point of Gx is contained with a multiplicity equal to the order of G_x, the *stabilizer of x in G*.

DEFINITION 8.4. We say that a group G acts properly discontinuously on X if the G-orbit of any point $x \in X$ is locally finite.

Since X is locally compact, a group G acts properly discontinuously on X if and only if each orbit has no accumulation point in X, and the order of the stabilizer of each point is finite. The first condition, however, is equivalent to the fact that each orbit of G is discrete. For, if $g_n(x) \to s \in X$, then for any $\varepsilon > 0$, $\rho(g_n(x), g_{n+1}(x)) < \varepsilon$ for sufficiently large n, but since g_n is an isometry, we have $\rho(g_n^{-1} g_{n+1}(x), x) < \varepsilon$, which implies that x is an accumulation point for its orbit Gx, i.e. Gx is not discrete. In fact, the discreteness of all orbits already implies the discreteness of the group (see Corollary 8.7 for subgroups of $PSL(2,\mathbb{R})$).

EXAMPLE B. Let us consider a group consisting of all transformations

$$z \to \frac{az+b}{cz+d} \; (a,b,c,d \in \mathbb{Z},\ ad-bc=1).$$

It is called the *modular group* and denoted by $PSL(2,\mathbb{Z}) \approx SL(2,\mathbb{Z})/\{\pm 1_2\}$.

It is clearly a discrete subgroup of $PSL(2,\mathbb{R})$ and hence a Fuchsian group.

Our next task is to show that a $\Gamma \subset PSL(2,\mathbb{R})$ is a Fuchsian group if and only if it acts properly discontinuously on $\mathcal{H}$.

LEMMA 8.5. *Let $z_0 \in \mathcal{H}$ be given and let K be a compact subset of $\mathcal{H}$. Then the set*

$$E = \{T \in PSL(2,\mathbb{R}) \mid T(z_0) \in K\}$$

is compact.

PROOF. $PSL(2,\mathbb{R})$ is topologized as a quotient space of $SL(2,\mathbb{R})$. Thus we have a continuous map $\psi : SL(2,\mathbb{R}) \to PSL(2,\mathbb{R})$ defined by

$$\psi \begin{bmatrix} a & b \\ c & d \end{bmatrix} = T, \text{ where } T(z) = \frac{az+b}{cz+d}.$$

If we show that

$$E_1 = \left\{ \begin{bmatrix} a & b \\ c & d \end{bmatrix} \in SL(2,\mathbb{R}) \mid \frac{az_0+b}{cz_0+d} \in K \right\}$$

is compact then it follows that $E = \psi(E_1)$ is compact. We prove that E_1 is compact by showing it is closed and bounded when regarded as a subset of $\mathbb{R}^4$ (identifying $\begin{bmatrix} a & b \\ c & d \end{bmatrix}$ with (a,b,c,d)). We have a continuous map $\beta : SL(2,\mathbb{R}) \to \mathcal{H}$ defined by $\beta(A) = \psi(A)(z_0)$. $E_1 = \beta^{-1}(K)$, thus it follows that E_1 is closed as the inverse image of the closed set K.

We now show that E_1 is bounded. As K is bounded there exists $M_1 > 0$ such that

$$\left| \frac{az_0+b}{cz_0+d} \right| < M_1,$$

for all $\begin{bmatrix} a & b \\ c & d \end{bmatrix} \in E_1$.

Also, as K is compact in $\mathcal{H}$, there exists $M_2 > 0$ such that

$$\operatorname{Im}\left(\frac{az_0+b}{cz_0+d}\right) \geq M_2.$$

(2.5) implies that the left-hand side of this inequality is $\operatorname{Im}(z_0)/|cz_0+d|^2$ so that

$$|cz_0+d| \leq \sqrt{\left(\frac{\operatorname{Im}(z_0)}{M_2}\right)},$$

and thus

$$|az_0+b| \leq M_1 \sqrt{\left(\frac{\operatorname{Im}(z_0)}{M_2}\right)},$$

and we deduce that a, b, c, d are bounded. □

THEOREM 8.6. *Let Γ be a subgroup of $PSL(2,\mathbb{R})$. Then Γ is a Fuchsian group if and only if Γ acts properly discontinuously on $\mathcal{H}$.*

PROOF. We first show that a Fuchsian group acts properly discontinuously on $\mathcal{H}$. Let $z \in \mathcal{H}$ and K be a compact subset of $\mathcal{H}$. We use Lemma 8.5 to see that $\{T \in \Gamma \mid T(z) \in K\} = \{T \in PSL(2,\mathbb{R}) \mid T(z) \in K\} \bigcap \Gamma$ is a finite set (it is the intersection of a compact and a discrete set), and hence Γ acts properly discontinuously. Conversely, suppose Γ acts properly discontinuously, but it is not a discrete subgroup of $PSL(2,\mathbb{R})$. Then there exists a sequence $\{T_k\}$ of distinct elements of Γ such that $T_k \to \mathrm{Id}$ as $k \to \infty$. Let $s \in \mathcal{H}$ be a point not fixed by any of T_k. Then $\{T_k(s)\}$ is a sequence of points distinct from s and $T_k(s) \to s$ as $k \to \infty$. Hence every closed hyperbolic disc centered at s contains infinitely many points of the Γ-orbit of s, i.e. Γ does not act properly discontinuously, a contradiction. □

COROLLARY 8.7. *Let Γ be a subgroup of $PSL(2,\mathbb{R})$. Then Γ acts properly discontinuously on $\mathcal{H}$ if and only if for all $z \in \mathcal{H}$, Γz, the Γ-orbit of z, is a discrete subset of $\mathcal{H}$.*

PROOF. Suppose Γ acts properly discontinuously on $\mathcal{H}$, hence each Γ-orbit is a locally finite family of points, hence a discrete set of $\mathcal{H}$. Conversely, suppose Γ does not act properly discontinuously on $\mathcal{H}$ and hence by Theorem 8.6 is not discrete. Repeating the argument in the proof of Theorem 8.6, we construct a sequence $\{T_k(s)\}$ of points distinct from s such that $T_k(s) \to s$, hence the Γ-orbit of the point s is not discrete. □

Corollary 8.7 implies the following: if $z \in \mathcal{H}$ and $\{T_n\}$ is a sequence of distinct elements in Γ such that $\{T_n(z)\}$ has a limit point $\alpha \in \mathbb{C} \cup \{\infty\}$, then $\alpha \in \mathbb{R} \cup \{\infty\}$.

9. Definition of a fundamental region

We are going to be concerned with fundamental regions of mainly Fuchsian groups, however it is convenient to give a definition in a slightly more general situation. As in §8, let X be a locally compact metric space, and Γ be a group of isometries acting properly discontinuously on X.

DEFINITION 9.1. A closed region $F \subset X$ (i.e. a closure of a non-empty open set $\mathring{F}$, called the interior of F) is defined to be a *fundamental region* for Γ if

(i) $\bigcup_{T\in\Gamma} T(F) = X$,

(ii) $\mathring{F} \cap T(\mathring{F}) = \emptyset$ for all $T \in \Gamma \setminus \{\mathrm{Id}\}$.

The set $\partial F = F \setminus \mathring{F}$ is called the *boundary* of F. The family $\{T(F) \mid T \in \Gamma\}$ is called the *tessellation* of X.

We shall prove in §10 that any Fuchsian group possesses a nice (connected and convex) fundamental region. Now we give an example in the simplest situation.

EXAMPLE C. Let Γ be the cyclic group generated by the transformation $z \to 2z$. Then the semi-annulus shown in Figure 9.1(a) is easily seen to be a fundamental region for Γ. It is already clear from this example that a fundamental region is not uniquely determined by the group: an arbitrary small perturbation of the lower semicircle determines a perturbation of the upper semicircle, and gives yet another fundamental region shown in Figure 9.1(b).

THEOREM 9.2. *Let F_1 and F_2 be two fundamental regions for a Fuchsian group Γ, and $\mu(F_1) < \infty$. Suppose that the boundaries of F_1 and F_2 have zero hyperbolic area. Then $\mu(F_2) = \mu(F_1)$.*

PROOF. We have $\mu(\mathring{F}_i) = \mu(F_i), i = 1, 2$. Now

$$F_1 \supseteq F_1 \cap (\bigcup_{T\in\Gamma} T(\mathring{F}_2)) = \bigcup_{T\in\Gamma} (F_1 \cap T(\mathring{F}_2)).$$

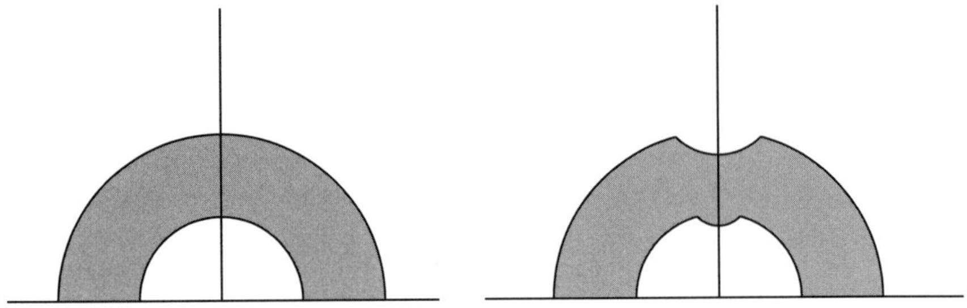

FIGURE 9.1. Fundamental domains for Example C

Since $\overset{\circ}{F}_2$ is the interior of a fundamental region, the sets $F_1 \cap T(\overset{\circ}{F}_2)$ are disjoint, and since μ is $PSL(2,\mathbb{R})$-invariant,

$$\mu(F_1) \geq \sum_{T\in\Gamma} \mu(F_1 \cap T(\overset{\circ}{F}_2)) = \sum_{T\in\Gamma} \mu(T^{-1}(F_1) \cap \overset{\circ}{F}_2) = \sum_{T\in\Gamma} \mu(T(F_1) \cap \overset{\circ}{F}_2).$$

Since F_1 is a fundamental region

$$\bigcup_{T\in\Gamma} T(F_1) = \mathcal{H},$$

and therefore

$$\bigcup_{T\in\Gamma} (T(F_1) \cap \overset{\circ}{F}_2) = \overset{\circ}{F}_2.$$

Hence

$$\sum_{T\in\Gamma} \mu(T(F_1) \cap \overset{\circ}{F}_2) \geq \mu(\bigcup_{T\in\Gamma} T(F_1) \cap \overset{\circ}{F}_2) = \mu(\overset{\circ}{F}_2) = \mu(F_2).$$

Interchanging F_1 and F_2, we obtain $\mu(F_2) \geq \mu(F_1)$. Hence $\mu(F_2) = \mu(F_1)$. □

Thus we have proved a very important fact: the area of a fundamental region, if it is finite, is a numerical invariant of the group. An example of a Fuchsian group with a fundamental region of infinite area is the group generated by $z \to z+1$ (see also Example C above). Obviously, a compact fundamental region has finite area. Non-compact regions also may have finite area. For example, for $\Gamma = PSL(2,\mathbb{Z})$ the fundamental region, which will be described in §10 (Example B), is a hyperbolic triangle with angles $\frac{\pi}{3}, \frac{\pi}{3}, 0$. By the Gauss-Bonnet formula (Theorem 5.4) its area is finite and is equal to $\pi - \frac{2\pi}{3} = \frac{\pi}{3}$.

THEOREM 9.3. *Let Γ be a discrete group of isometries of the upper half-plane $\mathcal{H}$, and Λ be a subgroup of Γ of index n. If*

$$\Gamma = \Lambda T_1 \cup \Lambda T_2 \cup \cdots \cup \Lambda T_n$$

is a decomposition of Γ into Λ-cosets and if F is a fundamental region for Γ then

(i) *$F_1 = T_1(F) \cup T_2(F) \cup \cdots \cup T_n(F)$is a fundamental region for Λ,*

(ii) *if $\mu(F)$ is finite and the hyperbolic area of the boundary of F is zero then $\mu(F_1) = n\mu(F)$.*

PROOF OF (i). Let $z \in \mathcal{H}$. Since F is a fundamental region for Γ, there exists $w \in F$ and $T \in \Gamma$ such that $z = T(w)$. We have $T = ST_i$ for some $S \in \Lambda$ and some $i, 1 \le i \le n$. Therefore

$$z = ST_i(w) = S(T_i(w)).$$

Since $T_i(w) \in F_1, z$ is in the Λ-orbit of some point of F_1. Hence the union of the Λ-images of F_1 is $\mathcal{H}$.

Now suppose that $z \in \mathring{F}_1$ and that $S(z) \in \mathring{F}_1$, for $S \in \Lambda$. We need to prove that $S = \text{Id}$. Let $\varepsilon > 0$ be so small that $B_\varepsilon(z)$ (the open hyperbolic disc of radius ε centered at z) is contained in $\mathring{F}_1$. Then $B_\varepsilon(z)$ has a non-empty intersection with exactly k of the images of $\mathring{F}$ under $T_1, \cdots, T_n$, where $1 \le k \le n$. Suppose these images are $T_{i_1}(\mathring{F}), \cdots, T_{i_k}(\mathring{F})$. Let $B_\varepsilon(S(z)) = S(B_\varepsilon(z))$ have a non-empty intersection with $T_j(\mathring{F})$ say, $1 \le j \le n$. It follows that $B_\varepsilon(z)$ has a non-empty intersection with $S^{-1}T_j(\mathring{F})$ so that $S^{-1}T_j = T_{i_\ell}$ where $1 \le \ell \le k$. Hence

$$\Lambda T_j = \Lambda S^{-1}T_j = \Lambda T_{i_\ell},$$

so that $T_j = T_{i_\ell}$ and $S = \text{Id}$. Hence $\mathring{F}_1$ contains precisely one point of each Λ-orbit. □

PROOF OF (ii). This follows immediately, as $\mu(T(F)) = \mu(F)$ for all $T \in PSL(2, \mathbb{R})$, and $\mu(T_i(F) \cap T_j(F)) = 0$ for $i \neq j$. □

10. The Dirichlet region

Let Γ be an arbitrary Fuchsian group and let $p \in \mathcal{H}$ be not fixed by any element of $\Gamma \setminus \{\text{Id}\}$. We define the *Dirichlet region for* Γ *centered at* p to be the set

(10.1) $$D_p(\Gamma) = \{z \in \mathcal{H} \mid \rho(z, p) \le \rho(z, T(p)) \text{ for all } T \in \Gamma\}.$$

By the invariance of the hyperbolic metric under $PSL(2, \mathbb{R})$ this region can also be defined as

(10.2) $$D_p(\Gamma) = \{z \in \mathcal{H} \mid \rho(z, p) \le \rho(T(z), p) \text{ for all } T \in \Gamma\}.$$

For each fixed $T_1 \in PSL(2, \mathbb{R})$,

(10.3) $$\{z \in \mathcal{H} \mid \rho(z, p) \le \rho(z, T_1(p))\}$$

is the set of points z which are closer in the hyperbolic metric to p than to $T_1(p)$. Clearly, $p \in D_p(\Gamma)$ and as the Γ-orbit of p is discrete (Corollary 8.7), $D_p(\Gamma)$ contains a neighborhood of p. In order to describe the set (10.3) we join the points p and $T_1(p)$ by a geodesic segment and construct a line given by the equation

$$\rho(z, p) = \rho(z, T_1(p)).$$

DEFINITION 10.1. A *perpendicular bisector* of the geodesic segment $[z_1, z_2]$ is the unique geodesic through w, the mid-point of $[z_1, z_2]$ orthogonal to $[z_1, z_2]$.

LEMMA 10.2. *A line given by the equation*

(10.4) $$\rho(z, z_1) = \rho(z, z_2)$$

is the perpendicular bisector of the geodesic segment $[z_1, z_2]$.

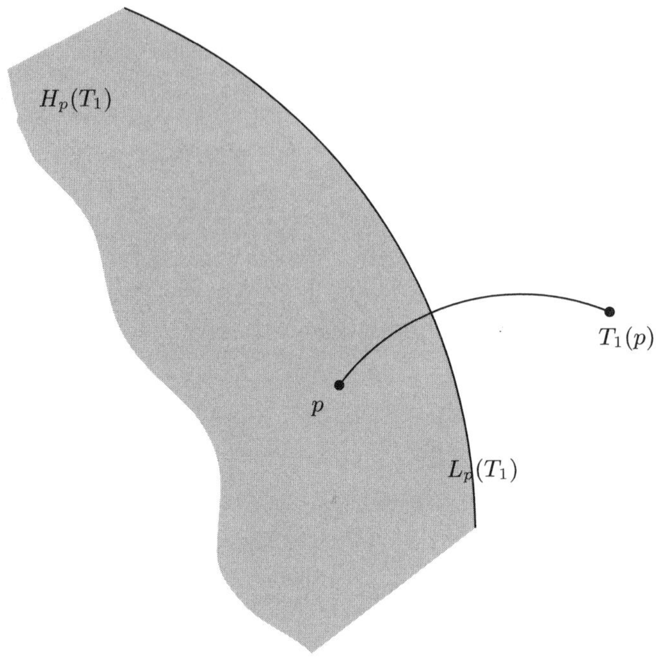

FIGURE 10.1. Construction of the Dirichlet region

PROOF. We may assume that $z_1 = i, z_2 = ir^2$ with $r > 0$: thus $w = ir$ and the perpendicular bisector is given by the equation $|z| = r$. On the other hand, by Theorem 3.5(b) (10.4) is equivalent to

$$\frac{|z - z_1|^2}{y} = \frac{|z - z_2|^2}{r^2 y}$$

which simplifies to $|z| = r$. □

We shall denote the perpendicular bisector of the geodesic segment $[p, T_1(p)]$ by $L_p(T_1)$, and the hyperbolic half-plane containing p described in (10.3) by $H_p(T_1)$ (see Figure 10.1). Thus $D_p(\Gamma)$ is the intersection of hyperbolic half-planes:

$$D_p(\Gamma) = \bigcap_{T \in \Gamma,\, T \neq \mathrm{Id}} H_p(T),$$

and thus is a *hyperbolically convex region.*

THEOREM 10.3. *If p is not fixed by any element of $\Gamma \setminus \{\mathrm{Id}\}$, then $D_p(\Gamma)$ is a connected fundamental region for Γ.*

PROOF. Let $z \in \mathcal{H}$, and Γz be its Γ-orbit. Since Γz is a discrete set, there exists $z_0 \in \Gamma z$ with the smallest $\rho(z_0, p)$. Then $\rho(z_0, p) \leq \rho(T(z_0), p)$ for all $T \in \Gamma$, and by (10.2) $z_0 \in D_p(\Gamma)$. Thus $D_p(\Gamma)$ contains at least one point from every Γ-orbit.

Next we show that if z_1, z_2 are in the interior of $D_p(\Gamma)$, they cannot lie in the same Γ-orbit. If $\rho(z, p) = \rho(T(z), p)$ for some $T \in \Gamma \setminus \{\mathrm{Id}\}$, then $\rho(z, p) = \rho(z, T^{-1}(p))$ and hence $z \in L_p(T^{-1})$. Then either $z \notin D_p(\Gamma)$ or z lies on the boundary of $D_p(\Gamma)$; hence if z is in the interior of $D_p(\Gamma)$, $\rho(z, p) < \rho(T(z), p)$

for all $T \in \Gamma \setminus \{\mathrm{Id}\}$. If two points z_1, z_2 lie in the same Γ-orbit, this implies $\rho(z_1,p) < \rho(z_2,p)$ and $\rho(z_2,p) < \rho(z_1,p)$, a contradiction. Thus the interior of $D_p(\Gamma)$ contains at most one point of each Γ-orbit. Being an intersection of closed half-planes, $D_p(\Gamma)$ is closed and convex. Thus $D_p(\Gamma)$ is path-connected, hence connected. □

EXAMPLE B. $\Gamma = PSL(2,\mathbb{Z})$. It is easily verified that ki ($k > 1$) is not fixed by any non-identity element of the modular group, so choose $p = ki$, where $k > 1$. We shall show that the region

$$F = \{z \in \mathcal{H} \mid |z| \geq 1, |Re(z)| \leq \frac{1}{2}\},$$

illustrated in Figure 10.2 is the Dirichlet region for Γ centered at p.

First, the isometries $T(z) = z+1$, $S(z) = -1/z$ are in Γ; and, as can be easily

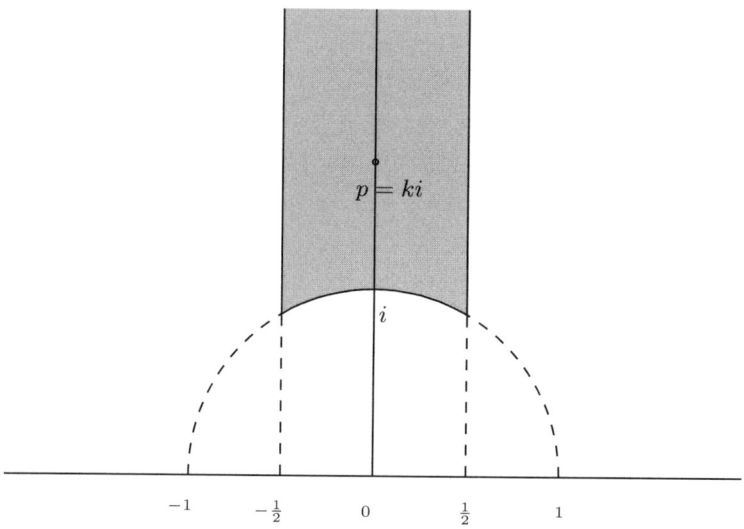

FIGURE 10.2. A Dirichlet region for $PSL(2,\mathbb{Z})$

verified, the three geodesic sides of F are $L_p(T), L_p(T^{-1})$ and $L_p(S)$. This shows that $D_p(\Gamma) \subset F$. If $D_p(\Gamma) \neq F$, there exists $z \in \mathring{F}$ and $h \in \Gamma$ such that $h(z) \in \mathring{F}$. We shall now show that this cannot happen. Suppose that

$$h(z) = \frac{az+b}{cz+d}, \quad (a,b,c,d \in \mathbb{Z}, ad-bc = 1).$$

Then

$$|cz+d|^2 = c^2|z|^2 + 2\,\mathrm{Re}(z)cd + d^2 > c^2 + d^2 - |cd| = (|c| - |d|)^2 + |cd|,$$

since $|z| > 1$ and $\mathrm{Re}(z) > -\frac{1}{2}$. This lower bound is an integer: it is non-negative and is not zero (this would be possible only if $c = d = 0$, which contradicts $ad - bc = 1$). Therefore it is at least 1 and $|cz+d| > 1$. Hence

$$\mathrm{Im}\, h(z) = \frac{\mathrm{Im}(z)}{|cz+d|^2} < \mathrm{Im}(z).$$

Exactly the same argument holds with z, h replaced by $h(z)$, h^{-1}, and a contradiction is reached: thus $D_p(\Gamma) = F$.

In the rest of this section, Γ will be a discrete group of orientation-preserving isometries of the unit disc $\mathcal{U}$ (sometimes also referred to as a Fuchsian group). We assume that 0 is not an elliptic fixed point, i.e. that $c \neq 0$ for all $T(z) = \frac{az+\bar{c}}{cz+\bar{a}}$ in the group Γ. We define

$$R_0 = \overline{\bigcap_{T\in\Gamma} \widehat{I}(T) \bigcap \mathcal{U}},$$

the closure of the set of points in $\mathcal{U}$ which are exterior to the isometric circles of all transformations in the group Γ. We shall prove that R_0 is a fundamental region for Γ, called the *Ford fundamental region.*

THEOREM 10.4. *R_0 is a fundamental region for Γ.*

PROOF. We shall prove that R_0 is a Dirichlet region $D_0(\Gamma)$, and the theorem will follow from Theorem 10.3. The perpendicular bisector I of the geodesic segment

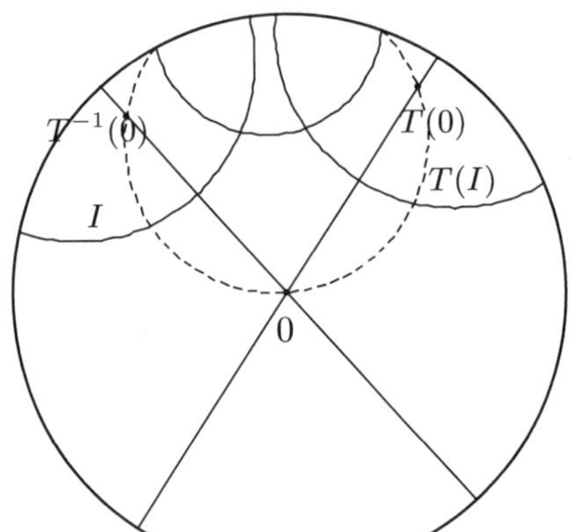

FIGURE 10.3. Ford region is Dirichlet region

$[T^{-1}(0), 0]$ is a geodesic in the unit disc model, hence the arc of an Euclidean circle orthogonal to the circle at infinity (see Figure 10.3). Since both geodesic segments $[T^{-1}(0), 0]$ and $[0, T(0)]$ are segments of the radii of the unit disc, $T(I)$ is the perpendicular bisector of $[0, T(0)]$ which is the arc of an Euclidean circle of the same radius. Thus the transformation T maps I to $T(I)$ without alteration of Euclidean lengths, and therefore the perpendicular bisector of $[0, T(0)]$ is the isometric circle $I(T^{-1})$. □

THEOREM 10.5. *Given any infinite sequence of distinct isometric circles $I_1, I_2, \cdots$ of transformations of a Fuchsian group Γ with radii $r_1, r_2, \cdots$ we have $\lim_{n\to\infty} r_n = 0$.*

PROOF. The transformations are of the form

$$T(z) = \frac{az + \bar{c}}{cz + \bar{a}} \; (a, c \in \mathbb{C}, |a|^2 - |c|^2 = 1). \tag{10.5}$$

Recall that the radius of $I(T)$ is equal to $\frac{1}{|c|}$. Let $\varepsilon > 0$ be given. There are only finitely many $T \in \Gamma$ with $|c| < 1/\varepsilon$. This follows from the discreteness of Γ and the relation $|a|^2 - |c|^2 = 1$. Hence there are only finitely many $T \in \Gamma$ with $I(T)$ of radius exceeding ε, and the theorem follows. □

11. Structure of a Dirichlet region

Dirichlet regions for Fuchsian groups can be quite complicated. They are bounded by geodesics in $\mathcal{H}$ and possibly by segments of the real axis. If two such geodesics intersect in $\mathcal{H}$, their point of intersection is called a *vertex* of the Dirichlet region. It will be shown that the vertices are isolated (see Proposition 11.3 below) so that a Dirichlet region is bounded by a union of (possibly, infinitely many) geodesics and possibly segments of the real axis (see Figure 11.1 for the unit disc model). We shall be interested in the tessellation of $\mathcal{H}$ formed by a Dirichlet

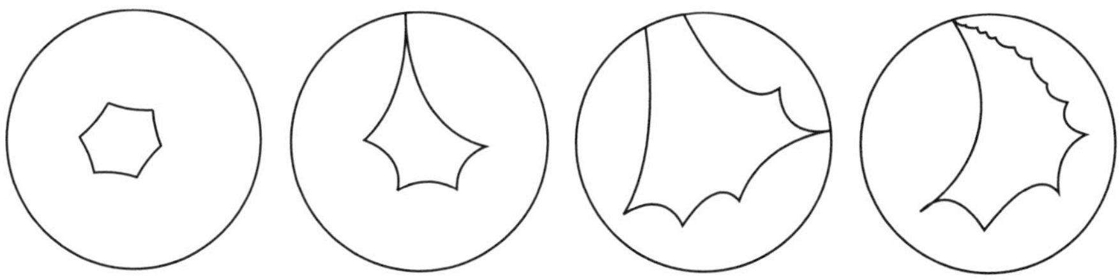

(i) (ii) (iii) (iv)

FIGURE 11.1. Dirichlet regions in the unit disc model

region F and all its images under Γ (called *faces*): $\{T(F) \mid T \in \Gamma\}$. This tessela-tion will be referred to as a *Dirichlet tessellation.* (See Figure 11.2 for a Dirichlet tessellation for the modular group.) The next theorem shows that the Dirichlet tessellation has nice local properties.

DEFINITION 11.1. A fundamental region F for a Fuchsian group Γ is called *locally finite* if the tessellation $\{T(F) \mid T \in \Gamma\}$ is locally finite (see the definition of locally finite family of subsets in §8).

THEOREM 11.2. *A Dirichlet region is locally finite.*

PROOF. Let $F = D_p(\Gamma)$, where p is not fixed by any element of $\Gamma \setminus \{\mathrm{Id}\}$. Let $a \in F$, and let $K \subset \mathcal{H}$ be a compact neighborhood of a. Suppose that $K \bigcap T_i(F) \neq \emptyset$ for some infinite sequence $T_1, T_2, \cdots$ of distinct elements of Γ. Let $\sigma = \sup_{z\in K} \rho(p, z)$. Since $\rho(p, z) \leq \rho(p, a) + \rho(a, z)$, for all $z \in K$, and K is bounded, σ is finite. Let $w_j \in K \bigcap T_j(F)$. Then $w_j = T_j(z_j)$ for $z_j \in F$, and by the triangle inequality,

$$\begin{aligned}
\rho(p, T_j(p)) &\leq \rho(p, w_j) + \rho(w_j, T_j(p)) \\
&= \rho(p, w_j) + \rho(z_j, p) \\
&\leq \rho(p, w_j) + \rho(w_j, p) \quad (\text{as } z_j \in D_p(\Gamma)) \\
&\leq 2\sigma
\end{aligned}$$

Thus the infinite set of points $T_1(p), T_2(p), \cdots$ belongs to a compact hyperbolic ball with center p and radius 2σ, but this contradicts the properly discontinuous action of Γ. □

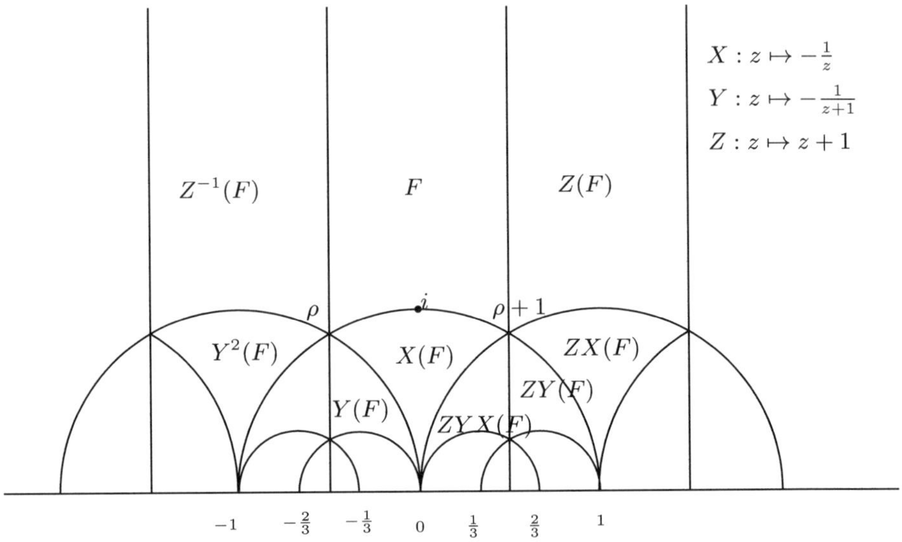

FIGURE 11.2. Dirichlet tessellation for the modular group

PROPOSITION 11.3. *The vertices of a Dirichlet region are* isolated, *that is every vertex of F has a neighborhood containing no other vertices of F.*

PROOF. If x lies on the side of the Dirichlet region $F = D_p(\Gamma)$, then there exists $T_x \in \Gamma$ such that $\rho(p, x) = \rho(T_x p, x)$, hence $\rho(p, x) = \rho(p, T_x^{-1} x)$. Now assume that a vertex $v \in \mathcal{H}$ is not isolated, i.e. there is a sequence of vertices $v_i \in F$ such that $v_i \to v$. According to the above remark, choose T_i such that $\rho(p, v_i) = \rho(p, T_i v_i)$. We have

$$\begin{aligned}\rho(v, T_i v_i) &\leq \rho(v, v_i) + \rho(v_i, T_i v_i) \leq \rho(v, v_i) \\ &+ \rho(v_i, p) + \rho(p, T_i v_i) = \rho(v, v_i) + 2\rho(v_i, p) \\ &\leq \rho(v, v_i) + 2\rho(v_i, v) + 2\rho(v, p).\end{aligned}$$

Hence for any $\varepsilon > 0$ $\rho(v, T_i v_i) < 2\rho(v, p) + \varepsilon$, for large i, which means that $T_i v_i \in K$ for all $i > N$ where K is a compact region in $\mathcal{H}$ which contradicts the local finiteness of F. □

COROLLARY 11.4. *A compact Dirichlet region has a finite number of vertices.*

We call two points $u, v \in \mathcal{H}$ *congruent* if they belong to the same Γ-orbit. First, notice that two points in a fundamental region F may be congruent only if they belong to the boundary of F. Suppose now that F is a Dirichlet region for Γ, and let us consider congruent vertices of F. The congruence is an equivalence relation on the vertices of F and the equivalence classes are called *cycles.* If u is fixed by an elliptic element S, then $v = Tu$ is fixed by the elliptic element TST^{-1}. Thus if one vertex of the cycle is fixed by an elliptic element, then all the vertices of that cycle are fixed by conjugate elliptic elements. Such a cycle is called an *elliptic cycle* and the vertices are called *elliptic vertices.* The number of elliptic cycles is equal to the number of non-congruent elliptic points in F.

Since the Dirichlet region F is a fundamental region, it is clear that every point $w \in \mathcal{H}$ fixed by an elliptic element S' of Γ lies on the boundary of $T(F)$ for some

$T \in \Gamma$. Hence $u = T^{-1}(w)$ lies on the boundary of F and is fixed by the elliptic element $S = T^{-1}S'T$. Since Γ is a Fuchsian group, S has a finite order k. Suppose first that $k \geq 3$: then as S is an isometry fixing u which maps geodesics to geodesics, u must be a vertex whose angle θ is at most $2\pi/k$. (See Figure 11.2 where the angle at the elliptic fixed point ρ of order 3 is $2\pi/6$.) The hyperbolically convex region F is bounded by a union of geodesics. The intersection of F with these geodesics is either a single point or a segment of a geodesic. These segments are called *sides* of F. If S has order 2, its fixed point might lie on the interior of a side of F. In this case, S interchanges the two segments of this side separated by the fixed point. We will include such elliptic fixed points as vertices of F, the angle at such vertex being π. Thus a *vertex* of F is a point of intersection in $\mathcal{H}$ of two bounding geodesics of F or a fixed point of an elliptic element of order 2. (All the previous definitions such as conjugate, elliptic cycles, etc. apply to this extended set of vertices.)

If a point in $\mathcal{H}$ has a nontrivial stabilizer in Γ, this stabilizer is a finite cyclic subgroup of Γ by Exercise 27; it is a *maximal finite cyclic subgroup* of Γ by Exercise 28. Conversely, every maximal finite cyclic subgroup of Γ is a stabilizer of a single point in $\mathcal{H}$. We can summarize the above as:

THEOREM 11.5. *There is a one-to-one correspondence between the elliptic cycles of F and the conjugacy classes of non-trivial maximal finite cyclic subgroups of Γ.*

EXAMPLE B. Let Γ be the modular group. The Dirichlet region F in Figure 9 has vertices in $\mathcal{H}$ at $\rho = \frac{-1+\sqrt{3}}{2}$, $\rho + 1 = \frac{1+\sqrt{3}}{2}$ and i. These are stabilized by the cyclic subgroups generated by $z \mapsto \frac{-z-1}{z}$, $z \mapsto \frac{z-1}{z}$, and $z \mapsto -\frac{1}{z}$, respectively. The vertices ρ and $\rho+1$ belong to the same cycle since they are congruent via $z \to z+1$. Each of them is fixed by an elliptic element of order 3. It is easy to check that these two vertices form an elliptic cycle. The point i is fixed by an elliptic element of order 2, and i is the only such point. Thus $\{i\}$ is an elliptic cycle consisting of just one vertex. By Theorem 11.5, the modular group has two conjugacy classes of maximal finite cyclic subgroups, one consisting of groups of order 2, the other consisting of groups of order 3.

DEFINITION 11.6. The orders of non-conjugate maximal finite cyclic subgroups of Γ are called the *periods* of Γ.

Each period is repeated as many times as there are conjugacy classes of maximal finite cyclic subgroups of that order. Thus the modular group has periods 2, 3.

A parabolic element can be considered as an elliptic element of infinite order; it has a unique fixed point in $\mathbb{R}\bigcup\{\infty\}$. Hence if a point in $\mathbb{R}\bigcup\{\infty\}$ has a non-trivial stabilizer in Γ all elements of which have only this fixed point, then this stabilizer is a *maximal (cyclic) parabolic subgroup* of Γ, and every maximal parabolic subgroup of Γ is a stabilizer of a single point in $\mathbb{R}\bigcup\{\infty\}$. Let F be a Dirichlet region for Γ with parabolic elements. It will be shown in §14 that in this case F is not compact (Theorem 14.2), and if additionally $\mu(F) < \infty$, then F has at least one *vertex at infinity*, i.e. two bounding geodesics of F meet there (Theorem 14.3). Moreover, each vertex at infinity is a parabolic fixed point for a maximal parabolic subgroup of Γ (Theorem 14.6), and non-congruent vertices at infinity of F are in a one-to-one correspondence with conjugasy classes of maximal parabolic subgroups of Γ (Corollary 14.7). If we allow infinite periods, the period ∞ will occur the same number of times as there are conjugacy classes of maximal parabolic subgroups.

This number is called the *parabolic class number of* Γ. It is easily calculated that in the modular group every parabolic element is conjugate to $z \to z + n$ for some $n \in \mathbb{Z}$, so that the modular group has periods 2, 3, ∞. The angle at a vertex at infinity is 0. With this convention, the Dirichlet region for the modular group described in §10 has a vertex at ∞ whose angle is $\frac{\pi}{\infty} = 0$.

The following result relates the sum of angles at all elliptic verticies belonging to an elliptic cycle with the order of that cycle.

THEOREM 11.7. *Let F be a Dirichlet region for Γ. Let $\theta_1, \theta_2, \cdots, \theta_t$ be the internal angles at all congruent vertices of F. Let m be the order of the stabilizer in Γ of one of these vertices. Then $\theta_1 + \cdots + \theta_t = 2\pi/m$.*

REMARKS. 1. As F is locally finite, there are only finitely many vertices in a congruent cycle.

2. As the stabilizers of two points in a congruent set are conjugate subgroups of Γ, they have the same order.

3. If a vertex is not a fixed point, we have $m = 1$ and $\theta_1 + \cdots + \theta_t = 2\pi$.

PROOF. Let $v_1, \ldots, v_t$ be the vertices of the congruent set, the internal angles being $\theta_1, \ldots, \theta_t$. Let

$$H = \{\mathrm{Id}, S, S^2, \ldots, S^{m-1}\}$$

be the stabilizer of v_1 in Γ. Then each $S^r(F)$ $(0 \leq r \leq m-1)$ has a vertex at v_1 whose angle is θ_1. Suppose $T_k(v_k) = v_1$ for some $T_k \in \Gamma$. Then the set of all elements which map v_k to v_1 is HT_k, a coset which has m elements, so the $S^r T_k(F)$ have v_1 as a vertex with an angle of θ_k. On the other hand, if a region $A(F)$ $(A \in \Gamma)$ has v_1 as a vertex, then $A^{-1}(v_1) \in F$, hence $A^{-1}(v_1) = v_i$ for some i, $1 \leq i \leq t$. Thus $A \in HT_i$, and $A(F)$ has been included in the above description. So we have mt regions surrounding v_1. These regions are distinct, for if $S^r T_k(F) = S^q T_l(F)$, then $S^r T_k = S^q T_l$, and hence $r = q$ and $k = l$. We conclude then that

$$m(\theta_1 + \cdots + \theta_t) = 2\pi.$$

□

We now consider the congruence of sides. Let s be a side of F, a Dirichlet region for a Fuchsian group Γ. If $T \in \Gamma \setminus \{\mathrm{Id}\}$ and $T(s)$ is a side of F, then s and $T(s)$ are called *congruent sides*. But $T(s)$ is also a side of $T(F)$ so that $T(s) \subseteq F \bigcap T(F)$. If a side of F has a fixed point of an elliptic element S of order 2 on it then S interchanges the two segments of this side. It is convenient to regard these two segments as distinct sides separated by a vertex. With this convention, one observes that for each side of F there exists another side of F congruent to it. There cannot be more than two sides in a congruent set. For, suppose that for some $T_1 \in \Gamma \setminus \{\mathrm{Id}\}$, $T_1(s)$ is also a side of F; then $T_1(s) = F \bigcap T_1(F)$. Thus $s = T_1^{-1}(F) \bigcap F = T^{-1}(F) \bigcap F$ so that $T_1^{-1}(F) \equiv T^{-1}(F)$ which implies $T_1 = T$. Thus the sides of F fall into congruent pairs. Hence if the number of sides of a Dirichlet region is finite, it is always even.

EXAMPLE B. The two vertical sides of the fundamental region for the modular group found in §10 (Figure 10.2) are congruent via the transformation $z \to z + 1$. The arc of the unit circle between ρ and $\rho + 1$ is the union of two sides: $[\rho, i]$ and $[i, \rho + 1]$, congruent via the elliptic transformation of order 2, $z \to -1/z$.

THEOREM 11.8. *Let $\{T_i\}$ be the subset of Γ consisting of those elements which pair the sides of some fixed Dirichlet region F. Then $\{T_i\}$ is a set of generators for Γ.*

PROOF. Let Λ be the subgroup generated by the set $\{T_i\}$. We have to show that $\Lambda = \Gamma$. Suppose that $S_1 \in \Lambda$, and that $S_2(F)$ is adjacent to $S_1(F)$, i.e. they share a side. Then $S_1^{-1}S_2(F)$ is adjacent to F. Hence $S_1^{-1}S_2 = T_k$ for some $T_k \in \{T_i\}$; and since $S_2 = S_1T_k$ we conclude that $S_2 \in \Lambda$. Suppose now $S_3(F)$ intersects $S_1(F)$ in a vertex v. Then $S_1^{-1}S_3(F)$ intersects F in a vertex $u = S_1^{-1}v$. By Theorem 11.2, there can only be finitely many faces with vertex u, and F can be "connected" with $S_1^{-1}S_3(F)$ by a finite chain of faces in such a way that each two consecutive ones share a side. Hence we can apply the above argument repeatedly to show that $S_3 \in \Lambda$. Let $X = \bigcup_{S\in\Lambda} S(F)$, $Y = \bigcup_{S\in\Gamma\setminus\Lambda} S(F)$. Then $X \bigcap Y = \emptyset$. Clearly $X \cup Y = \mathcal{H}$, so if we show that X and Y are closed subsets of $\mathcal{H}$, then as $\mathcal{H}$ is connected and $X \neq \emptyset$, we must have $X = \mathcal{H}$ and $Y = \emptyset$. This would show that $\Lambda = \Gamma$ and the result will follow.

We now show that any union $\bigcup V_j(F)$ of faces of the tessellation is closed. Suppose $\{z_i\}$ is an infinite sequence of points of $\bigcup V_j(F)$ which tends to some limit $z_0 \in \mathcal{H}$. Then $z_0 \in T(F)$ for some $T \in \Gamma$, and by Theorem 11.2, any neighborhood N of z_0 intersects only finitely many of the $V_j(F)$. Therefore, one face of this finite family, say $V_m(F)$, must contain a subsequence of $\{z_i\}$ tending to z_0. Since $V_m(F)$ is closed, $z_0 \in V_m(F) \subseteq \bigcup V_j(F)$. Thus $\bigcup V_j(F)$ is closed, and, in particular, X and Y are closed. □

EXAMPLE B. Theorem 11.8 implies that the modular group is generated by $z \to z+1$ and $z \to -1/z$.

12. Connection with Riemann surfaces and homogeneous spaces

Let Γ be a Fuchsian group acting on the upper half-plane $\mathcal{H}$, and F be a fundamental region for this action. The group Γ induces a natural projection (continuous and open) $\pi : \mathcal{H} \to \Gamma\backslash\mathcal{H}$, and the points of $\Gamma\backslash\mathcal{H}$ are the Γ-orbits. The restriction of π to F identifies the congruent points of F that necessarily belong to its boundary ∂F, and makes $\Gamma\backslash F$ into an oriented surface with possibly some *marked points* (which correspond to the elliptic cycles of F) and *cusps* (which correspond to non-congruent vertices at infinity of F), also known as an *orbifold*. Its topological type is determined by the number of cusps and by its *genus*—the number of handles if we view the surface as a sphere with handles. If F is locally finite, the quotient space $\Gamma\backslash\mathcal{H}$ is homeomorphic to $\Gamma\backslash F$ ([**5**], Theorem. 9.2.4), hence by choosing F to be a Dirichlet region which is locally finite by Theorem 11.2, we can find the topological type of $\Gamma\backslash\mathcal{H}$. We have seen in §9 (Theorem 9.2) that the area of a fundamental region (with nice boundary) is, if finite, a numerical invariant of the group Γ. Since the area on the quotient space $\Gamma\backslash\mathcal{H}$ is induced by the hyperbolic area on $\mathcal{H}$, the *hyperbolic area* of $\Gamma\backslash\mathcal{H}$, denoted by $\mu(\Gamma\backslash\mathcal{H})$, is well defined and equal to $\mu(F)$ for any fundamental region F. If Γ has a compact Dirichlet region F, then by Proposition 11.3, F has finitely many sides, and the quotient space $\Gamma\backslash\mathcal{H}$ is compact. We shall see in §14 (Corollary 14.4) that if one Dirichlet region for Γ is compact then all Dirichlet regions are compact. If, in addition, Γ acts on $\mathcal{H}$ without fixed points, $\Gamma\backslash\mathcal{H}$ is a compact *Riemann surface*—a 1-dimensional complex manifold—and its fundamental group is isomorphic to Γ [**30**].

Since Γ acts on $PSL(2,\mathbb{R})$ by left multiplication one can form the homogeneous space $\Gamma\backslash PSL(2,\mathbb{R})$. We have seen (Theorem 7.1) that $PSL(2,\mathbb{R})$ can be interpreted as the unit tangent bundle of the upper half-plane. It is easy to see (Exercise 24) that if F is a fundamental region for Γ in $\mathcal{H}$, SF is a fundamental region for Γ in $PSL(2,\mathbb{R})$. It also can be shown (see Exercise 25) that if Γ contains no elliptic elements, the homeomorphism described in Theorem 7.1 induces an homeomorphism of the corresponding quotient spaces. If Γ contains elliptic elements, an analogous result holds; however, the structure of the fibered bundle is violated in a finite number of marked points.

Since the fiber in $S(\Gamma\backslash\mathcal{H})$ over each point of $\Gamma\backslash\mathcal{H}$ is compact, $\Gamma\backslash\mathcal{H}$ is compact if and only if $S(\Gamma\backslash\mathcal{H})$ is compact.

13. Fuchsian groups of cofinite volume

THEOREM 13.1. (Siegel's Theorem) *If Γ is such that $\mu(\Gamma\backslash\mathcal{H}) < \infty$, then any Dirichlet region $F = D_p(\Gamma)$ has finitely many sides.*

PROOF. [**8**] Since the vertices of $D_p(\Gamma)$ are isolated (Proposition 11.3), any compact subset $K \subset \mathcal{H}$ contains only finitely many vertices. This takes care of the case in which F is compact. Now suppose that F is not compact.

The main ingredient of the proof is an estimation of the angles ω at vertices of the region F. More precisely, we are going to prove that

$$\sum_{\omega}(\pi - \omega) \leq \mu(F) + 2\pi, \tag{13.1}$$

where the sum is taken over all vertices of F lying in $\mathcal{H}$. We first notice that F is a star-like generalized polygon, and that the boundary of F, ∂F, is not necessarily connected. Let us connect all vertices of F with the point p by geodesics and

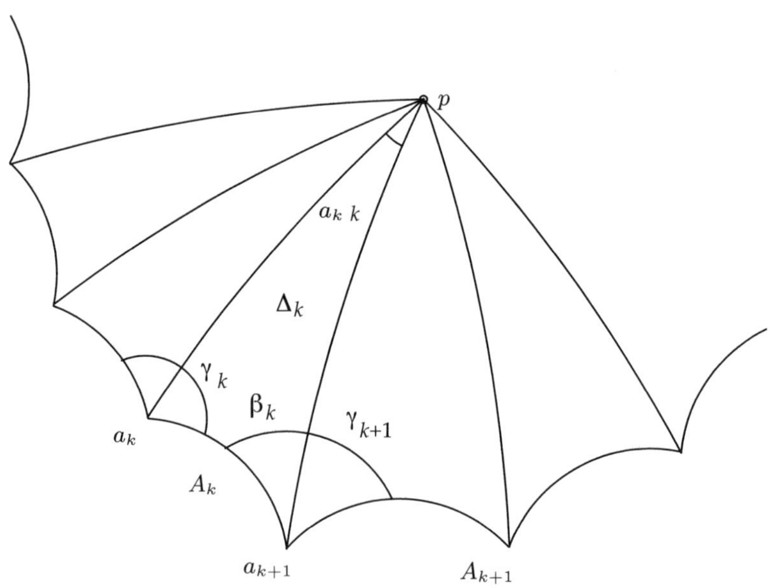

FIGURE 13.1. Proof of Siegel's Theorem

consider the triangles thus obtained. Let $\dots, A_m, A_{m+1}, \dots, A_n, \dots$ be a connected set of geodesic segments in ∂F with vertices $\dots, a_m, a_{m+1}, \dots, a_{n+1}, \dots$ (Figure 13.1).

We assume that this set is unbounded in both directions. We denote the triangle with the side A_k by Δ_k, its angles by $\alpha_k, \beta_k, \gamma_k$, and the angle between A_k and A_{k+1} by ω_k; thus we have

$$\omega_k = \beta_k + \gamma_{k+1}.$$

By the Gauss-Bonnet formula (Theorem 5.4) we have

$$\mu(\Delta_k) = \pi - \alpha_k - \beta_k - \gamma_k.$$

Thus

$$\sum_{k=m}^{n} \alpha_k + \sum_{k=m}^{n} \mu(\Delta_k) = \pi - \gamma_m - \beta_n + \sum_{k=m}^{n-1} (\pi - \omega_k). \tag{13.2}$$

The left-hand side of this equality is bounded since $\sum \alpha_k \le 2\pi$ and $\sum \mu(\Delta_k) \le \mu(F)$, hence the right-hand side is also bounded. It follows that $\sum(\pi - \omega_k)$ converges, and the following limits exist:

$$\lim_{m\to-\infty} \gamma_m = \gamma_\infty, \quad \lim_{n\to\infty} \beta_n = \beta_\infty.$$

Let us show that

$$\pi - \gamma_\infty - \beta_\infty \ge 0. \tag{13.3}$$

Since only finitely many segments $\{A_k\}$ may be a bounded distance from the point p, we have $a_k \to \infty$ as $k \to \infty$. Thus $\rho(p, a_{k+1}) > \rho(p, a_k)$ for infinitely many values of k, and for these values, as follows, for instance, from the Sine Rule (Theorem 6.1(i)), we have $\gamma_k > \beta_k$. On the other hand, $\beta_k + \gamma_k \le \pi$ and thus $\beta_k \le \pi/2$. Therefore $\beta_\infty \le \pi/2$. Similarly, $\gamma_\infty \le \pi/2$, and (13.3) follows.

Let $m \to -\infty, n \to \infty$. Taking into account (13.3) we obtain from 13.2 a limit inequality

$$\sum_{k=-\infty}^{\infty} \alpha_k + \sum_{k=-\infty}^{\infty} \mu(\Delta_k) \ge \sum_{k=-\infty}^{\infty} (\pi - \omega_k). \tag{13.4}$$

The inequality is obtained under the assumption that the connected set of segments $\{A_k\}$ is unbounded in both directions. Similar arguments apply in other cases when the connected set of segments is bounded at least in one direction. Adding up all these inequalities, we obtain a desired estimate

$$2\pi + \mu(F) \ge \sum_{\omega} (\pi - \omega), \tag{13.5}$$

where the sum is taken over all vertices of F which lie a finite hyperbolic distance from the point p, i.e. in $\mathcal{H}$.

Now we are going to prove, using this estimate, that the number of vertices which lie a finite distance from the point p is finite. Let a be a vertex and $a^{(1)} = a, a^{(2)}, \dots, a^{(n)}$ all vertices congruent to a. If we denote the angle at vertex $a^{(i)}$ by $\omega^{(i)}$, we have by Theorem 11.7

$$\omega^{(1)} + \omega^{(2)} + \dots + \omega^{(n)} = 2\pi, \tag{13.6}$$

if a is not a fixed point for any $T \in \Gamma - \{\mathrm{Id}\}$; and

$$\omega^{(1)} + \omega^{(2)} + \cdots + \omega^{(n)} = 2\pi/m, \tag{13.7}$$

if a is a fixed point of order m. Since F is convex, $\omega^{(i)} < \pi$, and for each cycle of the type (13.6), we have $n \geq 3$, and hence

$$\sum_{i=1}^{n}(\pi - \omega^{(i)}) = (n-2)\pi > \pi. \tag{13.8}$$

Comparing (13.8) with (13.5) we conclude that the number of cycles where a is not a fixed point for any $T \in \Gamma \setminus \{\mathrm{Id}\}$ is finite. For each cycle of the type (13.7) we have

$$\sum_{i=1}^{n}(\pi - \omega^{(i)}) = (n - \frac{2}{m})\pi > \frac{\pi}{3}. \tag{13.9}$$

Comparing (13.9) with (13.5) we conclude that the number of elliptic cycles of order ≥ 3 is finite. Finally, any elliptic fixed point of order 2 belongs to a segment of ∂F between two vertices which are not elliptic points of order 2, hence we see that the number of elliptic cycles of order 2 is also finite. Thus we have proved that there are only finitely many vertices at a finite distance from the point p.

It remains to show that the number of vertices at infinity is also finite. Let us take any N vertices at infinity: $B_1, \ldots, B_N$. It is obvious that there exists a hyperbolic polygon F_1 bounded by a finite number of geodesics and contained inside F such that its vertices at infinity are $B_1, \ldots, B_N$. An argument similar to that in the proof of (13.2) shows that the hyperbolic area of F_1 satisfies the following equation:

$$\sum_{\omega}(\pi - \omega) = 2\pi + \mu(F_1),$$

where ω are the angles at the vertices of F_1, and the sum is taken over all vertices of F_1. Since $\omega = 0$ for all vertices at infinity, we have

$$\pi N \leq 2\pi + \mu(F_1) \leq 2\pi + \mu(F).$$

Thus N is bounded from above, and the theorem follows. □

14. Cocompact Fuchsian groups

DEFINITION 14.1. A Fuchsian group is called *cocompact* if the quotient-space $\Gamma \backslash \mathcal{H}$ is compact.

The following results reveal the relationship between cocompactness of Γ and the absence of parabolic elements in Γ.

THEOREM 14.2. *If a Fuchsian group Γ has a compact Dirichlet region, then Γ contains no parabolic elements.*

PROOF. Let F be a compact Dirichlet region for Γ and

$$\eta(z) = \inf\{\rho(z, T(z)) \mid T \in \Gamma \setminus \{\mathrm{Id}\}, T \text{ not elliptic}\}.$$

Since the Γ-orbit of each $z \in \mathcal{H}$ is a discrete set (Corollary 8.7) and $T(z)$ is continuous, $\eta(z)$ is a continuous function of z and $\eta(z) > 0$. Therefore, as F is compact,

$\eta = \inf\{\eta(z) \mid z \in F\}$ is attained and $\eta > 0$. If $z \in \mathcal{H}$, there exists $S \in \Gamma$ such that $w = S(z) \in F$. Hence, if $T_0 \in \Gamma \setminus \{\mathrm{Id}\}$ is not elliptic,

$$\rho(z, T_0(z)) = \rho(S(z), S(T_0(z)) = \rho(w, ST_0S^{-1}(w)) \geq \eta,$$

and therefore

$$\inf\{\rho(z, T_0(z)) \mid z \in \mathcal{H},\ T_0 \text{ not elliptic}\} = \eta > 0.$$

Now suppose that Γ contains a parabolic element T_1. If for some $R \in PSL(2,\mathbb{R})$, $\Gamma_1 = R\Gamma R^{-1}$ then $R(F)$ will be a compact fundamental region for Γ_1. Thus by conjugating Γ in $PSL(2,\mathbb{R})$ we may assume that $T_1(z)$ or $T_1^{-1}(z)$ is the transformation $z \to z+1$. However, by Theorem 3.5(c), $\rho(z, z+1) \to 0$ as $\mathrm{Im}(z) \to \infty$, a contradiction. □

THEOREM 14.3.

(i) *If Γ has a non-compact Dirichlet region, then the quotient space $\Gamma\backslash\mathcal{H}$ is not compact.*

(ii) *If a Dirichlet region $F = D_p(\Gamma)$ for a Fuchsian group Γ has finite hyperbolic area but is not compact, then it has at least one vertex at infinity.*

PROOF. Let $F = D_p(\Gamma)$ be a non-compact Dirichlet region for Γ. We consider all oriented geodesic rays from the point p; each geodesic ray is uniquely determined by its direction l at the point p. Since F is a hyperbolically convex region, a geodesic ray in the direction l either intersects ∂F in a unique point or the whole geodesic ray lies inside F. Hence we can define a function $\tau(l)$ to be the length of a geodesic segment in the direction l inside F, $\tau(l)$ being equal to ∞ in the latter case. Obviously, $\tau(l)$ is a continuous function of l at the points where $\tau(l) < \infty$. Therefore if $\tau(l) < \infty$ for all l, the function $\tau(l)$ is bounded; hence the region F is compact. Thus if F is not compact, there are some directions l for which $\tau(l) = \infty$. After the identification of the congruent points of ∂F, we obtain a non-compact orbifold $\Gamma\backslash\mathcal{H}$ and (i) follows. To prove (ii), let us consider one such direction l_0. The intersection of the geodesic ray from p in the direction l_0 with the set of points at infinity belongs to $\partial_0 F$, the Euclidean boundary of F. By Theorem 13.1, F has finitely many sides, hence $\partial_0 F$ consists of finitely many free sides and vertices at infinity. Since $\mu(F) < \infty$, it is easy to see that $\partial_0 F$ cannot contain any free sides. Therefore this intersection is a vertex at infinity, and (ii) follows. □

COROLLARY 14.4. *The quotient space of a Fuchsian group Γ, $\Gamma\backslash\mathcal{H}$, is compact if and only if any Dirichlet region for Γ is compact.*

Let $p \in \mathcal{H}$ and $z(t)$, $0 \leq t < \infty$, be a geodesic ray from the point p. Let $B_t(p)$ be a hyperbolic circle centered at $z(t)$ and passing through the point p. Exercise 26 asserts that the limit of $B_t(p)$, as $t \to \infty$, exists. It is a Euclidean circle passing through p and through the end of the geodesic $z(t)$, s, corresponding to $t = \infty$. It is orthogonal to the geodesic $z(t)$ at s, hence is tangent to the real axis, and therefore is a horocycle (see §4, Exercise 12, and Figure 4.2). Since the geodesic ray through p is determined by its direction l, the horocycle depends on p and l and is denoted by $\omega(p, l)$ (see Figure 14.1). Notice that horocycles are not hyperbolic circles, however they may be considered as circles of infinite radius.

A horocycle through a point s at infinity is denoted by $\omega(s)$.

THEOREM 14.5. *Let S be a transformation in $PSL(2,\mathbb{R})$ fixing a point $s \in \mathbb{R}$. Then S is parabolic if and only if for each horocycle through $s, \omega(s)$, we have $S(\omega(s)) = \omega(s)$.*

PROOF. Suppose first that S is parabolic, and $R \in PSL(2,\mathbb{R})$ is such that $R(s) = \infty$. Then $S_0 = R \circ S \circ R^{-1}$ is a parabolic transformation fixing ∞, and therefore $S_0(z) = z + h$, $h \in \mathbb{R}$. Since S_0 is a Euclidean translation, it maps each horizontal line to itself. Since a linear fractional transformation maps circles and straight lines into circles and straight lines and preserves angles, we conclude that horocycles are mapped into horocycles. Thus $S(\omega(s)) = \omega(s)$.

Conversely, suppose S maps each horocycle $\omega(s)$ onto itself. Making the same conjugation as above, we move the fixed point s to ∞. Then $S(z) = az + b$. The condition that each horizontal line is mapped into itself implies that $a = 1$. Hence S is a parabolic element. □

THEOREM 14.6. *Suppose Γ has a non-compact Dirichlet region $F = D_p(\Gamma)$ with $\mu(F) < \infty$. Then*

(i) each vertex of F at infinity is a parabolic fixed point for some $T \in \Gamma$.

(ii) If ξ is a fixed point of some parabolic element in Γ, then there exists $T \in \Gamma$ s.t. $T(\xi) \in \partial_0(F)$.

PROOF OF (i). Let b be a vertex of F at infinity. Let us consider all images $S(F)$, $S \in \Gamma$, which have the point b as a vertex. Obviously, there are infinitely many of them. Let $b^{(1)} = b$, $b^{(2)}, \ldots, b^{(n)}$ be all vertices of F congruent to b:

$$b^{(k)} = T_k(b) \ (k = 1, \ldots, n).$$

We know from Theorem 13.1 that the number of such vertices is finite. Any image of F which has the point b as a vertex has a form

$$TT_i^{-1}(F) \ (i = 1, \ldots, n),$$

where T is any element of Γ which fixes the point b. Since there are infinitely many such images, and since T_i is only taken from a finite set of elements, we conclude that there are infinitely many elements $T \in \Gamma$ fixing b.

We shall show now that any such element T is a parabolic element. Suppose T is not parabolic. Let us consider a geodesic $z(t), 0 \le t \le \infty$, parametrized by its

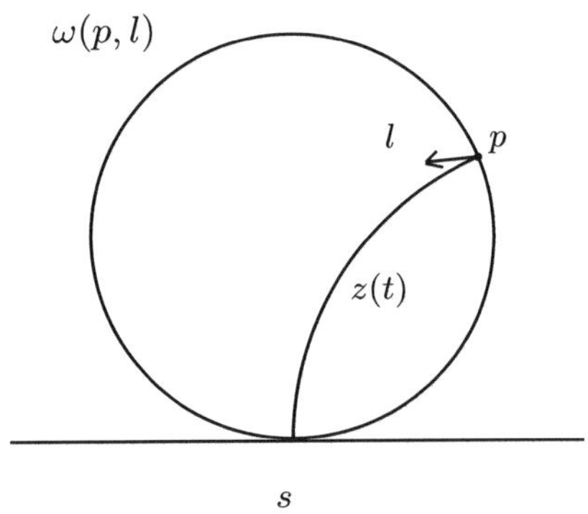

FIGURE 14.1. A horocycle

length, connecting the points p and $b(z(0) = p, z(\infty) = b)$. (See Figure 14.2.) Since F is a Dirichlet region the whole geodesic lies inside F and

$$\rho(p, z(t)) < \rho(T(p), z(t)), \quad 0 \le t < \infty. \tag{14.1}$$

Consider a horocycle $\omega(b)$ containing the point p. Since by our assumption T is

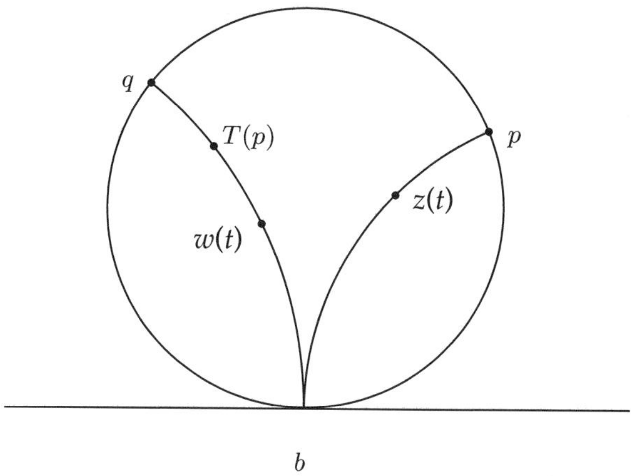

FIGURE 14.2. A vertex at infinity

not a parabolic transformation, $T(p)$ does not belong to $\omega(b)$. Then by Exercise 29 either $T(p)$ or $T^{-1}(p)$ lies inside $\omega(b)$. We may assume then that $T(p)$ lies inside $\omega(b)$. Let $w(t)$ be a geodesic passing through $T(p)$ and b. Let q be a second point of intersection of $\omega(b)$ and $w(t)$; we choose the parametrization of $w(t)$ by its length such that $w(0) = q$. We notice first that $\rho(z(t), w(t)) \to 0$ as $t \to \infty$. In order to see this, we conjugate Γ so that its action on $\mathcal{H}$ gives: $b = \infty$, $z(t) = a + it$, $w(t) = c + it$ $(t \ge t_0 > 0)$. Then using Theorem 3.5(c), we obtain

$$\sinh \left[\frac{1}{2}\rho(z(t), w(t))\right] = \frac{|a-c|}{2t} \to 0 \text{ as } t \to 0,$$

and the claim follows. We have

$$\begin{aligned} t = \rho(p, z(t)) = \rho(q, w(t)) &= \rho(q, T(p)) + \rho(T(p), w(t)) \\ &\ge \rho(q, T(p)) + \rho(T(p), z(t)) - \rho(z(t), w(t)), \end{aligned}$$

and hence for sufficiently large t, we have

$$\rho(p, z(t)) > \rho(T(p), z(t)),$$

contradiction with (14.1). □

PROOF OF (ii). See Exercise 30. □

We leave the proof of the following Corollary (Exercise 31).

COROLLARY 14.7. *There is a one-to-one correspondence between non-congruent vertices at infinity of a Dirichlet fundamental region for a non-cocompact Fuchsian group Γ with $\mu(\Gamma\backslash\mathcal{H}) < \infty$ and conjugacy classes of maximal parabolic subgroups of Γ.*

The following result is a direct consequence of Theorems 13.1, 14.3, and 14.6.

COROLLARY 14.8. *A Fuchsian group* Γ *is cocompact if and only if* $\mu(\Gamma\backslash\mathcal{H}) < \infty$ *and* Γ *contains no parabolic elements.*

15. The signature of a Fuchsian group

We now assume that Γ has a compact fundamental region F. By Corollary 11.4 F has finitely many sides, and hence finitely many vertices, finitely many elliptic cycles, and by Theorem 11.5, a finite number of periods, say $m_1, \dots, m_r$. As we have seen in §3.6 the quotient space $\Gamma\backslash\mathcal{H}$ is an orbifold, i.e. a compact, oriented surface of genus g with exactly r marked points. In this case we say that Γ has *signature* $(g; m_1, m_2, \dots, m_r)$.

THEOREM 15.1. *Let* Γ *have signature* $(g; m_1, \dots, m_r)$. *Then*

$$\mu(\Gamma\backslash\mathcal{H}) = 2\pi[(2g-2) + \sum_{i=1}^{r}\left(1 - \frac{1}{m_i}\right)].$$

PROOF. The area of the quotient space was defined in the beginning of §3.6: $\mu(\Gamma\backslash\mathcal{H}) = \mu(F)$ where F is a Dirichlet region. By Theorem 11.5 F has r elliptic cycles of vertices. (As described in §12, we include the interior point of a side fixed by an elliptic element of order 2 as a vertex whose angle is π, and then regard this side as being composed of two sides separated by this vertex.) By Theorem 11.7, the sum of angles at all elliptic vertices is $\sum_{i=1}^{4}\frac{2\pi}{m_i}$. Suppose there exist s other cycles of vertices. Since the order of the stabilizers of these vertices is equal to 1, the sum of angles at all these vertices is equal to $2\pi s$. Thus the sum of all angles of F is equal to

$$2\pi[\left(\sum_{i=1}^{r}\frac{1}{m_i}\right) + s].$$

The sides of F are matched up by elements of Γ. If we identify those matched sides, we obtain an orbifold of genus g. If F has n such sets of identified sides, we obtain a decomposition of $\Gamma\backslash\mathcal{H}$ into $(r+s)$ vertices, n edges, and 1 simply connected face. By the Euler formula,

$$2 - 2g = (r+s) - n + 1.$$

Exercise 32 gives a formula for the hyperbolic area of a hyperbolic polygon. Using it, we obtain

$$\mu(F) = (2n-2)\pi - 2\pi[\left(\sum_{i=1}^{r}\frac{1}{m_i}\right) + s] = 2\pi[(2g-2) + \sum_{i=1}^{r}\left(1 - \frac{1}{m_i}\right)].$$

□

It is quite surprising that the converse to Theorem 15.1 is also true, i.e. that there exists a Fuchsian group with a given signature. This first appeared in Poincaré's paper on Fuchsian groups [**27**], but the rigorous proof was given much later and is based on a more general result of Maskit [**24**].

THEOREM 15.2. (Poincaré's Theorem) *If* $g \geq 0$, $r \geq 0$, $m_i \geq 2$ $(1 \leq i \leq r)$ *are integers and if*

$$(2g-2)+\sum_{i=1}^{r}\left(1-\frac{1}{m_i}\right)>0,$$

then there exists a Fuchsian group with signature $(g; m_1, \ldots, m_r)$.

The sketch of the proof of Theorem 15.2 given in [**14**]; it is illustrated on the following example.

EXAMPLE D. Construction of a Fuchsian group with signature $(2; -)$. Since $r = 0$, a fundamental region is a regular hyperbolic octagon F_8 (see Figure 15.1) of hyperbolic area 4π. We call this group Γ_8. We suppose that t_0 is chosen such

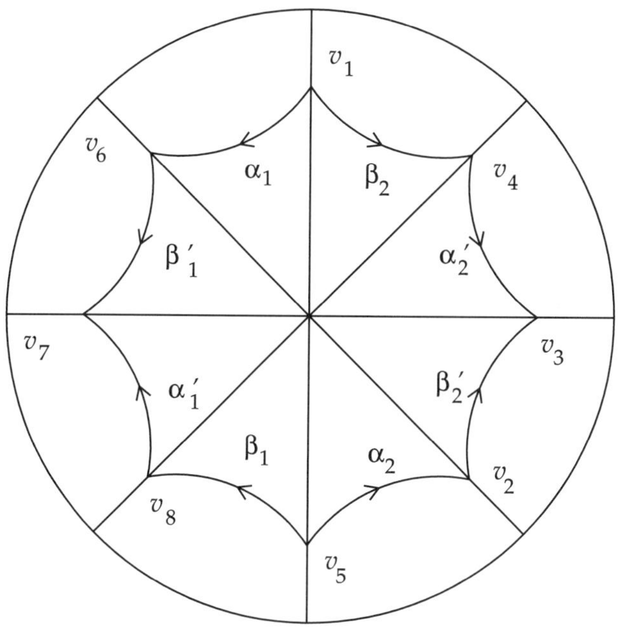

FIGURE 15.1. Fundamental region for the group Γ_8

that $\mu(N(t_0)) = 4\pi$. Then the area of each of the 8 isosceles hyperbolic triangles is equal to $\frac{\pi}{2}$; and since the angle at the origin is equal $\frac{\pi}{4}$, by the Gauss-Bonnet formula (Theorem 5.4) the two other angles are equal to $\frac{\pi}{8}$. The group Γ_8 is generated by 4 hyperbolic elements, A_1, A_2, B_1, and B_2, that identify the sides of F_8 as shown in Figure 15.1. Since all eight sides of F_8 are arcs of circles of the same Euclidean radius of equal Euclidean length, the sides identified by a generator must be isometric circles of this generator and its inverse. This allows us to use elementary geometry to explicitely write down those generators. Let

$$A_2 = \begin{bmatrix} a & c \\ \bar{c} & \bar{a} \end{bmatrix}, \tag{15.1}$$

then the isometric circle $I(A_2)$ is given by the equation $|\bar{c}z + \bar{a}| = 1$. By Exercise 17, A_2 maps $I(A_2)$ onto $I(A_2^{-1})$ in such a way that the center of $I(A_2)$, $-\frac{\bar{a}}{\bar{c}}$, is mapped onto the center of $I(A_2^{-1})$, $\frac{a}{\bar{c}}$. But from Figure 15.1 we see that $\frac{ia}{\bar{c}} = -\frac{\bar{a}}{\bar{c}}$, which implies $a = \pm|a|(\frac{1}{\sqrt{2}} + i\frac{1}{\sqrt{2}})$. Let the radius of $I(A_2) = R$, and the distance of the center of $I(A_2)$ from the origin be d. By elementary geometric arguments, we have $d = R(1+\sqrt{2})$. On the other hand, $|c| = \frac{1}{R}$, and $d = \frac{|a|}{|c|} = R|a|$, hence $|a| = 1+\sqrt{2}$; and since $|a|^2 - |c|^2 = 1$, we have $|c| = \sqrt{2+2\sqrt{2}}$. Now let us choose the + sign in the expression for a, i.e. $Arg(a) = \frac{\pi}{4}$. Since $Arg(-\frac{\bar{a}}{\bar{c}}) = \frac{\pi}{8}$, we obtain $Arg(c) = -\frac{5\pi}{8}$. Using the formulas $\cos\frac{5\pi}{8} = -\frac{\sqrt{2-\sqrt{2}}}{2}$ and $\sin\frac{5\pi}{8} = \frac{\sqrt{2+\sqrt{2}}}{2}$, we obtain the expressions for the numbers a and c in (15.1):

$$a = \frac{2+\sqrt{2}}{2}(1+i), c = -\frac{\sqrt[4]{2}}{2}(\sqrt{2}+i(2+\sqrt{2})).$$

Other generators of the group Γ_8 can also be expressed in terms of parameters a and c as follows: $A_1 = \begin{bmatrix} a & -c \\ -\bar{c} & \bar{a} \end{bmatrix}$, $B_1 = \begin{bmatrix} \bar{a} & -\bar{c} \\ -c & a \end{bmatrix}$, $B_2 = \begin{bmatrix} \bar{a} & \bar{c} \\ c & a \end{bmatrix}$.

Let $R : \mathcal{H} \to \mathcal{U}$ be a map given by $R(z) = \frac{zi+1}{z+i}$, see (4.4). Then $\Gamma = R^{-1}\Gamma_8 R$ be a subgroup of $PSL(2,\mathbb{R})$ whose generators are:

$$A_2 = \begin{bmatrix} \mathrm{Re}(a)+\mathrm{Im}(c) & \mathrm{Im}(a)+\mathrm{Re}(c) \\ -(\mathrm{Im}(a)-\mathrm{Re}(c)) & \mathrm{Re}(a)-\mathrm{Im}(c) \end{bmatrix},$$

$$A_1 = \begin{bmatrix} \mathrm{Re}(a)-\mathrm{Im}(c) & \mathrm{Im}(a)-\mathrm{Re}(c) \\ -(\mathrm{Im}(a)+\mathrm{Re}(c)) & \mathrm{Re}(a)+\mathrm{Im}(c) \end{bmatrix},$$

$$B_1 = \begin{bmatrix} \mathrm{Re}(a)+\mathrm{Im}(c) & -\mathrm{Im}(a)-\mathrm{Re}(c) \\ \mathrm{Im}(a)-\mathrm{Re}(c) & -\mathrm{Re}(a)-\mathrm{Im}(c) \end{bmatrix},$$

$$B_2 = \begin{bmatrix} \mathrm{Re}(a)-\mathrm{Im}(c) & -\mathrm{Im}(a)+\mathrm{Re}(c) \\ \mathrm{Im}(a)+\mathrm{Re}(c) & \mathrm{Re}(a)+\mathrm{Im}(c) \end{bmatrix}.$$

As elements of $PSL(2,\mathbb{R})$, the generators are:

$$A_2 = \begin{bmatrix} \frac{(2+\sqrt{2})(1-\sqrt[4]{2})}{2} & \frac{(2+\sqrt{2})-\sqrt[4]{2}\sqrt{2}}{2} \\ -\frac{(2+\sqrt{2})+\sqrt[4]{2}\sqrt{2}}{2} & \frac{(2+\sqrt{2})(1+\sqrt[4]{2})}{2} \end{bmatrix},$$

$$A_1 = \begin{bmatrix} \frac{(2+\sqrt{2})(1+\sqrt[4]{2})}{2} & \frac{(2+\sqrt{2})+\sqrt[4]{2}\sqrt{2}}{2} \\ -\frac{(2+\sqrt{2})-\sqrt[4]{2}\sqrt{2}}{2} & \frac{(2+\sqrt{2})(1-\sqrt[4]{2})}{2} \end{bmatrix},$$

$$B_1 = \begin{bmatrix} \frac{(2+\sqrt{2})(1-\sqrt[4]{2})}{2} & \frac{-(2+\sqrt{2})+\sqrt[4]{2}\sqrt{2}}{2} \\ -\frac{-(2+\sqrt{2})-\sqrt[4]{2}\sqrt{2}}{2} & \frac{(2+\sqrt{2})(1+\sqrt[4]{2})}{2} \end{bmatrix},$$

$$B_2 = \begin{bmatrix} \frac{(2+\sqrt{2})(1+\sqrt[4]{2})}{2} & \frac{-(2+\sqrt{2})-\sqrt[4]{2}\sqrt{2}}{2} \\ -\frac{-(2+\sqrt{2})+\sqrt[4]{2}\sqrt{2}}{2} & \frac{(2+\sqrt{2})(1-\sqrt[4]{2})}{2} \end{bmatrix}.$$

It can be shown that Γ_8 is derived from the quaternion algebra over $\mathbb{Q}(\sqrt{2})$.

Exercises

24. Prove that if F is a fundamental region for a Fuchsian group Γ on $\mathcal{H}$, then SF is a fundamental region for Γ on $S\mathcal{H}$.

25. Prove that if Γ is a Fuchsian group without elliptic elements, then $S(\Gamma\backslash\mathcal{H})$ is homeomorphic to $\Gamma\backslash PSL(2,\mathbb{R})$.

26. Show that the limit of $B_t(p)$ as $t \to \infty$ is a Euclidean circle passing through p and the end of the geodesic $z(t)$ corresponding to $t = \infty$, and orthogonal to the geodesic $z(t)$.

27. Prove that an elliptic subgroup of $PSL(2,\mathbb{Z})$ is a Fuchsian group if and only of it is finite.

28. If $ST = TS$ then S maps the fixed-point set of T to itself.

29. Let T be a non-parabolic transformation fixing a point b at infinity, $\omega(b)$ be a horocycle, $p \in \omega(b)$. Prove that either $T(p)$ or $T^{-1}(p)$ lies inside $\omega(b)$.

30. Let Γ be a non-elementary Fuchsian group, F a locally finite fundamental region for Γ, and ξ a fixed point of some parabolic element in Γ, then there exists $T \in \Gamma$ s.t. $T(\xi) \in \partial_0(F)$.

31. Give a careful proof of Corollary 14.7.

32. Prove the Gauss-Bonnet formula for an n-sided star-like hyperbolic polygon Π with angles $\alpha_1, \dots, \alpha_n$:

$$\mu(\Pi) = (n-2)\pi - \sum_{i=1}^{n} \alpha_i .$$

Lecture III. Geodesic flow

16. First properties

The geodesic flow $\{\tilde{\varphi}^t\}$ on $\mathcal{H}$ is defined as an $\mathbb{R}$-action on the unit tangent bundle $S\mathcal{H}$ which moves a tangent vector along the geodesic defined by this vector with unit speed. As was explained in §7, $S\mathcal{H}$ can be identified with $PSL(2,\mathbb{R})$, by sending v to the unique $g \in PSL(2,\mathbb{R})$ such that $z = g(i)$, $\zeta = g'(z)(\iota)$, where ι is the unit vector at the point i to the imaginary axis pointing upwards.

Under this identification the $PSL(2,\mathbb{R})$-action on $\mathcal{H}$ by Möbius transformations corresponds to left multiplications (Theorem 7.1), and the geodesic flow corresponds to the right multiplication by the one-parameter subgroup

$$a_t = \begin{pmatrix} e^{t/2} & 0 \\ 0 & e^{-t/2} \end{pmatrix} \text{ such that } \tilde{\varphi}^t(v) \leftrightarrow g a_t . \tag{16.1}$$

The orbit $\{ga_t\}$ projects to a geodesic through $g(i)$. The quotient space $\Gamma\backslash S\mathcal{H}$ can be identified with the unit tangent bundle of $M = \Gamma\backslash\mathcal{H}$, SM, although the structure of the fibered bundle is violated at elliptic fixed points and cusps (see §12 for details). The geodesic flow $\{\tilde{\varphi}^t\}$ on $\mathcal{H}$ descents to the *geodesic flow* $\{\varphi^t\}$ on the factor M via the projection $\pi : S\mathcal{H} \to SM$ of the unit tangent bundles (see e.g. [**16**, §5.3, 5.4] for more details).

In this chapter we will assume that $\mu(M) < \infty$.

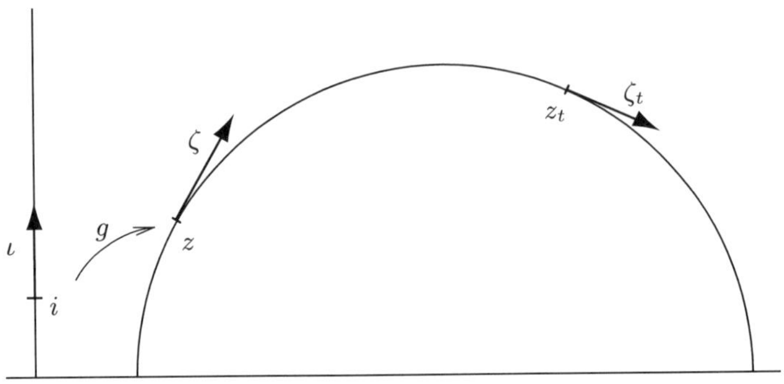

FIGURE 16.1. Geodesic flow on the upper half-plane $\mathcal{H}$

THEOREM 16.1. *The geodesic flow* $\{\varphi^t\} : SM \to SM$ *preserves the Liouville volume* $dv = d\mu d\theta$.

PROOF. In [**16**, Theorem 5.3.6] a more general statement for geodesic flows on Riemannian manifolds is a corollary from the fact that geodesic flows are Hamiltonian flows. In this case the theorem follows from the fact that the volume $dv = d\mu d\theta$ is the left-invariant Haar measure on $PSL(2,\mathbb{R})$ that is also right-invariant since $PSL(2,\mathbb{R})$ is an unimodular group. □

The following theorem is crucial in establishing further important properties of the geodesic flow.

THEOREM 16.2. *The geodesic flow* $\{\varphi^t\}$ *is Anosov, i.e. there exists a* C^∞ $\{\varphi^t\}$*-invariant decomposition of the tangent bundle to* SM, $T(SM) = E^0 \oplus E^+ \oplus E^-$ *such that*

(a) *The integral curves of* E^0 *are orbits of the geodesic flow.*
(b) *The integral curves of* E^+ *and* E^- *(we call them* stable *and* unstable *manifolds and denote them* W^+ *and* W^-*, respectively) are the unit normal vector fields to the horocycles orthogonal to the orbits of* $\{\varphi^t\}$.
(c) *There exist positive constants* C *and* λ *such that for any pair of points* $x_1, x_2 \in SM$ *lying on the same leaf of* W^+ *(or* W^-*),*

$$d^{W^\pm}(\varphi^t(x_1), \varphi^t(x_2)) \le Ce^{-\lambda|t|} d^{W^\pm}(x_1, x_2) \text{ for } t \ge 0, \ (t \le 0).$$

Here $d^{W^\pm}$ *is the distance on the corresponding stable or unstable manifold.*

PROOF. For each $v = (z, \zeta) \in SM$, let $z(t)$ be the geodesic through v with the fixed points $w = z(\infty)$ and $u = z(-\infty)$. As we saw in §14, there are two horocycles on $\mathcal{H}$ passing through the point z: one tangent to the real axis at w, another - at u. Let $W^+(v)$ be the unit vector field containing v and normal to the horocycle passing through w, and $W^-(v)$ be the unit vector field containing v and normal to the horocycle passing through u. In order to prove the estimates in (c), we "move" the geodesic $z(t)$ to the positive imaginary axis by a transformation $\gamma \in PSL(2,\mathbb{R})$ so that $\gamma(z) = i$ (this can be done by Exercise 4), and make the calculations for this particular case. The stable manifold W^+ will be mapped to the upward unit vector field normal to the horocycle $H = \mathbb{R} + i$, and the unstable manifold W^- will be mapped to the outward unit vector field normal to the horocycle passing

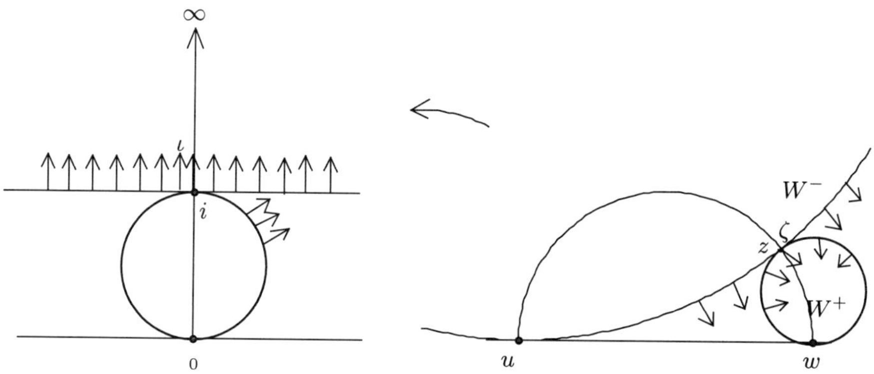

FIGURE 16.2. Stable and unstable manifolds

through i and 0 (see Figure 16.2). Let $x_1 = (i, \iota)$, where ι is the unit vector tangent to the imaginary axis pointed upwards, and $x_2 = (i + x, \iota)$. Then

$$d^{W^+}(x_1, x_2) = x,$$

and after the time $t > 0$,

$$d^{W^+}(\varphi^t(x_1), \varphi^t(x_2)) = xe^{-t},$$

so the estimates hold with $C = \lambda = 1$. The estimates for the unstable manifold are obtained by the change of direction of the flow. □

DEFINITION 16.3. A point $v \in SM$ is call *nonwandering* with respect to the flow $\{\varphi^t\}$ if for every open set $U \ni v$ there is T such that $\varphi^T(U) \cap U \neq \emptyset$.

REMARK. It follows from Poincaré Recurrence Theorem that since the geodesic flow is volume-preserving, every point of SM is nonwandering [**16**, Theorem 4.1.18].

17. Dynamics of the geodesic flow

Let $x \in SM$ and W_x^+ be the stable manifold containing x. Denote by D_x^+ the set of all points w in W_x^+ with $d^{W^+}(x, w) < \delta_0$, where δ_0 will be chosen later (see Figure 17.1). For any point $w \in D_x^+$ we will denote by W_w^- the unstable manifold containing w, and by D_w^- the set of all $z \in W_w^-$ with $d^{W^-}(z, w) < \delta_0$.

Let $S_x = \{z \mid z \in D_w^- \text{ for some } w \in D_x^+\}$. For small enough δ_0, S_x is a submanifold of dimension 2 transversal to the orbit of $\{\varphi^t\}$. This construction gives us a convenient way to parameterize SM locally by the coordinates (t, u, v): we parameterize D_x^+ by the length u measured over the stable leaf, and D_w^- by the length v measured over the unstable leaf. The coordinate t is the length on the orbit of the flow through x.

THEOREM 17.1 (Anosov Closing Lemma). *Suppose $x \in SM$ is such that $d(x, \varphi^T(x)) < \varepsilon$. Then*

(a) *there exists $x_0 \in SM$ whose orbit is closed, i.e. $\varphi^{T'}(x_0) = x_0$ and such that for some constant C, $d(x_0, x) < C\varepsilon$ and $|T' - T| < C\varepsilon$;*

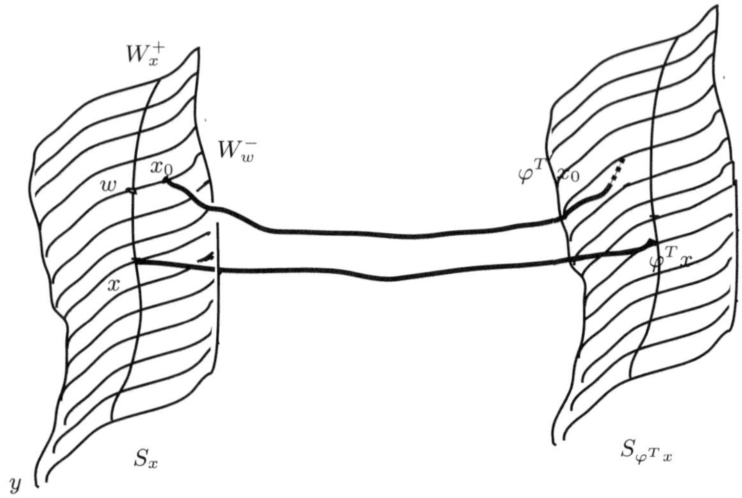

FIGURE 17.1. Dynamical parametrization of SM

(b) *for* $0 \le t \le T$ *and some other constant* C'

$$d(\varphi^t x_0, \varphi^t x) \le C' \varepsilon e^{-\min(t, T-t)}.$$

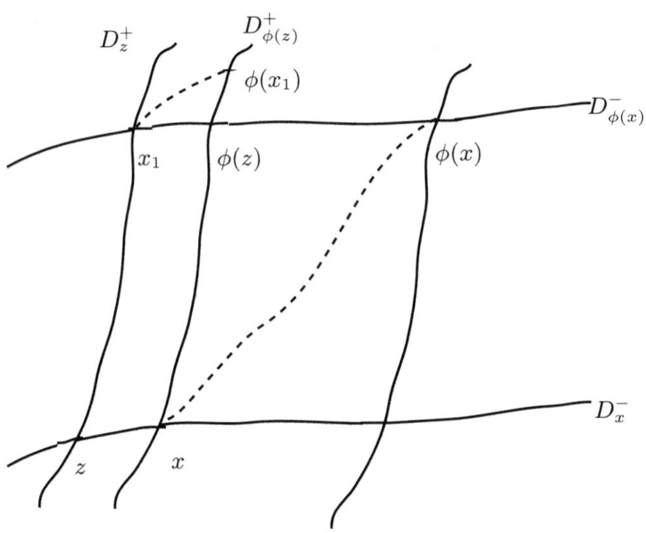

FIGURE 17.2. Proof of Anosov Closing Lemma

PROOF. Consider $S_x \ni x$, as in the beginning of this section, of some fixed size δ_0. Then there exists t_0 with $|t_0| < C_1\varepsilon$ for some constant C_1 and $\gamma \in \Gamma$ so that $\tilde{\varphi}^{T+t_0} S_x \ni d\gamma x$. Consider a map $\phi = (d\gamma)^{-1}\tilde{\varphi}^{T+t_0} : S_x \to S_x$. Since $d\gamma$ is an isometry, we have $d(x, \phi(x)) < C_1\varepsilon$. There exists $z \in \mathcal{D}_x^-$ such that $\phi(z) \in \mathcal{D}_x^+$ (see Figure 17.2). By property (c), $d^{W^-}(z, x) \le C_2 \varepsilon e^{-T}$ for some constant C_2. Take

$x_1 \in \mathcal{D}^-_{\phi(z)} \cap \mathcal{D}^+_z$, then $\phi(x_1) \in \mathcal{D}^+_{\phi(z)}$ and $d^{W^+}(\phi(x_1), \phi(z)) < C_3\varepsilon e^{-T}$ for some constant C_3. Therefore $d(x_1, \phi(x_1)) < C_4\varepsilon e^{-T}$. Continuing this process we get a fixed point $x_0 : \phi(x_0) = x_0$, i.e. $(d\gamma)x_0 = \varphi^{T'} x_0$ for some $T' : |T' - T| < C_1\varepsilon$.

We have $d(x, x_0) < C\varepsilon$. By construction $x_0 \in S_x$. Let $w \in D^+_x$ be a point such that $x_0 \in \mathcal{D}^-_w$. Then for some constant C_3, $d^{W^+}(x, w) < C_3\varepsilon$ and $d^{W^-}(w, x_0) < C_3\varepsilon$. For the same reason since $d(\varphi^T x, \varphi^{T'} x_0) < C\varepsilon$, we can conclude that for some t_1 with $|t_1| < C_2'\varepsilon$, $\varphi^{T'+t_1} x_0 \in S_{\varphi^T x}$ and that $d^{W^+}(\varphi^T x, \varphi^T w) < C_4\varepsilon$ and $d^{W^-}(\varphi^T w, \varphi^T x_0) < C_4\varepsilon$ for another constant C_4. Property (c) implies that $d^{W^+}(\varphi^t x, \varphi^t w) < C_5\varepsilon e^{-t}$ and $d^{W^-}(\varphi^t w, \varphi^t x_0) \le C_5\varepsilon e^{-(T-t)}$ and therefore $d(\varphi^t x, \varphi^t x_0) \le C_6\varepsilon e^{-\min(t, T-t)}$.

□

The following results are obtained by geometric considerations (see [**16**, §5.4] for cocompact case).

THEOREM 17.2. *Let $M = \mathcal{H}\backslash\Gamma$ and Γ be a Fuchsian group such that $\mu(M) < \infty$. Then the geodesic flow $\{\varphi^t\}$ has a dense orbit on SM, that is, it is* topologically transitive.

PROOF. We will prove that for any two nonempty open balls $U, V \in SM$ there is $t \in \mathbb{R}$ such that $\varphi^t(U) \cap V \neq \emptyset$, a property equivalent to topological transitivity [**16**, Lemma 1.4.2]. It is convenient to visualize this using the unit disc model $\mathcal{U}$ for the hyperbolic plane (see Figure 17.3).

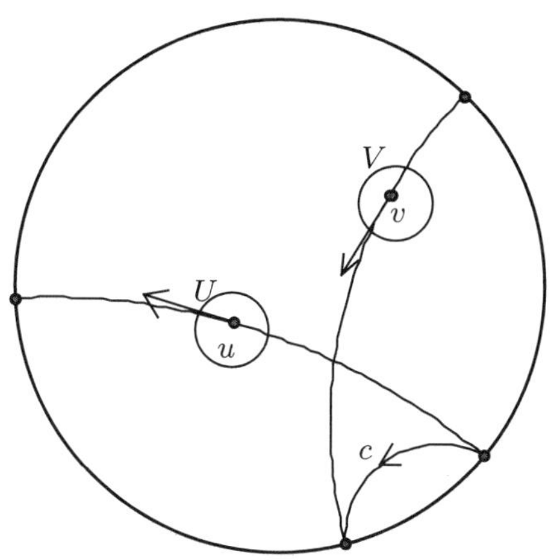

FIGURE 17.3. Topological transitivity of the geodesic flow

Using remark at the end of §16 and Theorem 17.1 we can find two periodic points, $u \in U$ and $v \in V$ (whose lifts to $S\mathcal{U}$ we also denote by u and v). Let c_u and c_v be geodesics in $\mathcal{U}$ such that $\dot{c_u} = u$ and $\dot{c_v} = v$. We may assume that $c_u(-\infty) \neq c_v(\infty)$, otherwise we replace c_u by $\gamma(c_u)$ for some $\gamma \in \Gamma$. Consider the geodesic c such that $c(\infty) = c_v(\infty)$ and $c(-\infty) = c_u(-\infty)$. By Theorem 16.2, for each $t \in \mathbb{R}$ we can find two numbers $g_u(t)$ and $g_v(t)$ such that

$$d(\dot{c_u}(g_u(t)), \dot{c}(t)) < e^{-t} \text{ as } t \to -\infty$$

and

$$d(\dot{c_v}(g_v(t)), \dot{c}(t)) < e^{-t} \text{ as } t \to \infty.$$

Since c_u and c_v project to closed geodesics on $\Gamma\backslash\mathcal{U}$, this shows that there exist t_1 and t_2 such that the projection of $\dot{c}(t_1)$ to SM is in U and the projection of $\dot{c}(t_2)$ to SM is in V. This yields the claim. □

The following important result follows immediately from Theorems 17.2 and 17.1:

COROLLARY 17.3. *Periodic orbits of the geodesic flow are dense in SM.*

THEOREM 17.4. *The Liouville measure $dv = d\mu d\phi$ on SM is ergodic under the geodesic flow.*

PROOF. The proof [**16**, Theorem 5.4.16] uses so-called "Hopf arguments", an important tool for hyperbolic dynamic. We will show that the ergodic average

$$f^+(x) = f_{\varphi^t}(x) := \lim_{T\to\infty} \frac{1}{T} \int_0^T f(\varphi^t(x))dt$$

is constant a.e. for every function $f \in L^1(SM, v)$, a property equivalent to ergodicity. It is sufficient to prove that for a continuous function f with compact support (hence uniformly continuous) since such functions are dense in $L^1(SM, v)$. Consider such an f. Then by Birkhoff Ergodic Theorem [**16**, Theorem 4.1.2]

$$\lim_{T\to\infty} \frac{1}{T} \int_0^T f(\varphi^t(x))dt$$

exists a.e. First we show that $f^+(x)$ is constant on stable leaves. Suppose the limit exists for some $p \in SM$. We will prove that it also exists for all $q \in W^+(p)$ and is independent on q: Given $\varepsilon > 0$, there exists T_0 such that

$$f(\varphi^t(p)) - f(\varphi^t(q)) < \varepsilon$$

for all $t > T_0$ by uniform continuity. But this means that

$$\left|\frac{1}{T} \int_o^T (f(\varphi^t(p)) - f(\varphi^t(q)))dt\right| < \varepsilon$$

for sufficiently large T, as required. Since existence and the value of the limit is φ^t-invariant, $f^+(x) = f_{\varphi^t}(x)$ is, in fact, constant on weak stable manifolds, the integral manifold of $E^0 \oplus E^+$. Consider also the negative time average

$$f^-(x) = \lim_{T\to\infty} \frac{1}{T} \int_{-T}^0 f(\varphi^t(x))dt.$$

It exists and is constant a.e. on unstable manifolds. Furthermore, by a corollary to Birkhoff Ergodic Theorem [**16**, Proposition 4.1.3] $f^+(x) = f^-(x)$ a.e. In terms of the local C^1 coordinates on SM (t, u, v) introduced in the proof of Theorem 17.1, on a small open set U, by Fubini's Theorem, for t in a set C of full measure $f^+(t, u, v) = f^-(t, u, v)$ for almost all (u, v). But then for any such t_1 and t_2 the corresponding sets of (u, v) intersect since they both have full measure. Therefore, $f^+(t_1, u, v) = f^+(t_2, u, v)$ for a.e. (u, v), hence $f^+(x)$ is constant on C, as required. □

18. Livshitz's Theorem

A. Livshitz in [**23**] proved his theorem for Anosov systems on any compact manifold and showed that the obtained function was C^1. Guillemin and Kazhdan in [**10**] gave a proof of Livshitz's theorem for Anosov flows, and in [**11**] they gave a proof for geodesic flows on any compact surface and showed that the function was C^∞.

Here we will prove a version of Livshitz's Theorem for the geodesic flow on SM of finite volume and possibly with cusps. In this case, it is natural to formulate it for the class $\mathcal{BL}$ of bounded Lipschitz functions.

THEOREM 18.1. *Let Γ be a discrete subgroup of $PSL(2,\mathbb{R})$, $M = \Gamma\backslash\mathcal{H}$, $\mu(M) < \infty$, $X = S(M)$, and let $\{\varphi^t\}$ be the geodesic flow acting on X. Let $f \in \mathcal{BL}(X)$ be a function having zero integrals over all periodic orbits of $\{\varphi^t\}$. Then there exists a function F on X satisfying a Lipschitz condition with some constant α and differentiable in the direction of the geodesic flow and such that $\mathcal{D}F = f$, where $\mathcal{D} = \frac{d}{dt}$ is the operator of differentiation along the orbits of the flow $\{\varphi^t\}$.*

PROOF. Consider a point $x_0 \in X$ with a dense orbit $\mathcal{O}(x_0)$ and define a map $F : \mathcal{O}(x_0) \to \mathbb{R}$ by the formula $F(x) = \int_0^s f(\varphi^t(x_0))dt$ at $x = \varphi^s(x_0)$. We want to prove that F extends to a function on X satisfying a Lipschitz condition and $\mathcal{D}F = f$. We claim that $F(x)$ satisfies a Lipschitz condition on the orbit $\mathcal{O}(x_0)$. Let $y_1 = \varphi^{t_1}(x_0)$, $y_2 = \varphi^{t_1+T}(x_0) \in \mathcal{O}(x_0)$ and $d(y_1, y_2) < \varepsilon$. Then

$$F(y_2) - F(y_1) = \int_0^T f(\varphi^t(y_1))dt.$$

By Theorem 17.1 we can find a point $x_1 \in X$ such that $\varphi^{T'}x_1 = x_1$ for some $T' : |T' - T| < C_1\varepsilon$ and such that for $0 \le t \le T$, $d(\varphi^t x_1, \varphi^t y_1) < C_2\varepsilon$ for some constant C_2. Since the orbit $\mathcal{O}(x_1)$ is periodic, we have

$$\int_0^{T'} f(\varphi^t(x_1))dt = 0.$$

Therefore, using the estimates of Theorem 17.1(b) we have

$$|F(y_2) - F(y_1)| = |\int_0^T f(\varphi^t(y_1))dt - \int_0^{T'} f(\varphi^t(x_1))dt|$$

$$\le |\int_0^T f(\varphi^t(y_1)) - f(\varphi^t(x_1))dt| + |\int_T^{T'} f(\varphi^t(x_0)dt| < \alpha\varepsilon$$

for some constant α. The Lipschitz property of the function f is used to estimate the first term, and the boundeness to estimate the second. This proves the claim. Therefore F can be extended from the dense set to a function satisfying a Lipschitz condition on X. Since $\frac{d}{dt}F = f$ on the dense set it follows that F is differentiable in the direction of the geodesic flow and $\frac{d}{dt}F = f$ on X. □

Exercises

33. Prove that if the function f is C^1, then the function F is also C^1. *Hint:* show that F is differentiable in the directions of W_x^+ and W_x^- using the transversality of these curves in SM, the fact that they depend continuously on x, and the uniform convergence of the integral expression for derivative.

Lecture IV. Symbolic coding of geodesics

19. Representation of the geodesic flow as a special flow

A *cross-section* C for the geodesic flow is a subset of the unit tangent bundle SM visited by (almost) every geodesic infinitely often both in the future and in the past. In other words, every $v \in C$ defines an oriented geodesic $\gamma(v)$ on M which will return to C infinitely often. The function $f : C \to \mathbb{R}$ giving the *time of the first return* to C is defined as follows: if $v \in C$ and t is the time of the first return of $\gamma(v)$ to C, then $f(v) = t$. The map $R : C \to C$ defined by $R(v) = \varphi^{f(v)}(v)$ is called the *first return map*. Thus $\{\varphi^t\}$ can be represented as a *special flow* on the space

$$C^f = \{(v, s) \mid v \in C,\ 0 \le s \le f(v)\},$$

given by the formula $\varphi^t(v, s) = (v, s+t)$ with the identification $(v, f(v)) = (R(v), 0)$.

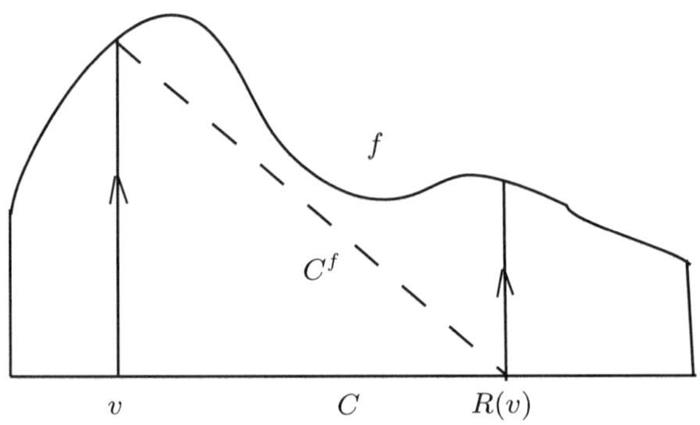

FIGURE 19.1. Geodesic flow is a special flow

Let $\mathcal{N}$ be a finite or countable alphabet, $\mathbb{N}^{\mathbb{Z}} = \{x = \{n_i\}_{i\in\mathbb{Z}} \mid n_i \in \mathbb{N}\}$ be the space of all bi-infinite sequences endowed with the Tikhonov (product) topology,

$$\sigma : \mathbb{N}^{\mathbb{Z}} \to \mathbb{N}^{\mathbb{Z}} \text{ defined by } \{\sigma x\}_i = n_{i+1}$$

be the left shift map, and $\Lambda \subset \mathbb{N}^{\mathbb{Z}}$ be a closed σ-invariant subset. Then (Λ, σ) is called a *symbolic dynamical system*. There are some important classes of such dynamical systems. The whole space $(\mathbb{N}^{\mathbb{Z}}, \sigma)$ is called the *Bernoulli shift*. If the space Λ is given by a set of simple transition rules which can be described with the help of a matrix consisting of zeros and ones, we say that (Λ, σ) is a *one-step topological Markov chain* or simply a *topological Markov chain* (sometimes (Λ, σ) is also called a *subshift of finite type*). Similarly, if the space Λ is determined by specifying which $(k+1)$-tuples of symbols are allowed, we say that (Λ, σ) is a *k-step topological Markov chain* (a precise definition is given in Section 25).

In order to represent the geodesic flow as a special flow over a symbolic dynamical system, one needs to choose an appropriate cross-section C and code it, i.e. to find an appropriate symbolic dynamical system (Λ, σ) and a continuous surjective map $\mathfrak{C} : \Lambda \to C$ (in some cases the actual domain of $\mathfrak{C}$ is Λ except a finite or countable set of excluded sequences) defined such that the diagram

$$\begin{array}{ccc} \Lambda & \xrightarrow{\sigma} & \Lambda \\ {\scriptstyle\mathfrak{C}}\downarrow & & \downarrow{\scriptstyle\mathfrak{C}} \\ C & \xrightarrow{R} & C \end{array}$$

is commutative. We can then talk about *coding sequences* for geodesics defined up to a shift which corresponds to a return of the geodesic to the cross-section C. Notice that usually the coding map is not injective but only finite-to-one (see e.g. [**1**, §3.2 and §5]).

There are two essentially different methods of coding geodesics on surfaces of constant negative curvature. The geometric code, with respect to a given fundamental region, is obtained by a construction universal for all Fuchsian groups. The second method is specific for the modular group and is of arithmetic nature: it uses continued fraction expansions of the end points of the geodesic at infinity and a so-called reduction theory. Bowen and Series [**6**] extended some of the ideas behind this type of coding to general Fuchsian groups by using the so-called "boundary expansions". In fact, a generalization of continued fractions discussed in §22 makes this connection even more promising: the arithmetic codings for the modular surface via (a, b)-continued fractions still can be viewed as boundary expansions by properly partitioning the real axis into three intervals labeled by T, T^{-1}, and S. Some other classes of continued fractions work for other arithmetic Fuchsian groups, in particular, for congruence subgroups [**17**, §7] and for Hecke triangle groups [**25**].

20. Geometric coding

The Morse method. We first describe the general method of coding geodesics on a surface of constant negative curvature by recording the sides of a given fundamental region cut by the geodesic. This method first appeared in a paper by Morse [**26**] in 1921. However, in a 1927 paper, Koebe [**22**] mentioned an unpublished work from 1917, where the same ideas were apparently used. Starting with [**29**] Series called this method *Koebe-Morse*, but since this earlier work by Koebe has not been traced, we think it is more appropriate to call this coding method the *Morse method.* We will follow [**15**] in describing the Morse method for a finitely generated Fuchsian group Γ of cofinite hyperbolic area.

A Dirichlet fundamental region $\mathcal{D}$ of Γ always has an even number of sides identified by generators of Γ and their inverses (Theorems 13.1 and 11.8); we denote this set by $\{g_i\}$. We label the sides of $\mathcal{D}$ (on the inside) by elements of the set $\{g_i\}$ as follows: if a side s is identified in $\mathcal{D}$ with the side $g_i(s)$, we label the side s by g_i. By labeling all the images of s under Γ by the same generator g_i we obtain the labeling of the whole net $\mathcal{S} = \Gamma(\partial\mathcal{D})$ of images of sides of $\mathcal{D}$, such that each side in $\mathcal{S}$ has two labels corresponding to the two images of $\mathcal{D}$ shared by this side. We assign to an oriented geodesic in $\mathcal{H}$ a bi-infinite sequence of elements of $\{g_i\}$ which label the successive sides of $\mathcal{S}$ this geodesic crosses.

We describe the *Morse coding sequence* of a geodesic in $\mathcal{H}$ under the assumption that it does not pass through any vertex of the net $\mathcal{S}$—we call such *general position geodesics.* (Morse called the coding sequences *admissible line elements*, and some authors [**29, 9**] referred to them as *cutting sequences.*) We assume that the geodesic intersects $\mathcal{D}$ and choose an initial point on it inside $\mathcal{D}$. After exiting $\mathcal{D}$, the geodesic enters a neighboring image of $\mathcal{D}$ through the side labeled, say, by g_1 (see Figure 20.1). Therefore this image is $g_1(\mathcal{D})$, and the first symbol in the code is g_1. If

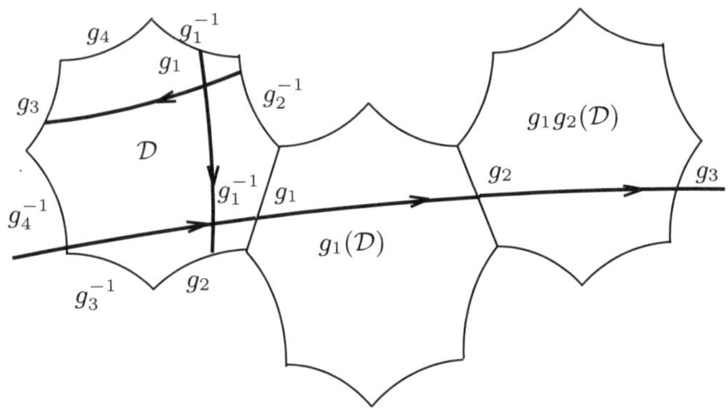

FIGURE 20.1. Morse coding

it enters the second image of $\mathcal{D}$ through the side labeled by g_2, the second image is $(g_1g_2g_1^{-1})(g_1(\mathcal{D})) = g_1g_2(\mathcal{D})$, and the second symbol in the code is g_2, and so on. Thus we obtain a sequence of all images of $\mathcal{D}$ crossed by our geodesic in the direction of its orientation: $\mathcal{D}$, $g_1(\mathcal{D})$, $g_1g_2(\mathcal{D}),\dots$, and a sequence of all images of $\mathcal{D}$ crossed by our geodesic in the opposite direction: $g_0^{-1}(\mathcal{D})$, $(g_0g_{-1})^{-1}(\mathcal{D}),\dots$. Thus, the Morse coding sequence is

$$[\dots g_{-1}, g_0, g_1, g_2, \dots].$$

By mapping the oriented geodesic segments between every two consecutive crossings of the net $\mathcal{S}$ back to $\mathcal{D}$ (as shown in Figure 20.1), we obtain a geodesic in $\mathcal{D}$. The coding sequence described above may be obtained by taking generators labeling the sides of $\mathcal{D}$ (on the outside) the geodesic hits consequently.

A geodesic on M is closed if and only if it is the projection of the axis of a hyperbolic element in Γ. For general position geodesics, a coding sequence is periodic if and only if the geodesic is closed. If a geodesic is the axis of a primitive hyperbolic element $g \in \Gamma$, i.e. a hyperbolic element which is not a power of another element in Γ, we have

$$g = g_1g_2\cdots g_n$$

for some n. In this case the sequence is periodic with the least period $[g_1, g_2, \dots, g_n]$.

An ambiguity in assigning a Morse code occurs whenever a geodesic passes though a vertex of $\mathcal{D}$: such geodesics have more than one code, and closed geodesics have non-periodic codes along with periodic ones (see [**9, 18**] for relevant discussions).

For free groups Γ with properly chosen fundamental regions, all reduced (here this simply means that a generator g_i does not follow or precede g_i^{-1}) bi-infinite sequences of elements from the generating set $\{g_i\}$ are realized as Morse coding sequences of geodesics on M (see [**29**]), but, in general, this is not the case. Even for the classical example of $\Gamma = PSL(2,\mathbb{Z})$ with the standard fundamental region F (Figure 10.2) no elegant description of admissible Morse coding sequences is known and probably does not exist. Important results in this direction were obtained in [9], where the admissible coding sequences were described in terms of forbidden blocks. The set of generating forbidden blocks found in [9] has an intricate structure attesting the complexity of the Morse code.

Geometric code for the modular surface. Let $\Gamma = PSL(2,\mathbb{Z})$, and $M = \Gamma\backslash\mathcal{H}$ be the modular surface. Recall that the generators of $PSL(2,\mathbb{Z})$ acting on $\mathcal{H}$ are $T(z) = z+1$ and $S(z) = -\frac{1}{z}$. The Morse code with respect to the standard fundamental region F can be assigned to any oriented geodesic γ in F (which does not go to the cusp of F in either direction), and can be described by a bi-infinite sequence of integers as follows. The boundary of F consists of four sides: left and right vertical, identified and labeled by T, and T^{-1}, respectively; left and right circular both identified and labeled by S (see Figure 20.2). It is clear

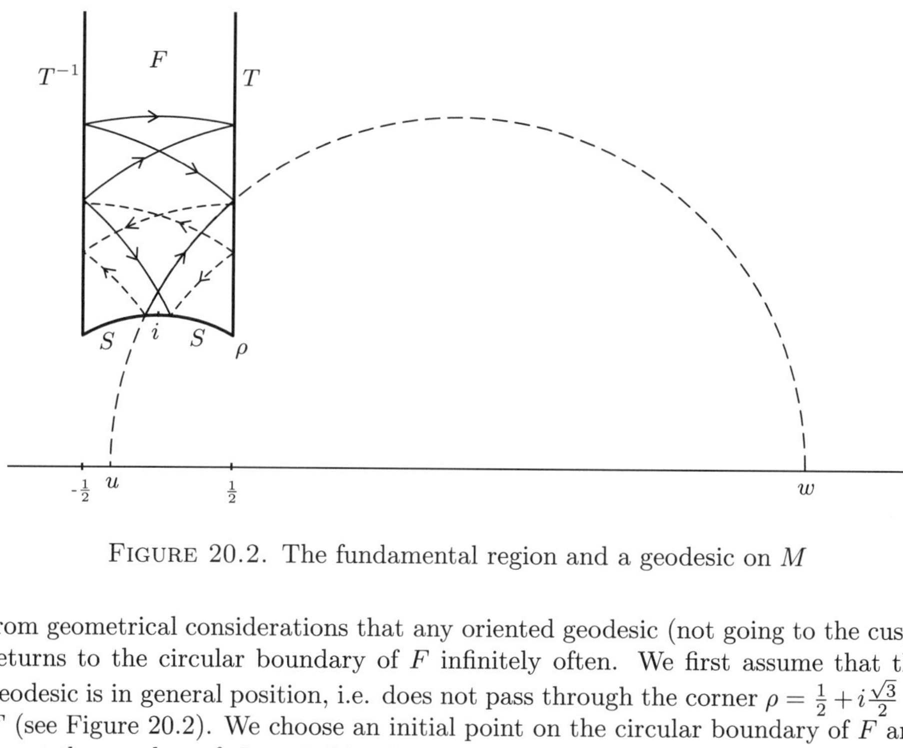

FIGURE 20.2. The fundamental region and a geodesic on M

from geometrical considerations that any oriented geodesic (not going to the cusp) returns to the circular boundary of F infinitely often. We first assume that the geodesic is in general position, i.e. does not pass through the corner $\rho = \frac{1}{2} + i\frac{\sqrt{3}}{2}$ of F (see Figure 20.2). We choose an initial point on the circular boundary of F and count the number of times it hits the vertical sides of the boundary of F moving in the direction of the geodesic. A positive integer is assigned to each block of hits of the right vertical side (or a block of T's in the Morse code), and a negative, to each block of hits of the left vertical side (or a block of T^{-1}'s). Moving the initial point in the opposite direction allows us to continue the sequence backwards. Thus we obtain a bi-infinite sequence of nonzero integers

$$[\gamma] = [\dots, n_{-2}, n_{-1}, n_0, n_1, \dots],$$

uniquely defined up to a shift, which is called the *geometric code* of γ. Moving the initial point in either direction until its return to one of the circular sides of F corresponds to a shift of the geometric coding sequence $[\gamma]$. Recall that a geodesic in general position is closed if and only if the coding sequence is periodic. We refer to the least period $[n_0, n_1, \dots, n_m]$ as its geometric code. For example, the geometric code of the closed geodesic on Figure 20.2 is $[4, -3]$.

A geodesic with geometric code $[\gamma]$ can be lifted to the upper half-plane $\mathcal{H}$ (by choosing the initial point appropriately) so that it intersects

$$T^{\pm 1}(F), \dots, T^{n_0}(F), T^{n_0}S(F), \dots, T^{n_0}ST^{n_1}S(F), \dots,$$

in the positive direction (the sign in the first group of terms is chosen in accordance with the sign of n_0, etc.) and

$$S(F), ST^{\mp 1}(F), \dots, ST^{-n_{-1}}(F), \dots, ST^{-n_{-1}}ST^{-n_{-2}}(F), \dots,$$

in the negative direction.

The case when a geodesic passes through the corner ρ of F was described to a great extent in [**9**, §7]. Such a geodesic has multiple codes obtained by approximating it by general position geodesics which pass near the corner ρ slightly higher or slightly lower. If a geodesic hits the corner only once it has exactly two codes. If a geodesic hits the corner at least twice, it hits it infinitely many times and is closed; if it hits the corner n times in its period, it has exactly $2n+2$ codes, i.e. shift-equivalence classes of coding sequences, some of which are not periodic. It is unknown, however, whether there is an upper bound on the number of shift-equivalence classes of coding sequences corresponding to closed geodesics [**9**, §9].

Canonical codes considered in [**15**] were obtained by the convention that a geodesic passing through ρ in the clockwise direction exits F through the right vertical side of F labeled by T (this corresponds to the approximation by geodesics which pass near the corner ρ slightly higher). According to this convention, the geometric codes of the axes of transformations $A_4 = T^4S$, $A_{3,6} = T^3ST^6S$ and $A_{6,3} = T^6ST^3S$ are $[4], [3,6]$ and $[6,3]$, respectively. However, all these geodesics have other codes. For example, the axis of A_4 has a code $[2,-1]$ obtained by approximation by geodesics which pass near the corner ρ slightly lower, and two non-periodic codes for the same closed geodesic are

$$[\dots, 4, 4, 3, -1, 2, -1, 2, -1, 2, \dots] \text{ and } [\dots, 2, -1, 2, -1, 2, -1, 3, 4, 4, \dots].$$

For more details, see [**9, 18**].

21. Symbolic representation of geodesics via geometric code.

Let

$$\mathbb{N}^{\mathbb{Z}} = \{x = \{n_i\}_i \in \mathbb{Z} \mid n_i \in \mathbb{N}\}$$

be the set of all bi-infinite sequences on the alphabet $\mathbb{N} = \{n \in \mathbb{Z} \mid n \neq 0\}$, endowed with the Tykhonov product topology, and $\sigma : \mathbb{N}^{\mathbb{Z}} \to \mathbb{N}^{\mathbb{Z}}$ be the left shift map given by $\{\sigma x\}_i = n_{i+1}$. Let X_0 be the set of admissible geometric coding sequences for general position geodesics in M, and X be its closure in the Tykhonov product topology. It was proved in [**9**, Theorem 7.2] that every sequence in X is a geometric code of a unique oriented geodesic in M, and every geodesic in M has at least one and at most finitely many codes (see examples above). Thus X is a closed σ-invariant subspace of $\mathbb{N}^{\mathbb{Z}}$.

The cross-section for the geometric code and its partition. Since every oriented geodesic that does not go to the cusp of F in either direction returns to the circular boundary of F infinitely often, the set $B \subset SM$ consisting of all unit vectors in SM with base points on the circular boundary of F and pointing inside F (see Figure 21.1) is a cross-section which captures the geometric code.

We parameterize the cross-section B by the coordinates (ϕ, θ), where $\phi \in [-\pi/6, \pi/6]$ parameterizes the arc and $\theta \in [-\phi, \pi - \phi]$ is the angle the unit vector makes with the positive horizontal axis in the clockwise direction. The elements of the partition of B are labeled by the symbols of the alphabet $\mathbb{N}$, $B = \cup_{n\in\mathbb{N}} C_n$, and are defined by the following condition: $C_n = \{v \in B \mid n_0(v) = n\}$, i.e. C_n consists of all tangent vectors v in B such that, for the coding sequence of the corresponding geodesic in $\mathcal{H}$, $n_0(x) = n$. Let $R : B \to B$ be the first return map. Since the first return to the cross-section exactly corresponds to the left shift of the coding sequence x associated to v, we have $n_0(R(v)) = n_1(v)$. The infinite geometric partition and its image under the return map R are sketched on Figure 21.2. Boundaries between the elements of the partition shown on Figure 21.2 correspond to geodesics going into the corner; the two vertical boundaries of the cross-section B are identified and correspond to geodesics emanating from the corner. They have more than one code. For example, the codes $[4]$ and $[\dots, 2, -1, 2, -1, 2, -1, 3, 4, 4, 4, \dots]$ correspond to the point on the right boundary of B between C_4 and C_3, and the codes $[2, -1]$ and $[\dots, 4, 4, 4, 4, 3, -1, 2, -1, 2, -1, 2, \dots]$ correspond to the point on the left boundary between C_2 and C_3 that are identified and are the four codes of the axis of A_4.

The coding map for the geometric code. It was proved in [**9**, Lemma 7.1] that if a sequence of general position geodesics is such that the sequence of their coding sequences converges in the product topology, then the sequence of these geodesics converges to a limiting geodesic uniformly. Since the tangent vectors in the cross-section B are determined by the intersection of the corresponding geodesics with the unit circle, we conclude that the sequence of images of the coding sequences under the map $\mathfrak{C} : X \to B$ converges to the image of the limiting coding sequence. This implies that the map $\mathfrak{C}$ is continuous.

Which geometric codes are realized? Not all bi-infinite sequences of nonzero integers are realized as geometric codes. For instance, the periodic sequence $\{\overline{8, 2}\}$ is not a geometric code since the geometric code of the axis of T^8ST^2S is $[6, -2]$, as can be seen on Figure 21.3 [**15**].

Figure 21.2 gives an insight into the complexity of the geometric code, where the elements C_n and their forward iterates $R(C_n)$ are shown. Each C_n is a curvilinear

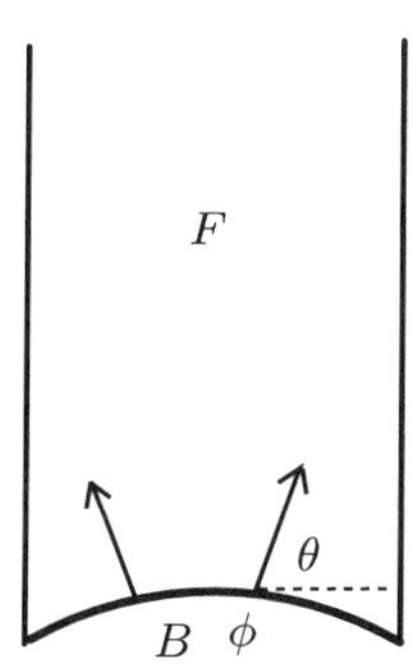

FIGURE 21.1. The cross-section B

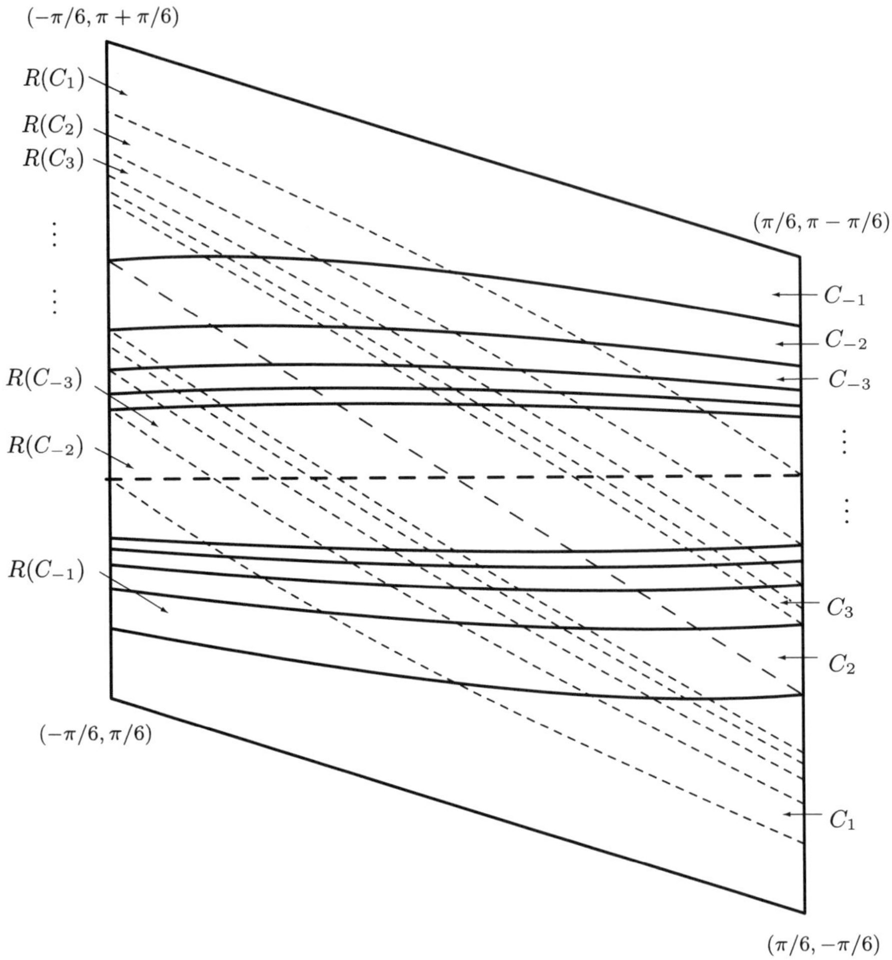

FIGURE 21.2. The infinite geometric partition and its image under the return map R

quadrilateral with two vertical and two "horizontal" sides, and each $R(C_n)$ is a curvilinear quadrilateral with two vertical and two "slanted" sides. The horizontal sides of C_n are mapped to vertical sides of $R(C_n)$, and the vertical sides of C_n are stretched across the parallelogram representing B and mapped to the "slanted" sides of $R(C_n)$.

If $n_0(v) = n$ and $n_1(v) = m$ for some vector $v \in B$, then $R(C_n) \cap C_m \neq \emptyset$. Therefore, as Figure 21.2 illustrates, the symbol 2 in a geometric code cannot be followed by 1, 2, 3, 4 and 5.

We say that C_m and $R(C_n)$ intersect "transversally" if their intersection is a curvilinear parallelogram with two "horizontal" sides belonging to the horizontal boundary of C_m and two "slanted" sides belonging to the slanted boundary of $R(C_n)$. Notice that for each transverse intersection $R(C_n) \cap C_m$ its forward iterate under R stretches to a strip inside $R(C_m)$ between its two vertical sides. Hence, the symbol m can follow symbol n in a coding sequence.

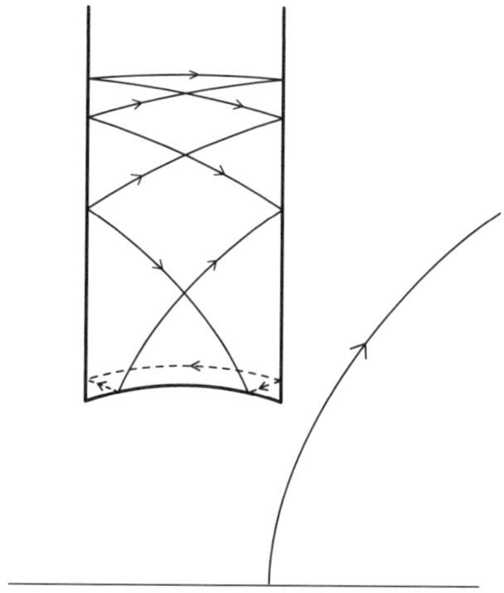

FIGURE 21.3. The geometric code of the axis of T^8ST^2S is $[6,-2]$

We also observe that the elements C_m and $R(C_n)$ intersect transversally if and only if $|n| \geq 2$, $|m| \geq 2$, and

$$|1/n + 1/m| \leq 1/2.$$

This is a flow-invariant subset which constitutes the essential part of the set of geometrically Markov codes; see Theorems 25.2 and 25.4 in §25.

22. Arithmetic codings

In this section we describe a method of constructing arithmetic codes for geodesics on the modular surface M using expansions of the end points of their lifts to $\mathcal{H}$ in what we call *generalized minus continued fractions.* Three classical continued fraction expansions described in **[19]** were defined using different integer-valued functions $(\cdot)$ (generalized "integer part" functions) that are included into a 2-parameter family of integer-valued functions suggested for consideration recently by Don Zagier,

$$(x)_{a,b} = \begin{cases} \lfloor x-a \rfloor & \text{if } x < a \\ 0 & \text{if } a \leq x < b \\ \lceil x-b \rceil & \text{if } x \geq b\,, \end{cases} \tag{22.1}$$

where $\lfloor x \rfloor$ denotes the integer part of x, $\lceil x \rceil = \lfloor x \rfloor + 1$.

If $(a,b) \in \mathcal{P} = \{(a,b) \in \mathbb{R}^2 \,|\, a \leq 0 \leq b,\, b-a \geq 1,\, -ab \leq 1\}$, any irrational number x can be expressed in a unique way as an infinite (a,b)-*continued fraction*

$$x = n_0 - \cfrac{1}{n_1 - \cfrac{1}{n_2 - \cfrac{1}{\ddots}}}$$

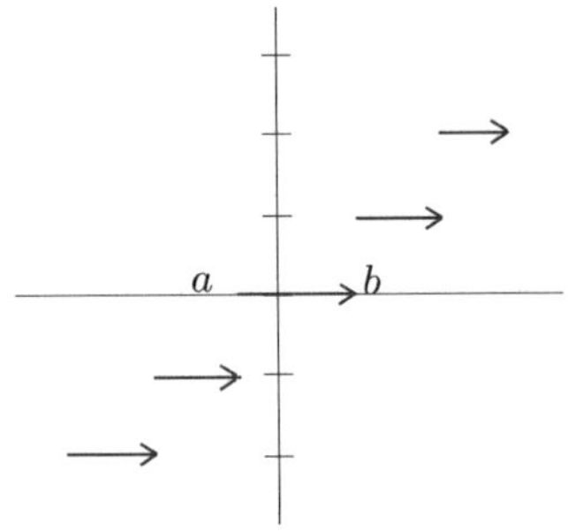

FIGURE 22.1. The function $(x)_{a,b}$

that we will denote by $(n_0, n_1, \cdots)$ for short. The "digits" n_i, $i \geq 1$, are non-zero integers determined recursively by

$$n_0 = (x)_{a,b},\ x_1 = -\frac{1}{x - n_0}, \text{ and } n_i = (x_i)_{a,b},\ x_{i+1} = -\frac{1}{x_i - n_i}. \tag{22.2}$$

The following theorem is a starting point of the study of what we call (a, b)-*continued fractions* in a joint paper of the author with I. Ugarcovici [**21**].

THEOREM 22.1. *Let $\{n_i\}$ be a sequence of integers defined by (22.2) and*

$$r_k = (n_0, n_1, \ldots, n_k) := n_0 - \cfrac{1}{n_1 - \cfrac{1}{n_2 - \cfrac{1}{\ddots - \cfrac{1}{n_k}}}}.$$

Then the sequence r_k converges to x.

The three classical continued fraction expansions can be now described as follows.

***G*-expansion** $(a = -1, b = 0)$. The function

$$(x)_{-1,0} = \lceil x \rceil = \lfloor x \rfloor + 1$$

(that differs for integers from the classical ceiling function) gives the *minus continued fraction expansion* described in [**31**] and used in [**15**] for coding closed geodesics. Since the coding procedure for closed geodesic is the same as the Gauss reduction theory for indefinite integral quadratic forms, we refer to this expansion as the *Gauss-* or *G-expansion* and call the corresponding code *G-code*. G-codes for oriented geodesics, not necessarily closed, were introduced in [**12**]. The digits $n_0, n_1, \ldots$ of a G-expansion satisfy the condition $n_i \geq 2$, if $i \geq 1$. Conversely, any infinite sequence of integers $n_0, n_1, n_2, \ldots$ with $n_i \geq 2$ for $i \geq 1$ defines a real number whose G-expansion is $\lceil n_0, n_1, n_2, \ldots \rceil$.

***A*-expansion** $(a = -1, b = 1)$. The function

$$(x)_{-1,1} = \lceil x \rfloor = \begin{cases} \lfloor x \rfloor & \text{if } x \geq 0 \\ \lceil x \rceil & \text{if } x < 0 \end{cases}$$

gives an expansion which was used in [**19**] to reinterpret the classical Artin code *(A-code)*. This expansion has digits of alternating signs, and we call it the *A-expansion.* Conversely, any infinite sequence of nonzero integers with alternating signs $n_0, n_1, n_2, \ldots$ defines a real number whose A-expansion is $\lceil n_0, n_1, n_2, \ldots \rfloor$.

The G- and A-expansions satisfy the following properties:

(1) Two irrationals x, y are $PSL(2,\mathbb{Z})$-equivalent $\Longleftrightarrow$ their expansions have the same tail, that is, if $x = (n_0, n_1, \ldots)$ and $y = (m_0, m_1, \ldots)$ then $n_{i+k} = m_{i+l}$ for some integers k, l and all $i \geq 0$;
(2) A real number x is a quadratic irrationality $\Longleftrightarrow$ $(n_0, n_1, \ldots)$ is eventually periodic;
(3) Let x and x' be conjugate quadratic irrationalities, i.e. the roots of a quadratic polynomial with integer coefficients. If $x = (\overline{n_0, n_1 \ldots, n_k})$, then $\frac{1}{x'} = (\overline{n_k, \ldots, n_1, n_0})$.

Let us remark that properties (2) and (3) are also valid for the regular continued fractions, while property (1) holds if one replaces $PSL(2,\mathbb{Z})$ by $PGL(2,\mathbb{Z})$.

***H*-expansion** $(a = -\frac{1}{2}, b = \frac{1}{2})$. This expansion is obtained using the function

$$(x)_{-\frac{1}{2},\frac{1}{2}} = \langle x \rangle$$

(the nearest integer to x). It was first used by Hurwitz [**13**] in order to establish a reduction theory for indefinite real quadratic forms, and we call it the *Hurwitz- or H-expansion.* The digits n_i $(i \geq 1)$ of an H-expansion satisfy $|n_i| \geq 2$, and if $|n_i| = 2$ then $n_i n_{i+1} < 0$. Conversely, any infinite sequence of integers $n_0, n_1, n_2, \ldots$ with the above property defines an irrational number whose H-expansion is $\langle n_0, n_1, n_2, \ldots \rangle$.

The H-expansion satisfies property (2), but not (1) and (3). There is a minor exception to property (1): it is possible for two irrationals not sharing the same tail to be $PSL(2,\mathbb{Z})$-equivalent, but this can happen if and only if one irrational has a tail of 3's in its H-expansion and the other one has a tail of -3's, i.e. the irrationals are equivalent to $r = (3-\sqrt{5})/2$ ([**13, 7**]). Property (3) is more serious. In order to construct a meaningful code, we need to use a different expansion for $1/u$ (introduced also by Hurwitz) so that a property similar to (3) is satisfied. It uses yet another integer-valued function which is a part of the (a,b)-family:

$$(x)_{r-1,1-r} = \langle\!\langle x \rangle\!\rangle = \begin{cases} \langle x \rangle - \operatorname{sgn}(x) & \text{if} \quad \operatorname{sgn}(x)(\langle x \rangle - x) > r, \\ \langle x \rangle & \text{otherwise,} \end{cases}$$

and is called the *H-dual expansion.* Now if $x = \langle \overline{n_0, n_1 \ldots, n_k} \rangle$, then $\frac{1}{x'}$ has a purely periodic H-dual expansion $\frac{1}{x'} = \langle\!\langle \overline{n_k, \ldots, n_1, n_0} \rangle\!\rangle$. The formula for $\langle\!\langle \cdot \rangle\!\rangle$ comes from the fact that if $x = \langle n_0, n_1, \ldots \rangle$ then the entries n_i satisfy the asymmetric restriction: if $|n_i| = 2$, then $n_i n_{i+1} < 0$. For more details, see [**13, 7, 19**]; for a general definition of a dual expansion see §23.

Convergents. If $x = (n_0, n_1, \ldots)$, then the *convergents* $r_k = (n_0, n_1, \ldots, n_k)$ can be written as p_k/q_k where p_k and q_k are obtained inductively as:

$$p_{-2} = 0\,,\; p_{-1} = 1\,;\; p_k = n_k p_{k-1} - p_{k-2} \;\text{ for } k \geq 0$$
$$q_{-2} = -1\,,\; q_{-1} = 0\,;\; q_k = n_k q_{k-1} - q_{k-2} \;\text{ for } k \geq 0\,.$$

The following properties are shared by all three expansions:

- $1 = q_0 \leq |q_1| < |q_2| < \ldots;$

- $p_{k-1}q_k - p_kq_{k-1} = 1$, for all $k \geq 0$.

The rates of convergence, however, are different. For the A- and H-expansions we have

$$\left| x - \frac{p_k}{q_k} \right| \leq \frac{1}{q_k^2}, \tag{22.3}$$

while for the G-expansion we only have

$$\left| x - \frac{p_k}{q_k} \right| \leq \frac{1}{q_k}. \tag{22.4}$$

A quadratic irrationality x has a purely periodic expansion if and only if x and x' satisfy certain *reduction inequalities*, which give us the notion of a *reduced geodesic* for each code.

DEFINITION 22.2. An oriented geodesic in $\mathcal{H}$ going from u to w is called

- *G-reduced* if $0 < u < 1$ and $w > 1$;
- *A-reduced* if $|w| > 1$ and $-1 < \operatorname{sgn}(w)u < 0$;
- *H-reduced* if $|w| > 2$ and $\operatorname{sgn}(w)u \in [r-1, r]$, $r = (3-\sqrt{5})/2$.

Now we can describe a reduction algorithm which works for each arithmetic α-code, where $\alpha = G, A, H$. For the H-code we consider only geodesics whose end point w is not equivalent to r.

Reduction algorithm. Let γ be an arbitrary geodesic on $\mathcal{H}$, with end points u and w, and $w = (n_0, n_1, n_2, \dots)$. We construct the sequence of real pairs $\{(u_k, w_k)\}$ $(k \geq 0)$ defined by $u_0 = u$, $w_0 = w$ and

$$w_{k+1} = ST^{-n_k} \dots ST^{-n_1}ST^{-n_0}w\,, \quad u_{k+1} = ST^{-n_k} \dots ST^{-n_1}ST^{-n_0}u\,.$$

Each geodesic with end points u_k and w_k is $PSL(2,\mathbb{Z})$-equivalent to γ by construction.

THEOREM 22.3. *The above algorithm produces in finitely many steps an α-reduced geodesic $PSL(2,\mathbb{Z})$-equivalent to γ, i.e. there exists a positive integer ℓ such that the geodesic with end points u_ℓ and w_ℓ is α-reduced.*

To an α-reduced geodesic γ, we associate a bi-infinite sequence of integers

$$(\gamma) = (\dots, n_{-2}, n_{-1}, n_0, n_1, n_2, \dots)$$

called its *arithmetic code*, by juxtaposing the α-expansions of $w = (n_0, n_1, n_2, \dots)$ and $1/u = (n_{-1}, n_{-2}, \dots)$ (for the H-code we need to use the dual H-expansion of $1/u$).

REMARK. Any further application of the reduction algorithm to an α-reduced geodesic yields α-reduced geodesics whose codes are left shifts of the code of the initial α-reduced geodesic.

The proof of Theorem 22.3 follows the same general scheme for each code, but the notion of reduced geodesic is different in each case, and so are the properties of the corresponding expansions and estimates.

Now we associate to any oriented geodesic γ in $\mathcal{H}$ the α-code of a reduced geodesic $PSL(2,\mathbb{Z})$-equivalent to γ, which is obtained by the reduction algorithm described above.

THEOREM 22.4. *Each geodesic γ in $\mathcal{H}$ is $PSL(2,\mathbb{Z})$-equivalent to an α-reduced geodesic ($\alpha = G, A, H$). Two reduced geodesics γ and γ' in $\mathcal{H}$ having arithmetic codes $(\gamma) = (n_i)_{i=-\infty}^{\infty}$ and $(\gamma') = (n_i')_{i=-\infty}^{\infty}$ are $PSL(2,\mathbb{Z})$-equivalent if and only if for some integer l and all integers i one has $n_i' = n_{i+l}$.*

EXAMPLE E. Let γ be a geodesic on $\mathcal{H}$ from $u = \sqrt{5}$ to $w = -\sqrt{3}$. The G-expansions are

$$w = \lceil -1, 2, \overline{2, 3} \rceil , \quad 1/u = \lceil 1, \overline{2, 6, 2, 2} \rceil .$$

First, we need to find an equivalent G-reduced geodesic. For this we use the reduction algorithm described above for G-expansions and construct the sequence (u_1, w_1), $(u_2, w_2), \ldots,$ until we obtain a G-reduced pair equivalent to (u, w). We have

$$\begin{aligned} w_1 &= ST(w) = (1+\sqrt{3})/2, \quad u_1 = ST(u) = (1-\sqrt{5})/4\,, \\ w_2 &= ST^{-2}(w_1) = 1 + 1/\sqrt{3},\ u_2 = ST^{-2}(u_1) = (7-\sqrt{5})/11 \end{aligned}$$

and the pair (u_2, w_2) is already G-reduced. The G-expansions of $1/u_2$ and w_2 are

$$w_2 = \lceil \overline{2, 3} \rceil , \quad 1/u_2 = \lceil 3, \overline{2, 2, 6, 2} \rceil ,$$

hence $\lceil \gamma \rceil = \lceil \overline{2, 6, 2, 2}, 3, \overline{2, 3} \rceil = \lceil \ldots, 2, 2, 6, 2, 2, 2, 6, 2, 2, 3, 2, 3, 2, 3, 2, 3 \ldots \rceil$.

23. Reduction theory and attractors

The notion of a "reduced" geodesic can be explained by studying a map on the boundary $\bar{\mathbb{R}} = \mathbb{R} \cup \{\infty\}$ associated with (a, b)-continued fraction expansion and its natural extension.

Let

$$f_{a,b}(x) = \begin{cases} T(x) = x + 1 & \text{if } x < a \\ S(x) = -\dfrac{1}{x} & \text{if } a \le x < b \\ T^{-1}(x) = x - 1 & \text{if } x \ge b\,. \end{cases}$$

The map $f_{a,b}$ induces a continued fraction algorithm if the orbit of any point returns to the interval $[a, b)$ infinitely often, and consists of blocks of T's and T^{-1}'s separated by S's, i.e. exactly if the parameters (a, b) belong to the set

$$\mathcal{P} = \{(a, b) \,|\, a \le 0 \le b,\ b - a \ge 1,\ -ab \le 1\}$$

indroduced in §22.

We define a two-dimensional natural extension map of $f_{a,b}$,

$$F_{a,b} : \bar{\mathbb{R}}^2 \setminus \Delta \to \bar{\mathbb{R}}^2 \setminus \Delta,$$

where $\Delta = \{(x, y) \in \bar{\mathbb{R}}^2 \mid x = y\}$ is the "diagonal", by

$$F_{a,b}(x, y) = \begin{cases} (x+1, y+1) & \text{if } y < a \\ \left(-\dfrac{1}{x}, -\dfrac{1}{y}\right) & \text{if } a \le y < b \\ (x-1, y-1) & \text{if } y \ge b\,. \end{cases} \tag{23.1}$$

Numerical experiments led Don Zagier to conjecture that such a map $F_{a,b}$ has several interesting properties for all parameter pairs $(a, b) \in \mathcal{P}$ that we list under **Reduction theory conjecture.**

(1) The map $F_{a,b}$ possesses a global attractor set $D_{a,b} = \cap_{n=0}^{\infty} F_{a,b}^{n}(\bar{\mathbb{R}}^2 \setminus \Delta)$ on which $F_{a,b}$ is essentially bijective.
(2) The set $D_{a,b}$ consists of two (or one, in degenerate cases) connected components each having *finite rectangular structure*, i.e. bounded by non-decreasing step-functions with a finite number of steps.
(3) Every point (x, y) of the plane $(x \neq y)$ is mapped to $D_{a,b}$ after finitely many iterations of $F_{a,b}$.

This conjecture is true for the classical cases whose attractors are shown on Figure 23.1, although for the H-expansion property (3) does not holds for some points (x, y) with y equivalent to $r = (3 - \sqrt{5})/2$. It has been proved recently in [**21**]

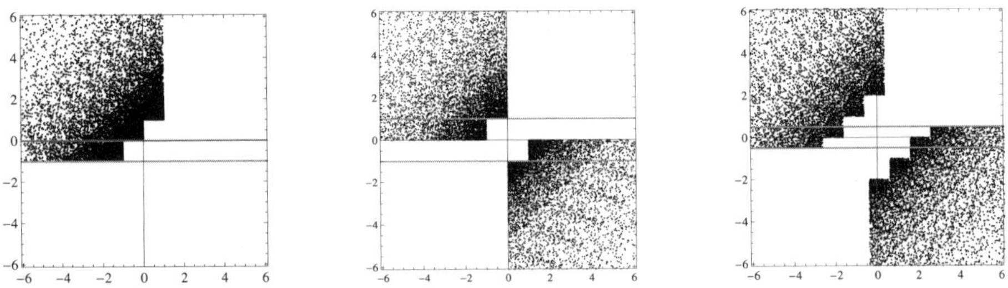

$D_{-1,0}$ G-expansion | $D_{-1,1}$ A-expansion | $D_{-1/2,1/2}$ H-expansion

FIGURE 23.1. Attractors for the classical cases

for an open dense subset of parameter pairs $(a, b) \in \mathcal{P}$. A typical attractor $D_{a,b}$ $(a = -\frac{4}{5}, b = \frac{2}{5})$ is shown on Figure 23.2. The main result of [**21**] is the following theorem:

THEOREM 23.1. *There exists an explicit one-dimensional Lebesgue measure* 0 *uncountable set* $\mathcal{E}$ *that lies on the diagonal boundary* $b = a + 1$ *of* $\mathcal{P}$ *such that:*

(a) *for all* $(a, b) \in \mathcal{P}\backslash\mathcal{E}$ *the map* $F_{a,b}$ *has an attractor* $D_{a,b}$ *satisfying properties (1) and (2) above;*
(b) *for an open and dense set in* $\mathcal{P} \setminus \mathcal{E}$ *property (3), and hence the Reduction theory conjecture, holds. For the rest of* $\mathcal{P}\backslash\mathcal{E}$ *property (3) holds for almost every point of the plane.*

If one identifies a geodesic on the upper half-plane with a pair of real numbers $(x, y) \in \bar{\mathbb{R}}^2$, $x \neq y$ — its end points, then $F_{a,b}$ maps a geodesic from x to y to a geodesic $PSL(2, \mathbb{Z})$-equivalent to it, and hence can be perceived as a *reduction map*.

Coding via (a, b)-continued fractions. In [**20**] we explained how Theorem 23.1 can be used to describe a reduction procedure for (almost) every geodesic in $\mathcal{H}$. In what follows we will denote the end points of geodesics by u and w, and whenever we refer to geodesics, we use (u, w) as coordinates on $D_{a,b}$.

First, we notice that the orbit of any point in $D_{a,b}$ returns to the subset $\Lambda_{a,b} = F_{a,b}(D_{a,b} \cap \{a \leq w \leq b\})$ infinitely often.

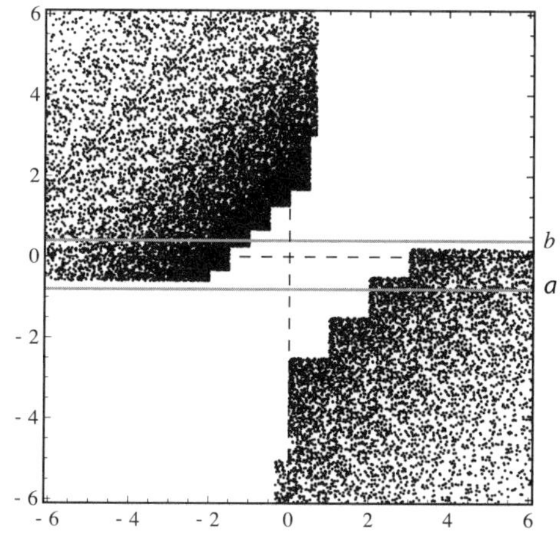

FIGURE 23.2. A typical attractor $D_{a,b}$ ($a = -\frac{4}{5}, b = \frac{2}{5}$).

DEFINITION 23.2. A geodesic in $\mathcal{H}$ from u to w is called (a, b)-*reduced* if $(u, w) \in \Lambda_{a,b}$.

It is easy to see from Figure 23.1 that the sets $\Lambda_{a,b}$ for the three classical cases are given by the inequalities of Definition 22.2.

In order to use (a, b)-expansions for coding geodesics we need the notion of a *dual expansion.*

DEFINITION 23.3. The (a, b)-expansion has a *dual* expansion if the reflection of $D_{a,b}$ about the line $y = -x$ is the attractor set of some (a', b')-expansion. If $(a', b') = (a, b)$, then the (a, b)-expansion is called *self-dual.*

It is shown in [**20**] that the parameter pairs $(a, b) \in \mathcal{P} \setminus \mathcal{E}$ that admit dual expansions form a discrete set in $\mathcal{D} \setminus \mathcal{E}$, where

$$\mathcal{D} = \{(a, b) \mid -1 \le a \le 0 \le b \le 1, b - a \ge 1\} \subset \mathcal{P},$$

and there are no parameter pairs (a, b) that admit dual expansions in the set $\mathcal{P} \setminus \mathcal{D}$.

It is obvious from Figure 23.1 that the attractors for Gauss and Artin codes are symmetric with respect to the line $y = -x$, hence these classical codes are self-dual. Figure 23.3 shows the attractors for the Hurwitz expansion and its dual.

The cross-section. In what follows we assume that $(a, b) \in \mathcal{D} \setminus \mathcal{E}$. Then every (a, b)-reduced geodesic from u to w intersects the unit half-circle. Let $C_{a,b} = P \cup Q_1 \cup Q_2$, where P consists of the unit vectors based on the circular boundary of the fundamental region F pointing inward such that the corresponding geodesic γ on the upper half-plane $\mathcal{H}$ is (a, b)-reduced, Q_1 consists of the unit vectors based on the right vertical boundary of F pointing inward such that $TS(\gamma)$ is (a, b)-reduced, and Q_2 consists of the unit vectors based on the left vertical boundary of F pointing inward such that $T^{-1}S(\gamma)$ is (a, b)-reduced. Then a.e. orbit of $\{\varphi^t\}$ returns to $C_{a,b}$, i.e. $C_{a,b}$ is a *cross-section* for $\{\varphi^t\}$, and $\Lambda_{a,b}$ is a parametrization of $C_{a,b}$, as shown on Figure 23.4. It is easy to see that for the G-code Q_2 and the left half of P are absent.

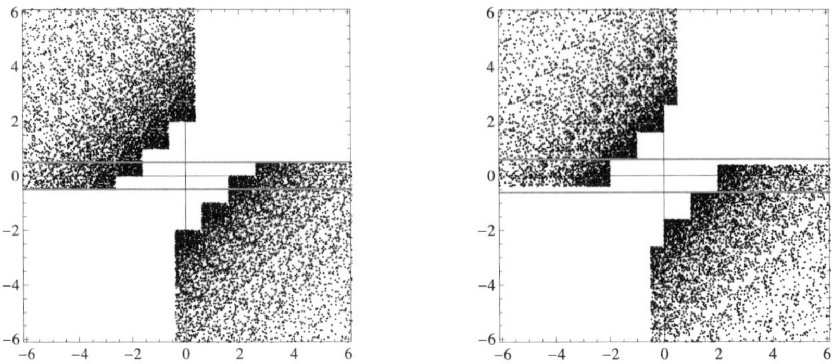

$(-\frac{1}{2}, \frac{1}{2})$ $(r-1, 1-r)$

FIGURE 23.3. Attractors for the Hurwitz expansion and its dual

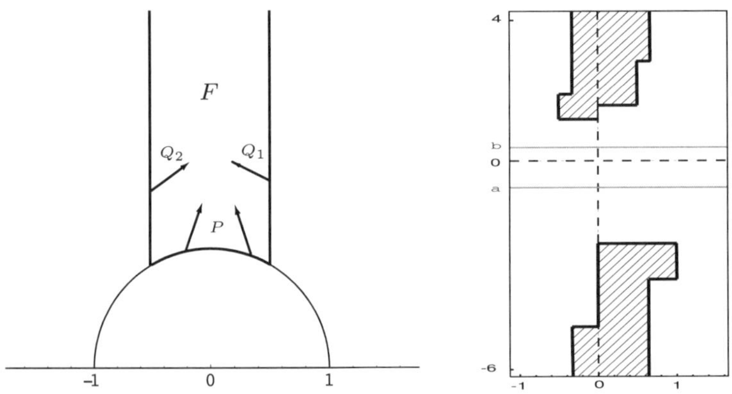

FIGURE 23.4. The cross-section $C_{a,b}$ (left) and its parametrization $\Lambda_{a,b}$ (right)

Reduction algorithm. Let us now assume that (a, b)-expansion has a dual (a', b'). Let γ be a geodesic on $\mathcal{H}$, from u and w, and $w = (n_0, n_1, \cdots)_{a,b}$. We construct the sequence of real pairs $\{(u_k, w_k)\}$ $(k \geq 0)$ defined by $u_0 = u$, $w_0 = w$ and $w_{k+1} = ST^{-n_k} w_k\,,\quad u_{k+1} = ST^{-n_k} u_k\,.$ Notice that $(u_k, w_k) = F_{a,b}^{n_k}(u, w)$ for the appropriate $n_k \geq 1$. Each geodesic with end points u_k and w_k is $PSL(2, \mathbb{Z})$-equivalent to γ by construction.

By Theorem 23.1 the reduction algorithm works in exactly the same way as that for the classical cases described in §22. For (almost) every geodesic in $\mathcal{H}$, the above algorithm produces in finitely many steps an (a, b)-reduced geodesic $PSL(2, \mathbb{Z})$-equivalent to γ, i.e. there exists a positive integer ℓ such that the geodesic with end points u_ℓ and w_ℓ is (a, b)-reduced. To an (a, b)-reduced geodesic γ, we associate a bi-infinite sequence of integers

$$(\gamma) = (\dots, n_{-2}, n_{-1}, n_0, n_1, n_2, \dots),$$

its coding sequence, by juxtaposing the (a, b)-expansion of $w = (n_0, n_1, \dots)_{a,b}$ and the dual (a', b')-expansion of $1/u = (n_{-1}, n_{-2}, \dots)_{a',b'}$.

If γ is a geodesic on $\mathcal{H}$, we denote by $\bar{\gamma}$ the canonical projection of γ on M. The geodesic $\bar{\gamma}$ on M can be represented as a bi-infinite sequence of geodesic segments between successive returns to the cross-section $C_{a,b}$. To each segment one can associate the corresponding (a,b)-reduced geodesic γ_i on $\mathcal{H}$. Thus we obtain a sequence of reduced geodesics $\{\gamma_i\}_{i=-\infty}^{\infty}$ representing the geodesic $\bar{\gamma}$. If one associates to γ_i (with end points u, w) its coding sequence, $(\gamma_i) = (\ldots, n_{-2}, n_{-1}, n_0, n_1, n_2, \ldots)$, then $\gamma_{i+1} = ST^{-n_0}(\gamma_i)$, because the map ST^{-n_0} gives the first return to the cross-section $C_{a,b}$. Thus all (a,b)-reduced geodesics γ_i in the sequence produce the same, up to a shift, coding sequence, which we call the *(a,b)-code* of γ and denote by (γ). The left shift of the sequence corresponds to the return of the geodesic to the cross-section $C_{a,b}$. We remark that if $\bar{\gamma}$ is a closed geodesic on M then its coding sequence is periodic $w = (n_0, n_1, \ldots, n_m)_{a,b}$, $\frac{1}{u} = (n_m, \ldots, n_1, n_0)_{a',b'}$.

24. Symbolic representation of geodesics via arithmetic codes

Let $\mathcal{N}_G^{\mathbb{Z}}$ be the Bernoulli space on the infinite alphabet $\mathcal{N}_G = \{n \in \mathbb{Z} \mid n \geq 2\}$. In §22 we proved that each oriented geodesic which does not go to the cusp of M in either direction admits a unique G-code, $\lceil \gamma \rceil \in \mathcal{N}_G^{\mathbb{Z}}$ which does not contain a tail of 2's. Taking the closure of the set of such G-codes we obtain the entire space $\mathcal{N}_G^{\mathbb{Z}}$. Now, each bi-infinite sequence $x \in \mathcal{N}_G^{\mathbb{Z}}$ produces a geodesic on $\mathcal{H}$ from $u(x)$ to $w(x)$, where

$$w(x) = \lceil n_0, n_1, \ldots \rceil, \quad \frac{1}{u(x)} = \lceil n_{-1}, n_{-2}, \ldots \rceil. \tag{24.1}$$

Notice that if a sequence has a tail of 2's then the oriented geodesic goes to the cusp. Thus the set of all oriented geodesics on M can be described symbolically as the Bernoulli space $X_G = \mathcal{N}_G^{\mathbb{Z}}$.

For the A-code, the set of all oriented geodesics (which do not go to the cusp) on M can be described symbolically as a countable one-step Markov chain $X_A \subset \mathcal{N}_A^{\mathbb{Z}}$ with the infinite alphabet $\mathcal{N}_A = \{n \in \mathbb{Z} \mid n \neq 0\}$ and transition matrix A,

$$A(n,m) = \begin{cases} 1 & \text{if } nm < 0, \\ 0 & \text{otherwise}. \end{cases} \tag{24.2}$$

Recall that for the H-code in the reduction algorithm and Theorem 22.4 we assumed that the end point w of a geodesic is not equivalent to $r = (3-\sqrt{5})/2$ since not all geodesics with w equivalent to r can be H-reduced. Taking the closure of the set of all such H-codes, we obtain a set X_H containing also the bi-infinite sequences with tails of 3's or -3's. These exceptional sequences are H-codes of some geodesics with end points w equivalent to r, but not of all such geodesics. Moreover, each geodesic with both end points u and w equivalent to r has two H-codes (see Figure 24.1 for the only closed such geodesic) [**13**]. The set X_H is a countable one-step Markov chain $X_H \subset \mathcal{N}_H^{\mathbb{Z}}$ with infinite alphabet $\mathcal{N}_H = \{n \in \mathbb{Z} \mid |n| \geq 2\}$ and transition matrix H,

$$H(n,m) = \begin{cases} 0 & \text{if } |n| = 2 \text{ and } nm > 0, \\ 1 & \text{otherwise}. \end{cases} \tag{24.3}$$

Therefore, for $\alpha = G, A, H$, the space X_α is a closed shift-invariant subset of $\mathcal{N}_\alpha^{\mathbb{Z}}$.

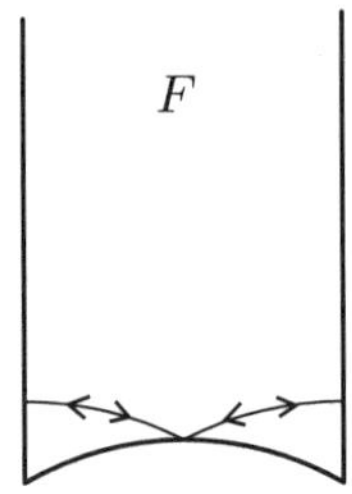

FIGURE 24.1. An exceptional geodesic with two H-codes, $\langle\overline{3}\rangle$ and $\langle\overline{-3}\rangle$

Coding maps for arithmetic codes. As shown above, the coding map for each arithmetic α-code ($\alpha = G, A$), $\mathfrak{C}_\alpha : X_\alpha \to C_\alpha$ is a bijection between the cross-section C_α and the symbolic space $X_\alpha \subset \mathcal{N}_\alpha^{\mathbb{Z}}$. The map $\mathfrak{C}_H : X_H \to C_H$ is surjective, and essentially one-to-one: the only exception is given by the H-codes corresponding to geodesics whose repelling end points are equivalent to r; for these exceptional H-codes the map is two-to-one.

The product topology on $\mathbb{N}_\alpha^{\mathbb{Z}}$ is induced by the distance function

$$d(x, x') = \frac{1}{m},$$

where $x = (n_i), x' = (n_i') \in \mathbb{N}_\alpha^{\mathbb{Z}}$, and $m = \max\{k \mid n_i = n_i' \text{ for } |i| \leq k\}$.

PROPOSITION 24.1. *The map $\mathfrak{C}_\alpha$ is continuous.*

PROOF. If $d(x, x') < \frac{1}{m}$, then the α-expansions of the attracting end points $w(x)$ and $w(x')$ of the corresponding geodesics given by (24.1) have the same first m digits. Hence the first m convergents of their α-expansions are the same, and by (22.4) and (22.3) $|w(x) - w(x')| < \frac{1}{m}$. Similarly, the first m digits of $\frac{1}{u(x)}$ and $\frac{1}{u(x')}$ are the same, and hence $|u(x) - u(x')| < \frac{u(x)u'(x)}{m} < \frac{1}{m}$. Therefore the geodesics are uniformly $\frac{1}{m}$-close. But the tangent vectors $v(x), v(x') \in C_\alpha$ are determined by the intersection of the corresponding geodesic with the unit circle. Hence, by making m large enough we can make $v(x')$ as close to $v(x)$ as we wish. □

The partition of the cross-section C_α. We parameterize the lift of the cross-section C_α to $S\mathcal{H}$, C_a by the coordinates (ϕ, θ), where $\phi \in [0, \pi]$ parameterizes the unit semicircle (counterclockwise) and $\theta \in [-\pi/2, (3\pi)/2]$ is the angle the unit vector makes with the positive horizontal axis (counterclockwise). The angle θ depends on ϕ and is determined by the condition that the corresponding geodesic is α-reduced.

The elements of the partition of C_a are labeled by the symbols of the corresponding alphabet $\mathcal{N}_\alpha$, $C_a = \bigcup_{n \in \mathbb{N}_\alpha} C_n$ and are defined by the following condition: C_n consists of all tangent vectors v in C_a such that for the coding sequence of the corresponding geodesic in $\mathcal{H}$, $n_0(x) = n$. The partitions of C_a (and therefore of C_α by projection) corresponding to the α-code ("the horizontal element") and

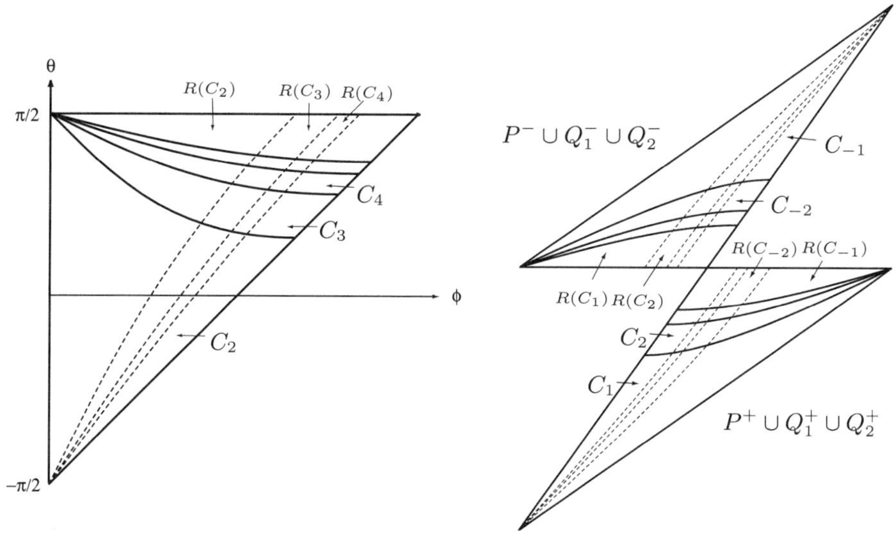

FIGURE 24.2. Infinite partition for the G-code (A-code, respectively) and its image under the return map R

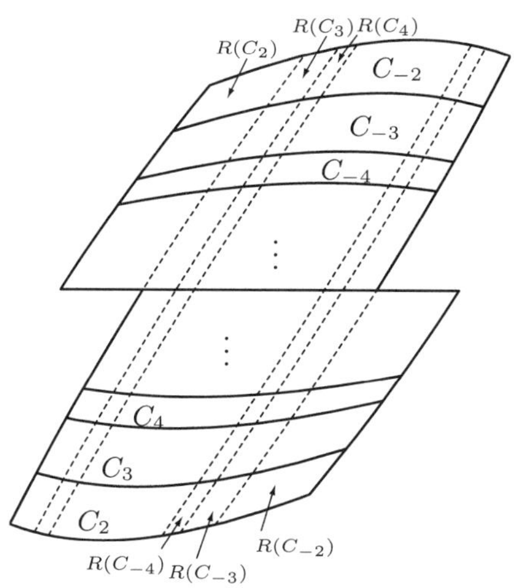

FIGURE 24.3. Infinite partition for the H-code and its image under the return map R

their iteration under the first return map R to the cross-section C_a ("the vertical element") were obtained in [**19**], and are shown on Figures 24.2 and 24.3.

If we parameterize the cross-section C_a by using the coordinates u, w and the inequalities given in Definition 22.2, as was explained in §23, the pictures become even simpler, each element of the partition is a rectangle. We have chosen the coordinates (ϕ, θ) here to be consistent with the parametrization of the cross-section associated to the geometric code in §21.

Some results of this section can be illustrated geometrically since the Markov property of the partition is equivalent to the Markov property of the shift space: the symbol m follows the symbol n in the coding sequence if and only if $R(C_n) \cap C_m \neq \emptyset$, and since all intersections are transversal, according to [**1**, Theorem 7.9], each partition is Markov.

Coding map for (a, b)**-continued fractions.** If an (a, b)-expansion has a dual and the attaractor $D_{a.b}$ has finite rectangular structure, one can code geodesics on the modular surface, as was explained in §23. The geodesic flow becomes a special flow over a symbolic dynamical system $(X_{a,b} \subset \mathbb{N}^{\mathbb{Z}}, \sigma)$, on the infinite alphabet $\mathbb{N} = \mathbb{Z} \setminus \{0\}$, where $X_{a,b}$ is the closure of the set of admissible sequences and σ is the left shift map. The coding map

$$\mathfrak{C}_{\mathrm{a,b}} : X_{a,b} \to C_{a,b}$$

defined by

$$\mathfrak{C}_{\mathrm{a,b}}((\ldots, n_{-2}, n_{-1}, n_0, n_1, n_2, \ldots)) = (1/(n_{-1}, n_{-2}, \ldots)_{a',b'}, (n_0, n_1, \ldots)_{a,b})$$

is continuous, surjective, and essentially one-to-one. However, it is not known whether the set of admissible coding sequences $X_{a,b} \subset \mathbb{N}^{\mathbb{Z}}$ is always Markov.

25. Complexity of the geometric code

Deciding which bi-infinite sequences of nonzero integers are admissible geometric codes is a nontrivial task. We present some known classes of such admissible sequences, and show that the space X of all geometric codes is not a topological Markov chain.

The arithmetic codes we considered in §24 provide partial results: by identifying certain classes of geometric codes which coincide with arithmetic codes we obtain classes of admissible geometric codes. The first result of this kind was obtained in [**12**]:

THEOREM 25.1. *A bi-infinite sequence of positive integers* $\{\ldots, n_{-1}, n_0, n_1, n_2, \ldots\}$ *is an admissible geometric code if and only if*

$$\frac{1}{n_i} + \frac{1}{n_{i+1}} \leq \frac{1}{2} \quad \text{for all } i \in \mathbb{Z}. \tag{25.1}$$

The corresponding geodesics are exactly those for which geometric codes coincide with G-codes.

The pairs forbidden by Theorem 25.1, $\{2, p\}$, $\{q, 2\}$, $\{3, 3\}$, $\{3, 4\}$, $\{4, 3\}$, $\{3, 5\}$, and $\{5, 3\}$—we call them *Platonic restrictions*—are of Markov type. More precisely, the set of all bi-infinite sequences satisfying relation (25.1) can be described as a one-step countable topological Markov chain $X_P \subset \mathbb{N}_G^{\mathbb{Z}}$, with the alphabet $\mathbb{N}_G$ and transition matrix P,

$$P(n, m) = \begin{cases} 1 & \text{if } 1/n + 1/m \leq 1/2\,, \\ 0 & \text{otherwise}\,. \end{cases} \tag{25.2}$$

Clearly, X_P is a shift-invariant subset of X.

The geodesics identified in Theorem 25.1 have the property that all their segments in F are positively (clockwise) oriented. Following [**12**] we call them *positive geodesics*, and the corresponding class of sequences *positive coding sequences*.

A wider class of admissible coding sequences, which includes the positive ones, has been identified in [**18**]:

THEOREM 25.2. *Any bi-infinite sequence of integers* $\{\dots, n_{-1}, n_0, n_1, n_2, \dots\}$ *such that*

$$\left|\frac{1}{n_i} + \frac{1}{n_{i+1}}\right| \le \frac{1}{2} \quad \text{for } i \in \mathbb{Z} \tag{25.3}$$

is realized as a geometric code of a geodesic on M.

The set of all bi-infinite sequences satisfying relation (25.3) can be described as a one-step countable topological Markov chain, with the alphabet $\mathbb{N} = \{n \in \mathbb{Z} \mid n \neq 0\}$ and transition matrix M,

$$M(n,m) = \begin{cases} 1 & \text{if } |1/n + 1/m| \le 1/2\,, \\ 0 & \text{otherwise}\,. \end{cases} \tag{25.4}$$

We denote the associated one-step Markov chain by X_M. Clearly, X_M is a closed shift-invariant subset of X.

Following [**18**] we call the admissible geometric coding sequences identified in Theorem 25.2 and the corresponding geodesics, *geometrically Markov*. In [**19**] we show that the H-code comes closest to the geometric code:

THEOREM 25.3. *For any geometrically Markov geodesic whose geometric code does not contain* 1*'s and* -1*'s, the* H*-code coincides with the geometric code.*

The set X_M is a σ-invariant subset strictly included in X. For example, $[5, 3, -2]$ is an admissible geometric code, obtained as the code of the closed geodesic corresponding to the axis of $T^5ST^3ST^{-2}S$ (see Figure 25.1), but it is not geometrically Markov. Moreover, the latter is also an example of a non-geometrically Markov geodesic for which geometric and H-codes coincide. A natural question would be to characterize completely the class of geodesics for which the two codes coincide.

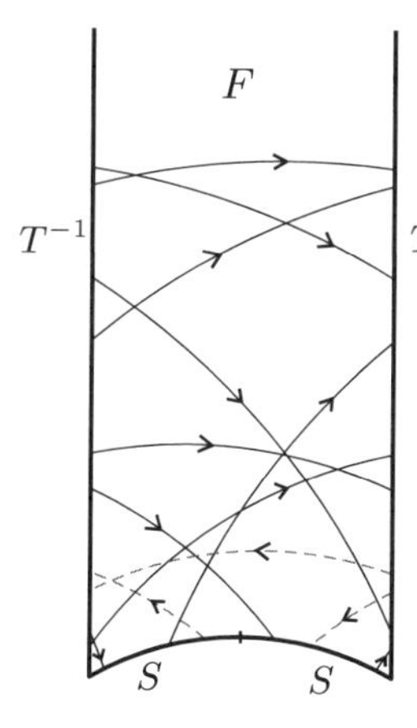

FIGURE 25.1. Geometric code $[5, 3, -2]$

The following theorems were proved in [**18**]:

THEOREM 25.4. *The set X_M is a maximal, transitive one-step countable topological Markov chain in the set of all geometric codes X.*

THEOREM 25.5. *The set X_M is the maximal symmetric (i.e. given by a symmetric transition matrix) one-step countable topological Markov chain in the set of all geometric codes X.*

The following result is an extension of a theorem proved in [**19**]:

THEOREM 25.6. *For any geometrically Markov geodesic whose geometric code consists of symbols with alternating signs, the A-code coincides with the geometric code.*

Unlike the spaces of admissible arithmetic codes X_G, X_A, and X_H which in §24 were proved to form topological Markov chains, the space of admissible geometric codes X is very complicated. In order to state the complexity result proved in [**18**] we recall the notion of a k-step topological Markov chain defined on the alphabet $\mathbb{N}$ (see [**16**, §1.9] for the finite alphabet definition):

DEFINITION 25.7. Given an integer $k \geq 1$ and a map $\tau : \mathbb{N}^{k+1} \to \{0,1\}$, the set

$$X_\tau = \{x \in \mathbb{N}^{\mathbb{Z}} \mid \tau(n_i, n_{i+1}, \dots, n_{i+k}) = 1 \; \forall \; i \in \mathbb{Z}\}$$

with the restriction of the left-shift map σ to X_τ is called the *k-step topological Markov chain with alphabet $\mathbb{N}$ and transition map τ.*

Without loss of generality we always assume that the map τ is *essential*, i.e. $\tau(n_1, n_2, \dots, n_{k+1}) = 1$ if and only if there exists a bi-infinite sequence in X_τ containing the $(k+1)$-block $\{n_1, n_2, \dots, n_{k+1}\}$.

THEOREM 25.8. *The space X of geometric codes is not a k-step topological Markov chain, for any integer $k \geq 1$.*

The proof of this result is contained in [**17**].

26. Applications of arithmetic codes

Calculation of the return time for special flows. In §21 and §24 we have constructed four continuous surjective coding maps. The map $\mathfrak{C} : X \to B$ for the geometric code and the map $\mathfrak{C}_H : X_H \to C_H$ (for the H-code) are essentially one-to-one, (and finite-to-one everywhere) while the maps for the other two arithmetic codes, $\mathfrak{C}_\alpha : X_\alpha \to C_\alpha$ ($\alpha = G, A$) are bijections. In all cases the first return to the cross-section corresponds to the left-shift of the coding sequence. This provides four symbolic representations of the geodesic flow $\{\varphi^t\}$ on SM as a special flow over (Λ, σ), where $\Lambda = X_G, X_A, X_H, X$, with the ceiling function f being the time of the first return to the cross-section $C = C_G, C_A, C_H, B$, i.e. four symbolic representations of the geodesic flow on the space

$$\Lambda^f = \{(x,y) : x \in \Lambda, 0 \leq y \leq f(x)\} \tag{26.1}$$

as explained in §19.

For $\Lambda = X_G, X_A, X_H, X$ and $C = C_G, C_A, C_H, B$, respectively, the ceiling function $f(x)$ on Λ is the time of the first return of the geodesic $\gamma(x)$ to the cross-section C. The following theorem was proved in [**12**] for the G-code, and appeared for other arithmetic codes in [**19**], and for the geometric code in [**18**]. The same formula holds for all (a, b)-codes as well. The proof for all codes is the same. A similar formula for Artin's original code has appeared earlier in [**28**].

THEOREM 26.1. *Let $x \in \Lambda$ and $w(x)$, $u(x)$ be the end points of the corresponding geodesic $\gamma(x)$. Then*

$$f(x) = 2 \log |w(x)| + \log g(x) - \log g(\sigma x),$$

where

$$g(x) = \frac{|w(x) - u(x)|\sqrt{w(x)^2 - 1}}{w(x)^2\sqrt{1 - u(x)^2}}.$$

This formula was used to obtain topological entropy estimates in [**12**] and [**18**].

Factor-maps associated with arithmetic codes and their invariant measures. Let $(u, w) \in \Lambda_{a,b}$. Making a change of variables $x = -\frac{1}{w}$, $y = u$ we obtain a compact region $D_{a,b} \subset [a, b) \times [-1, 1]$. The reduction map in these coordinates

$$G_{a,b} : D_{a,b} \to D_{a,b}$$

is given by the formula

$$G_{a,b}(x, y) = \left(-\frac{1}{x} - \left(-\frac{1}{x}\right)_{a,b}, -\frac{1}{y - \left(-\frac{1}{x}\right)_{a,b}} \right).$$

It may be considered a (natural) extension of the Gauss-type map $g_{a,b} : [a, b) \to [a, b)$,

$$g_{a,b}(x) = -\frac{1}{x} - \left(-\frac{1}{x}\right)_{a,b}; \; g_{a,b}(0) = 0.$$

One sees immediately that the following diagram

$$\begin{array}{ccc} D_{a,b} & \xrightarrow{G_{a,b}} & D_{a,b} \\ {\scriptstyle\pi}\downarrow & & \downarrow{\scriptstyle\pi} \\ [a, b) & \xrightarrow{g_{a,b}} & [a, b) \end{array}$$

is commutative if $\pi(x, y) = x$.

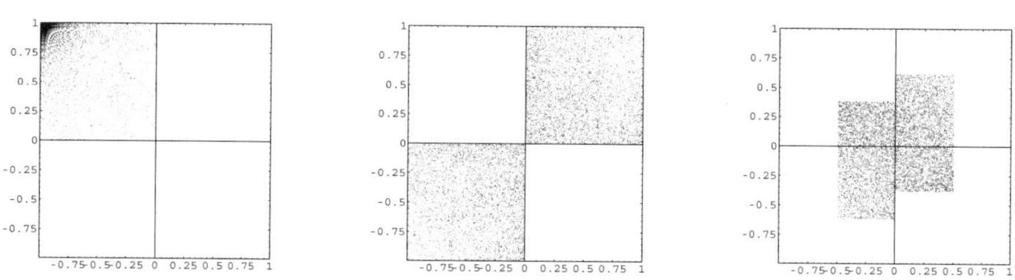

$D_{-1,0}$ $\quad$ $D_{-1,1}$ $\quad$ $D_{-\frac{1}{2},\frac{1}{2}}$

FIGURE 26.1. The three classical attractors in compact form

In order to calculate the invariant measure for the map $g_{a,b}$ we use the parametrization of $S\mathcal{H}$ by (u, w, s), considered in [**2**]: a vector in $S\mathcal{H}$ is identified with (u, w, s), where u, w are the end points of the associated geodesic in $\mathcal{H}$, and s is the distance to a predetermined point on the geodesic (for example, the midpoint). In this parametrization the geodesic flow has a particularly simple form:

$$\varphi^t : (u, w, s) \mapsto (u, w, s + t). \tag{26.2}$$

The Liouville measure $d\nu$ on $S\mathcal{H}$, introduced in §16, in these coordinates is given by the formula

$$d\nu = \frac{du\, dw\, ds}{(w-u)^2}, \tag{26.3}$$

and its invariance under $\{\varphi^t\}$ follows immediately from (26.2) and (26.3). The measure on the cross-section $\Lambda_{a,b}$ invariant for the first return map is obtained by dropping ds: $d\nu_{\Lambda_{a,b}} = \frac{du\, dw}{(w-u)^2}$, and, by the above change of variables, the invariant measure on $D_{a,b}$ is given by $d\bar{\nu}_{D_{a,b}} = \frac{dx\, dy}{(1+xy)^2}$. The invariant measure on $[a, b)$ is obtained by integrating $d\bar{\nu}_{D_{a,b}}$ with respect to dy as explained in [**2**]. Thus, if we know the exact shape of the set $D_{a,b}$, we can calculate the invariant measure precisely.

G-**code.** In this case $g_G : [-1, 0) \to [-1, 0)$ is given by $g_G(x) = -\frac{1}{x} - \lceil -\frac{1}{x} \rceil$, $D_{-1,0} = [-1, 0] \times [0, 1]$, and the invariant measure for g_G on $[-1, 0)$ is

$$d\mu_G = \frac{dx}{1+x}.$$

(See also [**3**] for a similar computation.)

A-**code.** In this case $g_A : [-1, 1) \to [-1, 1)$ is given by $g_A = -\frac{1}{x} - \lceil -\frac{1}{x} \rfloor$, $D_{-1,1} = \{[-1, 0] \times [-1, 0]\} \cup \{[0, 1] \times [0, 1]\}$, and the invariant measure is

$$d\mu_A = \left(\frac{\chi_{[-1,0]}}{1-x} + \frac{\chi_{[0,1]}}{1+x} \right) dx.$$

H-**code.** In this case $g_H = [-\frac{1}{2}, \frac{1}{2}) \to [-\frac{1}{2}, \frac{1}{2})$, $g_H = -\frac{1}{x} - \langle -\frac{1}{x} \rangle$, and the invariant measure is

$$d\mu_H = \left(\frac{\chi_{[-\frac{1}{2},0]}}{(1+rx)(1+(r-1)x)} + \frac{\chi_{[0,\frac{1}{2}]}}{(1-rx)(1+(1-r)x)} \right) dx.$$

The formulae for $d\mu_A$ and $d\mu_H$ rectify the formulae given in [**17**].

In [**20**] we derive the formulae for the invariant measure for a wider class of (a, b)-codes.

Classical results proved using arithmetic codes. Artin [**4**] used regular continued fractions to prove the topological transitivity of the geodesic flow on the modular surface (i.e. the existence of a dense geodesic) and the density of closed geodesics. In fact, any Markov (a, b)-code, in particular, any arithmetic α-code ($\alpha = G, A, H$) can be used for this purpose since the Markov property allows us to list all admissible periodic coding sequences.

Exercises

34. Let $A = \begin{pmatrix} 1 & 2 \\ 1 & 3 \end{pmatrix}$, $B = \begin{pmatrix} 2 & 1 \\ 3 & 2 \end{pmatrix}$, and $C = \begin{pmatrix} 0 & -1 \\ 1 & 4 \end{pmatrix}$. Use one of the classical continued fraction expansions to determine which of these matrices are conjugate in $PSL(2, \mathbb{Z})$.

35. Verify the formulae for the invariant measures $d\mu_G$, $d\mu_A$, $d\mu_H$.

References

[1] R. Adler, *Symbolic dynamics and Markov partitions*, Bull. Amer. Math. Soc. **35** (1998), no. 1, 1–56.

[2] R. Adler, L. Flatto, *Cross section maps for geodesic flows, I (The Modular surface)*, Birkhäuser, Progress in Mathematics (ed. A. Katok) (1982), 103–161.

[3] R. Adler and L. Flatto, *The backward continued fraction map and geodesic flow*, Ergod. Th. & Dynam. Sys. **4** (1984), 487–492.

[4] E. Artin, *Ein Mechanisches System mit quasiergodischen Bahnen*, Abh. Math. Sem. Univ. Hamburg **3** (1924), 170–175.

[5] A. Beardon, *The Geometry of Discrete Groups*, Springer-Verlag, 1983.

[6] R. Bowen, C. Series, *Markov maps associated with Fuchsian groups*, Inst. Hautes Études Sci. Publ. Math. No. 50 (1979), 153–170.

[7] D. Fried, *Reduction theory over quadratic imaginary fields*, J. Number Theory **110** (2005), no. 1, 44–74.

[8] I.M. Gelfand, M.I. Graev, I.I. Pyatetskii-Shapiro, *Representation Theory and Automorphic Functions* (English translation). W.B. Saunders, Philadelphia, 1969.

[9] D. J. Grabiner, J. C. Lagarias, *Cutting sequences for geodesic flow on the modular surface and continued fractions*, Monatsh. Math. **133** (2001), no. 4, 295–339.

[10] V. Guillemin and D. Kazhdan, *O*n the cohomology of certain dynamical systems, Topology, **19** (1980), 291–299.

[11] V. Guillemin and D. Kazhdan, *S*ome inverse spectral results for negatively curved 2-manifolds, Topology, **19** (1980), 301–312.

[12] B. Gurevich, S. Katok, *Arithmetic coding and entropy for the positive geodesic flow on the modular surface*, Moscow Math. J. **1** (2001), no. 4, 569–582.

[13] A. Hurwitz, *Über eine besondere Art der Kettenbruch-Entwicklung reeler Grossen*, Acta Math. **12** (1889) 367–405.

[14] S. Katok, Fuchsian groups, University of Chicago Press, 1992

[15] S. Katok, *Coding of closed geodesics after Gauss and Morse*, Geom. Dedicata **63** (1996), 123–145.

[16] A. Katok, B. Hasselblatt, *Introduction to the Modern Theory of Dynamical Systems*, Cambridge University Press, 1995.

[17] S. Katok and I. Ugarcovici, *Symbolic dynamics for the modular surface and beyond*, Bull. of the Amer. Math. Soc., **44**, no. 1 (2007), 87–132.

[18] S. Katok, I. Ugarcovici, *Geometrically Markov geodesics on the modular surface*, Moscow Math. J. **5** (2005), 135–151.

[19] S. Katok, I. Ugarcovici, *Arithmetic coding of geodesics on the modular surface via continued fractions*, 59–77, CWI Tract **135**, Math. Centrum, Centrum Wisk. Inform., Amsterdam, 2005.

[20] S. Katok, I. Ugarcovici, *Theory of (a,b)-continued fraction transformations and applications*, to appear, Electron. Res. Announc. Math. Sci. (2010).

[21] S. Katok, I. Ugarcovici, *Structure of attractors for (a,b)-continued fraction transformations*, preprint.

[22] P. Koebe, *Riemannsche Mannigfaltigkeiten und nicht euklidische Raumformen*, Sitzungsberichte der Preußischen Akademie der Wissenschaften, *I* (1927), 164–196; *II, III* (1928), 345–442; *IV* (1929), 414–557; *V, VI* (1930), 304–364, 504–541; *VII* (1931), 506–534.

[23] A. Livshitz, *S*ome homological properties of U -systems, Mat. Zametki **10** (1971), 555–564.

[24] B. Maskit, *On Poincaré's Theorem for fundamental polygons*, Adv. Math. **7** (1971) 219–230.

[25] D. Mayer, T. Mühlenbruch, *Nearest λ_q-multiple fractions*, Proceedings of the Spectrum and Dynamics workshop, CRM Proceedings and Lecture Notes Series, (CRM and AMS), Montreal 2008.

[26] M. Morse, *A one-to-one representation of geodesics on a surface of negative curvature*, Trans. Amer. Math. Soc. **22** (1921), 33–51.

[27] H. Poincaré, Théorie des groupes Fuchsiens, Acta Math. 1 (1882) 1–62.

[28] C. Series, *The modular surface and continued fractions*, J. London Math. Soc. (2) **31** (1985), 69–80.

[29] C. Series, *Geometrical Markov coding of geodesics on surfaces of constant negative curvature*, Ergod. Th. & Dynam. Sys. **6** (1986), 601–625.

[30] G. Springer, *Introduction to Riemann Surfaces*, 2nd ed., Chelsea Publ. Co., New York, 1981.

[31] D. Zagier, *Zetafunkionen und quadratische Körper: eine Einführung in die höhere Zahlentheorie*, Springer-Verlag, 1982.

Department of Mathematics, The Pennsylvania State University, University Park, PA 16802

E-mail address: katok_s@math.psu.edu

Clay Mathematics Proceedings
Volume 10, 2010

Chaoticity of the Teichmüller flow

Artur Avila

ABSTRACT. We consider the Teichmüller flow restricted to connected components of strata of the moduli space of Abelian differentials. Various aspects of "chaoticity" of this flow allow efficient stochastic modelling and permit to obtain a fine description of typical vertical flows on translation surfaces.

1. Introduction

A *translation surface* is a compact Riemann surface S together with the choice of a non-zero Abelian differential (holomorphic 1-form) ω.

Let $\Sigma \subset S$ be the set of zeros of ω, also called *singularities*. We let κ_p, $p \in \Sigma$ be the order of p as a zero. By the Gauss-Bonnet formula, $\sum \kappa_p = 2g - 2$ where g is the genus of S (and thus we must have $g \geq 1$).

Any non-singular point has a neighborhood where one can define a *regular chart* such that ω becomes dz. The family of such charts forms an atlas on $S \setminus \Sigma$, whose coordinate transitions are translations. On the other hand, for every $p \in \Sigma$ there exists a holomorphic chart in a neighborhood of p where ω becomes $z^{\kappa_p} dz$.

Alternatively, one can give the structure of a translation surface to a compact orientable surface S, with singularities at Σ (a finite subset of S), by fixing a maximal atlas on $S \setminus \Sigma$ whose coordinate transitions are translations, and such that each $p \in \Sigma$ has a punctured neighborhood isomorphic to a $(\kappa_p + 1)$-folded cover of a punctured neighborhood of 0 in $\mathbb{R}^2$, for some positive integer κ_p.

The translation structure yields a flat metric on S with conical singularities at Σ, which are responsible for carrying the negative curvature when $g \geq 2$: the total angle around $p \in \Sigma$ is $2\pi(\kappa_p + 1)$. The total area is given by $\int |\omega|^2 < \infty$. It is often convenient to restrict attention to *normalized* surfaces, of total area 1.

Translation surfaces carry a natural dynamical system, the *vertical flow* (or northbound flow), given by "going up" with unit speed. This flow is defined for all times outside the singularities and *vertical separatrices* stemming from them, and it is clearly area preserving.

1.1. Moduli spaces. Considering translation surfaces in genus g modulo isomorphism, one arrives at the moduli space of Abelian differentials $\mathcal{M}_g$. There is a natural action of $\mathrm{SL}(2, \mathbb{R})$ on $\mathcal{M}_g$, which is most easily seen at the level of regular charts, where it acts by postcomposition. The *Teichmüller flow* T_t is the diagonal flow $T_t(x) = \begin{pmatrix} e^t & 0 \\ 0 & e^{-t} \end{pmatrix} x$. The area is preserved by the $\mathrm{SL}(2,\mathbb{R})$ action.

By fixing the order of zeros, as an unordered list κ of positive integers (not necessarily distinct), one defines *strata* $\mathcal{M}_{g,\kappa} \subset \mathcal{M}_g$. Strata are invariant under the $\mathrm{SL}(2,\mathbb{R})$ action, but are not necessarily connected. They may have at most 3 connected components. Connected components of strata were classified in [**KZ**].

The moduli space $\mathcal{M}_g$ can be given a natural complex analytic structure. The strata are (not necessarily closed) complex analytic subvarieties of $\mathcal{M}_g$ of complex dimension $2g+n-1$, where n is the number of singularities. The strata can be given a finer complex affine structure, whose transition charts preserve the integer lattice, and hence also preserve the canonical volume in $\mathbb{C}^{2g+n-1}$. This structure is still preserved by the $\mathrm{SL}(2,\mathbb{R})$ action. The strata have infinite volume, but the set of translation structures of area *at most* 1 in some strata has a finite volume [**M, V1**], which was computed in [**EO**].

Since area is an invariant of the $\mathrm{SL}(2,\mathbb{R})$ action, we now restrict considerations to normalized surfaces and introduce the corresponding *real* analytic subvarieties $\mathcal{M}_g^1$ and $\mathcal{M}_{g,\kappa}^1$. For each connected component $\mathcal{C}$ of some $\mathcal{M}_{g,\kappa}^1$, there is a well-defined probability measure $\mu_{\mathcal{C}}$ in the Lebesgue measure class, coming (up to scaling) from the volume form on the set of translation structures of area at most 1.

We will basically be concerned with properties of *typical* translation surfaces, that is, those that hold $\mu_{\mathcal{C}}$-almost everywhere in some $\mathcal{C}$.

1.2. Renormalization. We have introduced above two classes of dynamical systems: the vertical flow on a normalized surface S and the Teichmüller flow on some connected component of strata $\mathcal{C}$. Both preserve natural probability measures, but the vertical flow is slow (zero entropy) and the Teichmüller flow is fast (entropy $2g+n-2$ [**V2**]).

The basic relation between them is that the Teichmüller flow can be seen as a renormalization flow on the space of vertical flows: The time-e^t map of the vertical flow parametrized by some $x \in \mathcal{C}$ is conjugate to the time-1 map of the vertical flow parametrized by $T_t(x)$. Accordingly, the investigation of the behavior of T_t, with respect to the invariant probability measure $\mu_{\mathcal{C}}$, often yields fundamental information on the dynamics of the vertical flow for typical surfaces $x \in \mathcal{C}$.

An early (and the most well-known) example of succesful application of this principle was the proof by Masur and Veech of unique ergodicity of vertical flows for typical surfaces, which was obtained in [**M, V1**] as a consequence of the Poincaré Recurrence Theorem applied to the Teichmüller flow.

The Poincaré Recurrence Theorem of course just expresses the invariance of the probability measure $\mu_{\mathcal{C}}$. In these notes, we will be basically concerned with the consequences for the vertical flow of subtler "chaotic" properties of the Teichmüller flow.

"Chaoticity" expresses lack of predictability and of additional invariant structures. While this might be thought to be bad at first, it turns out to be quite helpful, since it allows us to forget about the specifics of messy deterministic dynamics and replace it efficiently by stochastic models, which are much easier to analyze. The unpredictability of the Teichmüller flow has been an important topic of research on its own. The first achievement was the proof of ergodicity of T_t, showed in [**M, V1**]. In view of the existence of an underlying $\mathrm{SL}(2,\mathbb{R})$ action, ergodicity automatically implies the stronger property of *mixing*:

$$\int \phi \cdot (\psi \circ T_t)\, d\mu_{\mathcal{C}} \to 0 \text{ for every } \phi, \psi \in L^2_0(\mu_{\mathcal{C}}), \tag{1.1}$$

where $L^2_0(\mu_{\mathcal{C}})$ denotes the Hilbert space of L^2 observables $\phi : \mathcal{C} \to \mathbb{C}$ with zero average. This raises the problem of the speed of mixing, that is, how fast is the convergence in (1.1). As usual, no bounds can be obtained for general observables, but in view of the underlying $\mathrm{SL}(2,\mathbb{R})$ action there is a natural class $\mathcal{H} \subset L^2(\mu_{\mathcal{C}})$, containing compactly supported smooth observables, for which one can expect to show *exponential* mixing. In genus 1, the Teichmüller flow is just the geodesic flow of the modular surface, and hence exponential mixing was known for a long time. Exponential mixing was shown to also hold in the higher genus case in [**AGY**], and it is discussed in [**Y**], Section 14.

As for the dynamics of vertical flows, we will focus on two problems, where the role of chaoticity has been fundamental: the *Kontsevich-Zorich conjecture* and *typical weak mixing*. Actually, for those two problems we will not actually need to use exponential mixing of the Teichmüller flow itself, but some easier mixing properties of related *discrete renormalization dynamics*, which we will describe later, and which are much easier to show.

Acknowledgements: This research was partially conducted during the period the author served as a Clay Research Fellow.

2. The Kontsevich-Zorich conjecture

2.1. The Zorich phenomenon. Let us fix some connected component $\mathcal{C}$ of some $\mathcal{M}^1_{g,\kappa}$. Take a typical translation surface $S \in \mathcal{C}$, and choose an arbitrary point $x \in S$ for which the vertical flow $f_t(x)$ is defined for all $t \geq 0$. Unique ergodicity of the vertical flow, proved in [**M, V1**] (for typical S), means that the trajectory $\{f_t(x)\}_{t \geq 0}$ is equidistributed on S, that is, for every $\phi : M \to \mathbb{R}$ continuous, we have

$$\frac{1}{T} \int_0^T \phi(f_t(x))dx. \tag{2.1}$$

In particular, a long piece of trajectory is a Jordan arc which is very dense in S. How does this arc "winds around" the surface?

To give a precise sense to this question, let us denote by Γ_T the Jordan arc $\{f_t(x),\, 0 \leq t \leq T\}$, let γ_T be a curve (not necessarily simple), obtained by "closing" Γ_T with a small segment (say, whose length is at most the diameter of the surface) joining $f_T(x)$ and x. Then $[\gamma_T] \in H_1(S, \mathbb{Z})$ is well-defined *up to bounded error*, and we may ask about the asymptotics of $[\gamma_T]$ as T grows.

A classical result of Schwartzman [**S**] states that, due to unique ergodicity, there exists an *asymptotic cycle* $c \in H_1(M, \mathbb{R})$, such that

$$\frac{1}{T}[\gamma_T] \to c. \tag{2.2}$$

The asymptotic cycle can be explicitly related, by Poincaré duality, with $[\Re\omega] \in H^1(S, \mathbb{R})$, where ω is the Abelian differential giving the translation structure, and it is easily seen to be non-zero (for typical surfaces).

When $g = 1$, that is basically the whole story: it is easy to see that $[\gamma_T]$ differs from Tc by a bounded error. For higher genus however, there is much more that can be said about the asymptotics, as was first discovered numerically by Zorich [**Z1**]. He found out that if $g \geq 2$ the error term has a polynomial amplitude: indeed, if F_1 is the line $\mathbb{R}c$, we have

$$0 < \lambda_2 \equiv \limsup \frac{\ln \operatorname{dist}([\gamma_T], F_1)}{\ln T} < 1. \tag{2.3}$$

Moreover, Zorich found out that there exists an "asymptotic plane" $F_2 \supset F_1$ which captures most of the error term in the sense that

$$\limsup \frac{\ln \operatorname{dist}([\gamma_T], F_2)}{\ln T} < \lambda_2. \tag{2.4}$$

In genus 2, it turns out that the $[\gamma_T]$ stay at bounded distance from F_2. When $g \geq 3$ however, the oscillations about F_2 remain of polynomial amplitude. Thus

$$0 < \lambda_3 \equiv \limsup \frac{\ln \operatorname{dist}([\gamma_T], F_2)}{\ln T} < \lambda_2. \tag{2.5}$$

As before, the oscillations are again concentrated along a subspace F_3 with $\dim F_3 = 3$ and in genus 3 the $[\gamma_T]$ stay at bounded distance from F_3.

This process can be continued, yielding the following picture. For a typical translation surface in genus g, there is a "deviation spectrum" $1 > \lambda_2 > \cdots > \lambda_g > 0$, and an asymptotic flag $F_1 \subset \cdots \subset F_g \subset H_1(M, \mathbb{R})$, where $\dim F_i = i$, such that F_1 is spanned by the asymptotic cycle,

$$\lambda_{i+1} = \limsup \frac{\ln \operatorname{dist}([\gamma_T], F_i)}{\ln T}, \quad 1 \leq i \leq g-1, \tag{2.6}$$

and the distance from the $[\gamma_T]$ to F_g is uniformly bounded. Moreover, F_g is a Lagrangian subspace of $H_1(M, \mathbb{R})$ (endowed with the intersection form).

Notice that it is quite immediate that deviation exponents should be invariant under the Teichmüller flow, so ergodicity implies that they only depend on the connected component of strata where the surface is taken (recall we are only discussing the picture for typical surfaces).

How the Teichmüller flow can be used to understand the Zorich phenomenon? The basic idea is again that a long (time T) trajectory of the vertical flow becomes short (time 1) by application of the Teichmüller flow for time $\ln T$. In doing so, however, the translation structure becomes much deformed: for instance, the closed geodesics of moderate length for the original translation structure become much longer in the new translation structure. But since the Teichmüller flow is recurrent, one can expect that the new translation structure is not so deformed after all, *if viewed appropriately*: other geodesics may have become of moderate length after all, and one can look for a map which corresponds geodesics of moderate length for both structures. This *unwinding* map will have a complicated action on homology, which will be behind the rich behavior described by the Zorich phenomenon. We now turn to a more precise description.

2.2. The Kontsevich-Zorich conjecture. Consider the vector bundle $p : \tilde{\mathcal{C}} \to \mathcal{C}$ whose fiber $p^{-1}(x)$ is the cohomology $H^1(S, \mathbb{R})$ of the underlying surface S. Each fiber carries a well-defined lattice corresponding to $H^1(S, \mathbb{Z})$. In a small neighborhood of some $x \in \mathcal{C}$, there is a unique linear map between the fibers which preserves the lattice and is close to the identity. In particular, there is a unique lattice preserving continuous flow $\tilde{T}_t$ on $\tilde{\mathcal{C}}$, which is linear on the fibers and projects down to the Teichmüller flow. It is called the Kontsevich-Zorich cocycle. Notice that the Kontsevich-Zorich cocycle is symplectic with respect to the intersection form.

We let $B_t(x) : p^{-1}(x) \to p^{-1}(T_t(x))$ be the action on the fiber. It is possible to show that one can introduce a continuous inner product on $H^1(M, \mathbb{R})$ with respect to which

$$C^{-1} \leq e^{-t}\|B_t\| \leq C, \tag{2.7}$$

for every $x \in \mathcal{C}$, $t \geq 0$. By the Oseledets Theorem, it follows that for almost every x, $\lim_{t\to\infty}(B_t(x)^* B_t(x))^{1/t}$ converges to a positive operator $L(x)$. The numbers $\lambda_1 \geq \cdots \geq \lambda_{2g}$ such that $L(x)$ is orthogonally equivalent to the diagonal matrix with entries e^{λ_i} are almost surely independent of x, and are called the Lyapunov exponents of the Kontsevich-Zorich cocycle. The Kontsevich-Zorich conjecture states that all the Lyapunov exponents are distinct (or in other words, the Lyapunov spectrum is simple).

The Kontsevich-Zorich cocycle can be used to measure the homological behavior of the unwinding map we alluded before. It is thus no wonder that the main characteristic quantities of the cocycle, the Lyapunov exponents, are related to the asymptotic behavior in homology of the trajectories of the vertical flow. Indeed Zorich [**Z1**] showed that the full picture for this asymptotic behavior described above is equivalent to the Kontsevich-Zorich conjecture.

Since the Kontsevich-Zorich cocycle is symplectic, it easily follows that $\lambda_i = -\lambda_{2g-i+1}$, and by (2.7), we also conclude that $\lambda_1 = 1$. Zorich [**Z2**] (using the fundamental work of Veech [**V2**]) showed that $\lambda_1 > \lambda_2$. Forni [**F**] proved that $\lambda_g > \lambda_{g+1}$ (equivalently, $\lambda_g > 0$). The full conjecture was proved in [**AV1**]. We will describe a few ideas of the proof, especially the fundamental importance of "chaoticity".

2.3. Random products of matrices. It will be convenient to move our focus from the Teichmüller flow to a discrete analogous.

Choose a small transverse section Σ to the Teichmüller flow on $\mathcal{C}$, and let $f : \Sigma \to \Sigma$ be the Poincaré map, that is $f(x) = T_{r(x)}(x)$ where $r(x) = \min\{t > 0, T_t(x) \in \Sigma\}$. Let $\Sigma_r = \{(x,t),\, x \in \Sigma, 0 \leq t < r(x)\}$. There is a natural flow f_t on Σ_r: $f_s(x,t) = (f^k(x), w)$ where $k \geq 0$ and $0 \leq w \leq r(f^k(x))$ are unique such that $w = t + s - \sum_{i=0}^{k-1} r(f^i(x))$. The map $\operatorname{proj} : \Sigma_r \to \mathcal{C}$, $\operatorname{proj}(x,t) = T_t(x)$ is injective, and by ergodicity its image has full measure. It follows that there is a probability measure μ on Σ such that the probability measure μ_r on Σ_r given by $\mu_r(A \times [a,b]) = (\int r\, d\mu)^{-1}\mu(A)(b-a)$ is the pullback of $\mu_{\mathcal{C}}$.

Since Σ is small, the fibers $p^{-1}(x)$, $x \in \Sigma$ can be all identified with some fixed vector space H. We let $B(x) : H \to H$, be given (up to identification) by $B_{r(x)}(x)$.

We define a discrete time cocycle $B_n(x) = B(f^{n-1}(x)) \cdots B(x)$. The Lyapunov exponents $\theta_1 \geq \cdots \geq \theta_{2g}$ of this cocycle (logarithms of eigenvalues of $\lim(B_n(x)^* B_n(x))^{1/2n}$ for μ-almost every x) are easily related to the Kontsevich-Zorich exponents: $\theta_i = \int r d\mu \lambda_i$.

To understand the cocycle, it is convenient to think of the sequence

$$(B(f^k(x)))_{k=0}^{\infty}$$

as a sequence of random variables. The "stochastic process generating this sequence" is stationary: the probability that a particular block $(B_i)_{i=0}^{l-1}$ appears in position k (that is, $B(f^{k+j}(x)) = B_j$, $0 \leq j \leq l-1$) is independent of k.

Ideally, for the purpose of stochastic modelling, we would like to ask for the sequence of random variables to be independent: that the probability of appearance of a block $(B_i)_{i=0}^{k+l-1}$ is the product of the probabilities of appearance of $(B_i)_{i=0}^{k-1}$ and $(B_i)_{i=0}^{l-1}$. This is too much to ask in our case. Almost as good however, is the following form of "almost independence": the probability of appearance of a block $(B_i)_{i=0}^{k+l-1}$ is given by the product of the probabilities of appearance of $(B_i)_{i=0}^{k-1}$ and $(B_{i+k})_{i=0}^{l-1}$, up to a fixed multiplicative constant $C > 0$. It turns out that it is possible to choose Σ so that the resulting sequence of random variables displays almost independence.[1]

An important result of Raugi-Guivarch and Goldsheid-Margulis [**GR, GM**] gives a criterion for the simplicity of the Lyapunov spectrum for independent products of matrices. This was later extended to matrix products arising from dynamical systems [**BV, AV2**]. A version of this criterion was proved in [**AV1**] for almost independent products.

Let K be the *support* of the cocycle, that is, the set of all linear operators $B : H \to H$ such that for every neighborhood $B \in U$, $\mu\{x \in \Sigma,\ B(x) \in U\} > 0$. Let $\mathcal{B}$ be the monoid generated by K. We say that $\mathcal{B}$ is *twisting* if for any families $(F_i)_{i=1}^{k}$, $(G_i)_{i=1}^{k}$ of subspaces of H such that F_i and G_i have complementary dimensions, there exists $B \in \mathcal{B}$ such that $B \cdot F_i$ is transverse to G_i for $1 \leq i \leq k$. We say that $\mathcal{B}$ is *pinching* if for every $C > 0$, there exists $B \in \mathcal{B}$ such that the eigenvalues $e^{\delta_1} \geq \cdots \geq e^{\delta_{2g}}$ of B^*B satisfy $\delta_i \geq \delta_{i+1} + C$ (we say that B is C-pinched in this case).

THEOREM 2.1. *If $\mathcal{B}$ is pinching and twisting then the Lyapunov spectrum is simple.*

Thus, as long as sufficiently "rich" behavior (pinching and twisting) is seen at all by the Kontsevich-Zorich cocycle, and irrespective of any quantitative estimate about how frequently such behavior occurs, one may conclude that the Kontsevich-Zorich conjecture holds. Thus unpredictability, in the form of almost independence, reduces a quantitative question to a qualitative one.

2.4. Why pinching and twisting imply simplicity. We give a rough idea of why pinching and twisting of the monoid force simplicity of the Lyapunov spectrum.

Let us consider a model problem, which contains all the ideas. Consider two matrices $A^{(0)}$ and $A^{(1)}$ in $\mathrm{SL}(d, \mathbb{R})$ which generate a monoid which is pinching and twisting, and let us show that the Lyapunov exponents of an independent product (assigning a positive probability to each of them) of those two matrices are in fact distinct.

Let $\Sigma = \{0, 1\}^{\mathbb{Z}}$ and let $f : \Sigma \to \Sigma$ be the shift $f((x_n)_{n\in\mathbb{Z}}) = (x_{n+1})_{n\in\mathbb{Z}}$. Let $A : \Sigma \to \mathrm{SL}(n, \mathbb{R})$ be given by $A(x) = A^{(x_0)}$, and let $A_k(x) = A(f^{k-1}(x)) \cdots A(x)$ for $k \geq 0$. Let μ be the corresponding Bernoulli measure on Σ.

2.4.1. *u-states and s-states.* Fix $1 \leq i \leq d-1$ and let $G(i, n)$ be the Grassmanian of i-planes in $\mathbb{R}^n$. Define a map $F : \Sigma \times G(i, n) \to \Sigma \times G(i, n)$ given by $F(x, w) = (f(x), A(x) \cdot w)$. A probability measure ν on $\Sigma \times G(i, n)$, which projects to μ on the first factor, can be thought of as a μ-measurable function that associates

[1]To be precise, to achieve this one first passes to a finite cover of the Teichmüller flow, as will be described later, before choosing Σ.

to each $x \in \Sigma$ a probability measure ν_x on $G(i, n)$. It is called a u-state if it only depends on the past: $\nu_x = \nu_y$ if $x_j = y_j$, $j \geq 0$. It is called a s-state if it only depends on the future: $\nu_x = \nu_y$ if $x_j = y_j$, $j < 0$.

The set of u-states is compact with respect to the weak-$*$ topology and forward invariant under the action of F_*, so F-invariant u-states exist: take any u-state $\tilde{\nu}$, say $\tilde{\nu} = \mu \times \eta$ for some probability measure η in $G(i, n)$, and let ν be a Cesaro limit of $F_*^k \tilde{\nu}$.

Suppose that we knew that the Lyapunov spectrum is simple. Then, by the Oseledets Theorem, for μ-almost every $x \in \Sigma$, and for every $w \in G(i, n)$ which is transverse to the sum $s(x)$ of the Oseledets spaces (at x) corresponding to the $(n - i)$ smallest Lyapunov exponents, $A_n(x) \cdot w$ is exponentially close to the sum $u(f^n(x))$ of the Oseledets spaces (at $f^n(x)$) corresponding to the i largest Lyapunov exponents. It immediately follows that if we construct an invariant u-state ν as Cesaro limit of $F_*^k(\mu \times \text{Leb})$, the conditional measures ν_x are Dirac masses located at $u(x)$.

In the argument below we will somewhat reverse this reasoning. We will first show that pinching and twisting imply that the invariant u-states and s-states have the properties one would expect if the Lyapunov spectrum were simple, and then we will argue that those properties are enough to conclude simplicity.

2.4.2. *Analysis of conditional measures.* Fix an invariant u-state ν. We will first show that ν_x is a Dirac measure for μ-almost every x.

Since $x \mapsto \nu_x$ is measurable, μ-almost every $x \in \Sigma$ is a measurable continuity point. In other words, for every $\epsilon > 0$ there exists n_0 such that if $n \geq n_0$, the probability that ν_y is at distance at least ϵ of ν_x given that y is n-close to x (that is, $x_i = y_i$ if $|i| < n$) is at most ϵ.

Define ν_x^n to be the average over all y which are n-close to x of ν_y. By measurable continuity, we find out that $\nu_x^n \to \nu_x$ for μ-almost every x.

Notice that $F_*^k(\nu_x^n) = A_k(x)_* \nu_{f^k(x)}^{n-k}$, $0 \leq k \leq n$ by construction. Let η be the average of ν_x over all Σ (thus $\eta = \nu_x^0$ for μ-almost every x). We can conclude that $A_k(f^{-k}(x))_* \eta \to \nu_x$ for μ-almost every x.

The twisting condition implies that η is not supported on a small algebraic set. This means that fixing any finite set $W \subset G(n-i, n)$, the subset of $G(i, n)$ consisting of subspaces which are transverse to all elements of W has positive η-probability. Indeed, let z be any point in the support of some η. Thus there exists a positive probability that z belongs to the support of ν_x. By the twisting condition, there exists some x and $k \geq 1$ such that $B = A_k(x)$ is such that $B \cdot z$ is transverse to all element of W. But since ν_x only depend on the past, there is a positive probability that $A_k(x) = B$, given that z belongs to the support of ν_x. We conclude that there exists a positive probability that $B \cdot z$ belongs to the support of $\nu_{f^k(x)}$. This implies, by invariance, that $B \cdot z$ belongs to the support of η, as desired.

Let us now show that $A_k(f^{-k}(x))_* \eta$ converges to a Dirac measure for μ-almost every x. It is enough to prove this for a positive measure set of x, by ergodicity, and for a subsequence of k, by covergence.

Using the pinching and twisting condition, it is easy to see that there exists $k > 0$ and a matrix B with simple Lyapunov spectrum such that $B = A_k(x)$ with positive probability. From this it follows that $B_*^n \eta$ converges to a measure with finite support $Z \subset G(i, n)$ (contained in the set of i-planes spanned by eigenvectors of B).

Using the twisting condition, there exists $\tilde{k}$ and $\tilde{B}$ such that $\tilde{B} \cdot z$ is transverse to w for every $z \in G(i,n)$ and $w \in G(n-i,n)$ which are spanned by eigenvectors of B, and $\tilde{B} = A_{\tilde{k}}(x)$ with positive probability. This implies that $B^{(n)} = B^n \tilde{B} B^n$ is such that $B^{(n)}_* \eta$ converges to a Dirac measure, supported on some $z_0 \in G(i,n)$.

Fix n large. Since f is ergodic, for almost every x, there exists arbitrarily large m such that $A_{2kn+\tilde{k}}(f^{-m}(x)) = B^{(n)}$. Of course, it may still happen that $A_m(f^{-m}(x))_* \eta$ is not concentrated around $A_{m-2kn-\tilde{k}}(f^{-m+2kn+\tilde{k}}(x))_* z_0$. But we can ensure that this does not happen, with positive probability, by the following trick. One can show that the twisting condition implies that there is some $k_0 > 0$ such for every $z \in G(i,n)$, if ρ is any probability measure concentrated near z and $P \in \mathrm{SL}(n,\mathbb{R})$ is arbitrary, then $P_* A_{k_0}(x)_* \rho$ is concentrated near $PA_{k_0}(x) \cdot z$ with positive probability. Thus, by changing the coordinates of x between $-m+2kn+\tilde{k}+1$ and $-m+2kn+\tilde{k}+k_0$ (with positive probability conditioned on fixing the other coordinates) we get $A_m(f^{-m}(x))_* \eta$ concentrated near a Dirac mass.

We have now established that $\nu_x = \lim (A_n(f^{-n}(x))_* \eta$ is a Dirac measure for almost every x. Thus there exists a measurable function $x \mapsto u(x) \in G(i,n)$ such that $\nu_x = \delta_{u(x)}$. This function has a few key properties we will exploit: $A(x) \cdot u(x) = u(f(x))$, $u(x)$ only depends on the past, and moreover, for every $w \in G(n-i,n)$, $u(x)$ is transverse to w with positive probability (since η is not supported on a small algebraic set).

By a similar argument, there exists a function $x \mapsto s(x) \in G(n-i,n)$ such that $A(x) \cdot s(x) = s(f(x))$ and $s(x)$ only depends on the future. Notice that $u(x)$ is transverse to $s(x)$ for almost every x. Indeed, if this was not the case then by ergodicity $u(x)$ would not be transverse to $s(x)$, for almost every x, which would imply that for every w in the support of $s_* \mu$, $u(x)$ is not transverse to w for almost every x, a contradiction.

2.4.3. *Conclusion.* We will now use the functions u and s constructed before to prove that there is a gap between the i-th and the $(i+1)$-th Lyapunov exponents.

Since u is transverse to s almost everywhere, it follows that there exists $z \in G(i,n)$, $w \in G(n-i,n)$ with z transverse to w such that if V is a small neighborhood of z and W is a small neighborhood of w then $(u(x), s(x)) \in V \times W$ for a set U of x with positive measure. For almost every $x \in U$, we know that $A_n(f^{-n}(x))_* \eta \to \delta_{u(x)}$. Since η is not supported on a small algebraic set, it easily follows that for almost every $x \in U$, for every $C > 0$, if n is sufficiently large and $f^{-n}(x) \in U$ then

$$\ln \frac{\|A_n(f^{-n}(x)) \cdot a\| \|b\|}{\|A_n(f^{-n}(x)) \cdot b\| \|a\|} \geq C$$

for every $a \in u(x) \setminus \{0\}$ and $b \in s(x) \setminus \{0\}$. Replacing U by a smaller set, still with positive measure, we may assume that

$$\ln \frac{\|A_n(f^{-n}(x)) \cdot a\| \|b\|}{\|A_n(f^{-n}(x)) \cdot b\| \|a\|} \geq 1$$

for every $x \in U$ and $n \geq 1$ such that $f^{-n}(x) \in U$. By ergodicity, for almost every $x \in \Sigma$, for every $a \in u(x) \setminus \{0\}$ and $b \in s(x) \setminus \{0\}$,

$$\liminf \frac{1}{n} \ln \frac{\|A_n(f^{-n}(x)) \cdot a\|}{\|A_n(f^{-n}(x)) \cdot b\|} \geq \mu(U).$$

This implies that the smallest Lyapunov exponent along $u(x)$ is at least $\mu(U)$ larger than the largest Lyapunov exponent along $s(x)$. This gives the separation of the i-th and $(i+1)$-th exponents. Since $1 \leq i \leq d-1$ is arbitrary, simplicity of the spectrum follows.

2.5. Richness in the Kontsevich-Zorich cocycle. Verifying that the Kontsevich -Zorich cocycle is pinching and twisting means basically to show that "complicated" unwinding matrices must eventually appear, without being concerned with an estimate of their likelyhood.

The Teichmüller flow in each strata of large dimension is somewhat difficult to analyse directly. However, there are relations between the dynamics on different strata that allow us to proceed by "induction on the dimension", so that information can be passed from lower dimensional strata to higher dimensional ones.

Basically, what we attempt to exploit is the "compactification idea" that lower dimensional (connected components of) strata sort of embbed "in the boundary" of higher dimensional (connected components of) strata. This idea, very classical in the analysis of moduli spaces, is usually considered from the geometric point of view. A key achievement of [**AV1**] is to implement this from the dynamical point of view.

As one goes to the boundary of some strata, something degenerates: there exist *saddle connections*, straight segments joining singularities of Σ, of short length. For simplicity, assume that just one saddle connection is short.

Perhaps this is a saddle connection joining two distinct singularities. In this case, the degeneration corresponds to collapsing singularities. This is the simplest case to treat: indeed, the Teichmüller flow on some $\mathcal{C}$ is just the restriction of the Teichmüller flow on $\mathcal{M}_g$, and if there are at least two singularities then $\mathcal{C}$ is not closed in $\mathcal{M}_g$, and there is really some lower dimensional $\mathcal{C}'$ in the boundary of $\mathcal{C}$ in $\mathcal{M}_g$.

It is more difficult to treat the case where there is a single singularity, for in this case $\mathcal{C}$ is closed in $\mathcal{M}_g$. Collapsing a saddle connection then breaks the surface itself since it necessarily takes us into a lower genus situation (topologically one needs to cut the surface along an homologically non-trivial loop). Fortunately, it is enough for us to consider a particular way this degeneration can happen, the inverse procedure (increasing the genus) of which was described in [**KZ**].

Though the above description is quite intuitive, geometrically, we actually focus entirely on the combinatorics. To introduce combinatorics, it is convenient to work in a finite cover $\mathcal{C}^*$ of strata. The work of Veech [**V1**] introduces finitely many simply connected domains U_π, covering almost all of $\mathcal{C}^*$. Each U_π comes with a canonical trivialization of the relative homology bundle. As we move along the Teichmüller flow, one moves through the U_π. For each U_π, there are two ways to get in and two ways to get out (the interfaces being simply connected as well). In this way, we can draw a connected graph, whose vertices are the π and there are two oriented arrows entering and leaving each vertex. Since each U_π comes with a trivialization of the relative homology bundle, to each arrow there is a well-defined matrix in $\mathrm{SL}(d, \mathbb{R})$ relating the trivializations. The absolute homology bundle is identified with a subspace $H_\pi \subset \mathbb{R}^d$, which is preserved by the matrices.

The graph, called Rauzy diagram, is a good "topologically Markov" model of the Teichmüller flow dynamics. Indeed any finite concatenation of arrows in the Rauzy diagram actually is realized by orbits. One is tempted to consider it also as a probabilistic Markov model. However, this does not work too well because the expected number of transitions (necessary to move along the Teichmüller flow for a fixed positive length of time) is infinite.

To correct for this, we replace this nice finite model by an infinite one, which does not suffer however from this last defficiency. Given a finite sequence of arrows γ in the Rauzy diagram, let U_γ be the set of points whose initial sequence of transitions is given by γ. One first locates an appropriate γ such that U_γ is compact. We then look at the first return map to U_γ. As it turns out, one can return to U_γ by infinitely many distinctly combinatorial ways, represented by an infinite set of finite paths Γ in the Rauzy diagram. This time, we get a nice probabilistic Markov model: trajectories of the Teichmüller flow can be written as concatenation of paths of Γ, this induces a shift-invariant measure on $\Gamma^{\mathbb{Z}}$, and this measure has the nice "almost independence" property we described before.

We have just seen how to pass from the Rauzy diagram to the setting of the previous section. It remains to identify richness in the Rauzy diagram itself. Fix some vertex π in the Rauzy diagram, and let $\Pi(\pi)$ be the set of paths that start and end in π. Then $\Pi(\pi)$ is a monoid by concatenation. The set of matrices associated to elements of $\Pi(\pi)$ is a monoid as well, and it is this monoid that we show is pinching and twisting.

The induction on complexity idea is then realized as follows. Given a complicated Rauzy diagram $\mathcal{R}$, one finds a simpler one (by a procedure called "simple reduction") $\mathcal{R}'$, vertices $\pi \in \mathcal{R}$ and $\pi' \in \mathcal{R}'$ and an embedding of $\Pi(\pi') \to \Pi(\pi)$ that respects the monoid structure. We are however mostly concerned with relating the representation of $\Pi(\pi')$ on $\mathbb{R}^{d'}$ and of $\Pi(\pi)$ on $\mathbb{R}^d$ (actually, $d' = d-1$), and more precisely, their restrictions to $H_{\pi'}$ and H_π. It turns out that there exists an injective linear map $\mathbb{R}^{d'} \to \mathbb{R}^d$ which makes everything commute. If the genus of the surfaces corresponding to both Rauzy diagrams are the same, $g' = g$, then this map is actually an isomorphism $H_{\pi'} \to H_\pi$. This proves that pinching and twisting in $\mathcal{R}'$ implies pinching and twisting in $\mathcal{R}$.

In this way, we can reduce the complexity of the problem as much as possible, until we are forced to decrease the genus. Thus the simple reduction procedure gives $g' = g-1$. At this moment, pinching and twisting for $\mathcal{R}'$ certainly imply some richness for $\mathcal{R}$, but not as much as pinching and twisting. What happens in fact is that the image $\Pi' \subset \Pi(\pi)$ of the monoid $\Pi(\pi')$ stabilizes a non-zero vector in H_π. This certainly means that the representation of Π' is not pinching and twisting, but has an important consequence. Since the action of Π' on H_π is symplectic and stabilizes some direction $\lambda_0 \in \mathbb{P}H_\pi$, Π' acts on $H' = H_{\lambda_0} = H^{\lambda_0}/\lambda_0$, where H^{λ_0} is the symplectic orthogonal to λ_0. Now the space H' has the same dimension as $H_{\pi'}$, and in fact the representations of $\Pi(\pi')$ on $H_{\pi'}$ and of Π' on H' are isomorphic.

Consider now the Lagrangian flag space $\mathcal{L}$ on H_π. Elements of $\mathcal{L}$ are increasing sequences $F_1 \subset \cdots \subset F_g \subset H_\pi$ of isotropic subspaces such that $\dim F_i = i$. Let $\mathcal{L}_\lambda$ be the Lagrangian flag space on H_λ. Then $\mathcal{L}$ fibers over $\mathbb{P}H_\pi$, the fiber over λ being isomorphic to $\mathcal{L}_\lambda$.

We are now in position to prove, by induction, that the action of $\Pi(\pi)$ on $\mathcal{L}$ is minimal (which implies twisting). We have identified a submonoid Π' which fixes one of the fibers of the fibration of $\mathcal{L}$ over $\mathbb{P}H_\pi$ (the one over λ_0), and by induction it acts minimally over it. Thus it is enough to show that $\Pi(\pi)$ acts minimally on $\mathbb{P}H_\pi$. This is in turn a simple consequence of the usual theory of the Teichmüller flow, basically a translation of the fact that the Oseledets directions corresponding to the extremal Lyapunov exponent of the Teichmüller flow depend on the base point through an open analytic map, which is immediate from Veech's work [**V1**].

The proof of pinching is more subtle, and we just describe the strategy. By twisting and pinching for $\Pi(\pi')$, there is some $\gamma \in \Pi'$ which acts on H_π by an isomorphism fixing λ_0. Moreover, all Lyapunov exponents are simple except for the middle one which appears with multiplicity 2. A specific combinatorial construction (expected, though not shown, to correspond to adding an extra Dehn twist) gives another path γ' with the same Lyapunov exponents of γ, but ensures that 1 becomes an eigenvalue of algebraic multiplicity 1. It is easy to see that for every $C > 0$, large iterates of a matrix with those properties are C-pinched (the pinching growing logarithmically in the central space).

3. Weak mixing

Let $I \subset \mathbb{R}$ be an interval (all intervals will be assumed to be closed in the left and open in the right) and let $f : I \to I$ be an interval exchange transformation (see also [**Y**], Section 3). Thus f is a bijection with the property that I can be partitioned into $d \geq 2$ intervals I_α, labelled by an alphabet $\mathcal{A}$ on d letters, so that $f|I_\alpha$ is a translation. Define $\pi_t, \pi_b : \mathcal{A} \to \{1, ..., d\}$ so that π_t gives the ordering of the intervals I_α while π_b gives the ordering of the intervals $f(I_\alpha)$ and let $\lambda \in \mathbb{R}_+^{\mathcal{A}}$ be the vector of lengths of the I_α. We call $\pi = (\pi_t, \pi_b)$ the combinatorial data and λ the length data. We will assume that π is irreducible, so that $\pi_t^{-1}\{1, ..., k\} \neq \{1, ..., k\}$ for $1 \leq k \leq d-1$ (otherwise I can be partitioned into two intervals which are invariant under f).

We will say that π is a rotation if $\pi_t(\alpha) - \pi_b(\alpha) \bmod d$ is constant. In this case, f has a single actual discontinuity. If π is a rotation, the cyclic order of the interval is preserved: identifying the extremes of I, one gets a circle and f acts by a rigid rotation.

If π is not a rotation, then f does not preserve the natural cyclic order. Could it preserve some other, less obvious, one? More precisely, does there exist a measurable non-constant map $h : I \to S^1$ and $\alpha \in \mathbb{R}$ such that $h(f(x)) = e^{2\pi i \alpha} h(x)$? If this does not happen, one says that f is weak mixing. In [**AF**], we have shown that for every π that is not a rotation, f is weak mixing for a typical choice of the length data. This had been previously shown by Katok-Stepin [**KS**], and then by Veech [**V3**], for countably many distinct combinatorial data π, but several cases escaped considerations, for instance $d = 2g$, $\pi_t(\alpha) + \pi_b(\alpha) = 2g + 1$, $g \geq 2$.

3.1. Relation with translation surfaces. The relation between interval exchange transformations and translation surfaces is well known (for a detailed discussion, see [**Y**], Section 4). Fix an interval exchange transformation f, with combinatorial and length data (π, λ). Let $\tau \in \mathbb{R}^{\mathcal{A}}$ (the *suspension* data) be any vector such that $\sum_{\pi_t(\alpha) \leq k} \tau_\alpha > 0$, $1 \leq k \leq d-1$ and $\sum_{\pi_b(\alpha) \leq k} \tau_\alpha < 0$, $1 \leq k \leq d-1$ (for instance, $\tau_\alpha = \pi_b(\alpha) - \pi_t(\alpha)$ has this property). Then one associates to (π, λ, τ) a translation surface by the zippered rectangles construction of Veech [**V1**].

The basic idea of the construction is the following. Let $\zeta_\alpha = (\lambda_\alpha, \tau_\alpha) \in \mathbb{R}^2$. Consider two simple polygonal arcs in $\mathbb{R}^2$ obtained by concatenating, starting from 0, the vectors ζ_α, in the order given either by π_t or π_b. Notice that those two arcs have the same endpoints, so together they define a closed polygonal curve P. It is often the case that P is simple: in this case, starting from the closed domain bounded by P, we can identify (via translations) each of the d pairs of obviously parallel sides, so to obtain a compact surface with a natural translation structure. The general case involves some extra cutting and pasting which we will not detail here.

Reciprocally, consider a translation surface which has no vertical or horizontal connections. Take an initial segment of an eastbound separatrix (p, x), starting at p, such that there exists an initial segment of a vertical separatrix (q, x) which does not intersect (p, x). Then the first return map to (p, x) is basically an interval exchange transformation (except that the domain consists of open intervals).

3.2. The Rauzy algorithm. There is a natural dynamics in the space of interval exchange transformations, the Rauzy algorithm (see also [**Y**], Section 7). Let $\alpha = \pi_t^{-1}(d)$ and $\beta = \pi_b^{-1}(d)$. Assuming that $\lambda_\alpha \neq \lambda_\beta$, let $I' \subset I$ be obtained by cutting out at the right side of I an interval of size $\min\{\lambda_\alpha, \lambda_\beta\}$. Then the first return map f' to I' is again an interval exchange transformation. This new interval exchange transformation may have a different combinatorial data. To describe this, consider two operations on combinatorics. Think of π as a pair of rows, the top and the bottom, where the letters of $\mathcal{A}$ are ordered according to π_t and π_b. The top operation keeps the top (respectively, bottom) row unchanged, and takes the last letter of the bottom (respectively, top) row, β (respectively, α), and puts it back in the bottom (respectively, top) row just after the position taken by the letter α (respectively, β), which is last in the top (respectively, bottom) row. The combinatorial data of f' is obtained by applying the top or the bottom operation to π, according to whether $\lambda_\alpha > \lambda_\beta$ or $\lambda_\alpha < \lambda_\beta$, and the length data is modified by changing the larger of λ_α and λ_β to $|\lambda_\alpha - \lambda_\beta|$.

Let $\mathcal{R}$ be the set of all combinatorics that can be obtained from π by applying a sequence of tops and bottoms. Then the elements of $\mathcal{R}$ consist of irreducible combinatorial data, and the top and the bottom operations are invertible on $\mathcal{R}$. By considering the length vector projectively, this defines a map $R : \mathcal{R} \times \mathbb{P}\mathbb{R}_+^{\mathcal{A}} \to \mathcal{R} \times \mathbb{P}\mathbb{R}_+^{\mathcal{A}}$, defined outside a finite union of hyperplanes. It takes each $\{\pi\} \times \mathbb{P}\mathbb{R}_+^{\mathcal{A}}$, cuts it in 2 and takes each part onto some $\{\pi'\} \times \mathbb{P}\mathbb{R}_+^{\mathcal{A}}$ by a projective map.

The set $\mathcal{R}$ together with the top and bottom operations is precisely the Rauzy diagram that we have discussed previously. As mentioned before, to each vertex of the Rauzy diagram, one associates a vector space isomorphic to $\mathbb{R}^d$, in fact just $\mathbb{R}^{\mathcal{A}}$, and to each arrow an invertible linear map between the respective vector spaces. Thus if γ is an arrow starting at π, whose last letter in the top is α and in the bottom is β, B_γ is defined by $(B_\gamma^*)^{-1} \cdot \lambda = \lambda'$, where all coordinates of λ' are the same as λ except that $\lambda'_\alpha = \lambda_\alpha - \lambda_\beta$ if γ is a top and $\lambda'_\beta = \lambda_\beta - \lambda_\alpha$ if γ is a bottom.

As mentioned before, there is a canonical subspace $H_\pi \subset \mathbb{R}^{\mathcal{A}}$ such that if γ is an arrow starting at π and ending at π' then $B_\gamma \cdot H_\pi = H_{\pi'}$. This space can be defined as follows. If f is an interval exchange transformation, we have $f(x) = x + (\Omega_\pi \cdot \lambda)_\alpha$ for $x \in I_\alpha$, where $\Omega_\pi : \mathbb{R}^{\mathcal{A}} \to \mathbb{R}^{\mathcal{A}}$ is a matrix with entries $-1, 0, 1$: $(\Omega_\pi \cdot \lambda)_\alpha = \sum_{\pi_b(\beta) < \pi_b(\alpha)} \lambda_\beta - \sum_{\pi_t(\beta) < \pi_t(\alpha)} \lambda_\beta$. Thus Ω_π is an antisymmetric matrix, but not necessarily invertible. The image of Ω_π is H_π. Restricted to H_π, Ω_π induces a symplectic form, so H_π has even dimension. It is easy to check that H_π and its symplectic form are preserved by B_γ.

As for the problem of simplicity of the spectrum, it is convenient to work with an "acceleration" of the Rauzy algorithm. For $x \in \mathcal{R} \times \mathbb{P}\mathbb{R}_+^{\mathcal{A}}$, let $\Delta^{(n)}(x)$ be the connected component of the domain of R^n containing x. For Lebesgue almost every

x, there exists n such that $\Delta = \Delta^{(n)}(x)$ is compactly contained in $\mathcal{R} \times \mathbb{P}\mathbb{R}_+^{\mathcal{A}}$, and in fact we may require that B_γ has all entries positive, where γ is the concatenation of the first n arrows obtained by applying the Rauzy algorithm to x (we will call such Δ *nice*). Masur and Veech [**M, V1**] showed that the first return map R_Δ to Δ has full Lebesgue measure (Poincaré recurrence for the Teichmüller flow). This first return map has infinitely many connected components in its domain, all of them simplices, and restricted to each of them it is a projective map, onto Δ. Those two properties imply that there exists an absolutely continuous invariant probability measure invariant by R_Δ, whose density is continuous and positive on the closure.

3.3. The Veech criterium. Masur and Veech [**M, V1**] showed that almost every interval exchange transformation is ergodic. Assume that f is ergodic but not weak mixing. Then there is a non-trivial solution of $\phi \circ f = e^{2\pi i t}\phi$. Since ϕ is non-constant and f is ergodic, $t \notin \mathbb{Z}$.

Basically the only apriori fact about a measurable function such as ϕ is that almost every point is a measurable continuity point. Let us apply several times the Rauzy algorithm to f, arriving at an interval exchange transformation $f^{(n)}$ defined on a very small interval $I^{(n)}$. The functional equation gives rise to a new equation

$$\phi \circ f^{(n)} = e^{2\pi i h_\alpha^{(n)} t}\phi, \quad x \in I_\alpha^{(n)}, \tag{3.1}$$

where $h_\alpha^{(n)}$ is such that $f^{(n)}|I_\alpha^{(n)}$ is obtained by iterating f $h_\alpha^{(n)}$ times. Notice that $h^{(n)} = B_\gamma \cdot h$ where $h_\alpha = 1$, $\alpha \in \mathcal{A}$ and γ is an appropriate concatenation of n arrows in the Rauzy diagram, corresponding to the different steps of the Rauzy algorithm needed to go from f to $f^{(n)}$.

Let us hope that ϕ is close to a constant in $I^{(n)}$ and that all intervals $I_\alpha^{(n)}$ have approximately the same size (in the sense that the ration of the largest to the smallest of the intervals is bounded). Then (3.1) implies that $h^{(n)}t$ must be close to $\mathbb{Z}^{\mathcal{A}}$.

Let us now assume that f has combinatorial and length data (π, λ) belonging to some nice Δ. Let n_i be the sequence of times such that $R^{n_i}(\pi, [\lambda]) \in \Delta$. The precompactness of Δ immediately implies that the intervals $I_\alpha^{(n_i)}$ are all commensurable. Now assume that Δ is extra nice: Δ is a component of the domain of the first return map to some nice Δ'. Then it is easy to see that if $k_i \leq \min_\alpha h_\alpha^{(n_i)}$ is maximal such that $f^j|I^{(n_i)}$ is a translation for $1 \leq j \leq k_i$ then $\liminf k_i|I^{(n_i)}| > 0$. This easily implies that a measurable function is closer and closer to becoming constant on $I^{(n_i)}$.

This is Veech's criterium: if f is not weak mixing, then $th^{(n)} \to \mathbb{Z}^{\mathcal{A}}$, along a subsequence corresponding to extra nice renormalizations.

3.4. Veech's weak mixing Theorem. Veech showed that if $h \notin H_\pi$ then f is weak mixing for almost every λ. The argument uses a rather simple fact about the matrices B_γ. If $\gamma \in \Pi(\pi)$, that is, γ is a path in $\mathcal{R}$ starting and ending at π then B_γ preserves H_π. Thus B_γ^* preserves the orthogonal of H_π. It turns out that B_γ^* is the identity on the orthogonal of H_π.

Both H_π and its orthogonal complement intersect $\mathbb{Z}^{\mathcal{A}}$ in cocompact lattices whose sum is $\mathbb{Z}^{\mathcal{A}}$. It follows that the orthogonal projection h_0 of $h = (1, ..., 1)$ on the orthogonal complement of H_π is a primitive element of $\mathbb{Z}^{\mathcal{A}}$: thus for any non-integer t, $h_0 t \notin \mathbb{Z}^{\mathcal{A}}$.

Now assume that f is not weak mixing, but is otherwise typical. Consideration of an extra nice Δ yields a sequence of moments n_i such that $h^{(n_i)}t$ is close to the lattice. It follows that the orthogonal projection of $h^{(n_i)}t$, onto the orthogonal complement of H_π is also close to an integer, and since it is constant, it must be an integer. Since f is typical, it is ergodic so t is not an integer, contradiction.

3.5. Weak mixing for arbitrary combinatorics. In the proof of Veech's weak mixing Theorem, we used that the matrix B_γ acts very simply in the transverse direction to H_π. If $h \in H_\pi$, the situation is rather more complicated, since we have already seem that the monoid $\Pi(\pi)$ acts on H_π in a rich way (pinching and twisting). This richness is however central to the argument of weak mixing for such combinatorics, which involves the idea of chaoticity.

There are two aspects that follow from richness that we will use. The first is the simplicity of the Lyapunov spectrum. The second, a consequence of twisting is that if $E \subset H_\pi$ is an arbitrary 2-plane then it is in general position with respect to the stable Oseledets space, for almost every $[\lambda] \in \mathbb{R}_+^{\mathcal{A}}$. Recall that by simplicity and symplecticity, the stable space and the unstable space have dimension g where $\dim H_\pi = 2g$.

We will assume that $g \geq 2$, otherwise π is either a rotation or $h \notin H_\pi$. We setup the cocycle (R_Δ, A) where $A(x) = B_\gamma | H_\pi$, where γ is the appropriate sequence of arrows taken by x to return to Δ.

Using twisting one gets that a fixed line not passing through the origin "sees" the positive second Lyapunov exponent: it tends to be kicked further away from the origin under most iterations. By general arguments one gets the following finite formulation:

LEMMA 3.1. *There exists $\epsilon > 0$ such that for every $\delta > 0$ there exists n_0 such that if $n > n_0$ and $L \subset H_\pi$ is a line not passing by the origin then the probability (with respect to the invariant absolutely continuous probability measure μ on Δ) that $\frac{d(A_n(x)\cdot L,0)}{d(L,0)} < e^{\epsilon n}$ is at most δ (where d is the distance in $\mathbb{R}^{\mathcal{A}}$).*

The relevance of this lemma is the following. We are trying to show that non-integer points in the line $\mathbb{R}(1,...,1)$ can not converge to the integer lattice under cocycle iteration. The lemma guarantees that points near the origin get kicked away, on average, as we iterate the cocycle, and in fact this also work at the level of lines.

The problem is that the cocycle is unbounded, so when a point very near the origin is kicked away it may become close to another integer point. In a sense there is some simple local dynamics (of a hyperbolic fixed point) combined which may mix with "global transitions", in a picture reminiscent of Smale's horseshoe. Such complication is addressed by a probabilistic argument, which forms the core of [**AF**].

3.5.1. *Stochastic modelling.* We will consider first a somewhat simplified model. Assume that $\int \|A(x)\|^{1+\delta} d\mu(x) < \infty$ for some δ. This is certainly false in our setting, but allows us to focus already on a large part of the problem. We will also assume that the matrix product arising from (f, A) is independent, rather than almost independent. The ideas involved in getting to the almost independent setting turn out to be of only technical interest.

Assume that there exists a positive measure set of parameters for which the corresponding interval exchange transformation is not weak mixing. We recall that the iterates of the matrices $A(x)$ have all entries positive thus the direction of h is being expanded. By the Veech criterium, we know that $A_k(x) \cdot th$ is asymptotic to some sequence $p_k \subset H_\pi$ with integer coordinates, for some $t \in \mathbb{R} \setminus \mathbb{Z}$. Then $L_k(x) = A_k(x) \cdot \mathbb{R}h - p_k$ is a sequence of lines converging to the origin but not passing through it.

There are only countable many possibilities for the L_k. Thus for every $\delta > 0$ there exists some line L not passing through the origin, such that with positive probability there exists $z = z(x) \in L$ with $d(A_k(x) \cdot z(x), \mathbb{Z}^{\mathcal{A}}) < \delta$ for every $k \geq 0$. We will get to a contradiction by taking δ sufficiently small. Notice that L is parallel to the positive cone (this ensures that the cocycle expands along L as well).

Taking the first iterate of the cocycle basically means choosing one of the countably many possibilities for the matrix $A = A(x)$. Given A, we produce a finite set of lines L', called the *children* of L, by the following rule. For each $z \in L$ such that $d(z, 0) < \delta$ and $d(A \cdot z, p) < \delta$ for some $p \in \mathbb{Z}^{\mathcal{A}}$, $L' - p$ is a child of L. If $p = 0$ we will say that the child is trivial.

It is clear that by making $\delta \to 0$, the non-trivial children become more rare. In fact, a non-trivial child depends on the existence of some z which is close to 0 but $A \cdot z$ is close to a non-zero integer, so $\|A\|$ is at least of order δ^{-1}. This will be central in the argument, as it allows us to concentrate on the trivial child (at least under the simplified assumptions, otherwise non-trivial children are most annoying).

Applying successively the cocycle, we get a sequence of finite sets of lines not passing through the origin, and depending on x. Of course, for a given x this sequence may collapse to the empty set in finite time. We will show that this in fact happens for almost every x. This contradicts the assumption on L and concludes the proof.

Let $p_n(L)$ be the probability that the process continues for at least n steps. This is well-defined for all lines L (say, not passing through the origin and parallel to the positive cone), not just the special one we defined (but we only arrive at a contradiction if we prove that $p_n(L) \to 0$ for the special one).

We want to estimate $p_n(J)$, and show that $p_n(J) \to 0$ for all J. Of course, if J is a line that is very close to the origin, then $p_n(J)$ will be close to 1 for a long time. Thus the dependence on J is certainly important. We postulate that $p_n(J) \leq Ce^{-\kappa n} d(J, 0)^{-\rho}$ for appropriate parameters $\kappa, \rho > 0$. This is coherent with the idea that $d(J, 0)$ small implies large survival times.

We prove this estimate by induction. For each matrix A that may be applied, let J_A be the set of children. Then clearly

$$p_{n+1}(J) \leq \sum P(A) \sum_{J' \in J_A} p_n(J'),$$

where $p(A)$ is the probability of choosing A and we have used independence. Thus it is enough to prove that

$$\sum P(A) \sum_{J' \in J_A} d(J', 0)^{-\rho} d(J, 0)^{\rho} < e^{-\kappa}$$

for some $\kappa, \rho > 0$. We split the sum into two parts. The contribution of the trivial child is bounded by

$$\sum P(A) d(A \cdot J, 0)^{-\rho} d(J, 0)^{\rho}.$$

We want to show that this is less than 1 for small $\rho > 0$. For $\rho = 0$, this is actually 1. Using the hypothesis that $\int \|A\|^{\epsilon} d\mu < \infty$ for some $\epsilon > 0$ (which is satisfied in our setting [**AGY**]), one justifies taking the derivative with respect to ρ at $\rho = 0$. It has the form

$$-\sum P(A) \ln \frac{d(A \cdot J, 0)}{d(J, 0)}$$

which is basically what is estimated by the lemma. Thus the derivative is negative and for small ρ we have

$$\sum P(A) d(A \cdot J, 0)^{-\rho} d(J, 0)^{\rho} < e^{-2\kappa}$$

for some $\kappa > 0$.

We now consider the non-trivial children. It is easy to see that there are at most about $\delta \|A\|^{-1}$ of those. On the other hand, a simple geometric estimate shows that $d(J', 0)^{-1}$ is at most about $\|A^{-1}\|$, since some point which is δ-near 0 is taken $d(J', 0)$ near a non-zero integer after applying A which has integer coordinates. We can thus estimate the contribution of non-trivial children by

$$\sum P(A) \delta^{1-\rho} \|A\| \|A^{-1}\|^{\rho}.$$

Under the simplifying integrability assumption, this is $O(\delta^{1-\rho})$ in δ for small ρ, hence as $\delta \to 0$ the non-trivial children are asymptotically irrelevant and the result follows.

3.5.2. *Conditioning.* We now describe how to get rid of the simplifying assumption. Basically, nothing is assumed in the original [**AF**], but the argument is quite technical. Let us thus assume that $\|A\|^{\epsilon}$ is integrable for small $\epsilon > 0$, which was unknown when [**AF**] was written but is now a theorem [**AGY**].

The basic idea is to exploit knowledge of "spatial localization" of the exploding terms of the integral. This is explained by the following tale. Suppose that in the stock market there are assets which yield a return, after one unit of time, of plus 0.9 or minus 0.5, with probability 1/2. Two strategies are being considered. Either each time one chooses a single asset to invest, or each time one redistributes the money equally among the assets. The first strategy yields almost sure ruin, though the expectation of the fortune is actually exponentially large, but the second one makes the speculator exponentially rich almost surely. We have treated the problem so far under the framework of the second investment strategy, but there is determinism in the problem, so the modelling by the first strategy is more accurate. Indeed, given J, each choice of x initiates a civilization timeline, and at each unit of time the entire civilization is subject to the same matrix. This explains why one can hope for the almost sure collapse of the civilization started by L, even while the expected number of children is infinite: the expectation of the logarithm of the number of children is still finite.

Back to the problem, we take into account spatial localization by breaking the set of matrices into pieces. One piece consists of a large finite set, with large probability, Ω. The other pieces are formed by individual matrices outside Ω. Each point x has then a history, described by which piece is visited as we iterate the Rauzy algorithm.

We postulate that, by choosing Ω sufficiently large, for almost every x, the probability of survival up to time n given that the history up to time n is the same as for x, decreases exponentially according to a similar rule as described before:

denoting this conditional probability $p_n(J|x)$ we have

$$p_n(J|x) \leq C_n(x) d(J, 0)^{-\rho}$$

where $C_n(x) \to 0$ exponentially for almost every x. It is important to consider a large set of Ω, instead of just letting each matrix to be a piece, otherwise the modelling would become totally deterministic (probabilities are either 0 or 1). Notice that the stochastic modelling elliminates, for instance, the need to account for the precise position of the lines (and not merely the distance to 0), using the lemma.

The estimates are now mostly parallel to before. If $A(x) \in \Omega$, $p_{n+1}(J|x)$ becomes smaller than $p_n(J|R_\Delta(x))$ by a definite factor. Thus $C_n(x)$ involves (multiplicatively) a term basically $e^{-\kappa k}$ where k is the number of visits to Ω up to n, which is roughly of size n for large n since Ω has large probability. This provided δ is sufficiently small, so that there is no need to account for non-trivial children in Ω. When $A(x) \notin \Omega$, on the other hand, one accounts precisely for what happens, and one has a product of terms which are polynomial in the size of the matrices one visits outside Ω. By making Ω large, those visits become rare, and by integrability of the logarithm of $\|A\|$, this term becomes negligible compared to the first one.

References

[AF] Avila, Artur; Forni, Giovanni Weak mixing for interval exchange transformations and translation flows. Ann. of Math. (2) 165 (2007), no. 2, 637–664.

[AGY] Avila, Artur; Gouëzel, Sbastien; Yoccoz, Jean-Christophe Exponential mixing for the Teichmüller flow. Publ. Math. Inst. Hautes Études Sci. No. 104 (2006), 143–211.

[AV1] Avila, Artur; Viana, Marcelo Simplicity of Lyapunov spectra: proof of the Zorich-Kontsevich conjecture. Acta Math. 198 (2007), no. 1, 1–56.

[AV2] Avila, Artur; Viana, Marcelo Simplicity of Lyapunov spectra: a sufficient criterion. Port. Math. (N.S.) 64 (2007), no. 3, 311–376.

[BV] Bonatti, C.; Viana, M. Lyapunov exponents with multiplicity 1 for deterministic products of matrices. Ergodic Theory Dynam. Systems 24 (2004), no. 5, 1295–1330.

[EO] Eskin, Alex; Okounkov, Andrei Asymptotics of numbers of branched coverings of a torus and volumes of moduli spaces of holomorphic differentials. Invent. Math. 145 (2001), no. 1, 59–103.

[F] Forni, Giovanni Deviation of ergodic averages for area-preserving flows on surfaces of higher genus. Ann. of Math. (2) 155 (2002), no. 1, 1–103.

[GM] Goldsheid, I. Ya.; Margulis, G. A. Lyapunov exponents of a product of random matrices. Russian Math. Surveys 44 (1989), no. 5, 11–71.

[GR] Guivarch, Y.; Raugi, A. Products of random matrices: convergence theorems. Random matrices and their applications (Brunswick, Maine, 1984), 31–54, Contemp. Math., 50, Amer. Math. Soc., Providence, RI, 1986.

[KS] Katok, A. B.; Stepin, A. M. Approximations in ergodic theory. Uspehi Mat. Nauk 22 1967 no. 5 (137), 81–106.

[KZ] Kontsevich, Maxim; Zorich, Anton Connected components of the moduli spaces of Abelian differentials with prescribed singularities. Invent. Math. 153 (2003), no. 3, 631–678.

[M] Masur, Howard Interval exchange transformations and measured foliations. Ann. of Math. (2) 115 (1982), no. 1, 169–200.

[S] Schwartzman, Sol Asymptotic cycles. Ann. of Math. (2) 66 (1957), 270–284.

[V1] Veech, William A. Gauss measures for transformations on the space of interval exchange maps. Ann. of Math. (2) 115 (1982), no. 1, 201–242.

[V2] Veech, William A. The Teichmüller geodesic flow. Ann. of Math. (2) 124 (1986), no. 3, 441–530.

[V3] Veech, William A. The metric theory of interval exchange transformations. I. Generic spectral properties. Amer. J. Math. 106 (1984), no. 6, 1331–1359.

[Y] Yoccoz, J.-C. Interval exchange maps and translation surfaces. This volume.

[Z1] Zorich, Anton Asymptotic flag of an orientable measured foliation on a surface. Geometric study of foliations (Tokyo, 1993), 479–498, World Sci. Publishing, River Edge, NJ, 1994.

[Z2] Zorich, Anton Finite Gauss measure on the space of interval exchange transformations. Lyapunov exponents. Ann. Inst. Fourier (Grenoble) 46 (1996), no. 2, 325–370.

CNRS UMR 7599, Laboratoire de Probabilités et Modèles aléatoires, Université Pierre et Marie Curie–Boite courrier 188, 75252–Paris Cedex 05, France

Current address: IMPA, Estrada Dona Castorina 110, Rio de Janeiro, 22460-320, Brazil

E-mail address: artur@math.sunysb.edu

Clay Mathematics Proceedings
Volume 10, 2010

Orbital counting via Mixing and Unipotent flows

Hee Oh

Contents

This note is an expanded version of my lectures given at the Clay summer school in 2007.

1. Introduction: motivation

Let $\mathbf{X}$ be a projective algebraic variety defined over $\mathbb{Q}$, that is, $\mathbf{X}$ is the set of (equivalence classes of) zeros of homogeneous polynomials with coefficients in $\mathbb{Q}$. The set $\mathbf{X}(\mathbb{Q})$ of rational points in $\mathbf{X}$ consists of rational zeros of the polynomials.

The following is a classical question in number theory:

"understand the set $\mathbf{X}(\mathbb{Q})$ of rational points".

More detailed questions can be formulated as follows:

(1) Is $\mathbf{X}(\mathbb{Q})$ non-empty?
(2) If non-empty, is $\mathbf{X}(\mathbb{Q})$ infinite?
(3) If infinite, is $\mathbf{X}(\mathbb{Q})$ Zariski dense in $\mathbf{X}$?
(4) If Zariski dense, setting

$$\mathbf{X}_T := \{x \in \mathbf{X}(\mathbb{Q}) : \text{“ size”}(x) < T\},$$

what is the asymptotic growth rate of $\#\mathbf{X}_T$ as $T \to \infty$?

(5) Interpret the asymptotic growth rate of $\#\mathbf{X}_T$ in terms of geometric invariants of $\mathbf{X}$
(6) Describe the asymptotic distribution of $\mathbf{X}_T$ as $T \to \infty$?

A basic principle in studying these questions is that

(1.1) "the geometry of $\mathbf{X}$ governs the arithmetic of $\mathbf{X}$".

A good example demonstrating this philosophy is the Mordell conjecture proved by Faltings [**30**]:

THEOREM 1.2. *If $\mathbf{X}$ is a curve of genus at least 2, then $\mathbf{X}(\mathbb{Q})$ is finite.*

Note in the above theorem that a purely geometric property of $\mathbf{X}$ imposes a very strong restriction on the set $\mathbf{X}(\mathbb{Q})$.

Smooth projective varieties are classified roughly into three categories according to the ampleness of its canonical class $\mathrm{K}_\mathbf{X}$. For varieties of general type ($\mathrm{K}_\mathbf{X}$ ample), the following conjecture by Bombieri, Lang and Vojta is a higher dimensional analogue of the Mordell conjecture [**65**]:

CONJECTURE 1.3. *If $\mathbf{X}$ is a smooth projective variety of general type, then $\mathbf{X}(K)$ is not Zariski dense for any number field K.*

At the other extreme are Fano varieties ($-\mathrm{K}_\mathbf{X}$ ample). It is expected that a Fano variety should have a Zariski dense subset of rational points, at least after passing to a finite field extension of $\mathbb{Q}$. Moreover Manin formulated conjectures in 1987 on the questions (4) and (5) for some special "size functions", which were soon generalized by Batyrev and Manin for more general size functions [**1**], and Peyre made a conjecture on the question (6) [**50**].

In these lecture notes, we will focus on the question (4). First, we discuss the notion of the *size* of a rational point in $\mathbf{X}(\mathbb{Q})$. One way is simply to look at the Euclidean norm of the point. But this usually gives us infinitely many points of size less than T, and does not encode enough arithmetic information of the point.

Height function— A suitable notion of the size of a rational point is given by a height function. For a general variety $\mathbf{X}$ over $\mathbb{Q}$ and to every line bundle L of $\mathbf{X}$ over $\mathbb{Q}$, one can associate a height function H_L on $X(\mathbb{Q})$, unique up to multiplication by a bounded function, via Weil's height machine (cf. [**38**]). On the projective space $\mathbb{P}^n$, the height $\mathrm{H} = \mathrm{H}_{\mathcal{O}_{\mathbb{P}^n}(1)}$ associated to the line bundle $\mathcal{O}_{\mathbb{P}^n}(1)$ of a hyperplane is defined by:

$$\mathrm{H}(x) := \sqrt{x_0^2 + \cdots + x_n^2}$$

where $(x_0, \cdots, x_n)$ is a primitive integral vector representing $x \in \mathbb{P}^n(\mathbb{Q})$, which is unique up to sign.

It is clear that there are only finitely many rational points in $\mathbb{P}^n(\mathbb{Q})$ of height less than T, as there are only finitely many (primitive) integral vectors of Euclidean norm at most T. Schanuel [**57**] showed that as $T \to \infty$,

$$\#\{x \in \mathbb{P}^n(\mathbb{Q}) : \mathrm{H}_{\mathcal{O}_{\mathbb{P}^n(1)}}(x) < T\} \sim c \cdot T^{n-1}$$

for an explicit constant $c > 0$, and this is the simplest case of Manin's conjecture.

For a general variety $\mathbf{X}$ over $\mathbb{Q}$, a very ample line bundle L of $\mathbf{X}$ over $\mathbb{Q}$ defines a $\mathbb{Q}$-embedding $\psi_\mathrm{L} : \mathbf{X} \to \mathbb{P}^n$. Then a height function H_L is the pull back of the height function $H_{\mathcal{O}_{\mathbb{P}^n}(1)}$ to $\mathbf{X}(\mathbb{Q})$ via ψ_L. For an ample line bundle L, we set

$$H_L = H_{L^k}^{1/k}$$

for $k \in \mathbb{N}$ such that L^k is ample.

Note that for any ample line bundle L,

$$\#\{x \in \mathbf{X}(\mathbb{Q}) : H_L(x) < T\} < \infty.$$

For a subset U of $\mathbf{X}$ and $T > 0$, we define

$$N_U(L,T) := \#\{x \in \mathbf{X}(\mathbb{Q}) \cap U : H_L(x) < T\}.$$

Manin's conjecture— Manin's conjecture (or more generally the Batyrev-Manin conjecture) says the following (cf. [**1**]):

CONJECTURE 1.4. *Let* $\mathbf{X}$ *be a smooth Fano variety defined over* $\mathbb{Q}$. *For any ample line bundle* L *of* $\mathbf{X}$ *over* $\mathbb{Q}$, *there exists a Zariski open subset* U *of* $\mathbf{X}$ *such that (possibly after passing to a finite field extension), as* $T \to \infty$,

$$N_U(L,T) \sim c \cdot T^{a_L} \cdot (\log T)^{b_L - 1}$$

where $a_L \in \mathbb{Q}_{>0}$ *and* $b_L \in \mathbb{Z}_{\geq 1}$ *depend only on the geometric invariants of* L *and* $c = c(H_L) > 0$.

More precisely, the constants a_L and b_L are given by:

$a_L := \inf\{a : a[L] + [K_{\mathbf{X}}] \in \Lambda_{\text{eff}}(\mathbf{X})\}$;

$b_L :=$ the codimension of the face of $\Lambda_{\text{eff}}(\mathbf{X})$ containing $a_L[L] + [K_{\mathbf{X}}]$ in its interior

where $\text{Pic}(\mathbf{X})$ denotes the Picard group of $\mathbf{X}$ and $\Lambda_{\text{eff}}(\mathbf{X}) \subset \text{Pic}(\mathbf{X}) \otimes \mathbb{R}$ is the cone of effective divisors.

REMARK 1.5. *The restriction to a Zariski open subset is necessary in Conjecture 1.4: for the cubic surface* $\mathbf{X}$: $\sum_{i=0}^3 x_i^3 = 0$, *the above conjecture predicts the* $T(\log T)^3$ *order of rational points of height* $\sqrt{\sum_{i=0}^3 x_i^2} < T$, *but the curve* $\mathbf{Y}$ *given by the equations* $x_0 = -x_1$ *and* $x_2 = -x_3$ *contains the* T^2 *order of rational points of height bounded by* T.

Note that the asymptotic growth of the number of rational points of bounded height, which is arithmetic information on $\mathbf{X}$, is controlled only by the geometric invariants of $\mathbf{X}$; so this conjecture as well embodies the basic principle (1.1) for Fano varieties.

Conjecture 1.4 has been proved for smooth complete intersections of small degree [**6**], flag varieties [**31**], smooth toric varieties [**2**], smooth equivariant compactifications of horospherical varieties [**60**], smooth equivariant compactifications of vector groups [**11**], smooth bi-equivariant compactifications of unipotent groups [**59**] and wonderful compactifications of semisimple algebraic groups ([**58**] and [**33**]). We refer to survey articles ([**62**], [**63**], [**10**]) for more backgrounds.

In the first part of these notes, we will discuss a recent work of Gorodnik and the author [**35**] which solves new cases of Conjecture 1.4 for certain compactifications of homogeneous varieties. In contrast to most of the previous works which were based on the harmonic analysis on the corresponding adelic spaces in order to establish analytic properties of the associated height zeta function, our approach is to use the dynamics of flows on the homogeneous spaces of adele groups.

Approach—We will be interested in the projective variety $\mathbf{X}$ which is the compactification of an affine homogeneous variety $\mathbf{U}$, and try to understand the asymptotic of the number of rational points of $\mathbf{U}$ of height less than T (note that $\mathbf{U}$ is a Zariski open subset of $\mathbf{X}$ and hence it suffices to count rational points lying in $\mathbf{U}$.). More precisely, let $\mathbf{U}$ be an orbit $u_0\mathbf{G}$ where $\mathbf{G} \subset \mathrm{PGL}_{n+1}$ is an algebraic group defined over $\mathbb{Q}$ and $u_0 \in \mathbb{P}^n(\mathbb{Q})$. And let $\mathbf{X} \subset \mathbb{P}^n$ be the Zariski closure of $\mathbf{U}$, and consider the height function H on $\mathbf{X}(\mathbb{Q})$ obtained by the pull pack of $\mathrm{H}_{\mathcal{O}_{\mathbb{P}^n}(1)}$.

We attempt to forget about the ambient geometric space $\mathbf{X}$ for the time being and to focus on the rational points $\mathbf{U}(\mathbb{Q})$ of the affine homogeneous variety $\mathbf{U}$.

We would like to prove that

$$N_T := \#\{x \in \mathbf{U}(\mathbb{Q}) : \mathrm{H}(x) < T\} \sim c \cdot T^a \cdot (\log T)^{b-1} \tag{1.6}$$

for some $a, c > 0$ and $b \geq 1$. How does one prove such a result? Or where should the growth rate $T^a(\log T)^{b-1}$ come from?

Consider for a moment how to count integral vectors of Euclidean norm less than T in the plane. How does one know that the asymptotic of the number N_T of such integral vectors is of the form πT^2, as $T \to \infty$? It is because that one can show that N_T is asymptotic to the area of disc of radius T and compute that the area is πT^2, using calculus.

It turns out that one can follow the same basic strategy for counting rational points. We will first understand the set $\mathbf{U}(\mathbb{Q})$ of rational points as a discretely imbedded subset in certain ambient locally compact space and show that N_T is asymptotic to the volume of a suitably defined *height ball* in this ambient space.

What is this ambient space where the set $\mathbf{U}(\mathbb{Q})$ can be put as a discrete subset? In the real algebraic variety $\mathbf{U}(\mathbb{R})$, why is $\mathbf{U}(\mathbb{Q})$ not discrete? It is because the denominators of points can tend to infinity along prime numbers. The resolution of this issue can be found precisely using the language of *adeles*. In section 2, we define the adeles and discuss their basic properties.

Once we have defined the adelic space $\mathbf{U}(\mathbb{A})$ which contains $\mathbf{U}(\mathbb{Q})$ as a discrete subset, we will be extending the height function H of $\mathbf{X}(\mathbb{Q})$ to a continuous proper function on $\mathbf{U}(\mathbb{A})$, which we again denote by H, so that $B_T := \{x \in \mathbf{U}(\mathbb{A}) : \mathrm{H}(x) \leq T\}$ is a compact subset of $\mathbf{U}(\mathbb{A})$. Then

$$\{x \in \mathbf{U}(\mathbb{Q}) : \mathrm{H}(x) \leq T\} = \mathbf{U}(\mathbb{Q}) \cap B_T.$$

Our techniques based on the dynamical approach work for the *orbital* counting function. That is, we will be looking only at a single $\mathbf{G}(\mathbb{Q})$-orbit in $\mathbf{U}(\mathbb{Q})$ at a time; fixing a $\mathbf{G}(\mathbb{Q})$-orbit $\mathcal{O} := u_0\mathbf{G}(\mathbb{Q})$ for $u_0 \in \mathbf{U}(\mathbb{Q})$, what can we say about

$$N_T(\mathcal{O}) := \# u_0\mathbf{G}(\mathbb{Q}) \cap B_T?$$

Being able to count points in each orbit $u_0\mathbf{G}(\mathbb{Q})$ is good news and bad news at the same time; it gives finer information on the rational points $\mathbf{U}(\mathbb{Q})$, but does not quite say about the behavior of the total $\mathbf{U}(\mathbb{Q})$, since there are oftentimes infinitely many $\mathbf{G}(\mathbb{Q})$-orbits in $\mathbf{U}(\mathbb{Q})$.

Denote by $\mathbf{L}$ the stabilizer subgroup of u_0 in $\mathbf{G}$. To summarize, we will have a discrete $\mathbf{G}(\mathbb{Q})$-orbit $u_0\mathbf{G}(\mathbb{Q}) = \mathbf{L}(\mathbb{Q})\backslash\mathbf{G}(\mathbb{Q})$ in the homogeneous space $u_0\mathbf{G}(\mathbb{A}) = \mathbf{L}(\mathbb{A})\backslash\mathbf{G}(\mathbb{A}) \subset \mathbf{U}(\mathbb{A})$ and we would like to count points of the orbit $u_0\mathbf{G}(\mathbb{Q})$ in a growing sequence of compact subsets B_T of $\mathbf{L}(\mathbb{A})\backslash\mathbf{G}(\mathbb{A})$.

In section 3, we explain a general strategy due to Duke-Rudnick-Sarnak [**21**] on the orbital counting problem for $[x_0]\Gamma \cap B_T$ in a homogeneous space $L\backslash G$; here

Γ is a lattice in a second countable locally compact group G, $[x_0]\Gamma$ is discrete in $L\backslash G$ for $[x_0] = L$, and $\{B_T \subset L\backslash G\}$ is a family of compact subsets. This method reduces the counting problem for $N_T(\mathcal{O})$ into understanding

(i) the asymptotic behavior of adelic periods $\mathbf{L}(\mathbb{Q})\backslash\mathbf{L}(\mathbb{A})g_i$, as $g_i \to \infty$ in $\mathbf{L}(\mathbb{A})\backslash\mathbf{G}(\mathbb{A})$;
(ii) certain regularity problem for the volume of adelic height balls $B_T \subset \mathbf{L}(\mathbb{A})\backslash\mathbf{G}(\mathbb{A})$.

The techniques needed to establish (ii) are completely disjoint from those for (i), and this is where an input of algebraic/arithmetic geometry is needed. In section 4, we present the main ergodic result on the translates of semisimple periods in the adelic homogeneous space based on the study of unipotent flows and establish (i) when both $\mathbf{G}$ and $\mathbf{L}$ are connected semisimple and $\mathbf{L}$ is a maximal $\mathbb{Q}$-subgroup of G. In section 5, we explain how to deduce some cases of Manin's conjecture using this approach, which is the main result of [**35**]. We also explain the implications of this result on the rational points of *affine* varieties.

In section 6, we deduce mixing theorems for adelic groups as a special case of the results in section 4. We also explain in this section how the equidistribution of Hecke points is related to the adelic mixing theorem. In section 7, we interpret a quantitative adelic mixing theorem as a bound toward the Ramanujan conjecture on the automorphic spectrum. In section 8, we explain how the mixing theorems can be used to prove the equidistribution of symmetric periods, based on the special geometric property of an affine symmetric space, called the wavefront property. This approach gives an effective counting for the S-integral points on affine symmetric varieties. In the last section 9, we discuss a problem of Linnik on the representations of integers by an invariant polynomial, and extend the main result of [**29**] using the ergodic result presented in section 4.

We have not sought at all to state the most general statements. On the contrary, we will be speaking only on the simplest cases in many occasions, for instance, we stick to the field of *rational numbers* $\mathbb{Q}$, rather than a number field, and to the homogeneous spaces $\mathbf{L}\backslash\mathbf{G}$ with $\mathbf{L}$ being a *maximal* connected $\mathbb{Q}$-subgroup of $\mathbf{G}$.

Acknowledgement: I gratefully acknowledge the collaborations with Alex Gorodnik and Yves Benoist. In particular, the results presented in sections 4 and 5 are based on [**35**] and those in section 8 are based on [**3**]. I thank Peter Sarnak for helpful comments on the preliminary version of these notes. I thank my parents who took care of my baby Joy during the summer school. My research was partially supported by NSF grant 0629322).

2. Adeles: definition and basic properties

In this section, we define the notion of adeles, and state some of their basic properties as we would need (cf. [**66**], [**51**]).

2.1. Restricted topological product. Let X_p be a second countable locally compact topological space for each p in a given countable index set I. The goal is to construct a reasonable locally compact space which contains all X_p, $p \in I$. The first attempt will be simply taking the direct product $\prod_p X_p$. However the direct product topology does not yield a locally compact space unless either I is finite or almost all X_p are compact.

Hence we will do something else which will make the product a locally compact space and this can be achieved by taking the restricted topological product. Suppose that an open compact subset $K_p \subset X_p$ is given for each p in a co-finite subset I_0 of I.

Set

$$X_I := \prod_p (X_p : K_p) = \{(x_p)_{p\in I} \in \prod_p X_p : x_p \in K_p \text{ for almost all } p\}.$$

We endow on X_I the topology generated by subsets of the form $\prod_{p\in I} V_p$ where V_p is open in X_p and $V_p = K_p$ for almost all p. The space X with this topology is called the restricted topological product of X_p's with respect to the distinguished open subsets K_p's. In the case when I is a finite set, X_I is simply the direct product of X_p, $p \in I$.

We need the following basic properties of X_I:

FACTS 2.1. .

(1) *X_I is a second countable locally compact space.*
(2) *Any compact subset of X_I is contained in*

$$X^S := X_S \times \prod_{p\notin S} K_p$$

for some finite $S \subset I$.
(3) *If μ_p's are Borel measures on X_p's such that $\mu_p(K_p) = 1$ whenever $p \in I_0$, then the restricted product $\mu := \otimes^*_{p\in I}\mu_p$ on X_I is defined as follows: first for each finite subset $S \subset I$, define the measure μ^S on X^S to be simply the direct product $\prod_{p\in S}\mu_p \times \prod_{p\notin S}\mu_p|_{K_p}$.*

Now for any $f \in C_c(X_I)$ whose support is contained in X^S, set

$$\mu(f) := \mu^S(f).$$

It is easy to check that μ is well-defined since μ^S's are compatible with each other. Since X_I can be written as the union $\cup_S X^S$ over finite subsets $S \subset I$, μ defines a Borel measure on X_I by the Riesz representation theorem.
(4) *If each X_p is a group (resp. ring), X_I is a group (resp. ring) using the componentwise operations. If μ_p is a (resp. left invariant) Haar measure of X_p for each $p \in I$, then μ is a (resp. left-invariant) Haar measure of X_I.*

2.2. $\mathbb{Q}$ is a lattice in the adele group $\mathbb{A}$. Denote by $R = \{\infty, 2, 3, \cdots\}$ the set of all primes including the infinite prime ∞. For $p = \infty$, $|\cdot|_\infty$ denotes the usual absolute value on $\mathbb{Q}$ and for p finite, $|\cdot|_p$ denotes the normalized p-adic absolute value on $\mathbb{Q}$, i.e.,

$$\left|p^k \frac{a}{b}\right|_p = p^{-k}$$

if p does not divide ab. We obtain the locally compact fields $\mathbb{R} = \mathbb{Q}_\infty$ and $\mathbb{Q}_p$'s by taking the completions of $\mathbb{Q}$ with respect to $|\cdot|_p$'s. We set

$$\mathbb{Z}_p := \{x \in \mathbb{Q}_p : |x|_p \leq 1\}$$

for each $p \in R_f := R - \{\infty\}$.

DEFINITION 2.2. The adele ring $\mathbb{A}$ over $\mathbb{Q}$ is defined to be the restricted topological product of $\mathbb{Q}_\infty := \mathbb{R}$ and $(\mathbb{Q}_p, \mathbb{Z}_p)$'s for $p \in R_f$, that is,

$$\mathbb{A} := \prod_{p \in R} (\mathbb{Q}_p : \mathbb{Z}_p).$$

Since every element of $\mathbb{Q}$ belongs to $\mathbb{Z}_p$ for almost all p, $\mathbb{Q}$ can be considered as a subset of $\mathbb{A}$ under the diagonal embedding. That $\mathbb{Q}$ is a discrete subset of $\mathbb{A}$ follows from the following observation, which shows that 0 is an isolated point in $\mathbb{A}$:

$$\{x \in \mathbb{Q} : |x|_\infty < 0.5,\ x \in \mathbb{Z}_p \text{ for all } p \in R_f\} = \{0\}.$$

Moreover $\mathbb{Q}$ is a lattice in $\mathbb{A}$, that is, the quotient of $\mathbb{A}$ by $\mathbb{Q}$ has finite volume (with respect to a Haar measure of $\mathbb{A}$). To show this, we consider the ring $\mathbb{A}_f$ of finite adeles, i.e., the subring of $\mathbb{A}$ whose ∞-component is trivial. Note that $\mathbb{A}_f = \prod_{p \in R_f} (\mathbb{Q}_p : \mathbb{Z}_p)$, and that every element of $\mathbb{A}$ can be written as (x_∞, x_f) with $x_\infty \in \mathbb{R}$ and $x_f \in \mathbb{A}_f$. We set $\mathbb{Z}_f = \prod_{p \in R_f} \mathbb{Z}_p$.

LEMMA 2.3. .

(1) *$\mathbb{Q}$ is dense in $\mathbb{A}_f$.*
(2) $\mathbb{A} = \mathbb{Q} + ((0,1] \times \mathbb{Z}_f)$.
(3) *$\mathbb{Q}$ is a lattice in $\mathbb{A}$.*

PROOF. For (1), it suffices to show that $\mathbb{Z}$ is dense in $\mathbb{Z}_f$. Any open subset in $\mathbb{Z}_f$ contains a subset of the form $\prod_{p \in S}(a_p + p^{m_p}\mathbb{Z}_p) \times \prod_{p \notin S} \mathbb{Z}_p$ for some finite subset S. Hence (1) follows from the Chinese remainder theorem, which says that there exists $x \in \mathbb{Z}$ such that $x = a_p \bmod p^{m_p}$ for all $p \in S$.

(1) implies that $\mathbb{A}_f = \mathbb{Q} + \mathbb{Z}_f$, as $\mathbb{Z}_f$ is an open subgroup of $\mathbb{A}_f$. Now for $(x_\infty, x_f) \in \mathbb{A}$, we have $x_f = y_f + z \in \mathbb{Z}_f + \mathbb{Q}$. Hence $(x_\infty, x_f) = (x_\infty - z, y_f) + (z, z)$. Now $x_\infty - z = y_\infty + n$ for some $y_\infty \in (0,1]$ and $n \in \mathbb{Z}$. Hence

$$(x_\infty, x_f) = (y_\infty, y_f - n) + (n + z, n + z),$$

proving (2). (3) follows from (2). □

Let $\mathbf{U}$ be an affine variety defined over $\mathbb{Q}$. Choose any $\mathbb{Q}$-isomorphism α of $\mathbf{U}$ onto a Zariski closed $\mathbb{Q}$-subvariety $\mathbf{U}'$ of the N-dimensional affine space A^N for some N. Then the adele space $\mathbf{U}(\mathbb{A})$ corresponding to $\mathbf{U}$ is defined to be the restricted topological product of $\mathbf{U}(\mathbb{Q}_p)$'s with respect to $\alpha^{-1}(\mathbf{U}' \cap \mathbb{Z}_p^N)$'s. Note that this definition of $\mathbf{U}(\mathbb{A})$ does not depend on the choice of a $\mathbb{Q}$-isomorphism α. The set $\mathbf{U}(\mathbb{Q})$ imbeds into $\mathbf{U}(\mathbb{A})$ as a discrete subset.

The adele space $\mathbf{U}(\mathbb{A})$ for a general variety $\mathbf{U}$ is then defined using the open coverings of $\mathbf{U}$ by affine subsets.

2.3. $\mathbf{G}(\mathbb{Q})$ is a lattice in $\mathbf{G}(\mathbb{A})$ if $\mathbf{G}$ admits no non-trivial $\mathbb{Q}$-character. Let $\mathbf{G} \subset \mathrm{GL}_n$ be a connected $\mathbb{Q}$-group. For a commutative ring J, $\mathrm{GL}_n(J)$ is defined to be the matrices with entries in J and with determinant being a unit in J. We set

$$\mathbf{G}(J) = \mathbf{G} \cap \mathrm{GL}_n(J).$$

Note that $\mathbf{G}(\mathbb{Z}_p)$ is a compact open subgroup of $\mathbf{G}(\mathbb{Q}_p)$ for each finite p. We claim that the adele space $\mathbf{G}(\mathbb{A})$ associated to $\mathbf{G}$ coincides with the restricted topological product of $\mathbf{G}(\mathbb{R})$ and $\mathbf{G}(\mathbb{Q}_p)$'s with respect to $\mathbf{G}(\mathbb{Z}_p)$'s for $p \in R_f$.

To see this, we use the restriction, say α, to $\mathbf{G}$ of the map $\mathrm{GL}_n \to A^{n^2+1}$ given by $g \mapsto (g, \det(g)^{-1})$, where A^{n^2+1} is the n^2+1-dimensional affine space. Since $\alpha^{-1}(\alpha(\mathbf{G}) \cap \mathbb{Z}_p^{n^2+1}) = \mathbf{G}(\mathbb{Z}_p)$, this shows that $\mathbf{G}(\mathbb{A})$ defined at the end of the previous subsection is equal to $\prod_{p\in R}(\mathbf{G}(\mathbb{Q}_p) : \mathbf{G}(\mathbb{Z}_p))$.

We note that $\mathbf{G}(\mathbb{A})$, with the component-wise group operation, is a second countable locally compact group with a (left-invariant) Haar measure $\mu := \otimes^* \mu_p$ where μ_p is a (left-invariant) Haar measure on $\mathbf{G}(\mathbb{Q}_p)$ with $\mu_p(\mathbf{G}(\mathbb{Z}_p)) = 1$ for each finite $p \in R_f$.

THEOREM 2.4. [**7**] *If* $\mathbf{G}$ *admits no non-trivial* $\mathbb{Q}$*-character, then* $\mathbf{G}(\mathbb{Q})$ *is a lattice in* $\mathbf{G}(\mathbb{A})$.

We give an outline of the proof of this theorem for $\mathbf{G} = \mathrm{SL}_n$. We often write an element of $\mathbf{G}(\mathbb{A})$ as $(g_\infty, g_f) \in \mathbf{G}(\mathbb{R}) \times \mathbf{G}(\mathbb{A}_f)$, where $\mathbf{G}(\mathbb{A}_f)$ denotes the subgroup of finite adeles, i.e., with the trivial component at ∞.

THEOREM 2.5. *Let* $\mathbf{G} = \mathrm{SL}_n$.

(1) $\mathbf{G}(\mathbb{Q})$ *is dense in* $\mathbf{G}(\mathbb{A}_f)$.
(2) $\mathbf{G}(\mathbb{Q})\backslash\mathbf{G}(\mathbb{A}) = \Sigma_0 \times \prod_{p\in R_f} \mathbf{G}(\mathbb{Z}_p)$ *where* $\Sigma_0 = \mathbf{G}(\mathbb{Z})\backslash\mathbf{G}(\mathbb{R})$.
(3) $\mathbf{G}(\mathbb{Q})$ *is a lattice in* $\mathbf{G}(\mathbb{A})$.

PROOF. For $1 \le i, j \le n$, denote by $U_{ij}(\mathbb{Q}_p)$ (resp. $U_{ij}(\mathbb{Q})$) the unipotent one parameter subgroup $I_n + \mathbb{Q}_p E_{i,j}$ (resp. $I_n + \mathbb{Q}E_{i,j}$) where E_{ij} is the matrix whose only non-zero entry is 1 at the (i,j)-entry. Since $\mathbb{Q}$ is dense in $\mathbb{A}_f$ and $U_{ij}(\mathbb{Q}_p)$'s generate $\mathrm{SL}_n(\mathbb{Q}_p)$, $\mathrm{SL}_n(\mathbb{Q}_p)$ is contained in the closure of $\mathrm{SL}_n(\mathbb{Q})$ in $\mathrm{SL}_n(\mathbb{A}_f)$, and hence any finite product of $\mathrm{SL}_n(\mathbb{Q}_p)$'s is contained in the closure of $\mathrm{SL}_n(\mathbb{Q})$ in $\mathrm{SL}_n(\mathbb{A}_f)$. Since they form a dense subset in $\mathrm{SL}_n(\mathbb{A}_f)$, this proves (1). By (1), we have

$$\mathrm{SL}_n(\mathbb{A}_f) = \mathrm{SL}_n(\mathbb{Q}) \prod_{p\in R_f} \mathrm{SL}_n(\mathbb{Z}_p).$$

It is easy to deduce (2) and (3) from this, using the facts that $\mathrm{SL}_n(\mathbb{Z}) = \mathrm{SL}_n(\mathbb{Q}) \cap \prod_{p\in R_f} \mathrm{SL}_n(\mathbb{Z}_p)$ and that $\mathrm{SL}_n(\mathbb{Z})$ is a lattice in $\mathrm{SL}_n(\mathbb{R})$. □

3. General strategy on orbital counting

We now explain a general strategy for the orbital counting. This, at least in an explicit form, was first described and used in the work of Duke-Rudnick-Sarnak [**21**].

Let G be a locally compact second countable group and $L < G$ a closed subgroup. Let $\Gamma \subset G$ be a lattice such that the intersection $L \cap \Gamma$ is a lattice in L. This in particular implies that both G and L are unimodular, and that the orbit $x_0\Gamma$ is a discrete subset of $L\backslash G$ for $x_0 = [L]$. For a given sequence of growing compact subsets $B_n \subset L\backslash G$, we would like to understand the asymptotic behavior of $\#B_n \cap x_0\Gamma$ as $n \to \infty$. Heuristics suggest that

$$\#B_n \cap x_0\Gamma \sim \nu(B_n)$$

where ν is a measure on $L\backslash G$ determined as follows:

NOTATION 3.1. Let μ_G and μ_L be the Haar measures on G and L such that $\mu_G(\Gamma\backslash G) = 1 = \mu_L(\Gamma \cap L\backslash L)$. There exists the unique G-invariant measure ν on

$L\backslash G$ which is compatible with μ_G and μ_L, in the sense that for any $\psi \in C_c(G)$,

$$\int_G \psi \, d\mu_G = \int_{[g]\in L\backslash G} \int_{h\in L} \psi(hg) d\mu_L(h) d\nu([g]).$$

DEFINITION 3.2 (Counting function). Define the following counting function on $\Gamma\backslash G$:

$$F_n(g) = \sum_{\gamma\in\Gamma\cap L\backslash\Gamma} \chi_{B_n}(x_0\gamma g).$$

Noting that

$$F_n(e) = \#B_n \cap x_0\Gamma$$

we will present two conditions which guarantee that

$$F_n(e) \sim \nu(B_n) \quad \text{as } n \to \infty.$$

We denote by $\mathcal{P}(\Gamma\backslash G)$ the space of Borel probability measures on $\Gamma\backslash G$ and by $C_c(\Gamma\backslash G)$ the space of continuous functions on $\Gamma\backslash G$ with compact support. For a subgroup $K < G$, $C_c(\Gamma\backslash G)^K$ denotes a subset of $C_c(\Gamma\backslash G)$ consisting of right K-invariant functions.

Since $\Gamma\backslash\Gamma L = (\Gamma \cap L)\backslash L$ is a closed orbit in $\Gamma\backslash G$, we may consider μ_L as a probability measure in $\Gamma\backslash G$ supported in $\Gamma\backslash\Gamma L$.

DEFINITION 3.3. .

(1) For $\mu \in \mathcal{P}(\Gamma\backslash G)$ and $g \in G$, we denote by $g.\mu$ the translation of μ by g, i.e.,

$$g.\mu_L(E) := \mu_L(Eg^{-1})$$

for any Borel subset E of $\Gamma\backslash G$.

(2) For a subset $\mathcal{F} \subset C_c(\Gamma\backslash G)$ and a sequence $\nu_i \in \mathcal{P}(\Gamma\backslash G)$, we say that ν_i weakly converges to μ, as $i \to \infty$, relative to $\mathcal{F}$ if for all $\psi \in \mathcal{F}$,

$$\lim_{i\to\infty} \nu_i(\psi) = \mu(\psi).$$

(3) For a sequence $g_i \in L\backslash G$, the translate $\Gamma\backslash\Gamma L g_i$ is said to become equidistributed in $\Gamma\backslash G$, as $i \to \infty$, relative to the family $\mathcal{F}$ if the sequence $(g_i).\mu_L$ weakly converges to μ_G relative to $\mathcal{F}$.

(4) If $\mathcal{F} = C_c(\Gamma\backslash G)$, we omit the reference to $\mathcal{F}$ in (2) and (3).

DEFINITION 3.4. Let K be a compact subgroup of G. A family $\{B_n\}$ of K-invariant compact subsets of $L\backslash G$ is called K-well-rounded if there exists $c > 0$ such that for every $\epsilon > 0$, there exists a neighborhood U_ϵ of e in G satisfying

$$\nu(B_n U_\epsilon K - \cap_{g\in U_\epsilon K} B_n g) < c \cdot \epsilon \cdot \nu(B_n)$$

for all large n. For $K = \{e\}$, we simply say that B_n is well-rounded.

PROPOSITION 3.5. *Let K be a compact subgroup of G and $\{B_n \subset L\backslash G\}$ a sequence of K-invariant compact subsets with $\nu(B_n) \to \infty$ as $n \to \infty$. Assume that the following hold:*

(1) *For any sequence $g_i \to \infty$ in $L\backslash G$, the translate $\Gamma\backslash\Gamma L g_i$ becomes equidistributed in $\Gamma\backslash G$ relative to the family $C_c(\Gamma\backslash G)^K$;*

(2) *The sequence $\{B_n\}$ is K-well-rounded.*

Then as $n \to \infty$

$$\#x_0\Gamma \cap B_n \sim \nu(B_n).$$

PROOF. The proof consists of two steps. In the first step, we show that the condition (1) implies the weak-convergence of $\frac{1}{\nu(B_n)}F_n$, i.e., for all $\psi \in C_c(\Gamma\backslash G)^K$,

$$\lim_{n\to\infty} \frac{1}{\nu(B_n)} \int_{\Gamma\backslash G} F_n(x)\,\psi(x)\,d\mu_G(x) \;=\; \int_{\Gamma\backslash G} \psi(x)\,d\mu_G(x)\,. \tag{3.6}$$

Observe that

$$\begin{aligned}
\int_{\Gamma\backslash G} F_n\psi\,d\mu_G &= \int_{\Gamma\backslash G}\left(\sum_{\gamma\in\Gamma\cap L\backslash\Gamma} \chi_{B_n}(x_0\gamma g)\psi(g)\right) d\mu_G(g)\\
&= \int_{\Gamma\cap L\backslash G} \chi_{B_n}(x_0 g)\psi(g)\,d\mu_G(g)\\
&= \int_{L\backslash G}\int_{\Gamma\cap L\backslash L} \chi_{B_n}(x_0 g)\,\psi(hg)\,d\mu_L(h)d\nu(g)\\
&= \int_{x_0 g\in B_n}\left(\int_{h\in\Gamma\backslash\Gamma L} \psi(hg)d\mu_L(h)\right) d\nu(g)\\
&= \int_{x_0 g\in B_n}\left(\int_{\Gamma\backslash G} \psi(x)\,d(g.\mu_L)(x)\right) d\nu(g)
\end{aligned}$$

By (1), for any $\epsilon > 0$, there exists a compact subset $C_\epsilon \subset L\backslash G$ such that

$$\sup_{x_0 g\notin C_\epsilon} |\int \psi d(g.\mu_L) - \int \psi d\mu_G| \le \epsilon.$$

Hence

$$\begin{aligned}
|\int_{x_0 g\in B_n}\left(\int_{\Gamma\backslash G} \psi(x)\,d(g.\mu_L)(x)\right) d\nu(g) - \nu(B_n)\int \psi d\mu_G|\\
\le 2\|\psi\|_\infty \nu(C_\epsilon\cap B_n) + \epsilon\nu(B_n).
\end{aligned}$$

Since $\nu(B_n)\to\infty$, we deduce that

$$\limsup_n \left|\frac{1}{\nu(B_n)}\int_{x_0 g\in B_n}\left(\int_{\Gamma\backslash G} \psi(x)\,d(g.\mu_L)(x)\right) d\nu(g) - \int \psi d\mu_G\right| \le \epsilon.$$

Hence (3.6) follows from (1).

In order to deduce the pointwise convergence from the weak convergence, we will now use the assumption that $\{B_n\}$ is K-well rounded. Fix $\epsilon > 0$ and $U_\epsilon \subset G$ be a symmetric open neighborhood of e in G as in the definition of the K-well-roundedness of B_n. Define $F^+_{n,\epsilon}$ and $F^-_{n,\epsilon}$ similarly to F_n but using $B^+_{n,\epsilon} := B_n U_\epsilon K$ and $B^-_{n,\epsilon} := \cap_{u\in U_\epsilon K} B_n u$ respectively, in place of B_n.

It is easy to check that for all $g \in U_\epsilon K$,

$$F^-_{n,\epsilon}(g) \le F_n(e) \le F^+_{n,\epsilon}(g). \tag{3.7}$$

Choose a K-invariant non-negative continuous function ψ_ϵ on $\Gamma\backslash G$ with support in $\Gamma\backslash\Gamma U_\epsilon K$ and with the integral $\int_{\Gamma\backslash G}\psi_\epsilon d\mu_G$ one.

By integrating (3.7) against ψ_ϵ, we have

$$\langle F^-_{n,\epsilon}, \psi_\epsilon\rangle \le F_n(e) \le \langle F^+_{n,\epsilon}, \psi_\epsilon\rangle.$$

Applying (3.6) to $F^\pm_{n,\epsilon}$, which we may since $\nu(B^\pm_{n,\epsilon}) \to \infty$, we have

$$\langle F^\pm_{n,\epsilon}, \psi_\epsilon\rangle \sim_n \nu(B^\pm_{n,\epsilon}).$$

Therefore there are constants $c_1, c_2 > 0$ such that for any $\epsilon > 0$,

$$\limsup_n \frac{F_n(e)}{\nu(B_n)} \le (1 + c_1\epsilon)\cdot \limsup_n \frac{\nu(B^+_{n,\epsilon})}{\nu(B_n)} \le (1 + c_1\epsilon)(1 + c_2\epsilon).$$

Similarly,

$$(1 - c_1\epsilon)(1 - c_2\epsilon) \le \liminf_n \frac{F_n(e)}{\nu(B_n)}.$$

Since $\epsilon > 0$ is arbitrary, it follows that

$$\lim_{n\to\infty} \frac{F_n(e)}{\nu(B_n)} = 1.$$

□

REMARK 3.8. The above proposition was considered only for $K = \{e\}$ in [**21**] (also in [**26**]). In applications where G and L are (the identity components of) real Lie groups, this is usually sufficient. However when G and L are the adelic groups associated to non-simply connected semisimple $\mathbb{Q}$-groups, the equidistribution in Prop. 3.5 (1) does not usually hold for all of $C_c(\Gamma\backslash G)$ (cf. Theorem 4.3). Hence it is necessary to consider the above modification by introducing compact subgroups K.

Let $\mu_n \in \mathcal{P}(\Gamma\backslash G)$ denote the average of measures $x.\mu_L$, $x \in B_n$: for $\psi \in C_c(\Gamma\backslash G)$,

$$\mu_n(\psi) := \frac{1}{\nu(B_n)}\int_{x\in B_n} (x.\mu_L)(\psi)\, d\nu(x) = \frac{1}{\nu(B_n)}\int_{x\in B_n}\int_{\Gamma\cap L\backslash L} \psi(hx)\, d\mu_L(h)\, d\nu(x).$$

The proof of the above proposition yields the following stronger version:

PROPOSITION 3.9. *Let K be a compact subgroup of G and $\{B_n \subset L\backslash G\}$ a sequence of K-invariant compact subsets with $\nu(B_n) \to \infty$ as $n \to \infty$. Suppose:*

(1) *μ_n weakly converges to μ_G relative to $C_c(\Gamma\backslash G)^K$;*
(2) *The sequence $\{B_n\}$ is K-well-rounded.*

Then as $n \to \infty$,

$$\# x_0\Gamma \cap B_n \sim_n \nu(B_n).$$

4. Equidistribution of semisimple periods via unipotent flows

Let $\mathbf{G} \subset \mathrm{GL}_n$ be a connected semisimple algebraic $\mathbb{Q}$-group and $\mathbf{L}$ a connected semisimple $\mathbb{Q}$-subgroup of $\mathbf{G}$. Note that $\mathbf{G}(\mathbb{Q})\backslash\mathbf{G}(\mathbb{A})$ and $\mathbf{L}(\mathbb{Q})\backslash\mathbf{L}(\mathbb{A})$ have finite volumes by Theorem 2.4. We are interested in the asymptotic behavior of the translate

$$\mathbf{L}(\mathbb{Q})\backslash\mathbf{L}(\mathbb{A})g_i \subset \mathbf{G}(\mathbb{Q})\backslash\mathbf{G}(\mathbb{A})$$

as $g_i \to \infty$ in $\mathbf{L}(\mathbb{A})\backslash\mathbf{G}(\mathbb{A})$.

We hope that the translate $\mathbf{L}(\mathbb{Q})\backslash\mathbf{L}(\mathbb{A})g_i$ becomes equidistributed in the space $\mathbf{G}(\mathbb{Q})\backslash\mathbf{G}(\mathbb{A})$ as $g_i \to \infty$ in $\mathbf{L}(\mathbb{A})\backslash\mathbf{G}(\mathbb{A})$. One obvious obstruction is the existence of a proper $\mathbb{Q}$-subgroup $\mathbf{M}$ of $\mathbf{G}$ which contains $\mathbf{L}$ properly, since the sequence

$\mathbf{L}(\mathbb{Q})\backslash\mathbf{L}(\mathbb{A})g_i$ then remains entirely inside the closed subset $\mathbf{M}(\mathbb{Q})\backslash\mathbf{M}(\mathbb{A})$ for $g_i \in \mathbf{M}(\mathbb{A})$, and hence the desired equidistribution cannot happen for those sequences $g_i \in \mathbf{M}(\mathbb{A})$.

In the case when both $\mathbf{G}$ and $\mathbf{L}$ are simply connected, this is the only obstruction. In the rest of this section, we assume that $\mathbf{L}$ is a maximal connected $\mathbb{Q}$-subgroup of $\mathbf{G}$, unless mentioned otherwise. We closely follow the exposition in [**35**] to which we refer for details.

THEOREM 4.1. *Suppose that both* $\mathbf{G}$ *and* $\mathbf{L}$ *are simply connected. Then* $\mathbf{L}(\mathbb{Q})\backslash\mathbf{L}(\mathbb{A})g_i$ *becomes equidistributed in* $\mathbf{G}(\mathbb{Q})\backslash\mathbf{G}(\mathbb{A})$ *for any sequence* $g_i \to \infty$ *in* $\mathbf{L}(\mathbb{A})\backslash\mathbf{G}(\mathbb{A})$.

In the general case, when $\mathbf{G}$ and $\mathbf{L}$ are not necessarily simply connected, the above theorem does not hold, because there are many finite index subgroups of $\mathbf{G}(\mathbb{A})$ which contain $\mathbf{G}(\mathbb{Q})$ as a lattice and the entire dynamics may happen only in these smaller pieces of $\mathbf{G}(\mathbb{Q})\backslash\mathbf{G}(\mathbb{A})$.

To overcome this issue, we consider a simply connected covering $\pi : \tilde{\mathbf{G}} \to \mathbf{G}$ over $\mathbb{Q}$. The map π induces a map from $\tilde{\mathbf{G}}(\mathbb{A})$ to $\mathbf{G}(\mathbb{A})$, which we again denote by π by abuse of notation.

LEMMA 4.2. *For any compact open subgroup* W *of* $\mathbf{G}(\mathbb{A}_f)$, *the product*

$$G_W := \mathbf{G}(\mathbb{Q})\pi(\tilde{\mathbf{G}}(\mathbb{A}))W$$

is a normal subgroup of $\mathbf{G}(\mathbb{A})$ *with finite index.*

THEOREM 4.3. [**35**] *Fix a compact open subgroup* W *of* $\mathbf{G}(\mathbb{A}_f)$. *Then for any* $g_i \to \infty$ *in* $\mathbf{L}(\mathbb{A}) \cap G_W\backslash G_W$, *the translate* $\mathbf{L}(\mathbb{Q})\backslash(\mathbf{L}(\mathbb{A}) \cap G_W)g_i$ *becomes equidistributed in* $\mathbf{G}(\mathbb{Q})\backslash G_W$ *relative to the family* $C_c(\mathbf{G}(\mathbb{Q})\backslash G_W)^W$.

Analogous statement for $\mathbf{L}$ being a maximal $\mathbb{Q}$-anisotropic torus was proved in [**64**] for $\mathbf{G} = \mathrm{PGL}_2$ and in [**23**] for $\mathbf{G} = \mathrm{PGL}_3$.

In the rest of this section, we will outline the proof of Theorem 4.1; so we assume that both $\mathbf{G}$ and $\mathbf{L}$ are simply connected. The first step is to reduce the equidistribution problem in the homogeneous spaces of an adele group to the S-algebraic setting, using the strong approximation theorem.

It is convenient to define the following:

DEFINITION 4.4. .

(1) For a semisimple $\mathbb{Q}$-subgroup $\mathbf{L}$ of $\mathbf{G}$, an element $p \in R$ is called isotropic for $\mathbf{L}$ if $\mathbf{N}(\mathbb{Q}_\mathbf{p})$ is non-compact for any non-trivial normal $\mathbb{Q}$-subgroup $\mathbf{N}$ of $\mathbf{L}$.
(2) For a semisimple $\mathbb{Q}$-subgroup $\mathbf{L}$ of $\mathbf{G}$, an element $p \in R$ is called strongly isotropic for $\mathbf{L}$ if $\mathbf{N}(\mathbb{Q}_\mathbf{p})$ is non-compact for any non-trivial normal $\mathbb{Q}_p$-subgroup $\mathbf{N}$ of $\mathbf{L}$.
(3) A finite subset S of R is called strongly isotropic (resp. isotropic) for $\mathbf{L}$ if S contains a strongly isotropic (resp. isotropic) p for $\mathbf{L}$.

We fix a finite subset $S \subset R$ containing ∞ in the rest of this section. We set

$$\mathbf{G}_S := \prod_{p \in S} \mathbf{G}(\mathbb{Q}_p).$$

Denoting by $\mathbf{G}(\mathbb{A}_S)$ the subgroup of $\mathbf{G}(\mathbb{A})$ with trivial S-components, the group $\mathbf{G}(\mathbb{A})$ can be naturally identified with the product $\mathbf{G}_S \times \mathbf{G}(\mathbb{A}_S)$.

THEOREM 4.5 (Strong approximation property). *Let S be* $\mathbf{G}$*-isotropic. Then for any compact open subgroup* W_S *of* $\mathbf{G}(\mathbb{A}_S)$,

$$\mathbf{G}(\mathbb{Q})W_S = \mathbf{G}(\mathbb{A}_S).$$

See [**51**, 7.4].

Hence any element $g \in \mathbf{G}(\mathbb{A})$ can be written as

$$g = (\gamma_g, \gamma_g)(g_S, w)$$

where $\gamma_g \in \mathbf{G}(\mathbb{Q})$, $g_S \in \mathbf{G}_S$ and $w \in W_S$. Note that g_S is determined uniquely up to the left multiplication by an element of $\mathbf{G}(\mathbb{Q}) \cap W_S$.

DEFINITION 4.6. A subgroup Γ of $\mathbf{G}(\mathbb{Q})$ is called an S-congruence subgroup if $\Gamma = \mathbf{G}(\mathbb{Q}) \cap W_S$ for some compact open subgroup W_S of $\mathbf{G}(\mathbb{A}_S)$.

Note that an S-congruence subgroup is a lattice in $\mathbf{G}_S$, embedded diagonally.

EXAMPLE 4.7. If $S = \{\infty, p\}$ and $W := \prod_{q \neq p} \mathrm{SL}_n(\mathbb{Z}_q)$, then the diagonal embedding of $\Gamma = \mathrm{SL}_n(\mathbb{Q}) \cap W = \mathrm{SL}_n(\mathbb{Z}[1/p])$ is a lattice in $\mathrm{SL}_n(\mathbb{R}) \times \mathrm{SL}_n(\mathbb{Q}_p)$.

PROPOSITION 4.8. *Let S be isotropic for both* $\mathbf{G}$ *and* $\mathbf{L}$*, and W_S a compact open subgroup of* $\mathbf{G}(\mathbb{A}_S)$*. Set* $\Gamma := \mathbf{G}(\mathbb{Q}) \cap W_S$.

(1) *The map $g \mapsto g_S$ induces a $\mathbf{G}_S$-equivariant topological isomorphism*

$$\Phi : \mathbf{G}(\mathbb{Q})\backslash\mathbf{G}(\mathbb{A})/W_S \to \Gamma\backslash\mathbf{G}_S.$$

(2) *For any $g = (\gamma_g, \gamma_g)(g_S, w) \in \mathbf{G}(\mathbb{A})$, the map Φ, via the restriction, induces the topological isomorphism*

$$\mathbf{G}(\mathbb{Q})\backslash\mathbf{G}(\mathbb{Q})\mathbf{L}(\mathbb{A})gW_S/W_S \simeq \Gamma\backslash\Gamma\gamma_g^{-1}\mathbf{L}_S\gamma_g g_S.$$

PROOF. Since the map $g \mapsto (g_S, w)$ induces a $\mathbf{G}_S$-equivariant homeomorphism between $\mathbf{G}(\mathbb{Q})\backslash\mathbf{G}(\mathbb{A})$ and $\Gamma\backslash(\mathbf{G}_S \times W_S)$, (1) follows. The strong approximation theorem 4.5 applied to $\mathbf{L}$ implies

$$\mathbf{L}(\mathbb{A}) = \mathbf{L}(\mathbb{Q})\mathbf{L}_S(gW_Sg^{-1} \cap \mathbf{L}(\mathbb{A}_S)).$$

Hence any $x \in \mathbf{L}(\mathbb{A})$ can be written, modulo $\mathbf{L}(\mathbb{Q})$, as $(x_S, gw'g^{-1})$ for some $w' \in W_S$ and $x_S \in \mathbf{L}_S$. Now

$$xg = (\gamma_g, \gamma_g)(\gamma_g^{-1}x_S\gamma_g g_S, ww').$$

Hence $\Phi[xg]$ is represented by $\gamma_g^{-1}x_S\gamma_g g_S$ in $\Gamma\backslash\mathbf{G}_S$. □

The above proposition implies the following:

LEMMA 4.9 (Basic Lemma). *Let S be isotropic both for* $\mathbf{G}$ *and* $\mathbf{L}$*. For a sequence* $g \in \mathbf{L}(\mathbb{A})\backslash\mathbf{G}(\mathbb{A})$ *going to infinity, the following are equivalent:*

- *The translate* $\tilde{Y}_g := \mathbf{L}(\mathbb{Q})\backslash\mathbf{L}(\mathbb{A})g$ *becomes equidistributed in* $\tilde{X} := \mathbf{G}(\mathbb{Q})\backslash\mathbf{G}(\mathbb{A})$;
- *For any compact open subgroup W_S of* $\mathbf{G}(\mathbb{A}_S)$ *and the corresponding S-congruence subgroup* $\Gamma = \mathbf{G}(\mathbb{Q}) \cap W_S$*, the translate* $Y_g := \Gamma\backslash\Gamma\gamma_g^{-1}\mathbf{L}_S\gamma_g g_S$ *becomes equidistributed in* $X := \Gamma\backslash\mathbf{G}_S$.

PROOF. Let $\tilde{\mu}_g$, $\tilde{\mu}$, μ_g and μ be the invariant probability measures on $\tilde{Y}_g$, $\tilde{X}$, Y_g and X respectively. Note that $\bigcup_{W_S} C_c(\tilde{X})^{W_S}$, where the union is taken over all open compact subgroups W_S of $\mathbf{G}(\mathbb{A}_S)$, is dense in $C_c(\tilde{X})$, and that any function $\tilde{f} \in C_c(\tilde{X})^{W_S}$ corresponds to a unique function $f \in C_c((\mathbf{G}(\mathbb{Q}) \cap W_S)\backslash \mathbf{G}_S)$, and vice versa, by Proposition 4.8. Moreover

$$\tilde{\mu}_g(\tilde{f}) = \mu_g(f) \text{ and } \tilde{\mu}(\tilde{f}) = \mu(f).$$

Therefore the claim follows. □

In the following theorem, let $\mathbf{G}$ be a connected simply connected semisimple $\mathbb{Q}$-group and $\{\mathbf{L}_i\}$ a sequence of simply connected semisimple $\mathbb{Q}$-groups of $\mathbf{G}$. Suppose that S is *strongly* isotropic for all $\mathbf{L}_i$. Let Γ be an S-congruence subgroup of $\mathbf{G}(\mathbb{Q})$, and denote by μ_i the invariant probability measure supported on the closed orbit $\Gamma\backslash\Gamma\mathbf{L}_{i,S}$.

The following theorem is proved in [**35**], generalizing the works of Mozes-Shah [**46**], and of Eskin-Mozes-Shah ([**27**], [**25**]).

THEOREM 4.10. *Let $\{g_i \in \mathbf{G}_S\}$ be given.*

(1) *If the centralizer of each $\mathbf{L}_i$ is $\mathbb{Q}$-anisotropic, then $(g_i).\mu_i$ does not escape to infinity, that is, for any $\epsilon > 0$, there is a compact subset $\Omega \subset \Gamma\backslash\mathbf{G}_S$ such that*

$$(g_i).\mu_i(\Omega) > 1 - \epsilon \quad \textit{for all large } i.$$

(2) *If $\nu \in \mathcal{P}(\Gamma\backslash\mathbf{G}_S)$ is a weak-limit of $(g_i).\mu_i$, then the following hold:*
- $$\operatorname{supp}(\nu) = \Gamma\backslash\Gamma g\Lambda(\nu)$$
  where $\Lambda(\nu) := \{x \in \mathbf{G}_S : x.\nu = \nu\}$, $g \in \mathbf{G}_S$ and $g\Lambda(\nu)g^{-1}$ is a finite index subgroup of $\mathbf{M}_S$ for some connected $\mathbb{Q}$-group $\mathbf{M}$ with no non-trivial $\mathbb{Q}$-character.
- *For some $\gamma_i \in \Gamma$,*
  $$\gamma_i\mathbf{L}_i\gamma_i^{-1} \subset \mathbf{M}$$
  and for some $h_i \in \mathbf{L}_S$, the sequence $\gamma_i h_i g_i$ converges to g.

The proof of this theorem is based on the theory of unipotent flows on homogeneous spaces. To see which unipotent flows we use, pick $p \in S$ such that each $\mathbf{L}_i$ has no anisotropic $\mathbb{Q}_p$-factor. Since S is finite, we may assume the existence of such p, by passing to a subsequence if necessary. Then $\mathbf{L}_i(\mathbb{Q}_p)$ is generated by unipotent one-parameter subgroups in it and acts ergodically on each $\Gamma\backslash\Gamma\mathbf{L}_{i,S}$. From this, we deduce that there exists a one-parameter unipotent subgroup U_i in $\mathbf{L}_i(\mathbb{Q}_p)$ which acts ergodically on $\Gamma\backslash\Gamma\mathbf{L}_{i,S}$. Then $U_i' := g_i^{-1}U_ig_i$ acts ergodically on $\Gamma\backslash\Gamma\mathbf{L}_{i,S}g_i$ and the measure $(g_i).\mu_i$ is a U_i'-invariant ergodic measure. We then need to understand the limiting behavior of invariant ergodic measures on $\Gamma\backslash\mathbf{G}_S$ under unipotent one parameter subgroups.

The rest of the proof is then based on the generalization to the S-arithmetic setting of theorems of Mozes-Shah on limits of ergodic measures invariant under unipotent flows [**46**] and of Dani-Margulis [**16**] on the behavior of unipotent flows near cusps. Main ingredients of the proof are the measure classification theorem of Ratner [**52**], generalized in the S-arithmetic setting by Ratner [**53**] and independently by Margulis-Tomanov [**43**], and the linearization methods developed by Dani-Margulis [**17**].

Note that $g \to \infty$ in $\mathbf{L}(\mathbb{A})\backslash\mathbf{G}(\mathbb{A})$ if and only if $(g_S, \gamma_g) \to \infty$ in $\mathbf{L}_S\backslash\mathbf{G}_S \times \mathbf{L}(\mathbb{Q})\backslash\mathbf{G}(\mathbb{Q})/\Gamma$. Therefore, using the Basic lemma 4.9, Theorem 4.1 follows from the following corollary of Theorem 4.10:

COROLLARY 4.11. *Suppose that* $\mathbf{L}$ *is a maximal connected* $\mathbb{Q}$*-subgroup of* $\mathbf{G}$. *For* $(g_i, \delta_i) \to \infty$ *in* $\mathbf{L}_S\backslash\mathbf{G}_S \times \Gamma\backslash\mathbf{G}(\mathbb{Q})/\mathbf{L}(\mathbb{Q})$ *the sequence* $\Gamma\backslash\Gamma\delta_i\mathbf{L}_S g_i$ *becomes equidistributed in* $\Gamma\backslash\mathbf{G}_S$.

5. Well-rounded sequence and Counting rational points

Let $\mathbf{G}$ be a connected semisimple algebraic $\mathbb{Q}$-group and $\mathbf{L}$ a semisimple maximal connected $\mathbb{Q}$-subgroup of $\mathbf{G}$. Let $\mathbf{U} = \mathbf{L}\backslash\mathbf{G}$ and $u_0 = [\mathbf{L}]$.

Theorem 4.3 yields the following corollary by Proposition 3.5:

COROLLARY 5.1. *If* $\{B_T \subset \mathbf{L}(\mathbb{A})\backslash\mathbf{G}(\mathbb{A})\}$ *is a family of compact subsets which is* W*-well-rounded for some compact open subgroup* W *of* $\mathbf{G}(\mathbb{A}_f)$, *then*

$$\# u_0\mathbf{G}(\mathbb{Q}) \cap B_T \sim \nu(B_T \cap u_0 G_W)$$

provided $\nu(B_T \cap u_0 G_W) \to \infty$, *where* ν *is the invariant measure on* $u_0 G_W$ *which is compatible with invariant probability measures on* $\mathbf{G}(\mathbb{Q})\backslash G_W$ *and* $\mathbf{L}(\mathbb{Q})\backslash(G_W \cap \mathbf{L}(\mathbb{A}))$.

PROOF. In order to apply Proposition 3.5, we set $G = G_W$, $L = G_W \cap \mathbf{L}(\mathbb{A})$ for $\mathbf{L} = \mathrm{Stab}_{\mathbf{G}}(u_0)$ and $\Gamma = \mathbf{G}(\mathbb{Q})$. By Theorem 4.3, the translate $\Gamma\backslash\Gamma L x$ becomes equidistributed in $\Gamma\backslash G$ relative to $C_c(\Gamma\backslash G)^W$, as $x \to \infty$ in $L\backslash G$. Hence the claim follows from Proposition 3.5. □

5.1. Rational points on projective varieties. Let $\mathbf{G}$ be a connected semisimple algebraic $\mathbb{Q}$-group with a $\mathbb{Q}$-rational representation $\mathbf{G} \to \mathrm{GL}_{n+1}$. Let $\mathbf{U} := u_0\mathbf{G} \subset \mathbb{P}^n$ for some $u_0 \in \mathbb{P}^n(\mathbb{Q})$. Let $\mathbf{X} \subset \mathbb{P}^n$ denote the Zariski closure of $\mathbf{U}$, which is then a $\mathbf{G}$-equivariant compactification of $\mathbf{U}$. We assume that the stabilizer $\mathbf{L}$ in $\mathbf{G}$ of u_0 is connected and semisimple.

Let L be the line bundle of $\mathbf{X}$ given by the pull back of the line bundle $\mathcal{O}_{\mathbb{P}^n}(1)$. Then L is very ample and $\mathbf{G}$-linearized. Let $\mathrm{s}_0, \cdots, \mathrm{s}_n$ be the global sections of L obtained by pulling back the coordinate functions x_i's.

Recall that the height function H_{L} on $\mathbf{X}(\mathbb{Q})$ is then given as follows: for $x \in \mathbf{X}(\mathbb{Q})$,

$$\mathrm{H}_{\mathrm{L}}(x) := \mathrm{H}_{\mathcal{O}_{\mathbb{P}^n}(1)}(x) = \sqrt{x_0^2 + \cdots + x_n^2} \tag{5.2}$$

where $(x_0, \cdots, x_n)$ is a primitive integral vector proportional to $(\mathrm{s}_0(x), \cdots, \mathrm{s}_n(x))$. In order to extend H_{L} to $\mathbf{U}(\mathbb{A})$, we assume that there is a $\mathbf{G}$-invariant global section s of L such that $\mathbf{U} = \{\mathrm{s} \neq 0\}$.

DEFINITION 5.3. Define $\mathrm{H}_{\mathrm{L}} : \mathbf{U}(\mathbb{A}) \to \mathbb{R}_{>0}$ by

$$\mathrm{H}_{\mathrm{L}}(x) := \prod_{p \in R} \mathrm{H}_{\mathrm{L},p}(x_p) \quad \text{for } x = (x_p)$$

where

$$\mathrm{H}_{\mathrm{L},p}(x_p) = \begin{cases} \max_{0\le i\le n} \left|\frac{\mathrm{s}_i(x_p)}{\mathrm{s}(x_p)}\right|_p & \text{for } p \text{ finite} \\ \frac{\sqrt{\sum_i \mathrm{s}_i(x_\infty)^2}}{|\mathrm{s}(x_\infty)|_\infty} & \text{for } p=\infty \end{cases}.$$

Observe that this definition of H_L agrees with the one in (5.2) on $\mathbf{U}(\mathbb{Q})$, using the product formula $\prod_{p\in R}|\mathrm{s}(x)|_p = 1$ for $x\in\mathbf{U}(\mathbb{Q})$.

Set

$$B_T := \{x\in\mathbf{U}(\mathbb{A}) : \mathrm{H}_\mathrm{L}(x)\le T\};$$

$$W_\mathrm{L} := \{g\in\mathbf{G}(\mathbb{A}_f) : \mathrm{H}_\mathrm{L}(xg)=\mathrm{H}_\mathrm{L}(x) \quad \text{for all } x\in\mathbf{U}(\mathbb{A})\}.$$

Then W_L is an open compact subgroup of $\mathbf{G}(\mathbb{A}_f)$ under which B_T is invariant.

In view of Corollary 5.1, we would like to show that $B_T\cap u_0G_{W_\mathrm{L}}$ is W_L-well rounded. So far we have completely ignored any geometry of $\mathbf{X}$; here is the place where it enters into the counting problem of rational points.

In the following theorem, we suppose that there are only finitely many $\mathbf{G}(\mathbb{A})$-orbits in $\mathbf{U}(\mathbb{A})$. This finiteness condition is equivalent to that $\mathbf{G}(\mathbb{Q}_p)$ acts transitively on $\mathbf{U}(\mathbb{Q}_p)$ for almost all p, as well as to that there are only finitely many $\mathbf{G}(\mathbb{Q})$-orbits in $\mathbf{U}(\mathbb{Q})$. This is always satisfied if $\mathbf{L}$ is simply connected. Borovoi **[35]** classified symmetric spaces $\mathbf{L}\backslash\mathbf{G}$ with this finiteness property when $\mathbf{G}$ is absolutely simple.

THEOREM 5.4. *For any $x_0\in\mathbf{U}(\mathbb{A})$, we have*

(1) *the family $\{x_0G_{W_\mathrm{L}}\cap B_T\}$ is W_L-well rounded;*
(2) *for some $a\in\mathbb{Q}_{>0}$ and $b\in\mathbb{Z}_{\ge1}$ (explicitly given in terms of* $\operatorname{div}(\mathrm{s})$ *and the canonical class $\mathrm{K}_\mathbf{X}$),*

$$\operatorname{Vol}(x_0G_{W_\mathrm{L}}\cap B_T)\asymp T^a\log T^{b-1}$$

($\asymp$ means the ratio of the two sides is between bounded constants uniformly for all $T>1$).

To give some idea of the proof, consider the height zeta function on $x_0\mathbf{G}(\mathbb{A})$:

$$\eta(s) := \int_{x_0\mathbf{G}(\mathbb{A})} \mathrm{H}_\mathrm{L}(x)^{-s}\,d\nu(x).$$

Understanding the analytic properties of η provides the asymptotic growth of $\nu(x_0\mathbf{G}(\mathbb{A})\cap B_T)$ by Tauberian type argument. By the assumption on the finiteness, $x_0\mathbf{G}(\mathbb{Q}_p)=\mathbf{U}(\mathbb{Q}_p)$ for almost all p, and hence for some finite subset $S\subset R$,

$$\eta(s)=\prod_{p\in S}\int_{x_p\mathbf{G}(\mathbb{Q}_p)}\mathrm{H}_{\mathrm{L},p}^{-s}(y_p)\,d\nu_p(y_p)\cdot\prod_{p\notin S}\int_{\mathbf{U}(\mathbb{Q}_p)}\mathrm{H}_{\mathrm{L},p}^{-s}(y_p)\,d\nu_p(y_p)$$

where $\nu=\otimes^*\nu_p$ and $x=(x_p)$. Using the equivariant resolution of singularities and by passing to a finite field extension, we may assume that $\mathbf{X}$ is smooth with $\mathbf{X}\setminus\mathbf{U}$ a strict normal crossing divisor consisting of geometrically irreducible components:

$$\mathbf{X}\setminus\mathbf{U}=\cup_{\alpha\in\mathcal{A}}D_\alpha.$$

If ω denotes a nowhere zero differential form on $\mathbf{U}$ of top degree, we write

$$\operatorname{div}(\mathrm{s})=\sum_{\alpha\in\mathcal{A}}m_\alpha D_\alpha \quad\text{and}\quad -\operatorname{div}(\omega)=\sum_{\alpha\in\mathcal{A}}n_\alpha D_\alpha$$

for $m_\alpha \in \mathbb{N}$ and $n_\alpha \in \mathbb{Z}$. Each of the local integral $\int_{\mathbf{U}(\mathbb{Q}_p)} \mathrm{H}_{\mathrm{L},p}^{-s}(y_p)\, d\nu_p(y_p)$ admits a formula analogous to Denef's formula for Igusa zeta function. And putting these together, one can regularize $\eta(s)$ by the Dedekind zeta function and obtain that $\eta(s)$ has a meromorphic continuation to the half plane $\Re(s) \geq a - \epsilon$ for some $\epsilon > 0$ with a unique pole at $s = a$ of order b, where

$$a = \max_{\alpha \in \mathcal{A}} \frac{n_\alpha}{m_\alpha} \quad \text{and} \quad b = \#\{\alpha \in \mathcal{A} : \frac{n_\alpha}{m_\alpha} = a\}.$$

This argument has been carried out by Chambert-Loir and Tschinkel [**11**]. We have that $a > 0$ (see [**5**]) and by Tauberian argument that

$$\nu(x_0 \mathbf{G}(\mathbb{A}) \cap B_T) \sim c \cdot T^a (\log T)^{b-1}.$$

Note here that using the finiteness of $\mathbf{G}(\mathbb{A})$-orbits on $\mathbf{U}(\mathbb{A})$, the computation for the local integrals over $x_p \mathbf{G}(\mathbb{Q}_p)$ at almost all p becomes that over $\mathbf{X}(\mathbb{Q}_p)$ and hence a geometric problem. Without this assumption, one probably needs to use motivic integration.

Since $x_0 \mathbf{G}(\mathbb{A})$ can be covered by finitely many translates of $x_0 G_{W_{\mathrm{L}}}$, it is easy to deduce from here that $\nu(x_0 G_{W_{\mathrm{L}}} \cap B_T) \asymp T^a (\log T)^{b-1}$, although it does not yield the asymptotic equality. The W_{L}-well-roundedness of the sequence $x_0 G_{W_{\mathrm{L}}} \cap B_T$ does not immediately follow from this as well, but requires knowing a subtle Hölder property of local integral at ∞ (see Benoist-Oh [**3**], or Gorodnik-Nevo [**34**]).

THEOREM 5.5. *Assume that*

(i) $\mathbf{L}$ *is a maximal connected* $\mathbb{Q}$*-subgroup of* $\mathbf{G}$*;*
(ii) *there are only finitely many* $\mathbf{G}(\mathbb{A})$*-orbits in* $\mathbf{U}(\mathbb{A})$.

Then

(1) *for any* $u_0 \in \mathbf{U}(\mathbb{Q})$,

$$\#\{x \in u_0 \mathbf{G}(\mathbb{Q}) : \mathrm{H}_{\mathrm{L}}(x) < T\} \sim \nu(B_T \cap u_0 G_{W_{\mathrm{L}}})$$

where ν *is the invariant measure on* $u_0 G_{W_{\mathrm{L}}}$ *which is compatible with invariant probability measures on* $\mathbf{G}(\mathbb{Q}) \backslash G_{W_{\mathrm{L}}}$ *and* $\mathbf{L}(\mathbb{Q}) \backslash (G_{W_{\mathrm{L}}} \cap \mathbf{L}(\mathbb{A}))$.

(2) *there exist* $a \in \mathbb{Q}_{>0}$ *and* $b \in \mathbb{N}$ *(explicitly given in terms of* $\operatorname{div}(s)$ *and the canonical class of* $\mathbf{X}$*) such that*

$$\#\{x \in \mathbf{U}(\mathbb{Q}) : \mathrm{H}_{\mathrm{L}}(x) < T\} \asymp T^a \log T^{b-1}.$$

Generalizing the work of De Concini and Procesi [**18**] on the construction of the wonderful compactification of symmetric varieties, Luna introduced in [**42**] the notion of a wonderful variety: a smooth connected projective $\mathbf{G}$-variety $\mathbf{X}$ is called *wonderful* of rank l if (1) $\mathbf{X}$ contains l irreducible $\mathbf{G}$-invariant divisors with strict normal crossings, (2) $\mathbf{G}$ has exactly 2^l-orbits in $\mathbf{X}$. In particular, a wonderful variety is of Fano type. For a $\mathbf{G}$-homogeneous variety $\mathbf{U}$, a wonderful variety $\mathbf{X}$ is called the wonderful compactification of $\mathbf{U}$ if it is a $\mathbf{G}$-equivariant compactification of $\mathbf{U}$. Using the work of Brion on the computation of $\operatorname{Pic}(\mathbf{X})$ and $\Lambda_{\mathrm{eff}}(\mathbf{X})$ etc., we can verify that $a = a_{\mathrm{L}}$ and $b = b_{\mathrm{L}}$ as predicted by Manin and the height function H_{L} associated to a very ample line bundle L of $\mathbf{X}$ arises as described in the beginning of (5.1) provided $\mathbf{L}$ has finite index in its normalizer. Therefore we deduce the following special case of Manin's conjecture.

THEOREM 5.6. *Under the same assumption as in Theorem 5.5, let* $\mathbf{X}$ *be a wonderful variety. Then for any ample line bundle* L *of* $\mathbf{X}$ *over* $\mathbb{Q}$,

$$\#\{x \in \mathbf{U}(\mathbb{Q}) : \mathrm{H_L}(x) < T\} \asymp T^{a_{\mathrm{L}}} \log T^{b_{\mathrm{L}}-1}$$

where a_{L} *and* b_{L} *are as in Manin's conjecture. Moreover if* $\mathbf{G}$ *is simply connected, there exists* $c > 0$ *such that*

$$\#\{x \in \mathbf{U}(\mathbb{Q}) : \mathrm{H_L}(x) < T\} \sim c \cdot T^{a_{\mathrm{L}}} \log T^{b_{\mathrm{L}}-1}.$$

The above theorem applies to the wonderful compactification of a connected adjoint semisimple algebraic group $\mathbf{G}$, since $\mathbf{G}$ can be identified with $\Delta(\mathbf{G})\backslash\mathbf{G}\times\mathbf{G}$ where $\Delta(\mathbf{G})$ is the diagonal embedding of $\mathbf{G}$ into $\mathbf{G}\times\mathbf{G}$. This case was previously obtained in [**58**] with rate of convergence and an alternative approach was given in [**33**] (see [**61**] for the comparison of two methods).

5.2. Rational points on affine varieties. The main difference of the counting problem between an affine variety $\mathbf{V}$ and a Zariski open subset $\mathbf{U}$ of a projective variety $\mathbf{X}$ lies in the way of defining a height function. Recall that a height function on $\mathbf{U}$ is obtained by pulling back the height function on the projective space into which $\mathbf{X}$ is embedded. For an affine variety $\mathbf{V}$ in an affine n-space, one could also try to embed it into a projective space and use the height function there. However it is natural to ask if the following definition of a height works: for $x = (x_p) \in \mathbf{V}(\mathbb{A})$,

$$\mathrm{H}(x) := \prod_{p\in R} \|x_p\|_p,$$

where $\|\cdot\|_\infty$ is the Euclidean norm on $\mathbb{R}^n$, and $\|\cdot\|_p$ is the p-adic maximum norm on $\mathbb{Q}_p^n$ for each finite p.

For a general affine variety $\mathbf{V}$ (for instance, for the affine n-space), this may not be well-defined. We discuss the case of homogeneous affine varieties in the following.

Let $\mathbf{G}$ be a connected semisimple algebraic $\mathbb{Q}$-group with a $\mathbb{Q}$-rational representation $\mathbf{G} \to \mathrm{GL}_n$. Fix a non-zero vector $v_0 \in \mathbb{Q}^n$ such that the orbit $\mathbf{V} = v_0\mathbf{G}$ is an affine $\mathbb{Q}$-subvariety. We assume that the stabilizer $\mathbf{L}$ in $\mathbf{G}$ of v_0 is a semisimple maximal connected $\mathbb{Q}$-subgroup of $\mathbf{G}$.

LEMMA 5.7. *For almost all* p,

$$\delta_p := \min_{x\in\mathbf{V}(\mathbb{Q})} \mathrm{H}_p(x) := \|x\|_p \geq 1.$$

PROOF. It is well known that there exists a $\mathbf{G}$-invariant non-zero homogeneous polynomial f with integral coefficients, that is, $\mathbf{V} \subset \{f = r\}$ for some $r \in \mathbb{Q}\setminus\{0\}$. Now for any p coprime to r as well as to the coefficients of f, we claim that $\delta_p \geq 1$. Suppose not; for some $x \in \mathbf{V}(\mathbb{Q})$, p divides each coordinate of x. Write $x = p^k x'$ where $k \geq 1$ and the denominator of any coordinate of x' is divisible by p. Now if d is the degree of f, $f(x) = p^{kd} f(x') = r$ and $|f(x')|_p \leq 1$. Hence $|r|_p \leq p^{-kd}$, yielding contradiction. □

We write an element of $\mathbf{V}(\mathbb{Q})$ as $\left(\frac{x_1}{x_0}, \cdots, \frac{x_n}{x_0}\right)$ where $x_0, \cdots, x_n \in \mathbb{Z}$, $x_0 > 0$ and $\mathrm{g.c.d}(x_0, \cdots, x_n) = 1$.

LEMMA 5.8. *For* $x \in \mathbf{V}(\mathbb{Q})$,

$$\mathrm{H}(x) \asymp \sqrt{x_0^2 + \cdots + x_n^2},$$

in the sense that the ratio is between two bounded constants uniformly for all $x \in \mathbf{V}(\mathbb{Q})$.

PROOF. By the product formula,

$$\begin{aligned}
\mathrm{H}(x) &= \prod_p |x_0|_p \cdot \mathrm{H}(x) \\
&\leq \prod_p |x_0|_p \sqrt{\sum_{i=1}^n \frac{x_i^2}{x_0^2} + 1} \cdot \prod_{p \in R_f} \max_{1 \leq i \leq n} \{\frac{|x_i|_p}{|x_0|_p}, 1\} \\
&\leq \sqrt{\sum_{0 \leq i \leq n} x_i^2} \cdot \prod_{p \in R_f} \max_{0 \leq i \leq n} |x_i|_p = \sqrt{\sum_{0 \leq i \leq n} x_i^2}.
\end{aligned}$$

By Lemma 5.7 and its proof,

$$\|x\|_p = \max_i \{\frac{|x_i|_p}{|x_0|_p}, \delta_p\}$$

for some $0 < \delta_p \leq 1$ which is 1 for almost all p.

Using $\delta_\infty > 0$, we can show that there exists $0 < C < 1$ such that

$$\sum_{i=1}^n y_i^2 \geq C^2 (\sum_{i=1}^n y_i^2 + 1)$$

for any $(y_1, \cdots, y_n) \in \mathbf{V}(\mathbb{R})$. Hence

$$\begin{aligned}
\mathrm{H}(x) &\geq C \prod_p |x_0|_p \sqrt{\sum_{i=1}^n \frac{x_i^2}{x_0^2} + 1} \cdot \prod_{p \in R_f} \max_{1 \leq i \leq n} \{\frac{|x_i|_p}{|x_0|_p}, \delta_p\} \\
&\geq (C \prod_p \delta_p) \cdot \sqrt{\sum_{0 \leq i \leq n} x_i^2} \cdot \prod_{p \in R_f} \max_{0 \leq i \leq n} |x_i|_p \\
&= (C \prod_p \delta_p) \cdot \sqrt{\sum_{0 \leq i \leq n} x_i^2}.
\end{aligned}$$

This proves the claim. □

THEOREM 5.9. *Suppose that there are only finitely many* $\mathbf{G}(\mathbb{A})$*-orbits in* $\mathbf{V}(\mathbb{A})$. *Then for some* $a \in \mathbb{Q}_{>0}$ *and* $b \in \mathbb{Z}_{\geq 1}$,

$$\begin{aligned}
&\#\{x \in \mathbf{V}(\mathbb{Q}) : \mathrm{H}(x) < T\} \\
&\asymp \#\{\left(\frac{x_1}{x_0}, \cdots, \frac{x_n}{x_0}\right) \in \mathbf{V}(\mathbb{Q}) : \sqrt{x_0^2 + \cdots + x_n^2} < T\} \\
&\asymp T^a (\log T)^{b-1}.
\end{aligned}$$

PROOF. To deduce this from Theorem 5.5, consider the embedding of GL_n into GL_{n+1} by $A \mapsto \mathrm{diag}(A, 1)$, and of $\mathbf{V}$ into the projective space $\mathbb{P}^n$ by $(x_1, \cdots, x_n) \mapsto [x_1 : \cdots : x_n : 1]$. This identifies $\mathbf{V}$ with the orbit $\mathbf{U} := [(v_0 : 1)]\mathbf{G}$ and $\mathrm{s} = x_{n+1}$ is an invariant section of the line bundle L obtained by pulling back $\mathcal{O}_{\mathbb{P}^n}(1)$, satisfying $\mathbf{U} = \{\mathrm{s} \neq 0\}$. Note for $x \in \mathbf{V}(\mathbb{Q})$,

$$\mathrm{H}_{\mathrm{L}} \left(\frac{x_1}{x_0} : \cdots : \frac{x_n}{x_0} : 1\right) = \mathrm{H}_{\mathrm{L}}(x_1 : \cdots : x_n : x_0) = \sqrt{x_0^2 + \cdots + x_n^2}.$$

Therefore this theorem is a special case of Corollary 5.5. □

When $\mathbf{G}$ is simply connected, we can replace $\asymp$ with $\sim$ in Theorem 5.4, and hence obtain the following example. Since $\mathbf{V} = \{x \in \mathrm{SL}_{2n} : x^t = -x\}$ is a homogeneous variety $\mathrm{Sp}_{2n} \setminus \mathrm{SL}_{2n}$ for the action $v.g = g^t v g$, we have:

EXAMPLE 5.10. *Let $n \geq 2$. For some $a \in \mathbb{Q}^+$, $b \in \mathbb{Z}_{\geq 0}$ and $c > 0$, as $T \to \infty$,*

$$\#\{x \in \mathrm{SL}_{2n}(\mathbb{Q}) : x^t = -x, \max_{1 \leq i,j \leq 2n} \{|x_{ij}|, |x_0|\} < T\} \sim c \cdot T^a (\log T)^{b-1}.$$

where $x = \left(\frac{x_{ij}}{x_0}\right)$, $x_{ij} \in \mathbb{Z}$, $x_0 \in \mathbb{N}$ and $\mathrm{g.c.d}\{x_{ij}, x_0 : 1 \leq i,j \leq 2n\} = 1$.

S-integral points: We keep the same assumption on $\mathbf{V}$ from 5.2. Let S be a finite set of primes containing ∞, and consider the following S-height function on $\mathbf{V}_S := \prod_{p \in S} \mathbf{V}(\mathbb{Q}_p)$: for $x = (x_p)_{p \in S} \in \mathbf{V}_S$,

$$\mathrm{H}_S(x) := \prod_{p \in S} \|x_p\|_p, \tag{5.11}$$

where $\|\cdot\|_\infty$ is the Euclidean norm on $\mathbb{R}^n$, and $\|\cdot\|_p$ is the p-adic maximum norm on $\mathbb{Q}_p^n$.

Set

$$B_S(T) := \{x \in \mathbf{V}_S : \mathrm{H}_S(x) < T\}.$$

The following is obtained in [**3**, Prop. 8.11] for any $\mathbb{Q}$-algebraic group $\mathbf{G}$ and a closed orbit $\mathbf{V} = v_0 \mathbf{G}$.

THEOREM 5.12. *For any $v_0 \in \mathbf{V}_S$, the family $B_S(T) \cap v_0 \mathbf{G}_S$ is well-rounded and*

$$\mathrm{vol}(B_S(T) \cap v_0 \mathbf{G}_S) \asymp T^a (\log T)^b$$

for some $a \in \mathbb{Q}_{>0}$ and $b \in \mathbb{Z}_{\geq 0}$.

When $S = \{\infty\}$, we can replace $\asymp$ with $\sim$.

The notation $\mathbb{Z}_S$ is the subring of $\mathbb{Q}$ consisting of elements whose denominators are prime to all $p \notin S$. Hence if $S = \{\infty\}$, then $\mathbb{Z}_S = \mathbb{Z}$.

If $\Gamma \subset \mathbf{G}(\mathbb{Q})$ is an S-congruence subgroup which preserves $\mathbf{V}(\mathbb{Z}_S)$, there are only finitely many Γ-orbits in $\mathbf{V}(\mathbb{Z}_S)$, say, $v_1\Gamma, \cdots, v_l\Gamma$. Set $\mathbf{L}_i = \mathrm{Stab}_{\mathbf{G}}(v_i)$ and $\mathbf{L}_{i,S} = \prod_{p \in S} \mathbf{L}_i(\mathbb{Q}_p)$.

COROLLARY 5.13. *Suppose that S is strongly isotropic for each $\mathbf{L}_i$. Then*

$$\#\{x \in \mathbf{V}(\mathbb{Z}_S) : \mathrm{H}_S(x) < T\} \asymp T^a (\log T)^b.$$

PROOF. Since $\pi(\tilde{\mathbf{G}}_S)$ is a finite index normal subgroup of $\mathbf{G}_S$ for the simply connected cover $\pi : \tilde{\mathbf{G}} \to \mathbf{G}$, its proof immediately reduces to the case when both $\mathbf{G}$ and $\mathbf{L}$ are simply connected groups.

Then Proposition 3.5 and Corollary 4.11 imply that

$$\#\{x \in \mathbf{V}(\mathbb{Z}_S) : \mathrm{H}_S(x) < T\} \sim_T \sum_{i=1}^{l} \frac{\mathrm{vol}((\Gamma \cap \mathbf{L}_{i,S}) \backslash \mathbf{L}_{i,S})}{\mathrm{vol}(\Gamma \backslash \tilde{\mathbf{G}}_S)} \mathrm{vol}(B_S(T) \cap v_i \mathbf{G}_S). \tag{5.14}$$

Hence the claim follows from Theorem 5.12. □

The asymptotic (5.14) in the case $S = \{\infty\}$ (and hence the integral points case) was proved by Eskin-Mozes-Shah [**27**] using the unipotent flows. Their result is more general since they also deal with maximal reductive (non-semisimple) $\mathbb{Q}$-groups. We believe Corollary 5.13 should hold only assuming that the stabilizer of v_0 is a reductive maximal $\mathbb{Q}$-subgroup, extending the argument [**27**] to the S-arithmetic setting. When $\mathbf{V}$ is symmetric and $S = \{\infty\}$, (5.14) was proved even earlier by Duke-Rudnick-Sarnak [**21**] and by Eskin-McMullen [**26**] (see section 8 for more discussion.)

5.3. Equidistribution. Suppose that $\mathbf{G}$ and $\mathbf{L}$ are simply connected. Then the connected components of $\mathbf{V}(\mathbb{R})$ are precisely $\mathbf{G}(\mathbb{R})$-orbits and $\mathbf{G}(\mathbb{A}_f)$ acts transitively on $\mathbf{V}(\mathbb{A}_f)$. Fix a compact subset $\Omega \subset v_0\mathbf{G}(\mathbb{R})$ with smooth boundary.

COROLLARY 5.15. *As $m \to \infty$, subject to $b_m \neq \emptyset$,*

$$\#\{x \in \mathbf{V}(\mathbb{Q}) : x \in \Omega, \text{ denominator of } x \text{ is } m\} \sim \mathrm{vol}(\Omega) \times \mathrm{vol}(b_m)$$

where for $m = p_1^{k_1} \cdots p_r^{k_r}$

$$b_m = \{(x_p) \in \mathbf{V}(\mathbb{A}_f) : \|x_{p_i}\|_{p_i} = p_i^{k_i} \ \forall 1 \le i \le r, \ x_p \in \mathbf{V}(\mathbb{Z}_p) \quad \text{for all } p \neq p_i\}.$$

PROOF. The above follows from the observation that the family $\Omega \times b_m$ is W-well-rounded for any compact open subgroup W of $\mathbf{G}(\mathbb{A}_f)$ which preserves $\prod_{p \in R_f} \mathbf{V}(\mathbb{Z}_p)$ □

The above corollary implies that the rational points in $\mathbf{V}$ with denominator m are equidistributed on $\mathbf{V}(\mathbb{R})$ as $m \to \infty$.

6. Mixing and Hecke points

In this section, we will discuss an ergodic theoretic proof of the mixing of adelic groups as a special case of Theorem 4.1, and show that the adelic mixing is equivalent to the equidistribution of Hecke points together with the mixing of a finite product of corresponding local groups.

6.1. Mixing. We begin by recalling the notion of mixing in the homogeneous case. Let G be a locally compact second countable group and Γ a lattice in G. Let dx denote the probability invariant measure on $\Gamma\backslash G$. The group G acts on $L^2(\Gamma\backslash G)$ by $g\psi(x) = \psi(xg)$ for $g, x \in G, \psi \in L^2(\Gamma\backslash G)$.

DEFINITION 6.1. The right translation action of G on the space $\Gamma\backslash G$ is called *mixing* if for any $\psi, \phi \in L^2(\Gamma\backslash G)$,

$$\langle g_i\psi, \phi\rangle = \int_{\Gamma\backslash G} \psi(xg_i)\phi(x)\, dx \to \int \psi dx \cdot \int \phi\, dx$$

for any $g_i \in G$ going to infinity.

We need the following well known lemma: we denote by $\Delta(G)$ the diagonal embedding of G into $G \times G$.

LEMMA 6.2. *The following are equivalent:*

- *the right translation action of G on $\Gamma\backslash G$ is mixing;*
- *for any sequence $g \to \infty$ in G, the translate $\Delta(\Gamma)\backslash\Delta(G)(e, g)$ becomes equidistributed in $(\Gamma \times \Gamma)\backslash G \times G$.*

PROOF. Observe that for $\psi, \phi \in C_c(\Gamma\backslash G)$,

$$\langle g\psi, \phi\rangle = \int_{x\in\Gamma\backslash G} \psi(xg)\phi(x)dx = \int_{(x,x)\in\Delta(\Gamma)\backslash\Delta(G)} (\psi\otimes\phi)(xg, x)dx.$$

Since the set of finite linear combinations of $\psi\otimes\phi$, $\psi, \phi \in C_c(\Gamma\backslash G)$ is dense in $C_c((\Gamma\times\Gamma)\backslash(G\times G))$, the claim follows. □

6.2. Hecke orbits. Denote by $\mathrm{Comm}(\Gamma) < G$ the commensurator group of Γ, that is, $a\in \mathrm{Comm}(\Gamma)$ if and only if $a\Gamma a^{-1}\cap\Gamma$ has a finite index both in Γ and $a\Gamma a^{-1}$.

DEFINITION 6.3 (Hecke orbits). If $a\in\mathrm{Comm}(\Gamma)$,

$$\mathrm{T}_\Gamma(a) := \Gamma\backslash\Gamma a\Gamma$$

is called the Hecke orbit associated to a.

Using the bijection $\Gamma\backslash\Gamma a\Gamma = \Gamma\cap a^{-1}\Gamma a\backslash\Gamma$ given by $[a]\gamma\mapsto[\gamma]$, we have

$$\deg_\Gamma(a) := \#\,\mathrm{T}_\Gamma(a) = [\Gamma : \Gamma\cap a^{-1}\Gamma a].$$

EXAMPLE 6.4. For $\Gamma = \mathrm{SL}_2(\mathbb{Z})$ and $a = \mathrm{diag}(p, p^{-1})$, $\Gamma\backslash\Gamma a\Gamma$ is in bijection with $\mathrm{SL}_2(\mathbb{Z}_p)a^{-1}\,\mathrm{SL}_2(\mathbb{Z}_p)/\,\mathrm{SL}_2(\mathbb{Z}_p) = \mathrm{SL}_2(\mathbb{Z}_p)a\,\mathrm{SL}_2(\mathbb{Z}_p)/\,\mathrm{SL}_2(\mathbb{Z}_p)$, that is, the $\mathrm{SL}_2(\mathbb{Z}_p)$-orbit of a^{-1} in the Bruhat-Tits tree. Hence $\mathrm{T}_\Gamma(a)$ corresponds to the $p(p+1)$ vertices in the $p+1$-regular tree of distance 2 from the vertex $x_0 := \mathbb{Z}_p\oplus\mathbb{Z}_p$. For $a = (\sqrt{p}, \sqrt{p}^{-1})$, $\mathrm{T}_\Gamma(a)$ gives $p+1$ vertices of distance 1 from x_0.

The following observation was first made in a paper by Burger and Sarnak [**9**].

LEMMA 6.5. *For a sequence $a_i\in\mathrm{Comm}(\Gamma)$, the following are equivalent:*

- *the Hecke orbit $\mathrm{T}_\Gamma(a_i)$ is equidistributed in $\Gamma\backslash G$ as $i\to\infty$, that is, for any $\psi\in C_c(\Gamma\backslash G)$,*

$$\frac{1}{\deg_\Gamma(a_i)}\sum_{x\in\mathrm{T}_\Gamma(a_i)}\psi(x)\to\int_{\Gamma\backslash G}\psi\,dx.$$

- *the orbit $[(e, a_i^{-1})]\Delta(G)$ becomes equidistributed in $(\Gamma\times\Gamma)\backslash(G\times G)$ as $i\to\infty$.*

PROOF. We use the homeomorphism between the space of Γ-invariant probability measures on $\Gamma\backslash G$ and the space of $\Delta(G)$-invariant probability measures on $(\Gamma\times\Gamma)\backslash(G\times G)$ given by $\mu\mapsto\tilde\mu$ where

$$\tilde\mu(f) = \int_{\Gamma\backslash G}\int_{\Gamma\backslash G} f(y, xy)d\mu(x)dy$$

(cf. [**4**, Prop. 8.1]). If μ_a is the probability measure which is the average of the dirac measures of the Hecke point $\mathrm{T}_\Gamma(a)$, then $\tilde\mu_a$ is the $\Delta(G)$-invariant measure supported on the orbit $(\Gamma\times\Gamma)\backslash(e, a_i^{-1})\Delta(G)$. Hence the claim follows. □

Moreover,

$$\deg_\Gamma(a) = \mathrm{vol}(\Gamma\cap a^{-1}\Gamma a\backslash G) = \mathrm{vol}((\Gamma\times\Gamma)\backslash(1, a)\Delta(G))$$

where the volumes are induced by the Haar measure on G which gives volume 1 for $\Gamma\backslash G$.

6.3. Adelic mixing. Let $\mathbf{G}$ be a connected semisimple $\mathbb{Q}$-group. We will deduce the mixing of $\mathbf{G}(\mathbb{A})$ on $\mathbf{G}(\mathbb{Q})\backslash\mathbf{G}(\mathbb{A})$ from Theorem 4.1 for $\mathbf{G}$ simply connected and $\mathbb{Q}$-simple.

Fix a finite set S of primes, which contains ∞ if $\mathbf{G}(\mathbb{R})$ is non-compact.

Fixing an imbedding of $\mathbf{G}$ into GL_n, we set

$$\mathrm{H}_S(x) := \prod_{p\in S} \max |x(p)_{ij}|_p$$

where $x = (x(p))_{p\in S} \in \mathbf{G}_S$.

THEOREM 6.6. *Let $\mathbf{G}$ be connected simply connected and almost $\mathbb{Q}$-simple. The following equivalent statements hold for any $\mathbf{G}$-isotropic subset S:*

1. *The right translation action of $\mathbf{G}(\mathbb{A})$ on $\mathbf{G}(\mathbb{Q})\backslash\mathbf{G}(\mathbb{A})$ is mixing.*
2. *For any $g_i \to \infty$ in $\mathbf{G}(\mathbb{A})$, the translate*
$$[(e,e)]\Delta(\mathbf{G}(\mathbb{A}))(e,g_i)$$
becomes equidistributed in $\tilde{X}\times\tilde{X}$.
3. *For any S-congruence subgroup Γ of $\mathbf{G}(\mathbb{Q})$, $a_i \in \mathbf{G}(\mathbb{Q})$ and $x_i \in \mathbf{G}_S$ such that $\deg_\Gamma(a_i)\cdot \mathrm{H}_S(x_i) \to \infty$, the closed orbit*
$$[(e,a_i)]\Delta(\mathbf{G}_S)(e,x_i)$$
becomes equidistributed on $(\Gamma\times\Gamma)\backslash(\mathbf{G}_S\times\mathbf{G}_S)$.
4. *For any S-congruence subgroup Γ of $\mathbf{G}(\mathbb{Q})$, and $a_i \in \mathbf{G}(\mathbb{Q})$ with $\deg_\Gamma(a_i) \to \infty$,*
   - *the Hecke orbit $\mathrm{T}_\Gamma(a_i)$ becomes equidistributed in $\Gamma\backslash\mathbf{G}_S$;*
   - *the right translation action of $\mathbf{G}_S$ on $\Gamma\backslash\mathbf{G}_S$ is mixing.*

PROOF. The condition $\mathbf{G}$ being $\mathbb{Q}$-simple implies that the diagonal embedding of $\mathbf{G}$ into $\mathbf{G}\times\mathbf{G}$ is a maximal connected $\mathbb{Q}$-group. Hence (2) follows from Theorem 4.1.

The equivalence between (2) and (3) comes from the basic lemma 4.9. (1) and (2) are equivalent by Lemma 6.2. (3) and (4) are equivalent by Lemmas 6.2 and 6.5. □

REMARK 6.7. Since (2) follows from Theorem 4.1 which is proved using the unipotent flows on S-arithmetic setting, we have obtained by the equivalence of (1) and (2) an ergodic theoretic proof of the adelic mixing.

For S as above, the mixing of $\mathbf{G}_S$ on $\Gamma\backslash\mathbf{G}_S$ is a well-known consequence of the Howe-Moore theorem [**39**] on the decay of matrix coefficients for $\mathbf{G}(\mathbb{Q}_p)$'s, $p\in S$.

Hence the above theorem says that the adelic mixing is a consequence of the Howe-Moore theorem for $\mathbf{G}_S$ together with the equidistribution of Hecke points for *all* S-congruence subgroups for some fixed isotropic subset S (and hence for all isotropic S).

The equidistribution of Hecke points was obtained with a rate in [**14**] except for one case of some $\mathbb{Q}$-anisotropic form of a special unitary group. This last obstruction was removed by Clozel soon afterwards [**13**]. For $S=\{\infty\}$, a different proof for the equidistribution was given in [**28**] (without rates), using a theorem of Mozes-Shah [**46**] on unipotent flows.

The adelic mixing theorem can also be deduced from the property of the automorphic spectrum $L^2(\mathbf{G}(\mathbb{Q})\backslash\mathbf{G}(\mathbb{A}))$ based on the work of [**14**], [**48**], [**13**] and this approach is explained in the paper [**33**] and gives a rate of convergence for the mixing.

7. Bounds toward the Ramanujan conjecture on the automorphic spectrum

In this section, we discuss a quantitative adelic mixing statement and how they can be understood in view of the Ramanujan conjecture concerning the automorphic spectrum. We refer to [**55**] and [**12**] for the background on the Ramanujan conjecture.

We assume that $\mathbf{G}$ is a connected and absolutely simple $\mathbb{Q}$-group (e.g., $\mathbf{G} = \mathrm{SL}_n$ or PGL_n).

Note that the right translation action of $\mathbf{G}(\mathbb{A})$ on $\mathbf{G}(\mathbb{Q})\backslash\mathbf{G}(\mathbb{A})$ defines a unitary representation ρ of $\mathbf{G}(\mathbb{A})$ on the Hilbert space $L^2(\mathbf{G}(\mathbb{Q})\backslash\mathbf{G}(\mathbb{A}))$, and hence a unitary representation of $\mathbf{G}(\mathbb{Q}_p)$ for each $p \in R$. Roughly speaking, the automorphic dual $\hat{\mathrm{Aut}}(\mathbf{G})_p$ of $\mathbf{G}(\mathbb{Q}_p)$ is the closure in the unitary dual of $\mathbf{G}(\mathbb{Q}_p)$ of the subset consisting of all irreducible constituents of the unitary representation $\rho|_{\mathbf{G}(\mathbb{Q}_p)}$.

A most sophisticated form of the Ramanujan conjecture is an attempt to identify $\hat{\mathrm{Aut}}(\mathbf{G})_p$ in the unitary dual $\hat{\mathbf{G}}(\mathbb{Q}_p)$. In the case of $\mathbf{G} = \mathrm{PGL}_2$ over $\mathbb{Q}$, the Ramanujan conjecture says that for each p, any infinite dimensional irreducible representation in $\hat{\mathrm{Aut}}(\mathbf{G})_p$ is strongly $L^{2+\epsilon}(\mathrm{PGL}_2(\mathbb{Q}_p))$ for any $\epsilon > 0$.

Definition 7.1. .

- A unitary representation ρ of $\mathbf{G}(\mathbb{Q}_p)$ is strongly $L^{p+\epsilon}$ if there exists a dense subset of vectors v, w such that the matrix coefficient function $\mathbf{G}(\mathbb{Q}_p) \to \mathbb{C}$ defined by
$$g \mapsto \langle \rho(g)v, w\rangle$$
is $L^{p+\epsilon}(\mathbf{G}(\mathbb{Q}_p))$-integrable for all $\epsilon > 0$.
- A unitary representation ρ is tempered if ρ is strongly $L^{2+\epsilon}$.

By the classification of the unitary dual of $\mathrm{PGL}_2(\mathbb{Q}_p)$ we know that there exists an irreducible unitary representation ρ_p of $\mathrm{PGL}_2(\mathbb{Q}_p)$ which is not L^{m_p}-integrable for arbitrary large m_p. By forming a (restricted) tensor product $\otimes'_{p\in R}\rho_p$, one can construct an irreducible unitary representation of $\mathrm{PGL}_2(\mathbb{A})$. A point made by the Ramanujan conjecture is that such a unitary representation cannot arise as a nontrivial irreducible constituent of $L^2(\mathrm{PGL}_2(\mathbb{Q})\backslash \mathrm{PGL}_2(\mathbb{A}))$ if any of m_p is strictly larger than 2.

The following theorem says that in the case of $\mathbb{Q}$-rank at least 2, there is an obstruction of forming such a unitary representation of $\mathbf{G}(\mathbb{A})$ already in the level of unitary dual of $\mathbf{G}(\mathbb{Q}_p)$.

Theorem 7.2. *([**48**], [**47**]) Suppose that* $\mathrm{rank}_{\mathbb{Q}}(\mathbf{G}) \geq 2$ *(e.g.,* PGL_n, $n \geq 3$*). Then there exists a positive number* m *(explicit and independent of* p*) such that any infinite dimensional irreducible unitary representation of* $\mathbf{G}(\mathbb{Q}_p)$ *is strongly* L^m *for all* $p \in R$.

To each root α of a maximal $\mathbb{Q}_p$-split torus of G, we associate the algebraic subgroup H_α (isomorphic either to $\mathrm{PGL}_2(\mathbb{Q}_p)$ or to $\mathrm{SL}_2(\mathbb{Q}_p)$) generated by the one-dimensional root subgroups $N_{\pm\alpha}$. The key step of the proof is then to show that for any irreducible unitary representation ρ of $\mathbf{G}(\mathbb{Q}_p)$, the restriction to H_α

is tempered. In showing this, the main tool is Mackey's theory on the unitary representations of the semi-direct product $\mathrm{PGL}_2(\mathbb{Q}_p) \ltimes U_p$ (or $\mathrm{SL}_2(\mathbb{Q}_p) \times U_p$) for some non-trivial unipotent algebraic group U_p. In the case when $\mathbb{Q}_p$-rank is at least 2, one can always find such U_p so that H_α sits inside $H_\alpha \ltimes U_p \subset \mathbf{G}(\mathbb{Q}_p)$. Once we have a bound for those H_α's, we make use of the properties of tempered representations to extend the bound to the whole group $\mathbf{G}(\mathbb{Q}_p)$ [**48**].

The following is obtained in [**33**]: Choose a height function on $\mathbf{G}(\mathbb{A})$, for instance,

$$\mathrm{H}(g) := \prod_{p\in R} \max |(g_p)_{ij}|_p \quad \text{for } g = (g_p)_p \in \mathbf{G}(\mathbb{A})$$

using some $\mathbb{Q}$-embedding of $\mathbf{G} \to \mathrm{SL}_n$.

We write

$$L^2(\mathbf{G}(\mathbb{Q})\backslash\mathbf{G}(\mathbb{A})) = L^2_{00}(\mathbf{G}(\mathbb{Q})\backslash\mathbf{G}(\mathbb{A})) \oplus \hat{\bigoplus}_{\chi\in\Lambda} \mathbb{C}\chi$$

where Λ denotes the set of all automorphic characters. Hence the Hilbert space $L^2_{00}(\mathbf{G}(\mathbb{Q})\backslash\mathbf{G}(\mathbb{A}))$ denotes the orthogonal complement to the subspace spanned by automorphic characters. If $\mathbf{G}$ is simply connected, Λ has only the trivial character.

THEOREM 7.3 (Adelic mixing). [**33**] *Fix a maximal compact subgroup K of $\mathbf{G}(\mathbb{A})$. There exists $k > 0$ (explicit) and $c \geq 1$ such that for any K-invariant $\psi_1, \psi_2 \in L^2_{00}(\mathbf{G}(\mathbb{Q})\backslash\mathbf{G}(\mathbb{A}))$, we have*

$$|\langle g\psi_1, \psi_2\rangle| \leq c \cdot \mathrm{H}(g)^{-k}\|\psi_1\| \cdot \|\psi_2\| \quad \textit{for all } g \in \mathbf{G}(\mathbb{A}).$$

We stated the above theorem only for K-invariant functions for simplicity. However the same holds for smooth functions as well, provided the L^2-norms of ψ_i are replaced by suitable Sobolev norms of ψ_i (see [**33**, Thm. 3.25] for details.) In fact, in most applications, we need the smooth version of Theorem 7.3.

EXAMPLE 7.4. Let $n \geq 3$. Let ψ_1, ψ_2 be $K := \mathrm{SO}_n \times \mathrm{SL}_n(\mathbb{Z}_f)$-invariant functions in $L^2(\mathrm{SL}_n(\mathbb{Q})\backslash \mathrm{SL}_n(\mathbb{A}))$ with $\int \psi_i = 0$ and $\|\psi_i\| = 1$.

For any $\epsilon > 0$, there is $C = C_\epsilon > 0$ such that

$$|\langle g\psi_1, \psi_2\rangle| \leq C \cdot \prod_p \prod_{i=1}^{[n/2]} \left(\frac{a_{p,i}}{a_{p,n+1-i}}\right)^{-1/2+\epsilon} \leq C \cdot \mathrm{H}(g)^{-1/2+\epsilon}$$

where $g = (g_p)$ and $g_p = \mathrm{diag}(a_{p,1}, \cdots, a_{p,n})$ with $a_{p,1} \geq \cdots \geq a_{p,n} > 0$ and $a_{\infty,i} \in \mathbb{R}^+$ and $a_{p,i} \in p^{\mathbb{Z}}$ for finite p.

The Hilbert space $L^2_{00}(\mathbf{G}(\mathbb{Q})\backslash\mathbf{G}(\mathbb{A}))$ can be decomposed into the direct integral of irreducible unitary $\mathbf{G}(\mathbb{A})$-representations ρ and each ρ is of the form $\otimes'_{p\in R}\rho_p$ where ρ_p is an irreducible infinite dimensional unitary representation of $\mathbf{G}(\mathbb{Q}_p)$. Hence $\langle g\psi_1, \psi_2\rangle$ is the direct integral of $\prod_p \langle \rho_p(g_p)\psi_{1p}, \psi_{2p}\rangle$'s. Now we are reduced to understanding the matrix coefficients $\langle \rho_p(g_p)\psi_{1p}, \psi_{2p}\rangle$ for an infinite dimensional irreducible unitary representation ρ_p of $\mathbf{G}(\mathbb{Q}_p)$.

Although we know the decay phenomenon for the matrix coefficients of unitary representations over each local field $\mathbb{Q}_p$ by the work of Howe-Moore [**39**], we need to have some uniformity of the decay over all p's, namely a weak form of the Ramanujan conjecture for $\mathbf{G}$. In the case of the $\mathbb{Q}$ rank is at least 2 and hence the $\mathbb{Q}_p$-rank of $\mathbf{G}$ is at least 2 for each p, this is achieved in [**48**]. In the case when the $\mathbb{Q}$-rank of $\mathbf{G}$ is one, the $\mathbb{Q}_p$-rank may be one or higher. When its $\mathbb{Q}_p$-rank is higher, one uses again [**48**] and when the $\mathbb{Q}_p$-rank is one, one now has to use more than the fact that ρ_p is an infinite dimensional unitary representation of $\mathbf{G}(\mathbb{Q}_p)$. Using the fact that ρ_p is indeed an automorphic representation, one uses the lifting of automorphic bound of $\mathrm{SL}_2(\mathbb{Q}_p)$ to $\mathbf{G}(\mathbb{Q}_p)$ due to Burger-Sarnak [**9**] and Clozel-Ullmo [**15**]. Finally when the $\mathbb{Q}$-rank and the $\mathbb{Q}_p$-rank of $\mathbf{G}$ are 0 and 1 respectively, Clozel analyzed what kind of automorphic representations occur in this situation and obtained a necessary bound for the decay [**13**], based on Jacquet-Langlands correspondence, and base changes of Rogawski and Clozel.

REMARK 7.5. The bounds toward the Ramanujan conjecture we discuss in this section are very crude in many cases, not at all close to the optimal bounds (but they are optimal in the example 7.4 due to the continuous spectrum.) We point out that in recent applications to sieve, obtaining the Ramanujan bounds as close to optimal ones are very critical (see [**8**]).

We remark that using the volume computation for the adelic height balls made by Shalika, Takloo-Bighash and Tschinkel [**58**] one can deduce the following from 7.3:

COROLLARY 7.6. *The quasi-regular representation ρ of $\mathbf{G}(\mathbb{A})$ on $L^2_{00}(\mathbf{G}(\mathbb{Q})\backslash\mathbf{G}(\mathbb{A}))$ is strongly L^q for some explicit $q > 0$.*

We remark that the Ramanujan conjecture for PGL_2 implies that the representation $\rho := L^2_{00}(\mathrm{PGL}_2(\mathbb{Q})\backslash \mathrm{PGL}_2(\mathbb{A}))$ is strongly $L^{4+\epsilon}$: the conjecture implies that (cf. [**33**, proof of Thm 3.10]) for any $K := \mathrm{PO}_2 \times \prod_p \mathrm{PGL}_2(\mathbb{Z}_p)$-finite unit vectors v and w,

$$|\langle \rho(g)v, w\rangle| \le d_{v,w} \cdot \prod_{p\in R} \Xi_p(g_p) \quad \text{for } g = (g_p) \in \mathrm{PGL}_2(\mathbb{A})$$

where Ξ_p denotes the Harish-Chandra function of $\mathrm{PGL}_2(\mathbb{Q}_p)$ and $d_{v,w}$ depends only on the dimensions of K-span of v and w.

From

$$\Xi_p \begin{pmatrix} p^k & 0 \\ 0 & 1 \end{pmatrix} = p^{-k/2}\left(\frac{k(p-1)+(p+1)}{p+1}\right),$$

we can deduce that for any $\epsilon > 0$, there is $C_\epsilon > 0$ such that for $g = (g_p) \in \mathbf{G}(\mathbb{A})$,

$$\mathrm{H}^{-1/2}(g) \le \Xi(g) \le C_\epsilon\, \mathrm{H}(g)^{-1/2+\epsilon}$$

where $\Xi := \prod_p \Xi_p$, $\mathrm{H} = \prod_p \mathrm{H}_p$ and $\mathrm{H}_p(g_p)$ is the maximum p-adic norm of g_p.

Note that for any $\epsilon > 0$, there is $c_\epsilon > 0$ such that for any $\sigma > 0$,

$$\int_{\mathrm{PGL}_2(\mathbb{R})} \|g\|_\infty^{-\sigma/2} dg_\infty \le c_\epsilon \int_0^\infty e^{-t(\sigma/2)} e^{t(1+\epsilon)} dt.$$

Hence $\int_{\mathrm{PGL}_2(\mathbb{R})} \|g\|_\infty^{-s/2} dg_\infty$ absolutely converges for $\Re(s) > 2$.

Observe that

$$\begin{aligned}\int_{\mathrm{PGL}_2(\mathbb{Q}_p)} \mathrm{H}_p(g_p)^{-s/2} dg_p &= \sum_{k\geq 0} p^{-ks/2} \operatorname{vol}(\mathrm{PGL}_2(\mathbb{Z}_p) \begin{pmatrix} p^k & 0 \\ 0 & 1 \end{pmatrix} \mathrm{PGL}_2(\mathbb{Z}_p)) \\ &= 1 + \sum_{k\geq 1} p^{-ks/2} p^k (1+p^{-1}) \\ &= \left(1-p^{-(s/2-1)}\right)^{-1} (1+p^{-s/2}) \\ &= \zeta_p\left(\frac{s}{2}-1\right)(1+p^{-s/2}).\end{aligned}$$

Therefore

$$\int_{\mathrm{PGL}_2(\mathbb{A})} \mathrm{H}(g)^{-s/2} dg = \int_{\mathrm{PGL}_2(\mathbb{R})} \|g\|_\infty^{-s/2} dg_\infty \times \prod_p (1+p^{-s/2}) \cdot \zeta(\frac{s}{2}-1).$$

Since the Riemann zeta function $\zeta(s)$ has a pole at $s=1$ and $\prod_p(1+p^{-s/2})$ absolutely converges for $\Re(s) > 2$, the height zeta function

$$\mathcal{Z}(s) := \int_{\mathrm{PGL}_2(\mathbb{A})} \mathrm{H}(g)^{-s/2} dg$$

has a meromorphic continuation to $\Re(s) > 2$ with an isolated pole at $s = 4$. In particular, for any $\epsilon > 0$, $\mathcal{Z}(4+\epsilon) < \infty$ and hence $\int_{\mathrm{PGL}_2(\mathbb{A})} \Xi(g)^{4+\epsilon}\, dg < \infty$, proving the claim.

Remark: By the equivalence of Lemma 6.2, we can deduce from the quantitative mixing theorem the equidistribution of the closed $\Delta(G)$-orbits $X_a := [(e, a^{-1})]\Delta(G)$ with respect to the Haar measure with the rate given by $\operatorname{vol}(X_a)^{-k}$ for some $k > 0$. Analogous statement is true even in the positive characteristic case since the results in [**48**] are valid.

A much more general result in this direction (characteristic zero case) was recently obtained in [**22**].

8. Counting via mixing and the wavefront property

Let G be a locally compact and second countable group, Γ a lattice in G and L a closed subgroup of G such that $L \cap \Gamma$ is a lattice in L.

Can the mixing of G on $\Gamma\backslash G$ be used to count points in the Γ-orbit $[e]\Gamma$ on $L\backslash G$? There are two alternative methods developed by Duke-Rudnick-Sarnak [**21**] and by Eskin-McMullen [**26**] which show that the answer is yes. Although both papers are based eventually on the spectral gap property (or the mixing property) of the group actions, the ways to use the spectral gap property are different.

In this section, we will present the methods in [**26**] which make use of a certain geometric property of $L\backslash G$, called the wavefront property.

REMARK 8.1. We remark that the idea of using the mixing property in counting problems goes back to Margulis' 1970 thesis [**44**].

We give a slight variant of the wavefront property.

DEFINITION 8.2. Let G be a locally compact group and L a closed subgroup of G.

- For a (non-compact) Borel subset E of G, the triple $(G, L : E)$ is said to have the wavefront property if for every neighborhood U of e in G, there exists a neighborhood V of e in G such that

$$LVg \subset LgU \quad \text{for all } g \in E.$$

- We say (G, L) has the wavefront property if $(G, L : E)$ has the wavefront property for some Borel subset $E \subset G$ satisfying $G = LE$.

It is easy to observe that if $(G, L : E)$ has the wavefront property, so does $(G, L : EK)$ for any compact subgroup K of G.

This property means roughly that the g-translate of a small neighborhood of the base point $z_0 := [L]$ in $L\backslash G$ remains near $z_0 g$ uniformly over all $g \in E$.

We give two examples for $G = \mathrm{SL}_2(\mathbb{R})$ below. For $a_t := \mathrm{diag}(e^{t/2}, e^{-t/2})$. let

$$A^+ = \{a_t : t \geq 0\} \quad \text{and} \quad A^- = \{a_t : t \leq 0\},$$

and $A = A^+ \cup A^-$. Let

$$N = \{\begin{pmatrix} 1 & x \\ 0 & 1 \end{pmatrix} : x \in \mathbb{R}\} \quad \text{and} \quad N^- = \{\begin{pmatrix} 1 & 0 \\ x & 1 \end{pmatrix} : x \in \mathbb{R}\}.$$

Set $K = \mathrm{SO}_2 = \{x \in \mathrm{SL}_2(\mathbb{R}) : xx^t = e\}$.

EXAMPLE 8.3. The triple $(\mathrm{SL}_2(\mathbb{R}), N : A^- K)$ has the wavefront property: Let U be an ϵ-neighborhood of e in $\mathrm{SL}_2(\mathbb{R})$. Since NAN^- is a Zariski dense open subset of G, we may assume that $U = (N \cap U)(A \cap U)(N^- \cap U)$. For any $a_t, a \in A$, and $n^- \in N^-$, observe that

$$(an^-)a_t = a_t a(a_t^{-1} n^- a_t).$$

Since the conjugation by a_t^{-1} is a contracting automorphism of N^- for any $t < 0$, we have $a_t^{-1}(N^- \cap U)a_t \subset N^- \cap U$. Therefore for $a_t \in A^-$,

$$NUa_t = N(A \cap U)(N^- \cap U)a_t \subset Na_t(A \cap U)(N^- \cap U) \subset Na_t U.$$

Hence the claim is proved.

EXAMPLE 8.4. The pair $(\mathrm{SL}_2(\mathbb{R}), K)$ has the wavefront property: It suffices to prove that $(\mathrm{SL}_2(\mathbb{R}), K : A^+)$ has the wavefront property, since $G = KA^+K$ by the Cartan decomposition. Let U be an ϵ-neighborhood of e in $\mathrm{SL}_2(\mathbb{R})$. Using the Iwasawa decomposition $G = KAN$, we may assume that $U = (K \cap U)(A \cap U)(N \cap U)$. For any $a_t \in A^+$, since the conjugation by a_t is a contraction on N,

$$KUa_t = K(A \cap U)(N \cap U)a_t \subset Ka_t(A \cap U)a_t^{-1}(N \cap U)a_t \subset Ka_t U.$$

Hence the claim is proved.

PROPOSITION 8.5. *Let $L < G$ be locally compact groups as above and $E \subset G$. Suppose the following:*

(1) *The right translation action of G on $\Gamma\backslash G$ is mixing;*
(2) *The wavefront property holds for $(G, L : E)$.*

Then, for any sequence $g_i \in E$ tending to ∞ in $L\backslash G$, the translate $\Gamma\backslash\Gamma L g_i$ becomes equidistributed in $\Gamma\backslash G$.

PROOF. Let $Y = (\Gamma \cap L)\backslash L$ and $X = \Gamma\backslash G$. Denote by μ_L and μ_G the Haar measures on L and G which give one on Y and X respectively. For $\psi \in C_c(\Gamma\backslash G)$, we would like to show that

$$(8.6) \qquad I_g := \int_Y \psi(yg)\, d\mu_L(y) \to \int_X \psi\, d\mu_G \quad \text{as } g \in E \text{ goes to infinity in } L\backslash G.$$

Suppose first that Y is compact. Then we can choose a Borel subset W in G transversal to L, so that the multiplication $m : Y \times W \to YW$ is a bijection onto its image $YW \subset X$. By the wavefront property, for any small neighborhood U of e in G, there exists W so that YWg remains inside YgU for all $g \in E$. Hence by the uniform continuity of ψ, and by taking W small enough, we can assure that I_g is close to

$$\frac{1}{\mathrm{vol}(W)} \int_{YWg} \psi\, d\mu_G = \frac{1}{\mathrm{vol}(W)} \langle g\psi, \chi_{YW}\rangle$$

where χ_{YW} is the characteristic function of YW. It now follows from the mixing that

$$\frac{1}{\mathrm{vol}(W)} \langle g\psi, \chi_{YW}\rangle \sim \int_X \psi\, d\mu_G$$

as $g \to \infty$, and hence (8.6) holds. When Y is non-compact, such a W does not exist in general. In this case, we work with a big compact piece Y_ϵ of Y with co-volume less than ϵ. The above argument then gives that I_g is close to $\mu_L(Y_\epsilon)\int_X \psi d\mu_G$ for all large $g \in L\backslash E$. Since $\mu_L(Y \setminus Y_\epsilon) \le \epsilon$ and $\|\psi\|_\infty$ is bounded, we can deduce (8.6). We refer to [**26**] for more details. In the case when Y is non-compact, the above modification is explained in [**3**]. □

COROLLARY 8.7. *The sequence* $\mathrm{SL}_2(\mathbb{Z})\backslash \mathrm{SL}_2(\mathbb{Z})a_tN$ *becomes equidistributed in the space* $\mathrm{SL}_2(\mathbb{Z})\backslash \mathrm{SL}_2(\mathbb{R})$ *as* $t \to -\infty$.

PROOF. Since a_t normalizes N,

$$\mathrm{SL}_2(\mathbb{Z})\backslash \mathrm{SL}_2(\mathbb{Z})a_tN = \mathrm{SL}_2(\mathbb{Z})\backslash \mathrm{SL}_2(\mathbb{Z})Na_t.$$

Hence by the mixing of $\mathrm{SL}_2(\mathbb{R})$ on $\mathrm{SL}_2(\mathbb{Z})\backslash \mathrm{SL}_2(\mathbb{R})$ and Example 8.3, we can deduce the claim. □

This corollary can be interpreted as the equidistribution of long closed horocycles, since N-orbits are precisely horocycles in the identification of $\mathrm{SL}_2(\mathbb{Z})\backslash \mathrm{SL}_2(\mathbb{R})$ with the unit tangent bundle of the modular surface $\mathrm{SL}_2(\mathbb{Z})\backslash\mathbb{H}$. This result was first obtained by Sarnak [**54**] with rates of convergence. The approach explained here has been well known and can be made effective using the quantitative mixing.

Putting Propositions 8.5 and 3.5 together, we state:

COROLLARY 8.8. *Let* $L < G$ *be locally compact groups as above and* $E \subset G$ *a Borel set. Let* $\{B_n \subset L\backslash G\}$ *be a sequence of compact subsets whose volume tending to infinity. Suppose the following:*

(1) *The right translation action of* G *on* $\Gamma\backslash G$ *is mixing.*
(2) *The wavefront property holds for* $(G, L : E)$.
(3) $\mathrm{vol}\,(B_n) \sim_n \mathrm{vol}(B_n \cap L\backslash LE)$ *and* $\{B_n\}$ *is well-rounded.*

Then for $x_0 = [L]$, *as* $n \to \infty$,

$$\#x_0\Gamma \cap B_n \sim \mathrm{vol}(B_n).$$

Using the well known fact that the Riemannian balls are well-rounded, we deduce the following from Example 8.4 and the above corollary:

EXAMPLE 8.9. Let Γ be a lattice in $\mathrm{SL}_2(\mathbb{R})$. Set $x_0 = i \in \mathbb{H} = \{x+iy : y>0\}$ and let d denote the hyperbolic distance. Let B_t denote the ball of radius t centered at x_0. We then have

$$\{x \in x_0\Gamma : d(x, x_0) < t\} \sim \mathrm{vol}(B_t).$$

8.1. Affine symmetric variety. We now discuss more general examples of G and L for which Corollary 8.8 can be applied. These are affine symmetric pairs, which are generalizations of Riemannian symmetric pairs.

Let $\mathbf{G}$ be a connected semisimple $\mathbb{Q}$-group. A $\mathbb{Q}$-subgroup $\mathbf{L}$ is called symmetric if there is an involution σ of $\mathbf{G}$ defined over $\mathbb{Q}$ such that $\mathbf{L} = \{g \in \mathbf{G} : \sigma(g) = g\}$.

THEOREM 8.10. *Let $\mathbf{L}$ be a symmetric subgroup of $\mathbf{G}$. For any finite set of primes S, the pair $(\mathbf{G}_S, \mathbf{L}_S)$ satisfies the wavefront property.*

PROOF. The proof easily reduces to the case when S is a singleton. When $S = \{\infty\}$, this was proved by Eskin-McMullen [**26**]. Their proof was based on the Cartan decomposition for real symmetric spaces. The claim is obtained in [**3**] for $S = \{p\}$ based on the Cartan decomposition for p-adic symmetric spaces ([**5**] and [**19**]). □

Let $\mathbf{V}$ be an affine symmetric variety defined over $\mathbb{Q}$, i.e., $\mathbf{V} = v_0\mathbf{G}$ where $\mathbf{V} \subset \mathrm{SL}_n$ is a $\mathbb{Q}$-embedding and $v_0 \in \mathbb{Q}^n$ and the stabilizer $\mathbf{L}$ of v_0 is a symmetric subgroup of $\mathbf{G}$. Let S be a finite set of primes which contains ∞ if $\mathbf{G}(\mathbb{R})$ is non-compact, and consider the height function H_S on $\mathbf{V}_S = \prod_p \mathbf{V}(\mathbb{Q}_p)$ as in 5.11. Set

$$B_S(T) := \{x \in \mathbf{U}_S : \mathrm{H}_S(x) < T\}.$$

THEOREM 8.11. *Assume that $\mathbf{G}$ is $\mathbb{Q}$-simple. As $T \to \infty$,*

$$\#\{x \in \mathbf{V}(\mathbb{Z}_S) : \mathrm{H}_S(x) < T\} \sim \mathrm{vol}(B_S(T)) \asymp T^a(\log T)^b$$

for some $a \in \mathbb{Q}_{>0}$ and $b \in \mathbb{Z}_{\geq 0}$.

PROOF. Let Γ be an S-congruence subgroup preserving $\mathbf{V}(\mathbb{Z}_S)$. It suffices to obtain the asymptotic for $\# v_0\Gamma \cap B_S(T)$ assuming that $\mathbf{G}$ is simply connected (see the proof of Corollary 5.13). So we have the mixing of the right translation action of $\mathbf{G}_S$ on $\Gamma\backslash\mathbf{G}_S$. By Theorem 5.12, for each $v_0 \in \mathbf{V}_S$, the family $v_0\mathbf{G}_S \cap B_S(T)$ is well-rounded. Hence the claim follows from Theorems 8.8 and 8.10. □

We note that in this theorem there is no restriction on S and the stabilizer may not be semisimple, unlike Corollary 5.13. As mentioned in section 5, Theorem 8.11 in the case $S = \{\infty\}$ was proved in [**21**] and [**26**] without an explicit computation of the asymptotic growth of $\mathrm{vol}(B_\infty(T))$ in general. The asymptotic $\mathrm{vol}(B_\infty(T)) \sim cT^a(\log T)^b$ was obtained independently in [**45**] and [**36**] for group varieties, and [**37**] for general symmetric varieties.

The main advantage of using the mixing property in counting problems is its effectiveness. In fact, the above theorem is shown effectively in [**3**], by obtaining the effective versions of (2) and (3) in Theorem 8.8. We note that this was done in [**45**] in the case of group varieties (see also [**34**]).

Let $\Gamma \subset \mathbf{G}(\mathbb{Q})$ be an S-congruence subgroup. Set $X_S := \Gamma\backslash\mathbf{G}_S$ and $Y_S := \Gamma \cap \mathbf{L}_S\backslash\mathbf{L}_S$. Also set $S_f := S \setminus \{\infty\}$.

DEFINITION 8.12. We say that the translate $Y_S g$ becomes effectively equidistributed in X_S as $g \to \infty$ in $\mathbf{L}_S\backslash\mathbf{G}_S$ if there exist $m \in \mathbb{N}$ and $r > 0$ such that, for any compact open subgroup W of $\mathbf{G}_{S_f}$ and any compact subset C of X_S, there exists $c = c(W, C) > 0$ satisfying that for any $\psi \in C_c^\infty(X_S)^W$ with support in C, one has for all $g \in \mathbf{G}_S$

$$|\int_{Y_S} \psi(yg)\, d\mu_{Y_S}(y) - \int_{X_S} \psi\, d\mu_{X_S}| \le c \cdot \mathcal{S}_m(\psi)\, \mathrm{H}_S(v_0 g)^{-r}$$

where $\mathcal{S}_m(\psi)$ depends only on the L^2-Sobolev norm of ψ of order m at ∞.

DEFINITION 8.13. A sequence of Borel subsets B_n in $\mathbf{V}_S$ is said to be effectively well-rounded if

(1) it is invariant under a compact open subgroup W of $\mathbf{G}_{S_f}$,
(2) there exists $\kappa > 0$ such that, uniformly for all $n \ge 1$ and all $0 < \epsilon < 1$,

$$\mathrm{vol}(B_{n,\epsilon}^+ - B_{n,\epsilon}^-) = O(\epsilon^\kappa \,\mathrm{vol}(B_n))$$

where $B_{n,\epsilon}^+ = B_n U_\epsilon W$ and $B_{n,\epsilon}^- = \cap_{u \in U_\epsilon W} B_n u$, and U_ϵ denotes the ball of center e and radius ϵ in $\mathbf{G}(\mathbb{R})$.
(3) for any $k > 0$, there exists $\delta > 0$ such that, uniformly for all large $n > 1$ and all $0 < \epsilon < 1$, one has

$$\int_{B_{n,\epsilon}^+} \mathrm{H}_S^{-k}(z)\, dz = O(\mathrm{vol}(B_n)^{1-\delta}).$$

It is not hard to adapt the proof of Proposition 3.5 to prove the following:

PROPOSITION 8.14. *Suppose that*

(1) *the translate $Y_S g$ becomes effectively equidistributed in X_S as $g \to \infty$ in $\mathbf{L}_S \backslash \mathbf{G}_S$;*
(2) *a sequence $\{B_n \subset \mathbf{V}_S\}$ of Borel subsets is effectively well-rounded and $\mathrm{vol}(B_n) \to \infty$.*

Then there exists a constant $\delta > 0$ such that

$$\# v_0 \Gamma \cap B_n = \mathrm{vol}(B_n)(1 + O(\mathrm{vol}(B_n)^{-\delta})).$$

THEOREM 8.15. [**3**] *Let S be a finite set of primes including ∞. In the same setup as in Theorem 8.11, there exists $\delta > 0$ such that*

$$\#\{x \in \mathbf{V}(\mathbb{Z}_S) : \mathrm{H}_S(x) < T\} \sim_T \mathrm{vol}(B_S(T))(1 + \mathrm{vol}(B_S(T))^{-\delta}).$$

Besides the effective equidistribution of the translates $\Gamma \backslash \Gamma \mathbf{L}_S g$ for symmetric pairs $(\mathbf{G}, \mathbf{L})$, we need the effective well-roundedness of the height balls, which was obtained in the complete generality of a homogeneous variety in [**3**, Prop. 14.2]. This subtle property of the height balls was properly addressed first in [**45**] for group varieties.

9. A problem of Linnik: Representations of integers by an invariant polynomial II

Let f be an integral homogeneous polynomial of degree d in n-variables, and consider the level sets

$$\mathbf{V}_m := \{x \in \mathbb{C}^n : f(x) = m\} \quad \text{for } m \in \mathbb{N}.$$

Then $\mathbf{V}_m(\mathbb{Z}) := \mathbf{V}_m \cap \mathbb{Z}^n$ is precisely the set of integral vectors representing m by f. Linnik asked a question on whether the radial projection of $\mathbf{V}_m(\mathbb{Z})$ on $\mathbf{V}_1(\mathbb{R})$ becomes equidistributed as $m \to \infty$ [**40**]. In the case when $\mathbf{V}_m$ is a homogeneous space of a semisimple algebraic group, this question has been studied intensively in recent years (see for instance, [**56**], [**20**], [**14**], [**32**] [**28**], [**29**], [**49**], [**24**], [**23**], [**22**], etc.,)

In this section, we discuss a generalization of the main results of Eskin-Oh in [**29**], which explains the title of this section. To formulate our results, denote by $\mathrm{pr}_\infty : \mathbf{V}_m(\mathbb{R}) \to \mathbf{V}_1(\mathbb{R})$ the radial projection given by $\mathrm{pr}_\infty(x) = m^{-1/d}x$. For a subset Ω of $\mathbf{V}_1(\mathbb{R})$, set

$$N_m(f, \Omega) := \# \mathrm{pr}_\infty(\mathbf{V}_m(\mathbb{Z})) \cap \Omega. \tag{9.1}$$

Let $\mathbf{G}$ be a connected semisimple algebraic $\mathbb{Q}$-group with a given $\mathbb{Q}$-embedding $\mathbf{G} \subset \mathrm{GL}_n$ and a non-zero vector $v_0 \in \mathbb{Q}^n$ such that

$$v_0\mathbf{G} = V_1.$$

We assume that $\mathbf{L} := \mathrm{Stab}_{\mathbf{G}}(v_0)$ is a semisimple maximal connected $\mathbb{Q}$-group.

Connected components of $\mathbf{V}_1(\mathbb{R})$ are precisely the orbits of the identity component $\mathbf{G}(\mathbb{R})^\circ$. On each connected component $\mathcal{O}$, fix a $\mathbf{G}(\mathbb{R})^\circ$-invariant measure with respect to which the volumes of subsets of $\mathcal{O}$ are computed below.

THEOREM 9.2. *Fix a connected component $\mathcal{O}$ of $\mathbf{V}_1(\mathbb{R})$. As $m \to \infty$ along primes, the projection $\mathrm{pr}_\infty(\mathbf{V}_m(\mathbb{Z}))$ becomes equidistributed on $\mathcal{O}$, provided $N_m(f, \mathcal{O}) \neq 0$.*

The equidistribution in the above means that for any compact subsets $\Omega_1, \Omega_2 \subset \mathcal{O}$ of boundary measure zero and of non-empty interior, we have

$$\frac{N_m(f, \Omega_1)}{N_m(f, \Omega_2)} \sim \frac{\mathrm{vol}(\Omega_1)}{\mathrm{vol}(\Omega_2)}.$$

In the case when $\mathbf{L}$ has no compact factors over the reals, this theorem was obtained in [**29**] for any $m \to \infty$ provided $\mathrm{pr}_\infty(\mathbf{V}_m(\mathbb{Z}))$ has no constant infinite subsequence, which is clearly a necessary condition.

REMARK 9.3. Since we allow $m \to \infty$ only along the primes, the above theorem is weaker than what is desired. We think that our argument can be modified to obtain the equidistribution as long as the sequence m is co-prime to a fixed prime number, by proving a suitable generalization of [**17**, Thm. 3] in the S-arithmetic setting.

EXAMPLE 9.4. Fix $n \geq 3$. Let

$$\mathbf{V}_m := \{x \in \mathrm{M}_n : x = x^t, \quad \det(x) = m\}.$$

Then the projection of $\mathbf{V}_m(\mathbb{Z})$ to $\mathbf{V}_1(\mathbb{R})$ becomes equidistributed as $m \to \infty$ along primes.

In this example, $\mathbf{V}_1$ is a finite union of homogeneous spaces $\mathrm{SO}(p,q)\backslash \mathrm{SL}_n$ where $0 \leq p, q \leq n$ ranges over non-negative integers such that $p \leq q$ and $p + q = n$. Since $\mathrm{SO}(p,q)$'s are maximal connected subgroups of SL_n, and the sequence $\{\mathrm{pr}_\infty(\mathbf{V}_m(\mathbb{Z})) \cap \mathcal{O}\}$ is non-empty for each connected component $\mathcal{O}$ of $\mathbf{V}_1(\mathbb{R})$, the claim follows from Theorem 9.2.

We now discuss the proof of Theorem 9.2. The proof makes use the p-adic unipotent flows. Using the idea of dynamics in the homogeneous space of p-adic groups by extending the homogeneous space of $\mathbf{G}(\mathbb{R})$ to that of $\mathbf{G}(\mathbb{R}) \times \mathbf{G}(\mathbb{Q}_p)$ was already implicit in the work of Linnik [**41**], as pointed out in [**23**]. This idea was also used in [**24**] and [**23**].

Since $\mathbf{G}(\mathbb{R})^\circ$ is equal to $\pi(\tilde{\mathbf{G}}(\mathbb{R}))$ where $\pi : \tilde{\mathbf{G}} \to \mathbf{G}$ is the simply connected cover, we may assume without loss of generality that $\mathbf{G}$ is simply connected.

Choose p which is strongly isotropic for $\mathbf{L}$. We denote by $\mathbb{Z}_p^*$ the group of p-adic units. Then $\mathbb{Z}_p^*$ is the disjoint union $\cup_{i=1}^k u_i(\mathbb{Z}_p^*)^d$ where d is the degree of f and $\{x \in \mathbb{Z}_p^* : x^d = 1\} = \{u_1, \cdots, u_k\}$. For $m \in \mathbb{N}$ with $(m,p) = 1$, choose $\alpha_m \in \mathbb{Z}_p^*$ such that $u_i/m = \alpha_m^d$ for some $1 \le i \le k$. We then define a projection

$$\mathrm{pr} : \cup_{(m,p)=1} \mathbf{V}_m(\mathbb{Q}) \to \mathbf{V}_1(\mathbb{R}) \times (\cup_i \mathbf{V}_{u_i}(\mathbb{Q}_p))$$

by

$$\mathrm{pr}(x) = (\mathrm{pr}_\infty(x), \mathrm{pr}_p(x)) = (m^{-1/d}x, \alpha_m x).$$

Set $I_i := \{m \in \mathbb{N} : p \nmid m, m \in u_i(\mathbb{Z}_p^*)^d\}$ so that $m \in I_i$ means $\mathrm{pr}(\mathbf{V}_m(\mathbb{Q})) \subset \mathbf{V}_1(\mathbb{R}) \times \mathbf{V}_{u_i}(\mathbb{Q}_p)$.

Note that for $\Omega \subset \mathbf{V}_1(\mathbb{R})$ and for $m \in I_i$,

$$\#\,\mathrm{pr}_\infty(\mathbf{V}_m(\mathbb{Z})) \cap \Omega = \#\,\mathrm{pr}(\mathbf{V}_m(\mathbb{Z}[p^{-1}]) \cap (\Omega \times \mathbf{V}_{u_i}(\mathbb{Z}_p)).$$

Therefore Theorem 9.2 follows from the equidistribution of $\mathrm{pr}(\mathbf{V}_m(\mathbb{Z}[p^{-1}])$ on each $\mathbf{G}(\mathbb{R}) \times \mathbf{G}(\mathbb{Q}_p)$-orbit in $\cup_i(\mathbf{V}_1(\mathbb{R}) \times \mathbf{V}_{u_i}(\mathbb{Q}_p))$.

Set $S = \{\infty, p\}$, and let $\Gamma \subset \mathbf{G}(\mathbb{Z}[p^{-1}])$ be an S-congruence subgroup preserving $\mathbf{V}(\mathbb{Z}[p^{-1}])$. Let μ_G and μ_L be the invariant probability measures on $\Gamma\backslash\mathbf{G}_S$ and $\Gamma \cap \mathbf{L}_S\backslash\mathbf{L}_S$ respectively, and for each $\mathbf{G}_S$ orbit $\mathcal{O}_S$, let $\mu_{\mathcal{O}_S}$ denote the invariant measure on $\mathcal{O}_S$ compatible with μ_G and μ_L.

Fix $1 \le i \le k$ and for a $\mathbf{G}_S$-orbit $\mathcal{O}_S = \mathcal{O}_\infty \times \mathcal{O}_p$ in $\mathbf{V}_1(\mathbb{R}) \times \mathbf{V}_{u_i}(\mathbb{Q}_p)$ which contains $\mathrm{pr}(\xi_0)$ for some $\xi_0 \in \mathbf{V}(\mathbb{Q})$, we define for each $\xi \in \mathbf{V}_m(\mathbb{Q})$ with $\mathrm{pr}(\xi) \in \mathcal{O}_S$,

$$\omega(\xi) := \frac{\mu_L(g_\xi^{-1}\Gamma g_\xi \cap \mathbf{L}_S\backslash\mathbf{L}_S)}{\mu_G(\Gamma\backslash\mathbf{G}_S)}$$

where $g_\xi \in \mathbf{G}_S$ such that $\mathrm{pr}(\xi_0)g_\xi = \mathrm{pr}(\xi)$.

PROPOSITION 9.5. *For any sequence* $\mathrm{pr}(\xi_m) \in \mathrm{pr}(\mathbf{V}_m(\mathbb{Z}[p^{-1}])) \cap \mathcal{O}_S$ *and as* $m \to \infty$ *along* $(m,p) = 1$, *the sequence* $\mathrm{pr}(\xi_m)\Gamma$ *is equidistributed on* $\mathcal{O}_S$ *unless it contains a constant infinite subsequence.*

In fact, for any $\psi \in C_c(\mathcal{O}_S)$,

$$\sum_{x \in \mathrm{pr}(\xi_m)\Gamma} \psi(x) \sim \omega(\xi_m) \cdot \int \psi \, d\mu_{\mathcal{O}_S}. \tag{9.6}$$

PROOF. For simplicity, let $g_m = g_{\xi_m}$. By the duality [**29**] it suffices to prove that the closed orbit $\Gamma\backslash\Gamma g_m \mathbf{L}_S$ becomes equidistributed in $\Gamma\backslash\mathbf{G}_S$. Set $\mathbf{L}_m$ to be the stabilizer of ξ_m in $\mathbf{G}$. Then

$$\mathbf{L}_m(\mathbb{R}) \times \mathbf{L}_m(\mathbb{Q}_p) = g_m(\mathbf{L}(\mathbb{R}) \times \mathbf{L}(\mathbb{Q}_p))g_m^{-1}.$$

In particular, $\mathbf{L}_m(\mathbb{Q}_p)$ is conjugate to $\mathbf{L}(\mathbb{Q}_p)$ by an element of $\mathbf{G}(\mathbb{Q}_p)$. Therefore p is strongly isotropic for all $\mathbf{L}_m$. It follows from Theorem 4.10 that if the equidistribution (9.6) we desire does not hold, then by passing to a subsequence, there exist m_0 and $\{\delta_m \in \Gamma : m \ge m_0\}$ such that

$$\mathbf{L}_m = \delta_m^{-1} \mathbf{L}_{m_0} \delta_m.$$

Since $\mathbf{L}$ has finite index in its normalizer, this means, by passing to a subsequence, the existence of $\gamma'_m \in \Gamma$ such that

$$g_l^{-1}\gamma'_m g_m \in \mathbf{L}(\mathbb{R}) \times \mathbf{L}(\mathbb{Q}_p) \quad \text{for all large } m, l$$

and hence $\mathrm{pr}(\xi_m)\Gamma = \mathrm{pr}(\xi_l)(\Gamma)$ for all large m and l in I_i. This is contradiction. □

For each connected component $\mathcal{O}_\infty$ of $\mathbf{V}_1(\mathbb{R})$, and $m \in I_i$, we define

$$\omega_m(\mathcal{O}_\infty) := \sum_{\xi_m \in \Omega_m} \omega(\xi_m) \cdot \mu_{\mathcal{O}_p}(\mathbf{V}_{u_i}(\mathbb{Z}_p) \cap \mathcal{O}_p)$$

where Ω_m is the set of representatives of Γ-orbits in $\mathbf{V}_m(\mathbb{Z}[p^{-1}])$,

$$\mathcal{O}_p = \mathrm{pr}_p(\xi_m)\mathbf{G}(\mathbb{Q}_p)$$

and the measure $\mu_{\mathcal{O}_p}$ is determined so that its product with $\mu_{\mathcal{O}_\infty}$ is compatible with μ_G and μ_L.

Theorem 9.2 follows from:

THEOREM 9.7. *For any connected component $\mathcal{O}_\infty$ of $\mathbf{V}_1(\mathbb{R})$, and for any compact subset $\Omega \subset \mathcal{O}_\infty$ of boundary measure zero, we have*

$$N_m(f, \Omega) \sim \omega_m(\mathcal{O}_\infty) \cdot \mu_{\mathcal{O}_\infty}(\Omega)$$

if $m \to \infty$ along primes, and $N_m(f, \mathcal{O}_\infty) \neq 0$.

PROOF. First note that $\mathrm{pr}(\mathbf{V}_m(\mathbb{Q})) \cap \mathrm{pr}(\mathbf{V}_l(\mathbb{Q})) = \emptyset$ for any primes m and l.

Therefore Proposition 9.5 implies that for any compact subset $\Omega \subset \mathcal{O}_\infty$ of smooth boundary, as $m \to \infty$ in I_i along primes,

$$\begin{aligned}\#\,\mathrm{pr}(\xi_m)\Gamma \cap (\Omega \times \mathbf{V}_{u_i}(\mathbb{Z}_p)) &\sim \omega(\xi_m)\mu_{\mathcal{O}_S}(\Omega \times (\mathbf{V}_{u_i}(\mathbb{Z}_p) \cap \mathcal{O}_p)) \\ &= \omega(\xi_m) \cdot \mu_{\mathcal{O}_\infty}(\Omega) \cdot \mu_{\mathcal{O}_p}(\mathbf{V}_{u_i}(\mathbb{Z}_p) \cap \mathcal{O}_p).\end{aligned}$$

Since for $m \in I_i$

$$\#\,\mathrm{pr}(\mathbf{V}_m(\mathbb{Z})) \cap \Omega = \sum_{\xi_m \in \Omega_m} \#\,(\mathrm{pr}(\xi_m)\Gamma \cap (\Omega \times \mathbf{V}_{u_i}(\mathbb{Z}_p)))$$

the claim follows.

□

References

[1] V. V. Batyrev and Yu. I. Manin. Sur le nombre des points rationnels de hauteur borné des variétés algébriques. *Math. Ann.*, 286(1-3):27–43, 1990.
[2] V. V. Batyrev and Yu. Tschinkel. Rational points on toric varieties. In *Number theory (Halifax, NS, 1994)*, volume 15 of *CMS Conf. Proc.*, pages 39–48. Amer. Math. Soc., Providence, RI, 1995.
[3] Yves Benoist and Hee Oh. Effective equidistribution of S-integral points on symmetric varieties, 2007. Preprint.
[4] Yves Benoist and Hee Oh. Equidistribution of rational matrices in their conjugacy classes. *Geom. Funct. Anal.*, 17(1):1–32, 2007.
[5] Yves Benoist and Hee Oh. Polar decomposition for p-adic symmetric spaces. *Int. Math. Res. Not. IMRN*, (24):Art. ID rnm121, 20, 2007.
[6] B. J. Birch. Forms in many variables. *Proc. Roy. Soc. Ser. A*, 265:245–263, 1961/1962.
[7] Armand Borel. Some finiteness properties of adele groups over number fields. *Inst. Hautes Études Sci. Publ. Math.*, (16):5–30, 1963.
[8] Jean Bourgain, Alex Gamburd, and Peter Sarnak. Sieving and expanders. *C. R. Math. Acad. Sci. Paris*, 343(3):155–159, 2006.
[9] M. Burger and P. Sarnak. Ramanujan duals. II. *Invent. Math.*, 106(1):1–11, 1991.

[10] Antoine Chambert-Loir. Lecture on height zeta functions:At the confluence of algebraic geometry, algebraic number theory and analysis arXive:0812.0947, 2009
[11] Antoine Chambert-Loir and Yuri Tschinkel. On the distribution of points of bounded height on equivariant compactifications of vector groups. *Invent. Math.*, 148(2):421–452, 2002.
[12] L. Clozel. Spectral theory of automorphic forms. In *Automorphic forms and applications*, volume 12 of *IAS/Park City Math. Ser.*, pages 43–93. Amer. Math. Soc., Providence, RI, 2007.
[13] Laurent Clozel. Démonstration de la conjecture τ. *Invent. Math.*, 151(2):297–328, 2003.
[14] Laurent Clozel, Hee Oh, and Emmanuel Ullmo. Hecke operators and equidistribution of Hecke points. *Invent. Math.*, 144(2):327–351, 2001.
[15] Laurent Clozel and Emmanuel Ullmo. Équidistribution des points de Hecke. In *Contributions to automorphic forms, geometry, and number theory*, pages 193–254. Johns Hopkins Univ. Press, Baltimore, MD, 2004.
[16] S. G. Dani and G. A. Margulis. Asymptotic behaviour of trajectories of unipotent flows on homogeneous spaces. *Proc. Indian Acad. Sci. Math. Sci.*, 101(1):1–17, 1991.
[17] S. G. Dani and G. A. Margulis. Limit distributions of orbits of unipotent flows and values of quadratic forms. In *I. M. Gelfand Seminar*, volume 16 of *Adv. Soviet Math.*, pages 91–137. Amer. Math. Soc., Providence, RI, 1993.
[18] C. De Concini and C. Procesi. Complete symmetric varieties. In *Invariant theory (Montecatini, 1982)*, volume 996 of *Lecture Notes in Math.*, pages 1–44. Springer, Berlin, 1983.
[19] P. Delorme and V. Se'cherre. An analogue of the cartan decomposition for p-adic symmetric spaces. *preprint arXiv: math/0612545*, 2007.
[20] W. Duke. Hyperbolic distribution problems and half-integral weight Maass forms. *Invent. Math.*, 92(1):73–90, 1988.
[21] W. Duke, Z. Rudnick, and P. Sarnak. Density of integer points on affine homogeneous varieties. *Duke Math. J.*, 71(1):143–179, 1993.
[22] M. Einsiedler, G. Margulis, and A. Venkatesh. Effective equidistribution for closed orbits of semisimple groups on homogeneous spaces. *Invent. Math.*, 177(1):137–212, 2009.
[23] Manfred Einsiedler, Elon Lindenstrauss, Philippe Michel, and Akshay Venkatesh. The distribution of periodic torus orbits and Duke's theorem for cubic fields. *Preprint*, 2007.
[24] J. Ellenberg and A. Venkatesh. Local-global principles for representations of quadratic forms. *Invent. Math.*, 171(2):257–279, 2008.
[25] A. Eskin, S. Mozes, and N. Shah. Non-divergence of translates of certain algebraic measures. *Geom. Funct. Anal.*, 7(1):48–80, 1997.
[26] Alex Eskin and Curt McMullen. Mixing, counting, and equidistribution in Lie groups. *Duke Math. J.*, 71(1):181–209, 1993.
[27] Alex Eskin, Shahar Mozes, and Nimish Shah. Unipotent flows and counting lattice points on homogeneous varieties. *Ann. of Math. (2)*, 143(2):253–299, 1996.
[28] Alex Eskin and Hee Oh. Ergodic theoretic proof of equidistribution of Hecke points. *Ergodic Theory Dynam. Systems*, 26(1):163–167, 2006.
[29] Alex Eskin and Hee Oh. Representations of integers by an invariant polynomial and unipotent flows. *Duke Math. J.*, 135(3):481–506, 2006.
[30] Gerd Faltings. Diophantine approximation on abelian varieties. *Ann. of Math. (2)*, 133(3):549–576, 1991.
[31] Jens Franke, Yuri I. Manin, and Yuri Tschinkel. Rational points of bounded height on Fano varieties. *Invent. Math.*, 95(2):421–435, 1989.
[32] Wee Teck Gan and Hee Oh. Equidistribution of integer points on a family of homogeneous varieties: a problem of Linnik. *Compositio Math.*, 136(3):323–352, 2003.
[33] Alex Gorodnik, François Maucourant, and Hee Oh. Manin's and Peyre's conjectures on rational points and adelic mixing. *Ann. Sci. Éc. Norm. Supér. (4)*, 41(3):383–435, 2008.
[34] Alex Gorodnik and Amos Nevo. Ergodic theory of lattice subgroups. 2007. Preprint.
[35] Alex Gorodnik and Hee Oh. Rational points on homogeneous varieties and equidistribution of adelic periods, with an appendix by M. Borovoi, 2008. Preprint.
[36] Alex Gorodnik and Barak Weiss. Distribution of lattice orbits on homogeneous varieties. *Geom. Funct. Anal.*, 17(1):58–115, 2007.
[37] Alexander Gorodnik, Hee Oh, and Nimish Shah. Integral points on symmetric varieties and Satake compactifications. *Amer. J. Math.*, 131(1):1–57, 2009.

[38] M. Hindry and J. Silverman. *Diophantine geometry*, volume 201 of *Graduate Texts in Mathematics*. Springer-Verlag, New York, 2000.

[39] Roger E. Howe and Calvin C. Moore. Asymptotic properties of unitary representations. *J. Funct. Anal.*, 32(1):72–96, 1979.

[40] Ju. V. Linnik. Additive problems and eigenvalues of the modular operators. In *Proc. Internat. Congr. Mathematicians (Stockholm, 1962)*, pages 270–284. Inst. Mittag-Leffler, Djursholm, 1963.

[41] Yu. V. Linnik. *Ergodic properties of algebraic fields*. Translated from the Russian by M. S. Keane. Ergebnisse der Mathematik und ihrer Grenzgebiete, Band 45. Springer-Verlag New York Inc., New York, 1968.

[42] D. Luna. Toute variété magnifique est sphérique. *Transform. Groups*, 1(3):249–258, 1996.

[43] G. A. Margulis and G. M. Tomanov. Invariant measures for actions of unipotent groups over local fields on homogeneous spaces. *Invent. Math.*, 116(1-3):347–392, 1994.

[44] Grigoriy A. Margulis. *On some aspects of the theory of Anosov systems*. Springer Monographs in Mathematics. Springer-Verlag, Berlin, 2004. With a survey by Richard Sharp: Periodic orbits of hyperbolic flows, Translated from the Russian by Valentina Vladimirovna Szulikowska.

[45] F. Maucourant. Homogeneous asymptotic limits of Haar measures of semisimple linear groups and their lattices. *Duke Math. J.*, 136(2):357–399, 2007.

[46] Shahar Mozes and Nimish Shah. On the space of ergodic invariant measures of unipotent flows. *Ergodic Theory Dynam. Systems*, 15(1):149–159, 1995.

[47] Hee Oh. Harmonic analysis and Hecke operators. In *Rigidity in dynamics and geometry (Cambridge, 2000)*, pages 363–378. Springer, Berlin, 2002.

[48] Hee Oh. Uniform pointwise bounds for matrix coefficients of unitary representations and applications to Kazhdan constants. *Duke Math. J.*, 113(1):133–192, 2002.

[49] Hee Oh. Hardy-Littlewood system and representations of integers by an invariant polynomial. *Geom. Funct. Anal.*, 14(4):791–809, 2004.

[50] Emmanuel Peyre. Hauteurs et mesures de Tamagawa sur les variétés de Fano. *Duke Math. J.*, 79(1):101–218, 1995.

[51] Vladimir Platonov and Andrei Rapinchuk. *Algebraic groups and number theory*, volume 139 of *Pure and Applied Mathematics*. Academic Press Inc., Boston, MA, 1994. Translated from the 1991 Russian original by Rachel Rowen.

[52] Marina Ratner. On Raghunathan's measure conjecture. *Ann. of Math. (2)*, 134(3):545–607, 1991.

[53] Marina Ratner. Raghunathan's conjectures for Cartesian products of real and p-adic Lie groups. *Duke Math. J.*, 77(2):275–382, 1995.

[54] Peter Sarnak. Asymptotic behavior of periodic orbits of the horocycle flow and eisenstein series. *Comm. Pure Appl. Math.*, 34(6):719–739, 1981.

[55] Peter Sarnak. Notes on the generalized Ramanujan conjectures. In *Harmonic analysis, the trace formula, and Shimura varieties*, volume 4 of *Clay Math. Proc.*, pages 659–685. Amer. Math. Soc., Providence, RI, 2005.

[56] Peter C. Sarnak. Diophantine problems and linear groups. In *Proceedings of the International Congress of Mathematicians, Vol. I, II (Kyoto, 1990)*, pages 459–471, Tokyo, 1991. Math. Soc. Japan.

[57] S. Schanuel. On heights in number fields. *Bull. Amer. Math. Soc.*, 70:262–263, 1964.

[58] Joseph Shalika, Ramin Takloo-Bighash, and Yuri Tschinkel. Rational points on compactifications of semi-simple groups. *J. Amer. Math. Soc.*, 20(4):1135–1186 (electronic), 2007.

[59] Joseph A. Shalika and Yuri Tschinkel. Height zeta functions of equivariant compactifications of the Heisenberg group. In *Contributions to automorphic forms, geometry, and number theory*, pages 743–771. Johns Hopkins Univ. Press, Baltimore, MD, 2004.

[60] Matthias Strauch and Yuri Tschinkel. Height zeta functions of toric bundles over flag varieties. *Selecta Math. (N.S.)*, 5(3):325–396, 1999.

[61] Ramin Takloo-Bighash. Bounds for matrix coefficients and arithmetic applications. In *Eisenstein series and applications*, volume 258 of *Progr. Math.*, pages 295–314. Birkhäuser Boston, Boston, MA, 2008.

[62] Yuri Tschinkel. Fujita's program and rational points. In *Higher dimensional varieties and rational points (Budapest, 2001)*, volume 12 of *Bolyai Soc. Math. Stud.*, pages 283–310. Springer, Berlin, 2003.

[63] Yuri Tschinkel. Geometry over nonclosed fields. In *International Congress of Mathematicians. Vol. II*, pages 637–651. Eur. Math. Soc., Zürich, 2006.
[64] A. Venkatesh. Sparse equidistribution problem, period bounds, and subconvexity, 2007. To appear in Annals of Math.
[65] Paul Vojta. *Diophantine approximations and value distribution theory*, volume 1239 of *Lecture Notes in Mathematics*. Springer-Verlag, Berlin, 1987.
[66] André Weil. *Adeles and algebraic groups*, volume 23 of *Progress in Mathematics*. Birkhäuser Boston, Mass., 1982. With appendices by M. Demazure and Takashi Ono.

MATHEMATICS DEPARTMENT, BROWN UNIVERSITY, PROVIDENCE, RI AND KOREA INSTITUTE FOR ADVANCED STUDY, SEOUL, KOREA
E-mail address: heeoh@math.brown.edu

Clay Mathematics Proceedings
Volume 10, 2010

Equidistribution on the modular surface and L-functions

Gergely Harcos

ABSTRACT. These are notes for two lectures given at the 2007 summer school "Homogeneous Flows, Moduli Spaces and Arithmetic" in Pisa, Italy. The first lecture introduces Heegner points and closed geodesics on the modular surface $\mathrm{SL}_2(\mathbb{Z})\backslash\mathcal{H}$ and highlights some of their arithmetic significance. The second lecture discusses how subconvex bounds for certain automorphic L-functions yield quantitative equidistribution results for Heegner points and closed geodesics.

1. Lecture One

Let us start the discussion with the equivalence of integral binary quadratic forms. The concept was introduced by Lagrange [**15**] and studied by Gauss [**9**] in a systematic fashion.

An *integral binary quadratic form* is a homogeneous polynomial

$$\langle a,b,c\rangle := ax^2+bxy+cy^2 \in \mathbb{Z}[x,y]$$

with associated *discriminant*

$$d := b^2-4ac \in \mathbb{Z}.$$

The possible discriminants are the integers congruent to 0 or 1 mod 4. We shall assume that the form $\langle a,b,c\rangle$ is not a product of linear factors in $\mathbb{Z}[x,y]$, then d is not a square, hence $ac\neq 0$. If $d<0$ then $ac>0$ and we shall assume that we are in the *positive definite* case $a,c>0$. Furthermore, we shall assume that d is a *fundamental discriminant* which means that it cannot be written as $d'e^2$ for some smaller discriminant d'. Then $\langle a,b,c\rangle$ is a *primitive* form which means that a,b,c are relatively prime. The possible fundamental discriminants are the square-free numbers congruent to 1 mod 4 and 4 times the square-free numbers congruent to 2 or 3 mod 4.

EXAMPLE 1. The first few negative fundamental discriminants are: -3, -4, -7, -8, -11, -15, -19, -20, -23, -24. The first few positive fundamental discriminants are: 5, 8, 12, 13, 17, 21, 24, 28, 29, 33.

The author was supported by European Community grant MEIF-CT-2006-040371 under the Sixth Framework Programme.

Lagrange [**15**] discovered that every form $\langle a,b,c\rangle$ with a given discriminant d can be reduced by some integral unimodular substitution

$$(x,y)\mapsto(\alpha x+\beta y,\gamma x+\delta y),\qquad \begin{pmatrix}\alpha & \beta\\ \gamma & \delta\end{pmatrix}\in \mathrm{SL}_2(\mathbb{Z}),$$

to some form with the same discriminant that lies in a finite set depending only on d. Forms that are connected by such a substitution are called *equivalent.* It is easiest to understand this reduction by looking at the simple substitutions

$$(x,y)\overset{T}{\mapsto}(x-y,y)\qquad\text{and}\qquad (x,y)\overset{S}{\mapsto}(-y,x). \tag{1}$$

The induced actions on forms are given by

$$\langle a,b,c\rangle\overset{T}{\mapsto}\langle a,b-2a,c+a-b\rangle\qquad\text{and}\qquad \langle a,b,c\rangle\overset{S}{\mapsto}\langle c,-b,a\rangle.$$

Now a given form $\langle a,b,c\rangle$ can always be taken to some $\langle a,b',c'\rangle$ with $|b'|\leqslant|a|$ by applying T or T^{-1} a few times. If $|a|\leqslant|c'|$ then we stop our reduction. Otherwise we apply S to get some $\langle a'',b'',c''\rangle$ with $|a''|<|a|$ and we start over with this form. In this algorithm we cannot apply S infinitely many times because $|a|$ decreases at each such step. Hence in a finite number of steps we arrive at an equivalent form $\langle a,b,c\rangle$ whose coefficients satisfy

$$|b|\leqslant|a|\leqslant|c|,\qquad b^2-4ac=d. \tag{2}$$

These constraints are satisfied by finitely many triples (a,b,c). Indeed, we have

$$|d|=|b^2-4ac|\geqslant 4|ac|-b^2\geqslant 3b^2, \tag{3}$$

so there are only $\ll|d|^{1/2}$ choices for b and for each such choice there are only $\ll_\varepsilon d^\varepsilon$ choices for a and c since the product ac is determined by b. We have shown that the number of equivalence classes of integral binary quadratic forms of fundamental discriminant d, denoted $h(d)$, satisfies the inequality

$$h(d)\ll_\varepsilon |d|^{1/2+\varepsilon}. \tag{4}$$

In the case $d<0$ it is straightforward to compile a maximal list of inequivalent forms satisfying (2). There is an algorithm for $d>0$ as well but it is less straightforward. In fact the subsequent findings of this lecture can be turned into an algorithm for all d. Note that for $d>0$ (3) implies $4ac=b^2-d<0$, hence by an extra application of S we can always arrange for a reduced form $\langle a,b,c\rangle$ with $a>0$.

EXAMPLE 2. The equivalence classes for $d=-23$ are represented by the forms $\langle 1,1,6\rangle$, $\langle 2,\pm1,3\rangle$. Hence $h(-23)=3$. The equivalence classes for $d=21$ are represented by the forms $\langle 1,1,-5\rangle$, $\langle -1,1,5\rangle$. Hence $h(21)=2$.

To obtain a geometric picture of equivalence classes of forms we shall think of $\mathbb{Q}(\sqrt{d})$ as embedded in $\mathbb{C}$ such that $\sqrt{d}/i>0$ for $d<0$ and $\sqrt{d}>0$ for $d>0$. For $q_1,q_2\in\mathbb{Q}$ we shall consider the conjugation

$$\overline{q_1+q_2\sqrt{d}}:=q_1-q_2\sqrt{d}.$$

Each form $\langle a,b,c\rangle$ decomposes as

$$ax^2+bxy+cy^2=a(x-zy)(x-\bar{z}y),$$

where

$$z := \frac{-b+\sqrt{d}}{2a}, \qquad \bar{z} := \frac{-b-\sqrt{d}}{2a}. \tag{5}$$

Using (1) we can see that the action of $\mathrm{SL}_2(\mathbb{Z})$ on z and $\bar{z}$ is the usual one given by fractional linear transformations:

$$z \overset{T}{\mapsto} z+1 \qquad \text{and} \qquad z \overset{S}{\mapsto} -1/z.$$

Therefore in fact we are looking at the standard action of $\mathrm{SL}_2(\mathbb{Z})$ on certain conjugate pairs of points of $\mathbb{Q}(\sqrt{d})$ embedded in $\mathbb{C}$. For $d<0$ we consider the points $z \in \mathcal{H}$ and obtain $h(d)$ points on $\mathrm{SL}_2(\mathbb{Z})\backslash\mathcal{H}$. These are the *Heegner points* of discriminant $d<0$. For $d>0$ we consider the geodesics $G_{\bar{z},z} \subset \mathcal{H}$ connecting the real points $\{\bar{z}, z\}$ and obtain $h(d)$ geodesics on $\mathrm{SL}_2(\mathbb{Z})\backslash\mathcal{H}$.

It is a remarkable fact that for $d>0$ any geodesic $G_{\bar{z},z}$ as above becomes closed when projected to $\mathrm{SL}_2(\mathbb{Z})\backslash\mathcal{H}$, and its length is an important arithmetic quantity associated with the number field $\mathbb{Q}(\sqrt{d})$. To see this take any matrix $M \in \mathrm{GL}_2^+(\mathbb{R})$ which takes 0 to $\bar{z}$ and ∞ to z, for example[1]

$$M := \begin{pmatrix} z & \bar{z} \\ 1 & 1 \end{pmatrix}, \tag{6}$$

then M takes the positive real axis (resp. geodesic) connecting $\{0,\infty\}$ to the real segment (resp. geodesic) connecting $\{\bar{z}, z\}$. In particular, using that M is a conformal automorphism of the Riemann sphere, we see that $G_{\bar{z},z}$ is the semicircle above the real segment $[\bar{z}, z]$, parametrized as

$$G_{\bar{z},z} = \{g(\lambda)i : \lambda > 0\}, \qquad \text{where} \qquad g(\lambda) := M\begin{pmatrix} \lambda & 0 \\ 0 & \lambda^{-1} \end{pmatrix}.$$

Moreover, the unique isometry of $\mathcal{H}$ fixing the geodesic $G_{\bar{z},z}$ and taking $g(1)i$ to $g(\lambda)i$ is given by the matrix

$$M\begin{pmatrix} \lambda & 0 \\ 0 & \lambda^{-1} \end{pmatrix} M^{-1} \in \mathrm{SL}_2(\mathbb{R}). \tag{7}$$

Therefore we want to see that for some $\lambda > 1$ the matrix

$$M\begin{pmatrix} \lambda & 0 \\ 0 & \lambda^{-1} \end{pmatrix} M^{-1} = \frac{1}{z-\bar{z}}\begin{pmatrix} z\lambda - \bar{z}\lambda^{-1} & z\bar{z}(\lambda^{-1}-\lambda) \\ \lambda - \lambda^{-1} & z\lambda^{-1} - \bar{z}\lambda \end{pmatrix} \tag{8}$$

is in $\mathrm{SL}_2(\mathbb{Z})$, and then the projection of $G_{\bar{z},z}$ to $\mathrm{SL}_2(\mathbb{Z})\backslash\mathcal{H}$ has length

$$\int_1^{\lambda^2} \frac{dy}{y} = 2\ln(\lambda)$$

for the smallest such $\lambda > 1$. A necessary condition for λ is that the sum and difference of diagonal elements of the matrix (8) are integers and so are the anti-diagonal elements as well. Using that

$$z - \bar{z} = \frac{\sqrt{d}}{a}, \qquad z + \bar{z} = \frac{-b}{a}, \qquad z\bar{z} = \frac{c}{a}$$

this is equivalent to:

$$\lambda + \lambda^{-1} \in \mathbb{Z}, \qquad \{a,b,c\}\frac{\lambda - \lambda^{-1}}{\sqrt{d}} \subset \mathbb{Z}.$$

---

[1] we assume here that $a>0$ which is legitimate as we have seen

As $\gcd(a,b,c)=1$ we can simplify this to

$$\lambda+\lambda^{-1}\in\mathbb{Z},\qquad\text{and}\qquad\frac{\lambda-\lambda^{-1}}{\sqrt{d}}\in\mathbb{Z}.$$

In other words, there are integers m,n such that

$$\lambda=\frac{m+n\sqrt{d}}{2}\qquad\text{and}\qquad\lambda^{-1}=\frac{m-n\sqrt{d}}{2}. \tag{9}$$

As $\lambda>1$ the integers m,n are positive and they satisfy the diophantine equation

$$m^2-dn^2=4. \tag{10}$$

The equations (9)–(10) are not only necessary but also sufficient for (8) to lie in $\mathrm{SL}_2(\mathbb{Z})$. Namely, (8)–(10) imply that

$$M\begin{pmatrix}\lambda & 0\\ 0 & \lambda^{-1}\end{pmatrix}M^{-1}=\begin{pmatrix}\frac{m-bn}{2} & -nc\\ na & \frac{m+bn}{2}\end{pmatrix}\in\mathrm{SL}_2(\mathbb{Z}) \tag{11}$$

since

$$m\pm bn\equiv m^2-dn^2\equiv 0\pmod 2.$$

The λ's given by (9)–(10) are exactly the totally positive[2] units in the ring of integers $\mathcal{O}_d$ of $\mathbb{Q}(\sqrt{d})$. These units form a group isomorphic to $\mathbb{Z}$ by Dirichlet's theorem, therefore there is a smallest $\lambda=\lambda_d>1$ among them (which generates the group). In other words, the sought $\lambda=\lambda_d>1$ exists and comes from the smallest positive solution of (10). In classical language, the matrices (11) are the *automorphs* of the form $\langle a,b,c\rangle$.

To summarize, the $\mathrm{SL}_2(\mathbb{Z})$-orbits of forms $\langle a,b,c\rangle$ with given fundamental discriminant d give rise to $h(d)$ Heegner points on $\mathrm{SL}_2(\mathbb{Z})\backslash\mathcal{H}$ for $d<0$ and $h(d)$ closed geodesics of length $2\ln(\lambda_d)$ for $d>0$ where $\lambda_d=(m+n\sqrt{d})/2$ is the smallest totally positive unit of $\mathcal{O}_d$ greater than 1. This geometric picture is even more interesting in the light of the following refinement of (4) which is a consequence of Dirichlet's class number formula and Siegel's theorem (see [**5**, Chapters 6 and 21]):

$$\begin{aligned}|d|^{1/2-\varepsilon}\ll_\varepsilon h(d)\ll_\varepsilon |d|^{1/2+\varepsilon},&\qquad d<0,\\ d^{1/2-\varepsilon}\ll_\varepsilon h(d)\ln(\lambda_d)\ll_\varepsilon d^{1/2+\varepsilon},&\qquad d>0.\end{aligned} \tag{12}$$

This shows that the set of Heegner points of discriminant $d<0$ has cardinality about $|d|^{1/2}$, while the set of closed geodesics of discriminant $d>0$ has total length about $d^{1/2}$.

2. Lecture Two

In the light of (12) the natural question arises if the set Λ_d of Heegner points (resp. closed geodesics) of fundamental discriminant d becomes equidistributed in $\mathrm{SL}_2(\mathbb{Z})\backslash\mathcal{H}$ as $d\to-\infty$ (resp. $d\to+\infty$). That is, given a smooth and compactly supported weight function $g:\mathrm{SL}_2(\mathbb{Z})\backslash\mathcal{H}\to\mathbb{C}$ do we have

$$\begin{aligned}\frac{1}{h(d)}\sum_{z\in\Lambda_d}g(z)&\to\int_{\mathrm{SL}_2(\mathbb{Z})\backslash\mathcal{H}}g(z)\,d\mu(z),\qquad d\to-\infty,\\ \frac{1}{h(d)\,2\ln(\lambda_d)}\sum_{G\in\Lambda_d}\int_G g(z)\,ds(z)&\to\int_{\mathrm{SL}_2(\mathbb{Z})\backslash\mathcal{H}}g(z)\,d\mu(z),\qquad d\to+\infty,\end{aligned} \tag{13}$$

[2]i.e. positive under both embeddings $\mathbb{Q}(\sqrt{d})\hookrightarrow\mathbb{R}$

where $d\mu(z)$ abbreviates the $\mathrm{SL}_2(\mathbb{R})$-invariant probability measure on $\mathrm{SL}_2(\mathbb{Z})\backslash\mathcal{H}$ and $ds(z)$ abbreviates the hyperbolic arc length? Duke [**6**] proved that the answer is yes in the sharper form that the difference of the two sides is $\ll_g |d|^{-\delta}$ for some fixed $\delta > 0$. Earlier Linnik [**16**] established the above limits with error term $\ll_g (\log|d|)^{-A}$ for all $A > 0$ under the condition that $\left(\frac{d}{p}\right) = 1$ for a fixed odd prime p.

We shall discuss Duke's quantitative result and a refinement of it from the modern perspective of subconvex bounds for automorphic L-functions. Our first step is to decompose spectrally the weight function considered in (13) as

$$g(z) = \langle g, 1\rangle + \sum_{j=1}^{\infty} \langle g, u_j\rangle u_j(z) + \frac{1}{4\pi}\int_{-\infty}^{\infty} \langle g, E(\cdot, \tfrac{1}{2}+it)\rangle E(z, \tfrac{1}{2}+it)\, dt,$$

where

$$\langle f_1, f_2\rangle := \int_{\mathrm{SL}_2(\mathbb{Z})\backslash\mathcal{H}} f_1(z)\overline{f_2(z)}\, d\mu(z),$$

the $\{u_j\}$ are Hecke–Maass cusp forms on $\mathrm{SL}_2(\mathbb{Z})\backslash\mathcal{H}$ with $\langle u_j, u_j\rangle = 1$, and the Eisenstein series $E(z, \frac{1}{2}+it)$ are obtained by meromorphic continuation from

$$E(z,s) := \frac{1}{2}\sum_{\substack{m,n\in\mathbb{Z}\\ \gcd(m,n)=1}} \frac{\Im z^s}{|mz+n|^{2s}}, \qquad \Re s > 1.$$

The above decomposition converges in $L^2(\mathrm{SL}_2(\mathbb{Z})\backslash\mathcal{H})$ and also pointwise absolutely and uniformly on compact sets, see [**14**, Theorem 7.3]. If

$$\Delta := -y^2\left(\frac{\partial^2}{\partial x^2} + \frac{\partial^2}{\partial y^2}\right)$$

denotes the hyperbolic Laplacian and we use the notation and fact

$$\Delta u_j(z) = \left(\tfrac{1}{4}+t_j^2\right) u_j(z), \qquad \Delta E(z, \tfrac{1}{2}+it) = \left(\tfrac{1}{4}+t^2\right) E(z, \tfrac{1}{2}+it),$$

then for any smooth and compactly supported $g(z)$ and for any $B > 0$ we have

$$(14) \qquad \langle g, u_j\rangle \ll_{g,B} (1+|t_j|)^{-B}, \qquad \langle g, E(\cdot, \tfrac{1}{2}+it)\rangle \ll_{g,B} (1+|t|)^{-B}.$$

Therefore in order to establish Duke's theorem with an error term $\ll_g |d|^{-\delta'}$ it suffices to show that if g is a Hecke–Maass cusp form with $\langle g, g\rangle = 1$ or a standard Eisenstein series $E(\cdot, \frac{1}{2}+it)$ then for some fixed $\delta > 0$ and $A > 0$ the sums considered in (13) satisfy

$$(15) \qquad \sum_{\Lambda_d} \cdots \ll (1+|t|)^A |d|^{\frac{1}{2}-\delta},$$

where $t = t_g$ is the spectral parameter of g, i.e.

$$\Delta g(z) = \left(\tfrac{1}{4}+t^2\right) g(z).$$

At this point we remark that any such g has a Fourier decomposition of the form

$$g(x+iy) = c_1 y^{\frac{1}{2}+it} + c_2 y^{\frac{1}{2}-it} + \sqrt{y}\sum_{n\neq 0} \rho_g(n) K_{it}(2\pi|n|y) e^{2\pi i n x},$$

where $c_{1,2}$ are some constants[3] and K_{it} is a Bessel function. The Fourier coefficients $\rho_g(n)$ are proportional to the Hecke eigenvalues of g, and by a result of Hoffstein–Lockhart [**10**] we have the uniform bound (see also [**14**, (3.25)])

$$|\rho_g(1)| \ll_\varepsilon (1+|t|)^\varepsilon e^{\frac{\pi}{2}|t|}. \tag{16}$$

We note that for t bounded away from zero we have a similar lower bound, with exponent $-\varepsilon$ in place of ε, as proved by Iwaniec [**13**] (see also [**14**, Theorem 8.3]).

Now we state a formula which can be attributed to several people[4] and relates the sums in (15) to central values of automorphic L-functions:

$$\left|\sum_{\Lambda_d}\cdots\right|^2 = c_d\,|d|^{\frac{1}{2}}\,|\rho_g(1)|^2\,\Lambda\left(\tfrac{1}{2},g\right)\Lambda\left(\tfrac{1}{2},g\otimes\left(\tfrac{d}{\cdot}\right)\right), \tag{17}$$

where the factor c_d is positive and takes only finitely many different values. In this formula $\Lambda(s,\Pi)$ denotes the *completed* L-function; the finite part $L(s,\Pi)$ of the L-function is defined in terms of Hecke eigenvalues; the infinite part of the L-function is a product of exponential and gamma factors whose contribution in (17) is $\ll (1+|t|)e^{-\pi|t|}$ by Stirling's approximation. Using also (16) we conclude that (15) follows by a subconvex bound of the form

$$L\left(\tfrac{1}{2},g\otimes\left(\tfrac{d}{\cdot}\right)\right) \ll (1+|t|)^A\,|d|^{\frac{1}{2}-\delta},$$

where $\delta>0$ and $A>0$ are some fixed constants (different from those in (15)). In the case when g is a cusp form such a bound was proved by Duke–Friedlander–Iwaniec [**7**] for any $\delta<\frac{1}{22}$, by Bykovskiĭ [**3**] for any $\delta<\frac{1}{8}$, and by Conrey–Iwaniec [**4**] for any $\delta<\frac{1}{6}$. In the case when g is an Eisenstein series $E(\cdot,\frac{1}{2}+it)$ the above becomes

$$\left|L\left(\tfrac{1}{2}+it,\left(\tfrac{d}{\cdot}\right)\right)\right|^2 \ll (1+|t|)^A\,|d|^{\frac{1}{2}-\delta},$$

and this was established by Burgess [**2**] for any $\delta<\frac{1}{8}$, and by Conrey–Iwaniec [**4**] for any $\delta<\frac{1}{6}$.

We shall now formulate a refinement of (13) using the natural action of the narrow ideal class group H_d of $\mathbb{Q}(\sqrt{d})$ on Λ_d. This action comes from the natural bijection $H_d\leftrightarrow\Lambda_d$ which we describe in the Appendix. Note in particular that $|H_d|=h(d)$ by this bijection. Given some $z_0\in\Lambda_d$ when $d<0$ and some $G_0\in\Lambda_d$ when $d>0$, and given some subgroup $H\leqslant H_d$ one can ask if

$$\begin{aligned}\frac{1}{|H|}\sum_{\sigma\in H} g(z_0^\sigma) &\to \int_{\mathrm{SL}_2(\mathbb{Z})\backslash\mathcal{H}} g(z)\,d\mu(z), \qquad d\to-\infty,\\ \frac{1}{|H|\,2\ln(\lambda_d)}\sum_{\sigma\in H}\int_{G_0^\sigma} g(z)\,ds(z) &\to \int_{\mathrm{SL}_2(\mathbb{Z})\backslash\mathcal{H}} g(z)\,d\mu(z), \qquad d\to+\infty.\end{aligned} \tag{18}$$

Using characters of the abelian group H_d we can decompose the sums over H into twisted sums over H_d:

$$\sum_{\sigma\in H}\cdots = \sum_{\sigma\in H_d}\frac{1}{(H_d:H)}\sum_{\substack{\psi\in\widehat{H}_d\\ \psi_{|H}\equiv 1}}\psi(\sigma)\ldots = \frac{|H|}{|H_d|}\sum_{\substack{\psi\in\widehat{H}_d\\ \psi_{|H}\equiv 1}}\sum_{\sigma\in H_d}\psi(\sigma)\ldots$$

[3] $c_1=c_2=0$ if g is a cusp form, $c_1=|c_2|=1$ if g is an Eisenstein series $E(\cdot,\frac{1}{2}+it)$

[4] Dirichlet, Hecke, Siegel, Maass, Shimura, Waldspurger, Kohnen–Zagier, Duke, Katok–Sarnak, Guo, Zhang, Popa; see the references for (20) of which (17) is a special case

Note that the number of characters of H_d restricting to the identity character on H is $(H_d : H)$. Therefore if we have, uniformly for all characters $\psi : H_d \to \mathbb{C}^\times$ and for all L^2-normalized Hecke–Maass cusp forms or standard Eisenstein series in the role of g,

$$\begin{aligned} &\sum_{\sigma \in H_d} \psi(\sigma) g(z_0^\sigma) \ll (1+|t|)^A |d|^{\frac{1}{2}-\delta}, && d<0, \\ &\sum_{\sigma \in H_d} \psi(\sigma) \int_{G_0^\sigma} g(z)\, ds(z) \ll (1+|t|)^A |d|^{\frac{1}{2}-\delta}, && d>0, \end{aligned} \tag{19}$$

where $\delta > 0$ and $A > 0$ are fixed constants, then by the same discussion as above, the limits (18) follow with a strong error term $\ll_g |d|^{-\delta'}$ as long as

$$(H_d : H) \ll |d|^\eta$$

for any fixed constant $0 < \eta < \delta$.

The twisted sums in (19) can be related to central automorphic L-values similarly as in (17). The formula is based on the deep work of Waldspurger [**21**] and was carefully derived by Zhang [**22**] when $d < 0$ and by Popa [**19**] when $d > 0$:

$$\left| \sum_{\sigma \in H_d} \overline{\psi(\sigma)} \dots \right|^2 = c_d |d|^{\frac{1}{2}} |\rho_g(1)|^2 \Lambda\left(\tfrac{1}{2}, g \otimes f_\psi\right). \tag{20}$$

Here f_ψ is the so-called Jacquet–Langlands lift of ψ, discovered by Hecke [**12**] and Maass [**17**] in this special case: it is a modular form on $\mathcal{H}$ of level $|d|$ and nebentypus $\left(\frac{d}{\cdot}\right)$ with the same completed L-function as ψ. In particular, when g is an Eisenstein series $E(\cdot, \frac{1}{2}+it)$ the identity (20) follows from [**20**, pp. 70 and 88] and [**14**, (3.25)].

If the character $\psi : H_d \to \mathbb{C}^\times$ is real-valued then it is one of the genus characters discovered by Gauss [**9**]. In this case, as observed by Kronecker [**20**, p. 62],

$$\Lambda(s,\psi) = \Lambda(s, (\tfrac{d_1}{\cdot}))\Lambda(s, (\tfrac{d_2}{\cdot})),$$

where $d = d_1 d_2$ is a factorization of d into fundamental discriminants d_1 and d_2, whence (20) simplifies to

$$\left| \sum_{\sigma \in H_d} \overline{\psi(\sigma)} \dots \right|^2 = c_d |d|^{\frac{1}{2}} |\rho_g(1)|^2 \Lambda\left(\tfrac{1}{2}, g \otimes (\tfrac{d_1}{\cdot})\right) \Lambda\left(\tfrac{1}{2}, g \otimes (\tfrac{d_2}{\cdot})\right).$$

In fact (17) is the special case of this formula when ψ is the trivial character ($d_1 = 1$, $d_2 = d$). The necessary estimate (19) follows by the subconvex bounds discussed before:

$$L\left(\tfrac{1}{2}, g \otimes (\tfrac{d_i}{\cdot})\right) \ll (1+|t|)^A |d_i|^{\frac{1}{2}-\delta}, \qquad i = 1, 2.$$

If the character $\psi : H_d \to \mathbb{C}^\times$ is not real-valued then f_ψ is a cusp form of level $|d|$ and nebentypus $\left(\frac{d}{\cdot}\right)$, and we need a subconvex bound of the form

$$L\left(\tfrac{1}{2}, g \otimes f_\psi\right) \ll (1+|t|)^A |d|^{\frac{1}{2}-\delta}.$$

In the case when g is a cusp form such a bound was proved by Harcos–Michel [**11**] with $\delta = \frac{1}{3000}$. In the case when g is an Eisenstein series $E(\cdot, \frac{1}{2}+it)$ the above becomes

$$\left| L\left(\tfrac{1}{2}+it, \psi\right) \right|^2 \ll (1+|t|)^A |d|^{\frac{1}{2}-\delta},$$

and this was established by Duke–Friedlander–Iwaniec [**8**] with $\delta = \frac{1}{12000}$ and by Blomer–Harcos–Michel [**1**] with $\delta = \frac{1}{1000}$.

Finally we remark that the above ideas have been greatly extended by several researchers. The interested reader should consult the excellent survey of Michel–Venkatesh [**18**].

3. Appendix

In this Appendix we consider an arbitrary fundamental discriminant d and regard $\sqrt{d}$ as a complex number which lies on the positive real axis or positive imaginary axis depending on the sign of d. We show that the equivalence classes of forms of fundamental discriminant d can be mapped bijectively to narrow ideal classes of the quadratic number field $\mathbb{Q}(\sqrt{d})$ in a natural fashion. As the latter classes form an abelian group under multiplication this will exhibit a natural multiplication law on the equivalence classes of forms. This law, discovered by Gauss [**9**], is called *composition* in the classical theory.

Recall that a fractional ideal of $\mathbb{Q}(\sqrt{d})$ is a finitely generated $\mathcal{O}_d$-module contained in $\mathbb{Q}(\sqrt{d})$ and two nonzero fractional ideals are equivalent (in the narrow sense) if their quotient is a principal fractional ideal generated by a totally positive element of $\mathbb{Q}(\sqrt{d})$. Here "totally positive element" can clearly be changed to "element of positive norm" where the norm of $\mu \in \mathbb{Q}(\sqrt{d})$ is given by $N(\mu) = \mu\bar{\mu}$. Recall also that we can represent equivalence classes of forms of fundamental discriminant d by some

$$Q_i(x,y) = a_i x^2 + b_i xy + c_i y^2 = a_i(x - z_i y)(x - \bar{z}_i y), \qquad i = 1, \dots, h(d),$$

with

$$a_i > 0, \qquad z_i := \frac{-b_i + \sqrt{d}}{2a_i}, \qquad \bar{z}_i := \frac{-b_i - \sqrt{d}}{2a_i}.$$

It will suffice to show that each fractional ideal I of $\mathbb{Q}(\sqrt{d})$ is equivalent to some fractional ideal

$$I_i := \mathbb{Z} + \mathbb{Z}z_i, \qquad i = 1, \dots, h(d),$$

and that the fractional ideals I_i are pairwise inequivalent.

Any fractional ideal I can be written as

$$I = \mathbb{Z}\omega_1 + \mathbb{Z}\omega_2 \qquad \text{with} \qquad \frac{\bar{\omega}_1\omega_2 - \omega_1\bar{\omega}_2}{\sqrt{d}} > 0.$$

We associate to I (and ω_1, ω_2) the binary quadratic form

$$Q_I(x,y) := \frac{(x\omega_1 - y\omega_2)(x\bar{\omega}_1 - y\bar{\omega}_2)}{N(I)},$$

where $N(I) > 0$ is the absolute norm of I, i.e. the multiplicative function that agrees with $(\mathcal{O}_d : I)$ for integral ideals I. We claim first that $Q_I(x,y)$ has integral coefficients and discriminant d. To see the claim we can assume that I is an integral ideal since $Q_I(x,y)$ does not change if we replace I by nI (and ω_i by $n\omega_i$) for some positive integer n. Then ω_1, ω_2 and their conjugates are in $\mathcal{O}_d$ and the claim amounts to:

- $N(I) \mid \omega_1\bar{\omega}_1,\ \omega_1\bar{\omega}_2 + \bar{\omega}_1\omega_2,\ \omega_2\bar{\omega}_2$;
- $(\omega_1\bar{\omega}_2 - \bar{\omega}_1\omega_2)^2 = N(I)^2 d$.

The first statement follows from the fact that $\omega_1, \omega_2, \omega_1 + \omega_2$ are elements of I, hence their norms are divisible by $N(I)$. The second statement follows by writing $\mathcal{O}_d$ as $\mathbb{Z} + \mathbb{Z}\omega$ and then noting that

$$\begin{vmatrix} \omega_1 & \bar{\omega}_1 \\ \omega_2 & \bar{\omega}_2 \end{vmatrix}^2 = (\mathcal{O}_d : I)^2 \begin{vmatrix} 1 & 1 \\ \omega & \bar{\omega} \end{vmatrix}^2 = N(I)^2 d.$$

The claim implies that there is a unique i and a unique $\begin{pmatrix} \alpha & \beta \\ \gamma & \delta \end{pmatrix} \in \mathrm{SL}_2(\mathbb{Z})$ such that

$$Q_I(\alpha x + \beta y, \gamma x + \delta y) = Q_i(x, y).$$

We can write this as

$$\frac{N(\alpha\omega_1 - \gamma\omega_2)}{N(I)}(x - zy)(x - \bar{z}y) = a_i(x - z_i y)(x - \bar{z}_i y),$$

where

$$z := \frac{-\beta\omega_1 + \delta\omega_2}{\alpha\omega_1 - \gamma\omega_2}. \tag{21}$$

This implies immediately that

$$N(\alpha\omega_1 - \gamma\omega_2) = a_i N(I) > 0. \tag{22}$$

Then a straightforward calculation yields

$$\frac{z - \bar{z}}{\sqrt{d}} = \frac{\alpha\delta - \beta\gamma}{N(\alpha\omega_1 - \gamma\omega_2)} \frac{\bar{\omega}_1\omega_2 - \omega_1\bar{\omega}_2}{\sqrt{d}} > 0$$

which by

$$\frac{z_i - \bar{z}_i}{\sqrt{d}} = \frac{1}{a_i} > 0$$

forces that $z = z_i$. But then (21)–(22) imply that

$$I = \mathbb{Z}\omega_1 + \mathbb{Z}\omega_2 = \mathbb{Z}(\alpha\omega_1 - \gamma\omega_2) + \mathbb{Z}(-\beta\omega_1 + \delta\omega_2)$$

is equivalent to

$$\mathbb{Z} + \mathbb{Z}z = \mathbb{Z} + \mathbb{Z}z_i = I_i.$$

Now assume that I_i and I_j are equivalent, i.e. there is some $\mu \in \mathbb{Q}(\sqrt{d})$ such that

$$\mu(\mathbb{Z} + \mathbb{Z}z_i) = \mathbb{Z} + \mathbb{Z}z_j, \qquad N(\mu) > 0.$$

Then we certainly have some $\begin{pmatrix} \alpha & \beta \\ \gamma & \delta \end{pmatrix} \in \mathrm{GL}_2(\mathbb{Z})$ such that

$$\mu = \alpha + \beta z_j, \qquad \mu z_i = \gamma + \delta z_j.$$

In particular,

$$z_i = \frac{\gamma + \delta z_j}{\alpha + \beta z_j} \qquad \text{with} \qquad N(\alpha + \beta z_j) > 0.$$

By a straightforward calculation as before,

$$\frac{z_i - \bar{z}_i}{\sqrt{d}} = \frac{\alpha\delta - \beta\gamma}{N(\alpha + \beta z_j)} \frac{z_j - \bar{z}_j}{\sqrt{d}},$$

which shows that

$$\alpha\delta - \beta\gamma = 1 \qquad \text{and} \qquad N(\alpha + \beta z_j) = \frac{z_j - \bar{z}_j}{z_i - \bar{z}_i} = \frac{a_i}{a_j}.$$

Now we obtain

$$a_i(x - z_i y)(x - \bar{z}_i y) = a_j\big((\alpha + \beta z_j)x - (\gamma + \delta z_j)y\big)\big((\alpha + \beta \bar{z}_j)x - (\gamma + \delta \bar{z}_j)y\big),$$

i.e.

$$Q_i(x, y) = Q_j(\alpha x - \gamma y, -\beta x + \delta y), \qquad \begin{pmatrix} \alpha & -\gamma \\ -\beta & \delta \end{pmatrix} \in \mathrm{SL}_2(\mathbb{Z}).$$

This clearly implies that $i = j$, since otherwise the forms Q_i and Q_j are inequivalent.

Incidentally, we see that the equivalence class of the associated form $Q_I(x, y)$ only depends on the narrow class of I (in particular, it is independent of the choice of ordered basis of I) and two fractional ideals I and J are in the same narrow class if and only if $Q_I(x, y)$ and $Q_J(x, y)$ are equivalent.

References

[1] V. Blomer, G. Harcos, P. Michel, *Bounds for modular L-functions in the level aspect*, Ann. Sci. École Norm. Sup. **40** (2007), 697–740.

[2] D. A. Burgess, *On character sums and L-series. II*, Proc. Lond. Math. Soc. **13** (1963), 524–536.

[3] V. A. Bykovskiĭ, *A trace formula for the scalar product of Hecke series and its applications*, translated in J. Math. Sci (New York) **89** (1998), 915–932.

[4] B. Conrey, H. Iwaniec, *The cubic moment of central values of automorphic L-functions*, Ann. of Math. **151** (2000), 1175–1216.

[5] H. Davenport, *Multiplicative number theory [revised and with a preface by H. L. Montgomery]*, 3rd edition, Graduate Texts in Mathematics 74, Springer-Verlag, New York, 2000.

[6] W. Duke, *Hyperbolic distribution problems and half-integral weight Maass forms*, Invent. Math. **92** (1988), 73–90.

[7] W. Duke, J. Friedlander, H. Iwaniec, *Bounds for automorphic L-functions*, Invent. Math. **112** (1993), 1–8.

[8] W. Duke, J. Friedlander, H. Iwaniec, *The subconvexity problem for Artin L-functions*, Invent. Math. **149** (2002), 489–577.

[9] C. F. Gauss, *Disquisitiones arithmeticae [translated by A. A. Clarke and revised by W. C. Waterhouse, C. Greither and A. W. Grootendorst]*, Springer-Verlag, New York, 1986.

[10] J. Hoffstein, P. Lockhart, *Coefficients of Maass forms and the Siegel zero (with an appendix by D. Goldfeld, J. Hoffstein and D. Lieman)*, Ann. of Math. **140** (1994), 161–181.

[11] G. Harcos, P. Michel, *The subconvexity problem for Rankin–Selberg L-functions and equidistribution of Heegner points. II*, Invent. Math. **163** (2006), 581–655.

[12] E. Hecke, *Über Modulfunktionen und die Dirichletschen Reihen mit Eulerscher Produktentwicklung. I.*, Math. Ann. **114** (1937), 1–28; *II.*, ibid. **114** (1937), 316–351.

[13] H. Iwaniec, *Small eigenvalues of Laplacian for $\Gamma_0(N)$*, Acta Arith. **56** (1990), 65–82.

[14] H. Iwaniec, *Spectral methods of automorphic forms*, 2nd edition, Graduate Studies in Mathematics 53, American Mathematical Society, Providence, RI; Revista Matemática Iberoamericana, Madrid, 2002.

[15] J. L. Lagrange, *Recherche d'arithmétique*, Nouv. Mém. Acad. Berlin (1773), 265–312.

[16] Y. V. Linnik, *Ergodic properties of algebraic fields [translated from the Russian by M. S. Keane]*, Ergebnisse der Mathematik und ihrer Grenzgebiete, Band 45, Springer-Verlag, New York, 1968.

[17] H. Maass, *Über eine neue Art von nichtanalytischen automorphen Funktionen und die Bestimmung Dirichletscher Reihen durch Funktionalgleichungen*, Math. Ann. **121** (1949), 141–183.

[18] P. Michel, A. Venkatesh, *Equidistribution, L-functions and ergodic theory: on some problems of Yu. Linnik*, International Congress of Mathematicians. Vol. II, 421–457, Eur. Math. Soc., Zürich, 2006.

[19] A. Popa, *Central values of Rankin L-series over real quadratic fields*, Compos. Math. **142** (2006), 811–866.

[20] C. L. Siegel, *Advanced analytic number theory*, 2nd edition, Tata Institute of Fundamental Research Studies in Mathematics, 9, Tata Institute of Fundamental Research, Bombay, 1980.

[21] J.-L. Waldspurger, *Sur les valeurs de certaines fonctions L automorphes en leur centre de symétrie*, Compos. Math. **54** (1985), 173–242.
[22] S. Zhang, *Gross–Zagier formula for* GL_2, Asian J. Math. **5** (2001), 183–290.

Alfréd Rényi Institute of Mathematics, Hungarian Academy of Sciences, POB 127, Budapest H-1364, Hungary
E-mail address: gharcos@renyi.hu

Clay Mathematics Proceedings
Volume 10, 2010

Eigenfunctions of the laplacian on negatively curved manifolds : a semiclassical approach

Nalini Anantharaman

Contents

An introduction to semiclassical analysis.

1. Mechanics.

1.1. Three approaches to classical mechanics. The variational approach. The Maupertuis or Euler principle [**M1744, E1744**] is the mechanical analogue of the Fermat principle in optics: a solid of mass $m = 1$, moving under

Key words and phrases. Quantum chaos, Schrödinger equation, quantum unique ergodicity.

the effect of a force $F = -\mathrm{grad}\, V$, with a total energy E, follows a trajectory γ which minimizes the action

$$S(\gamma) = \int \sqrt{2(E - V(\gamma))}\|d\gamma\| \tag{1.1}$$

among all curves with the same endpoints, and under the constraint that $\frac{\|\dot\gamma(t)\|^2}{2} + V(\gamma(t)) = E$ for all t. More precisely, we should look for *critical points* of S, among all paths with given endpoints, and constant total energy E. In these notes, we work on a riemannian manifold (X, g), and $\|.\|_x$ is the norm defined on T_xX by the riemannian metric : $\|v\|_x^2 = g_x(v, v)$. In other words, the Maupertuis principle says that the trajectories of energy E are geodesics for a new, degenerate metric, $2(E - V(x))g_x$.

The dual formulation, due to Lagrange [**L1788**], is to find the extrema of the functional

$$A(\gamma) = \int_0^T \left(\frac{\|\dot\gamma(t)\|^2_{\gamma(t)}}{2} - V(\gamma(t)) \right) dt \tag{1.2}$$

among all curves going from x to y in a given time T. Let us introduce the lagrangian $L(x, v) = \frac{\|v\|_x^2}{2} - V(x)$; the movement is described by the Euler-Lagrange equation

$$\frac{d}{dt}\left(\frac{\partial L}{\partial v}(\gamma, \dot\gamma)\right) = \frac{\partial L}{\partial x}(\gamma, \dot\gamma), \tag{1.3}$$

or more explicitly $D_{\dot\gamma}\dot\gamma = -\mathrm{grad}\, V(\gamma)$. This second order equation defines a local flow (ϕ^t_{EL}) on the tangent bundle TX, called the Euler-Lagrange flow.

Hamiltonian point of view. The hamiltonian is the Fenchel–Legendre transform of L with respect to the variable v :

$$H(x, \xi) = \xi.v - L(x, v)$$

with $\xi = \frac{\partial L}{\partial v}$; we are in a nice situation where the Legendre transformation

$$\mathcal{L}\mathrm{eg} : (x, v) \mapsto \left(x, \frac{\partial L}{\partial v}\right)$$

defines a diffeomorphism between the tangent bundle TX and the cotangent bundle T^*X. Its inverse is $\mathcal{L}\mathrm{eg}^{-1} : (x, \xi) \mapsto \left(x, \frac{\partial H}{\partial \xi}\right)$. In fact, in our case, $\mathcal{L}\mathrm{eg}$ is nothing else than the natural identification between TX and T^*X, provided by the riemannian metric. We can define a scalar product g^x on T_x^*X by $g^x(\xi, \xi) = g_x(v, v) = \|v\|_x^2$, with $\xi = \frac{\partial L}{\partial v}$. The vector ξ is called the momentum, and $H(x, \xi) = \frac{g^x(\xi,\xi)}{2} + V(x)$ is the total energy of the system. We shall also denote $g^x(\xi, \xi) = \|\xi\|_x^2$, but the reader should not confuse the norms $\|.\|_x$ on T_x^*X and T_xX.

The Euler-Lagrange equation (1.3) is equivalent to Hamilton's system of equations,

$$\begin{cases} \dot x = \frac{\partial H}{\partial \xi} \\ \dot \xi = -\frac{\partial H}{\partial x}, \end{cases} \tag{1.4}$$

which define a local flow (ϕ^t_H) on T^*X, called the hamiltonian flow. This flow is conjugate to (ϕ^t_{EL}) via the diffeomorphism $\mathcal{L}\mathrm{eg}$. It preserves the energy H, in the sense that $H(x(t), \xi(t))$ is constant for any trajectory of the flow $(x(t), \xi(t))$. The hamiltonian flow also preserves the Liouville measure $dx\, d\xi$.

If a is a function on T^*X (an "observable quantity" in the language of Heisenberg), and if we denote $a_t = a \circ \phi_H^t$, we have

$$\frac{da}{dt} = \{H, a\},$$

where $\{.\,,.\}$ denotes the Poisson bracket, $\{H, a\} = \sum \partial_{\xi_j} H\, \partial_{x_j} a - \partial_{\xi_j} a\, \partial_{x_j} H$.

A more intrinsic way of writing the Hamilton equations (1.4) would be to note that the vector field on the right hand side is the symplectic gradient of H, with respect to the *canonical symplectic form* on T^*X. Let us define the Liouville 1-form on the cotangent bundle, defined by

$$\alpha_{(x,\xi)}(P) = \xi . d\pi(P) \text{ for all } P \in T_{(x,\xi)}(T^*X),$$

where $\pi : T^*X \longrightarrow X$ is the usual projection, and $d\pi$ its tangent map. The cotangent bundle T^*X can be endowed with the symplectic form

$$\omega = -d\alpha. \tag{1.5}$$

In local coordinates, $\alpha = p.dq$ and $\omega = dq \wedge dp$, if p and q denote respectively the "momentum" and "position" functions, $p(x, \xi) = \xi$, $q(x, \xi) = x$. The reader can check that the right hand side of (1.4) is the expression in local coordinates of the symplectic gradient X_H of H, defined by $dH = \omega(X_H, .)$. The Poisson bracket is given by $\{f, g\} = -\omega(X_f, X_g) = dg(X_f)$, for any two functions f, g on T^*X.

One can show that the flow ϕ_H preserves the symplectic form ω. In the language of symplectic geometry, a (local) diffeomorphism of T^*X which preserves ω is called a canonical transformation.

Hamilton–Jacobi equation, generating functions. This third point of view, called the Hamilton–Jacobi approach, meets many technical difficulties, but it is the key tool to understand the semiclassical analysis of the Schrödinger equation.

Around 1830, Hamilton introduced a new formalism, in which the action is seen as a function of the endpoints x and y [**H1830, H1834**]. Let $\gamma : [0, T] \longrightarrow X$ be a solution of the Euler–Lagrange equation, joining x to y in time $T > 0$. To simplify the discussion, we consider here the nice, but usually unrealistic situation, where such a trajectory is unique. We can then consider the lagrangian action $A(x, y; T) = \int_0^T L(\gamma, \dot\gamma) dt$ as a function of x, y, T, and check that

$$\frac{\partial A}{\partial x} = -\dot\gamma(0); \; \frac{\partial A}{\partial y} = \dot\gamma(T), \tag{1.6}$$

and

$$\frac{\partial A}{\partial T} = -E$$

where E is the energy $E = \frac{\|\dot\gamma\|^2}{2} + V(\gamma)$, constant along the trajectory γ. If we freeze the variable y (thus fixing an initial or rather "final" condition) and see A as a function of $x \in X$ and $T > 0$, we have

$$\frac{\partial A}{\partial T} + H(x, \partial_x A) = 0. \tag{1.7}$$

Hamilton then argues that being able to integrate the hamiltonian vector field (1.4) is equivalent to finding the generating function A, solution of the Hamilton–Jacobi equation (1.7) for any initial condition (or a large enough family of initial conditions). By this procedure, the ordinary differential equations (1.3) or (1.4) have been replaced by a single PDE. Quoting Hamilton, *"even if it should be thought that no practical facility is gained, yet an intellectual pleasure may result from the reduction of [...] all researches respecting the forces and motions of body, to the study of one characteristic function"*.

Let us also consider the Legendre transform of $A(x, y; T)$ with respect to the variable T,

$$S(x, y; E) = ET + A(x, y; T) \tag{1.8}$$

where T and E are related by

$$\frac{\partial A}{\partial T} = -E,$$

which implies that

$$\frac{\partial S}{\partial E} = T.$$

The function S is nothing else but the Maupertuis action (1.1) of the trajectory γ joining x to y with energy E:

$$S(x, y; E) = \int \sqrt{2(E - V(\gamma))}\|\dot{\gamma}\|dt = \int_0^T \|\dot{\gamma}\|^2 dt.$$

We still have

$$\frac{\partial S}{\partial x} = -\dot{\gamma}(0); \quad \frac{\partial S}{\partial y} = \dot{\gamma}(T). \tag{1.9}$$

If we freeze the final state y, the function S solves the stationary Hamilton–Jacobi equation,

$$H(x, \partial_x S) = E. \tag{1.10}$$

The solutions of the time-dependent Hamilton–Jacobi equation (1.7) and of the stationary equation (1.10) are related by the Legendre transform (1.8).

The Hamilton–Jacobi equation (1.7) has a simple geometrical intepretation. Consider a subset of the cotangent bundle T^*X, of the form $\mathcal{L}_0 = \{(x, d_x A_0), x \in \Omega_0\}$, with Ω_0 an open subset of X. This is a particular case of a *lagrangian submanifold* in T^*X (see Definition 12.1). Let $\mathcal{L}_0$ evolve under the hamiltonian flow, and consider $\mathcal{L}_t = \phi_H^t \mathcal{L}_0$: because ϕ_H^t preserves the symplectic form ω, $\mathcal{L}_t$ is still a lagrangian manifold. Let us assume that, for $t \in [0, T]$, $\mathcal{L}_t$ still projects diffeomorphically to an open subset of $\Omega_t \subset X$. This means exactly that $\mathcal{L}_t$ is of the form $\mathcal{L}_t = \{(x, d_x A_t), x \in \Omega_t\}$ for some smooth function A_t. It can be shown that the relation $\mathcal{L}_t = \phi_H^t \mathcal{L}_0$ is equivalent to A_t solving the Hamilton–Jacobi equation (1.7), with the condition that Ω_t is the image of $\mathcal{L}_0$ under the "exponential" map associated with $\mathcal{L}_0$:

$$\exp^t_{\mathcal{L}_0} : \mathcal{L}_0 \longrightarrow X, \tag{1.11}$$

$$\xi \mapsto \pi(\phi_H^t \xi) \tag{1.12}$$

(the notation π denotes the projection $T^*X \longrightarrow X$).

This approach suffers from the notorious problem of caustics (Figure 1). Usually, the exponential map will only be a diffeomorphism if the energy H is bounded on $\mathcal{L}_0$, and if t is small enough. For large times two kinds of problems arise,

– exp is not injective (two trajectories starting in $\mathcal{L}_0$ land at the same point in X)

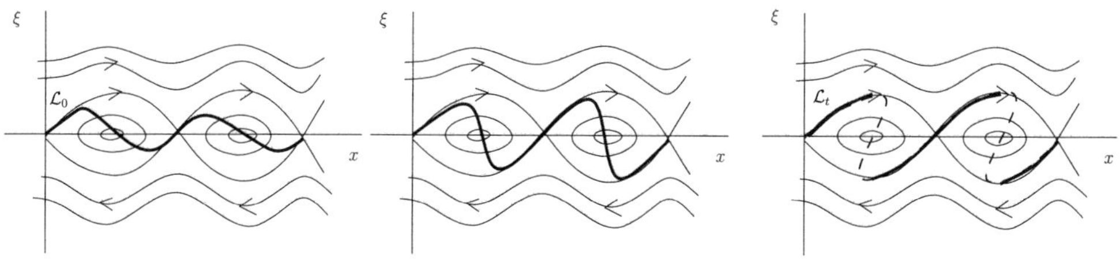

FIGURE 1. Appearance of caustics for large times.

– the tangent map $d\exp$ is not injective (focal points, conjugate points).

Geometrically, this means that after some time $\mathcal{L}_t$ will cease to project diffeomorphically to X. From a PDE point of view, this means that the equation (1.7) does not, in general, have globally defined smooth solutions.

Although the problem of caustics makes the Hamilton–Jacobi equation rather difficult to work with, it is, nevertheless, the key tool to understand Schrödinger's equation and its semiclassical analysis. Semiclassical methods often break down with the appearance of caustics, or a little after.

We now review Schrödinger's view of mechanics, but also the work of Born, Heisenberg and Jordan, which lead to the idea of quantization.

1.2. Quantum/wave mechanics. At the beginning of the twentieth century, it became clear that classical mechanics was not applicable to certain problems, like the study of energy radiation in atoms. People started looking for new physical laws, but it was not until 1925 that theories judged as satisfactory were elaborated. These theories involve Planck's constant $h = 2\pi\hbar = 6.626068 \times 10^{-34} m^2.\text{kg}/s$ (the "action quantum"), and one is supposed to recover classical mechanics when letting h tend to 0 in the equations.

Quantenmechanik. In 1925, Heisenberg, Born and Jordan gave some new laws of mechanics, supposed to replace the old Hamilton equations (1.4). Consider a hamiltonian system with d degrees of freedom, meaning that the manifold X has dimension d. In fact let us take $X = \mathbb{R}^d$ as in the paper [**BHJ25-II**]. In classical mechanics the time evolution is given by equation (1.4), defining a symplectic flow on the phase space T^*X. According to the quantum mechanics of [**BHJ25-II**], the time evolution of the system is ruled by the five following principles :

(0) The "phase space" is a Hilbert space $\mathcal{H}$.

(1) The "observable quantities" are described by linear operators (= infinite matrices). Heisenberg, Born and Jordan used a boldface letter $\mathbf{a}$ to denote the quantum observable corresponding to the classical observable a; if a is a real–valued function on T^*X then the corresponding operator $\mathbf{a}$ is hermitian.

(2) Main rules : We consider, in particular, an algebra of operators generated by the momentum and position observables, $\mathbf{p} = (\mathbf{p}_1, \mathbf{p}_2, ..., \mathbf{p}_d)$ and $\mathbf{q} =$

$(\mathbf{q}_1, ..., \mathbf{q}_d)$. These operators must obey the following commutation rules :

$$[\mathbf{p}_k, \mathbf{q}_l] = \frac{\hbar}{i}\delta_{kl}\, I, \tag{1.13}$$

$$[\mathbf{p}_k, \mathbf{p}_l] = 0, \tag{1.14}$$

$$[\mathbf{q}_k, \mathbf{q}_l] = 0. \tag{1.15}$$

Consider now a classical observable f defined by a power series

$$f(p,q) = \sum \alpha_{sr} p^s q^r.$$

Then the quantum observable $\mathbf{f}$ should be defined by

$$\mathbf{f}(\mathbf{p},\mathbf{q}) = \sum \alpha_{sr} \frac{1}{s+1} \sum_{l=0}^{s} \mathbf{p}^{s-l}\mathbf{q}^r\mathbf{p}^l.$$

This prehistoric "quantization rule" can be applied, in particular, to define the hamiltonian operator $\mathbf{H}$.

(3) A canonical transformation is a transformation that sends the observables $(\mathbf{p}, \mathbf{q})$ to new observables $(\mathbf{P}, \mathbf{Q})$ satisfying the same commutation relations. We ask that a canonical transformation preserve hermitian operators, and sends an observable of the form $\mathbf{f}(\mathbf{p}, \mathbf{q})$ to $\mathbf{f}(\mathbf{P}, \mathbf{Q})$. Such a transformation is of the form $\mathbf{P} = \mathbf{S}\mathbf{p}\mathbf{S}^{-1}$, $\mathbf{Q} = \mathbf{S}\mathbf{q}\mathbf{S}^{-1}$, where $\mathbf{S}$ is a unitary operator.

(4) The equations of motion are

$$\begin{cases} \dot{\mathbf{p}} = -\frac{\partial \mathbf{H}}{\partial \mathbf{q}} \\ \\ \dot{\mathbf{q}} = \frac{\partial \mathbf{H}}{\partial \mathbf{p}}, \end{cases} \tag{1.16}$$

where we define

$$\frac{\partial \mathbf{f}}{\partial \mathbf{x}_1} = \lim_{\varepsilon \longrightarrow 0} \frac{1}{\varepsilon} \left(\mathbf{f}(\mathbf{x}_1 + \varepsilon I, \mathbf{x}_2, ..., \mathbf{x}_s) - \mathbf{f}(\mathbf{x}_1, \mathbf{x}_2, ..., \mathbf{x}_s)\right)$$

for $\mathbf{f}(\mathbf{x}_1, \mathbf{x}_2, ..., \mathbf{x}_s)$ a power series in the s observables $\mathbf{x}_1, \mathbf{x}_2, ..., \mathbf{x}_s$ (I is the identity operator).

It can be shown from formula (1.13) that we have the identity

$$[\mathbf{f}, \mathbf{g}] = \frac{\hbar}{i}\left(\frac{\partial \mathbf{f}}{\partial \mathbf{p}}\frac{\partial \mathbf{g}}{\partial \mathbf{q}} - \frac{\partial \mathbf{f}}{\partial \mathbf{q}}\frac{\partial \mathbf{g}}{\partial \mathbf{p}}\right)$$

holding for $\mathbf{f}, \mathbf{g}$ power series in the operators $\mathbf{p}$ and $\mathbf{q}$.

In particular, the equations (1.16) can be reexpressed as

$$\dot{\mathbf{f}} = \frac{i}{\hbar}[\mathbf{H}, \mathbf{f}]$$

for any observable $\mathbf{f}$.

(5) To integrate the equation of motion, we must find a unitary operator $\mathbf{S}$ such that

$$\mathbf{S}\mathbf{H}\mathbf{S}^{-1} = \mathbf{W} \tag{1.17}$$

is diagonal. In other words, we look for a canonical transformation which allows to express the solutions of (1.16) as a superposition of periodic motions[1].

In a basis where $\mathbf{H}$ is diagonal, we find, for any observable $\mathbf{f}$, that the matrix elements evolve according to

$$\mathbf{f}_{nm}(t) = \mathbf{f}_{nm}(0)e^{2i\pi\nu_{nm}t} \tag{1.18}$$

where the radiation sprectrum ν_{nm} ("physical spectrum") is related to the eigenvalues (E_n) of $\mathbf{H}$ ("mathematical spectrum") by

$$\nu_{nm} = \frac{E_n - E_m}{h}.$$

Wellenmechanik. In 1926, Erwin Schrödinger, independently of the work of Heisenberg, Born and Jordan, proposed a new equation, supposed to describe the state of our system submitted to a force field $-\mathrm{grad}\, V$, when the value of the energy E is given : the "stationary" Schrödinger equation is a second order elliptic PDE,

$$-\frac{\hbar^2}{2}\Delta\psi + V\psi = E\psi, \tag{1.19}$$

where E is the energy. As we shall see, this equation is closely related to the stationary Hamilton–Jacobi equation (1.10). The corresponding evolution equation reads

$$i\hbar\frac{\partial\phi}{\partial t} = \left(-\frac{\hbar^2}{2}\Delta + V\right)\phi. \tag{1.20}$$

These two forms of the equation are related by a time/energy Fourier transform $\phi(t) = \int e^{-iEt/\hbar}\psi_E dE$, which recalls the relation (1.8).

According to Schrödinger's theory, the energy spectrum can be computed by finding the values of E for which equation (1.19) admits solutions which are *"single–valued, finite, and continuous throughout configuration space"*.

Schrödinger, motivated by the works of De Broglie, gives an interpretation of ψ as a "wave function". *"The true mechanical process is realised or represented in a fitting way by the* wave processes *in q–space, and not by the motion of* image

[1]The analogy with the theory of classical hamiltonian systems can be pushed further. In fact, equation (1.16) is a linear hamiltonian flow, in an infinite dimensional space. Such systems are completely integrable, due to the fact that a unitary transformation diagonalizing $\mathbf{H}$ always exists. To be more explicit, let $(\mathcal{H}, \langle .,.\rangle)$ be a complex Hilbert space, seen as a real vector space endowed with the symplectic form $\omega(\phi,\psi) = \Im m\langle\phi,\psi\rangle$. If we use an orthonormal basis (e_n) to define coordinates, $\phi = \sum_n (x_n + i\xi_n)e_n$, then (x_n, ξ_n) are Darboux coordinates, meaning that $\omega = \sum_n dx_n \wedge d\xi_n$.

Let $\mathbf{H}$ be a self–adjoint operator; it can be used to define a quadratic hamiltonian $H(\psi) = \frac{1}{2}\langle\psi, \mathbf{H}\psi\rangle$. If we consider quadratic observables, $f(\psi) = \frac{1}{2}\langle\psi, \mathbf{f}\psi\rangle$, then the Poisson bracket defined by ω correspond to the usual commutator bracket,

$$\{f,g\}(\psi) = \frac{1}{2}\langle\psi, i[\mathbf{f},\mathbf{g}]\psi\rangle.$$

The Hamilton equations defined by H read $\frac{d\psi}{dt} = -i\mathbf{H}\psi$. Finally, linear transformations preserving ω are of the form $\psi \mapsto \mathbf{S}\psi$ where $\mathbf{S}$ is unitary.

Thus, finding a unitary $\mathbf{S}$ such that $\mathbf{S}^{-1}\mathbf{H}\mathbf{S}$ is diagonal amounts to finding a linear canonical transformation $\psi \mapsto \mathbf{S}\psi$ which transforms the hamiltonian H into $H(\mathbf{S}\psi) = \frac{1}{2}\sum(2\pi\nu_n)^2(x_n^2 + \xi_n^2)$. This means that we can integrate the equation of motion by decomposing it into a superposition of infinitely many independent harmonic oscillators.

points *in this space. The study of the motion of image points, which is the object of classical mechanics, is only an approximate treatment, and has, as such, just as much justification as geometrical or "ray" optics has, compared with the true optical process".* This approximation is only justified when the dimensions of the system are very large compared to the wave length : *"we inevitably became involved in irremovable contradictions if we tried, as was very natural, to maintain also the idea of* paths of systems *in these processes; just as we find the tracing of the course of* a light ray *to be meaningless, in the neighbourhood of a diffraction phenomenon".*

It is particularly interesting for us to note that Schrödinger derived the form of his equation by a heuristic argument, based on the desired asymptotic behaviour of the solutions when $\hbar \longrightarrow 0$:

Assume that our mechanical phenomenon is described by a wave function ψ, and assume that this wave has the particular form : $\psi(x,0) = \exp\left(i\frac{A(x,0)}{\hbar} + C\right)$ at $t=0$. Assume also that for $t>0$ the wave ψ looks like

$$\psi(x,t) \sim \exp\left(i\frac{A(x,t)}{\hbar} + C\right) + \text{ small error.} \tag{1.21}$$

To find the form of the equation satisfied by ψ, Schrödinger postulates that the phase A must approximately satisfy the Hamilton-Jacobi equation (1.7), when the wave length is very small (semiclassical approximation). In other words, we must (almost) see the wave move according to the classical motion. Thus, the point is to find an equation, looking like a wave equation, and such that (1.21) is an approximate solution if $\frac{\partial A}{\partial t} + H(x, d_x A) = 0$ (equation (1.7)) and $\hbar \longrightarrow 0$.

To find such an equation, Schrödinger actually works with the stationary formulation : this means that $A(x,t)$ is of the form $A(x,t) = -Et + S(x)$ where S solves $H(x, d_x S) = E$ (equation (1.10)). If S satisfies (1.10), then the local speed of propagation of ψ is

$$u(x) = \frac{-\frac{\partial A}{\partial t}}{|\nabla A|} = \frac{E}{\sqrt{2(E - V(x))}}$$

and the wave length is $\lambda(x) = \frac{\hbar}{\sqrt{2(E-V(x))}}$. This encourages Schrödinger to propose the equation

$$\frac{\partial^2 \psi}{\partial t^2} = u^2 \Delta\psi.$$

From the expression of u, and since the formula (1.21) is supposed to give an approximate solution when $\lambda \longrightarrow 0$, we find $-\frac{\hbar^2}{2}\Delta\psi + V\psi = E\psi$.

Let us now consider the propagation of an arbitrary wave ψ, and let us try to put Schrödinger's discussion into mathematical words. At time $t=0$, any initial state ψ can be written as

$$\psi(x) \sim \int a(x,\theta) \exp\left(\frac{i}{\hbar} A(x;\theta)\right) d\theta, \tag{1.22}$$

where θ varies in an open set of $\mathbb{R}^d$, $\exp\left(\frac{i}{\hbar}A(x;\theta)\right)$ is a generating family parametrized by θ, and a is a distribution. In $\mathbb{R}^d$, we can for instance take the plane waves $\exp\left(\frac{i}{\hbar}A(x;\theta)\right) = \exp\left(\frac{i}{\hbar}\langle x, \theta\rangle\right)$, and the decomposition (1.22) is the Fourier decomposition. By linearity of the Schrödinger equation, and by the approximate form of the solutions (1.21), after time t the wave looks like

$$\psi(t,x) \sim \int a(x,\theta) \exp\left(\frac{i}{\hbar} A(t,x;\theta)\right) d\theta, \tag{1.23}$$

where $A(t,x;\theta)$ is the solution of (1.7) with initial condition $A(x;\theta)$. If the oscillations are very rapid (λ small) we expect all these waves to interfer destructively, except at those points x where the phase has a stationary point,

$$\partial_\theta A(t,x,\theta_0) = 0$$

(for some θ_0). At such a point, we see essentially the wave $\exp\left(\frac{i}{\hbar} A(t,x;\theta_0)\right)$, with the frequency vector $\xi = \partial_x A(t,x,\theta_0)$. Thus, the wave front at time t can be represented by the subset of the cotangent space

$$\mathcal{L}(t) = \{(x,\xi), \text{ there exists } \theta_0, \partial_\theta A(t,x,\theta_0) = 0, \ \xi = \partial_x A(t,x,\theta_0)\}. \tag{1.24}$$

Assuming each $A(.,.,\theta)$ satisfies the Hamilton–Jacobi equation, one can check that $\mathcal{L}(t)$ is precisely the image of $\mathcal{L}(0)$ under the hamiltonian flow (1.4) at time t (see Exercise 12.9). In other words, the wave front is propagated according to the classical hamiltonian flow.

"The point of phase agreement for certain infinitesimal manifolds of wave systems, containing n parameters, moves according to the same laws as the image point of the mechanical system" [**Schr26-II**].

Recall that this is an approximation, valid when the wave length λ is very small; for mathematicians, this is the same as letting $\hbar$ tend to 0, and this is called the semiclassical limit. Schrödinger's heuristic discussion already contain the seeds of semiclassical analysis. Classical mechanics is obtained as a limiting case of wave mechanics by a phenomenon of constructive or destructive interferences.

Schrödinger writes : *"I consider it a very difficult task to give an exact proof that the superposition of these wave systems really produces a noticeable disturbance in only a relatively small region surrounding the point of phase agreement, and that everywhere else they practically destroy one another through interference"* [**Schr26-II**]. As we shall see in Section 4.2, this problem can in fact be handled by the stationary phase method, if we impose strong smoothness conditions on the distribution a.

2. Weyl quantization.

In [**Schr26-III**], Schrödinger realizes, in the case of $X = \mathbb{R}^d$, that his "wave mechanics" is equivalent to the "quantum mechanics" introduced by Born, Heisenberg and Jordan. The equivalence comes from the existence of an explicit *quantization procedure* , that is, a way to associate, to every function on the classical phase space $T^*X = \mathbb{R}^d \times \mathbb{R}^d$, an operator on the Hilbert space $\mathcal{H} = L^2(\mathbb{R}^d)$, so that the commutation rules (1.13) hold. Schrödinger's remark is that we can take the operators $\mathbf{q}_k =$ (multiplication by q_k) to the coordinate function q_k, and the operator $\mathbf{p}_k = \frac{\hbar}{i}\frac{\partial}{\partial q_k}$ to the function p_k.

One must then decide of a convention to define the operator $\mathbf{a}(\mathbf{q},\mathbf{p})$ associated to an arbitrary function $a(q,p)$ of (q,p). For instance, the function $p_k q_k$ could be represented by the operator $\mathbf{p}_k\mathbf{q}_k$ or by $\mathbf{q}_k\mathbf{p}_k$. Schrödinger leaves the issue open for general a, but recommends to quantize a hamiltonian of the form

$$H(q,p) = \frac{\|p\|^2}{2} + V(q),$$

where $\|.\|$ is a riemannian metric, by the operator $\mathbf{H} = -\frac{\hbar^2}{2}\triangle + V$, where $\triangle$ is the laplacian associated to the metric. In this representation, Heisenberg's equation (1.17), requiring to diagonalize the operator $\mathbf{H}$, can be written $-\frac{\hbar^2}{2}\triangle\psi + V\psi = E\psi$, which is exactly Schrödinger's equation (1.19). Thus, the two theories will give the same values of the energy spectrum. Schrödinger suggests, however, that two theories can be mathematically equivalent without being physically equivalent.

Weyl quantization. Hermann Weyl [**Weyl27**] gave a quantization rule which defines $\mathbf{a}(\mathbf{q},\mathbf{p})$ for any $a(q,p)$. He first proposed to quantize the observable $U_{p_0,q_0}(q,p) = e^{\frac{i}{\hbar}(p_0.q - q_0.p)}$ (with $q_0, p_0 \in \mathbb{R}^d$) by the operator $\mathbf{U}_{p_0,q_0}(\mathbf{q},\mathbf{p}) = e^{\frac{i}{\hbar}(p_0.\mathbf{q} - q_0.\mathbf{p})}$ (where $(\mathbf{q},\mathbf{p})$ are defined by Schrödinger's prescriptions). Then, the Fourier transform allows to quantize any observable : if a is decomposed into

$$a(q,p) = \int e^{\frac{i}{\hbar}(p_0.q - q_0.p)} \check{a}_\hbar(q_0,p_0)\frac{dq_0 dp_0}{(2\pi\hbar)^d},$$

then the Weyl quantization is

$$\mathbf{a}(\mathbf{q},\mathbf{p}) = \int e^{\frac{i}{\hbar}(p_0.\mathbf{q} - q_0.\mathbf{p})} \check{a}_\hbar(q_0,p_0)\frac{dq_0\, dp_0}{(2\pi\hbar)^d} =: \mathrm{Op}_\hbar^W(a).$$

We used the "symplectic" Fourier transform,

$$\check{a}_\hbar(q,p) = \int e^{\frac{-i}{\hbar}(p_0.q - q_0.p)} a(q,p)\frac{dqdp}{(2\pi\hbar)^d}.$$

One can also check that the following expression holds [**Foll**],

$$\mathrm{Op}_\hbar^W(a)f(x) = \frac{1}{(2\pi\hbar)^d}\int a\left(\frac{x+y}{2},\xi\right) e^{\frac{i}{\hbar}\xi.(x-y)} f(y)dy\, d\xi.$$

The Schrödinger representation. For $p = 0$, we have $U_{0,q}u(x) = u(x-q)$, so that $U_{0,q}$ corresponds to the translation of vector q in the position variable. Similarly, $U_{p,0}$ translates the Fourier transform $\mathcal{F}_\hbar u$ (defined in Section 4.1) by the vector p, and $U_{p,0}$ is interpreted as the translation of vector p in the momentum variable. To interpolate between these two cases, one usually says that $U_{p,q}$ is the operator corresponding to "translation of vector (q,p) in the phase space $\mathbb{R}^d \times \mathbb{R}^d$". The caveat is that $U_{0,q}$ and $U_{p,0}$ do not commute, in fact the operators $\mathbf{U}_{p,q}$ obey the following composition rule,

$$\mathbf{U}_{p,q}.\mathbf{U}_{p',q'} = \mathbf{U}_{p+p',q+q'} e^{\frac{i}{\hbar}\frac{1}{2}(pq'-q'p)}. \tag{2.1}$$

Consider the Heisenberg group $\mathbf{H}_d$ with d degrees of freedom, defined as $\mathbb{R}^{2d+1}$ endowed with the composition rule

$$(p,q,t).(p',q',t') = \left(p+p', q+q', t+t' + \frac{1}{2}(pq' - qp')\right), \quad (p,p',q,q' \in \mathbb{R}^d, t,t' \in \mathbb{R}).$$

Its Lie algebra is generated by $P_1, \ldots, P_d, Q_1, \ldots, Q_d, T$ with the relations

$$[P_j, P_k] = [Q_j, Q_k] = [P_j, T] = [Q_j, T] = 0; \;\; [P_j, Q_k] = \delta_{jk}T.$$

The identity (2.1) can be reinterpreted by saying that

$$\rho_h(p,q,t) = e^{\frac{it}{\hbar}}\mathbf{U}_{p,q}$$

defines a unitary representation from $\mathbf{H}_d$ into $L^2(\mathbb{R}^d)$, called the Schrödinger representation of parameter h. The associated infinitesimal representation is $P_k \mapsto \frac{\partial}{\partial q_k} = \frac{i}{\hbar}\mathbf{p}_k$, $Q_k \mapsto \frac{i}{\hbar}\mathbf{q}_k$, $T \mapsto \frac{i}{\hbar}I$.

THEOREM 2.1. *(Stone–von Neumann 1930* [**St30, vN31**]*, see* [**Foll**]*) Every irreducible unitary representation of* $\mathbf{H}_d$ *is equivalent to exactly one of the following representations :*

(a) ρ_h *(*$h \in \mathbb{R} \setminus \{0\}$*) acting on* $L^2(\mathbb{R}^d)$*;*

(b) $\sigma_{ab}(p,q,t) = e^{2\pi i(ap+bq)}$*, (*$a, b \in \mathbb{R}^d$*) acting on* $\mathbb{C}$*.*

3. Born's probabilitic interpretation of the Schrödinger equation.

Born discovered that the square modulus $|\psi|^2$ of the wave functions (satisfying the Schrödinger equation) could be used to predict the probability of where the "particle" would be found. More precisely, if ψ is normalized so that $\int |\psi(t,x)|^2 dx = 1$, then $|\psi(t,x)|^2$ gives the probability density of finding, in an experiment, the particle at x (at time t). This was the beginning of a tense philosophical (or physical) debate on the correct interpretation of the wave/particle duality.

"Let me say at the outset, that in this discourse, I am opposing not a few special statements of quantum physics held today (1950s), I am opposing as it were the whole of it, I am opposing its basic views that have been shaped 25 years ago, when Max Born put forward his probability interpretation, which was accepted by almost everybody. (E. Schrödinger, The Interpretation of Quantum Physics. Ox Bow Press, Woodbridge, CN, 1995).

"I don't like it, and I'm sorry I ever had anything to do with it" (Erwin Schrödinger talking about quantum physics).

4. The semiclassical limit.

Let us now turn to much more recent mathematical preoccupations. The main subject of these notes is to try to describe the localization of the probability density $|\psi(x)|^2dx$, for a Schrödinger eigenfunction, in the semiclassical limit $\hbar \longrightarrow 0$. The quantum/classical correspondence tells us, intuitively, that the eigenfunctions, which are stationary solutions of the Schrödinger equation, should look like invariant probability measures of the classical hamiltonian flow. In this section we give a quick survey (without proofs) of the mathematical tools used to study this question.

It is not really satisfactory, and usually practically impossible, to study the density $|\psi(x)|^2dx$ itself. This is because, when taking the modulus of ψ, we lose some precious information on the frequency vector of ψ (related to its phase, or complex argument). We need to study simultaneously the Fourier transform of ψ. Of course, rigourously speaking, one cannot study at the same time the *local* properties of a function and of its Fourier transform around some point $(x,\xi) \in T^*X$. This is expressed by Heisenberg's uncertainty principle, saying that one cannot localize a function around the point x without perturbing a lot the momentum (and vice-versa). Microlocal analysis[2] is a collection of mathematical techniques allowing to study the joint localization of a function and its Fourier transform; because of the uncertainty principle, this can only be meaningful asymptotically, in the limit $\hbar \longrightarrow 0$.

[2]More precisely, we will present here its $\hbar$-dependent version, also called semiclassical analysis, or "microlocal analysis with a small parameter".

4.1. Fourier transform. The Fourier transform

$$\mathcal{F}_\hbar(u)(\xi) = \hat{u}_\hbar(\xi) = (2\pi\hbar)^{-d/2} \int_{\mathbb{R}^d} e^{-\frac{i}{\hbar}\xi.x} u(x) \mathrm{d}x$$

allows to analyze a signal u in terms of its frequencies, at the scale $\hbar$. For $u \in C_o^\infty$, we have the decomposition

$$u(x) = (2\pi\hbar)^{-d/2} \int_{\mathbb{R}^d} e^{\frac{i}{\hbar}\xi.x} \hat{u}_\hbar(\xi) \mathrm{d}\xi \ .$$

4.2. The stationary phase method. This is a result describing the asymptotic behaviour, as $\hbar \to 0$, of an integral of the form :

$$I(\hbar) = \int_{\mathbb{R}^D} e^{\frac{i}{\hbar}S(x)} a(x) \, \mathrm{d}x$$

where $a \in C_o^\infty(\mathbb{R}^D)$ and $S \in C^\infty(\mathbb{R}^D, \mathbb{R})$.

The interferences between the different terms $e^{\frac{i}{\hbar}S(x)}$ are destructive, except at the stationary points of the phase S. The precise statement is :

- If S has no critical/stationary point in the support of a, then $I(\hbar) = O(\hbar^\infty)$ (this notation means that, for all $N > 0$, we have $I(\hbar) = O_N(\hbar^N)$).
- If S has a unique critical point x_0, supposed to be non–degenerate, in the support of a, then there is an asymptotic development in powers of $\hbar$, up to any order,

$$I(\hbar) \sim (2\pi\hbar)^{D/2} \frac{e^{i\sigma\pi/4}}{|\det S''(x_0)|^{\frac{1}{2}}} e^{iS(x_0)/\hbar} \left(\sum_{j=0}^{\infty} \hbar^j a_j \right) \tag{4.1}$$

where $S''(x_0)$ is the hessian matrix of S at x_0, $\sigma = n_+ - n_-$ is the index of $S''(x_0)$ (the difference between the number of positive and negative eigenvalues), and $a_0 = a(x_0)$. More generally, a_j can be expressed in terms of the derivatives of a up to order $2j$, at the point x_0.

For technical developments, one usually needs to work with functions a which are not necessarily compactly supported, but have a well behaved behaviour at infinity, and can be allowed to depend on $\hbar$. The choice of a class of "symbols" is a technical issue, which depends on the aims, but also on the tastes of the authors. For the sake of completeness we give an example of a convenient class of symbols. However, it is not required to understand all the technical issues to read the next chapters.

Symbol spaces. Let $D, d > 0$ be two integers, and let U be an open subset of $\mathbb{R}^D$. Let us define *symbols of order* m (independent of $\hbar$) :

$$\begin{aligned}\Sigma^m(U \times \mathbb{R}^d) := \big\{ a \in C^\infty(U \times \mathbb{R}^d; \mathbb{C}) / \\ \text{for every compact } K \subset U, \text{ for every } \alpha, \beta \text{ there exists } C \text{ such that} \\ |D_z^\alpha D_\xi^\beta a(z,\xi))| \leq C(1+|\xi|)^{m-|\beta|} \text{ for all } (z,\xi) \in K \times \mathbb{R}^d \big\}.\end{aligned}$$

For instance, this class contains functions which are homogeneous in a neighbourhood of infinity. We denote $\Sigma^{-\infty} = \cap_{m\in\mathbb{Z}} \Sigma^m$ — this class contains the smooth compactly supported functions $C_o^\infty(U \times \mathbb{R}^d)$.

We also define *semiclassical symbols of order m and degree l* — thus called because they depend on a parameter $\hbar$:

$$\Sigma^{m,l} = \{a_\hbar(z,\xi) = \hbar^l \sum_{j=0}^{\infty} \hbar^j a_j(z,\xi),\ a_j \in \Sigma^{m-j}\} \tag{4.2}$$

This means that $a_\hbar(x,\xi)$ has an asymptotic development in powers of $\hbar$; in the sense that

$$a - \hbar^l \sum_{j=0}^{N-1} \hbar^j a_j \in \hbar^{l+N}\Sigma^{m-N}$$

for all N, uniformly in $\hbar$. In this context, we denote $\Sigma^{-\infty,+\infty} = \cap_{m\geq 0}\Sigma^{-m,m}$.

In these definitions, $U \times \mathbb{R}^d$ can be replaced by a fiber bundle of rank d on a D–dimensional manifold.

Fresnel integrals, generalized stationary phase method. We can now describe the asymptotic behaviour, as $\hbar \to 0$, of the integral :

$$I_\hbar^S(a) = \int_{U\times\mathbb{R}^d} e^{\frac{i}{\hbar}S(z,\xi)} a(z,\xi)\, dzd\xi$$

where S is smooth, homogeneous of degree $n > 0$ near infinity with respect to ξ, and without critical points outside a compact subset of $U \times \mathbb{R}^d$. The integral $I_\hbar^S(a)$ is defined for $a \in \Sigma_o^{m,l}$, by continuous extension of the case $a \in C_o^\infty$. Here, the index $*_o$ in $\Sigma_o^{m,l}$ means that a is compactly supported with respect to z, with support independent of $\hbar$. Such non absolutely convergent oscillatory integrals are sometimes called Fresnel integrals, a well-known example is $\int_{\mathbb{R}^d} e^{\frac{i}{2\hbar}\|\xi\|^2} d\xi = (2\pi\hbar)^{d/2} e^{id\pi/4}$.

The previous asymptotic behaviour still holds in this setting.

4.3. Pseudodifferential operators. As we have seen, a quantization procedure is a way to associate an operator to a classical observable $a(p,q)$. Recall Schrödinger's prescriptions, $\mathbf{q}_k =$ (multiplication by q_k), and $\mathbf{p}_k = \frac{\hbar}{i}\frac{\partial}{\partial q_k}$, compatible with Heisenberg's commutation relations (1.13). To extend this definition to an arbitrary function of (p,q), we meet an obvious problem : to quantize the function $p_k q_k^2$, for instance, we could propose any of the operators $\mathbf{p}_k\mathbf{q}_k^2$, $\mathbf{q}_k^2\mathbf{p}_k$, or $\mathbf{q}_k\mathbf{p}_k\mathbf{q}_k$. There are many quantization procedures. We already met the Weyl quantization, which combines several remarkable features, like the fact that it associates a symmetric operator to a real symbol. Later on, we shall also define the anti-Wick positive quantization, which associates a nonnegative operator to a nonnegative symbol.

The theory of pseudodifferential operators with small parameter allows to describe the passage from the quantum theory to the classical theory when $\hbar \longrightarrow 0$. This is also called $\hbar$-dependent microlocal analysis, microlocal analysis with small parameter, or semiclassical analysis. Pseudodifferential operators were first developed by Hörmander [**Ho, Ho79**] for the study of the regularizing properties of partial differential equations (without any small parameter). Pseudodifferential operators with small parameter, manipulated by Maslov [**Masl65**] in the framework of semiclassical analysis, developed by Voros in mathematical physics [**Vor, Vor78**], were perfectioned by Sjöstrand, Robert, Helffer, [**DimSjo, Rob**]... I advise to read

[**Helffer1**] for a history of the first years of this theory in the seventies and an exhaustive bibliography; see also [**Helffer2**] for a survey of applications.

Symbol spaces depend on authors, and can be extremely sophisticated. Hörmander's definition has no $\hbar$ and involves symbols which are homogeneous near infinity, allowing to describe the regularizing properties of operators. The semiclassical symbol classes of [**DimSjo**] are rather aimed at describing the behaviour of operators when $\hbar \longrightarrow 0$, say in L^2 norm. The symbols we use here combine both approaches (after an idea of Y. Colin de Verdière) : taking $\hbar = 1$ we would find (one of) Hörmander's symbol spaces.

Pseudodifferential operators. Let Ω be an open subset of $\mathbb{R}^d$, and let $a = a_\hbar(x, y; \xi) \in \Sigma_o^{m,l}(\Omega \times \Omega \times \mathbb{R}^d)$. Here the index $_o$ means that for every compact $K \subset \Omega$, there exists a compact K' such that $a(x, y, \xi) = 0$ for $x \in K, y \notin K'$, $\xi \in \mathbb{R}^d$. Let u be a smooth function. We define :

$$\mathrm{OP}_\hbar(a)u(x) = (2\pi\hbar)^{-d} \int e^{\frac{i}{\hbar}\xi.(x-y)} a(x, y, \xi) u(y)\, \mathrm{d}y \mathrm{d}\xi,$$

the integral being well defined as a Fresnel integral. We denote $\Psi\mathrm{DO}^{m,l}(\Omega)$ these operators, called (proper) pseudodifferential operators of *degree* l and *order* m, on Ω. The intersection $\Psi\mathrm{DO}^{-\infty,\infty}$ of all the $\Psi\mathrm{DO}^{m,l}(\Omega)$ are the negligible operators : they are the operators with a smooth kernel $K_\hbar$, and such that all derivatives of $K_\hbar$ are $O(\hbar^\infty)$ uniformly on compact sets.[3]

The class of pseudodifferential operators includes differential operators (corresponding to a symbol which is polynomial in ξ), but has the advantage of being stable under inversion, or more general smooth functional calculus.

Note that several symbols $a(x, y, \xi)$ can give the same operator $\mathrm{OP}_\hbar(a)$. As a simple example, we note that $a(x, y, \xi) = V(x)$ and $a(x, y, \xi) = V(y)$ both give the operator of multiplication by V. It is often convenient to choose special representatives :

Weyl quantization. Left and right quantizations.

Here $\Omega = \mathbb{R}^d$.

We already met the Weyl quantization[4], $\mathrm{Op}_\hbar^W(a) = \mathrm{OP}_\hbar\left(a(\frac{x+y}{2}, \xi)\right)$. If $a \in \Sigma_o^{m,l}(\mathbb{R}^d \times \mathbb{R}^d)$ is compactly supported with respect to the first variable, then $\mathrm{Op}_\hbar^W(a) \in \Psi\mathrm{DO}^{m,l}$.

The inverse of Weyl quantization is explicit, given by the Wigner transform : if $K(x, y)$ is the kernel of the operator A, we let :

$$W_A(x, \xi) = (2\pi\hbar)^{-d/2} \int e^{\frac{iv\xi}{\hbar}} K\left(x + \frac{v}{2}, x - \frac{v}{2}\right) dv\ .$$

Then $A = \mathrm{Op}_\hbar^W(W_A)$. In particular, the Weyl symbol of an operator is unique.

Two other common quantizations are, the *left quantization*, defined by $\mathrm{Op}_\hbar^L(a) = \mathrm{OP}_\hbar\left(a(x, \xi)\right)$ where $a \in \Sigma_o^{m,l}(\mathbb{R}^d \times \mathbb{R}^d)$ and the *right quantization*, $\mathrm{Op}_\hbar^R(a) = \mathrm{OP}_\hbar\left(a(y, \xi)\right)$. The left and right symbols are both uniquely determined by the operator (there are explicit inversion formulas, too).

---

[3]Usually, in this theory, all the assertions about operators hold *modulo negligible operators.* Likewise, the assertions about functions hold *modulo negligible functions.* These are the smooth functions $u_\hbar(x)$ such that all derivatives are $O(\hbar^\infty)$ uniformly on compact sets of X.

[4]I try to stick to the notation OP for symbols $a \in \Sigma(\Omega \times \Omega \times \mathbb{R}^d)$, and Op for symbols $a \in \Sigma(\Omega \times \mathbb{R}^d)$.

EXAMPLE 4.1. To quantize the observable $a(p,q) = pq^2$, the left quantization chooses $\mathbf{q}^2\mathbf{p}$, the right quantization chooses $\mathbf{p}\mathbf{q}^2$, and the Weyl quantization forms the combination $\frac{1}{4}(\mathbf{p}\mathbf{q}^2 + 2\mathbf{q}\mathbf{p}\mathbf{q} + \mathbf{q}^2\mathbf{p}) = \frac{1}{2}(\mathbf{p}\mathbf{q}^2 + \mathbf{q}^2\mathbf{p})$.

EXERCISE 4.2. On $\mathbb{R}^d \times \mathbb{R}^d$, consider a lagrangian $L(x,v)$ defined by a riemannian metric,

$$L(x,v) = \frac{1}{2}g_x(v,v) = \frac{1}{2}\sum_{i,j=1}^{d} g_{ij}(x)v_iv_j.$$

Check that the corresponding hamiltonian is

$$H(x,\xi) = \frac{1}{2}g^x(\xi,\xi) = \frac{1}{2}\sum_{i,j=1}^{d} g^{ij}(x)\xi_i\xi_j,$$

where $(g^{ij}(x))$ is the inverse of the matrix $(g_{ij}(x))$.

Write the explicit expression of the laplacian $\triangle$ associated to the metric g.

Choose a quantization procedure $\mathrm{Op}_\hbar = \mathrm{Op}_\hbar^W, \mathrm{Op}_\hbar^L$ or $\mathrm{Op}_\hbar^R$.

Show that

$$\mathrm{Op}_\hbar(H) = -\frac{1}{2}\hbar^2 \triangle + \hbar\sum_{j=1}^{d} b_j(x)\frac{\hbar}{i}\frac{\partial}{\partial x_j} + \hbar^2 c(x)$$

for certain functions b_j, c, the expression of which depends on the choice of $\mathrm{Op}_\hbar$.

Show that there are functions $\tilde{b}_j, \tilde{c}$ such that

$$-\frac{1}{2}\hbar^2\triangle = \mathrm{Op}_\hbar\left(H(x,\xi) + \hbar\sum_j \tilde{b}_j(x)\xi_j + \hbar^2\tilde{c}(x)\right). \tag{4.3}$$

Compare with (4.2) to find the order and the degree of $-\hbar^2\triangle$ (of course, differential operators are pseudodifferential operators !).

The expression of b_j, c, d depends on the choice of $\mathrm{Op}_\hbar$. The first term $H(x,\xi)$ does not, it is called the principal symbol of $-\frac{1}{2}\hbar^2\triangle$.

Principal symbol. Let $a_\hbar \in \Sigma_o^{m,0}(\Omega \times \Omega \times \mathbb{R}^d)$. Applying the operator $A_\hbar = \mathrm{OP}_\hbar(a_\hbar) \in \Psi\mathrm{DO}^{m,0}$ to a function of the form $u(x)e^{iS(x)/\hbar}$, where u and S are smooth[5], the method of stationary phase gives the following asymptotics :

$$A_\hbar\left(u(x)e^{iS(x)/\hbar}\right) = a_0\left(x,x,S'(x)\right)u(x)e^{iS(x)/\hbar} + O(\hbar).$$

This shows that the function $a_0(x,x,\xi)$ on $\mathbb{R}^d \times \mathbb{R}^d = T^*\mathbb{R}^d$ does not depend on the choice of the symbol $a_\hbar(x,y,\xi)$, but only on the operator $A_\hbar$. It is called the principal symbol of $A_\hbar$, denoted $\sigma^0(A_\hbar)$. If $\sigma^0(A_\hbar) = 0$, then $A_\hbar$ actually belongs to $\Psi\mathrm{DO}^{m-1,1}$ (and conversely).

REMARK 4.3. For $a \in \Sigma_o^{m,0}$, we note that $\mathrm{Op}_\hbar^W(a)$, $\mathrm{Op}_\hbar^L(a)$, $\mathrm{Op}_\hbar^R(a)$ all have the same principal symbol $a_0(x,\xi)$. In other words,

$$\mathrm{Op}_\hbar^W(a) - \mathrm{Op}_\hbar^{R/L}(a) \in \Psi\mathrm{DO}^{m-1,+1}.$$

[5]Such a function is called a *WKB state*, see Section 14

Product. If $A_\hbar \in \Psi\mathrm{DO}^{m_1,0}$ and $B_\hbar \in \Psi\mathrm{DO}^{m_2,0}$, then the product $A_\hbar B_\hbar$ belongs to $\Psi\mathrm{DO}_o^{m_1+m_2,0}$, and the principal symbols are multiplied : $\sigma^0(A_\hbar B_\hbar) = \sigma^0(A_\hbar)\sigma^0(B_\hbar)$. This is proved by the stationary phase method.

An equivalent statement : if $a \in \Sigma_o^{m_1,0}(\mathbb{R}^d \times \mathbb{R}^d)$ and $b \in \Sigma_o^{m_2,0}(\mathbb{R}^d \times \mathbb{R}^d)$, then $\mathrm{Op}_\hbar(a)\,\mathrm{Op}_\hbar(b) \in \Psi\mathrm{DO}^{m_1+m_2,0}(\mathbb{R}^d)$, and

$$\mathrm{Op}_\hbar(ab) - \mathrm{Op}_\hbar(a)\,\mathrm{Op}_\hbar(b) \in \Psi\mathrm{DO}^{m_1+m_2-1,1}(\mathbb{R}^d). \tag{4.4}$$

Thanks to Remark 4.3, this statement does not depend on the choice of Op^W, Op^L or Op^R.

Brackets. If $A_\hbar \in \Psi\mathrm{DO}^{m_1,0}$ and $B_\hbar \in \Psi\mathrm{DO}^{m_2,0}$, then the bracket $[A_\hbar, B_\hbar]$ belongs to $\Psi\mathrm{DO}^{m_1+m_2-1,1}$, and

$$\sigma^0\left(\hbar^{-1}[A_\hbar, B_\hbar]\right) = \frac{1}{i}\left\{\sigma^0(A_\hbar), \sigma^0(B_\hbar)\right\};$$

where $\{.,.\}$ is the Poisson bracket.

Equivalently : if $a \in \Sigma_o^{m_1,0}(\mathbb{R}^d \times \mathbb{R}^d)$ and $b \in \Sigma_o^{m_2,0}(\mathbb{R}^d \times \mathbb{R}^d)$, we have

$$[\mathrm{Op}_\hbar(a), \mathrm{Op}_\hbar(b)] - \mathrm{Op}_\hbar\left(\frac{\hbar}{i}\{a,b\}\right) \in \Psi\mathrm{DO}^{m_1+m_2-2,2} \tag{4.5}$$

and again this statement does not depend on the choice of Op^W, Op^L or Op^R.

REMARK 4.4. There is also an integrated version of this result, called the Egorov Theorem. We will use it in the following form : assume the pseudodifferential operator $A_\hbar$ is self–adjoint. Define the Schrödinger flow $(U_\hbar^t) = (\exp -\frac{it}{\hbar}A_\hbar)$.

Let $a \in C_c^\infty(T^*\mathbb{R}^d)$. Then, for any given t in $\mathbb{R}$,

$$U_\hbar^{-t}\,\mathrm{Op}_\hbar(a)U_\hbar^t - \mathrm{Op}_\hbar(a \circ \phi^t_{\sigma^0(A_\hbar)}) \in \Psi\mathrm{DO}^{-\infty,1}. \tag{4.6}$$

Here $\phi^t_{\sigma^0(A_\hbar)}$ is the hamiltonian flow defined by the hamiltonian $\sigma^0(A_\hbar)$. The estimate is usually not uniform in t, so that one cannot invert the limits $\hbar \longrightarrow 0$ and $t \longrightarrow \infty$. This is a notorious source of problems when one tries to use the semiclassical approximation to understand the large time behaviour of solutions of the Schrödinger equation.

Pseudodifferential operators on a compact manifold. Let X be a compact C^∞ manifold of dimension d. Let (Ω_i, φ_i) be a finite atlas of X ($X = \cup\Omega_i$, $\varphi_i : \Omega_i \longrightarrow \mathbb{R}^d$). We use the φ_i to define local coordinates $\Phi_i : T^*\Omega_i \longrightarrow \mathbb{R}^d \times \mathbb{R}^d$ on T^*X as follows :

$$\Phi_i(x,p) = (\varphi_i(x), (d\varphi_i(x))^{-1}p).$$

These are symplectic (Darboux) coordinates on T^*X, i.e. the canonical symplectic form reads $\omega = \sum dx_j \wedge dp_j$ in these coordinates. Introduce a finite partition of unity $\chi_i \in C_o^\infty(\Omega_i)$ such that $\sum \chi_j^2 = 1$. For $a \in \Sigma_o^{m,l}(T^*X)$, we let :

$$\mathrm{Op}_\hbar(a)u = \sum_i \chi_i\left[\mathrm{OP}_\hbar\left(a \circ \Phi_i^{-1}\right)\left(\chi_i u \circ \varphi_i^{-1}\right)\right] \circ \varphi_i\,. \tag{4.7}$$

The map $a \mapsto \mathrm{Op}_\hbar(a)$ thus defined depends on the partition of unity and on the local coordinates; but its range does not, modulo negligible operators. The algebra $\Psi\mathrm{DO}^{m,l}(X)$ of pseudodifferential operators on X (modulo negligible operators) is thus well defined.

All the properties stated above can be extended to this case.

Continuity. Trace class and Hilbert-Schmidt operators. The domain of an operator in $\Psi\mathrm{DO}^{m,0}(\Omega)$ depends a lot on m, that is, on the growth of the symbol when $\xi \longrightarrow \infty$. When m decreases, the regularizing properties of the operator are improved. Without proof, let us mention that an operator in $\Psi\mathrm{DO}^{0,0}(\Omega)$ is bounded from $L^2(\Omega)$ to $L^2_{loc}(\Omega)$, uniformly with respect to $\hbar$. On a compact manifold X, an operator in $\Psi\mathrm{DO}^{m,0}$ is

- Hilbert-Schmidt if $m < -d/2$
- trace class if $m < -d$

In this latter case, the trace of $\mathrm{Op}(a)$ is given by the convergent integral,

$$\mathrm{Tr}\,\mathrm{Op}(a) = (2\pi\hbar)^{-d} \int_{T^*X} a(x,\xi) \mathrm{d}x \mathrm{d}\xi \ . \tag{4.8}$$

5. Semiclassical measures, microlocal lifts.

A quantization procedure Op is said to be *nonnegative* if $\mathrm{Op}(a)$ is a nonnegative operator as soon as a is a nonnegative function. The usual quantization procedures do not have this property.

Positive quantization on $\mathbb{R}^d$.

DEFINITION 5.1. (Coherent states) The coherent state (of size $\hbar$) centered at (x_0, ξ_0) is defined as the normalized gaussian state

$$e_{x_0,\xi_0}(x) = \frac{1}{(\pi\hbar)^{d/4}} e^{\frac{i}{\hbar}\xi_0 . x} \exp\left(-\frac{\|x - x_0\|^2}{2\hbar}\right)$$

For $(x,\xi) \in \mathbb{R}^d \times \mathbb{R}^d$, we shall denote $\Pi_{(x,\xi)}$ the orthogonal projector onto $\mathbb{C}e_{(x,\xi)}$.

THEOREM 5.2. *Let $a \in C^\infty_o(T^*\mathbb{R}^d)$. The operator defined by*

$$\mathrm{Op}^+(a) = (2\pi\hbar)^{-d} \int a(x,\xi)\, \Pi_{x,\xi}\, \mathrm{d}x \mathrm{d}\xi$$

belongs to the class $\Psi\mathrm{DO}^{-\infty,0}$, it is self–adjoint if a is real valued, and non-negative if a is non-negative. Its principal symbol is $a(x,\xi)$.

We have $\mathrm{Op}^+(1) = I$, which allows to extend the definition of Op^+ to the case when a is constant in a neighbourhood of infinity in T^*X.

This quantization is called the anti–Wick quantization.

To define a positive quantization procedure on a compact manifold X, we choose an atlas of X and a non-negative subordinate partition of unity, $\sum \chi_j^2 = 1$. For $a \in C^\infty_o(T^*X)$, we let $\mathrm{Op}^+_X(a) = \sum_j \chi_j \,\mathrm{Op}^+_{\mathbb{R}^d}(a)\chi_j$ — where $\mathrm{Op}^+_{\mathbb{R}^d}(a)$ is defined using local coordinates in the support of χ_j (see (4.7)). We can extend this definition to the case when a is constant in a neighbourhood of infinity in T^*X, by letting $\mathrm{Op}^+_X(1) = I$.

Semiclassical measures. Let X be a compact riemannian manifold; we denote Vol the riemannian volume on X. To a family $(u_\hbar)$ of normalized elements of $L^2(X, \mathrm{Vol})$, we can associate a family of distributions $\mu_\hbar$ by the formula $\mu_\hbar(a) = \langle u_\hbar, \mathrm{Op}^+_\hbar(a) u_\hbar\rangle_{L^2(X,\mathrm{Vol})}$. They are in fact probability measures on T^*X. To be able to take weak limits when $\hbar \longrightarrow 0$, we see them as probability measures on the compactification $\overline{T^*X}$ of T^*X obtained by adding a sphere bundle at infinity. We will call the measures $\mu_\hbar$ the *Husimi measures* associated to the family $(u_\hbar)$. The term *Wigner transform* will be exclusively used in the case $X = \mathbb{R}^d$, for the distributions $a \mapsto \langle u_\hbar, \mathrm{Op}_\hbar^W(a) u_\hbar \rangle$ defined thanks to the Weyl quantization. These distributions are also called *microlocal lifts* of the probability measures $|u_\hbar(x)|^2 d\mathrm{Vol}(x)$. This means that their projection down to X is $|u_\hbar(x)|^2 d\mathrm{Vol}(x) + O(\hbar)$.

Due to the uncertainty principle, these objects are not really meaningful for fixed $\hbar > 0$. In fact, their definition depends on a certain number of arbitrary choices, coming into play in the definition of Op : local coordinates, partition of unity, choice of the quantization procedure... However, the semiclassical limits of these distributions do not depend on all these arbitrary conventions : if $a \in \Sigma_o^{0,0}(T^*X)$, two definitions of $\mathrm{Op}(a)$ only differ by $O(\hbar)$ in L^2 operator norm.

We shall call any limit point of the sequence $(\mu_\hbar)$ in the weak topology a *semiclassical measure* associated to the family $(u_\hbar)$.

EXAMPLE 5.3. (Coherent states)

$$u_\hbar(x) = e_{x_0,\xi_0}(x) = \frac{1}{(\pi\hbar)^{d/4}} e^{\frac{i}{\hbar}\xi_0 \cdot x} \exp\left(-\frac{\|x - x_0\|^2}{2\hbar}\right)$$

Then there is a unique semiclassical measure, the Dirac mass at (x_0, ξ_0).

EXAMPLE 5.4. (Lagrangian states/WKB states) Let $u_\hbar(x) = b(x) e^{\frac{i}{\hbar} S(x)}$ where b and S are of class C^∞. In Section 14, we will call such functions *lagrangian states* associated to the lagrangian manifold $\mathcal{L} = \{(x, dS(x))\}$.

There is a unique semiclassical measure associated to $(u_\hbar)$, it is carried by the lagrangian $\mathcal{L}$ and projects to X as the measure $|b(x)|^2 \, d\mathrm{Vol}(x)$.

EXERCISE 5.5. You have noted that we sometimes omit to indicate the dependence on $\hbar$ in the definition of Op (which should be denoted $\mathrm{Op}_\hbar$). The choice of scaling is, nevertheless, very important, and the properties observed vary a lot according to the scaling.

In the previous example, show that the measures defined by

$$\mu_{\hbar,\alpha}(a) = \langle u_\hbar, \mathrm{Op}_{\hbar^\alpha}^+(a) u_\hbar \rangle$$

concentrate to the 0-section in T^*X if $\alpha > 1$, but concentrate to the sphere bundle at infinity $\overline{T^*X} \setminus T^*X$ if $\alpha < 1$.

When the $u_\hbar$ are the eigenfunctions of a hamiltonian operator as in (1.19), one can apply the following theorem :

THEOREM 5.6. *Let P be a self-adjoint pseudodifferential operator, denote p_0 its principal symbol. Let $(u_\hbar)$ be a family of tamed*[6] *smooth functions, such that $Pu_\hbar = O(\hbar^\infty)$ and $\|u_\hbar\|_{L^2} = 1$. Let $\mu_\hbar$ be the Husimi measures associated to $(u_\hbar)$. Then, every weak limit μ_0 of the measures $\mu_\hbar$ on $\overline{T^*X}$*

1. *is a probability measure on $\overline{T^*X}$.*
2. *projects on X to a weak limit of the measures $|u_\hbar(x)|^2 d\mathrm{Vol}(x)$.*
3. *is invariant under the hamiltonian flow of p_0.*
4. *its restriction to T^*X is carried by the energy level $\{p_0 = 0\}$.*
5. *If p_0 is elliptic at infinity, then μ_0 is carried by T^*X.*

[6]meaning that, for all $N \in \mathbb{N}$, for any compact K, there exists $k \in \mathbb{N}$ such that the C^N norm of $u_\hbar$ on K is $O(\hbar^{-k})$

The first two items have already been explained.

EXERCISE 5.7. Prove the third item by using the relation $\sigma^0\left(\hbar^{-1}[P,\mathrm{Op}(a)]\right) = -i\{p_0, a\}$ to show that $\int \{p_0, a\} d\mu_0 = 0$ for any $a \in C_o^\infty(T^*X)$.

Prove the fourth item by using the relation $\sigma^0\left(\mathrm{Op}(a)P\right) = a\, p_0$ to show that $\int a\, p_0\, d\mu_0 = 0$ for any $a \in C_o^\infty(T^*X)$.

We do not give here the precise definition of "elliptic at infinity". It implies that P is invertible in a neighbourhood of infinity in the class of pseudodifferential operators. More precisely, there exists a smooth a, taking the constant value 1 in a neighbourhood of infinity in T^*X, and a pseudodifferential operator $\mathrm{Op}(b)$ such that

$$\mathrm{Op}(a) = \mathrm{Op}(b)P + R$$

where $R \in \Psi\mathrm{DO}^{-\infty,\infty}$ is a negligible operator. From this fact, the last item follows easily. The ellipticity criterion is satisfied by the Schrödinger operator $\left(-\frac{\hbar^2\triangle}{2} + V - E\right)$ on a compact manifold X.

Eigenfunctions of the laplacian. Let (X, g) be a compact riemannian manifold, and $\triangle$ the laplacian on X associated to the metric. If $(-\hbar^2 \triangle - 1)u_\hbar = 0$, and if we denote $\mu_\hbar$ the corresponding Husimi measures, then every limit point of the family $(\mu_\hbar)_{\hbar\longrightarrow 0}$ is a probability measure μ_0 carried by the unit cotangent bundle S^*X, invariant under the geodesic flow (apply Theorem 5.6 and remember Exercise 4.2). It is a widely open problem to find all the possible limits μ_0 among the invariant measures on S^*X.

In the case of the round sphere or a flat torus, it is easy to construct families of eigenfunctions $(u_\hbar)$ for which $\mu_\hbar$ converges to the uniform measure on any given invariant lagrangian torus. On the flat torus $\mathbb{T}^d = \mathbb{R}^d/\mathbb{Z}^d$ for instance, the family $(e^{\frac{i}{\hbar}\xi_0 . x})$, where ξ_0 is a unit vector (and of course $\frac{\xi_0}{\hbar} \in 2\pi\mathbb{Z}^d$), has a unique semi-classical measure, the uniform measure on the lagrangian torus $\{(x, \xi_0), x \in \mathbb{T}^d\}$. More generally, for a completely integrable system, one can use WKB methods [**Brill26, Kr26, Wtz26, Kell58, Masl65**] to build quasimodes, in other words solutions of $\|(-\hbar^2\triangle - 1)u_\hbar\| = O(\hbar^\infty)$, the Husimi measures of which concentrate to any given invariant torus[7]. Historically, the case of completely integrable systems, or perturbations thereof, was the most important, since it is related to the study of small atoms and ions. The "opposite" case of chaotic systems has been studied only more recently [**Berr77, Vor77, Bo91**], but the question of the localization of stationary motions in ergodic systems was already asked explicitly by Einstein [**Ein17**].

In these notes, we shall focus on the case where the geodesic flow has a very chaotic behaviour. When the geodesic flow is ergodic, the semiclassical measures are essentially described by the Snirelman theorem [**Sn74, Ze87, CdV85**] (see Section 6). Let X be a compact riemannian manifold; call $0 < \lambda_1 \leq \lambda_2 \leq \cdots$ the eigenvalues of the laplacian, and let (ψ_j) be an orthonormal basis of eigenfunctions : $-\triangle \psi_j = \lambda_j \psi_j$. Denote μ_j the corresponding Husimi measures (the semiclassical parameter is $\hbar = \lambda_j^{-1/2}$). We shall call $(L_E)_E$ the disintegration of the Liouville measure $\mathrm{d}x\,\mathrm{d}\xi$ with respect to the value E of the hamiltonian $\frac{\|\xi\|^2}{2}$. We normalize

[7]Note that $\|(-\hbar^2 \triangle - 1)u_\hbar\| \leq \varepsilon \|u_\hbar\|$ implies that 1 is an ε-neighbourhood of the spectrum of $-\hbar^2\triangle$, but does not imply that $u_\hbar$ is close to an eigenfunction of the laplacian.

L_E to be a probability measure on the energy layer $\{\frac{\|\xi\|^2}{2} = E\}$. If the geodesic flow on S^*X is ergodic with respect to $L_{\frac{1}{2}}$, then there is a "density 1" subsequence of the family (μ_j) converging to $L_{\frac{1}{2}}$:

THEOREM 5.8 (Snirelman theorem). [**Sn74, Ze87, CdV85**] *Assume that the action of S^*X is ergodic, with respect to the Liouville measure $L_{\frac{1}{2}}$. Then, there exists a subset $\mathcal{S} \subset \mathbb{N}$ of density* 1, *such that*

$$\mu_j \underset{j\longrightarrow+\infty, j\in\mathcal{S}}{\longrightarrow} L_{\frac{1}{2}}.$$

In specific examples, what we would like to know is whether the whole sequence μ_j converges to the Liouville measure, or if there can be exceptional subsequences converging to other invariant measures. In the case of nonpositively curved surfaces with flat cylinders, it is believed that certain sequences of eigenfunctions concentrate asymptotically on these cylinders. But in (strictly) negative curvature, it was conjectured by Rudnick and Sarnak [**RudSa94**] that the Liouville measure is the unique limit point of the μ_js. It would imply, in particular, that the sequence of probability measures $|\psi_j(x)|^2 d\mathrm{Vol}(x)$ on X converges weakly to the riemannian volume measure Vol.

ENTROPY AND LOCALIZATION OF EIGENFUNCTIONS.

6. Motivations.

The field of *quantum chaos* tries to understand how the chaotic behaviour of a classical hamiltonian system is translated in quantum mechanics. For instance, let X be a compact riemannian C^∞ manifold, with negative sectional curvature. The geodesic flow has the Anosov property, which is considered as the ideal chaotic behaviour in the theory of dynamical systems. The corresponding quantum dynamics is the unitary flow generated by the Laplace-Beltrami operator on $L^2(X)$. One expects that the chaotic features of the geodesic flow can be seen in the spectral properties of the laplacian. The Random Matrix conjecture [**Bo91**] asserts that the large eigenvalues should, after proper renormalization, statistically resemble those of a large random matrix, at least for a generic Anosov metric. The Quantum Unique Ergodicity conjecture [**RudSa94**] (see also [**Berr77, Vor77**]) deals with the corresponding eigenfunctions ψ: it claims that the probability density $|\psi(x)|^2 dx$ should approach (in a weak sense) the riemannian volume, when the eigenvalue tends to infinity. In fact, a corresponding (stronger) property should hold for the microlocal lift of this measure to the cotangent bundle T^*X, which describes the distribution of the wave function ψ on the classical phase space (position and momentum).

To describe the problem, we will adopt a semiclassical point of view, that is, consider the eigenstates of eigenvalue 1 of the semiclassical laplacian $-\hbar^2\triangle$, in the semiclassical limit $\hbar \to 0$. We denote by $(\psi_k)_{k\in\mathbb{N}}$ an orthonormal basis of $L^2(X)$ made of eigenfunctions of the laplacian, and by $(-\frac{1}{\hbar_k^2})_{k\in\mathbb{N}}$ the corresponding eigenvalues:

$$-\hbar_k^2 \triangle\psi_k = \psi_k, \quad \text{with} \quad \hbar_{k+1} \leq \hbar_k\,. \tag{6.1}$$

We are interested in the high-energy eigenfunctions of $-\triangle$, in other words the semiclassical limit $\hbar_k \to 0$.

To an eigenfunction ψ_k corresponds a distribution on T^*X defined by

$$\mu_k(a) = \langle \psi_k, \mathrm{Op}_{\hbar_k}(a)\psi_k \rangle_{L^2(X)}, \qquad a \in C_o^\infty(T^*X).$$

Here $\mathrm{Op}_{\hbar_k}$ is a quantization procedure, set at the scale $\hbar_k$, which associates a bounded operator on $L^2(X)$ to any smooth phase space function a with nice behaviour at infinity. If a is a function on the manifold X, we have $\mu_k(a) = \int_X a(x)|\psi_k(x)|^2 dx + O(\hbar)$: the distribution μ_k is a *microlocal lift* of the probability measure $|\psi_k(x)|^2 dx$ into a phase space distribution. It contains the information about the frequency vector of ψ_k (in other words, the momentum), in addition to the position distribution $|\psi_k(x)|^2 dx$. The definition of μ_k is not canonical, it depends on a certain number of choices, like the choice of local coordinates, or of the quantization procedure (Weyl, anti-Wick, "right" or "left" quantization...); this somehow reflects the fact that, for $\hbar > 0$, it does not really make sense to study simultaneously the position and frequency of a wave. Mathematically speaking, one cannot study simultaneously the local properties of a function and of its Fourier transform around some point $(x, \xi) \in T^*X$. But the asymptotic behaviour of μ_k when $\hbar_k \longrightarrow 0$ does not depend on the arbitrary conventions involved in its definition. We saw that it is possible to construct $\mathrm{Op}^+_{\hbar_k}$ so that the μ_k are probability measures, in which case we call them *Husimi measures* associated to the eigenfunctions ψ_k. We call *semiclassical measures* the limit points of the sequence $(\mu_k)_{k\in\mathbb{N}}$, in the distribution topology.

The quantum hamiltonian $-\frac{\hbar^2\triangle}{2}$ generates the Schrödinger flow

$$(U_\hbar^t) = (\exp(it\hbar\frac{\triangle}{2}))$$

acting unitarily on $L^2(X)$. A solution of (6.1) is an invariant state of the flow $(U_\hbar^t)$, corresponding to the energy $\frac{1}{2}$ of the hamiltonian. In the semiclassical limit $\hbar \longrightarrow 0$, "quantum mechanics converges to classical mechanics". We will denote $|\cdot|_x$ the norm on T_x^*M given by the metric. The geodesic flow $(g^t)_{t\in\mathbb{R}}$ is the hamiltonian flow on T^*X generated by the hamiltonian $H(x, \xi) = \frac{|\xi|_x^2}{2}$. In the previous chapter we saw the following :

PROPOSITION 6.1. *Any semiclassical measure is a probability measure carried on the energy layer $H^{-1}(\frac{1}{2})$, that is, the unit cotangent bundle S^*X. This measure is invariant under the geodesic flow.*

If the geodesic flow has the Anosov property — for instance if X has negative sectional curvature — then there exist many invariant probability measures on S^*X, in addition to the Liouville measure. The geodesic flow has countably many periodic orbits, each of them carrying an invariant probability measure. There are still many others, like the equilibrium states obtained by variational principles [**KH**].

For manifolds with an ergodic geodesic flow (with respect to the Liouville measure), it has been known for some time that *almost all* eigenfunctions become uniformly distributed over S^*X, in the semiclassical limit. This property is dubbed as Quantum Ergodicity :

THEOREM 6.2. [**Sn74, Ze87, CdV85**] *Let X be a compact riemannian manifold, and assume that the action of the geodesic flow on S^*X is ergodic with respect to the Liouville measure $L_{\frac{1}{2}}$. Let $(\psi_k)_{k\in\mathbb{N}}$ be an orthonormal basis of $L^2(X)$*

consisting of eigenfunctions of the laplacian (6.1), *and let* (μ_k) *be the associated distributions on* T^*X.

Then, there exists a subset $\mathcal{S} \subset \mathbb{N}$ *of density* 1, *such that*

$$\mu_k \underset{k\longrightarrow\infty, k\in\mathcal{S}}{\longrightarrow} L_{\frac{1}{2}}.$$

Proof. Let us give the main lines of the argument, and see where the ergodicity comes into play. For all $a \in C_o^\infty(T^*X)$, one first shows, without using any assumption on the dynamics, that

$$\sum_{j,\,\lambda_j\leq E} \int a\, d\mu_j \underset{E\longrightarrow+\infty}{\sim} \frac{b_d}{(2\pi)^d} Vol(X) \int_{S^*X} a\, dL_{\frac{1}{2}} \times E^{d/2}. \tag{6.2}$$

The constant b_d is the volume of the euclidean d-dimensional ball. The idea is to express in two different ways the trace of $\mathrm{Op}_{\sqrt{E}}(a)$: the trace can be expressed either as a spectral sum $\sum_k \langle \psi_k, \mathrm{Op}(a)\psi_k\rangle$ or as the integral of the kernel on the diagonal (4.8). There are some technical details that we skip here.

From (6.2) one can deduce the Weyl asymptotics :

$$N(E) = \sharp\{j, \lambda_j \leq E\} \sim \frac{b_d}{(2\pi)^d} Vol(X)\, E^{d/2}$$

Thus, we have a Cesaro convergence :

$$\frac{1}{N(E)} \sum_{j,\,\lambda_j\leq E} \int a\, d\mu_j \underset{E\longrightarrow+\infty}{\longrightarrow} \int_{S^*X} a\, dL_{\frac{1}{2}}. \tag{6.3}$$

Using the ergodicity assumption, one can do better :

$$\frac{1}{N(E)} \sum_{j,\,\lambda_j\leq E} \left| \int a\, d\mu_j - \int_{S^*X} a\, dL_{\frac{1}{2}} \right|^2 \underset{E\longrightarrow+\infty}{\longrightarrow} 0. \tag{6.4}$$

Here is how. We know from Theorem 5.6 (3) that

$$|\int a\, d\mu_j - \int a\circ g^t\, d\mu_j| \longrightarrow 0$$

as $j \longrightarrow +\infty$, for any fixed t. Thus, we can write, for any given T,

$$\begin{aligned}
&\limsup_{E\longrightarrow\infty} \frac{1}{N(E)} \sum_{j,\,\lambda_j\leq E} \left| \int a\, d\mu_j - \int_{S^*X} a\, dL_{\frac{1}{2}} \right|^2 \\
&= \limsup \frac{1}{N(E)} \sum_{j,\,\lambda_j\leq E} \left| \int M^T a\, d\mu_j - \int_{S^*X} a\, dL_{\frac{1}{2}} \right|^2 \\
&\leq \limsup \frac{1}{N(E)} \sum_{j,\,\lambda_j\leq E} \mu_j \left((M^T a - \int_{S^*X} a\, dL_{\frac{1}{2}})^2 \right) \\
&= L_{\frac{1}{2}} \left((M^T a - \int_{S^*X} a dL_{\frac{1}{2}})^2 \right).
\end{aligned}$$

We denoted $M^T a = T^{-1}\int_0^T a\circ g^t dt$ the time average of a on the interval $[0, T]$. We used the Cauchy-Schwartz inequality, which requires to know that the μ_j can be assumed to be *probability* measures (see §5; this was the missing argument in Snirelman's original paper). In the last line, we used the Cesaro convergence (6.3) of the sequence (μ_j). Letting at the end T tend to $+\infty$, the ergodicity assumption means that

$$L_{\frac{1}{2}}\left((M^T a - \int_{S^*X} a\, dL_{\frac{1}{2}})^2\right) \underset{T\to\infty}{\longrightarrow} 0;$$

which proves (6.4).

Finally, the Snirelman theorem results from the classical lemma :

LEMMA 6.3. *Let (u_n) be a sequence of nonnegative numbers. If*

$$\frac{1}{n}\sum_{k=0}^{n} u_k \longrightarrow 0$$

then there exists $\mathcal{S} \subset \mathbb{N}$ of density 1 *such that $u_n \underset{n\in\mathcal{S}}{\longrightarrow} 0$.*

For each $a \in C_o^\infty(T^*X)$, the lemma yields the Snirelman theorem for some density one set $\mathcal{S} \subset \mathbb{N}$, possibly depending on a. Using the fact that $C_o^\infty(T^*X)$ has a countable dense subset, one can find some $\mathcal{S} \subset \mathbb{N}$ that works for *all* $a \in C_o^\infty(T^*X)$. □

REMARK 6.4. The result was subsequently extended to more general hamiltonians [**HelMR87**], to ergodic billiards [**GL93, ZeZw96**], and to certain discrete time symplectic dynamical systems.

The question of knowing, in particular cases, if there can exist "exceptional" subsequences with a different behaviour is widely open. On a negatively curved manifold, the geodesic flow satisfies the ergodicity assumption, and in fact much stronger properties : mixing, K–property,... In this case, the Quantum Unique Ergodicity conjecture [**RudSa94**] expresses the belief that there exists a unique semiclassical measure, namely the Liouville measure on S^*X : the whole sequence (μ_k) converges to $L_{\frac{1}{2}}$. In other words, in the semiclassical régime all eigenfunctions should become uniformly distributed over S^*X.

So far the most precise results on this question were obtained for manifolds X with constant negative curvature and *arithmetic* properties: see Rudnick–Sarnak [**RudSa94**], Wolpert [**Wol01**]. In that very particular situation, there exists a countable commutative family of self–adjoint operators commuting with the laplacian : the Hecke operators. One may thus decide to restrict the attention to bases of common eigenfunctions, often called "arithmetic" eigenstates, or Hecke eigenstates. A few years ago, Lindenstrauss [**Li06**] proved that the arithmetic eigenstates become asymptotically equidistributed (Arithmetic Quantum Unique Ergodicity). If there is some degeneracy in the spectrum of the laplacian, it could be possible that the Quantum Unique Ergodicity conjectured by Rudnick and Sarnak holds for one orthonormal basis but not for another. In the arithmetic case, it is believed that the spectrum of the laplacian has bounded multiplicity, in which case it would be a harmless assumption to consider only Hecke eigenstates.

Nevertheless, one may be less optimistic about the general conjecture. Faure–Nonnenmacher–De Bièvre exhibited in [**FNDB03**] a simple example of a symplectic Anosov dynamical system, namely the action of the linear hyperbolic automorphism $\begin{pmatrix} 2 & 1 \\ 1 & 1 \end{pmatrix}$ on the 2-torus, the Weyl–quantization of which does not satisfy the Quantum Unique Ergodicity conjecture. In this model, it is known [**KurRud00**] that there is one orthonormal family of eigenfunctions satisfying Quantum Unique

Ergodicity, but, due to high degeneracies in the spectrum, one can also construct eigenfunctions with a different behaviour. Precisely, [**FNDB03**] construct a family of eigenstates for which the semiclassical measure consists in two ergodic components: half of it is the Liouville measure, while the other half is a Dirac peak on a single unstable periodic orbit. It was also shown that this half-localization on a periodic orbit is *maximal* for this model [**FN04**] : a semiclassical measure cannot have more than half the mass carried by a finite union of closed orbits. Another type of semiclassical measure was recently obtained by Kelmer for a quantized automorphism on a higher-dimensional torus [**Kelm05**]: it consists in the Lebesgue measure on some invariant co-isotropic subspace of the torus. For these torus automorphisms, the existence of exceptional eigenstates seems to be due to some nongeneric algebraic properties of the classical and quantized systems. It has been believed for a while that any perturbation of the system which lifts the degeneracies in the spectrum will also destroy the counterexamples to Quantum Unique Ergodicity. However, Kelmer has recently disproved this belief : he showed that a non-linear perturbation of his previous construction [**Kelm05**], if it is done so as to preserve the invariant co-isotropic subspaces, will still contradict Quantum Unique Ergodicity [**Kelm06**], in the same way as before the perturbation. Moreover, the spectrum of the perturbed quantum model is simple [**Kelm08**].

7. Main result.

We wish to consider the *Kolmogorov–Sinai* entropy of semiclassical measures. We work on a compact manifold X of arbitrary dimension, and assume that the geodesic flow has the Anosov property. In fact, our method is very general, and can without any doubt be adapted to more general Anosov hamiltonian systems.

The Kolmogorov–Sinai entropy, also called metric entropy, of a (g^t)-invariant probability measure μ is a nonnegative number $h_{KS}(\mu)$ that describes, in some sense, the complexity of a μ-typical orbit of the flow. The precise definition will be given later, but for the moment let us just give a few facts. A measure carried on a closed geodesic has zero entropy. In constant curvature, the entropy is known to be maximal for the Liouville measure. More generally, an upper bound on the entropy is given by the Ruelle inequality: since the geodesic flow has the Anosov property, the energy layer S^*X is foliated into unstable manifolds of the flow, and for any invariant probability measure μ one has

$$h_{KS}(\mu) \leq \left| \int_{S^*X} \log J^u(\rho) d\mu(\rho) \right| . \tag{7.1}$$

In this inequality, $J^u(\rho)$ is the *unstable Jacobian* of the flow at the point $\rho \in S^*X$, defined as the Jacobian of the map g^{-1} restricted to the unstable manifold at the point $g^1\rho$. The average of $\log J^u$ over any invariant measure is negative. In fact, if μ is an invariant probability measure,

$$\int_{S^*X} \log J^u(\rho) d\mu(\rho) = - \int_{S^*X} \sum \lambda_j^+(\rho) d\mu(\rho)$$

where $\lambda_j^+(\rho)$ are the positive Lyapunov exponents of ρ. If X has dimension d and has constant sectional curvature -1, (7.1) just reads $h_{KS}(\mu) \leq d - 1$. Equality holds in (7.1) if and only if μ is the Liouville measure on S^*X [**LY85**].

Let μ be a (g^t)–invariant probability measure on S^*X. According to the Birkhoff ergodic theorem, for μ–almost every $\rho \in S^*X$, the weak limit

$$\mu^\rho = \lim_{|t|\longrightarrow\infty} \frac{1}{t}\int_0^t \delta_{g^s\rho} ds$$

exists, and is an ergodic probability measure. We can then write

$$\mu = \int_{S^*X} \mu^\rho d\mu(\rho),$$

which is called the ergodic decomposition of μ. One can prove that the ergodic probability measures are the extremal points of the compact convex set of (g^t)–invariant probability measures.

To understand the connection of our results with the previous discussion, it is important to know that the entropy if an *affine* functional on the convex set of (g^t)–invariant probability measures :

$$h_{KS}(\mu) = \int_{S^*X} h_{KS}(\mu^\rho) d\mu(\rho).$$

In what follows, we consider a certain subsequence of eigenstates $(\psi_{k_j})_{j\in\mathbb{N}}$ of the laplacian, such that the corresponding sequence (μ_{k_j}) converges to a certain semiclassical measure μ (see the discussion preceding Proposition 6.1). The subsequence (ψ_{k_j}) will simply be denoted by $(\psi_\hbar)_{\hbar\to 0}$, using the slightly abusive notation $\psi_\hbar = \psi_{\hbar_{k_j}}$ for the eigenstate ψ_{k_j}. Each state $\psi_\hbar$ satisfies

$$(-\hbar^2 \triangle - 1)\psi_\hbar = 0\,. \tag{7.2}$$

It is proved in [**A05**] that the entropy of any semiclassical measure associated with eigenfunctions of the laplacian is strictly positive. In [**AN07**] more explicit lower bounds were obtained. We shall prove here the following lower bound :

THEOREM 7.1. *Let μ be a semiclassical measure associated to the eigenfunctions of the laplacian on X. Then its metric entropy satisfies*

$$h_{KS}(\mu) \geq \left| \int_{S^*X} \log J^u(\rho) d\mu(\rho) \right| - \frac{(d-1)}{2}\lambda_{\max}\,, \tag{7.3}$$

where $d = \dim M$ and $\lambda_{\max} = \lim_{t\to\pm\infty} \frac{1}{t} \log \sup_{\rho\in S^*X} |dg^t_\rho|$ is the maximal expansion rate of the geodesic flow on S^*X.

In particular, if X has constant sectional curvature -1, this means that

$$h_{KS}(\mu) \geq \frac{d-1}{2}. \tag{7.4}$$

The bound (7.4) in the above theorem is much sharper than the bound proved in [**A05**] in the case of constant curvature. On the other hand, if the curvature varies a lot (still being negative everywhere), the right hand side of (7.3) may be negative, in which case the above bound is trivial and the result of [**A05**] is better. We believe this to be but a technical shortcoming of our method, and would actually expect the following bound to hold:

$$h_{KS}(\mu) \geq \frac{1}{2}\left| \int_{S^*X} \log J^u(\rho) d\mu(\rho) \right|\,. \tag{7.5}$$

Our result is compatible with the kind of counter-examples obtained by Faure–Nonnenmacher–De Bièvre [**FNDB03**]. It allows certain ergodic components to be carried by closed geodesics, but says that others must have positive entropy. Compare with the much stronger result obtained in the arithmetic case by Bourgain and Lindenstrauss :

THEOREM 7.2. [**BLi03**] *Let X be a congruence arithmetic surface, and (ψ_j) an orthonormal basis of eigenfunctions for the laplacian and the Hecke operators.*

Let μ be a corresponding semiclassical measure, with ergodic decomposition $\mu = \int_{S^*X} \mu^\rho d\mu(\rho)$. Then for almost all ergodic components we have $h_{KS}(\mu^\rho) \geq \frac{1}{9}$.

Quantum Unique Ergodicity would mean that $h_{KS}(\mu) = \left|\int_{S^*X} \log J^u(\rho)\, d\mu(\rho)\right|$ [**LY85**]. We believe however that (7.5) is the optimal result that can be obtained without using more precise information, like for instance upper bounds on the multiplicities of eigenvalues. Indeed, in the above mentioned examples of Anosov systems where the Quantum Unique Ergodicity conjecture is wrong, the bound (7.5) is actually *sharp* [**FNDB03, Kelm05, AN06**]. In those examples, the spectrum has very high degeneracies, which allows for much freedom to select the eigenstates, and could be responsible for the failure of Quantum Unique Ergodicity. Such high degeneracies are not expected to happen in the case of the laplacian on a negatively curved manifold. For the moment, however, there is no clear understanding of the precise relation between spectral degeneracies and failure of Quantum Unique Ergodicity. As explained above, at the time of revision of these notes, Kelmer had found a model (a non-linear symplectic diffeomorphism of $\mathbb{T}^{2d}$, $d \geq 2$ and its quantization), for which the spectrum is simple but Quantum Unique Ergodicity does not hold [**Kelm08**].

8. Definition of entropy, and main idea of the proof.

Let μ be a probability measure on T^*X. Let $(P_1, \ldots, P_K)$ be a finite measurable partition of the unit tangent bundle : $T^*X = P_1 \sqcup \ldots \sqcup P_K$. The Shannon entropy of μ with respect to the partition P is

$$h_P(\mu) = -\sum_{k=1}^{K} \mu(P_k) \log \mu(P_k). \tag{8.1}$$

Assume now that μ is (g^t)–invariant. For any integer n, denote $P^{\vee n}$ the partition formed by the sets $P_{\alpha_0} \cap g^{-1} P_{\alpha_1} \ldots \cap g^{-n+1} P_{\alpha_{n-1}}$. Denote

$$\begin{aligned}(8.2)\quad h_n(\mu, P) &= h_{P^{\vee n}}(\mu) \\ &= -\sum_{(\alpha_j)\in\{1,\ldots,K\}^n} \mu(P_{\alpha_0} \cap g^{-1} P_{\alpha_1} \ldots \cap g^{-n+1} P_{\alpha_{n-1}}) \log \mu(P_{\alpha_0} \cap g^{-1} P_{\alpha_1} \ldots \cap g^{-n+1} P_{\alpha_{n-1}}).\end{aligned}$$

If μ is (g^t)–invariant, it follows from the concavity of $x \mapsto -x \log x$ that

$$h_{n+m}(\mu, P) \leq h_n(\mu, P) + h_m(\mu, P), \tag{8.3}$$

in other words the sequence $(h_n(\mu, P))_{n\in\mathbb{N}}$ is subadditive. The entropy of μ with respect to the action of geodesic flow and to the partition P is defined by

$$h_{KS}(\mu, P) = \lim_{n\longrightarrow+\infty} \frac{h_n(\mu, P)}{n} = \inf_{n\in\mathbb{N}} \frac{h_n(\mu, P)}{n}. \tag{8.4}$$

The existence of the limit, and the fact that it coincides with the infimum, follow from a standard subadditivity argument. Note that $\mu(P_{\alpha_0} \cap g^{-1} P_{\alpha_1} \ldots \cap g^{-n+1} P_{\alpha_{n-1}})$ measures the μ–probability to visit successively P_{α_0}, P_{α_1},..., $P_{\alpha_{n-1}}$ at times 1, 2,...,

$n-1$ of the geodesic flow. The entropy measures the average exponential decay of these probabilities when n gets large. In particular, if there is a uniform exponential decay, that is, if there exists $C, \beta \geq 0$ such that $\mu(P_{\alpha_0} \cap g^{-1}P_{\alpha_1} ... \cap g^{-n}P_{\alpha_n}) \leq Ce^{-\beta n}$, for all n and all $\alpha_0, \ldots, \alpha_n$, then it is easy to see that $h_{KS}(\mu, P) \geq \beta$.

The entropy of μ with respect to the action of the geodesic flow is defined as

$$h_{KS}(\mu) = \sup_P h_{KS}(\mu, P), \tag{8.5}$$

the supremum running over all finite measurable partitions P. Assume μ is carried on the energy layer S^*X. Due to the Anosov property of the geodesic flow on S^*X, it is known that the supremum (8.5) is reached as soon as the maximum diameter of the sets $P_k \cap S^*X$ is small enough.

We will restrict our attention to partitions P which are actually partitions of the base X (lifted to T^*X) : $X = \sqcup_{k=1}^K P_k$. This choice is not crucial, but it simplifies certain aspects of the analysis.

The fact that the limit in (8.4) coincides with the infimum has a crucial consequence, the upper semicontinuity property of $h_{KS}(., P)$: if (μ_k) is a sequence of (g^t)–invariant probability measures converging weakly to μ, then

$$h_{KS}(\mu, P) \geq \limsup_k h_{KS}(\mu_k, P) \tag{8.6}$$

(provided μ does not charge the boundary of P).

Since our semiclassical measure μ is defined as a limit of Husimi measures associated to $\psi_\hbar$, a naive idea would be to estimate from below the entropy of $\psi_\hbar$ and then take the limit.

A first issue is to decide how to define the $\psi_\hbar$–probability to visit successively P_{α_0}, P_{α_1},..., $P_{\alpha_{n-1}}$ at times 1, 2,..., $n-1$.

From the definition of the Husimi measures, a first idea could be to consider

$$\left\langle \psi_\hbar, \mathrm{Op}_\hbar\left((\mathbb{1}_{P_{\alpha_0}})(\mathbb{1}_{P_{\alpha_1}} \circ g^1) \ldots (\mathbb{1}_{P_{\alpha_{n-1}}} \circ g^{n-1}) \right) \psi_\hbar \right\rangle. \tag{8.7}$$

To avoid dealing with characteristic functions (which are not quantized to pseudodifferential operators), we can smooth them by convolution and try replacing $\mathbb{1}_{P_k}$ by a smooth $\mathbb{1}^{sm}_{P_k}$. Even so, studying the large–n behaviour of (8.7) is very problematic. In fact, the derivatives of $(\mathbb{1}^{sm}_{P_{\alpha_0}})(\mathbb{1}^{sm}_{P_{\alpha_1}} \circ g^1) \ldots (\mathbb{1}^{sm}_{P_{\alpha_{n-1}}} \circ g^{n-1})$ grow like e^n, so that when n reaches the size $|\log \hbar|$ this function no longer belongs to any reasonable symbol space (the operator is not a pseudodifferential operator).

We also note that an overlap of the form (8.7) is a *hybrid* expression: this is a *quantum* matrix element, but the operator is defined in terms of the *classical* flow ! From the point of view of quantum mechanics, it is more natural to consider, instead, the operator obtained as the product of Heisenberg-evolved quantized functions, namely

$$\hat{P}_{\alpha_{n-1}}(n-1)\hat{P}_{\alpha_{n-2}}(n-2) \ldots \hat{P}_{\alpha_1}(1)\hat{P}_{\alpha_0} . \tag{8.8}$$

Here we used the shorthand notation $\hat{P}_k \stackrel{\text{def}}{=} \mathrm{Op}(\mathbb{1}^{sm}_{P_k})$, $k \in [1, K]$, and $\hat{P}_k(t) = U_\hbar^{-t}\hat{P}_k U_\hbar^t$. Instead of (8.7), a second idea is to consider

$$\left\langle \psi_\hbar, \hat{P}_{\alpha_{n-1}}(n-1) \ldots \hat{P}_{\alpha_1}(1)\hat{P}_{\alpha_0}\psi_\hbar \right\rangle. \tag{8.9}$$

as the $\psi_\hbar$–probability to visit successively P_{α_0}, P_{α_1},..., $P_{\alpha_{n-1}}$ at times 1, 2,..., $n-1$. However, the scalar product is a complex number, and can not be directly manipulated as a probability.

Our third and final try is to consider

$$\|\hat{P}_{\alpha_{n-1}}(n-1)\dots\hat{P}_{\alpha_1}(1)\hat{P}_{\alpha_0}\psi_\hbar\|^2. \tag{8.10}$$

In fact, if we do the smoothing of $\mathbb{1}_{P_k}$ so that

$$\sum_k (\mathbb{1}_{P_k}^{sm})^2 \equiv 1$$

then the norms (8.10) can actually be manipulated like probability measures :

$$\sum_{\alpha_0,\dots,\alpha_{n-1}} \|\hat{P}_{\alpha_{n-1}}(n-1)\hat{P}_{\alpha_{n-2}}(n-2)\dots\hat{P}_{\alpha_1}(1)\hat{P}_{\alpha_0}\psi_\hbar\|^2 = 1, \tag{8.11}$$

and

$$\sum_{\alpha_{n-1}} \|\hat{P}_{\alpha_{n-1}}(n-1)\hat{P}_{\alpha_{n-2}}(n-2)\dots\hat{P}_{\alpha_1}(1)\hat{P}_{\alpha_0}\psi_\hbar\|^2$$
$$= \|\hat{P}_{\alpha_{n-2}}(n-2)\dots\hat{P}_{\alpha_1}(1)\hat{P}_{\alpha_0}\psi_\hbar\|^2.$$

Finally, using the Egorov theorem (4.6), we see that, for fixed n,

$$\|\hat{P}_{\alpha_{n-1}}(n-1)\dots\hat{P}_{\alpha_1}(1)\hat{P}_{\alpha_0}\psi_\hbar\|^2 \underset{\hbar\longrightarrow 0}{\longrightarrow} \mu\left((\mathbb{1}_{P_{\alpha_0}}^{sm})^2\,(\mathbb{1}_{P_{\alpha_1}}^{sm}\circ g^1)^2\dots(\mathbb{1}_{P_{\alpha_{n-1}}}^{sm}\circ g^{n-1})^2\right)$$

if the Husimi measures of $\psi_\hbar$ converge to μ. Apart from the smoothing, this is the quantity we are interested in when computing entropy of μ (8.2).

It is proved in [**A05**] that

$$\|\hat{P}_{\alpha_{n-1}}(n-1)\dots\hat{P}_{\alpha_1}(1)\hat{P}_{\alpha_0}\psi_\hbar\|^2 \leq \frac{C}{\hbar^d}e^{-(d-1)n},$$

say, in dimension d and constant curvature -1, and assuming the diameter of the P_k is small enough[8]. From this, it would be tempting to deduce that the entropy of the $\psi_\hbar$–Husimi measures is bounded below by $d-1$, then use the semicontinuity property (8.6) to deduce that $h_{KS}(\mu) \geq d-1$ (thus proving quantum unique ergodicity).

Of course, we can not apply (8.6), since we are not in the situation of a sequence (μ_k) of g^t–invariant probability measures converging to μ. To use (8.6) we need to know if a similar property holds in our quantum framework, using expressions such as (8.10) to evaluate entropies. This is, in fact, NOT the case : a factor of 2 is lost somewhere in the proof, and we will end up proving

$$h_{KS}(\mu) \geq \frac{d-1}{2}.$$

As we shall see, this is due to the fact that the operators $\hat{P}_\alpha(t)$ appearing in (8.10) do not commute : for non commuting operators, the interpretation of $\|\hat{P}_{\alpha_{n-1}}(n-1)\dots\hat{P}_{\alpha_1}(1)\hat{P}_{\alpha_0}\psi_\hbar\|^2$ as "the probability to visit successively P_{α_0}, P_{α_1},..., $P_{\alpha_{n-1}}$ at times 1, 2,..., $n-1$" is not allowed. It is only acceptable for a time n small enough so that the operators almost commute (up to some small error). This restriction of the time range is responsible for the loss of a factor 2.

[8]To prove this estimate, we assume, without any loss of generality, that the injectivity radius of X is larger than 1.

THE ENTROPIC UNCERTAINTY PRINCIPLE.

In this chapter, we give the steps of the proof of our main result, inequality (7.3). To simplify the presentation we restrict ourselves to the case of constant curvature $\equiv -1$.

We start with a functional inequality called the "entropic uncertainty principle".

9. The abstract result...

We consider a complex Hilbert space $(\mathcal{H}, \langle .,. \rangle)$, and denote $\|\psi\| = \sqrt{\langle \psi, \psi \rangle}$ the associated norm. The same notation $\|.\|$ will also be used for the operator norm on $\mathcal{L}(\mathcal{H})$.

We define the following family of l_p norms on $\mathcal{H}^N$: for $\Psi = (\Psi_1, \dots, \Psi_N) \in \mathcal{H}^N$, we let

$$(9.1) \qquad \|\Psi\|_p \stackrel{\text{def}}{=} \left(\sum_{k=1}^{N} \|\Psi_k\|^p \right)^{1/p}, \quad 1 \leq p < \infty, \text{ and } \quad \|\Psi\|_\infty \stackrel{\text{def}}{=} \max_k \|\Psi_k\| .$$

For $p = 2$, this norm coincides with the Hilbert norm deriving from the scalar product

$$\langle \Psi, \Phi \rangle_{\mathcal{H}^N} = \sum_{k=1}^{N} \langle \Psi_k, \Phi_k \rangle_{\mathcal{H}}.$$

We can define similarly a family of l_p norms on $\mathcal{H}^M \ni \Phi = (\Phi_1, \dots, \Phi_M)$:

$$(9.2) \qquad \|\Phi\|_p \stackrel{\text{def}}{=} \left(\sum_{j=1}^{M} \|\Phi_j\|^p \right)^{1/p}, \quad 1 \leq p < \infty, \text{ and } \quad \|\Phi\|_\infty \stackrel{\text{def}}{=} \max_j \|\Phi_j\| .$$

For $\Psi \in \mathcal{H}^N$ with $\|\Psi\|_2 = 1$, we define its entropy,

$$h(\Psi) = - \sum_{k=1}^{N} \|\Psi_k\|^2 \log \|\Psi_k\|^2 ;$$

(and we define similarly the entropy of a normalized vector $\Phi \in \mathcal{H}^M$). The entropy is related to the l_p norms by the fact that $-\frac{1}{4}h(\Psi)$ is the derivative of $\|\Psi\|_p$ at $p = 2$.

Consider the action of a bounded operator $T : \mathcal{H}^N \to \mathcal{H}^M$, which we present as a $M \times N$ matrix $(T_{j\,k})$ of bounded operators on $\mathcal{H}$. We denote $\|T\|_{p,q}$ the norm of T from $l_p(\mathcal{H}^N)$ to $l_q(\mathcal{H}^M)$, for $1 \leq p, q \leq \infty$.

THEOREM 9.1 (Riesz interpolation theorem). [**DunSchw**, Section VI.10] *The function* $\log \|T\|_{1/a,1/b}$ *is a convex function of* (a, b) *in the square* $0 \leq a, b \leq 1$.

From this theorem, Maassen and Uffink derived a new form of uncertainty relations [**MaaUff88**].

THEOREM 9.2. *Assume that* $\|T\|_{2,2} = 1$, *which implies in particular that* $\|T_{jk}\| \leq 1$ *for all* j, k. *Introduce the real number* $c(T) = \max_{j,k} \|T_{jk}\|$, *where the norm is the operator norm in* $\mathcal{L}(\mathcal{H})$.

For all $\Psi \in \mathcal{H}$ *such that* $\|\Psi\|_2 = 1$ *and* $\|T\Psi\|_2 = 1$, *we have*

$$h(T\Psi) + h(\Psi) \geq -2 \log c(T) .$$

REMARK 9.3. The theorem proved in [**MaaUff88**] deals with $\mathcal{H} = \mathbb{C}$. The operator T is then simply represented by an $M \times N$ complex matrix, and $c(T)$ is the supremum of its entries. According to the theorem, if a matrix has small entries, then either $h(T\Psi)$ or $h(\Psi)$ must be large. Here $h(\Psi) = -\sum |\Psi_k|^2 \log |\Psi_k|^2$ measures the entropy of the probability vector $(|\Psi_k|^2)$, given by the coordinates of Ψ in the canonical basis of $\mathbb{C}^N$.

Proof: In the case $a = 1, b = 0$, we have for any Ψ,

$$\|T\Psi\|_\infty = \sup_j \|(T\Psi)_j\| \leq \sup_{j,k} \|T_{j,k}\| \sum_{k'} \|\Psi_{k'}\| = \sup_{j,k} \|T_{j,k}\| \, \|\Psi\|_1 ,$$

which can be written as $\|T\|_{1,\infty} \leq \sup_{j,k} \|T_{j,k}\| \stackrel{\text{def}}{=} c(T)$.

Let us assume that T is contracting on l_2 : $\|T\|_{2,2} \leq 1$. We take $t \in [0,1]$ and $a_t = \frac{1+t}{2}$, $b_t = \frac{1-t}{2}$ to interpolate between $(1/2, 1/2)$ and $(1, 0)$; Theorem 9.1 implies that

$$\|T\|_{1/a_t, 1/b_t} \leq c(T)^t .$$

COROLLARY 9.4. *Let the operator $T : \mathcal{H}^N \longrightarrow \mathcal{H}^M$ satisfy $\|T\|_{2,2} \leq 1$ and call $c(T) \stackrel{\text{def}}{=} \sup_{j,k} \|T_{j,k}\|$. Then, for all $t \in [0,1]$, for all $\Psi \in \mathcal{H}^N$,*

$$\|T\Psi\|_{\frac{2}{1-t}} \leq c(T)^t \, \|\Psi\|_{\frac{2}{1+t}} .$$

We define for any $r > 0$ or $-1 < r < 0$ the "moments"

$$M_r(\Psi) \stackrel{\text{def}}{=} \left(\sum_j \|\Psi_j\|^{2+2r} \right)^{1/r} .$$

Corollary 9.4 leads to the following family of "uncertainty relations":

$$\forall t \in (0,1),\ \forall \Psi \in \mathbb{C}^N, \qquad M_{\frac{t}{1-t}}(T\Psi)\, M_{\frac{-t}{1+t}}(\Psi) \leq c(T)^2 . \tag{9.3}$$

In the case $\|\Psi\|_2 = 1$, we notice that the moments converge to the same value when $r \to 0$ from above or below:

$$\lim_{r \to 0} M_r(\Psi) = e^{-h(\Psi)}, \quad \text{where} \quad h(\Psi) = -\sum_j \|\Psi_j\|^2 \log \|\Psi_j\|^2 .$$

If, furthermore, $\|T\Psi\|_2 = 1$, then the limit $t \to 0$ of the inequalities (9.3) yield the Entropic Uncertainty Principle stated in Theorem 9.2. □

We shall use Theorem 9.2 in the following particular case :

EXAMPLE 9.5. Suppose we have two partitions of unity $(\pi_k)_{k=1}^N$ and $(\tau_j)_{j=1}^M$, that is, two families of operators on $\mathcal{H}$ such that

$$\sum_{k=1}^N \pi_k \pi_k^* = Id, \qquad \sum_{j=1}^M \tau_j \tau_j^* = Id. \tag{9.4}$$

Let U be a unitary operator on $\mathcal{H}$. We can take $T_{j\,k} = \tau_j^* U \pi_k$.

LEMMA 9.6. *Let $T_{jk} = \tau_j^* U \pi_k$, for some bounded operator $U : \mathcal{H} \to \mathcal{H}$. Then we have the identity*

$$\|T\|_{2,2} = \|U\|_{\mathcal{L}(\mathcal{H})} .$$

PROOF. The operator T may be described as follows. Consider the following row and column vectors of operators on $\mathcal{H}$:

$$L \stackrel{\text{def}}{=} (\pi_1, \dots, \pi_N), \quad \text{as well as } C = \begin{pmatrix} \tau_1^* \\ \dots \\ \tau_M^* \end{pmatrix}.$$

We can write $T = CUL$. We insert this formula in the identity

$$\|T\|_{2,2}^2 = \|T^*T\|_{\mathcal{L}(\mathcal{H}^N)} = \left\|L^\dagger U^* C^\dagger CUL\right\|_{\mathcal{L}(\mathcal{H}^N)}$$

Using (9.4) for the τ_j, we notice that $C^\dagger C = Id_{\mathcal{H}}$, so that the norm above reads

$$\left\|L^\dagger U^* UL\right\|_{\mathcal{L}(\mathcal{H}^N)}.$$

Then, we use the identities

$$\left\|(UL)^\dagger(UL)\right\|_{\mathcal{L}(\mathcal{H}^N)} = \left\|(UL)(UL)^\dagger\right\|_{\mathcal{L}(\mathcal{H})} = \left\|(UL)L^\dagger U^*\right\|_{\mathcal{L}(\mathcal{H})} = \|UU^*\|_{\mathcal{L}(\mathcal{H})},$$

where we used (9.4) for the π_k. □

Therefore, if U is contracting (resp. $\|U\|_{\mathcal{L}(\mathcal{H})} = 1$) one has $\|T\|_{2,2} \le 1$ (resp. $\|T\|_{2,2} = 1$).

We also specify the vector Ψ by taking $\Psi_k = \pi_k^*\psi$ for some normalized $\psi \in \mathcal{H}$. From (9.4), we check that $\|\Psi\|_2 = \|\psi\|$, and also that $(T\Psi)_j = \tau_j^* U\psi$. Thus, if $\|U\psi\| = 1$, the relation (9.4) also implies $\|T\Psi\|_2 = \|U\psi\| = 1$. With this choice for T and Ψ, Theorem 9.2 reads as follows:

THEOREM 9.7. *Let U be an isometry on $\mathcal{H}$, and let π, τ be two quantum partitions of unity as in* (9.4).
Define $c_{\tau,\pi}(U) \stackrel{\text{def}}{=} \sup_{j,k} \|\tau_j^*\, U\, \pi_k\|_{\mathcal{L}(\mathcal{H})}$.
Then, for any normalized $\psi \in \mathcal{H}$, we have

$$h_\tau(U\psi) + h_\pi(\psi) \ge -2 \log c_{\tau,\pi}(U)$$

where $h_\pi(\psi) = -\sum_{k=1}^N \|\pi_k^*\psi\|^2 \log \|\pi_k^*\psi\|^2$ and $h_\tau(\psi) = -\sum_{j=1}^M \|\tau_j^*\psi\|^2 \log \|\tau_j^*\psi\|^2$.

Note that the definition $h_\pi(\psi) = -\sum_{k=1}^N \|\pi_k^*\psi\|^2 \log \|\pi_k^*\psi\|^2$ is somewhat analogous to (8.1), π playing the role of the partition P and ψ the role of the measure μ. The quantity $h_\pi(\psi)$ may be called the "Shannon entropy" of the state ψ with respect to the partition π.

10. ... applied to eigenfunctions of the laplacian...

In this section we define the data to input into Theorem 9.7, in order to obtain information on the eigenstates $\psi_\hbar$ (7.2) and the semiclassical measures μ considered in the previous chapters. Only the Hilbert space is fixed, $\mathcal{H} \stackrel{\text{def}}{=} L^2(X)$. All other data depend on the semiclassical parameter $\hbar$: the quantum partitions π, τ, the unitary operator U. Besides, we will need yet another technical variant of Theorem 9.7.

10.1. Smooth partition of unity. To evaluate the Kolmogorov–Sinai entropy of a semiclassical measure, we must start by decomposing T^*X into a finite partition. We actually specify the form of the partition we want to use. We work with a measurable partition $(P_k)_{k=1,\ldots,K}$ of the base $X : X = \sqcup P_k$, that we lift to a partition of the phase space T^*X.

For semiclassical methods we actually need to work with smooth functions, and so we introduce a smooth partition of unity $(\mathbb{1}^{sm}_{P_k})$, obtained by smoothing the characteristic functions $(\mathbb{1}_{P_k})$ with a convolution kernel. We require that the smoothing be done so that $\sum_{k=1}^{K}(\mathbb{1}^{sm}_{P_k})^2 \equiv 1$.

We finally denote $\hat{P}_k = \mathrm{Op}(\mathbb{1}^{sm}_{P_k})$: it is simply the operator of multiplication by $\mathbb{1}^{sm}_{P_k}$. We have

$$\sum_{k=1}^{K} \hat{P}_k^2 = I\,, \tag{10.1}$$

which means that these operators form a quantum partition of unity as in (9.4), which we will call $\mathcal{P}^{(0)}$.

10.2. Refinement of the partition under the Schrödinger flow. We denote by $U^t = \exp(it\hbar\triangle/2)$ the quantum propagator. With no loss of generality, we will assume that the injectivity radius of X is much greater than 1, and work with the propagator at time one, $U = U^1$. This propagator quantizes the geodesic flow at time one, g^1. The $\hbar$-dependence of U will be implicit in our notations.

As one does to compute the Kolmogorov–Sinai entropy of an invariant measure, we define a new quantum partition of unity by evolving and refining the initial partition $\mathcal{P}^{(0)}$ under the quantum evolution. For each time $n \in \mathbb{N}$ and any sequence of symbols $\boldsymbol{\alpha} = (\alpha_0 \cdots \alpha_{n-1})$, $\alpha_i \in [1, K]$ (we say that the sequence $\boldsymbol{\alpha}$ is of *length* $|\boldsymbol{\alpha}| = n$), we define the operators

$$\hat{P}_{\boldsymbol{\alpha}} = \hat{P}_{\alpha_{n-1}}(n-1)\hat{P}_{\alpha_{n-2}}(n-2)\ldots\hat{P}_{\alpha_0}\,. \tag{10.2}$$

We keep using the notation $A(t) = U^{-t}AU^t$ for the quantum evolution of an operator A. From (10.1) and the unitarity of U, the family of operators $\left\{\hat{P}_{\boldsymbol{\alpha}}\right\}_{|\boldsymbol{\alpha}|=n}$ obviously satisfies the relation $\sum_{|\boldsymbol{\alpha}|=n}\hat{P}_{\boldsymbol{\alpha}}\,\hat{P}^*_{\boldsymbol{\alpha}} = Id_{L^2}$, and therefore forms a quantum partition which we call $\mathcal{P}^{(n)}$. We also have $\sum_{|\boldsymbol{\alpha}|=n}\hat{P}^*_{\boldsymbol{\alpha}}\,\hat{P}_{\boldsymbol{\alpha}} = Id_{L^2}$, and we denote $\mathcal{T}^{(n)}$ the partition of unity given by the family of operators $\left\{\hat{P}^*_{\boldsymbol{\alpha}}\right\}_{|\boldsymbol{\alpha}|=n}$.

10.3. In the entropic uncertainty principle, Theorem 9.7, we shall input the following data :

- the quantum partition $\pi = \mathcal{P}^{(n)}$ is given by the family of operators $\{\hat{P}_{\boldsymbol{\alpha}},\ |\boldsymbol{\alpha}| = n\}$. The quantum partition $\tau = \mathcal{T}^{(n)}$ is given by the family of operators $\{\hat{P}^*_{\boldsymbol{\alpha}},\ |\boldsymbol{\alpha}| = n\}$. The integer n will always be of order $\mathcal{K}|\log\hbar|$, where $\mathcal{K}$ will be determined later.
- the isometry will be $\mathcal{U} = U^n$.

To apply Theorem 9.7 we will need an upper bound on

$$c_{\mathcal{T}^{(n)},\mathcal{P}^{(n)}}(\mathcal{U}) = \max_{|\boldsymbol{\alpha}|=|\boldsymbol{\alpha}'|=n} \|\hat{P}_{\boldsymbol{\alpha}'}\,U^n\,\hat{P}_{\boldsymbol{\alpha}}\|.$$

We remark that $\hat{P}_{\boldsymbol{\alpha}'} U^n \hat{P}_{\boldsymbol{\alpha}}$ can be developed as

$$U^{-n+1} \hat{P}_{\alpha'_{n-1}} U \cdots U \hat{P}_{\alpha'_1} U \hat{P}_{\alpha'_0} U \hat{P}_{\alpha_{n-1}} \cdots U \hat{P}_{\alpha_1} U \hat{P}_{\alpha_0}$$

or equivalently

$$U^n \hat{P}_{\alpha'_{n-1}}(2n-1) \cdots \hat{P}_{\alpha'_1}(n+1) \hat{P}_{\alpha'_0}(n) \hat{P}_{\alpha_{n-1}}(n-1) \cdots \hat{P}_{\alpha_1}(1) \hat{P}_{\alpha_0}.$$

10.4. The main estimate. Let us assume (without loss of generality) that the injectivity radius of X is greater than 1; and that the diameter of each P_k is small enough so that, for every j, k, for every $x, y \in P_j, P_k$, there is at most one unit speed geodesic joining x and y in time 1.

The estimate essentially proven in [**A05**] is :

THEOREM 10.1. *Let χ be an energy cut-off, that is, a smooth compactly supported function vanishing outside $H^{-1}([1/2-\varepsilon, 1/2+\varepsilon])$.*

Given $\mathcal{K} > 0$ and a partition $\mathcal{P}^{(0)}$, there exists $\hbar_{\mathcal{K},\mathcal{P}^{(0)},\chi}$ such that, for any $\hbar \leq \hbar_{\mathcal{K},\mathcal{P}^{(0)},\chi}$, for any positive integer $n \leq \mathcal{K}|\log \hbar|$, and any pair of sequences $\boldsymbol{\alpha}$, $\boldsymbol{\alpha}'$ of length n,

$$\|\hat{P}_{\boldsymbol{\alpha}'} U^n \hat{P}_{\boldsymbol{\alpha}} \operatorname{Op}(\chi)\| \leq C \hbar^{-\frac{d}{2}} e^{-(d-1)n} (1+O(\varepsilon))^n . \tag{10.3}$$

The constant C only depends on the riemannian manifold.

REMARK 10.2. The best bound we can hope for on the norm of the operators $\hat{P}_{\boldsymbol{\alpha}'} U^n \hat{P}_{\boldsymbol{\alpha}}$ is certainly the trivial one : $c_{\mathcal{T}^{(n)},\mathcal{P}^{(n)}}(\mathcal{U}) \leq 1$. We can only improve this bound if we insert the energy cut-off $\operatorname{Op}_\hbar(\chi)$, which has the effect of restricting our operators to functions oscillating at a certain (high) frequency. In fact, if we take χ of the form $f((2H-1))$, where f is a smooth function on $\mathbb{R}$, supported in $[1-2\varepsilon, 1+2\varepsilon]$, we can take $\operatorname{Op}(\chi)$ to be the corresponding function of the laplacian, obtained by functional calculus, $f((-\hbar^2 \triangle - 1))$. Thus, $\operatorname{Op}(\chi)$ is a smoothed version of the projection to the spectral window $[\frac{1-2\varepsilon}{\hbar^2}, \frac{1+2\varepsilon}{\hbar^2}]$ of the laplacian.

Since we have no good estimate on $\|\hat{P}_{\boldsymbol{\alpha}'} U^n \hat{P}_{\boldsymbol{\alpha}}\|$, but only on $\|\hat{P}_{\boldsymbol{\alpha}'} U^n \hat{P}_{\boldsymbol{\alpha}} \operatorname{Op}(\chi)\|$, we will need to modify accordingly the statement of the entropic uncertainty principle : see later Theorem 10.5

REMARK 10.3. If we were in variable curvature, instead of the exponent $d-1$ we would have a variable exponent depending on the local Lyapunov exponents.

The proof of Theorem 10.1 will be given in Section 15. The idea is rather simple, although the technicalities may seem difficult. We first show that any state in the image of $\operatorname{Op}(\chi)$ may be decomposed as a superposition of essentially $\hbar^{-\frac{d}{2}}$ normalized lagrangian states, supported on lagrangian manifolds transverse to the stable foliation. The lagrangian states we work with are truncated δ–functions, supported on spheres $S_z^* X$. The action of the operator $U^{n-1} \hat{P}_{\boldsymbol{\alpha}'} U^n \hat{P}_{\boldsymbol{\alpha}} = \hat{P}_{\alpha'_{n-1}} U \cdots U \hat{P}_{\alpha'_0} U \hat{P}_{\alpha_{n-1}} \cdots U \hat{P}_{\alpha_0}$ on such lagrangian states is translated, by the theory of Fourier integral operators (WKB methods), into a classical picture : an application of U corresponds, classically, to applying g^1, so that it stretches the lagrangian in the unstable direction. Each multiplication by $\hat{P}_\alpha$ corresponds to cutting out a small piece of lagrangian. This iteration of stretching and cutting is responsible for the exponential decay.

In [**AN07**] the estimate of Theorem 10.1 is modified by optimizing the shape of the cutoff χ. We want to take into account the fact that we are dealing with eigenfunctions of the laplacian, and not merely spectral packets in the window $[\frac{1-2\varepsilon}{\hbar^2}, \frac{1+2\varepsilon}{\hbar^2}]$. But we also want $\mathrm{Op}(\chi)$ to stay in a reasonable class of pseudodifferential operators. We consider a smooth function $\chi \in C^\infty(\mathbb{R};[0,1])$, with $\chi(t)=1$ for $|t|\leq 1$ and $\chi(t)=0$ for $|t|\geq 2$. Then, for some fixed $\delta\in(0,1)$, we rescale that function to obtain an $\hbar$-dependent cutoff near S^*X:

$$\text{(10.4)}\quad \forall \hbar\in(0,1),\ \forall n\in\mathbb{N},\ \forall \rho\in T^*X, \qquad \chi_\delta(\rho;\hbar) \stackrel{\mathrm{def}}{=} \chi\big(\hbar^{-1+\delta}(H(\rho)-1/2)\big)\,.$$

The cutoff χ_δ is localized in a tubular neighbourhood of S^*X of width $2\hbar^{1-\delta}$

THEOREM 10.4. [**AN07**] *Given $\mathcal{K}>0$ a partition $\mathcal{P}^{(0)}$ and $\delta>0$ small enough, there exists $\hbar_{\mathcal{K},\mathcal{P}^{(0)},\delta}$ such that, for any $\hbar\leq \hbar_{\mathcal{K},\mathcal{P}^{(0)},\delta}$, for any positive integer $n\leq \mathcal{K}|\log\hbar|$, and any pair of sequences $\boldsymbol{\alpha}$, $\boldsymbol{\alpha}'$ of length n,*

$$\text{(10.5)}\qquad \|\hat{P}_{\boldsymbol{\alpha}'}\,U^n\,\hat{P}_{\boldsymbol{\alpha}}\,\mathrm{Op}(\chi_\delta)\| \leq C\,\hbar^{-\frac{d-1}{2}-\delta}\,e^{-(d-1)n}(1+O(\hbar^\delta))^n\,.$$

The constant C only depends on the riemannian manifold (M,g).

Theorem 10.4 essentially improves the prefactor $\hbar^{-\frac{d}{2}}$ of Theorem 10.1. Its proof is similar, the main difficulty being to define $\mathrm{Op}(\chi_\delta)$ — the function χ_δ does not fall into one of the usual "nice" classes of symbols, since its derivatives explode quite fast when $\hbar\longrightarrow 0$. To define $\mathrm{Op}(\chi_\delta)$ would be much beyond the scope of these notes (see [**SZ99, AN07**]). We shall admit Theorem 10.4, and only prove the simpler version of Theorem 10.1.

10.5. Technical variant of the entropic uncertainty principle. As explained in Remark 10.2, we cannot apply Theorem 9.7 directly, because we need to insert our energy cut-off $\mathrm{Op}(\chi)$. On the other hand, this frequency cut–off does not really bother us, since it hardly modifies our eigenfunctions.

We generalize the statement of Theorem 9.7 by introducing an auxiliary operator $\mathcal{O}$.

THEOREM 10.5. [**AN07**] *Let $\mathcal{O}$ be a bounded operator on $\mathcal{H}$. Let $\mathcal{U}$ be an isometry on $\mathcal{H}$.*
Define $c_{\mathcal{O}}^{\tau,\pi}(\mathcal{U}) \stackrel{\mathrm{def}}{=} \sup_{j,k}\|\tau_j^*\,\mathcal{U}\,\pi_k\,\mathcal{O}\|_{\mathcal{L}(\mathcal{H})}$.

Then, for any $\theta\geq 0$, for any normalized $\psi\in\mathcal{H}$ satisfying

$$\forall k=1,\ldots,\mathcal{N},\qquad \|(Id-\mathcal{O})\pi_k^*\psi\|\leq\theta\,,$$

the entropies $h_\tau(\mathcal{U}\psi)$, $h_\pi(\psi)$ satisfy

$$h_\tau(\mathcal{U}\psi)+h_\pi(\psi)\geq -2\log\big(c_{\mathcal{O}}^{\pi,\tau}(\mathcal{U})+\mathcal{N}\,\theta\big).$$

10.6. Applying the entropic uncertainty principle. We now precise all the data we will use in the entropic uncertainty principle, Theorem 10.5:

- the quantum partition $\pi=\mathcal{P}^{(n)}$, $\tau=\mathcal{T}^{(n)}$ have already been defined in 10.3. The integer n will be of order $\mathcal{K}|\log\hbar|$, where the choice of $\mathcal{K}$ will be determined later. In the semiclassical limit, these partitions have cardinality $\mathcal{N}=K^n\asymp\hbar^{-K_0}$ for some fixed $K_0>0$.
- the isometry will be $\mathcal{U}=U^n$.
- the operator $\mathcal{O}$ is $\mathcal{O}=\mathrm{Op}(\chi_\delta)$. Since we are, at the end, interested in eigenfunctions of the laplacian, we need to know that this operator hardly

modifies them. In fact, for any $L > 0$, there exists $\hbar_L$ such that, for any $\hbar \leq \hbar_L$, any solution of $(-\hbar^2 \triangle - I)\psi_\hbar = 0$ satisfies

$$\forall \boldsymbol{\alpha}, |\boldsymbol{\alpha}| = n \leq \mathcal{K}|\log \hbar|, \ \|\big(\operatorname{Op}(\chi_\delta) - Id\big)\hat{P}^*_{\boldsymbol{\alpha}}\psi_\hbar\| \leq \hbar^L \|\psi_\hbar\|\,. \tag{10.6}$$

This means that, for an eigenfunction $\psi_\hbar$, all the states $\hat{P}^*_{\boldsymbol{\alpha}}\psi_\hbar$ are very sharply microlocalized near the energy layer S^*X.

- $\theta = \hbar^L$, and L will be chosen very large.

  All these quantities are defined for $n = \mathcal{K}|\log \hbar|$, $\mathcal{K}$ will be determined later, but fixed.

The entropies associated with a state $\psi \in \mathcal{H}$ are given by

$$h_{\mathcal{P}^{(n)}}(\psi) = - \sum_{|\boldsymbol{\alpha}|=n} \|\hat{P}^*_{\boldsymbol{\alpha}}\psi\|^2 \log\big(\|\hat{P}^*_{\boldsymbol{\alpha}}\psi\|^2\big)$$

and

$$h_{\mathcal{T}^{(n)}}(\psi) = - \sum_{|\boldsymbol{\alpha}|=n} \|\hat{P}_{\boldsymbol{\alpha}}\psi\|^2 \log\big(\|\hat{P}_{\boldsymbol{\alpha}}\psi\|^2\big).$$

We may apply Theorem 10.5 to any sequence of states satisfying (10.6).

COROLLARY 10.6. *Define*

$$c_{\operatorname{Op}\chi_\delta}(U^n) \stackrel{\text{def}}{=} \max_{|\boldsymbol{\alpha}|=|\boldsymbol{\alpha}'|=n} \|\hat{P}_{\boldsymbol{\alpha}'}\, U^n\, \hat{P}_{\boldsymbol{\alpha}} \operatorname{Op}(\chi_\delta)\|\,. \tag{10.7}$$

Then for any normalized state ϕ satisfying (10.6),

$$h_{\mathcal{T}^{(n)}}(U^n \phi) + h_{\mathcal{P}^{(n)}}(\phi) \geq -2\log\big(c_{\operatorname{Op}\chi_\delta}(U^n) + h^{L-K_0}\big)\,.$$

We now apply Corollary 10.6 to the particular case of the eigenstates $\psi_\hbar$. The estimate (10.5) can be rewritten as

$$c_{\operatorname{Op}\chi_\delta}(U^n) \leq C\,\hbar^{-\frac{d-1}{2}-\delta} e^{-(d-1)n}(1 + O(\hbar^\delta))^n\,.$$

We choose L large enough such that $\hbar^{L-K_0}$ is negligible in comparison with $\hbar^{-\frac{d-1}{2}-\delta}e^{-(d-1)n}$.

PROPOSITION 10.7. *Let $(\psi_\hbar)_{\hbar\to 0}$ be any sequence of eigenstates* (7.2). *Then, in the semiclassical limit, we have*

$$h_{\mathcal{T}^{(n)}}(\psi_\hbar) + h_{\mathcal{P}^{(n)}}(\psi_\hbar) \geq 2(d-1)n + (d-1+2\delta)\log\hbar + \mathcal{O}(1)\,. \tag{10.8}$$

This holds for $n \leq \mathcal{K}|\log \hbar|$ ($\mathcal{K}$ arbitrary) and $\hbar \leq \hbar_{\mathcal{K},\mathcal{P}^{(0)},\delta}$.

11. ...and the conclusion.

Before taking the limit $\hbar \to 0$, we prove that a similar lower bound holds if we replace $n \asymp |\log \hbar|$ by some fixed n_o, and $\mathcal{P}^{(n)}$ by the corresponding partition $\mathcal{P}^{(n_o)}$. Proposition 11.1 below is the semiclassical analogue of the classical subadditivity (8.3) of entropy for invariant measures.

We introduce the Ehrenfest time $n_E(\hbar) = \frac{(1-\delta')|\log\hbar|}{\lambda_{\max}}$ (δ' fixed, arbitrarily small). In constant curvature -1, the maximal expansion rate of the geodesic flow on S^*X is $\lambda_{\max} = 1$. The Ehrenfest time is the main limitation to use semiclassical methods to understand the large time behaviour of the Schrödinger flow : roughly speaking, we have $U^{-t}\operatorname{Op}_\hbar(a)U^t \sim \operatorname{Op}(a\circ g^t)$ for $|t| \leq \frac{n_E(\hbar)}{2}$, but for larger t we can no longer refer to the classical dynamics to understand $U^{-t}\operatorname{Op}_\hbar(a)U^t$.

PROPOSITION 11.1 (Subadditivity). *Let $\delta' > 0$. There is a function $R(n_o, \hbar)$ such that, for all integer n_o,*

$$\lim_{\hbar \longrightarrow 0} |R(n_o, \hbar)| = 0$$

and such that, for all n_o, $n \in \mathbb{N}$ with $n_o + n \leq n_E(\hbar)$, for any $(\psi_\hbar)$ normalized eigenstates satisfying (7.2), *the following inequality holds:*

$$h_{\mathcal{P}^{(n_o+n)}}(\psi_\hbar) \leq h_{\mathcal{P}^{(n_o)}}(\psi_\hbar) + h_{\mathcal{P}^{(n)}}(\psi_\hbar) + R(n_o, \hbar) .$$

We do not prove this proposition in these notes, but just make a few more comments. The non–commutative dynamical system formed by (U^t) acting on pseudodifferential operators is (approximately) commutative on time intervals of length $n_E(\hbar)$:

$$\| [\mathrm{Op}_\hbar(a)(t), \mathrm{Op}_\hbar(b)(-t)] \|_{L^2(X)} = \mathcal{O}(\hbar^{c\delta'}),$$

for any time $|t| \leq \frac{n_E(\hbar)}{2}$, or equivalently (using the unitarity of U^t)

$$\| [\mathrm{Op}_\hbar(a)(t), \mathrm{Op}_\hbar(b)] \|_{L^2(X)} = \mathcal{O}(\hbar^{c\delta'}),$$

for any time $|t| \leq n_E(\hbar)$. On such a time interval, we almost have a commutative dynamical system, up to small errors tending to 0 with $\hbar$. This roughly explains why the quantum entropy $h_{\mathcal{P}^{(n_o+n)}}(\psi_\hbar)$ has the same subadditivity property as the classical entropy (8.3), up to small errors, as long as $n_o + n$ remains bounded by the Ehrenfest time.

Thanks to this subadditivity, we may finish the proof of Theorem 7.1. Although Proposition 10.7 holds for $n \leq \mathcal{K}|\log \hbar|$ and $\mathcal{K}$ arbitrary, we are now limited by Proposition 11.1 to $\mathcal{K} = \frac{1-\delta'}{\lambda_{\max}}$. For $n = n_E(\hbar)$, Proposition 10.7 can be written

$$h_{\mathcal{P}^{(n)}}(\psi_\hbar) + h_{\mathcal{T}^{(n)}}(\psi_\hbar) \geq 2(d-1)n - \frac{(d-1+2\delta)\lambda_{\max}}{(1-\delta')} n + \mathcal{O}(1) . \tag{11.1}$$

Let $n_o \in \mathbb{N}$ be fixed and $n = n_E(\hbar)$. Using the Euclidean division $n = q n_o + r$ (with $r \leq n_o$), Proposition 11.1 implies that for $\hbar$ small enough,

$$\frac{h_{\mathcal{P}^{(n)}}(\psi_\hbar)}{n} \leq \frac{h_{\mathcal{P}^{(n_o)}}(\psi_\hbar)}{n_o} + \frac{h_{\mathcal{P}^{(r)}}(\psi_\hbar)}{n} + \frac{R(n_o, \hbar)}{n_o} .$$

A similar inequality holds with $\mathcal{P}$ replaced by $\mathcal{T}$.

Using (10.8) and the fact that $h_{\mathcal{P}^{(r)}}(\psi_\hbar)$ stays uniformly bounded (by a quantity depending on n_o) when $\hbar \to 0$, we find

$$\begin{aligned} \frac{1}{2}\left[\frac{h_{\mathcal{P}^{(n_o)}}(\psi_\hbar)}{n_o} + \frac{h_{\mathcal{T}^{(n_o)}}(\psi_\hbar)}{n_o}\right] \geq (d-1) - \frac{(d-1+2\delta)\lambda_{\max}}{2(1-\delta')} n \\ + \mathcal{O}(1) - \frac{R(n_o, \hbar)}{n_o} + \mathcal{O}_{n_o}(1/n) . \end{aligned} \tag{11.2}$$

We are now dealing with the partition $\mathcal{P}^{(n_o)}$, n_0 being fixed.

11.1. End of the proof. Let us take a subsequence of $(\psi_{\hbar_k})$ such that the Husimi measures $\mu_k = \mu_{\psi_{\hbar_k}}$ converge to a semiclassical measure μ on S^*X, invariant under the geodesic flow (see Prop. 6.1). We may take the limit $\hbar_k \to 0$ (so that $n \to \infty$) in the expression above. The norms appearing in the definition of $h_{\mathcal{P}^{(n_o)}}(\psi_{\hbar_k})$ and $h_{\mathcal{T}^{(n_o)}}(\psi_{\hbar_k})$can be written as

$$\|\hat{P}_{\boldsymbol{\alpha}}\,\psi_{\hbar_k}\| = \|\hat{P}_{\alpha_{n_o}}(n_o)\cdots\hat{P}_{\alpha_1}(1)\hat{P}_{\alpha_0}\,\psi_{\hbar_k}\| \tag{11.3}$$

$$\|\hat{P}^*_{\boldsymbol{\alpha}}\,\psi_{\hbar_k}\| = \|\hat{P}_{\alpha_0}\hat{P}_{\alpha_1}(1)\cdots\hat{P}_{\alpha_{n_o}}(n_o)\,\psi_{\hbar_k}\|\,. \tag{11.4}$$

For any sequence $\boldsymbol{\alpha}$ of length n_o, the laws of pseudodifferential calculus imply the convergence of $\|\hat{P}^*_{\boldsymbol{\alpha}}\,\psi_{\hbar_k}\|^2$ and $\|\hat{P}_{\boldsymbol{\alpha}}\,\psi_{\hbar_k}\|^2$ to the same quantity $\mu(\{\boldsymbol{\alpha}\})$, where $\{\boldsymbol{\alpha}\}$ is the function $(\mathbb{1}^{sm}_{P_{\alpha_0}})^2\,(\mathbb{1}^{sm}_{P_{\alpha_1}})^2\circ g^1\ldots(\mathbb{1}^{sm}_{P_{\alpha_{n_o-1}}})^2\circ g^{n_o-1}$ on T^*X. Thus $h_{\mathcal{P}^{(n_o)}}(\psi_{\hbar_k})$ and $h_{\mathcal{T}^{(n_o)}}(\psi_{\hbar_k})$ both semiclassically converge to the classical entropy

$$h_{n_o}(\mu) = h_{n_o}(\mu, (\mathbb{1}^{sm\,2}_{P_k})) = -\sum_{|\boldsymbol{\alpha}|=n_o}\mu(\{\boldsymbol{\alpha}\})\log\mu(\{\boldsymbol{\alpha}\})\,.$$

We have thus obtained the lower bound

$$\frac{h_{n_o}(\mu)}{n_o} \geq (d-1) - \frac{(d-1+2\delta)\lambda_{\max}}{2(1-\delta')}\,. \tag{11.5}$$

δ and δ' could be taken arbitrarily small, and at this stage they can be let vanish. Remember also that $\lambda_{\max} = 1$ for constant sectional curvature -1.

The Kolmogorov–Sinai entropy of μ (with respect to the partition $X = \sqcup P_k$) is by definition the limit of the first term $\frac{h_{n_o}(\mu)}{n_o}$ when n_o goes to infinity (8.2) (8.4), with the notable difference that the smooth functions $(\mathbb{1}^{sm}_{P_k})^2$ should be replaced by the characteristic functions $(\mathbb{1}_{P_k})$. We note, however, that the lower bound (11.5) does not depend on the derivatives of $(\mathbb{1}^{sm}_{P_k})^2$: as a result, the same bound carries over to the characteristic functions $(\mathbb{1}_{P_k})$.

We can finally let n_o tend to $+\infty$, to obtain (7.4).

□

The proof is finished, save for Theorem 10.1.

WKB METHODS.

To prove our main estimate (Theorem 10.1), we need to describe the action of the operator $U^t = \exp(it\hbar\frac{\triangle}{2})$ on "rapidly oscillating" functions, in the limit $\hbar \longrightarrow 0$. The idea, already used by Schrödinger, is to describe the action of $\exp(it\hbar\frac{\triangle}{2})$ on functions of the form $e^{\frac{i}{\hbar}S(x)}$, called WKB functions or lagrangian functions; and to use the fact that all the functions we consider can be represented as integral combinations of lagrangian functions.

12. Lagrangian submanifolds of T^*X and generating functions.

We have seen that T^*X is endowed with a "canonical" symplectic form ω, defined as follows. Let $\Omega \subset X$ be an open subset of X, endowed with a coordinate chart $\phi : X \longrightarrow \mathbb{R}^d$. Then $T^*\Omega \subset T^*X$ can be endowed with the coordinate chart

$$\Phi : T^*\Omega \longrightarrow \mathbb{R}^d \times (\mathbb{R}^d)^* \tag{12.1}$$

$$(x,p) \mapsto (\phi(x), (d\phi^*_x)^{-1}p). \tag{12.2}$$

On $T^*\Omega$, ω is defined as the pullback by Φ of the symplectic form $\sum_{i=1}^d dq_i \wedge dp_i$ of $\mathbb{R}^d \times (\mathbb{R}^d)^*$. We leave it to the reader to check that this definition does not depend on the choice of local coordinates. Thus, by choosing an atlas of X, one can define

ω on T^*X, and the definition does not depend on the atlas. In fact, ω can also be defined in an intrinsic way, as was done in Section 1 by formula (1.5).

DEFINITION 12.1. In the $2d$-dimensional symplectic manifold (T^*X, ω), a lagrangian submanifold is a d-dimensional submanifold on which the restriction of ω vanishes.

Equivalently, a submanifold $\mathcal{L} \subset T^*X$ is lagrangian if and only if, for all $\rho \in \mathcal{L}$, $T_\rho \mathcal{L}$ is its own ω-orthogonal in $T_\rho(T^*X)$.

EXAMPLE 12.2. On $T^*\mathbb{R}^d = \mathbb{R}^d \times (\mathbb{R}^d)^*$ endowed with the symplectic form $\sum_{i=1}^d dq_i \wedge dp_i$, affine subspaces of the form $\mathbb{R}^d \times \{\xi_0\}$ or $\{x_0\} \times \mathbb{R}^d$ are examples of lagrangian submanifolds. More generally, for any manifold X, the zero section $\{(x, 0), x \in X\} \subset T^*X$ is a lagrangian submanifold of T^*X. For any $x \in X$, the fiber T_x^*X is also lagrangian.

Generating functions.

EXERCISE 12.3. In $\mathbb{R}^d \times (\mathbb{R}^d)^*$ endowed with the symplectic form $\sum_{i=1}^d dq_i \wedge dp_i$, consider an linear subspace of the form Graph$A = \{(x, Ax)\}$ where A is a linear operator from $\mathbb{R}^d$ to itself. Show that Graph$A$ is lagrangian if and only if A is symmetric for the canonical euclidean structure on $\mathbb{R}^d$: $\langle Ax, y\rangle = \langle x, Ay\rangle$.

The following gives us many examples of lagrangian submanifolds.

EXERCISE 12.4. Let X be a smooth manifold and consider T^*X endowed with its usual symplectic structure. Let $\Omega \subset X$ be an open subset of X, and let a be a smooth 1–form on Ω. Consider the graph Graph $a = \{(x, a_x)\} \subset T^*\Omega$. Show that Graph a is lagrangian if and only if the 1–form a is closed : $da = 0$.

In particular, if Ω is simply connected, this implies the existence of a smooth function $S : \Omega \longrightarrow \mathbb{R}$ such that $a = dS$. The function S is called a generating function of the lagrangian manifold Graph a.

We denote $\pi : T^*X \longrightarrow X$ the canonical projection.

DEFINITION 12.5. Let $\mathcal{L} \subset T^*X$ be a lagrangian submanifold. The *caustic* of $\mathcal{L}$ is the set of points $\rho \in \mathcal{L}$ such that the restriction of π to $\mathcal{L}$ is not a local diffeomorphism at ρ.

In Example 12.2, the zero section in T^*X has empty caustic, whereas the case $\mathcal{L} = T_x^*X$ gives an example of a lagrangian submanifold for which the caustic is all of $\mathcal{L}$.

If ρ does not belong to the caustic, Exercise 12.4 shows there is a neighbourhood of ρ in which $\mathcal{L}$ is the graph of the differential dS, for some function S defined locally up to an additive constant. We say S is a generating function of $\mathcal{L}$ near ρ.

What happens on the caustic ?

Let $S(x, \theta)$ be a real–valued function on $\Omega_X \times \Omega_{\mathbb{R}^N}$ where Ω_X is an open subset of X and $\Omega_{\mathbb{R}^N}$ an open subset of $\mathbb{R}^N$. Let $C_S = \left\{(x, \theta), \frac{\partial S}{\partial \theta} = 0\right\}$. On C_S we assume that all the differentials $d_{(x,\theta)} \frac{\partial S}{\partial \theta_i}$ $(i = 1, \ldots, N)$ are linearly independent : then C_S is a smooth d–dimensional submanifold of $\Omega_X \times \Omega_{\mathbb{R}^N}$ (recall $d = dim\, X$). Define $j_S : C_S \longrightarrow T^*X$ by $j_S(x, \theta) = (x, \partial_x S(x, \theta))$.

PROPOSITION 12.6. *The map j_S is an immersion. Its image,*

$$\mathcal{L}_S = \{(x, \xi) \in T^*X, \text{ there exists } \theta \text{ with } \partial_\theta S(x, \theta) = 0 \text{ and } \xi = \partial_x S(x, \theta)\}$$

is a lagrangian submanifold of T^*X.

One calls $S(.,\theta)$ a generating family (or generating function) of $\mathcal{L}_S$. Compare with (1.24).

THEOREM 12.7. *Every lagrangian submanifold* $\mathcal{L}$ *of* T^*X *admits, locally, a generating family. More precisely : for* $\rho_0 \in \mathcal{L}$*, let* $N = \dim Ker\, d\pi_{\rho_0}$*. Then there is a neighbourhood* Ω *of* ρ_0 *in* T^*X*, there are open subsets* $\Omega_X \subset X$ *and* $\Omega_{\mathbb{R}^N} \subset \mathbb{R}^N$*, and a function* $S : \Omega_X \times \Omega_{\mathbb{R}^N} \longrightarrow \mathbb{R}$ *satisfying all the required conditions, such that*

$$\mathcal{L} \cap \Omega = \mathcal{L}_S.$$

Proof: Using (12.1) we see it is enough to consider the case $X = \mathbb{R}^d$. Let $\rho_0 = (x_0, \xi_0) \in T^*\mathbb{R}^d = \mathbb{R}^d \times (\mathbb{R}^d)^*$ and let $L = T_{\rho_0}\mathcal{L}$. It is a lagrangian linear subspace of $\mathbb{R}^d \times (\mathbb{R}^d)^*$. Let $\pi : \mathbb{R}^d \times (\mathbb{R}^d)^* \longrightarrow \mathbb{R}^d$ be the projection on the first coordinate. By assumption, $F = \pi(L)$ is a linear subspace of $\mathbb{R}^d$ of dimension $d - N$. Let G be a supplementary subspace of F in $\mathbb{R}^d$: $\mathbb{R}^d = F \oplus G$. We have a corresponding decomposition of the dual space, $\mathbb{R}^*_d = G^\circ \oplus F^\circ$, where F° is the space of linear forms vanishing on F, and similarly for G°. We leave it to the reader to check that the projection $P : L \longrightarrow F \times F^\circ$ is an isomorphism.

Since L is tangent to $\mathcal{L}$ at ρ_0, there is a neighbourhood Ω of ρ_0 such that $P : \mathcal{L} \longrightarrow F \times F^\circ$ is a diffeomorphism. In other words, there is a smooth map $\varphi : (F \times F^\circ) \cap \Omega \longrightarrow G \times G^\circ$ such that $\mathcal{L} \cap \Omega$ is the graph of φ. Writing $\varphi = (f, g)$ we have

$$\mathcal{L} \cap \Omega = \{(x_F, f(x_F, \xi_{F^\circ}), g(x_F, \xi_{F^\circ}), \xi_{F^\circ}), x_F \in F, \xi_{F^\circ} \in F^\circ\}.$$

For $\mathcal{L}$ to be lagrangian we must have $df \wedge d\xi_{F^\circ} + dx_F \wedge dg = 0$, in other words $d(fd\xi_{F^\circ} - gdx_F) = 0$. This means there exists, in a neighbourhood of ρ_0, a function $S(x_F, \xi_{F^\circ})$ such that $dS = fd\xi_{F^\circ} - gdx_F$ (equivalently, $f = \partial_{\xi_{F^\circ}} S, g = -\partial_{x_F} S$). Consider the function

$$\mathcal{S}(x_F, x_G, \xi_{F^\circ}) = \xi_{F^\circ}.x_G - S(x_F, \xi_{F^\circ})$$

defined on an open subset of $F \times G \times F^\circ = \mathbb{R}^d \times F^\circ$. It is now straightforward to check that this is a generating function of $\mathcal{L} \cap \Omega$, and $\dim F^\circ = N$ as announced. □

EXAMPLE 12.8. Here is a fundamental example : in $T^*\mathbb{R}^d = \mathbb{R}^d \times (\mathbb{R}^d)^*$ endowed with the canonical symplectic form $\sum_{i=1}^d dq_i \wedge dp_i$, a generating function for $T^*_x\mathbb{R}^d = \{x\} \times (\mathbb{R}^d)^*$ is $S(y, \theta) = \sum_{i=1}^d \theta_i(y_i - x_i) = \langle\theta, y - x\rangle$ (here $N = d$ and θ varies in $\mathbb{R}^d$).

EXERCISE 12.9. A crucial thing : (i) Show that a (connected) lagrangian submanifold $\mathcal{L}$ is invariant under the hamiltonian flow of H if and only if it is contained in some fixed energy layer $\{H = E\}$.

(ii) As a particular case, deduce that if $H(x, d_xS(x, \theta)) = E$ for any (x, θ), then the lagrangian manifold $\mathcal{L}_S$ generated by S is invariant under the hamiltonian flow (ϕ^t_H).

(iii) Assume now that S is a smooth function of (t, x, θ), and assume that

$$\frac{\partial S}{\partial t} + H(x, d_xS) = 0$$

for all (t, x, θ). Denote $S_t(x, \theta) = S(t, x, \theta)$. Show that the lagrangian manifold $\mathcal{L}_{S_t}$ is the image of $\mathcal{L}_{S_0}$ under ϕ^t_H. *Hint : reduce the problem to the previous one by considering the hamiltonian* $\overline{H}(x, t, \xi, E) = H(x, \xi) + E$ *on* $T^*(X \times \mathbb{R})$.

13. Lagrangian distributions.

Let $\mathcal{L}$ be a lagrangian submanifold of T^*X. For our applications we shall only be interested in the case where $\mathcal{L}$ is relatively compact and where it has a global generating function $S : \mathcal{L} = \mathcal{L}_S$, S being defined on $\Omega_X \times \Omega_{\mathbb{R}^N}$. In this case we define the (semiclassical) notion of a lagrangian function associated to $\mathcal{L}$ as follows :

DEFINITION 13.1. We denote $O^m(X, \mathcal{L}_S)$ the space of functions of the form

$$u_\hbar(x) = \frac{e^{i\alpha(\hbar)}}{(2\pi\hbar)^{N/2}} \int_{\Omega_{\mathbb{R}^N}} e^{i\frac{S(x,\theta)}{\hbar}} a_\hbar(x, \theta) d\theta$$

where

– $\alpha(\hbar)$ is a real number that may depend on $\hbar$,

– the function a defined on $\Omega_X \times \Omega_{\mathbb{R}^N}$ is smooth and has an asymptotic development when $\hbar \longrightarrow 0$,

$$a \sim \sum_{j=0}^{\infty} \hbar^{j+m} a_{j+m},$$

the asymptotic development holds in all C^k–norms on compact subsets,

– we assume that a is compactly supported with respect to the variable θ.

As usual, the class $O^m(X, \mathcal{L}_S)$ should actually be defined *modulo negligible functions,* which, we recall, are smooth functions $u_\hbar$ for which all the C^k–norms on compact sets are $\mathcal{O}(\hbar^\infty)$. Then, one can prove **[GS94]** that the definition of $O^m(X, \mathcal{L}_S)$ does not depend on the choice of the generating function S :

THEOREM 13.2. *If* $\mathcal{L}_S = \mathcal{L}_{S'}$ *then* $O^m(X, \mathcal{L}_S) = O^m(X, \mathcal{L}_{S'})$.

EXAMPLE 13.3. On $X = \mathbb{R}^d$, the Dirac mass at x

$$\delta_x(y) = \frac{1}{(2\pi\hbar)^d} \int_{\mathbb{R}^d} e^{\frac{\langle \xi, y-x \rangle}{\hbar}} d\xi$$

can be seen as a lagrangian distribution associated with the lagrangian submanifold $T^*_x\mathbb{R}^d$ ($S(y, \xi) = \langle \xi, y - x \rangle$), save for the fact that the symbol $a(y, \xi) \equiv 1$ is not compactly supported. Let $\chi(y, \xi)$ be a smooth, positive, compactly supported function, call Ω a bounded open set containing the support of χ. Then the function

$$\delta^\chi_x(y) = \frac{1}{(2\pi\hbar)^d} \int e^{\frac{\langle \xi, y-x \rangle}{\hbar}} \chi(y, \xi) d\xi$$

falls into the class $O^{-d/2}(\mathbb{R}^d, T^*_x\mathbb{R}^d \cap \Omega)$. Assume χ takes the constant value 1 in a neighbourhood of a certain compact subset $\mathcal{E} \subset T^*\mathbb{R}^d$. Then δ^χ_x is often called a "delta–function truncated away from $\mathcal{E}$" : it is a Dirac mass whose frequencies near $\mathcal{E}$ have not been touched, while the frequencies out of Ω have been suppressed.

14. WKB description of the operator $U^t = \exp(it\hbar\frac{\triangle}{2})$.

REMARK 14.1. The initials WKB stand for Wentzel, Kramers and Brillouin, who independently proposed this method to find approximate solutions of a 1-d *stationary* Schrödinger equation — in other words, to find approximate eigenfunctions [**Wtz26, Kr26, Brill26**]. The method was later generalized by Keller and Maslov to find approximate eigenfunctions (quasimodes) of higher dimensional, completely integrable systems [**Kell58, Masl65**].

Here we present the WKB method applied to the *evolutive* Schrödinger equation. It was first used by Van Vleck [**VV28**].

Recall that we are interested in the hamiltonian flow generated by $H(x,\xi) = \frac{\|\xi\|_x^2}{2}$, namely the geodesic flow, denoted g^t in Section 6. We wish to study the Schrödinger flow $U^t = e^{it\hbar\frac{\triangle}{2}}$ and to relate it to the geodesic flow as $\hbar \longrightarrow 0$.

Consider an initial state $u(0)$ of the form $u(0,x) = a_\hbar(0,x)\,\mathrm{e}^{\frac{i}{\hbar}S(0,x)}$, where $S(0,\bullet)$, $a_\hbar(0,\bullet)$ are smooth functions defined on a subset of $\Omega \subset X$, $a_\hbar$ has a fixed compact support in Ω and has an asymptotic development $a_\hbar \sim \sum_k \hbar^k\, a_k$, valid in all C^n–norms. This represents a WKB (or lagrangian) state, supported on the lagrangian manifold $\mathcal{L}(0) = \{(x, d_xS(0,x)),\ x \in \Omega\}$.

The WKB method consists in looking for an approximate expression[9] for the state $\tilde{u}(t) \stackrel{\text{def}}{=} U^t u(0)$, in the form

$$u(t,x) = \mathrm{e}^{\frac{iS(t,x)}{\hbar}}\, a_\hbar(t,x) = \mathrm{e}^{\frac{iS(t,x)}{\hbar}} \sum_{k=0}^{N-1} \hbar^k a_k(t,x) \tag{14.1}$$

where N is a fixed, arbitrarily large integer. We want $u(t)$ to solve $\frac{\partial u}{\partial t} = i\hbar\frac{\triangle_x u}{2}$ up to a remainder of order $\hbar^N$. Computing explicitly both sides of the equation, and identifying the successive powers of $\hbar$, we see that the functions S and a_k must satisfy the following partial differential equations:

$$\begin{cases} \frac{\partial S}{\partial t} + H(x, d_xS) = 0 \quad \text{(Hamilton-Jacobi equation)} \\[2ex] \frac{\partial a_0}{\partial t} = -\langle d_x a_0, d_xS(t,x)\rangle - a_0\frac{\triangle_x S(t,x)}{2} \quad \text{(0-th transport equation)}\,, \\[2ex] \frac{\partial a_k}{\partial t} = \frac{i\triangle a_{k-1}}{2} - \langle da_k, dS\rangle - a_k\frac{\triangle S}{2} \quad \text{(k-th transport equation)}\,. \end{cases} \tag{14.2}$$

Assume that, on a certain time interval — say $s \in [0,1]$ — the above equations have a well defined smooth solution $S(s,x)$, meaning that the transported lagrangian manifold $\mathcal{L}(s) = \phi_H^s\mathcal{L}(0)$ is of the form $\mathcal{L}(s) = \{(x, d_xS(s,x))\}$, where $S(s)$ is a smooth function defined on the open set $\pi\mathcal{L}(s)$. Under these conditions, we denote as follows the induced flow on X:

$$G_s^t : x \in \pi\mathcal{L}(s) \mapsto \pi g^{t-s}\big(x, d_xS(s,x)\big) \in \pi\mathcal{L}(t)\,, \tag{14.3}$$

In the first chapter we introduced the *exponential map* associated to $\mathcal{L}_s$: we have $G_s^t = \exp_{\mathcal{L}_s}^{t-s} \circ \pi^{-1}$. Note that $G_s^s = I$ and that the following composition rule holds : $G_{t_1}^{t_2} \circ G_{t_0}^{t_1} = G_{t_0}^{t_2}$.

[9]often called an Ansatz

We then introduce the following unitary operator T_s^t, which transports functions on $\pi\mathcal{L}(s)$ into functions on $\pi\mathcal{L}(t)$:

$$T_s^t(a)(x) = a \circ G_t^s(x)\, J_t^s(x)^{1/2}\,. \tag{14.4}$$

Here $J_t^s(x)$ is the Jacobian of the map G_t^s at the point x, measured with respect to the riemannian volume on X. It is given by

$$J_s^t(x) = \exp\left\{ \int_s^t \triangle S\big(\tau, G_s^\tau(x)\big)\, d\tau \right\}. \tag{14.5}$$

We leave it as an exercise to check this formula, and to deduce that the 0-th transport equation in (14.2) is explicitly solved by

$$a_0(t) = T_0^t\, a_0\,, \quad t \in [0,1]\,. \tag{14.6}$$

The higher-order terms $k \geq 1$ are given by

$$a_k(t) = T_0^t a_k + \int_0^t T_s^t\left(\frac{i \triangle a_{k-1}}{2}(s)\right) ds\,. \tag{14.7}$$

The function $u(t,x)$ defined by (14.1) satisfies the approximate equation

$$\frac{\partial u}{\partial t} = i\hbar \frac{\triangle u}{2} - i\hbar^N \mathrm{e}^{\frac{i}{\hbar}S(t,x)} \frac{\triangle a_{N-1}}{2}(t,x)\,.$$

By Duhamel's principle, the difference between $u(t)$ and the exact solution $\tilde{u}(t)$ is

$$u(t) - \tilde{u}(t) = -i\hbar^N \int_0^t U^{t-s}\left(\mathrm{e}^{\frac{i}{\hbar}S(s,x)} \frac{\triangle a_{N-1}}{2}(s,x)\right) ds,$$

and from the unitarity of U^t, this is bounded, for $t \in [0,1]$, by

$$\|u(t) - \tilde{u}(t)\|_{L^2} \leq \frac{\hbar^N}{2} \int_0^t \|\triangle a_{N-1}(s)\|_{L^2}\, ds \leq C\, t\, \hbar^N \Big(\sum_{k=0}^{N-1} \|a_k(0)\|_{C^{2(N-k)}} \Big)\,. \tag{14.8}$$

The constant C is controlled by the volumes of the sets $\pi\mathcal{L}(s)$ $(0 \leq s \leq t \leq 1)$, and by a certain number of derivatives of the flow G_t^s $(0 \leq s \leq t \leq 1)$.

REMARK 14.2. Elaborating on these methods, one proves the following : if u is a lagrangian state in $O^m(X,\mathcal{L})$, then $U^t u$ is a lagrangian state in $O^m(X, g^t\mathcal{L})$. We have proved it in the particular case when $g^t\mathcal{L}$ is a graph over X for all t. The operator U^t is called a *Fourier Integral Operator* associated with the transformation g^t.

This is the property Schrödinger had looked for when introducing his equation. We have, in addition, found the explicit formula for all the $a_k(t)$. For $k = 0$, equation (14.6) is called the Van Vleck formula.

15. Proof of the main estimate.

15.1. Decomposition of $\mathrm{Op}(\chi)u$ **into truncated delta–functions.** We can now prove Theorem 10.1, which estimates the norm of the operator

$$\hat{P}_{\alpha_n}(n)\hat{P}_{\alpha_{n-1}}(n-1)...\hat{P}_{\alpha_0}\,\mathrm{Op}(\chi) = U^{-n}\hat{P}_{\alpha_n}U\hat{P}_{\alpha_{n-1}}...U\hat{P}_{\alpha_0}\,\mathrm{Op}(\chi)$$

(where we denote $U^t = \exp(ith\frac{\triangle}{2})$ and $U = U^1$). Since U^t is unitary, the norm of this operator is also the same as the norm of $\hat{P}_{\alpha_n}U\hat{P}_{\alpha_{n-1}}...U\hat{P}_{\alpha_0}\,\mathrm{Op}(\chi)$.

The pseudo-differential operator $\mathrm{Op}(\chi)$ is defined in §4.3 :

$$\mathrm{Op}(\chi) = \sum_l \varphi_l \, \mathrm{OP}(\chi) \, \varphi_l$$

where (φ_l) is an auxiliary partition of unity on X (i.e. $\sum_l \varphi_l(x)^2 \equiv 1$) such that the support of each φ_l is endowed with local coordinates in $\mathbb{R}^d$. In local coordinates in the support of φ_l, $\mathrm{OP}(\chi)$ is then defined by the usual formula,

$$\mathrm{OP}(\chi)u(x) = (2\pi\hbar)^{-d} \int u(z) e^{i\frac{\langle \xi, x-z\rangle}{\hbar}} \chi(z,\xi) dz d\xi. \tag{15.1}$$

The function χ will be chosen of the form $\chi(z,\xi) = \chi_1(|\xi|_z)$ where χ_1 is a smooth function on $\mathbb{R}_+$ supported in $[1-\varepsilon/2, 1+\varepsilon/2]$ with $\chi_1 \equiv 1$ in a neighbourhood of 1. For $x \in \Omega_{\alpha_0}$, we can write

$$\mathrm{Op}(\chi)u(x) = \sum_l \int u(z) \delta_z^l(x) dz, \tag{15.2}$$

where we denote δ_z^l the truncated δ-function

$$\delta_z^l(x) = \varphi_l(x)\varphi_l(z) \int e^{\frac{i\langle \xi, x-z\rangle}{\hbar}} \chi(z,\xi) \frac{d\xi}{(2\pi\hbar)^d}. \tag{15.3}$$

Each δ_z^l is a lagrangian state associated with the lagrangian manifold $T_z^*X \cap H^{-1}\left((\frac{1}{2}-\varepsilon, \frac{1}{2}+\varepsilon)\right)$. Equation (15.2) means that every state in the image of $\mathrm{Op}(\chi)$ can be decomposed as an integral combination of the lagrangian states δ_z^l. We shall first estimate the norm of $\hat{P}_{\alpha_n} U \hat{P}_{\alpha_{n-1}} ... U \hat{P}_{\alpha_0} \delta_z^l$ for any z, and then use (15.2) to write, for an arbitrary function u,

$$\begin{aligned} \|\hat{P}_{\alpha_n} U \hat{P}_{\alpha_{n-1}} ... U \hat{P}_{\alpha_0} \mathrm{Op}(\chi) u\| &\le \sum_l \sup_z \|\hat{P}_{\alpha_n} U \hat{P}_{\alpha_{n-1}} ... U \hat{P}_{\alpha_0} \delta_z^l\| \int_X |u(y)| dy \\ &\le \sum_l \sup_z \|\hat{P}_{\alpha_n} U \hat{P}_{\alpha_{n-1}} ... U \hat{P}_{\alpha_0} \delta_z^l\| \sqrt{\mathrm{Vol}\, X} \|u\|_{L^2(X)} \end{aligned}$$

The estimates will be done by induction on n: we will propose an Ansatz – that is, an approximate expression – for $\hat{P}_{\alpha_n} U \hat{P}_{\alpha_{n-1}} ... U \hat{P}_{\alpha_0} \delta_z^l$, valid for "large" n. In what follows we omit the ^{l} superscript and just write δ_z.

15.2. The Ansatz for $n = 1$. At $n = 0$ we know that $\hat{P}_{\alpha_0} \delta_z(x)$ is a lagrangian state associated with the lagrangian manifold $\mathcal{L}^0 = T_z^*X \cap H^{-1}\left((\frac{1}{2}-\varepsilon, \frac{1}{2}+\varepsilon)\right)$, a union of spheres $H^{-1}(\frac{1}{2}+\eta) \cap T_z^*X$.

From Remark 14.2, we know that $U^t \hat{P}_{\alpha_0} \delta_z$ is a lagrangian state associated to

$$\mathcal{L}^0(t) = g^t \left(T_z^*X \cap H^{-1}((\frac{1}{2}-\varepsilon, \frac{1}{2}+\varepsilon)) \right).$$

If we assume that the injectivity radius of X is greater than $1 + 100\varepsilon$, then this is a graph over X for $0 < t < 1+\varepsilon$. This is just saying that the exponential map $\exp_z^t$ is a diffeormorphism from $T_z^*X \cap H^{-1}\left((\frac{1}{2}-\varepsilon, \frac{1}{2}+\varepsilon)\right)$ onto its image, for $0 < t < 1+\varepsilon$.

This means we have an Ansatz

$$U^t \hat{P}_{\alpha_0} \delta_z \sim (2\pi\hbar)^{-d/2} e^{\frac{iS^0(t,x|z)}{\hbar}} \Big(\sum_{k=0}^{\infty} \hbar^k b_k^0(t, x|z) \Big), \tag{15.4}$$

where the function $S^0(t,x|z)$ is a generating function of the lagrangian manifold $\mathcal{L}^0(t)$.

We denote

$$v^0(t;x|z) = e^{\frac{iS^0(t,x|z)}{\hbar}} b^0_\hbar(t,x|z), \tag{15.5}$$

$$b^0_\hbar(t,x|z) \stackrel{\text{def}}{=} \left(\sum_{k=0}^{N-1} \hbar^k b^0_k(t,x|z)\right) \tag{15.6}$$

For $t=1$, the function $v^0(1;x|z)$ gives us an approximation to $U\hat{P}_{\alpha_0}\delta_z$, the difference being bounded in L^2–norm by $\mathcal{O}(\hbar^{N-\frac{d}{2}})$.

15.3. Iteration of the WKB Ansätze. In this section we will obtain an approximate Ansatz for $\hat{P}_{\alpha_n}\dots U\hat{P}_{\alpha_1}U\hat{P}_{\alpha_0}\delta_z$. Above we have already performed the first step, obtaining an approximation $v^0(1,.|z)$ of $U\hat{P}_{\alpha_0}\delta_z$. Until Lemma 15.1 we will fix the base point z, and omit it in our notations when no confusion may arise; at the end we will obtain an estimate which is uniform in z.

Applying the multiplication operator $\hat{P}_{\alpha_1}$ to the state $v^0(1,x) := v^0(1,x|z)$, we obtain another WKB state which we denote as follows:

$$v^1(0,x) = b^1_\hbar(0,x)\, e^{\frac{i}{\hbar}S^1(0,x)}, \qquad \text{with} \qquad \begin{cases} S^1(0,x) := S^0(1,x|z)\,, \\ b^1_\hbar(0,x) := \hat{P}_{\alpha_1}(x)\, b^0_\hbar(1,x|z)\,. \end{cases}$$

This state is associated with the lagrangian manifold

$$\mathcal{L}^1(0) = \mathcal{L}^0(1)\cap T^*\Omega_{\alpha_1}\,.$$

If this intersection is empty, then $v^1(0)=0$, which means that $\hat{P}_{\alpha_1}v^0(1) = \mathcal{O}(\hbar^N)$ in L^2 norm. In the opposite case, we can evolve $v^1(0)$ following the procedure described in §14. For $t\in[0,1]$, and up to an error $\mathcal{O}_{L^2}(\hbar^N)$, the evolved state $U^t v^1(0)$ is given by the WKB Ansatz

$$v^1(t,x) = b^1_\hbar(t,x)\, e^{\frac{i}{\hbar}S^1(t,x)}, \qquad b^1_\hbar(t) = \sum_{k=0}^{N-1} b^1_k(t)\,.$$

The state $v^1(t)$ is associated with the lagrangian $\mathcal{L}^1(t) = g^t\,\mathcal{L}^1(0)$, and the function $b^1_\hbar(t)$ is supported inside $\pi\mathcal{L}^1(t)$.

15.3.1. *Evolved lagrangians.* We can iterate this procedure, obtaining a sequence of approximations

$$v^j(t) = U^t\,\hat{P}_{\alpha_j}v^{j-1}(1) + \mathcal{O}(\hbar^N), \quad \text{where } v^j(t,x) = v^j(t,x|z) = b^j_\hbar(t,x|z)\, e^{\frac{i}{\hbar}S^j(t,x|z)}\,. \tag{15.7}$$

(Again, the initial position z is fixed for the moment, and we do not always indicate in the notations the z-dependence). To show that this procedure is consistent, we must check that the lagrangian manifold $\mathcal{L}^j(t)$ supporting $v^j(t)$ does not develop caustics through the evolution ($t\in[0,1]$), and that the projection $\pi:\mathcal{L}^j(t)\to X$ remains injective. These were the two conditions required to apply the method of §14. We now show that these properties hold, due to our assumption that the curvature is negative (in fact, it is enough to assume that the geodesic flow has the Anosov property).

The manifolds $\mathcal{L}^j(t)$ are obtained by the following procedure. Knowing $\mathcal{L}^{j-1}(1)$, which is generated by the phase function $S^{j-1}(1)$, we take for $\mathcal{L}^j(0)$ the intersection

$$\mathcal{L}^j(0) = \mathcal{L}^{j-1}(1)\cap T^*\Omega_{\alpha_j}\,.$$

If this set is empty, then we stop the construction. Otherwise, this lagrangian is evolved into $\mathcal{L}^j(t) = g^t\mathcal{L}^j(0)$ for $t \in [0,1]$. Notice that the lagrangian $\mathcal{L}^j(t)$ corresponds to evolution at time $j+t$ of a piece of $\mathcal{L}^0(0)$: it is made up of the image under the geodesic flow of a compact piece of the fiber T_z^*X. If the geodesic flow is Anosov, the geodesic flow has no conjugate points — by a result of Klingenberg [**Kl74**]. This means precisely that $g^t\mathcal{L}^0(0)$ will not develop caustics.

On a negatively curved manifold, there cannot be two homotopic geodesics joining two points x and z. As a consequence, there cannot be two geodesics joining x and z in time $j+t$ and which fall in the same Ω_{α_k} for all integer times k (this holds if the injectivity radius is larger than 1 and if the diameter of the Ω_{α_k} is small enough). This means that, for any $j \geq 1$, $0 \leq t \leq 1$, the manifold $\mathcal{L}^j(t)$ projects injectively to $\pi\mathcal{L}^j$.

Finally, we recall that $\mathcal{L}^0(0)$ was obtained by propagating a piece of $T_z^*X \cap H^{-1}(\frac{1}{2}-\varepsilon, \frac{1}{2}+\varepsilon)$. Since the geodesic flow on each energy layer $H^{-1}(1/2+\eta)$ is Anosov, the sphere bundle $H^{-1}(1/2+\eta) \cap T_z^*X$ is uniformly transverse to the stable foliation in $H^{-1}(1/2+\eta)$ — also a result of [**Kl74**]. As a consequence, the action of the geodesic flow carries a piece of sphere $H^{-1}(1/2+\eta) \cap T_z^*X$ exponentially close to a piece of unstable leaf of $H^{-1}(1/2+\eta)$ when $t \to +\infty$. This transversality of spheres with the stable foliation is crucial in our choice of the "basis" δ_z.

15.3.2. *Exponential decay.* We now analyze the behaviour of the symbols $b_\hbar^j(t,x)$ appearing in (15.7), when $j \to \infty$. These symbols are constructed iteratively: starting from the function $b_\hbar^{j-1}(1) = \sum_{k=0}^{N-1} b_k^{j-1}(1)$ supported inside $\pi\mathcal{L}^{j-1}(0)$, we define

$$b_\hbar^j(0,x) = \hat{P}_{\alpha_j}(x)\, b_\hbar^{j-1}(1,x)\,, \qquad x \in \pi\mathcal{L}^j(0)\,. \tag{15.8}$$

The WKB procedure of §14 shows that for any $t \in [0,1]$,

$$U^t\, v^j(0) = v^j(t) + R_N^j(t)\,, \tag{15.9}$$

where the transported symbol $b_\hbar^{j-1}(t) = \sum_{k=0}^{N-1} \hbar^k\, b_k^{j-1}(t)$ is supported inside $\pi\mathcal{L}^j(t)$. The remainder satisfies

$$\|R_N^j(t)\| \leq C\, t\, \hbar^N \Big(\sum_{k=0}^{N-1} \|b_k^j(0)\|_{C^{2(N-k)}} \Big)\,. \tag{15.10}$$

To control this remainder when $j \to \infty$, we need to bound from above the derivatives of $b_\hbar^j$. Lemma 15.1 below shows that all terms $b_k^j(t)$ and their derivatives decay exponentially when $j \to \infty$, due to the Jacobian appearing in (14.4).

To understand the reasons of the decay, we first look at the principal symbol $b_0^j(1,x)$. It satisfies the following recurrence:

$$b_0^j(1,x) = T_j^{j+1}(\hat{P}_{\alpha_j} \times b_0^{j-1}(1))(x) = (\hat{P}_{\alpha_j} \times b_0^{j-1}(1)) \circ G_{j+1}^j(x)\sqrt{J_{j+1}^j(x)}\,; \tag{15.11}$$

using similar notations as above, the transport map $G_{s,t}$ is defined, for $j \leq s,\ t \leq j+1$, by $G_{s,t} := \exp^{t-s}_{\mathcal{L}^j(s-j)} \circ \pi^{-1}$, and maps $\pi\mathcal{L}^j(s-j)$ to $\pi\mathcal{L}^j(t-j)$. We denote J_s^t the jacobian of G_s^t. We recall that $G_{n-1}^n G_{n-2}^{n-1} \dots G_1^2 = G_1^n$, where both sides are defined.

Iterating this expression, and using the fact that $0 \leq \hat{P}_{\alpha_j} \leq 1$, we get at time n and for any $x \in \pi\mathcal{L}^n(0)$:

$$|b_0^n(0,x)| \leq |b_0^0(1, G_n^1(x))| \times \Big(J_n^{n-1}(x) J_{n-1}^{n-2}(G_n^{n-1}x) \dots J_2^1(G_3^2(x)) \Big)^{1/2}. \tag{15.12}$$

By the chain rule, this product of jacobians is simply $J^1_n(x)^{1/2} = J^n_1(G^1_n(x))^{-1/2}$.

Recall that $\mathcal{L}^0(0)$ intersected with each energy layer $S^{1+\eta}X := \{\xi \in T^*X, \|\xi\| = 1+\eta\}$ is just a piece of the sphere $S^{1+\eta}_z X$. Thus, if $d(x,z) = 1+\eta$, the jacobian $J^n_1(G^1_n(x))$ measures the expansion rate of the sphere $g^n(S^{1+\eta}_z X)$: in dimension d and curvature $\equiv -1$, it grows asymptotically like $e^{(d-1)(1+\eta)n}$ when $n \to \infty$. If $x \in \pi\mathcal{L}^1(0)$ (and if this last set is non-empty) we have $d(x,z) \geq 1-\varepsilon$ (because $\mathcal{L}^0(0)$ is contained in $H^{-1}(\frac{1}{2}-\varepsilon, \frac{1}{2}+\varepsilon)$). We obtain the following estimate on the principal symbol $b^n_0(0)$:

$$\forall n \geq 1 \quad \|b^n_0(0)\|_\infty \leq \|b^0_0(1)\|_\infty \,[\exp(-n(d-1)(1-\varepsilon)]^{1/2} \tag{15.13}$$

The following lemma, which we shall not prove here, shows that the upper bound extends to the full symbol $b^n_\hbar(0,x)$ and its derivatives.

LEMMA 15.1. *Take any index $0 \leq k \leq N$ and $m \leq 2(N-k)$. Then there exists a constant $C(k,m)$ such that*

$$\forall n \geq 1, \quad \forall x \in \pi\mathcal{L}^n(0), \qquad |d^m b^n_k(0,x|z)| \leq C(k,m)\, n^{m+3k}\, [\exp(-n(d-1)(1-\varepsilon))]^{1/2}. \tag{15.14}$$

This bound is uniform with respect to the initial point z. For $(k,m) \neq (0,0)$, the constant $C(k,m)$ depends on the partition $\mathcal{P}^{(0)}$, while $C(0,0)$ does not.

Taking into account the fact that the remainders $R^j_N(1)$ are dominated by the derivatives of the b^j_k (see (15.10)), the above statement translates into

$$\forall j \geq 1, \qquad \|R^j_N(1)\|_{L^2} \leq C(N)\, j^{3N}\, [\exp(-n(d-1)(1-\varepsilon))]^{1/2}\, \hbar^N.$$

A crucial fact for us is that the above bound also holds for the propagated remainder $\hat{P}_{\alpha_n} U \cdots U \hat{P}_{\alpha_{j+1}} R^j_N(1)$, due to the fact that the operators $\hat{P}_{\alpha_j} U$ have norms less than 1. As a result, the total error at time n is bounded from above by the sum of the errors $\|R^j_N(1)\|$. We obtain the following estimate for any $n > 0$:

$$\|\hat{P}_{\alpha_n} U \hat{P}_{\alpha_{n-1}} \cdots \hat{P}_{\alpha_1} v^0(1|z) - v^n(0|z)\| \leq C(N)\,\hbar^N \sum_{j=0}^{n} j^{3N}\, [\exp(-n(d-1)(1-\varepsilon))]^{1/2}. \tag{15.15}$$

The last term is bounded by $C(N)\hbar^N$. This bound is uniform with respect to the initial point z.

15.4. Conclusion. From (15.15), we see that we can use our Ansatz $v^n(0|z)$ to estimate the norm of $\hat{P}_{\alpha_n} U \hat{P}_{\alpha_{n-1}} \cdots \hat{P}_{\alpha_1} v^0(1|z)$, up to an error $\mathcal{O}(\hbar^N)$. From (15.14) and the definition (15.7) of $v^n(0|z)$, we have

$$\|v^n(0|z)\|_{L^2(X)} \leq [\exp(-n(d-1)(1-\varepsilon))]^{1/2} \sum_{k=0}^{N-1} C(k,0)\hbar^k n^{3k}. \tag{15.16}$$

As required in Theorem 10.1, let us now take an arbitrary large $\mathcal{K}$, and $n = \mathcal{K}|\log\hbar|$. In the inequalities (15.13) and (15.16), the right hand term is bounded *below* by a fixed power of $\hbar$ (more precisely, $\hbar^{-\frac{1}{2}\mathcal{K}(d-1)}$). Thus, we will choose N, the order of our WKB expansion, large enough so that the remainder (15.15) is negligible compared to $\hbar^{-\frac{1}{2}\mathcal{K}(d-1)}$.

Now, remember the relation (15.4) between $v^0(1|z)$ and $U\hat{P}_{\alpha_0}\delta_z$: note in particular the normalization factor $(2\pi\hbar)^{-d/2}$. The combination of (15.15) and (15.16) gives us

$$(15.17)\quad \|\hat{P}_{\alpha_n}U\hat{P}_{\alpha_{n-1}}\cdots\hat{P}_{\alpha_1}U\,\hat{P}_{\alpha_0}\delta_z\|_{L^2(X)} \leq \frac{2}{(2\pi\hbar)^{d/2}}[\exp(-n(d-1)(1-\varepsilon))]^{1/2}$$

for $n = \mathcal{K}|\log\hbar|$ and $\hbar \leq \hbar_{\mathcal{K}}$.

Combined with (15.2) and the subsequent discussion, we find

$$\|\hat{P}_{\alpha_n}U\hat{P}_{\alpha_{n-1}}...U\hat{P}_{\alpha_0}\operatorname{Op}(\chi)u\| \leq \frac{2l\sqrt{\operatorname{Vol}X}}{(2\pi\hbar)^{d/2}}\|u\|_{L^2(X)}[\exp(-n(d-1)(1-\varepsilon))]^{1/2}$$

which is the announced result.

References

[AF94] R. Alicki, M. Fannes, *Defining quantum dynamical entropy*, Lett. Math. Phys. **32** no. 1, 75–82 (1994).

[A05] N. Anantharaman, *Entropy and the localization of eigenfunctions,* à paraître dans Ann. Math.

[AN06] N. Anantharaman, S. Nonnenmacher, *Semi-classical entropy of the Walsh-quantized baker's map,* à paraître dans Ann. I.H.P.

[AN07] N. Anantharaman, S. Nonnenmacher, *Half-delocalization of eigenfunctions of the laplacian on an Anosov manifold,* prépublication.

[Bera77] P. Bérard, *On the wave equation on a compact Riemannian manifold without conjugate points.* Math. Z. **155** no. 3, 249–276 (1977).

[Berr77] M.V. Berry, *Regular and irregular semiclassical wave functions*, J.Phys. **A 10**, 2083-2091 (1977).

[Bo91] O. Bohigas, *Random matrix theory and chaotic dynamics*, in M.J. Giannoni, A. Voros and J. Zinn-Justin eds., *Chaos et physique quantique*, (École d'été des Houches, Session LII, 1989), North Holland, 1991.

[BDB03] F. Bonechi, S. De Bièvre, *Controlling strong scarring for quantized ergodic toral automorphisms*, Duke Math. J. **117** no. 3, 571–587 (2003).

[BHJ25-I] M. Born, W. Heisenberg, P. Jordan, *Zur Quantenmechanik,* Zeitschrift f. Physik **34**, 858–888 (1925).

[BHJ25-II] M. Born, W. Heisenberg, P. Jordan, *Zur Quantenmechanik II,* Zeitschrift f. Physik **35**, 557–615 (1925).

[BLi03] J. Bourgain, E. Lindenstrauss, *Entropy of quantum limits.* Comm. Math. Phys. **233** no. 1, 153–171 (2003).

[Brill26] L. Brillouin, *La mécanique ondulatoire de Schrödinger; une méthode générale de résolution par approximations successives,* C. R. A. S. **183**, 24–26 (1926).

[Broglie24] L. de Broglie, Annales de Physique (10) **3** 1925, p. 22 (Thèse, 1924).

[CdV85] Y. Colin de Verdière, *Ergodicité et fonctions propres du laplacien*, Comm. Math. Phys **102** no. 3, 497–502 (1985).

[CNT87] A. Connes, H. Narnhofer, W. Thirring, *Dynamical entropy of C^* algebras and von Neumann algebras*, Comm. Math. Phys. **112** no. 4, 691–719 (1987).

[DimSjo] M. Dimassi, J. Sjöstrand, *Spectral asymptotics in the semi-classical limit.* London Math. Soc. Lecture Notes Series 268, Cambridge University Press (1999).

[Dirac33] P. A. M Dirac, *The principles of quantum mechanics,* The Clarendon Press, Oxford (1935), second edition; also, Physik. Zeits. Sowjetunion **3**, 64 (1933).

[DH72] J.J. Duistermaat, L. Hörmander, *Fourier integral operators. II*, Acta Math. **128** no. 3-4, 183–269 (1972).

[DunSchw] N. Dunford and J.T. Schwartz, *Linear Operators, Part I*, Interscience, New York, 1958.

[Ein17] A. Einstein, *Zum Quantensatz von Sommerfeld und Epstein,* Verhandl. deut. physik. Ges. (1917).

[E1744] L. Euler, *De Motu Projectorum in medio non resistente, per methodum maximorum ac minimorum determinando,* Opera Omnia, Seria Prima Vol. XXIV Bernae 1952, Additamentum II 298–308 (1744).

[FNDB03] F. Faure, N, Nonnenmacher, S. De Bièvre, *Scarred eigenstates for quantum cat maps of minimal periods*, Comm. Math. Phys. **239** no. 3, 449–492, (2003).

[FN04] F. Faure, N, Nonnenmacher, *On the maximal scarring for quantum cat map eigenstates*, Comm. Math. Phys. **245** no. 1, 201–214 (2004).

[Foll] G. B. Folland, *Harmonic analysis in phase space,* Princeton University Press 1989.

[GL93] P. Gérard and E. Leichtnam, *Ergodic properties of eigenfunctions for the Dirichlet problem*, Duke Math. J. **71**(2), 559–607 (1993).

[GS94] A. Grigis, A. and J. Sjöstrand, Microlocal Analysis for differential Operators: an Introduction. London Math. Soc. Lecture Notes, 1994.

[Gutz] M. C. Gutzwiller, *Chaos in classical and quantum mechanics,* Springer-Verlag New York, 1990.

[H1830] W. R. Hamilton, *Theory of Systems of Rays,* Transactions of the Royal Irish Academy Vol. XV, 69–174 (1828). Supplement Vol. XVI Part I, 4–62 (1830). Second Supplement Vol. XVI Part II, 93–125 (1831).

[H1834] W. R. Hamilton, *On a General Method in Dynamics; by which the Study of the Motions of all free Systems of attracting or repelling Points is reduced to the Search and Differentiation of one central Relation, or characteristic Function,* Philosophical transactions of the Royal Society of London, 247–308 (1834).

[H25] W. Heisenberg, *Über quantentheoretische Umdeutung kinematischer und mechanischer Beziehungen,* Zeitschrift f. Physik **33**, 879–893 (1925).

[Helffer1] B. Helffer, *30 ans d'analyse semi-classique : bibliographie commentée,* notes disponibles sur la page personnelle de Bernard Helffer.

[Helffer2] B. Helffer, *On h-pseudodifferential operators and applications,* à paraître dans Encyclopedia of Mathematical Physics.

[HelMR87] B. Helffer, A. Martinez, D. Robert *Ergodicité et limite semi-classique*, Comm. Math. Phys. **109** no. 2, 313–326 (1987).

[Hell89] E. J. Heller, in: Chaos and Quantum Physics, Les Houches 1989. Ed. M.J. Giannoni, A. Voros, J. Zinn-Justin, Amsterdam, North-Holland 549–661 (1991).

[HisSig] P. D. Hislop, I. M. Sigal, *Introduction to spectral theory. With application to Schrödinger operators.* Applied Mathematical Science 113, Springer Verlag New York, 1996.

[Ho71] L. Hörmander, *Fourier integral operators. I,* Acta Math. 127 no. 1-2, 79–183 (1971).

[Ho79] L. Hörmander, *The Weyl Calculus of pseudodifferential operators,* Comm. Pure Appl. Math. 32, p. 359-443 (1979).

[Ho] L. Hörmander, *The Analysis of linear Partial Differential Operators.* Grundlehren der Mathematischen Wissenschaften, Springer Verlag (1984).

[KH] A.B. Katok, B. Hasselblatt, *Introduction to the modern theory of dynamical systems,* Encyclopedia of Mathematics and its applications **54**. Cambridge University Press 1995.

[Kell58] J. B. Keller, *Corrected Bohr-Sommerfeld Quantum conditions for Nonseparable Systems,* Ann. Physics **4**, 180–188 (1958).

[Kelm05] D. Kelmer, *Arithmetic quantum unique ergodicity for symplectic linear maps of the multidimensional torus*, prépublication 2005.

[Kelm06] D. Kelmer, *Scarring on invariant manifolds for perturbed quantized hyperbolic toral automorphisms*, http://arxiv.org/abs/math-ph/0607033

[Kelm08] D. Kelmer, *Scarring for quantum maps with simple spectrum,* http://arxiv.org/abs/0801.2493

[Kl74] W.P.A. Klingenberg, *Riemannian manifolds with geodesic flows of Anosov type,* Ann. of Math. (2) **99**, 1–13 (1974)

[Kr26] H. A. Kramers, *Wellenmechanik und halbzahlige quantisierung,* Zeitschrift f. Physik **39**, 828–840 (1926).

[Kraus87] K. Kraus, *Complementary observables and uncertainty relations*, Phys. Rev. **D 35**, 3070–3075 (1987).

[KurRud00] P. Kurlberg and Z. Rudnick, *Hecke theory and equidistribution for the quantization of linear maps of the torus*, Duke Math. J. **103**, 47–77 (2000).

[L1788] J.-L. Lagrange, *Mécanique analytique* éd. A. Blanchard, Paris 1965 (1788).

[LY85] F. Ledrappier, L.-S. Young, *The metric entropy of diffeomorphisms. I. Characterization of measures satisfying Pesin's entropy formula*, Ann. of Math. (2) **122** no. 3, 509–539 (1985).

[Leray] J. Leray, *Lagrangian analysis and quantum mechanics. A mathematical structure related to asymptotic expansions and the Maslov index.* English transl. by Carolyn Schroeder. Cambridge, Massachusetts; London: The MIT Press (1981).

[Li06] E. Lindenstrauss, *Invariant measures and arithmetic quantum unique ergodicity*, Ann. of. Math. (2) **163** no. 1, 165–219 (2006).

[MaaUff88] H. Maassen and J. B. M. Uffink, *Generalized entropic uncertainty relations*, Phys. Rev. Lett. **60**, 1103–1106 (1988).

[Masl65] V. P. Maslov, *Théorie des perturbations et méthodes asymptotiques,* suivi de deux notes complémentaires de V. I. Arnol'd et V. C. Bouslaev, préface de J. Leray, Dunod Paris (1965).

[M1744] P. L. Moreau de Maupertuis, Histoire de l'Académie Royale des Sciences MDCCXLVIII, Paris, Imprimerie Royale, 417–426 (1744).

[Rob] D. Robert, *Autour de l'approximation semi-classique.* Progress in Mathematics no. 68, Birkhäuser (1987).

[RudSa94] Z. Rudnick, P. Sarnak, *The behaviour of eigenstates of arithmetic hyperbolic manifolds*, Comm. Math. Phys **161** no. 1, 195–213 (1994).

[Sa95] P. Sarnak, *Arithmetic quantum chaos*, The Schur lectures (1992) (Tel Aviv), 183–236, Israel Math. Conf. Proc., 8, Bar-Ilan Univ., Ramat Gan, 1995.

[Sa03] P. Sarnak, *Spectra of hyperbolic surfaces*, Bull. A.M.S. **40** no. 4, 441–478 (2003).

[Sn74] A. I. Snirelman, *Ergodic properties of eigenfunctions (Russe)*, Uspehi Mat. Nauk. **29** no. 6 (180), 181–182 (1974).

[Schr26-I] E. Schrödinger, *Quantisierung als Eigenwertproblem (erste Mitteilung),* Annalen der Physik (4) **79** 361–376 (1926), translated into English in *Collected papers on Wave Mechanics*, Chelsea Publishing Company, NY, 1978.

[Schr26-II] E. Schrödinger, *Quantisierung als Eigenwertproblem (zweite Mitteilung),* Annalen der Physik (4) **79** 489–527 (1926), translated into English in *Collected papers on Wave Mechanics*, Chelsea Publishing Company, NY, 1978.

[Schr26-III] E. Schrödinger, *Über das Verhältnis der Heisenberg–Born–Jordanschen Quantenmechanik zu der meinen,* Annalen der Physik (4) **79** 734–756 (1926), translated into English in *Collected papers on Wave Mechanics*, Chelsea Publishing Company, NY, 1978.

[SZ99] J. Sjöstrand and M. Zworski, *Asymptotic distribution of resonances for convex obstacles*, Acta Math. **183**, 191–253 (1999)

[Somm] A. Sommerfeld, *Atombau und Spektrallinien,* Dritte Auflage, Braunschweig (1922), traduction française *La constitution de l'atome et les raies spectrales*, A. Blanchard (1923).

[St30] M. H. Stone, *Linear transformations in Hilbert space III: operational methods and group theory,* Proc. Nat. Acad. Sci. USA **16**, 172–175 (1930).

[VV28] J. H. Van Vleck, *The correspondence principle in the statistical interpretation of quantum mechanics,* Proc. Nat. Acad. Sci. USA **14**, 178–188 (1928).

[Vo95] D. Voiculescu, *Dynamical approximation entropies and topological entropy in operator algebras*, Comm. Math. Phys **170** 249–281 (1995).

[vN31] J. von Neumann, *Die Eindeutigkeit des Schrödingerschen Operatoren,* Math. Ann. **104**, 570–578 (1931).

[Vor] A. Voros, *Développements semi-classiques,* Thèse d'état (1977).

[Vor77] A. Voros, *Semiclassical ergodicity of quantum eigenstates in the Wigner representation, Stochastic Behavior in Classical and Quantum Hamiltonian Systems,* G. Casati, J. Ford, eds., in: Proceedings of the Volta Memorial Conference, Como, Italy, 1977, Lect. Notes Phys. Springer-Verlag, Berlin **93**, 326-333 (1979) .

[Vor78] A. Voros, *An algebra of pseudodifferential operators and the asymptotics of quantum mechanics,* J. Funct. An. **29**, 104–132 (1978).

[Weh79] A. Wehrl, *On the relation between classical and quantum-mechanical entropy*, Rept. Math. Phys. **16**, 353–358 (1979).

[Wtz26] G. Wentzel, *Eine Verallgemeinerung der quantenbedingungen für die Zwecke der Wellenmechanik,* Zeitschrift f. Physik **38**, 518–529 (1926).

[Weyl27] H. Weyl, *Quantenmechanik und Gruppentheorie,* Z. Physik **46**, 1–46 (1927).

[Wol01] S.A. Wolpert, *The modulus of continuity for* $\Gamma_0(m)/\mathbb{H}$ *semi-classical limits*, Commun. Math. Phys. **216**, 313–323 (2001).

[Ze86] S. Zelditch, *Pseudodifferential analysis on hyperbolic surfaces*, J. Funct. Anal. **68** no. 1, 72–105 (1986).

[Ze87] S. Zelditch, *Uniform distribution of eigenfunctions on compact hyperbolic surfaces*, Duke Math. J. **55** no. 4, 919–941 (1987).

[Ze96] S. Zelditch, *Quantum ergodicity of* C^* *dynamical systems*, Commun. Math. Phys, **177**, 507–528 (1996).

[ZeZw96] S. Zelditch and M. Zworski, *Ergodicity of eigenfunctions for ergodic billiards*, Commun. Math. Phys, **175**, 673–682 (1996).

C.M.L.S, École Polytechnique, F-91128 Palaiseau Cedex

Current address: Département de Mathématiques, Bâtiment 425, Faculté des Sciences d'Orsay, Université Paris-Sud 11 F-91405 Orsay Cedex

E-mail address: `Nalini.Anantharaman@math.u-psud.fr`